# Experimental Innovations in Surface Science

John T. Yates, Jr.

# Experimental Innovations in Surface Science

## A Guide to Practical Laboratory Methods and Instruments

AIP PRESS

Springer

John T. Yates, Jr.
Department of Chemistry and Department of Physics and Astronomy
University of Pittsburgh
Pittsburgh, PA 15260
USA

Library of Congress Cataloging-in-Publication Data
Yates, John T., Jr. 1935–
Experimental innovations in surface science : a guide to practical laboratory methods and instruments / John T. Yates, Jr.
p. cm.
Includes index.
ISBN 0-387-98332-5 (alk. paper)
1. Surface (Physics)—Laboratory manuals. I. Title.
QC173.4.S94Y38 1997
530.4'17'078—dc21 97-33269
CIP

Printed on acid-free paper.



Production managed by Victoria Evarretta; manufacturing supervised by Johanna Tschebull.
Photocomposed pages prepared from author's Microsoft Word files.
Printed and bound by Hamilton Printing Co., Rensselaer, NY.
Printed in the United States of America.

9 8 7 6 5 4 3 2 1

ISBN 0-387-98332-5 Springer-Verlag New York Berlin Heidelberg SPIN 10642715

# Preface

The modern era of surface science, developed in the last 30 years, features hundreds of new physical and chemical measurement methods such as low-energy electron diffraction and Auger electron spectroscopy, which form a vital foundation for the field. These measurement methods constitute the basis for a number of books used by beginning students as their introduction to experimental research.

This book is not about this particular experimental core of the field. Instead, it is about those practical laboratory methods used throughout the field, but seldom communicated coherently in writing and drawings to beginning students. These practical methods of design, construction, and measurement also form their own foundation for making possible significant research. They constitute a gold mine of clever ideas and "tricks of the trade" about how to use basic materials and components effectively for research and measurement in surface science. Approximately 250 practical methods are described in the book in enough detail that one could design a new experimental device from the book alone.

This book will be useful to beginning research workers in surface science as well as to those who work with similar methods in fields such as semiconductor technology, materials science, and physical electronics.

The contents of the book have been selected from unpublished information supplied by more than 100 workers in the field as well as from the literature, dating back to the mid-1960s. I have attempted to communicate the core ideas from published and unpublished accounts of research in easy-to-comprehend pictorial form, believing that pictures are the most effective mode for communication of information of this type. Thus, the reader can learn much just by looking at the pictures. The text associated with each drawing gives more details. In many cases, the text and its citations indicate the originators and subsequent contributors to the innovation, but in other cases the origin of the key ideas is unknown, making accurate historical citations impossible.

There are some topics that have not been addressed in the book, primarily because they are currently highly specialized and fast-moving activities that are best learned by advanced research students who have previously mastered some of the topics covered in this book. Untreated topics include many details of the scanning tunneling microscope (STM), complex molecular beam techniques, and complex laser-based measurements.

Updated information on United States and international suppliers of materials and components used in surface science research is featured at the end of the book. Here the reader will find a guide to product sources as well as addresses and other communications information for all companies mentioned in the text footnotes as well as many other company listings. I thank Dr. Jun Yoshinobu and Dr. Joachim Ahner for help with these compilations.

I very much appreciate the involvement of many friends and colleagues who have contributed to this book by sending me accounts of their innovative methods. This includes many of my own students and postdoctorals who have developed some of the experimental methods described here. I also thank the publishers of the Journal of Vacuum Science and Technology for permission to use many published contributions dating back to 1964. These were organized for me by Professor Michael Bozack and Mr. Kenneth W. Bryant at Auburn University, to whom I am greatly indebted, and were invaluable in writing a portion of this book. I thank Dr. Fred Dylla and Mrs. Maria Taylor for their encouragement during work with AIP and with Springer-Verlag as the book progressed. Mrs. Marge Augenstein and Mrs. Mary Beth Merenick supplied accurate assistance in organizing the collection and verification of information for the book at the University of Pittsburgh, and in checking the accuracy of the citations. Ms. Hana Novak took my hand-drawn mechanical drawings and skillfully converted them to computer graphics. The book would not have been produced without the financial assistance of the Alexander von Humboldt Stiftung, which made possible my stay in the Department of Physics, Philipps University, Marburg, Germany. I am indebted also to the University of Pittsburgh for a coincident sabbatical leave to write the book. I wish to thank my many German friends for their hospitality, particularly Dr. Heinz J. Jaensch and Professor Dieter Fick for their friendship during my seven months in Marburg. I also thank my parents, Mr. and Mrs. J. Thomas Yates, Sr., for their hospitality in Hagerstown, Maryland, during the completion of the book. And finally, and most importantly, I thank my wife, Kerin, for all of her support. Without her understanding and constructive criticism, writing and illustrating the book would have been impossible.

John T. Yates, Jr.
R.K. Mellon Professor
Department of Chemistry and
Department of Physics and Astronomy
Director, Surface Science Center
University of Pittsburgh

March 1997

# Contents


**Preface** v

**I. Vacuum System Technology** 1

**A. Motion in Vacuum** 3

1. Crystal Manipulator with Six Degrees of Motional Freedom 4
2. Rotation in Ultrahigh Vacuum 8
3. Crystal Positioning Using Capacitance 16
4. UHV Rotary Manipulator with Arcsec Resolution 20
5. Solenoid Driven Linear Motion Device 24
6. An Inexpensive Linear Motion Device 26
7. Translating Auger Spectrometer 28
8. Simple Multiple Motion Manipulator 30
9. Simple Device for Small Rotations in Ultrahigh Vacuum—Grating Application 32
10. Turntable Rotation in Ultrahigh Vacuum 36
11. Small Motions in UHV Systems Using Shape Memory Effect Alloys 38
12. Rotary Shutter Device Driven by Magnetic Eddy Current 42
13. High-Speed Ultrahigh Vacuum Motor 44
14. High-Speed Rotary Feedthrough for UHV Operation 46
15. Self-Lubricating Bearings in UHV 48
16. Lubrication for Heavy Sliding Loads in Ultrahigh Vacuum 50
17. Simple UHV Bearings for High-Speed Shafts 54
18. Flange-Mounted UHV Variable Aperture 56
19. Linear Motion Platform (LMP) 58
20. Thermally Compensated STM with Repeatable Sample Positioning 60

**B. Sample Transfer** 65

21. Sample Transfer from High Pressure to Ultrahigh Vacuum 66
22. Novel Sample Transfer from UHV Chamber to External Cell 76
23. UHV Sample Transfer Device with Low-Temperature Capability 80
24. Internal Cup High-Pressure Cell 84
25. High-Pressure Transfer Cell 86
26. Sample Transfer with Disconnect-I 90
27. Sample Transfer with Disconnect-II 94
28. UHV Access Port 96

C. Electrical Connections 99

29. Tungsten Ribbon—Attachment to Power Leads 100
30. Wrap Connection—Tungsten to Tungsten 102
31. Cold-Formed Wire Connector 104
32. Ultrahigh Resistance Vacuum Feedthroughs 106
33. Sliding Metal Electrical Contact 108
34. Low-Profile Electrical Lead Clamp 110

D. Pumping and Trapping 113

35. Water Aspirator/Sorption Pump Combination for Efficient UHV System Evacuation 114
36. Ballast Pumping—Vibration Free 116
37. The Use of Appendage Titanium Sublimation Pumps for Pumping Low Gas Loads in Ultrahigh Vacuum 118
38. Pressure-Equalizing Device in Complex Systems 120
39. Diode Ion Pump Performance in He Pumping 122
40. Vacuum Applications of Metal Foams 124
41. Cleaning of Ion Pumps by Chemical Etching 126
42. An Efficient Liquid Nitrogen Trap 128

E. Bakeout 131

43. Bakeout of Metal Ultrahigh Vacuum Systems 132

F. Behavior of UHV Systems 135

44. Wall Passivation in Stainless Steel Ultrahigh Vacuum Systems 136
45. Minimizing Wall Reactions in Ultrahigh Vacuum Systems—Gas Dosers 138

G. Mechanical Action on Samples 143

46. Simple UHV Crystal Cleaver 144
47. Piezoelectric Fatigue Apparatus for UHV Operation 146

H. Gasket Seals 149

48. Gasket Seals for Ultrahigh Vacuum Systems 150
49. Cryogenic Gasket Seals 154
50. Copper Gasket Removal Devices 156
51. Unconventional Compression Seals for Ultrahigh Vacuum 158

I. Leak Repairs and Detection 161

52. Coaxial Pumped He Leak Detection Probe 162
53. Temporary Leak Sealing of Welded Bellows 164
54. Atmospheric Permeation—Viton O-Ring 166

J. Specialized UHV Systems 171

55. Aluminum Ultrahigh Vacuum System 172
56. Surface Electrochemistry Apparatus 174

## II. Mechanical Fabrication Techniques 177

**A. Grids** 179

57. Hemispherical Grids—Formation and Piercing 180
58. Making Flat Mesh Grids of Large Size 184
59. Grid Fabrication Techniques—Conical Grid 186

**B. Conductive Coatings** 189

60. Deposition of Electrically Conductive $SnO_2$ Films 190

**C. Phosphor Screens** 193

61. Sedimentation Method for Depositing Phosphor Screens 194
62. Dusting Method for Coating Phosphor Screens 198

**D. Thermionic Emitters** 201

63. Thoriated Thermionic Emitters 202
64. Lanthanum Hexaboride Thermionic Emitters—Deposition on Metals 206
65. Directly Heated Lathanum Hexaboride Thermionic Emitters 210
66. Thermionic Emitter Mounting 214
67. Indirectly Heated Cathodes for High-Temperature Operation 216
68. Replacing Filaments in Glass Bayard–Alpert Gauges 218

**E. Shielding** 221

69. Magnetic Shielding in Ultrahigh Vacuum 222
70. Electrical Isolation of UHV Components 226

**F. Single Crystal Fabrication/Orientation** 229

71. Single Crystal Orientation, Grinding, and Polishing 230
72. A Simple Goniometer for Cutting Single Crystals 236
73. Ion Beam Polishing of Crystals to Subnanometer Roughness 238
74. Crystal Optical Alignment 240
75. Measurement of the Angles of Incidence in a LEED Experiment 244
76. Measurement of Angle of Incidence of an Electron Beam on a Single Crystal 246

**G. Tip Fabrication** 249

77. Fabrication of Metal Tips Using Zone Electropolishing 250
78. Etching STM Tips Reproducibly 254
79. Sharpening Single Crystal Metal Tips by Ion Bombardment 256

**H. Spot-Welding** 259

80. Spot-Welding Difficult Junctions 260

**I. Mechanical Procedures** 265

81. Stretching Thin Metal Foils to a Wrinkle-Free Condition 266
82. Making Micron-Size Holes for Supersonic Nozzles 268
83. Straightening and Braiding Wires 270

III. Measurement Methods 273

A. Electrons 275

84. Electron Gun Design and Behavior 276
85. Low-Energy Electron Gun for Broad Beam Irradiation 280
86. Low-Energy Electron Gun for Broad Beam Irradiation—Cylindrical Symmetry 282
87. Electron Energy Analyzers 288
88. Electron Spectrometer Calibration Using Graphite 292
89. Measurement of Electron Beam Angular Divergence 294
90. Spot Photometer for LEED Intensity Measurements 298
91. Modified Faraday Cup for Electron Current Measurement 300
92. Rejuvenation of Cu/Be Electron Multipliers 302
93. Geiger-Müller Counter for Inverse Photoemission 304

B. Ions 307

94. Ion Sputter Gun with Low Gas Emission 308
95. Alternate Ion Bombardment Sources 310
96. Ion Gun Based on Bayard–Alpert Gauge 314
97. Sputter Ion Gun—Cold Cathode 318
98. Broad Ion Sources—Improved Stability with Oxygen 320
99. Focusing of Low-Energy Ions by Magnetic Field 324
100. Measuring Ion Beam Dimensions 326
101. Beam Position Monitor—Charged Particles 328
102. Pulse-Counting LEED/ESDIAD Analyzer—Using Microchannel Plate Detection 330
103. Time-of-Flight Detection for Laser-Ionized Neutrals 334

C. Photons (UV) 337

104. Hollow Cathode Resonance Lamp for He(II) (40.8eV) Photoemission 338
105. He(I) and He(II) Ultraviolet Resonance Source 344
106. Photochemistry on Surfaces Using an Ultraviolet Lamp 348
107. Use of Polarized Ultraviolet Light for Photochemistry 352

D. Kinetics of Adsorption/Adsorbate Coverages 357

108. Absolute Adsorption Uptake Measurements from a Calibrated Effusive Molecular Beam 358
109. Measurement of Absolute Sticking Coefficient at Desorption Temperature 362

E. Mass Spectrometry 365

110. Shielded Quadrupole Mass Spectrometers—Temperature-Programmed Desorption 366
111. Enhanced Signal-to-Noise in Quadrupole Mass Spectrometers 372
112. Elimination of Spurious Electron Emission from a Quadrupole Mass Spectrometer 376
113. Ta Mass Spectrometer Filament 380
114. Mass Spectrometer Calibration by Molecular Decomposition 382

**F. Temperature-Programmed Desorption/Reaction** **385**

115. Thermal Desorption—Angular Resolved 386
116. Magic-Angle Thermal Desorption Mass Spectroscopy 390
117. High-Temperature Thermal Desorption Spectroscopy 394
118. Scanning Kinetic Spectroscopy—A Survey Method for Investigation of Surface Reaction Processes 398
119. Temperature-Programmed Reaction Spectroscopy 404

**G. Gas Chromatography—Enhanced Sensitivity** **407**

120. Gas Chromatography—Sensitivity Enhancement 408

**H. Work Function** **411**

121. Work Function Changes Using Retarding Diode Method 412
122. Work Function Measurements Using an Electron Gun 416
123. Work Function and Electron Reflection Coefficient Measurements with the Shelton Diode 420
124. Improved Piezoelectric Drive Kelvin Probe 424
125. Improved Electromagnet Drive Kelvin Probe 428
126. Kelvin Probe—High Temperature Flow Reactor 432

**I. Radioactive Adsorbates** **435**

127. Radiotracer Techniques for Surface Studies 436

**J. Thin Film Deposition** **439**

128. Thin Film Deposition to Absolute Surface Coverages 440
129. Ionization Gauge Measurement of Film Evaporation Rate 446
130. Reusable Quartz Crystals for Film Thickness Measurement 448
131. Hybrid Thin Film Deposition Processes—Stable and Metastable Films 450
132. Sputter-Coating Inner Walls of Tubes 452

**K. Infrared Spectroscopy** **455**

133. Infrared Reflection Absorption Spectrometer 456
134. IRAS Measurements at High Pressure 462
135. Internal Reflection IR Spectroscopy—Silicon 466
136. Internal Reflection IR Spectroscopy—Gallium Arsenide 470
137. Infrared Spectroscopy—Continuous Pumping of Detector Dewar 472

**L. Calorimeteric Heats of Adsorption—Single Crystals** **475**

138. Measurement of the Calorimetric Heat of Adsorption on Ultrathin Metal Single Crystals 476

**M. Surface Debye Temperature** **479**

139. Measuring the Surface Debye Temperature of a Single Crystal 480

**N. Friction** **485**

140. UHV Tribometer for Measuring the Coefficient of Friction 486
141. Pin-on-Disk Measurement—UHV 488

# IV. Thermal Control 491

## A. Heating Samples 493

142. Electronic Temperature Programmer—Metal Crystals 494
143. Temperature Programming of Metal and Semiconductor Crystals 498
144. Mounting Nonweldable Crystals for Resistive Heating and Cooling 504
145. Strain Free Single Crystal Mounting—Azimuthal Motion 506
146. Heating Silicon Crystals and Temperature Measurements 508
147. Avoiding Extraneous Electrical Effects Owing to Heating Crystals 512
148. Electron Bombardment Crystal Heating 516
149. Indirect Sample Heating 520
150. Indirect Heating of Compound Semiconductors 524
151. Heating Design—Insulator Crystals 526
152. Cement Mounting—Semiconductor Crystals 528
153. Low-Stress Mount for Fragile Semiconductor Crystals 530
154. Radiation Heating of Crystals through Glass Windows 532
155. Pyrolytic Graphite Heating Element for UHV 536
156. Heater—Paint-on Type 538
157. Mounting High-Temperature Tungsten Filaments 540

## B. Cooling Samples 543

158. 6 K—Cryogenic Crystal Holder 544
159. Cooling Crystals by Contact with a Thermal Reservoir 548
160. Cooling Samples on Manipulators—Thermal Conductivity Issues 550
161. Vacuum-Jacketed Cryogenic Manipulator Rod 554
162. Manipulator Component with Cryogenic Azimuthal Rotation 558
163. Cooling Using Direct Contact 560
164. Cooling Reservoir—Horizontal Manipulator 562
165. Cryogenic Connections 564
166. Sample Cooling Using Thermal Braid 566
167. Flexible Cooling Reservoir 570
168. Enhanced Cooling Using Liquid Nitrogen 572
169. Level Alarm—Liquid Nitrogen 574
170. Gas Thermal Switch 576

## C. Temperature Measurements 579

171. Tungsten-Rhenium Thermocouples—Calibration Over a Wide Temperature Range 580
172. Measuring Temperatures of Hot Filaments 584
173. Pyrometric Measurement of Temperatures 588
174. Pyrometer Measurements Through Windows Coated Inadvertently with Variable-Thickness Thin Films 592
175. Calibration Point for Optical Pyrometer Used for Silicon 596
176. Control of Silicon Temperature Using Resistivity 598

V. Delivery of Adsorbates to Surfaces 601

A. Gases 603

177. Design and Performance of Microcapillary Array Beam Doser 604
178. Gas Handling System for Array Beam Doser 608
179. Calibration and Use of Array Beam Doser 610
180. Beam Doser Shutoff—Condensable Gases 614
181. Neutral Beam Intensity Measurement Using a Stagnation Detector 616
182. Low Deadspace Aperture Valve for Repetitive Gas Dosing 620
183. Gas Flow Regulation—Squeeze Valve 624

B. Evaporation Sources 627

184. Degassing Evaporation Sources 628
185. Metal Evaporation Sources for Downward Evaporation 632
186. Ultralow Coverage Metal Evaporation 636
187. Congruent Evaporation of Multicomponent Materials 638
188. Evaporator for Eutectic-Forming Metals 640
189. Simple Lithium Metal Evaporation Source 642
190. Aluminum Evaporation Sources—Long Lived 644
191. Cr Evaporator Sources of Compact Design 648
192. Electron Beam Evaporator for Refractory Materials 652
193. Small Electron Beam Evaporator for Refractory Metals 656
194. Electron Beam Evaporator 658
195. Break-Seal Ampoule Doser 660
196. Cadmium Sulfide Evaporation Source 662
197. Arsenic Atom Source 664
198. Vapor Deposition of Organic Thin Films 668

C. Ions 671

199. Solid State Cesium Ion Gun Source 672
200. Alkaline Earth Metal Ion Sources 678

D. Active Gases and Species 681

201. High Flux Atomic Hydrogen Sources—Thermal 682
202. Atomic Hydrogen Beam Source—Radio Frequency Driven 688
203. Atomic Fluorine Source 690
204. Alkali Metal Sources 694
205. Thermal Source of Methyl Free Radicals 698
206. Production, Storage, and Use of Ozone 702
207. Detection of Ozone or Atomic Oxygen by Oxidation of Silver Films 706

E. Electrochemical Sources 709

208. Electrochemical Sulfur Source 710
209. Electrochemical Halogen Sources 712

F. Gas Purification 717

210. Gas Purification for Experiments at High Pressures 718
211. Production and Purity Measurement for Extremely Pure He 722

212. Purification of Gases by Cryogenic or Gettering Methods 726
213. Gas Purification by Permeation Through Metals 730
214. $O_2$ Source—Solid State 734
215. Reusable Sample Holder for Hydrogen Permeation Studies 736

**VI. UHV Windows** **739**

**A. Spectroscopic Windows** **741**

216. Mounting Infrared Windows on UHV Systems 742
217. Windows for Vacuum UV Transmission into Ultrahigh Vacuum Systems 748
218. Construction of Be Windows for Ultrahigh Vacuum Use 752
219. Bakeable Aluminum Ultrahigh Vacuum Window 756
220. Metal Thin Foil Windows—Welding 758
221. Preparation of Large-Area Si Single Crystal Windows 760
222. Strain-Free Optical Window Mounting 764

**B. Observation Windows** **767**

223. Shielded Observation Window 768
224. Heated Internal Window for Prevention of Film Deposition 770
225. UHV Glass Window Seal—Low Optical Distortion 772
226. Zero Length Window Assembly 774

**VII. Surface Preparation Methods** **777**

**A. Cleaning Metal and Semiconductor Crystals—Examples** **779**

227. Surface Segregation and the Cleaning of Metal Single Crystals 780
228. Chemical Method to Measure and Remove Carbon from Pd 784
229. Surface Cleaning Procedures—Silicon 790

**VIII. High-Area Solids** **793**

**A. Infrared Cells** **795**

230. Infrared Cell for Adsorption Studies on Supported Catalysts 796
231. Wide-Temperature-Range IR Cell for High-Area Solids 800

**B. Adsorption/Desorption—Thermal** **807**

232. Adsorption and Desorption from High-Area Solids 808
233. Microcalorimetric Studies of Adsorption Heat on Powders 812

**IX. Safety** **815**

**A. Protecting the Vacuum System** **817**

234. Delay Circuit for Turbopumping Protection Against Vacuum Loss from Power Interruptions 818
235. Fast-Closing Beam Valve for UHV Chamber Protection 822
236. Safety System for Oil Diffusion-Pumped UHV Systems 824

237. Folding Linear Magnetic Translator 828

**B. Protecting Personnel 831**

238. Electrical Shocks in the Laboratory 832
239. Accidental Electrical Charging from Ionization Gauge 836

## Appendixes 839

**Appendix A. Useful Books—Experimental Surface Science Methods 841**

**Appendix B. Manufacturers 844**

I. United States, United Kingdom, and Canada 844
II. Germany and Europe 877
III. Japan 892

## Index 895

# I. Vacuum System Technology


A. Motion in Vacuum . . . . . . . . . . 3
B. Sample Transfer . . . . . . . . . . 65
C. Electrical Connections . . . . . . . . . . 99
D. Pumping and Trapping . . . . . . . . . . 113
E. Bakeout . . . . . . . . . . 131
F. Behavior of UHV Systems . . . . . . . . . . 135
G. Mechanical Action on Samples . . . . . . . . . . 143
H. Gasket Seals . . . . . . . . . . 149
I. Leak Repairs and Detection . . . . . . . . . . 161
J. Specialized UHV Systems . . . . . . . . . . 171

# A. Motion in Vacuum

# 1

# Crystal Manipulator with Six Degrees of Motional Freedom

This design [1] is shown in two parts to separate the azimuthal motion (Fig. 1A) device from the tilt motion device (Fig. 1B). A central rod extends from a typical *x,y,z*-plus rotary motion manipulator. The azimuthal and tilt motion (pushing and pulling) devices are connected to the same central rod and are free to slide up and down on this rod. The central rod is slotted to keep the crystal mount linking fittings from rotating with respect to the rod. The two push-and-pull actuators permit the central rod that passes through them to rotate in two ball-bearing (BB) mounts. Azimuthal crystal motion is transmitted by the push-and-pull actuator, which, through two coupled universal joints, turns the crystal mounting platform on its own ball bearing mount behind. Tilt motion is transmitted by the push-and-pull actuator, which tilts the crystal mounting platform about the axis of the hexagonal nut shown in the Tilt Motion diagram. The degrees of freedom for motion at various points are shown by double-headed arrows. This design does not result in a decoupling of tilt motion from other degrees of freedom, which is a disadvantage in some cases, but the design has been used for more than ten years with success. A more recent design uses a circular slot to guide a pin in the crystal tilt mechanism that eliminates the crystal translational motion associated with the tilt motion. It is of interest to note that in this design, ordinary unlubricated ball bearings are used in ultrahigh vacuum with success and without seizure or binding.

The crystal is mounted on electrical and thermal isolation standoffs on the crystal mounting platform and may be heated by electron bombardment or by ohmic heating. Various thermocouples may be connected to the crystal, and cooling to about 40 K is achieved using a copper braid connected between the crystal mounting assembly and a liquid He cryostat. Appropriate provisions are made to shield the mounting assembly from emitted radiation generated during crystal heating so that the desorption of impurities from the mounting is minimized.

Because of the complexity of the design of a device of this sort, it is recommended that a working model first be constructed before final design parameters are set.

Many other designs allowing cooled crystals to be azimuthally rotated in ultrahigh vacuum have been published [2-12] and some are treated in this book.

## *References*


[1] Professor Peter Toennies, Max Planck Institut Stroemungsf, Bunsenstr. 20, D-37073 Göttingen, Germany (private communication).

[2] D. Venus, Rev. Sci. Instrum. **62,** 1361 (1991).
[3] Y. Dai, H. Li, and F. Jona, Rev. Sci. Instrum. **61,** 1724 (1990).
[4] T. Engel, D. Braid, and E.H. Conrad, Rev. Sci. Instrum. **57,** 487 (1986).
[5] J.J. Zinck and W.H. Weinberg, Rev. Sci. Instrum. **56,** 1285 (1985).
[6] K.D. Jamison and F.B. Dunning, Rev. Sci. Instrum. **55,** 1509 (1984).
[7] N.J. Wu and A. Ignatiev, Rev. Sci. Instrum. **56,** 752 (1985).
[8] J.A. Stroscio and W. Ho, Rev. Sci. Instrum. **55,** 1672 (1984).
[9] J. Larscheid and J. Kirschner, Rev. Sci. Instrum. **49,** 1486 (1978).
[10] K.D. Jamison and F.B. Dunning, Rev. Sci. Instrum. **55,** 1509 (1984).
[11] H. Durr, Th. Fauster, and R. Schneider, J. Vac. Sci. Technol. **A8,** 145 (1990).
[12] M. Michaud, P. Cloutier, and L. Sanche, Rev. Sci. Instrum. **66,** 2661 (1995).

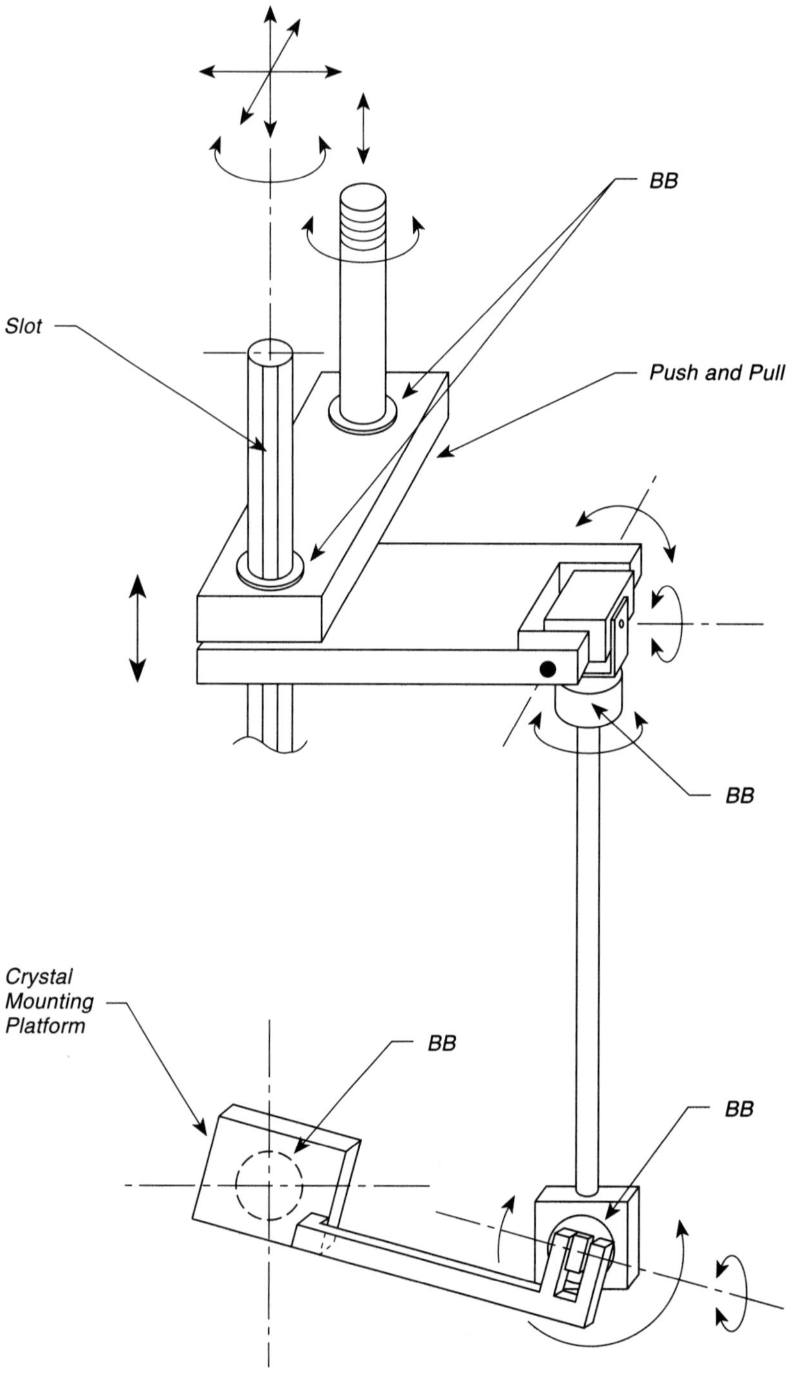


**Fig. 1A. Azimuthal Motion**

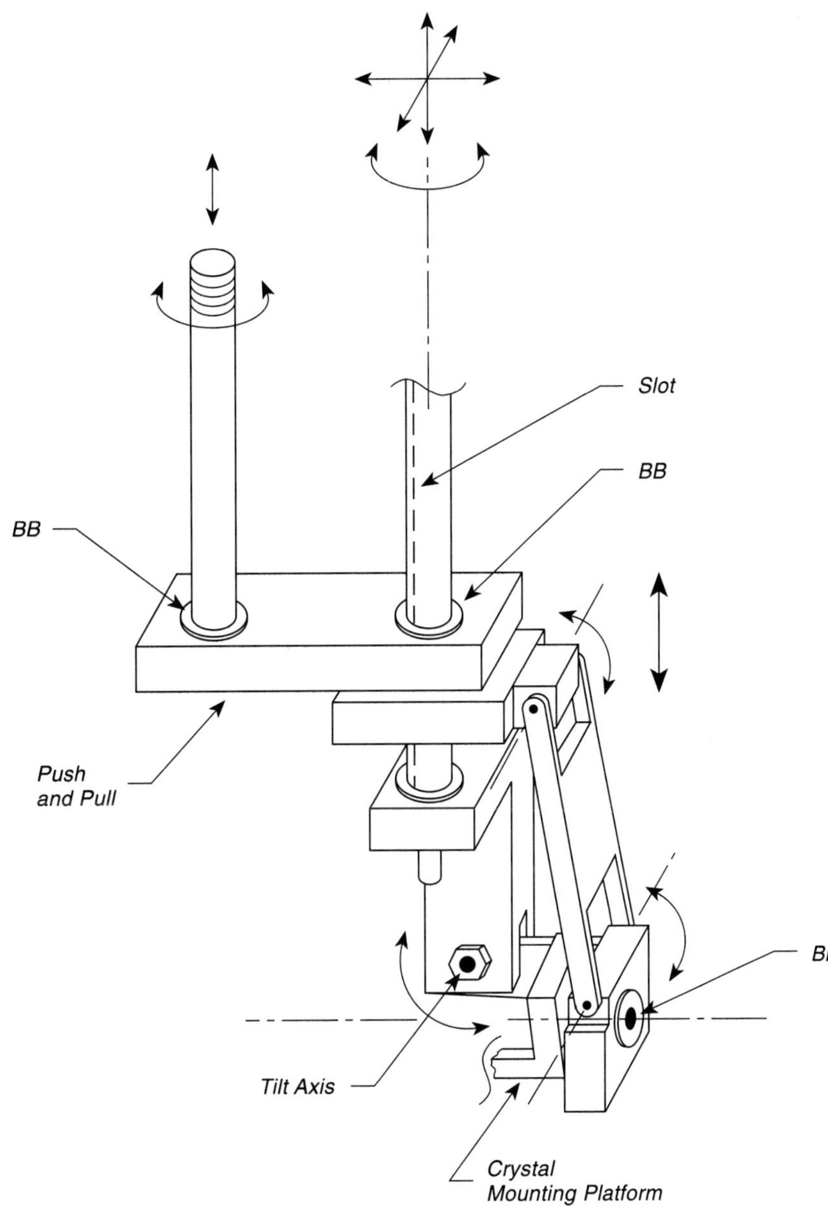


***Fig. 1B. Tilt Motion***

# 2

# Rotation in Ultrahigh Vacuum

The provision for rotation of samples in ultrahigh vacuum (UHV) is centrally important in most investigations, since the analytical facilities in most research chambers are located on different azimuths from the center line of the vacuum chamber. It is often desired to provide other motions, as well as heating and cooling of the crystal, but this section is concerned only with axial rotation. Historically, the problem was addressed in the first sixty inch cyclotron by Wilson [1], long before ultrahigh vacuum was achieved in metal systems. It is of interest to note that the differential pumping method employed by Wilson is still fundamentally important to UHV technology.

Figure 2A shows the Wilson seal as currently used in ultrahigh vacuum to rotate a long cylindrical cryostat on which the crystal is mounted using ceramic feedthroughs [2]. Two Viton gaskets, with a hole cut smaller than the rotary seal tube, are held in place in a brass housing that bolts onto a specially shaped lower Conflat flange having an upward-facing conical section to push the Viton gasket upward as shown. An assembly plate, below the upper Viton seal, also has an upward-directed conical support section. These two Viton seals make a strained contact with the polished-mirror-finished (600 grit) outer portion of the rotary seal, and provide the vacuum tight joint. To prevent significant leakage through the lower Viton seal gasket, the region between the two gaskets is differentially pumped with a rotary pump or a small ion pump. The cryostat tube, welded to a mini-Conflat flange is bolted to a mini-Conflat knife-edge sealing surface on top of the rotary seal. Two stabilizer bearings prevent wobbling of the central components; these may be made of phosphor bronze [2], Teflon, or they may even be ball bearings. Wires for heating the sample electrically, and thermocouple wires conveniently thread down the cryostat tube and exit from the bottom. As is obvious from the drawing, the bottom Conflat flange must be bolted on from below, meaning that a flanged extension tube from the chamber flange must be employed under the device. A handle added to the flange of the rotary seal (not shown) is helpful in applying torque, and the upper section of the brass housing is divided into degrees of arc for accurate angular positioning, using the pointer. Bakeout temperatures of 150°C are nondamaging to the Viton seals. For higher temperature bakeout, KALREZ perfluoro elastomer might be used [3]. The gaskets can most conveniently be made by use of a homemade punching jig to accurately locate and cut the center hole to size as well as to place the six radial holes through which pass the bolts clamping the assembly plates down on the Viton. A pressure rise of about $1 \times 10^{-10}$ Torr was observed on slow rotation.

The seal in Figure 2A is designed for liquid nitrogen cooling, and in order to achieve long cryogenic storage time, a 1-L storage Dewar with a hole in its bottom was fabricated [4]. To make a liquid nitrogen seal each day with the cryostat tube, a platform support for the Dewar is covered with some cotton wool that is saturated with water; sufficient ice forms to make an excellent seal. Ice formation under the support platform by condensation from the atmosphere is a problem that can be reduced by using a heater tape around the brass housing and possibly a fan to evaporate liquid water runoff. The apparatus can be inefficiently used with the gaseous discharge from a liquid He Dewar to achieve temperatures near 30 K, and the momentary transfer of liquid He brings the temperature to 15 K [2,4]. Another rotary seal based on this principle has been reported [5].

Figure 2B shows a second design for a rotary seal in which O-rings are employed as the sealing elements, with one stage of differential pumping between the O-rings [6]. A polished rotary-seal tube rotates against the slightly compressed Viton O-rings, which are held in place by tightening the flange on the compression pieces, which are made of copper and which slide in between the outer wall and the rotary seal. These two compression pieces contain a number of holes to provide a pumping opportunity between the inside and outside of the cylindrical pieces. The O-rings are lubricated with a trace of graphite powder. A different lubricant reported to be satisfactory for Viton O-rings consists of a mixture of graphite powder and Santovac 5 diffusion pump oil (a polyfluoroether) [7]. Differential pumping with a rotary pump or a small ion pump is used for the inner section holding the lower compression piece. A Teflon spacer at the top supplies a bearing surface for the rotary seal. Inside the rotary seal, a cryostat tube similar to that in Figure 2A may be inserted. A pressure rise of about $3 \times 10^{-11}$ Torr was observed on rotation [6].

Rotation using large differentially pumped seals is often employed to move heavy instruments inside the vacuum system, with a large-diameter rotating plate being moved through 360°. Auerbach [8] and his colleagues first reported the use of spring-loaded Teflon seals in this role in ultrahigh vacuum systems. Ordinary Teflon has a high thermal expansion coefficient, poor creep characteristics, and high permeability in static systems, so that differential pumping is required, along with a mechanical trick in which a spring is embedded inside the Teflon ring [9] to compensate for creep. These seals can also be obtained with Teflon filled with $MoS_2$ lubricant particles for ultimate lubrication of an already lubricious material [8]. However, the filled polymer has a higher gas permeability than pure Teflon. In Figure 2C, a diagram of a double-stage differentially pumped Teflon seal is shown [8]. The large-diameter rotating plate is centered with two ball bearings, which have been split at the inner and outer races to prevent distortion and binding owing to differences in expansion coefficient between the bearings and the stainless steel. Beneath the ball bearings are three spring-loaded Teflon seals. The space between the upper two seals is pumped with a rotary pump, while a small ion pump [11 L/s] pumps the space above the inner seal and the center seal. The system is baked to 200°C, and the pressure rise upon rotation is less than $10^{-10}$ Torr and usually unobservable. The pressure at the small ion pump is in the $10^{-8}$ Torr range. The estimated gas leakage under static conditions is $10^{-13}$ Torr L/s [8]. Seals based on this principle are now commercially available. Others have reported the use of this type of seal for rotary manipulators [10–12].

Sometimes rapid rotation in ultrahigh vacuum is desired, where rather massive components must turn in bearings. Figure 2D shows a magnetically driven rotation device that drives an inner cylinder suspended in vacuum. Two vacuum ball bearings are used at either end of the inner cylinder. These are ordinary ball bearings that are unlubricated, and they contain large cross section ball races. These bearings are routinely baked at 220°C, and proved satisfactory for 18 months, having undergone $5 \times 10^6$ rotations without any degradation in performance. The evolution of gases is proportional to the rotation speed as shown in the lower part of Figure 2D, and the rise in total pressure is about $2 \times 10^{-11}$ mbar at 100 rpm. Most of the gas evolution was due to $H_2O$ and $H_2$ liberation [13].

Other designs for achieving independent rotational motions around two axes are described in the literature [14-16]. Reference [14] describes an electrical readout for the angular motions accurate to 0.5° and based on motion of a wire-wound potentiometer.

## *References*


[1] R.R. Wilson, Rev. Sci. Inst. **12,** 91 (1941).
[2] V.R. Unwin, K. Horn, and P. Geng, Vakuum Tek **29,** 149 (1980).
[3] M.F. Zabielski, and P.R. Blaszuk, J. Vac. Sci. Tech. **13,** 644 (1976).
[4] J.T. Yates, Jr. (unpublished).
[5] J.J. Zinck and W.H. Weinberg, Rev. Sci. Inst. **56,** 1285 (1985).
[6] A. Pararas, S.T. Ceyer, and J.T. Yates, Jr. J. Vac. Sci. Tech. **21,** 1031 (1982).
[7] E. Puckrin, J.K. Fowler, and A.J. Slavin, J. Vac. Sci. Tech. **A7,** 2818 (1989).
[8] D.J. Auerbach, C.A. Becker, J.P. Cowin, and L. Wharton, Rev. Sci. Inst. **49,** 1518 (1978).
[9] The spring-loaded Teflon seals are available in many sizes from the Fluorocarbon Corporation, 10871 Kyle Street, P.O. Box 520, Los Alamitos, CA 90720. Their descriptive literature gives many details about the seals and the preparation of the metal surfaces on which they ride.
[10] S.M. George, J. Vac. Sci. Tech. **A4,** 2394 (1986).
[11] J.A. Stroscio and W. Ho, Rev. Sci. Inst. **55,** 1672 (1984).
[12] A particularly nice design of a manipulator with an easily demountable cold finger end for crystal replacement can be found in P. A. Thiel and J.W. Anderegg, Rev. Sci. Instrum. **55,** 1669 (1984).
[13] R.A. Kubiak, P. Stonestreet, S.M. Newstead, E.H.C. Parker, T. Whall, and N. Naylor, J. Vac. Sci. Technol. **A9,** 2797 (1991).
[14] J. Larscheid and J. Kirschner, Rev. Sci. Instrum. **49,** 1486 (1978).
[15] K.D. Jamison and F.B. Dunning, Rev. Sci. Instrum. **55,** 1509 (1984).
[16] H. Durr, Th. Fauster, and R. Schneider, J. Vac. Sci. Technol. **A8,** 145 (1990).

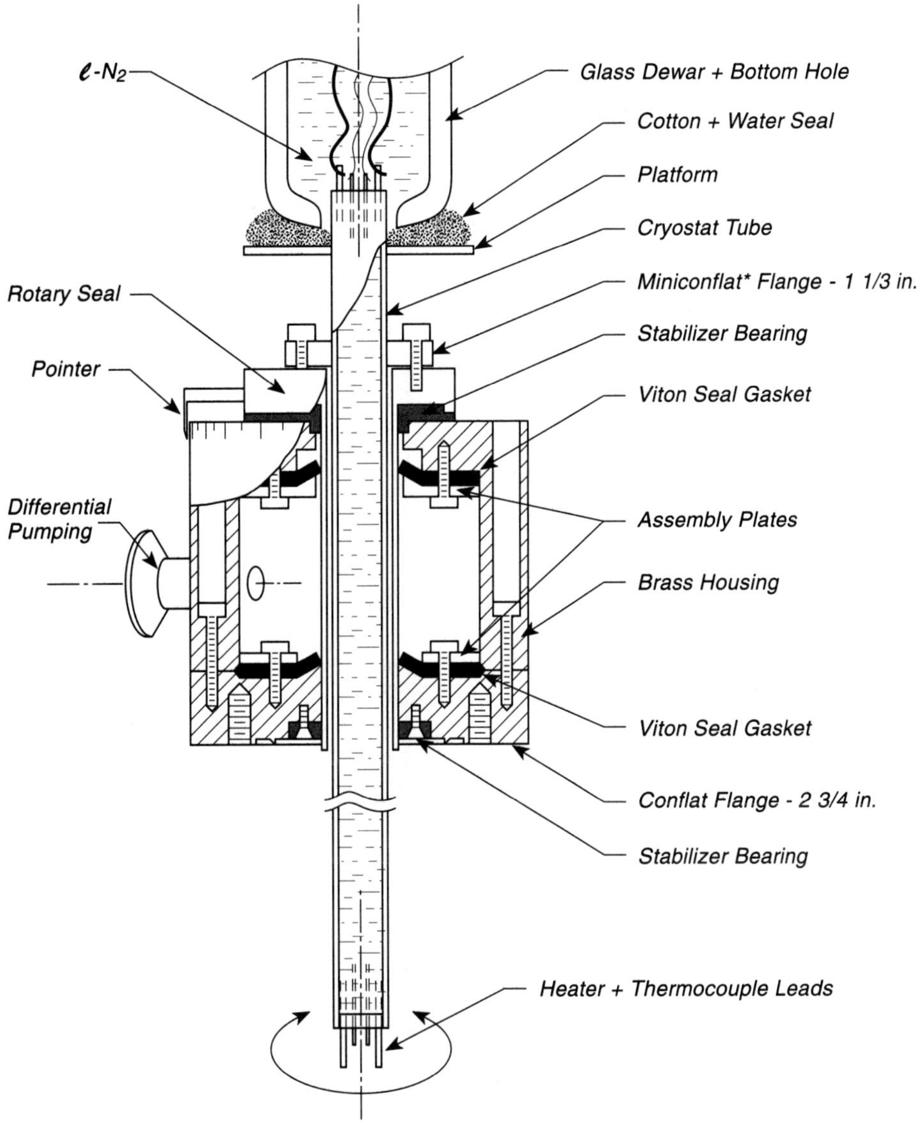


***Fig. 2A. Differentially Pumped Rotary Seal—Flat Viton Gaskets***

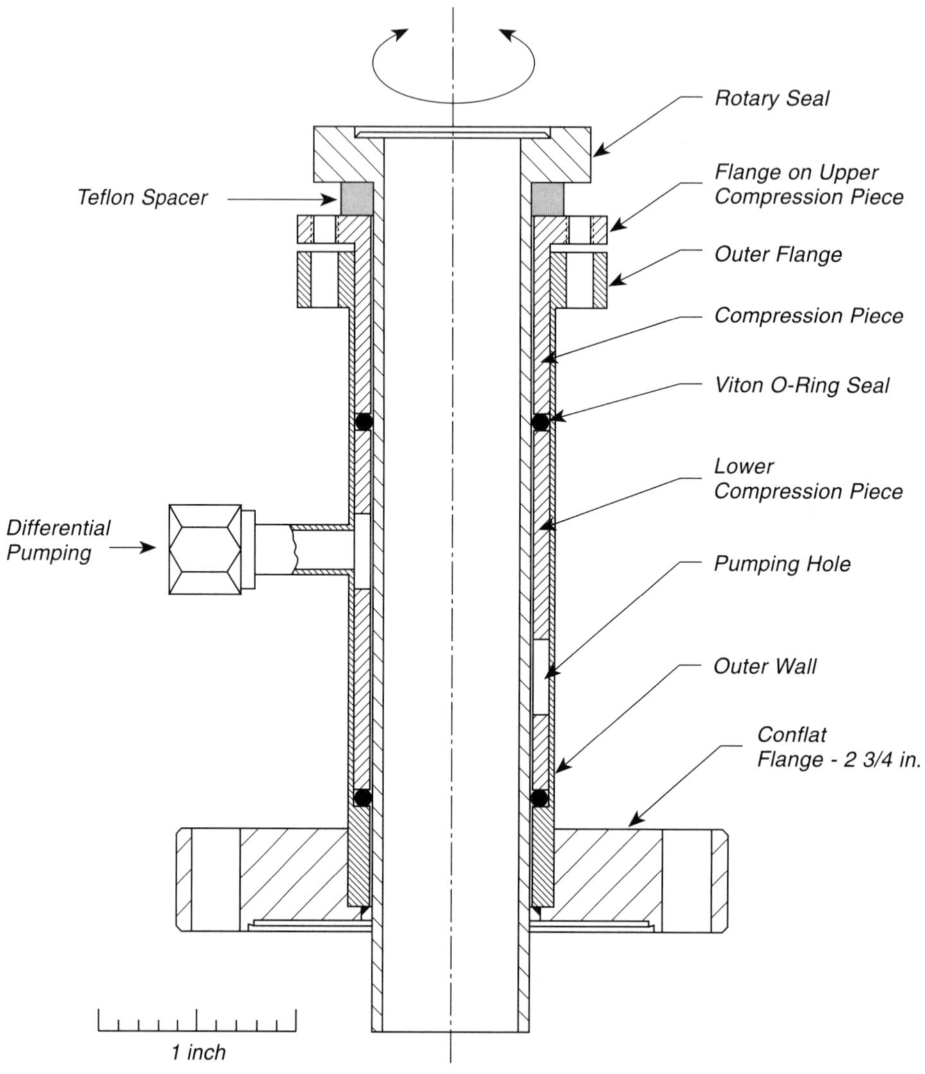


***Fig. 2B. Differentially Pumped Rotary Seal — Viton O-Ring***

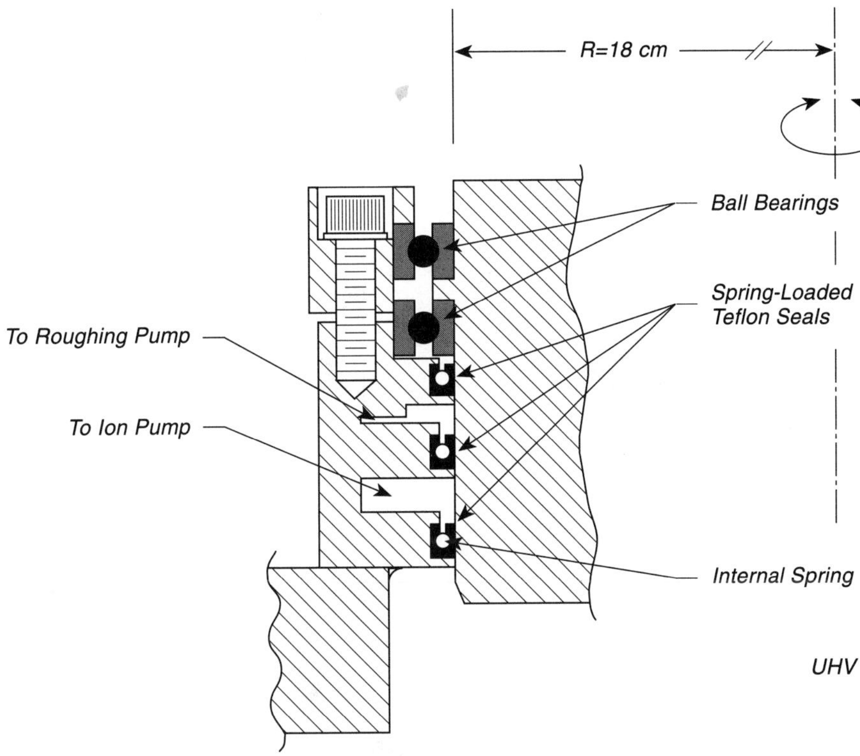


***Fig. 2C. Differentially Pumped Rotary Seal—Spring-Loaded Teflon Seals***

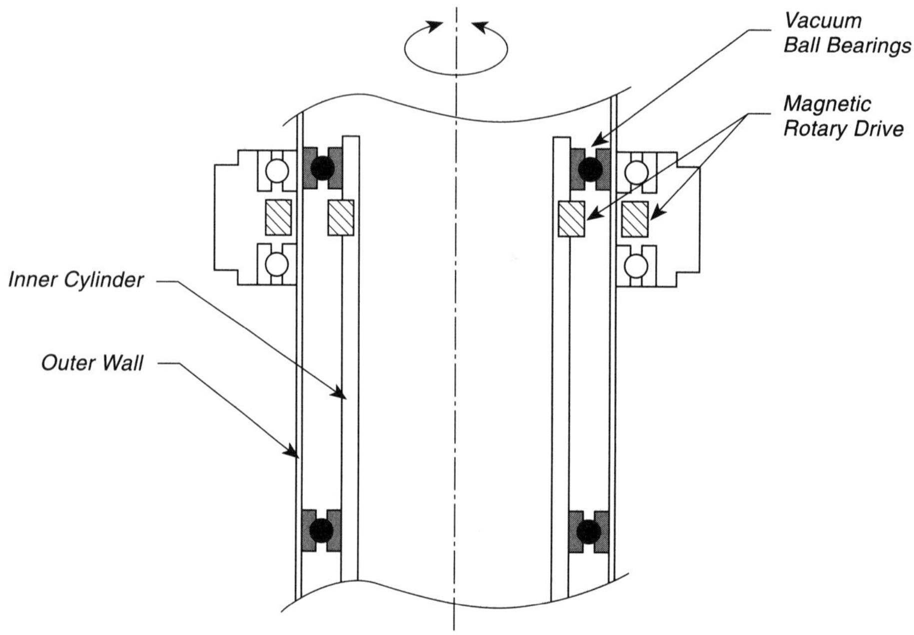


**Fig. 2D. Internal Ball Bearing**

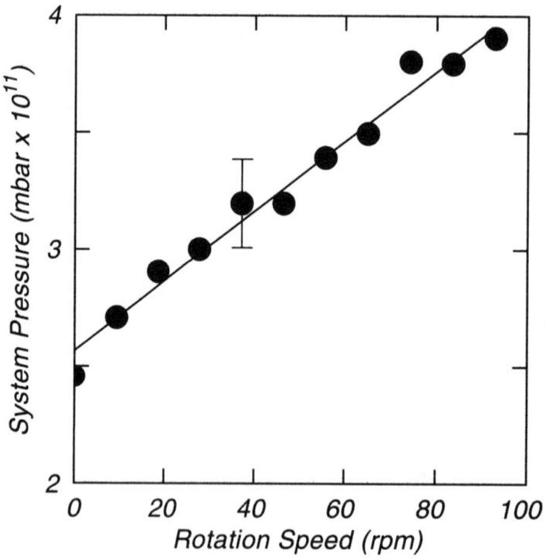

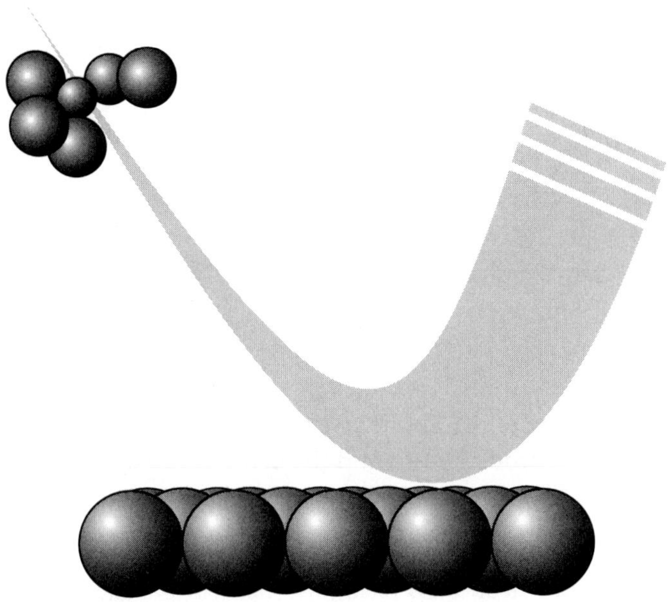

# 3

# Crystal Positioning Using Capacitance

The ability to position a single crystal precisely can be useful in many measurement procedures in ultrahigh vacuum. This is a problem when the positioning must be done at accuracies beyond the capability of micrometer drives on manipulators. Factors such as thermal expansion effects, backlash in precision screw drives, and vibration interfere with accurate positioning methods.

The topic of capacitive positioning of a single crystal is treated in the context of locating the crystal accurately in front of a mass spectrometer that is housed in an apertured shield for accurate and reproducible temperature programmed desorption measurements [1]. This connects to the topic on page 366.

Figure 3A shows a glass shield [1,2] mounted over a quadrupole mass spectrometer ionization source. A Ta ring is rigidly supported near the aperture of the shield and acts as the capacitive pickup terminal. The capacitance is of the order of 0.1 pF for a 1-mm spacing between the Ta ring and the crystal. The capacitance measuring circuit, to be described below, can measure capacitance changes in the range of 0.1%, meaning that the crystal can be positioned to 1 μm reproducibility, corresponding to an error in the measurement of the desorption rate from the crystal of only about 0.05%. Small tilting of the crystal will reduce this reproducibility, and the demonstrated reproducibility of actual measurements was better than 1% [2].

Figure 3B shows a slight variation on the design. Here, an Ag ring is employed as the capacitance pickup terminal, and it is fixed on the end of the glass shield as shown. The Ag plate, (12 mm OD; 6 mm ID, 1 mm thick) is held in place by use of 3 Ta wires that are spotwelded to the plate and wound around three glass eyelets on the shield. The Ag plate is connected through a Kapton-insulated and shielded lead to a BNC connector that exits the vacuum system. To avoid static charging, the glass shield is surrounded by a very thin grounded metal mesh [3]. An Aquadag [4] (colloidal graphite) coating can also be used for this purpose, or a metal shield like that shown on page 366 may be used. The spacing employed between the crystal and the aperture is reported to be only 10 μm in this design, and this may be reproduced to 1 μm [3].

The capacitance measuring circuit [2,3] is shown in schematic block diagram form in Figure 3C. It is a phase-shifted bridge circuit, working at 20 kHz [2]. A sinusoidal signal (20 Vpp) is applied to the crystal and the capacitive pickup is made to the ring aperture, since the crystal is often connected to unshielded leads and electronics outside of the vacuum system. The detection system is used as a simple lock-in amplifier.

## *References*

[1] P. Feulner and D. Menzel, J. Vac. Sci. Technol. **17**, 662 (1980).

[2] H. Schlichting and D. Menzel, Surface Sci. **285**, 209 (1993).

[3] H. Schlichting (private communication).

[4] Obtainable from Acheson Colloids, 1607 Washington Avenue, P.O. Drawer 611747, Port Huron, MI 48061-1747.

**Fig. 3. Crystal Positioning Using Capacitance**

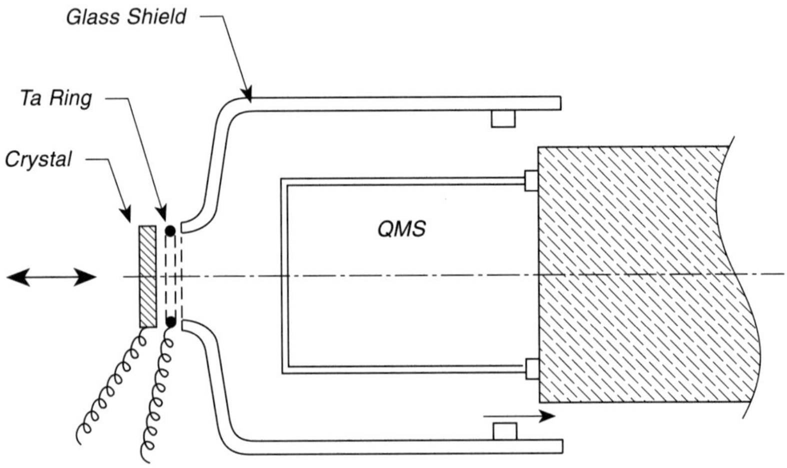


**A. Ring Capacitance**

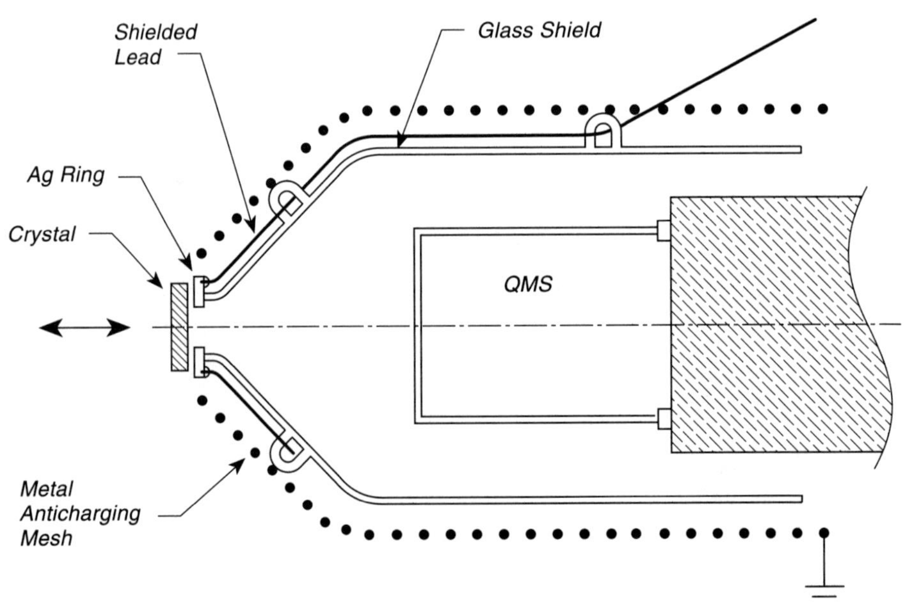


**B. Affixed Capacitance**

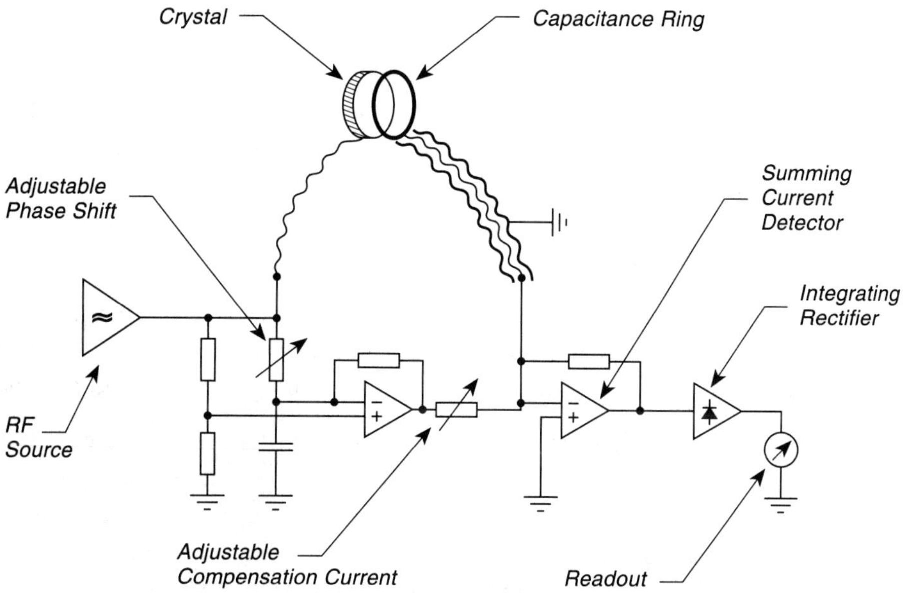


***C. Capacitance Measurement—Block Diagram***

# 4

# UHV Rotary Manipulator with Arcsec Resolution

In some cases, the use of a commercial crystal manipulator is inappropriate because of the poor resolution and reproducibility inherent in these designs. For certain cases, such as in X-ray standing wave spectroscopy (XSW), a high-precision rotary stage is needed [1].

Figure 4A shows a side view of an ultrahigh vacuum system containing both a coarse and a fine adjustment for rotating a single crystal about the horizontal axis. The crystal holder is mounted on a platform such that its surface is in a plane that contains the axis of the shaft that holds the crystal holder. This shaft, made of precision-ground 304 stainless steel, rotates and translates inside of two linear/rotary ball bearings that are mounted on the top flange. With a conventional rotary feedthrough and rotary manipulator contacted to the shaft by a U-joint, coarse sample manipulation in 7 arcsec increments can be achieved with a stepping motor/speed reducer.

Figure 4B shows the principle of the high-precision rotary device, where the shaft may be engaged by a lever, and where precise variations in the position of the top of the shaft may be used to make azimuthal rotations to better than arcsec accuracy and precision.

Figure 4C shows the detailed structure of the high-precision rotary device coupled to the upper flange by a bellows. The rotary feedthrough drives an actuating lever that pushes a pushrod and the lower clutch body against two springs. When the force on the pushrod is released, the upper and lower clutch bodies grab the shaft. The rotation of the shaft about its axis is determined by the setting of the upper micrometer which is driven by a stepping motor. A weight is used to maintain tension and contact between the upper micrometer and the movable flange, and incremental motion of 0.88 arcsec is achieved with a lever arm of 292 mm length and a micrometer displacement of $1.25 \times 10^{-3}$ mm [1].

## *References*


[1] T. Gog and S.M. Durbin, Rev. Sci. Instrum. **60,** 3030 (1989).

**Fig. 4. Rotary Motion—Arcsec Resolution**

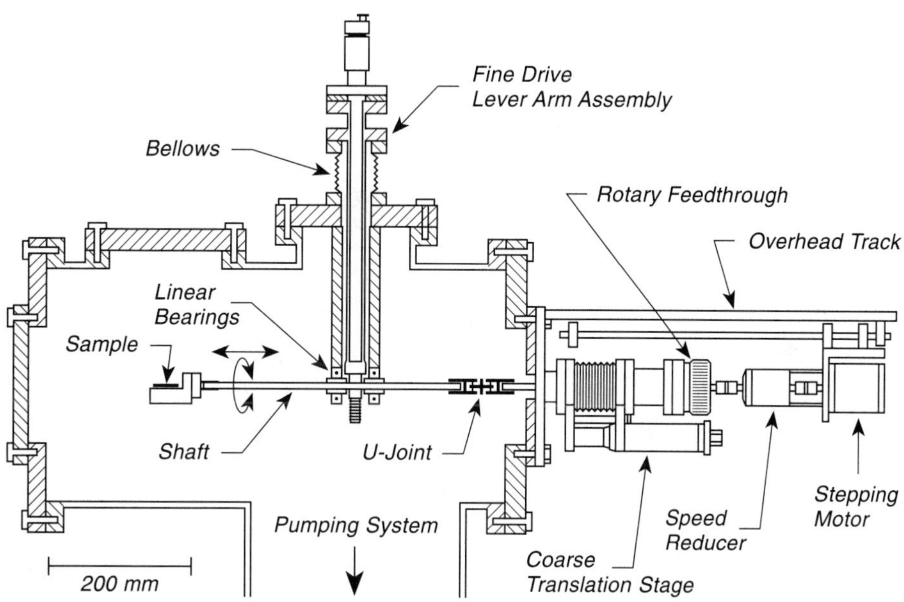


**A. Side View—Coarse and Fine Adjustments**

**Fig. 4. Rotary Motion—Arcsec Resolution**

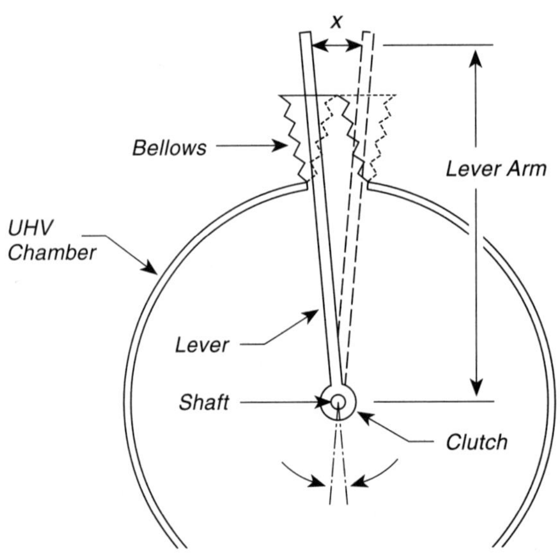


**B. Schematic - Fine Angular Adjustment**

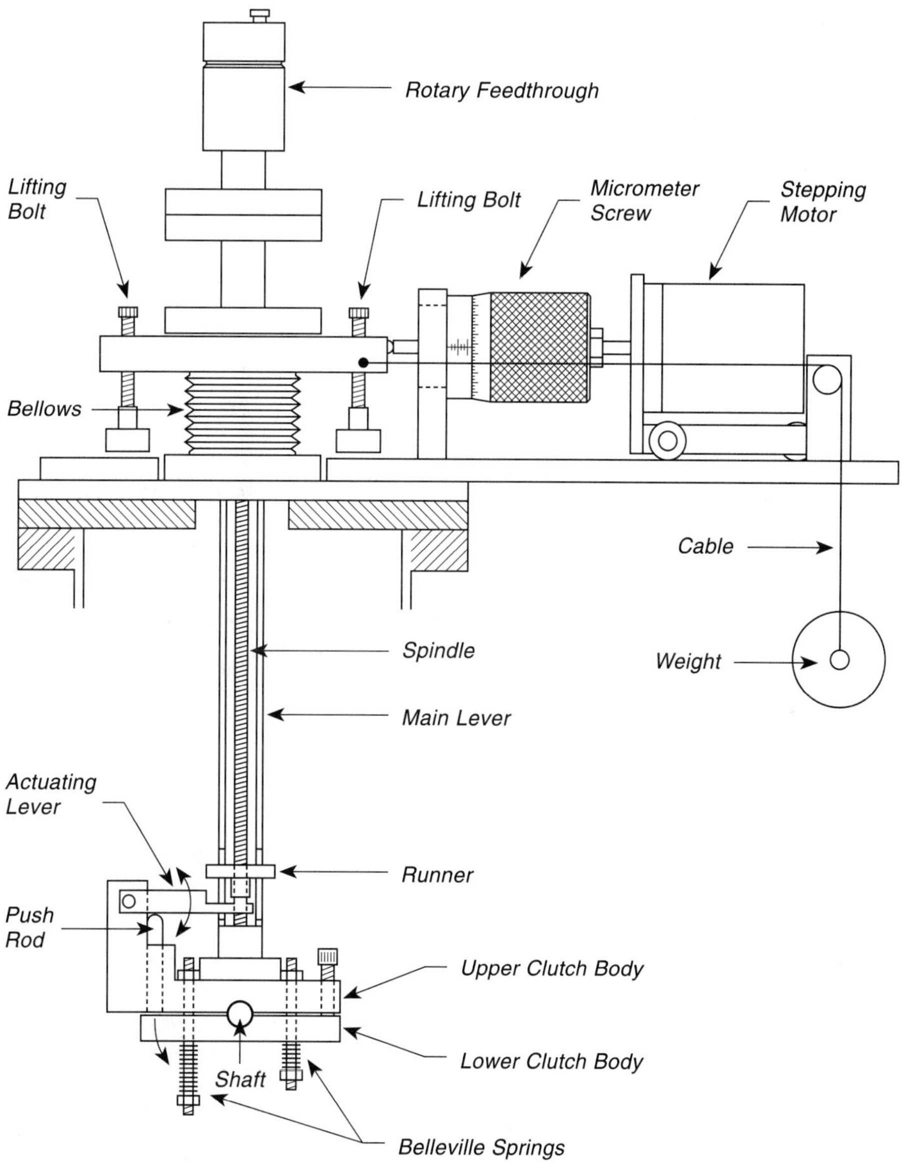


***C. Detail—Fine Angular Adjustment***

# 5

# Solenoid Driven Linear Motion Device

A rapid linear motion inside an ultrahigh vacuum system may be achieved by using a solenoid-power-driven actuator like that shown in Figure 5. All moving parts are inside the vacuum system, and the external solenoids do not require vacuum encapsulation. The actuator is a barrel-shaped plunger made of high-purity iron sized to make a loose fit inside the stainless steel tube. In this device a stroke of 3 cm was achieved. Each solenoid consists of 1400 turns of #24 magnet wire wound on an aluminum spool and giving a dc resistance of 14 ohms. Short electrical pulses are used on either solenoid to move the actuator in or out; the current is then turned off to prevent overheating or stray magnetic fields. In operation, the actuator will tilt as shown in the drawing when it moves inward. Measurements made on the actuator show that a torque of 500 g cm can be achieved at the internal pivot point. The closing time was measured to be 50 ms. The original paper [1] also provides a design for either an automatic or manual actuator circuit.

The solenoids are removed for system bakeout by releasing the lock ring on the back end. A similar magnetically driven shutter assembly using a Nd Fe-B magnetic encapsulated slug and an external ceramic-ring magnet has also been described [2].

## *References*


[1] R.W. Bertram, J. Vac. Sci. Tech. 7, 612 (1970).

[2] C.M. Rouleau and R.M. Park, J. Vac. Sci. Tech. **A11**, 464 (1993).

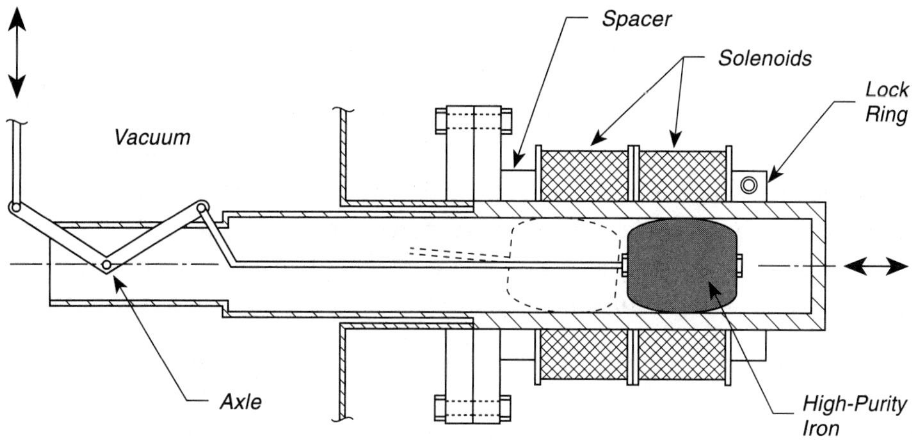


***Fig. 5. Linear Motion Actuator***

# 6

# An Inexpensive Linear Motion Device

A homemade linear motion device may be constructed quite easily from an inexpensive machinist's device as shown in Figure 6. The vise is made of high-grade gray iron and the jaws as received from the manufacturer are precision ground for squareness and parallelism to ±0.0001 cm per cm. The hardened jaw pieces are removed from the vise and replaced with aluminum clamps that have an inner lip machined in them to prevent the Conflat flanges from slipping through. These clamps are mounted on the vise and then tightened around the flanges holding the welded bellows. Machinist's vises giving up to 7.5 cm linear travel are available [1] at a cost of about $200–$300, and this device, while heavy, gives smooth and repeatable translation. The precise position can be determined using a vernier caliper between the clamps, or repeatable positions can be established by using simple spacer bars. The overall price is about a quarter that of a comparable commercial bellows manipulator [2].

## *References*


[1] MSC, 151 Sunnyside Boulevard, Plainview, Long Island, New York, 11803.

[2] H. Gutleben and J.T. Yates, Jr., J. Vac.Sci. Tech. **A9**, 170 (1991).

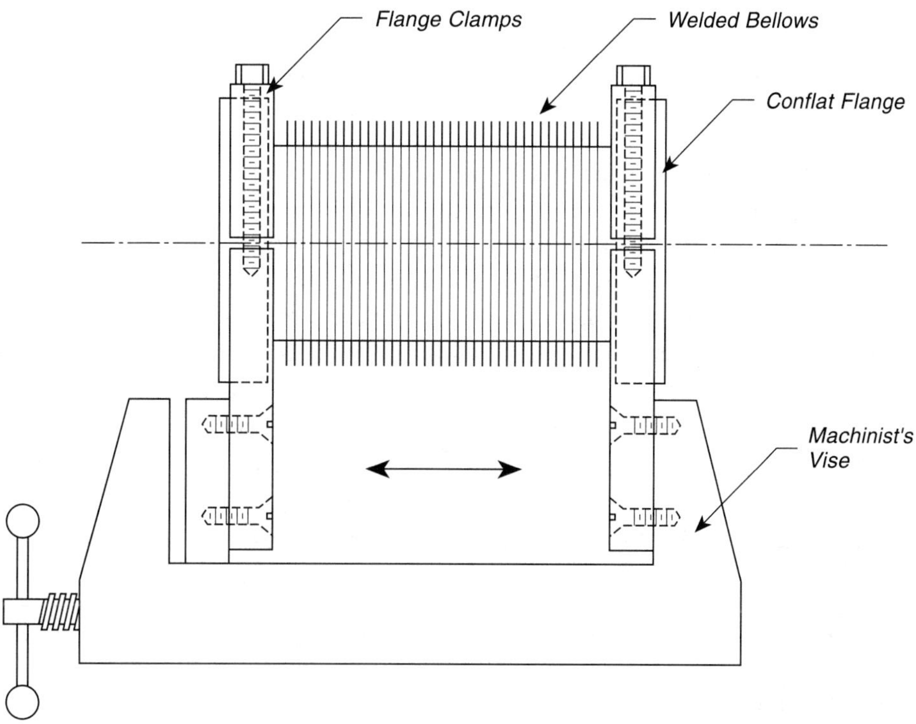


***Fig. 6. Linear Translation Device***

# 7

# Translating Auger Spectrometer

It is sometimes appropriate to move an Auger spectrometer or other large instrument over a large distance. This may be done in a conventional fashion by using a large-bellows translation device, which is both heavy and expensive. An alternate method for achieving the same result is to employ internal linear ball bearings and to translate with a small bellows-type linear translator. This requires that the electrical connections to the instrument be flexible over rather long extension distances.

Figure 7 shows a design [1] for translating a cylindrical mirror analyzer (CMA) Auger spectrometer over a 14-cm distance. This is done using a 1/2-in.-diameter linear ball bearing [2] that slides on a polished stainless support rod at the top, and is guided along a 1/4-in.-diameter rod-linear bearing assembly on the bottom. The use of the smaller rod on the bottom allows the system to have a little flexibility to prevent seizure due to slight misalignment. The first upper support bushing is clamped with a stainless steel ring to the μ-metal shield of the CMA spectrometer using three set screws at 120° to each other. The second support bearing is attached to a baseplate that rigidly supports the CMA by rods. The entire assembly is mounted on an 8-in.-diameter Conflat flange. Translation is achieved using a small linear translator, mounted on a small flange, which pushes and pulls on the baseplate at a point close to the upper support rod to avoid torque on the upper bushing during motion.

The flexible electrical connecting wires are looped to give the needed extension and are made of Teflon-coated coaxial cable [3]. The coaxial cable is made UHV compatible by stripping off the outer cover, leaving the cable shield, and contributes no significant outgassing problem.

Another design for a translating Auger spectrometer may be found in reference [4]. In this case, the miniature instrument (electron optics 30 mm diameter; 45 mm length) is held on a single linear translator mounted on a 2 3/4-in.-diameter Conflat flange. The resolution of this instrument is $R = \Delta E/E_o = 1.5\%$ [4].

## *References*


[1] M.C. Reuter and R.M. Tromp, J. Vac. Sci. Technol. **A6,** 2575 (1988).

[2] 1/2 in. -Thompson A-81420-SS; 1/4 in. - Thompson A-4812-SS.

[3] Custom Cable Corp., 242 Butler Street, P.O. 1050, Westbury, NY 11590-3193. Cable type: RG 316/V.

[4] D.N. McIlroy, P.A. Dowben, A. Knop, and E. Rühl, J. Vac. Sci. Technol. **B13,** 2142 (1995).

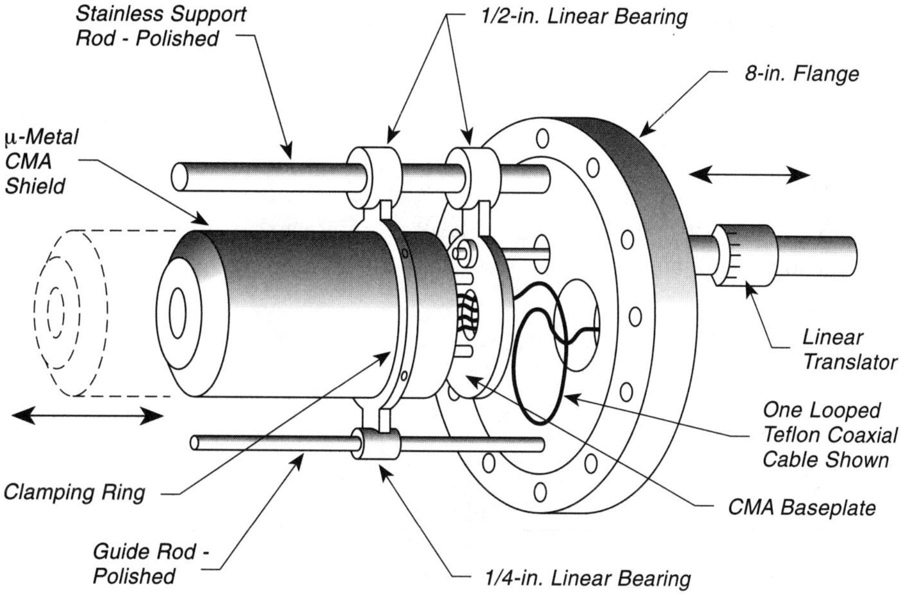


**Fig. 7. Translating CMA Auger Spectrometer**

# 8

# Simple Multiple Motion Manipulator

The provision of various degrees of motional freedom for samples, apertures, etc., in ultrahigh vacuum systems is often accomplished with complex manipulators such as that shown on page 4. When complex motion is required and the measurement of the position of the sample is not required, more simple devices are possible. In Figure 8, a device involving the combination of a port aligner and a linear translator is used to give multiple degrees of freedom for a rod. The port aligner allows $\pm 10^\circ$ motion within a solid cone, which is schematically shaded. The linear translator permits motion back and forth within this solid cone [1].

## *References*

[1] Dr. Oliver Schaff, Fritz Haber Institut der Max Planck Gessellschaft, Faraday Weg 4-6, D-14195 Berlin (private communication).

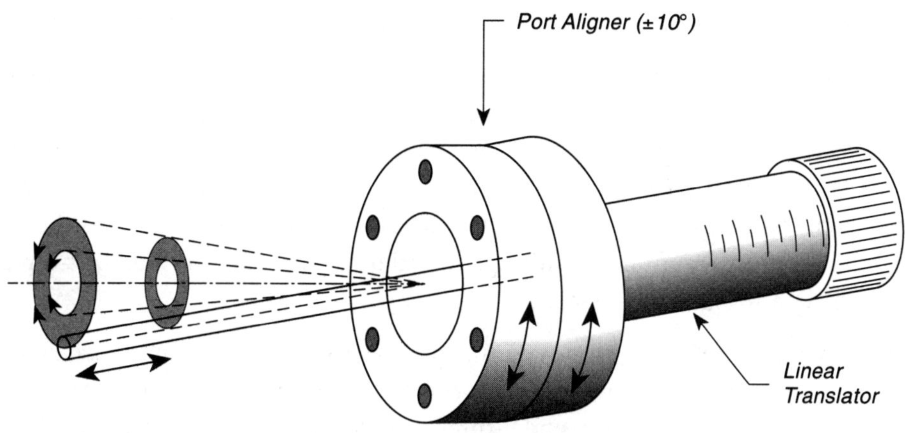


Fig. 8. Simple Multiple Motion

# 9

# Simple Device for Small Rotations in Ultrahigh Vacuum—Grating Application

The rotation of gratings and other optical components through small angles in ultrahigh vacuum for the purpose of alignment is not straightforward. This device permits the adjustment of a grating through an angular range of less than 1° with ease [1]. Adjustments in one direction or in two are possible, depending on the structure used.

Figure 9A shows a side view of a device permitting one-dimensional rotation about the *x*-axis. Two planar broad-leaf springs are connected to a base support, and to a movable platform. A plane extrapolation of the two springs intercepts at the origin, *o*, which is the center of the grating. Here the *x*-, *y*-, and *z*-axes are defined. An adjustment rod is connected to the base support, and to an external adjustment screw. Bellows, as shown, isolate the mounting system from the atmosphere. Adjustment of the screw in the *z*-direction rotates the grating about the *x*-axis.

The principle can be extended to rotation about two axes, as shown in the schematic (Fig. 9B). Here, two sets of orthogonal leaf springs are used, and screws for rotation of the adjustment rod operate in the *z*- and *x*-directions, leading to grating rotation about the x- and z-axis.

Figure 9C shows the use of the two-axis device for optimization of the resolution of a UHV spectrometer. In this case, an angular resolution of less than 0.1° was achieved for optimum spectrometer resolution adjustment [1].

## *References*

[1] W.A.M. van Bers, A.J.J. Franken, and P.K. Larsen, Rev. Sci. Instrum. **54**, 637 (1983).

**Fig. 9. Small Rotations in UHV**

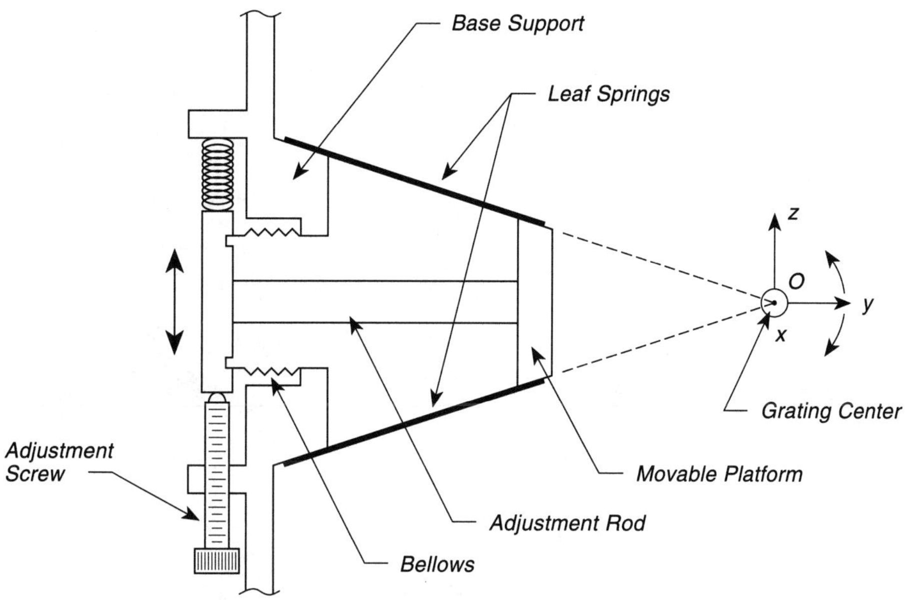


**A. One-Axis Device**

***Fig. 9. Small Rotations in UHV***

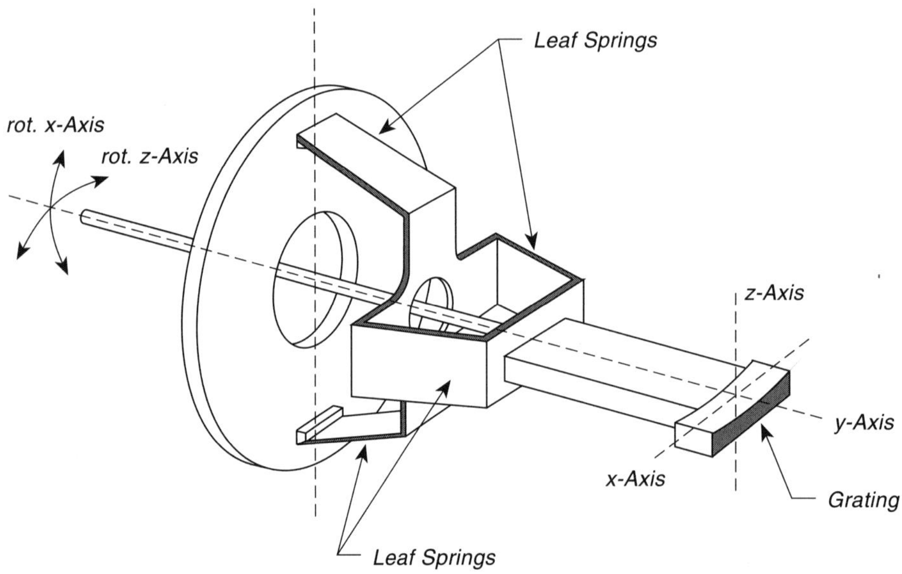


***B. Two-Axis Device***

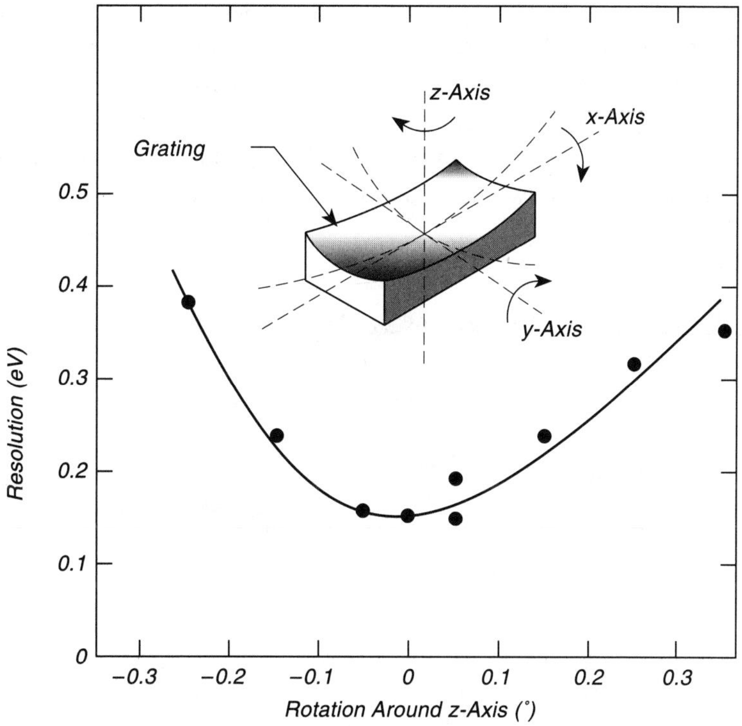


**C. Result—z Rotation**

# 10

# Turntable Rotation in Ultrahigh Vacuum

Figure 10 shows a turntable for the rotation of heavy equipment in UHVwith high angular precision and repeatability, and without the need for coated or lubricated bearings. Rotation is achieved using a rotary feedthrough to move the axle of the turntable. The flat bottom of the turntable rests on three spheres made of a metal carbide mounted in a V-groove. The spheres are of 10-mm diameter and the three define the plane on which the turnable rotates. The mechanical stability is excellent, and several tens of kilograms may be supported. In addition, friction is low in ultrahigh vacuum. The danger of the spheres coming close to each other is small, especially when forward and backward motion is used alternatively. The system is uninfluenced by bakeouts at 200°C [1].

## *References*

[1] W. Stocker and K.H. Rieder, J. Vac. Sci. and Tech. **A2,** 1606 (1984).

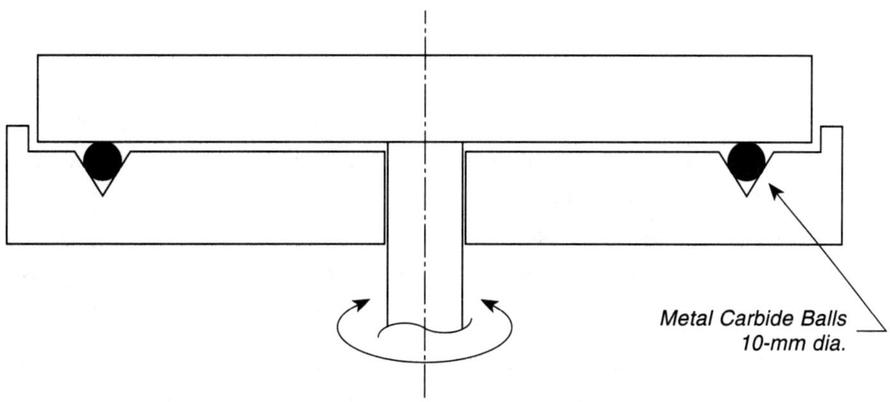


*Fig. 10. Rotary Platform*

# 11

# Small Motions in UHV Systems Using Shape Memory Effect Alloys

An unusual phenomenon present in certain alloys may be used to make small motions in ultrahigh vacuum systems by the application of a small heating current to a helix of the alloy material. The shape memory effect (SME) is so named because shape memory alloy (SMA) wire helixes can be made to change from the deformed "cold" shape to the "preferred" hot shape reversibly. The restoring force is quite large and may be used to achieve small motions as illustrated in the two examples shown here. There are numerous alloys that exhibit the SME, and a favorite SMA for UHV applications is NiTi. This material is easily formed into the hot shape and has a convenient transformation temperature. The shape change is caused by a transformation from a martensitic to an austenitic phase, with the high temperature austenitic material being produced at about 60°C, while the reverse transformation occurs at about 30°C. Deformations producing less that 2% local strain in the martensitic state can be reversed by heating through the transformation temperature. By setting the hot shape in a helical wire it is possible to get extensions of several hundred percent without exceeding this strain limit.

Figure 11A shows a shutter arrangement made with a 0.45-mm-diameter NiTi wire [1]. The hot shape of the left-hand helix was set by winding it on a cylindrical mandrel and then heating to 500°C for 1 h in a vacuum annealing oven to prevent oxidation. The helix was then stretched between two stainless steel posts and pinch clamped to each. The helix was therefore stretched from its preferred hot shape. Upon heating with a current of 2.5 A, the left segment of the helix contracts to its preferred hot shape, moving the attached shutter to the left in a period of less than 0.5 s. This operation stretches the right-hand cold helix beyond its plastic limit, which means that it has little restoring force. Thus, the heating current can be turned off after the shutter has moved to the left. To restore the shutter to its original position, one heats the right-hand portion of the helix.

The effect of bakeouts at 215°C on a severely stretched (300%) helix was tested over a 24-h period, and it was found that this treatment did not affect the hot shape memory [1].

A second example in which the SME is used to open and close a clamp on a manipulator is shown in Figure 11B [2]. Here, the jaws of the clamp are opened when the NiTi SMA wire is heated. The particular wire [3] used here is of small diameter (0.15 mm) and has been treated in manufacture to exhibit a shorter length at high temperature and a longer length at low temperature. For strong clamping, four such wires were braided together and connected to a copper wire by a simple overhand knot above the ceramic insulator tube, which passes through the back end of the stationary upper jaw. The other end of the wire bundle is grounded through the jaw. A restoring force to close the jaws when the heating current is off is supplied by a spring mounted on a screw that is fixed in the upper jaw but that does not touch the lower jaw. When the lower jaw opens, this spring is compressed. Electrical heating to a wire temperature of 115°C causes the jaws of the clamp to open up by about 2 mm when 0.21 A is passed through the four wires in vacuum. If the current is taken too high (0.23 A in vacuum), the annealing temperature will be exceeded, and the SMA wires will lose their pulling ability. The reader is referred to texts on the SME for more details [4]. Also engineering studies of SME test structures are described in reference [5].

## *References*


[1] A.P. Jardine, M. Ahmad, R.J. McClelland, and J.M. Blakely, J. Vac. Sci. Tech. **A6,** 3017 (1988); This idea is based on an earlier paper which also characterizes the behavior of a similar device as a function of repeated cycling; see B. J. Mulder, Vacuum **26,** 31 (1976).

[2] W.F. Smith, E.E. Ehrichs, and A.L. de Lozanne, J. Vac. Sci. Tech. **A10,** 576 (1992). This article contains additional references to other devices constructed with SME alloys.

[3] The thicker SME wire used in Figure 11A was obtained from RayChem, 300 Constitution Drive, Menlo Park, CA 94025.

[4] *Shape Memory Effects in Alloys,* edited by J. Perkins (Plenum, New York, 1975); also see T.W. Duerig and K.N. Melton, Proc. Mater. Res. Soc., Intern. Meeting on Advanced Mater. **9,** 581 (1989).

[5] M. Nishikawa, M. Kawai, T. Yokoyama, T. Hoshiya, M. Naganuma, M. Kondo, K. Yoshikawa, and K. Watanabe, J. Nucl. Matls. **179–181,** 1115 (1991).

***Fig. 11. Small Motions Using Shape Memory Effect***

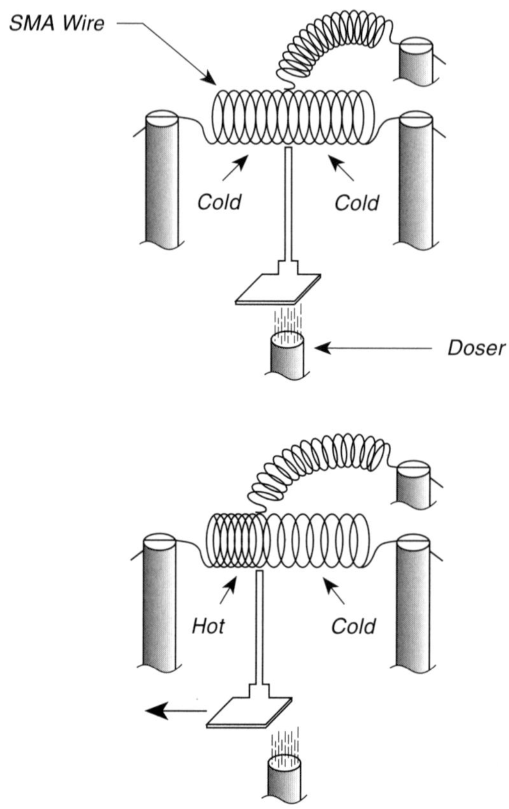


***A. Shape Memory Effect—Shutter***

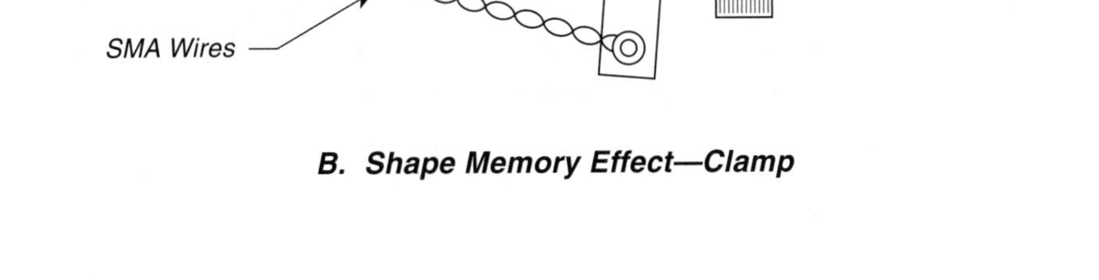


**B. Shape Memory Effect—Clamp**

# 12

# Rotary Shutter Device Driven by Magnetic Eddy Current

The use of magnetic eddy currents in an ultrahigh vacuum system to drive a low friction shutter is illustrated in Figure 12. The eddy current magnet, shown without its housing in the upper part of the Figure, drives the stepped copper rotor at 300 rpm when ball bearings are used, and 60-Hz power is applied to the coil. The copper disk is 0.8 mm in thickness. The stepped openings on the perimeter permit the operator to vary the open duty cycle for evaporations through the openings from an evaporation source mounted above the disk; operation near the perimeter gives larger fractional transmission than with the substrate placed more toward the axis.

The lower end of the iron cylinder [5 cm dia. x 7.5 cm long] contains a semicircular piece of high-purity copper, 0.6 cm thick. It is wrapped with 150 turns of #14 [1.6 mm dia.] magnet wire with Formvar insulation. This is housed in a stainless steel can to prevent outgassing, and the can is welded shut using the ceramic feedthrough for electrical contact to the coil inside.

The device works on the basis of the lag of the magnetic field at position *b* from that at position *a*. This means that when the magnetic field is maximized during the cycle at *a*, the eddy currents centered at *b* will also be maximized. A force on the volume element of the copper disk near *a* will be in the direction from *a* to *b*, and the disk will move. One quarter cycle later, the magnetic field at *b* will interact with the eddy currents centered at *a* and the force is again in the direction from *a* to *b*. During the next half cycle, both the magnetic field and eddy current directions are reversed, again giving a force in the same direction. A mechanical brake is employed to stop the rotation.

The shutter device is used to reduce the flux of evaporating material arriving at the substrate by a precise geometrical method involving only the dimensions of the transmission slots relative to the circumference of the sector disk. Thus, thick films deposited near the outer edge of the disk are proportionately reduced to thinner films deposited at a known radius under the disk. Measurements of the film thickness deposited above the shutter can be converted to sector thicknesses below the shutter [1]. A similar device, driven conventionally, for extending the range of a quartz microbalance has been described previously [2].

## *References*

[1] J.S. Brooks, R. Meservey, and R.K. MacNabb, J. Vac. Sci. Tech. **20,** 243 (1982).
[2] F.C. Wilson, D.B. Sandstrom, and H.B. Serreze, Rev. Sci. Instrum. **46,** 1593 (1975).

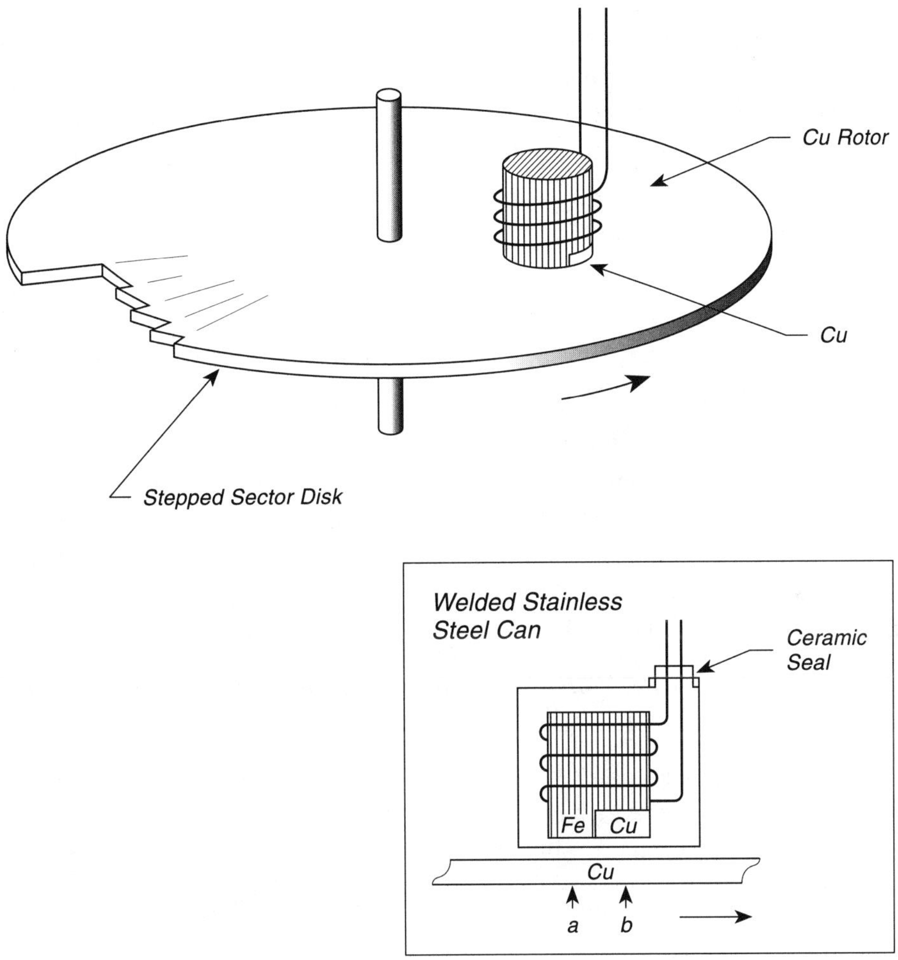


***Fig. 12. Eddy Current Rotor***

# 13

# High-Speed Ultrahigh Vacuum Motor

A modified hysteresis-synchronous motor was designed to eliminate most of the outgassing that occurs during motor operation [1]. There are two sources of gas evolution in motors working in vacuum. The first is outgassing of the bearings. The second is outgassing of the stator winding insulation owing to heating during operation.

The design shown in Figure 13 eliminates both of these effects to a high degree. The bearings [2] in the motor [3] were lubricated with a dry lubricant. The stator was isolated from the vacuum system and the rotor by a container sealed with two O-ring seals at either end. This container is filled with nitrogen to 1 atm to facilitate heat conduction away from the motor. Cooling coils that are brazed to the outside of the isolation container maintain low temperatures at the motor.

The main modification in the motor involved increasing the gap between rotor and stator to allow the isolation container wall (0.010 in. thick) to fit. This may be done in two ways. Either the inside diameter of the stator is increased, in which case the isolation container wall must be of magnetic stainless steel (this was done), or if the outer diameter of the rotor is decreased, the isolation container wall between rotor and stator must by made of nonmagnetic stainless steel.

Motors modified in this way have been working for more than one year at 200 Hz and exhibit a pressure rise of less than $1 \times 10^{-9}$ Torr [1].

To reduce the loss of gas still further, the motor housing could be filled with nitrogen gas at about 1 Torr pressure, which is still adequate for heat transfer. Alternatively, UHV sealants might be used to seal the motor [4].

## *References*


[1] T. Engel and D. Braid, Rev. Sci. Instrum. **56**, 1668 (1985).

[2] Barden Bartemp bearings, SR 188 SSTB.

[3] Globe motor, model 75 A 1003-2.

[4] Professor Thomas Engel, Department of Chemistry, BG-10 University of Washington, Seattle, WA 98195 (private communication).

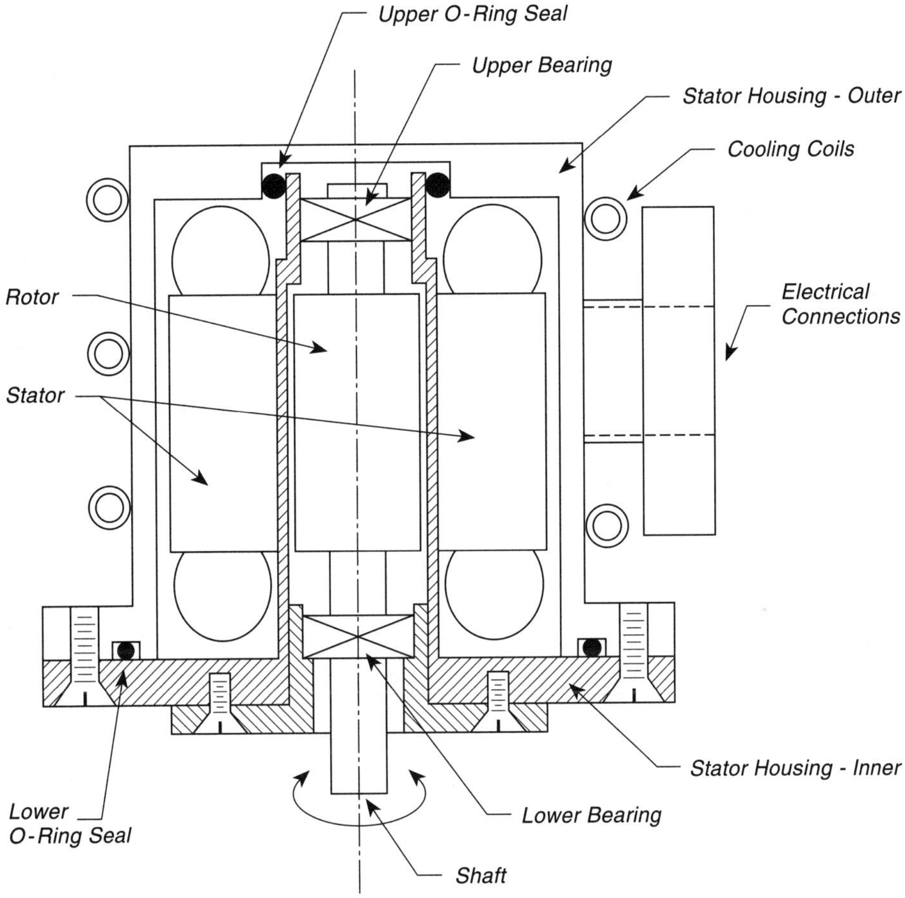


**Fig. 13. High-Speed UHV Motor**

# 14

# High-Speed Rotary Feedthrough for UHV Operation

A simple rotary feedthrough, suitable for high-speed rotation in an ultrahigh vacuum system, is shown in Figure 14. The device uses a hysteresis motor, in which the stator is externally mounted, for driving a shaft at speeds from 1200 to 15,000 rpm [1]. The pressure rise during operation is less than 1 x $10^{-10}$ Torr, since the stator, which generates gas on heating, is isolated from the vacuum system.

In the Figure, a shaft, held between two unlubricated bearings [2], is attached to a rotor and both pieces are ground together to assure good dynamic balance. Two spring washers eliminate axial play. The stator is held in place by plastic supports. The torque delivered is limited by the motor size chosen [3] and by the gap between stator and rotor, which is 1.5 mm here. After 1200 h of intermittent operation, no need for bearing replacement was encountered.

## *References*


[1] T. Engel, Rev. Sci. Instrum. 52, 301 (1981).

[2] The bearings were from Barden Bartemp, Type SR 188 SSTB.

[3] The rotor and stator were obtained from KaVo Electrotechnisches Werk GmbH D-7970 Leutkirch, Germany, Type EV 48-15.

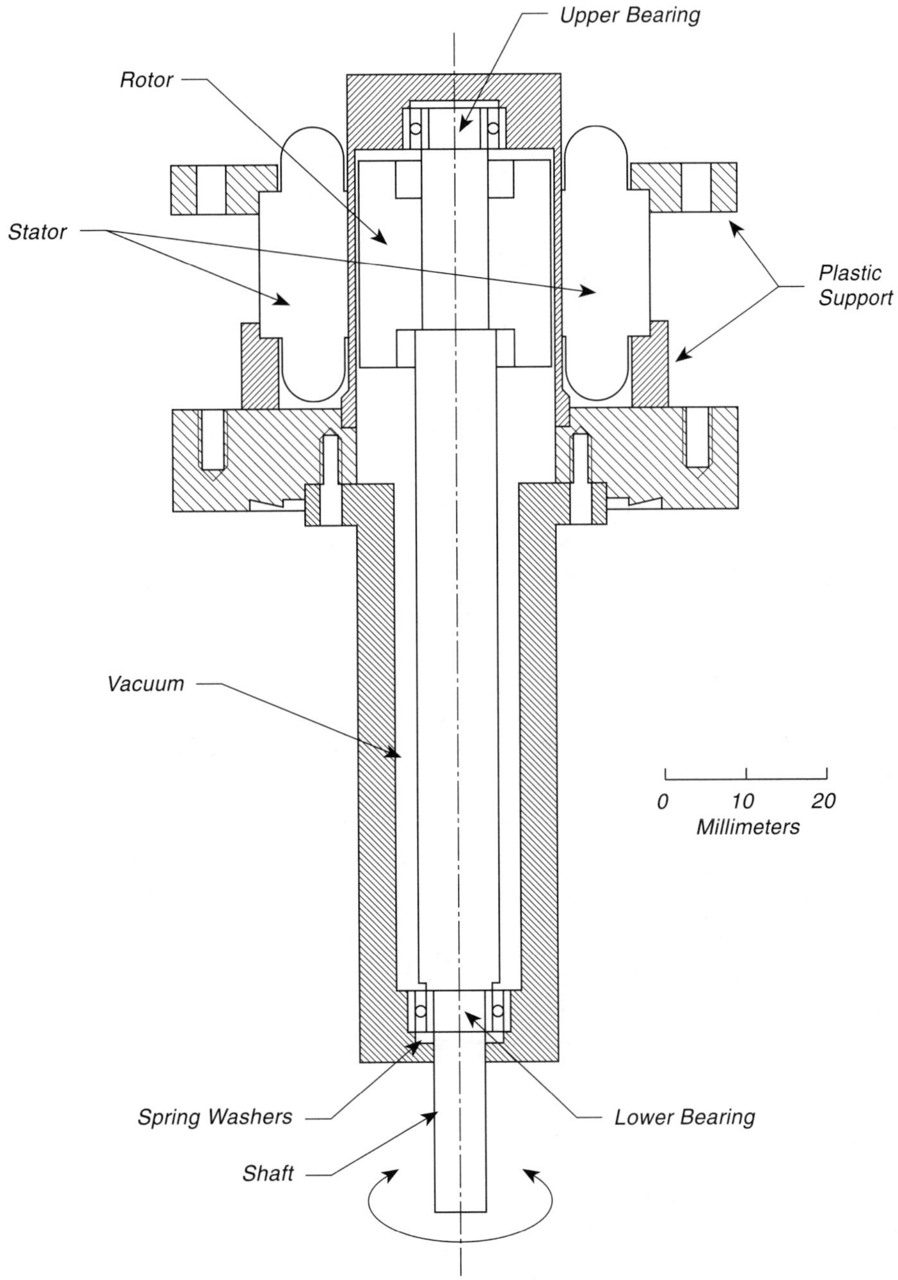


***Fig. 14. Very High Speed UHV Motor***

# 15

# Self-Lubricating Bearings in UHV

The use of boron nitride, BN, as a bearing lubricant has been demonstrated for light loads in a combination rotary-translator bearing [1]. This is especially convenient for magnetically driven shafts where only small forces may be applied.

Figure 15 shows an exploded view of the bearing assembly. The rotary and translation shaft, made of stainless steel, is supported at some distance from the bearing by V-blocks that can be moved into contact with the shaft, and these are not shown here. The shaft is held by WC bearings (0.48 cm diameter) which themselves are held in a BN ball separator cylinder containing holes that accommodate the WC balls. This BN ball separator provides continuous lubrication to the WC balls and the rod, since BN is a dry lubricant. The balls are held in their place by a type 6061 Al alloy retainer which slips over the top of the balls keeping them from touching each other. An outer-bearing sleeve then slips over all of the components and is stabilized by contact with the inside of the vacuum wall tubulation. The bearing assembly is loosely retained between the magnetic drive mechanism and a stop on its other end, not shown.

The system can be repeatedly baked at 275°C. Since BN is hygroscopic, when the chamber is open to atmosphere, the bearing is kept at 80–100°C using heating tape on the outside of the tubulation. No contamination of samples by BN has been detected using various surface analysis methods.

## *References*

[1] G.J. Collet, E.L. Garwin, and R.E. Kirby, Rev. Sci. Instrum. **58,** 479 (1987).

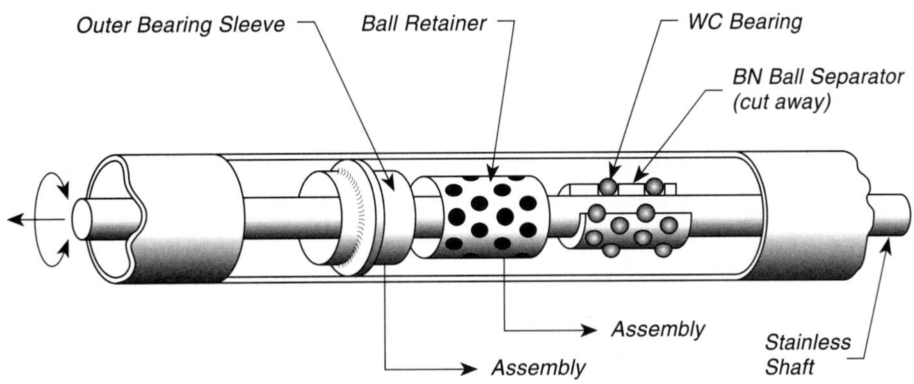


**Fig. 15. Exploded View—Self-Lubricating Translation-Rotation Bearing**

# 16

# Lubrication for Heavy Sliding Loads in Ultrahigh Vacuum

The lubrication of metal/metal interfaces for sliding motion in ultrahigh vacuum is a special problem related to the fact that practical surfaces that often slide together with low friction in air will seize in vacuum because of the absence of adsorbed molecular layers of water lubricant or other materials.

Figure 16A shows a test device used to evaluate the lifetime of a special coating on stainless steel surfaces that was able to lubricate bearing surfaces successfully for over 200,000 cycles. The comparable unlubricated surfaces failed after 1740 cycles [1].

A ballscrew that translates over spherical bearings in a race was installed, along with a parallel identical partner ballscrew behind the device plane (not shown), in a device which drove a 2-kg mass up and down in ultrahigh vacuum of the order of $10^{-10}$ Torr. The vacuum condition was achieved after a bakeout at 140°C. The motion drove the mass upward and downward 45 mm, at a speed of 2.25 mm/s. The screws and nuts of the ballscrews are made of 440C stainless steel, finished using electrochemical buffing to a roughness of about 0.1 μm, and then coated as described below.

The ballscrew was coated with a TiC film, 1.5 μm in thickness, and with a micro-Vickers hardness of about Hv 1100. An Ag coating of 0.3-μm thickness was then applied. The ball bearings mating to the ballscrew were also made of 440C stainless steel, coated with an Ag film of 0.3-μm thickness. The unlubricated surface, used as a control, was not treated with either TiC or Ag.

Figure 16B shows the wear track observed visually after 200,000 cycles. The thin-film Ag coating is still retained on sections of the screw that did not undergo contact with the balls. The TiC underlayer is visible in the contact regions and was not worn through. Surprisingly, when the device was operated in air, the untreated ballscrew performed well but the coated ballscrew did not move smoothly.

Plasma-coated Ag films on stainless steel screws for use in UHV have been found to resist cold welding [2]. Also, ordinary Ag plating of either the bolt threads or the nut threads of stainless steel bolts used internally or externally in baked UHV systems is reported to be an excellent lubricant using coatings of 10-μm thickness [3]. It has been found that coating metals such as Ta, Cu, Mo, and stainless steel with an oxide film, by heating to redness in air, will provide protection against galling during motion in UHV [4].

A good review on tribology and lubrication in the vacuum environment in space is found in reference [5].

## *References*

[1] H. Saeki, J. Ikeda, I. Kohzu, and H. Ishimaru, J. Vac. Sci. Technol. **A8,** 3360 (1990).

[2] R.A. MacGill, R.A. Castro, X.Y. Yao, and I.G. Brown, J. Vac. Sci. Technol. **A12,** 601 (1994).

[3] A. Contaldo, Rev. Sci. Instrum. **36,** 1510 (1965).

[4] Dr. W.F. Egelhoff, National Institute of Standards and Technology, Gaithersburg, MD 20899-0001 (private communication).

[5] E.W. Roberts and M.J. Todd, Wear **136,** 157 (1990).

Fig. 16. Heavy Load Lubrication in UHV

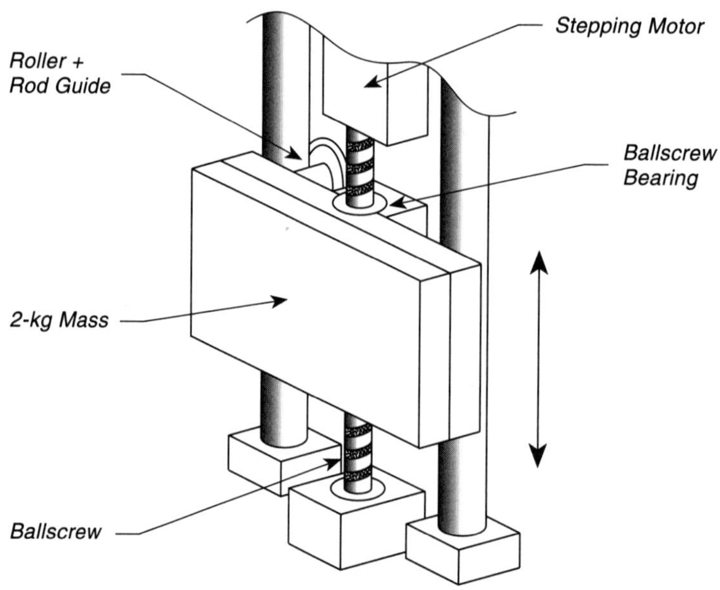


A. Ballscrew Test Device

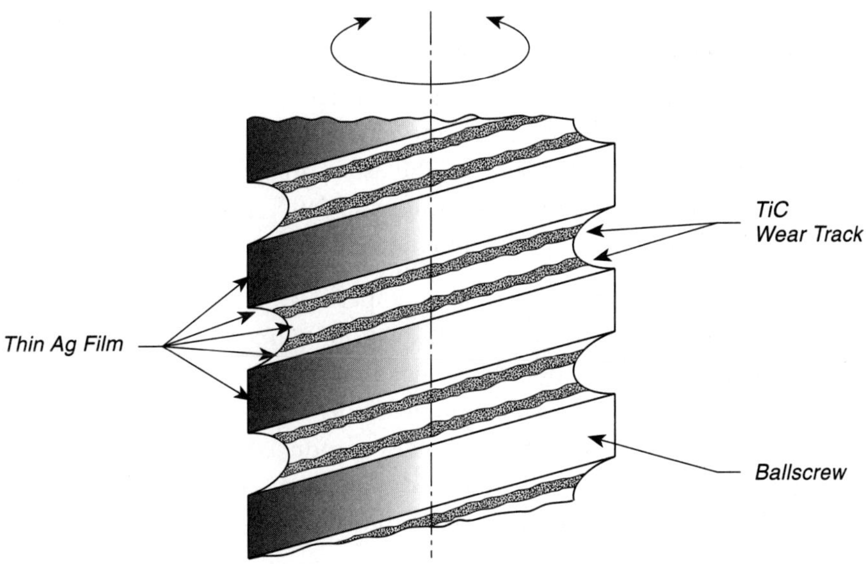


**B. Wear Track Patterns**

# 17

# Simple UHV Bearing for High-Speed Shafts

Steel ball bearings are often used unlubricated in ultrahigh vacuum. At high rotational speeds, over time, these bearings tend to become sticky before final seizure. The development shown in Figure 17 uses polytetrafluoroethylene (PTFE or Teflon) as a bearing material for continuous rotation at 10–1000 rpm. The bearing is machined using sharp oil-free tools, and care is taken not to introduce metallic particles into the PTFE. Cleaning is carried out in organic solvents, and traces of metal and oil may be removed by use of hot nitric acid solution, followed by soaking in distilled water.

The design has three essential components. The first is the use of relief slots made by a small, round file at 90° intervals around the inside of the bearing. Removal of about one-third of the contact area was found to be essential. The second feature is the placing of pumping holes in the PTFE bearing, and more of these can be supplied on the upper housing. The third feature of the design is the use of an 0.25–0.50-mm-diameter wire squeezed between the bearing and the housing as a friction point to prevent the bearing from rotating when torque is supplied from the shaft.

Bakeout below 200°C is appropriate, and temperatures above about 250°C are to be avoided. The bearing should be operated near room temperature and the mechanical admission to the bearing surface of flakes of metal from thin film deposition in the chamber should be avoided. Small partial pressures of fluorine and chlorine were seen initially, and these decreased with time. Initial use also produced small partial pressures of methane and carbon monoxide which soon disappeared. The rotary bearing was used with a 400 rpm shaft for a million rotations without any apparent degradation [1].

## *References*

[1] R.A.A. Kubiak, P. Driscoll, V. Manning, and R. Houghton, J. Vac. Sci. Tech. **A4,** 1951 (1986).

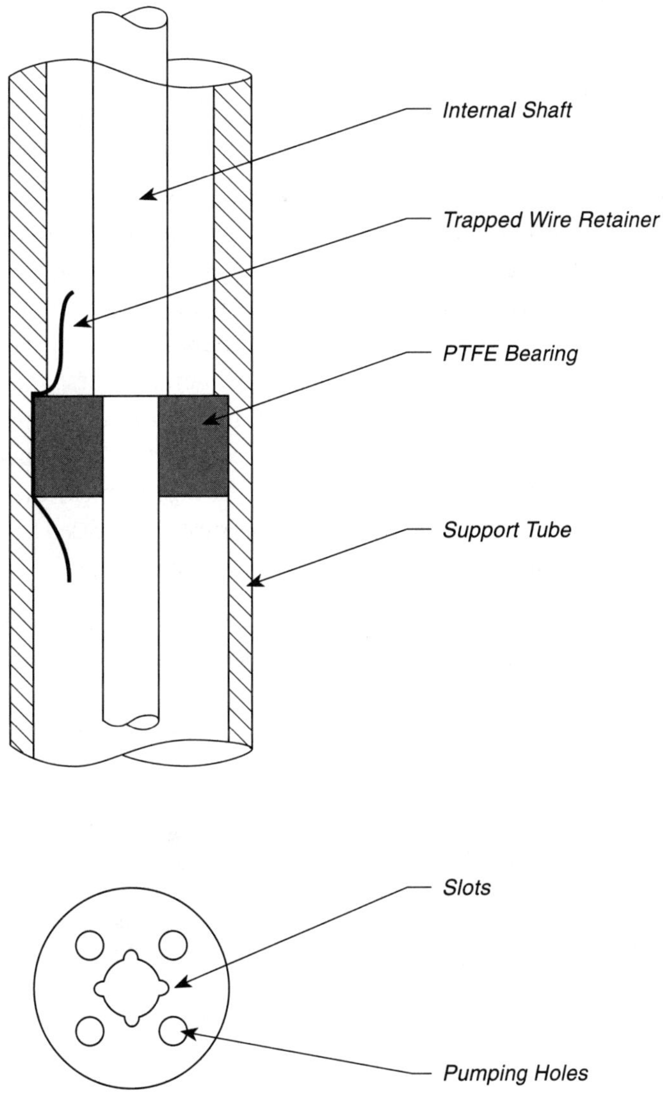


**Fig. 17. Internal Rotary Bearing**

# 18

# Flange-Mounted UHV Variable Aperture

The need for an internal aperture that can be adjusted in ultrahigh vacuum exists both in optical experiments and also in vacuum systems where variable pumping speed control is desired. An inexpensive aperture device is shown in Figure 18. It consists of a variable iris aperture [1] that may be purchased as a component, mounted inside a 2 3/4-in.-diameter Conflat flange. The aperture is mounted on an adapter ring that is tack welded into the flange to avoid distortion. The aperture is fixed from rotation by mounting pins that fit into the adapter ring and through the wall of the aperture housing. The aperture is tightened into place by a threaded ring containing holes for a spanner wrench. The original Conflat flange has had an inside slot milled around its inner circumference for the motion of the aperture control. This is accomplished by a linear actuator coupled to the aperture control lever by a Cu-Be strip (0.3 mm thick) that extends around the aperture housing inside the milled slot. The aperture is degreased before use and relubricated with a tiny amount of $MoS_2$ applied lightly with a cloth.

The iris opening ranges from 1 to 25 mm and can be reproduced to a precision of 2% if one is careful to avoid backlash by moving always in the same direction for final adjustment [2].

## *References*

[1] Edmund Scientific Company, 101 East Gloucester Pike, Barrington, NJ 08007-1380.
[2] F. Homan, J. Vac. Sci. Tech. 17, 664 (1980).

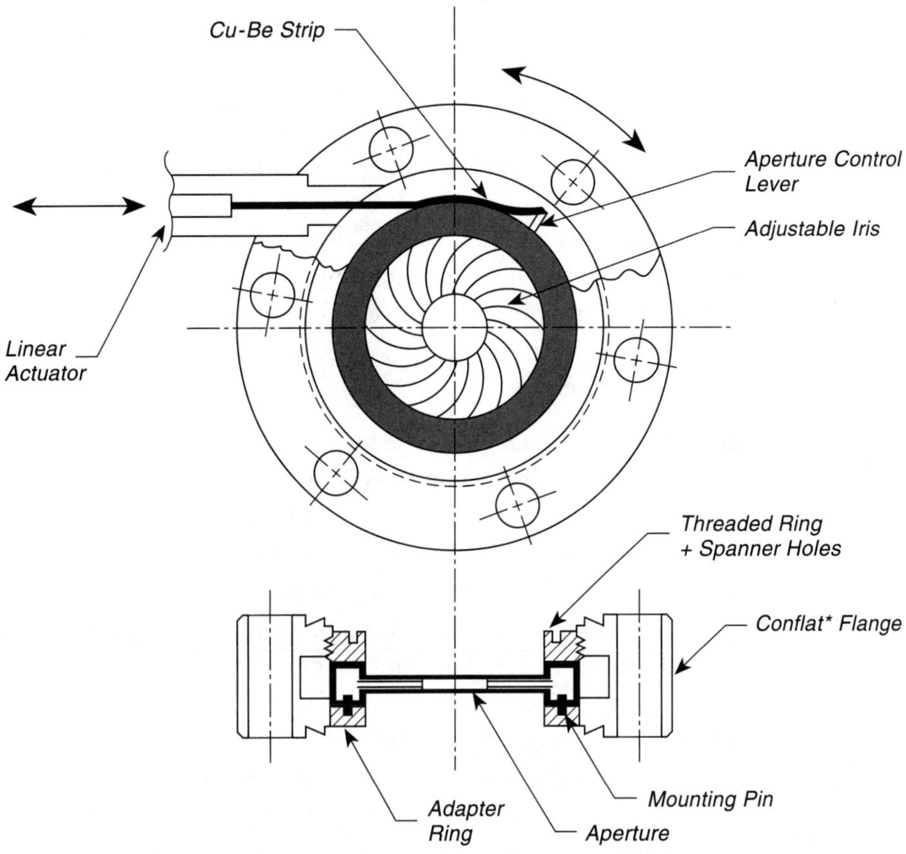


***Fig. 18. Adjustable UHV Aperture***

# 19

# Linear Motion Platform (LMP)

A clever mechanical design to convert a linear screw motion into a linear platform motion with reduced amplitude has been described by Dr. H.P. Rust and Mr. L. Cramer [1]. The design is used as an approach mechanism for an STM tip in a low-temperature instrument [2].

The device shown in Figures 19A,B is made entirely from a single piece of bronze, which, after fabrication, is rigidly attached to a base plate also made of bronze. The active LMP part is milled with three orthogonal slots in the form of a C, and then 6 "milled hinge" cavities are also produced as shown in Figure 19B by dashed lines. These milled hinge cavities have only a thin section of bronze remaining at the top of each cavity, giving excellent bending response through a few degrees. The LMP is then rigidly attached to the two projecting flats from the baseplate, and a screw motion, schematically shown as a thumbscrew, is added on the left.

The device works as follows. As the screw is moved up or down, the plate section above it moves up or down, and we will assign a length of motion *a* as shown in Figure 19A. This motion is reduced at point *b* by virtue of the relative length of the arms involved as the milled hinge is bent near the point *b*. The section of the LMP device labeled *c* is rigidly connected to the outer plate section containing the *a* and *b* motions, and there is therefore an equivalent *c* motion as shown, i.e., $b = c$. As section *c* moves up, a milled hinge, midway between *c* and c′ flexes as a result of the forces it experiences. This causes a symmetrical distortion in the bridged section *c*–*c*′, and leads to equal angular displacement on both ends of the platform, which occurs as a result of distortions in the region of the pair of milled hinges on each end of the platform. The small motion of the platform is schematically indicated by *d*. Because of the symmetry, the motion of the platform up and down will be in the normal direction without any tilting, which is the desired result.

The function of the device can best be understood by remembering that the light gray region, shown in Figure 19B, and containing *a, b,* and *c* and the symmetrically similar regions, is rigid and undergoes identical angular distortion away from the baseplate when the screw is advanced. The mechanical advantage is given by the ratio of the lever arms, $l_1$ and $l_2$.

## *References*


[1] Registered design: Dr. H.-P. Rust and L. Cramer, Deutsches Patentamt Gebrauchsmuster No. G 94 12 112.5, (27 July 1994).

[2] H-P. Rust, J. Buisset, E. K. Schweizer, and L. Cramer, Rev. Sci. Instrum. **68** (1) January 1997.

***Fig. 19. Linear Movable Platform***

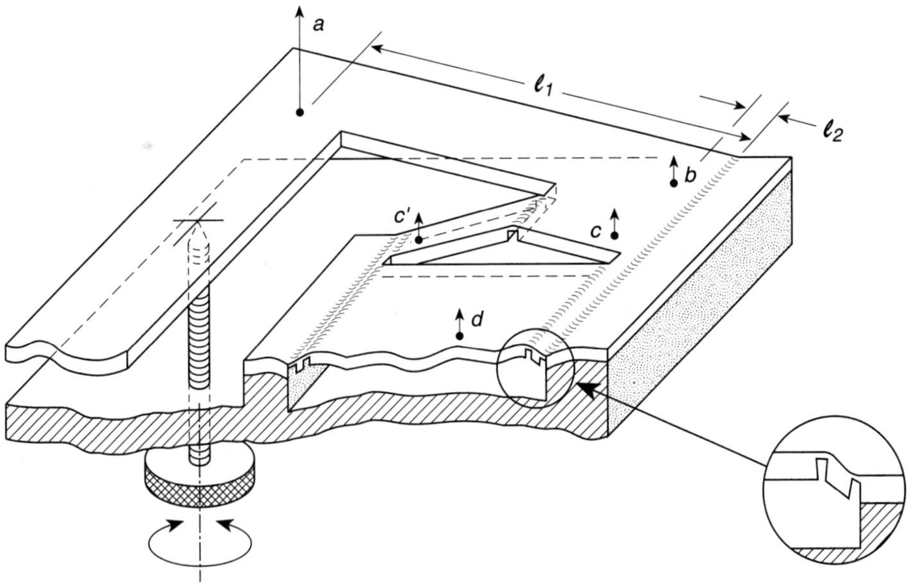


***A. 3D-Sectional View of Portion of LMP***

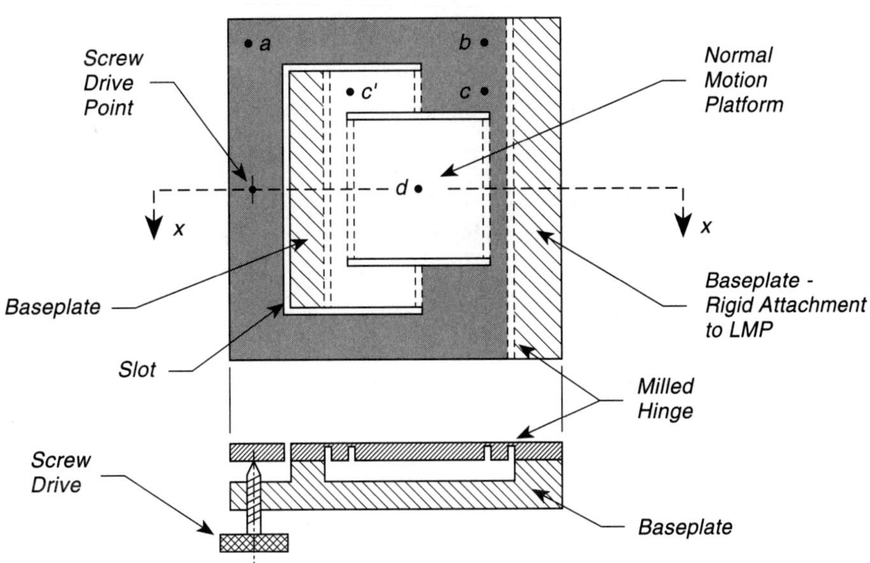


***B. Top & Sectional View (x-x)***

# 20

# Thermally Compensated STM with Repeatable Sample Positioning

Two problems common to most STM designs have to do with the thermal drift of the relative position of the STM tip and the sample and the inability to replace a sample in the same position so that the STM tip can return to the same image point. Both of these problems have been solved by a design that addresses these issues [1].

Figure 20A shows a side view of the STM design. Two piezotubes are mounted on the same base. The inner tube (scan piezotube) is used for scanning the tip, while the outer tube (translation piezotube) connects to the sample support assembly and is used to support and translate this assembly. Both of these piezotubes are of exactly the same length, so that thermal drifts are compensated in the $z$-direction; their circular symmetry eliminates thermal drifts in the $x,y$-directions. The outer translation piezotube is surrounded by a thermal shield, made of oxygen-free high conductivity (OFHC) Cu to eliminate temperature gradients, which would defeat the principle of having two identical-length piezotubes for compensation of thermal drift effects. The outer tube is connected to the sample holder through a quartz mounting assembly that consists of an outer quartz tube and a quartz or metal/ceramic sample holder. The sample holder slides (by means of five spherical bearing contacts) on two metallized quartz rails that are mounted inside the outer quartz tube as shown in Figure 20B. Operation of the translator piezodrive causes the sample to jump toward the tip as a result of the friction and inertia effects in the sample holder, and adjustment of the piezo activation waveform can be used to produce jumps from about $10^4$ Å to less than 5 Å. By working in the kHz frequency domain, translation rates up to 1 mm/s are possible.

The sample holder may be removed from the outer quartz tube containing the rods and the sample may be transferred and heated in a furnace or in a load lock. Upon return of the sample holder to the quartz rods, the design permits repositioning to an accuracy of several thousand Å, depending upon how much the sample is bumped during transfer. This is accomplished by using gravity to align the sample on the rails, which are tilted with respect to the horizontal plane. The repositioning scheme, using gravity to set the sample on the rails in the same position, is based on kinematic design principles that are well described elsewhere [2].

The details of the construction of the probe translator are shown in Figure 20C.

The thermal drift experienced by this design is about 1Å/h.

A piezo-driven translator for the tip, which moves in the *x,y*-directions over several millimeters by a stepping and gripping action, thereby eliminating mechanical positioning devices, has been described as a modification of this device [3].

## *References*


[1] J.W. Lyding, S. Skala, J.S. Hubacek, R. Brockenbrough, and G. Gammie, Rev. Sci. Instrum. **59**, 1897 (1988); also, U.S. Patent 4,841,148.

[2] E.B. Wilson, *An Introduction to Scientific Research,* 1st Ed., McGraw Hill, NY (1952).

[3] R.T. Brockenbrough and J.W. Lyding, Rev. Sci. Instrum. **64,** 2225 (1993).

## Fig. 20. Thermally Compensated STM—Repeatable Sample Positioning

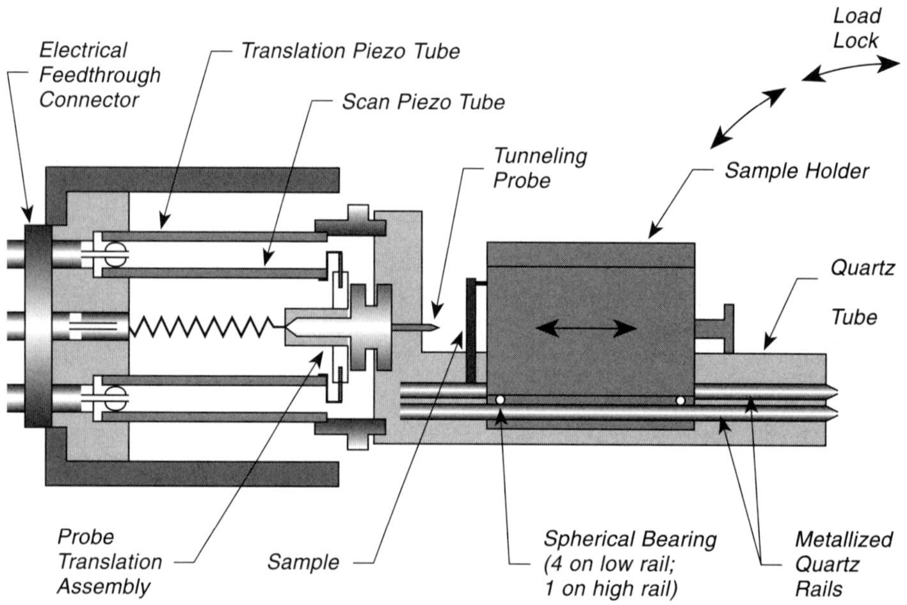


**A. Side View**

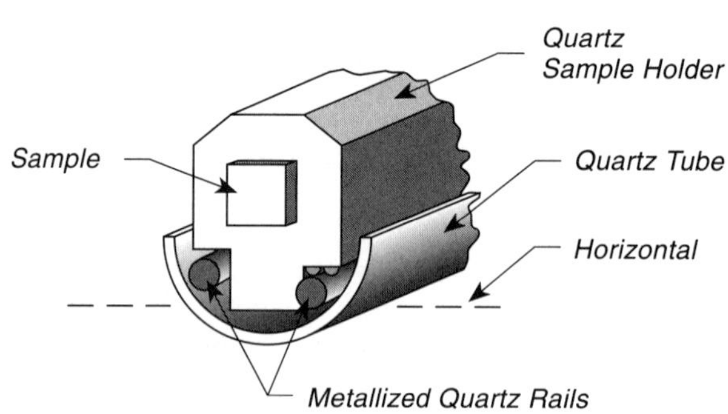


**B. Sample Holder**

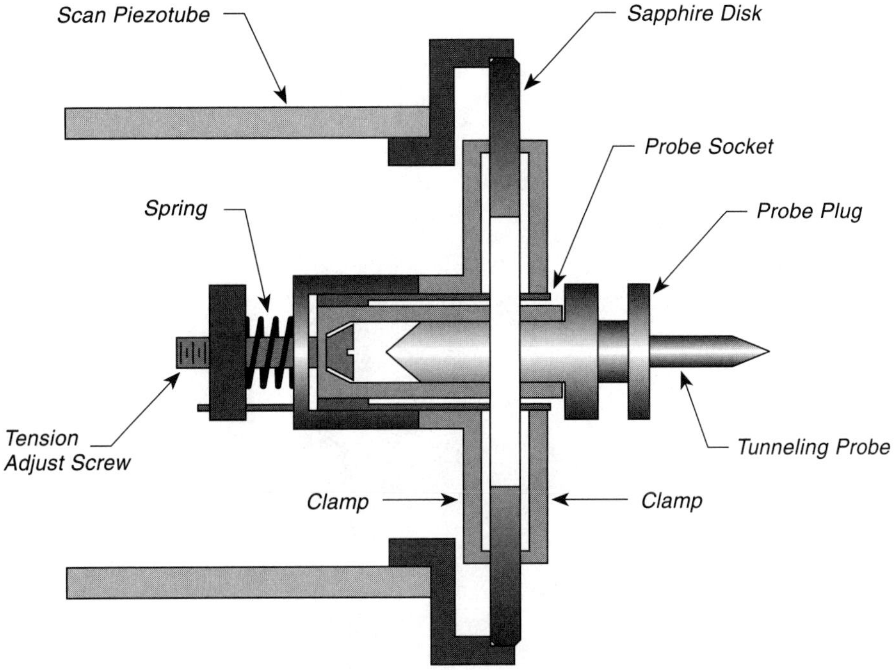


**C. Detail, Probe Translator**

# B. Sample Transfer

# 21

# Sample Transfer from High Pressure to Ultrahigh Vacuum

The ability to transfer samples from high-pressure reaction conditions into ultrahigh vacuum for surface analysis or for crystal preparation has its origin in the work of the Somorjai group at Berkeley [1], who devised apparatus to do this for studies of the kinetics of catalytic reactions on metal single crystal surfaces. Figure 21A shows an early version of this scheme. A single crystal mounted on a flange is surrounded by a movable reactor cup that seals the crystal into a small cavity capable of being pressurized. The reactor cup is sealed under strong force by a drive rod that operates inside a stainless steel welded bellows. Because of the strong sealing forces involved, the body of the ultrahigh vacuum chamber, shown cross-hatched in Figure 21A, is made of extra thick stainless steel to avoid significant distortion when the sealing force is applied. The crystal can be heated electrically in the reactor cup and its temperature measured with a thermocouple welded to the crystal edge. In some cases the inside of the high-pressure chamber is coated with a gold film to reduce wall reactivity. Gas circulation pumps cause mixing of the high-pressure gas in order to mimic stirred flow conditions. The design shown in Figure 21A would not permit Auger analysis of the crystal surface because of its fixed geometry, but later work permitted the crystal itself to be moved inside the ultrahigh vacuum system after the reactor cup was retracted [2,3]

Figure 21B shows an alternate design that has been widely used [4]. The crystal is mounted on a ceramic feedthrough at the end of a transfer tube housed inside a welded stainless steel bellows. The bellows conveniently moves on an external translation carriage containing two parallel stainless rods that accommodate linear bearings connected to the outer flange (not shown) of the retraction device. When retracted, the crystal sits in a reaction chamber which can be pressurized to about two atmospheres. The reaction chamber is closed off by two all metal sealed valves, one leading to the surface analysis/preparation chamber, and the second leading to the high-pressure chamber pumps. As in Figure 21A, the entire apparatus can be brought into ultrahigh vacuum condition by means of bakeout. The crystal can be translated into the analysis/surface preparation chamber where surface analysis and gas dosing can be carried out under well-controlled conditions. In the high-pressure reactor chamber, gas analysis is made by extraction of a small quantity of the gas either into the sampling loop of a flame ionization gas chromatograph, or into the mass spectrometer of the main ultrahigh vacuum chamber. For many analyses, the gas chromatograph is superior in sensitivity and selectivity to the mass spectrometer. Compared to the reactor cup design, the internal volume of the high-pressure cell shown in Figure 21B is large and wall effects may in principle be larger. For certain reactions, the wall effect has been measured and found to be small [5]. Viton O-ring sealing valves are likely to give poorer results because of absorption and liberation of gases from the Viton. Often, wall cooling is necessary when the single crystal operates at high pressure and at elevated temperature because of efficient heat transport to the walls through the dense gas. It may be necessary to cool the outer wall of the reactor chamber in these cases. Such cooling is not easy to achieve in the reactor cup design of Figure 21A. High-pressure chambers are often useful in chemical cleaning of crystals because of the high gas density achievable, let us say for oxygen or hydrogen treatment of a metal single crystal. There is a danger with all stainless steel reactor chambers when operated at high pressure with carbon monoxide gas. The carbon monoxide will produce nickel and iron carbonyls from the stainless walls that transfer their metal atoms to the hot metal surface (see pages 138 and 718).

Another type of sample transfer device was marketed by Leybold-Heraeus and is schematically shown in Figure 21C. A 2-cm-diameter polished stainless steel rod is arranged to move laterally by sliding through a number of differentially pumped O-ring seals that permit rod translation rapidly from a high-pressure reactor into the ultrahigh vacuum system. Differential pumping occurs by means of a mechanical pump in the first stage between the reactor cell and the ultrahigh vacuum system, and high-pressure gases used in the reactor cell would be carried away here. The second stage is pumped with a turbomolecular pump, and achieves a pressure of the order of $10^{-8}$ mbar [6,7]. This design was used by others [8,9] with minor modifications, and in the lower section of Figure 21C can be seen the manner in which a metal single crystal is mounted [8]. The high-pressure reactor is only a few cubic centimeters in volume and was entirely made of stainless steel and gold-plated pan bronze, except for Viton/Teflon seals. The small volume allows an enhanced sensitivity for product gases evolved in catalytic reactions at high pressures [8,9]. Leaks due to poor O-ring lubrication or to scratching of the stainless rod is avoided to some degree by providing Teflon liners, in the form of tubes which slide on the rod between the O-rings [10].

A more recent type of arrangement for sample transfer between an ultrahigh vacuum system and a high-pressure chamber is called the "rod-through-chamber" geometry. As shown in Figure 21D, a single crystal is mounted on a long feedthrough that contains a special conical shoulder that mates to a Teflon gasket when the crystal has been advanced into the high-pressure chamber. The conical contour shoulder is polished to a mirror finish, and acts like a plunger when closing the apparatus to isolate the two sections. This Teflon closure can seal against 1 atm pressure with a leak rate less than $10^{-9}$ Torr L/s. When the high-pressure chamber is evacuated and the probe is withdrawn, a gate valve is closed to completely isolate the ultrahigh vacuum system from the high-pressure chamber. While bakeout temperature limits are not specified in the original paper [11], temperatures in the 150–200°C range would not be expected to disturb the Teflon. This type of seal requires that a rather large force be applied to the rod in order to make a gas tight junction with the Teflon gasket.

In some cases, it is not appropriate to employ high forces along the rod to make a metal-to-Teflon seal. Figure 21E shows a method for making a high-force seal using another approach, on a crystal holder that moves vertically and that is not so robust as that shown in Figure 21D. Here the crystal is mounted on electrical leads that are cooled in a reservoir containing liquid nitrogen supplied through a tube entering from the left-hand side. The gas tight seal is again made by contact between a polished stainless steel surface on the bottom of the plunger, which is of spherical contour mating up with a square-edge Teflon gasket. The upper portion of the plunger is also of spherical contour, and when the plunger is seated properly on the Teflon gasket, tapered steel wedges are able to be brought in from the side to mate with the upper spherical surface. This gives a strong downward force, making the lower Teflon–stainless steel junction gas tight. The steel wedges are driven laterally by two separate linear motion feedthroughs. They enter through heavy guide blocks to provide strength needed to make the compression seal. In order to provide the nonrigid character needed to accommodate the downward motion produced by the tapered wedges, the plunger assembly is coupled to the liquid nitrogen reservoir by means of an internal bellows. This seal also has a leak rate less that $10^{-9}$ Torr L/s [11].

A number of other designs for transferring samples between vacuum systems have been devised and are discussed in references [12–17].

A novel design for a continuous stirred flow reactor for use in studying surface reaction kinetics at high pressure, but which cannot be used for placing the catalyst surface in a position for surface analysis, is discussed in reference [18]. The criteria for determining that the reactor is behaving as a stirred flow reactor are given in reference [18]. Another catalytic reactor, which permits the measurement of the spatial distribution of reactant and product gases, has been described in reference [19]. This reactor also does not permit the analysis of the catalyst surface composition by electron spectroscopies. In this design, by measuring the leak rate for the reactor through a capillary, the temperature of the gas in the vicinity of the catalyst surface can be measured [19,20].

## *References*


[1] D.R. Kahn, E.E. Petersen, and G.A. Somorjai, J. Catal. **34,** 294 (1974).

[2] D.W. Blakely, E.I. Kozak, B.A. Sexton and G.A. Somorjai, J. Vac. Sci. Tech. **13,** 1091 (1976).

[3] G.A. Somorjai, Surf. Sci., 89, 496 (1979); A.L. Cabrera, N.D. Spencer, E. Kozak, P.W Davies, and G.A. Somorjai, Rev. Sci. Instrum. **53,** 1888 (1982); T.G. Rucker, K. Franck, D. Colomb, M.A. Logan, and G.A. Somorjai, Rev. Sci. Instrum. **58,** 2292 (1987).

[4] D.W. Goodman, R.D. Kelley, T.E. Madey, and J.T. Yates, Jr., J. Catal. **63,** 226 (1980); see also, D.W. Goodman, J. Vac. Sci. Technol. **20,** 522 (1982).

[5] D.W. Goodman and J.T. Yates, Jr., J. Catal. **82,** 255 (1983).

[6] H.J. Krebs, H.P. Bonzel, and G. Gafner, Surf. Sci. **88,** 269 (1979).

[7] H.P. Bonzel, G. Broden and H.J. Krebs, Appl. Surf. Sci. **16,** 373 (1983).

[8] C.T. Campbell and M.T. Paffett, Surf. Sci. **139,** 396 (1984).

[9] C.T. Campbell, J. Vac. Sci. Tech. **A2,** 1024 (1984).

[10] Dr. Renato Zenobi, Analytical Chemistry, Universitätstr. 16, ETH Zentrum CHN C59, CH-8092 Zurich, Switzerland (private communication).
[11] T. Yamada, T. Misono, K-I. Tanaka, and Y. Murata, J. Vac. Sci. Tech. **A7**, 2808 (1989).
[12] C.A. Crider, G. Cisneros, P. Mark, and J.D. Levine, J. Vac. Sci. Technol. **13**, 1202 (1976).
[13] J.P. Hobson and E.V. Kornelsen, J. Vac. Sci. Technol. **16**, 701 (1979).
[14] R.E. Clausing, L. Heatherly, and L.C. Emerson, J. Vac. Sci. Technol. **16**, 708 (1979).
[15] J. Klebanoff, V.H. Ritz, and R.E. Thomas, J. Vac. Sci. Technol. **A2**, 1396 (1984).
[16] G.S. Chottiner, W.D. Jennings, and K.I. Pandya, J. Vac. Sci. Technol. **A5**, 2970 (1987).
[17] R. Raval, M.A. Harrison, D.A. King, and G. Caine, J. Vac. Sci. Technol. **A9**, 345 (1991).
[18] J.J. Vajo, W. Tsai, and W.H. Weinberg, Rev. Sci. Instrum. **56**, 1439 (1985).
[19] S. Lundgren, K.-E. Keck, and B. Kasemo, Rev. Sci. Instrum. **65**, 2696 (1994).
[20] B. Kasemo, Rev. Sci. Instrum. **50**, 1602 (1979).

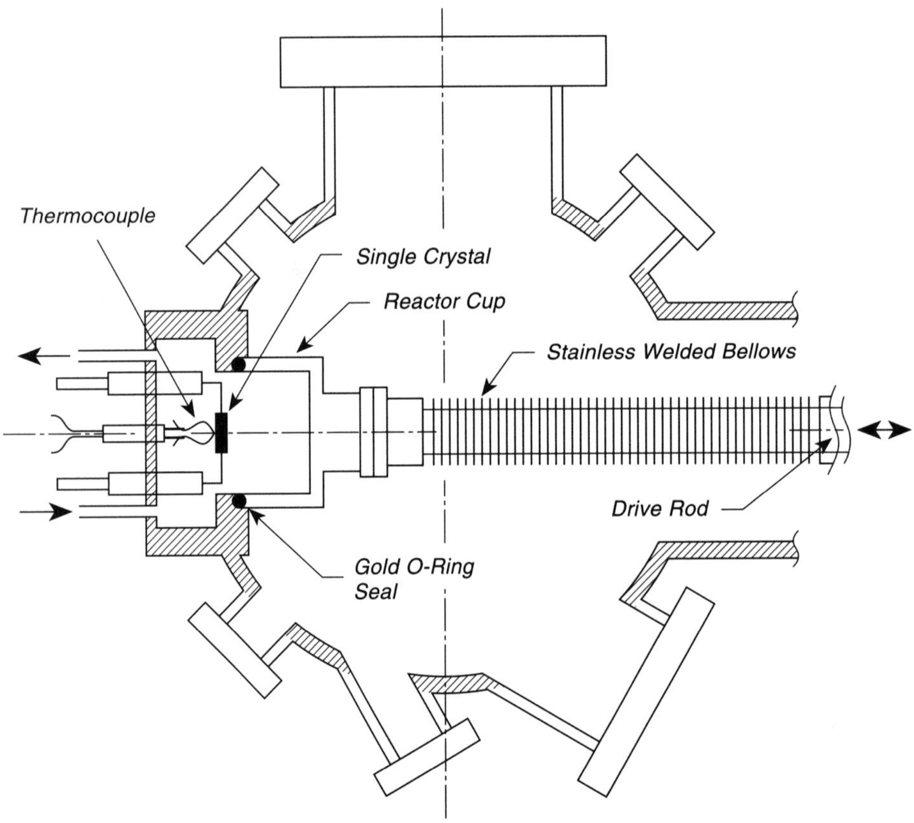


***Fig. 21A. Movable Reactor Cup—Stationary Crystal***

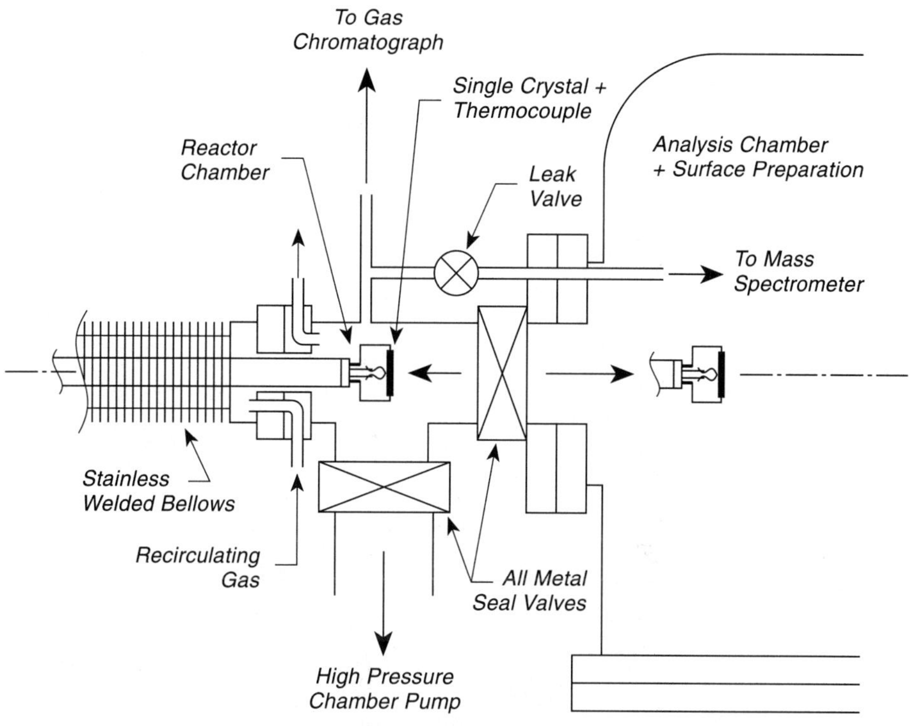


**Fig. 21B. High Pressure Cell—Movable Crystal**

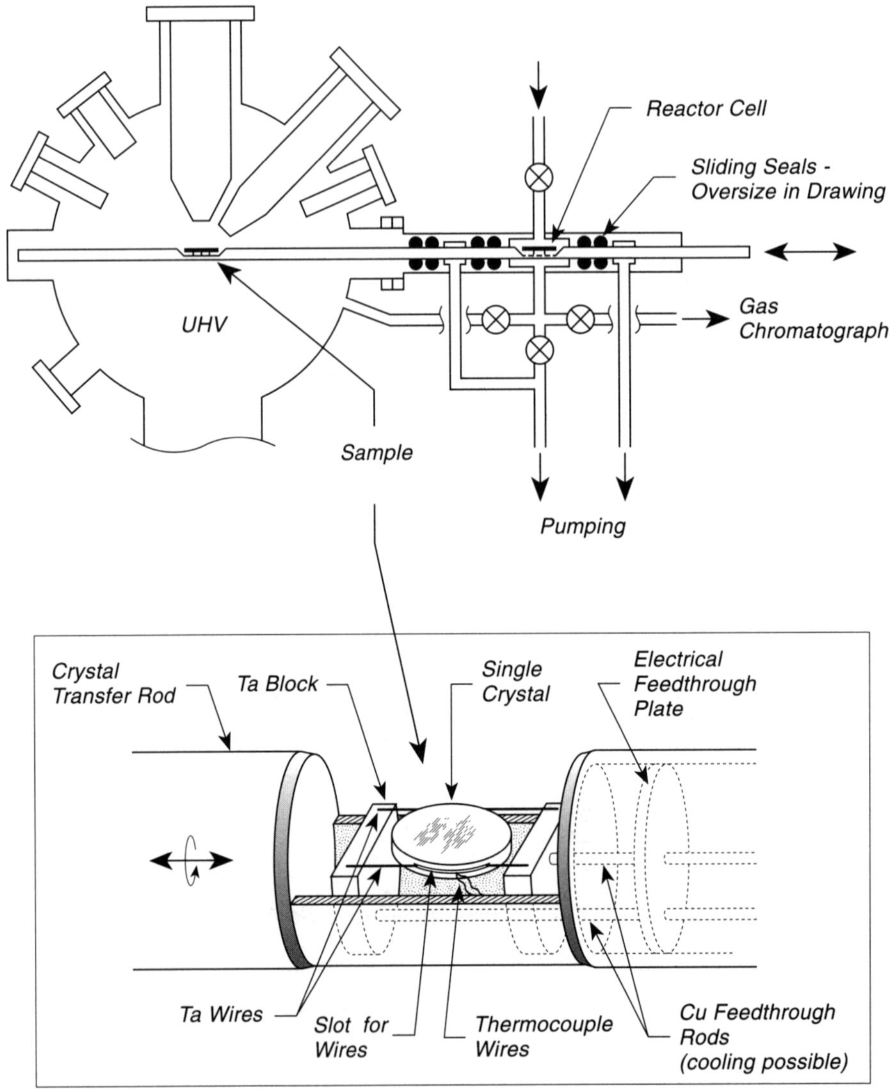


***Fig. 21C. Sliding Rod Seal***

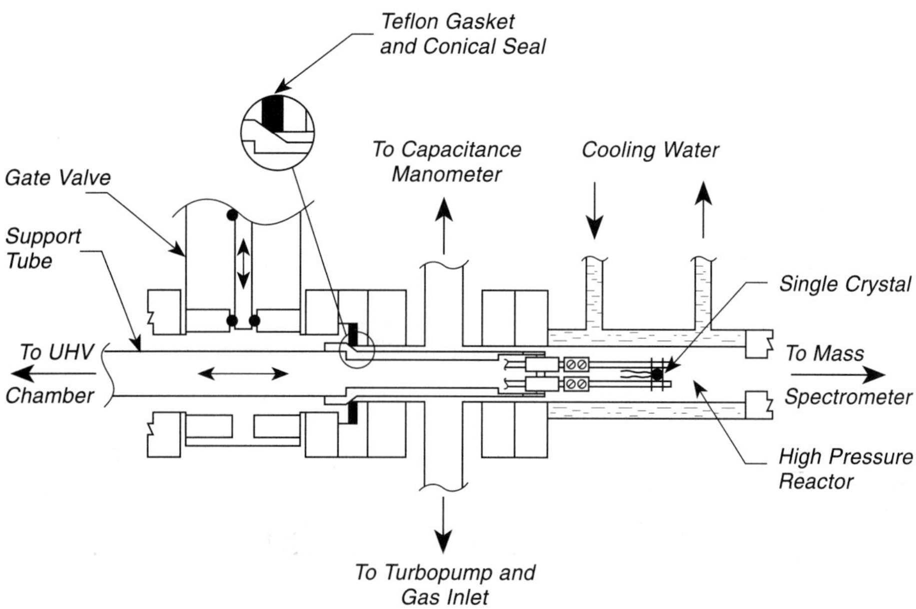


***Fig. 21D. Rod-Through-Chamber***

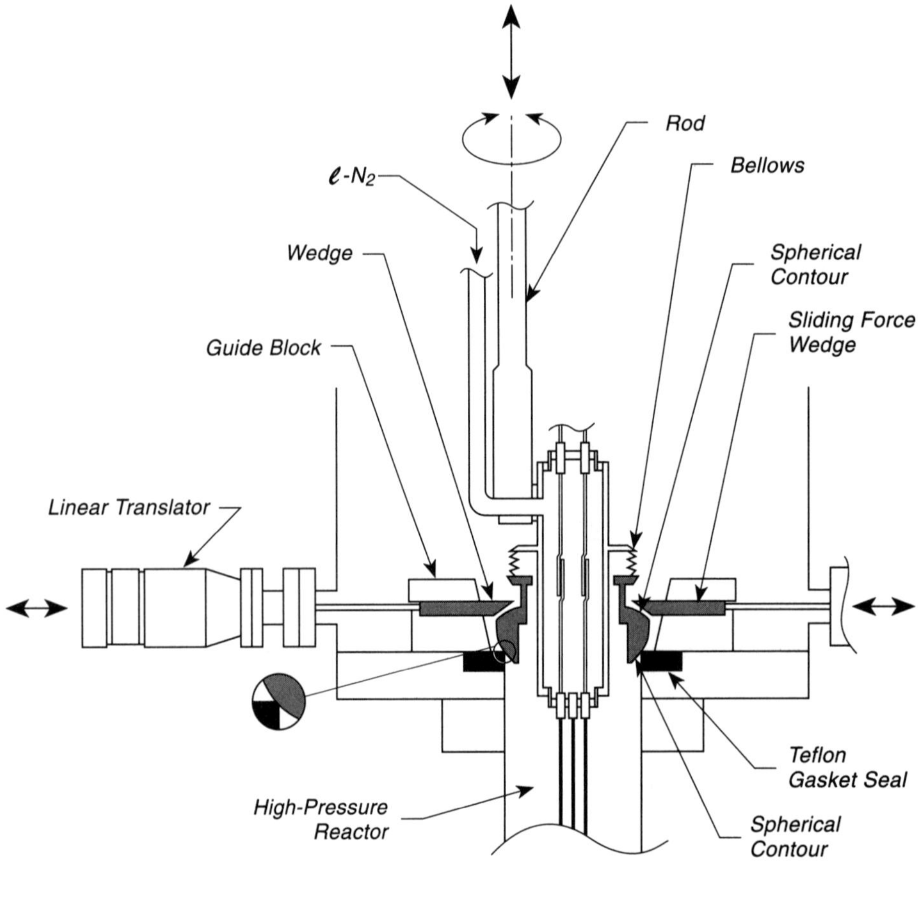


***Fig. 21E. Rod-Through-Chamber Seal—Local Sealing Force***

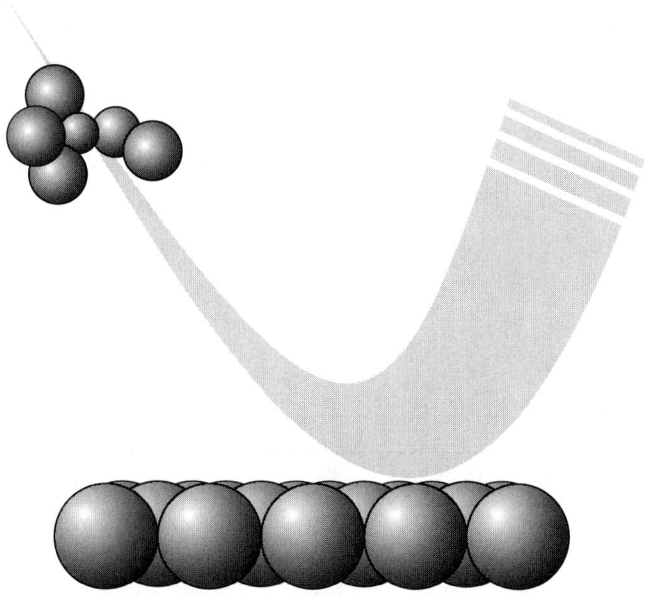

# 22

# Novel Sample Transfer from UHV Chamber to External Cell

A novel design of an ultrahigh vacuum system permitting the rotation of a manipulator containing a single crystal sample to several analytical ports, while also conveniently allowing sample access to an external cell, has been described [1]. In Figure 22A, an octagonal UHV chamber is shown in side view. The different ports on the octagonal faces contain analytical capabilities of various sorts. The manipulator is mounted on a doubly differentially pumped rotary seal on a 14-in.-diameter flange (see page 13 for one design), and its center line is 5 in. off center, allowing the sample manipulator to be rotated through 360° on a 5-in. radius circle to face the various analytical probes mounted on the octagonal faces of the chamber. The bellows manipulator has 18 in. of motion in the vertical direction, and is attached to a standard *x,y,z*-translating stage. The hollow tube holding the crystal may be rotated through 360° also, allowing axial rotation of the sample, and this rotational motion is achieved with a differentially pumped rotary seal at the top of the translating stage (see pages 11 and 12).

The unique access to an external chamber occurs by pushing the sample through three spring-loaded Teflon seals [2] that are doubly differentially pumped and that act as a sliding seal, as in the design shown on page 72. Accurate adjustment of the position of the vertical manipulator with the *x-y*-stage allows this to be done without strain. The Teflon seal section is separated from the large chamber above it with a small gate valve. The external cell is located below the sliding seal section and has a small volume. In various cells, high-pressure surface reactions, infrared reflection absorption spectroscopy, and electrochemistry [3] may be carried out, or if the cell is removed, repairs to the exposed crystal and its mounting may be made. This permits insertion of new crystal samples into the UHV chamber without requiring bakeout of the chamber, and the design is therefore a tremendous time saver.

The bottom portion of the feedthrough is polished to allow easy motion through the spring-loaded Teflon seals and the achievement of a vacuum seal as shown in Figure 22B. When the sample has been translated back into the UHV chamber, the gate valve may be closed to prevent outgassing from the external cell. A pressure of 2 atm in the external cell causes no measurable degradation of vacuum in the UHV chamber (typical pressure = $2 \times 10^{-10}$ Torr).

## *References*

[1] (a) R.A. Campbell and D.W. Goodman, Rev. Sci. Instrum. **63,** 172 (1992). (b) J. Szanyi and D.W. Goodman, Rev. Sci. Instrum. **64,** 2350 (1993). These articles also contain useful references to other external high-pressure cell designs.

[2] Furon Co., 3336 E. Lapaima Ave., Anaheim, CA 92806.

[3] L.-W. H. Leung, T.W. Gregg, and D.W. Goodman, Rev. Sci. Instrum. **62,** 1857 (1991).

***Fig. 22. Novel Sample Transfer to External Cell***

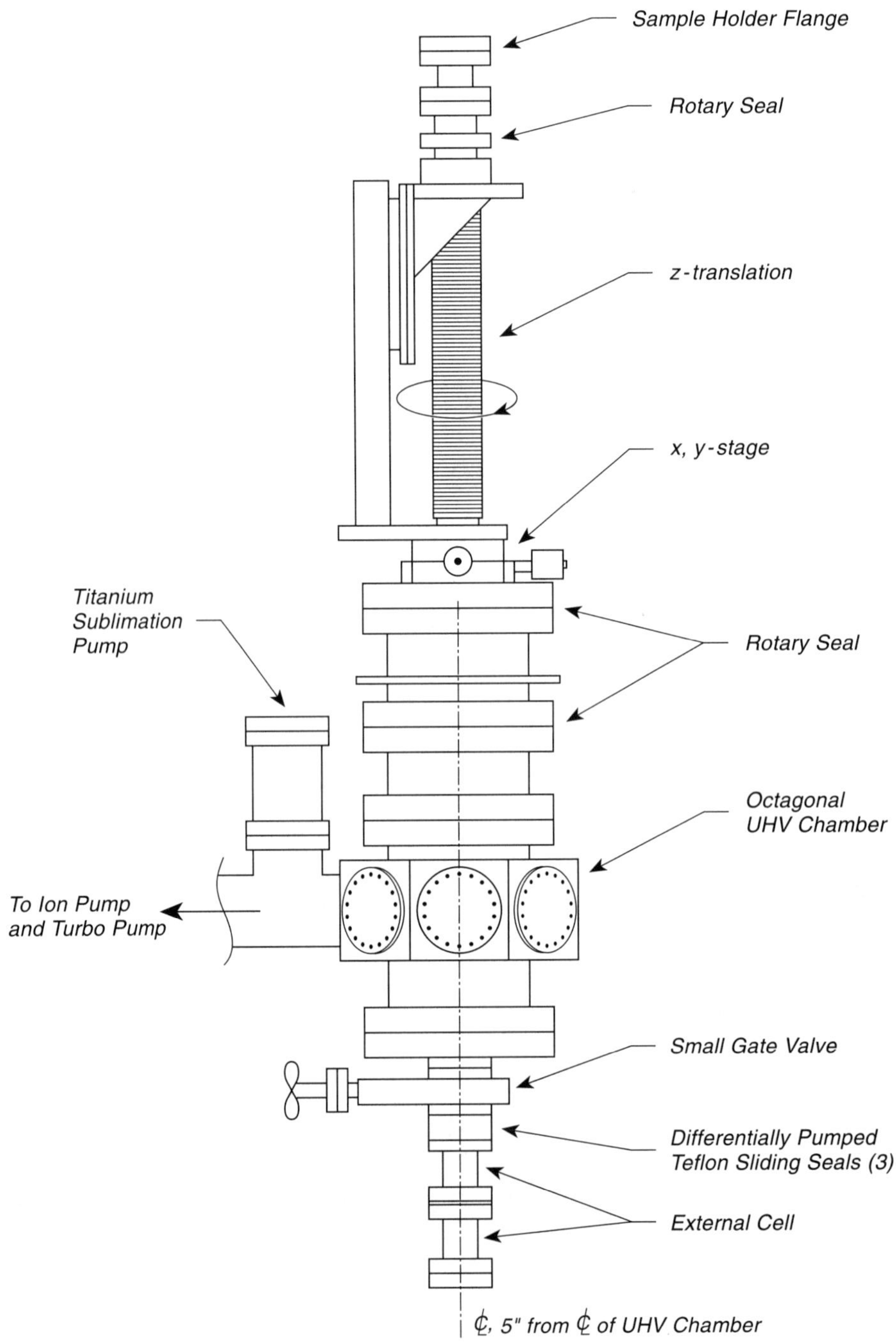


***A. UHV Chamber***

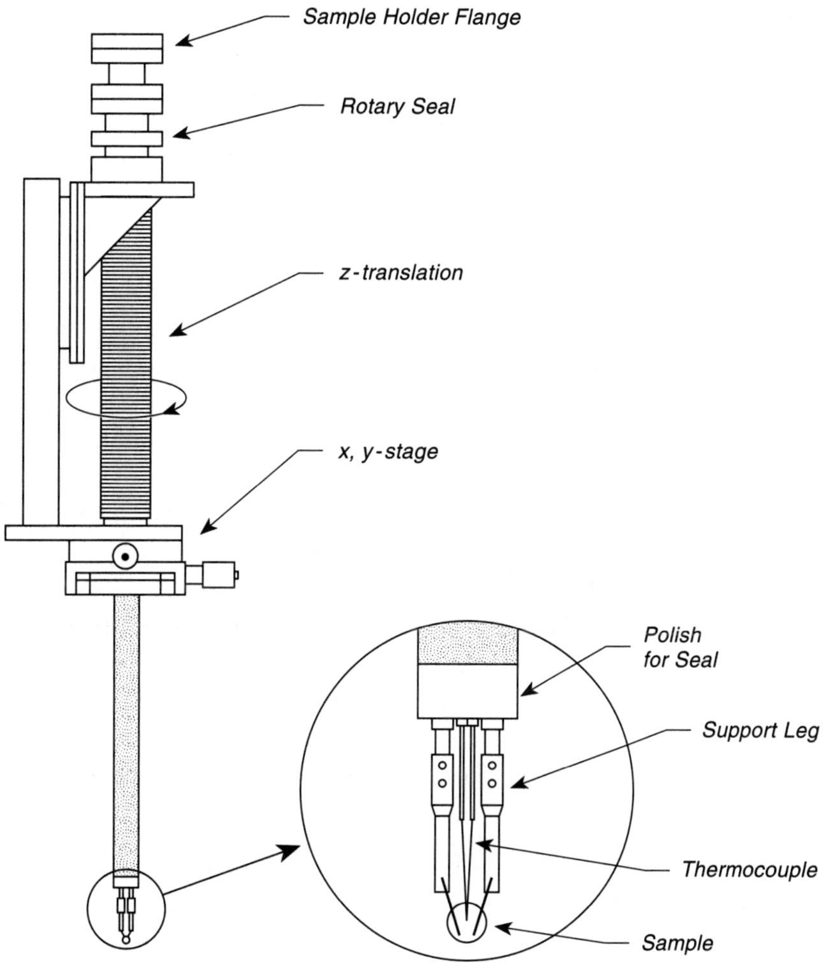


**B. Sample Transport Device**

# 23

# UHV Sample Transfer Device with Low-Temperature Capability

The transfer of a sample from a manipulator in a preparation chamber to a manipulator in an ultrahigh vacuum is a classic problem, and this has been dealt with on pages 90 and 94. However, when excellent cooling is required to temperatures below about 100 K, special designs must be used to permit rapid heat transfer through a demountable joint. Various designs to do this are found starting on page 548. A particularly appealing design, because of its simplicity, is shown in Figure 23 [1].

Figure 23A shows the basic vacuum system with a sample preparation chamber separated from the UHV chamber by a straight-through valve. The sample, initially held at the end of a horizontal manipulator in the sample preparation chamber, can be moved through the valve into the ultrahigh vacuum chamber, where it is transferred to the vertically oriented manipulator. This is done by transfer of a sample slide wedge, shown in Figure 23B. This wedge is made of Mo and is highly polished on its bottom contact surface to avoid cold welding when assembled. The sample is attached to a flat pad (0.3 x 0.4 in.$^2$) on the top, using spring clips, screws, or welding. A long Mo screw, fixed in the sample slide wedge, and rounded on the threaded end, is tightened by rotation of the transfer rod drawing the wedge shaped surfaces on the slide and the base together. The base is also made of Mo to avoid differential thermal expansion, and its upper surface facing out of the drawing is also polished to receive the slide wedge contact surface. Two screws at the top and bottom of the heating/cooling base act as stops for the sample slide wedge. With the correct adjustment of dimensions, there is actually no sliding of the wedge against the base contact surface; i.e., when the surfaces come together there is no room left for any sliding.

The heating/cooling base is connected by three screws to a copper cooling finger, which may be refrigerated with liquid He. A pancake radiation heater (see page 522) is attached by a Mo screw to the back of the base [2]. The pancake heater is surrounded by a radiation shield for maximum efficiency in heating. The cutout section of the base between the sample region and the copper block provides sufficiently poor thermal contact when heating to allow a high temperature to be reached, yet also sufficient thermal conduction for cooling the sample.

Transfer of the sample slide wedge is carried out using the inverted screwdriver, which rotates the Mo screw, drawing the slide wedge to the base. Two retaining pins help to align the sample slide wedge and then disengage when the screw is tightened and the transfer rod is retracted. A clutch spring permits rotation of the inverted screwdriver while the pins are engaged with the slide wedge.

The temperature range available was shown to be from 32 K to about 1270 K using 100 W of indirect heating. It was found that Aquadag [3] was needed to prevent galling of the contact surfaces other than the polished Mo wedge contact surfaces. A disadvantage of this design is that attachment of a thermocouple to the sample and transfer of the thermocouple is not provided.

Other transfer systems incorporating low-temperature capabilities for work at cryogenic temperatures are given in references [4-5].

## *References*


[1] N.J. DiNardo, J.E. Demuth, W.A. Thompson, and P.G. Ledermann, Rev. Sci. Instrum. **55,** 1492 (1984).

[2] Custom as well as standard heaters are available from Spectra-Mat, Inc., 100 Westgate Drive, Watsonville, CA 95076. The version used here has W/Re alloy wires embedded in an alumina matrix and housed in a Mo can.

[3] Aquadag is a colloidal graphite paint, which can be brushed onto a surface and then outgassed by bakeout. It provides lubrication and is electrically conductive. Obtain Dag 154 from Acheson Colloids, 1607 Washington Avenue, P.O. Drawer 611747, Port Huron, MI 48061-1747.

[4] G.S. Chottiner, W.D. Jennings, and K.I. Pandya, J. Vac. Sci. Technol. **A5,** 2970 (1987).

[5] R. Raval, M.A. Harrison, D.A. King and G. Caine, J. Vac. Sci. Technol. **A9,** 345 (1991).

***Fig. 23. UHV Transfer—Low-Temperature Capability***

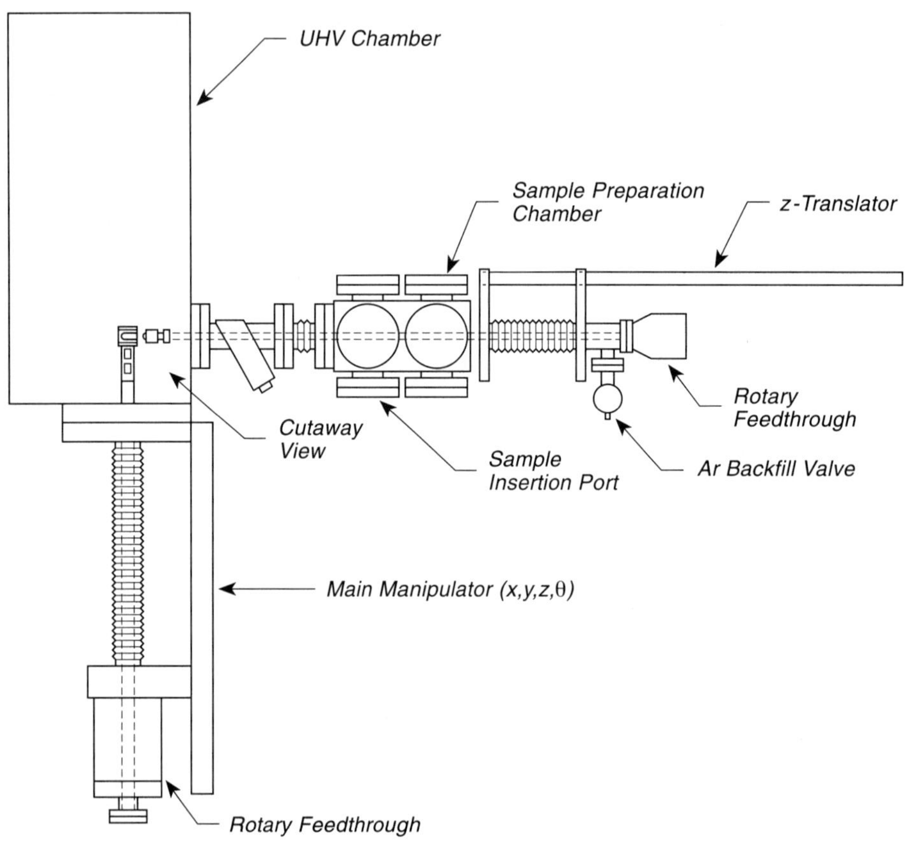


***A. System Configuration***

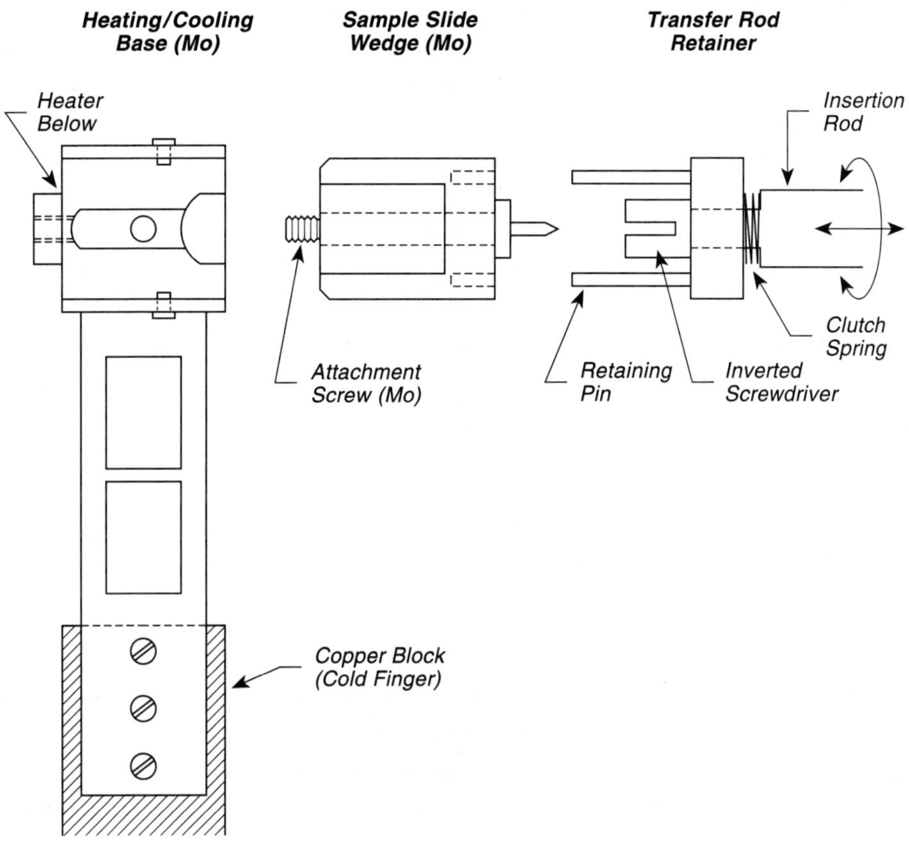


**B. Transfer Devices**

# 24

# Internal Cup High-Pressure Cell

Here is described the provision of an internal cup high-pressure cell that accommodates a sample mounted to move in a normal fashion to various analytical ports in an ultrahigh vacuum chamber [1]. A schematic diagram of the apparatus is shown in Figure 24A, where a sample, mounted in a special cup, can rotate to various analytical ports. One of its positions is opposite a reactor cup assembly that can be moved laterally by hydraulic pressure. The sample cup is rigidly held by a retainer assembly, connected to the reactor cup assembly, which prevents the application of large forces to the manipulator shaft when the seal is made using hydraulic pressure. Figure 24B shows an enlarged schematic view of the reactor cup assembly and its hydraulic cylinder. When the sample has been rotated into its retainer assembly, the reactor cup assembly is pressed against the sample cup, and an annealed Cu gasket produces the seal at a mini-Conflat sealing surface. This operation is assisted by an internally tapered guiding surface on the outer section of the reactor cup assembly. This method of sealing allows repetitive use of the same annealed gasket. The background pressure in the UHV system was found to rise into the low $10^{-9}$ Torr range when the cell was pressurized to 100–2500 Torr. The reactor cell (volume 45 $cm^3$), could be evacuated through gas lines passing through the heavy stainless steel reactor cup assembly, or more rapid evacuation could be achieved by opening the pressurized cell inside the UHV system, causing a pressure surge that decreased into the $10^{-9}$ Torr range in 1–2 min. The original paper [1] contains a brief review of other methods for building high-pressure cells.

A different design, permitting a single crystal catalyst sample to be moved into a small reactor cup, sealed with a Viton O-ring, has been described previously [2–4]. In this design also, the single crystal may be moved in an ordinary fashion to various positions to face different analytical probes mounted radially in an ultrahigh vacuum system. The design in reference [3] cleverly permits the reactor cup to be sealed off from the UHV system when the sample is removed, preventing outgassing effects from the inner walls from contaminating the UHV system.

## *References*


[1] Y-N Wang, R. McAllister, R.G. Herman, G.W. Simmons and K. Klier, Rev. Sci. Instrum. **63,** 5767 (1992).

[2] J.J. Weimer and F.A. Putnam, Rev. Sci. Instrum. **55,** 238 (1984).

[3] K.-E Keck, B. Kasemo, and A. Hoglund, Rev. Sci. Instrum. **54,** 574 (1983).

[4] M.J. Mummey and L.D. Schmidt, Surf. Sci. **91,** 301 (1980).

***Fig. 24. Internal Cup High-Pressure Cell***

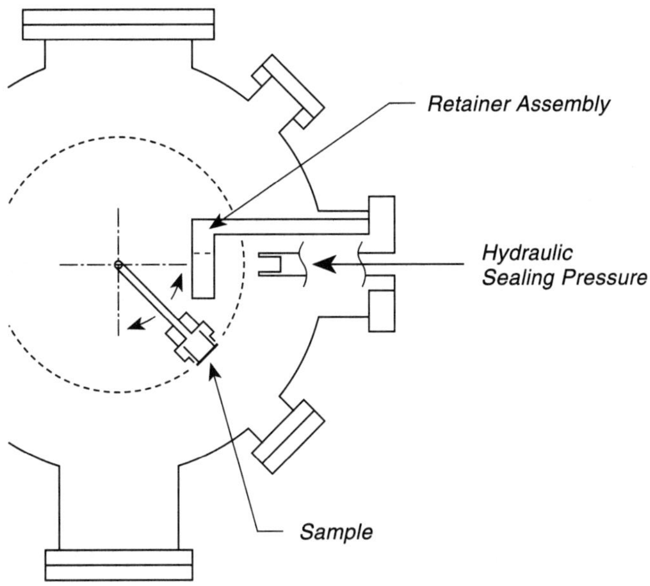


***A. Schematic of Rotating Sample + Cup Assembly***

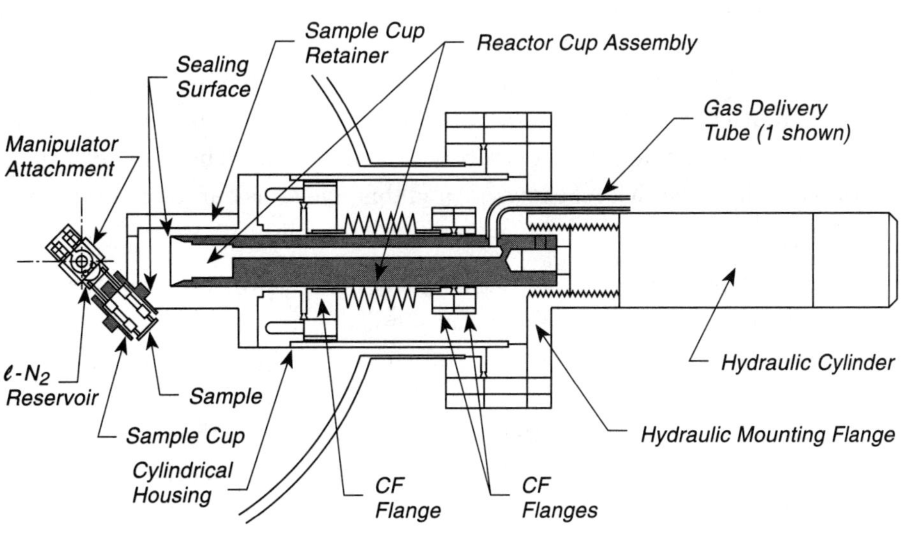


***B. Schematic of High-Pressure Reactor Cell***

# 25

# High-Pressure Transfer Cell

A high-pressure transfer cell that is designed to have a small (30 $cm^3$) volume and provision for sample cooling is shown in Figures 25A and B. Pressures as high as 15 bar are able to be sustained in the high-pressure cell. Sealing is accomplished by a stainless steel/Teflon seat. Impurities on the crystal that might possibly come from the Teflon under compression are eliminated in this design [1].

Figure 25A shows a schematic design for the transfer rod. The sample is electrically mounted on Mo connector rods with W, Ta, or Mo wires. A thermocouple is also welded to the sample. The transfer rod enters the chamber through differentially pumped spring loaded Teflon seals (not shown) such as those mentioned on pages 8–13 [2]. These seals permit both translation and rotation of the sample when it is in the UHV chamber. The transfer rod is self aligning as it enters the Teflon sealing position. The compression of a spring permits the force on the Teflon seal to be measured and reproduced at about 400 lb. Heating of the sample in vacuum to 1300 K is possible by resistive heating. Heating of the sample to about 700 K is possible in 10 bar of $H_2$ if cooling is achieved by using flowing air rather than liquid nitrogen. Cooling to about 130 K can be achieved with liquid nitrogen coolant. In order to prevent extreme cooling of the transfer rod when liquid nitrogen is used, a U-shaped Cu tube is inserted inside the transfer rod wall so that it is held compressively against the inner wall, and warm, dry air is continuously circulated through this tube when liquid nitrogen is used. This prevents failure of the sliding seal mechanism owing to cooling. Teflon spacers are inserted equidistantly inside the transfer rod using a pusher rod with calibrated markings. The outside of the transfer rod is polished to the correct rms roughness following the directions of the spring-loaded Teflon seal manufacturer [2]. The inside of the transfer rod is evacuated with a dual-stage mechanical pump when liquid nitrogen cooling is employed. The outside of the rod is protected with a hypalon/nylon bellows [3] and this bellows is purged continuously with nitrogen gas.

Figure 25B shows the transfer rod sealed into the high-pressure chamber. An alignment insert, H, is brazed between the vacuum housing and the high-pressure cell. It is made of a bakeable and UHV-compatible soft aluminized bronze [4] so that it does not scratch or damage the transfer rod. The annular space between the Teflon seat and the high-pressure chamber is of the order of 0.05 cm, and its length is of the order of 2.5 cm. Thus, at high pressure, possible diffusion of impurities from the Teflon to the sample is minimized by the long diffusion time. Calculations of the diffusion time indicate that at 10 bar of gas pressure, diffusion times of greater than 30 min are involved over the 6.5 cm distance from the Teflon to the sample.

The high-pressure cell is directly attached to a gas handling manifold that serves to introduce gases as well as to extract them for gas chromatographic analysis. In addition, direct pumping of the high-pressure cell by a small turbopump may be carried out.

Proof of the performance was made in two ways. First it was shown that a pressure of 10 bar of $H_2$ led to an increase of pressure of only $1 \times 10^{-9}$ Torr in the UHV chamber. Secondly, storage of an atomically clean Cu crystal for up to 30 min in the evacuated high-pressure cell does not lead to measurable contamination as measured by XPS or TPD methods. Also, long reaction periods at high pressure occur with constant reaction kinetics on a Cu single crystal surface [1].

## *References*

[1] A. Ludviksson, J. Yoshihara, and C.T. Campbell, Rev. Sci. Instrum. **66**, 4370 (1995). This paper also contains references to many other transfer devices in the literature.

[2] Obtained from: Fluorocarbon Co., 10871 Kyle Street, P.O. Box 520, Los Alamitos, CA 90720.

[3] Gortiflex Protector Bellows.

[4] No. UNS C95400, Copper and Brass Sales, 17401 Ten Mile, East Detroit, MI 48021.

## Fig. 25. High-Pressure Transfer Cell

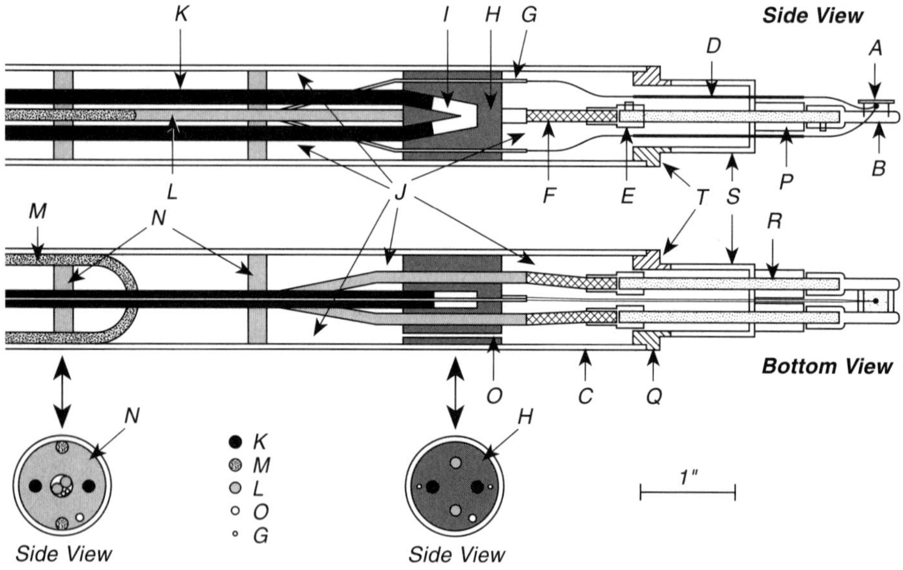


| | |
|---|---|
| A. Sample | K. Stainless Steel Coolant Line |
| B. Mo Connector / Rod | L. Teflon Insulated Power Wire |
| C. Stainless Steel Tube | M. Cu Air Line |
| D. Thermocouple Feedthrough | N. Teflon Spacers |
| E. Copper Connector | O. Evacuation Hole |
| F. Bare Power Wire | P. Ceramic Electrical Feedthrough |
| G. Teflon-Insulated Thermocouple Wire | Q. Stainless Steel Weld Adaptor |
| H. Cu Liquid Nitrogen Reservoir | R. Cu Power Feedthrough |
| I. Liquid Nitrogen Reservoir | S. Skirt for Ceramic Electrical Feedthrough |
| J. Internal Vacuum | T. Sealing Surface for Teflon Seat, D |

### A. Transfer Rod for UHV

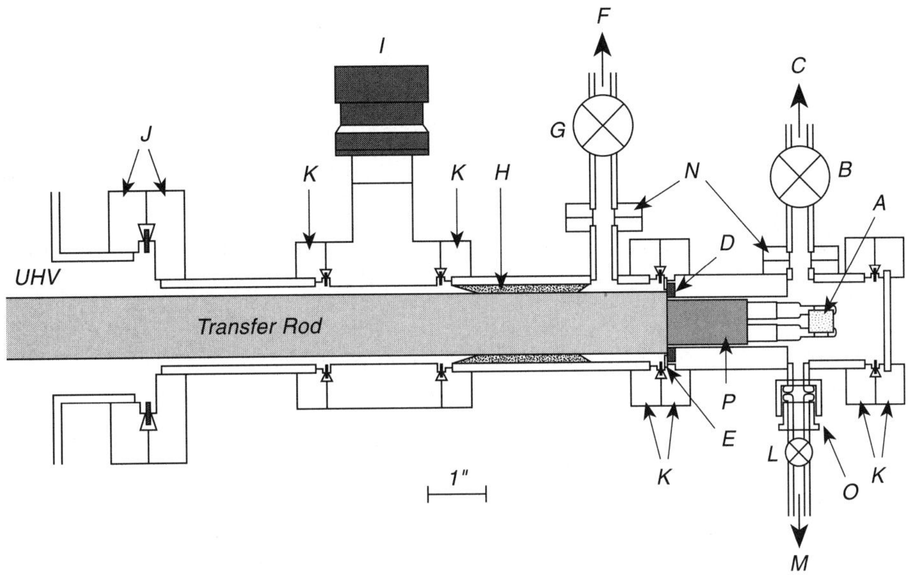


A. Sample
B. Valve to Turbopump
C. Turbopump
D. Teflon Seat
E. Retainer Ring for Teflon Seat
F. Turbopump
G. Valve to Turbopump
H. Aluminized Bronze Insert-Rod Alignment
I. Gate Valve
J. Conflat Flange on UHV Chamber
K. Conflat Flange on Gate Valve
L. Valve to Gas Handling Line + Gas Analysis
M. To Gas Handling Line + Gas Analysis
N. Conflat Flange
O. VCR Fitting + Cu Gasket
P. Ceramic Electrical Feedthrough

**B. High-Pressure Cell**

# 26

# Sample Transfer with Disconnect–I

## Transfer in Standard UHV Systems

Often it may be desirable to transfer a sample from one chamber to another and to disconnect the sample from the transfer device after transfer. In Figure 26 one may see that transfer from a reaction chamber to an ultrahigh vacuum chamber is accomplished by use of a magnetic transfer rod, which is commercially available. The sample, mounted on a ceramic sample holder ring at the end of the transfer rod, sits between three tines on the fork at the end of the transfer arm. The hooks of the tines lock into the sample holder ring by a slight twist, which may be accomplished with the transfer magnet. The electrical connecting pins are machined from 6-32 screws as shown in the central Figure, and these pins insert into homemade sockets made if necessary from alloys designed for thermocouple connections, etc. It is stated that the electrical connections do not have to be particularly tight and that ordinary misalignments in the pins supply the friction level needed [1]. The ceramic in this case was cast from machinable ceramic [2] and fired in a hobbyist's ceramic kiln.

The sample is transferred to a manipulator that has a central translating rod used to drive the tilt mechanism. This tilting action is carried out by moving the central manipulator rod up and down, causing the base holding the sample to tilt backward and forward as desired. A locking nut on this rod prevents overdriving the tilting mechanism. In the bottom portion of Figure 26, two types of samples are shown. On the left, a metallic foil substrate is welded to the screwhead ends of the mounting pins in the base. On the right, a single crystal of a metal is welded to a circular piece of foil which is welded to the mounting pins. A circular hole in the backing foil will permit electron bombardment of the back of the crystal. One might provide a radiation shield between the emitter filament and the ceramic base to reduce heating of the ceramic and outgassing.

The proper location of viewports for monitoring the sample transfer is recommended, or else two persons may be needed to transfer the sample.

## Transfer in STM Systems

The special considerations needed for the transfer of samples being used for scanning tunneling microscopy (STM) studies as well as other more conventional surface science studies have been reported by Sullivan and Kummel [3]. Here, a design coupling the STM sample to a cold bath, electron beam heating, two degrees of rotation, and electrical isolation, along with sample transfer are described, and a review of other manipulator designs is given [3]. Noteworthy is a simple design for transferring sample holders between forklike manipulators [4]. (See also page 80.)

## *References*

[1] A. Garcia and M. E. Kordesch, J. Vac. Sci. Tech. **A11**, 461 (1993).
[2] Aremco, P.O. Box 429, 23 Snowden Avenue, Ossining, NY 10562-0429.
[3] D.J.D. Sullivan and A.C. Kummel, Rev. Sci. Instrum. **63**, 4285 (1992).
[4] B.J. Mulder, J. Phys. **E12**, 908 (1979).

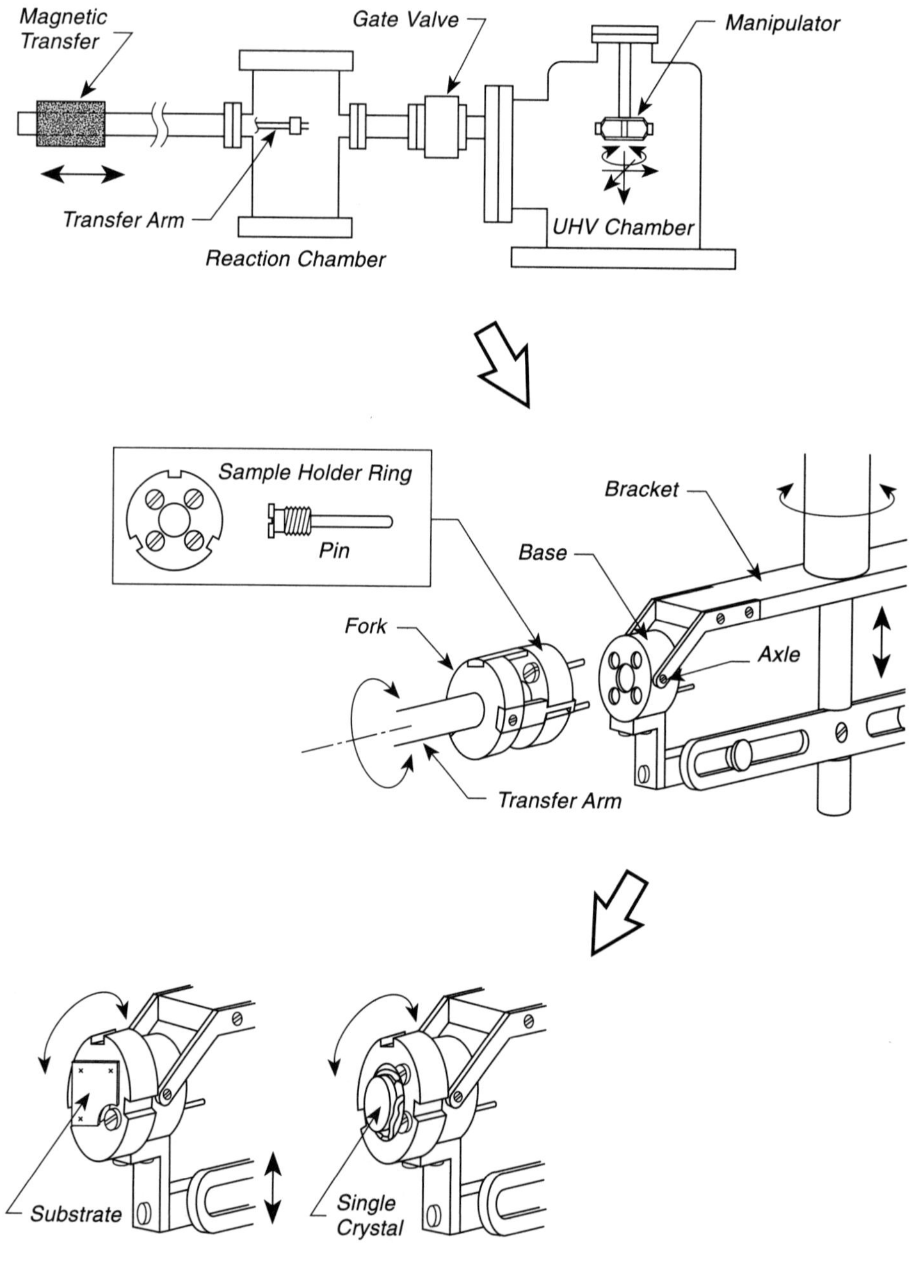


***Fig. 26. Longitudinal Sample Transfer—Disconnect***

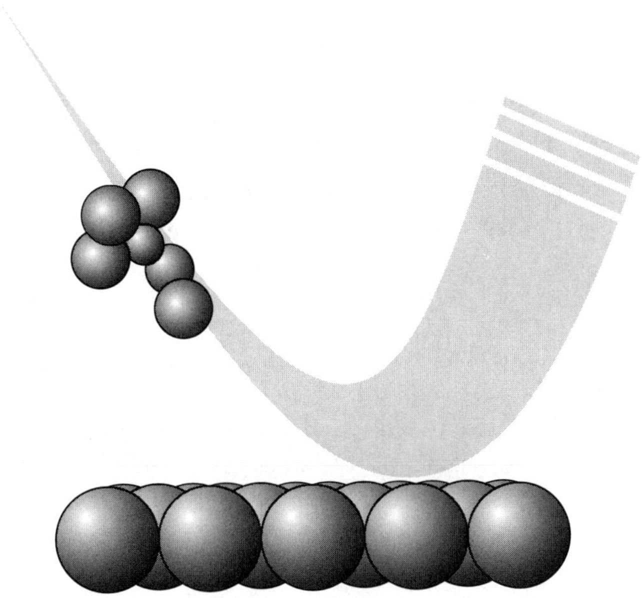

# 27

# Sample Transfer with Disconnect–II

A simple and inexpensive method to transfer a sample with disconnect [1] is shown in Figure 27. The sample is mounted on stainless steel or other metal posts using Ta clamps that bolt into the base of the holder as shown. This assembly is placed on a transfer screw rod in the the loadlock chamber, and then, after loadlock chamber evacuation, the valve between chambers is opened and the sample assembly is passed into the UHV chamber and plugged into the female banana plug receptacles. These receptacles are mounted on a rod that is connected to the rotary manipulator in the UHV chamber. Electrical power is supplied through Cu braid (see page 564) as shown. The sample transfer screw rod is then unscrewed and withdrawn. The UHV banana plugs [2] tend to allow the sample to sag a little, and if this is unacceptable, then a mechanical modification to the design near the banana plugs can be added.

In cases where ohmic heating of the sample is inappropriate, it may be backed with a section of a Si wafer (approximately 0.1 $\Omega$ cm ) for indirect resistive heating.

## *References*

[1] Dr. Ray Kaplan, Mail Code 6863, Naval Research Laboratory, Washington, DC 20375 (private communication).

[2] UHV banana plugs (Model Nos. BPBPM and BPBPF) can be obtained from: Surface/Interface Inc., 110 Pioneer Way, Suite H, Mountain View, CA 94041.

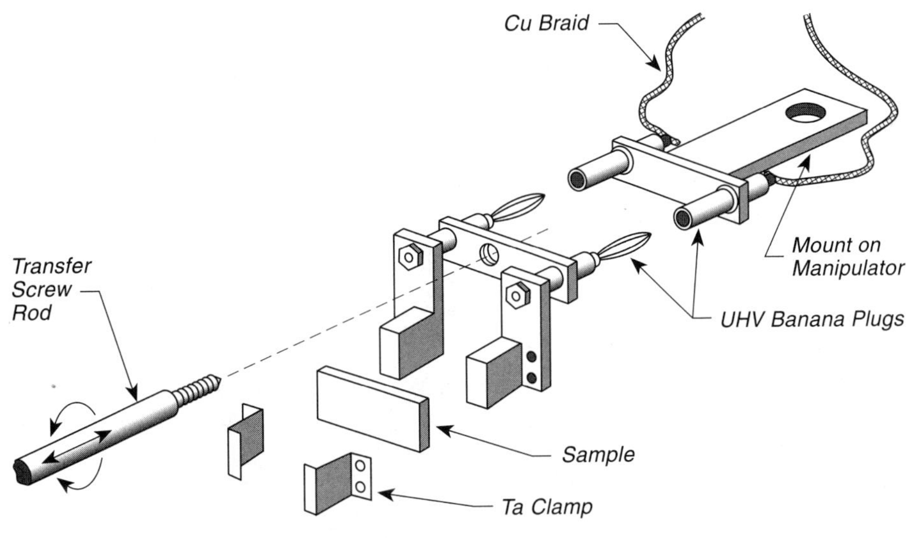


***Fig. 27. Sample Transfer with Disconnect***

# 28

# UHV Access Port

The entry into an ultrahigh vacuum chamber can often be achieved through a flanged entry port that contains a standard copper gasket. For large ports that permit access to the entire chamber, one wall of the system may be designed to be reversibly removable without a need for gasket replacement [1].

The design for a removable wall for a rectangular access port is shown in Figure 28. The port contains spring-loaded Teflon [2] seals in separate wells. The regions between the Teflon seals are differentially pumped to prevent small leakage from influencing the vacuum conditions in the chamber. Thus, an ion pump is used on the pumping region closest to the chamber, and a roughing pump is used on the pumping region closest to the atmosphere. The access port is arranged, by means of a recessed section, to be pulled down tightly against the mating surface, providing some further sealing at the very inner section of the seal.

## *References*


[1] D.J. Auerbach, C.A. Becker, J.P. Cowin, and L. Wharton, Rev. Sci. Instrum. **49,** 1518 (1978).

[2] The spring-loaded Teflon seals are available from the Fluorocarbon Co., 10871 Kyle Street, P.O. Box 520, Los Alamitos, CA 90720. Many sizes of the seals are available, and instructions for their use are given by the company.

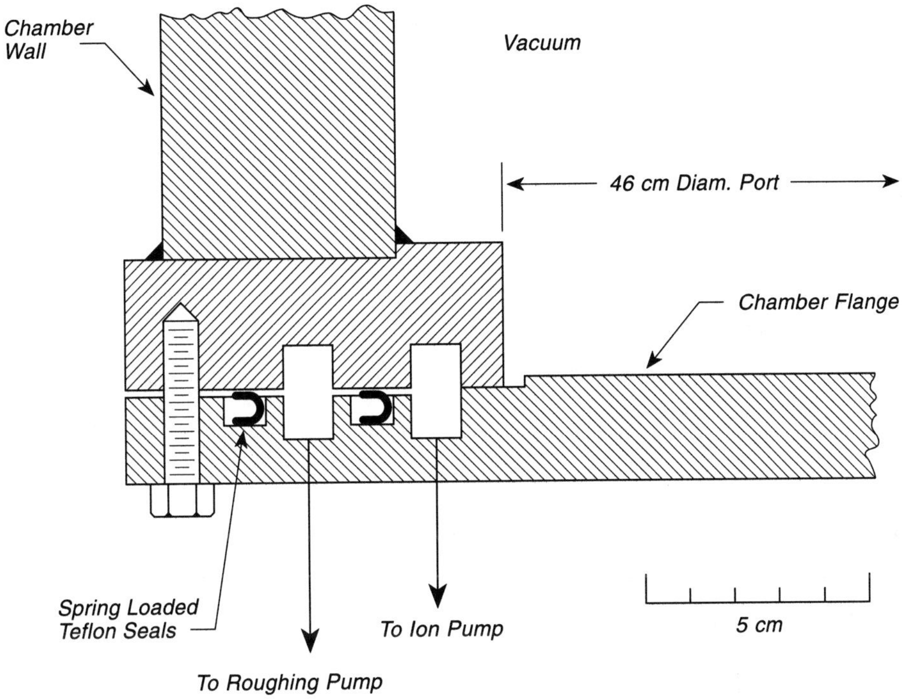


***Fig. 28. UHV Access Port***

# C. Electrical Connections

# 29

# Tungsten Ribbon—Attachment to Power Leads

Heated tungsten ribbons are used in many experiments as radiation sources, metal deposition sources, electron bombardment sources, and molecular cracking sources. Most commonly, the ribbons are attached to heating leads by spot welding. This method requires that the mounted power leads be removed from the vacuum system for ribbon replacement, a time-consuming procedure. An alternative method for ribbon attachment shown in Figure 29 may be carried out inside the vacuum system without welding [1].

A Ta feedthrough lead has a slot milled into it as shown in Figure 29(a). The length of the slot is about 1.5 times the ribbon width, and the width of the slot is about 1.5 times the ribbon thickness. A small cap is machined from Ta, for example, as shown in Figure 29(b). The inside diameter of the cap is larger than that of the feedthrough lead by one thickness of the ribbon. The cap length is three to four times that of the ribbon width, and the depth of the bore is 2.5 times the ribbon width, leaving plenty of material in the slotted upper region. The cutout on the cap is centered on the cap so as to allow the ribbon to pass as shown in Figure 29(c). The end of the lead extends internally above the cutout by about 0.5 mm, as shown by the dotted lines in Figure 29(c), preventing the cap from tilting when rotated.

To assemble the device, the ribbon is passed through the slot in the lead with the cap in place, and extends about one-half the lead diameter beyond the cap. To secure the ribbon, the cap is rotated one-half turn as shown, wrapping the ribbon around the lead under the cap, and locking it in place.

This method will work on very thin W ribbon and has been extended to 0.010-mm-thick ribbon material, 1.2 mm wide; this particular mounting was done on 1.5-mm-diameter Ta leads. To keep the ribbon straight after it has been used, it is common to strain the two leads toward each other before making the second cap attachment. The strain in the two supporting leads will place tension on the ribbon and keep it straight.

## *References*


[1] Mr. Gerald L. Fowler, 6816 Kelly Ave., N.E., Albuquerque, NM 87109 (private communication).

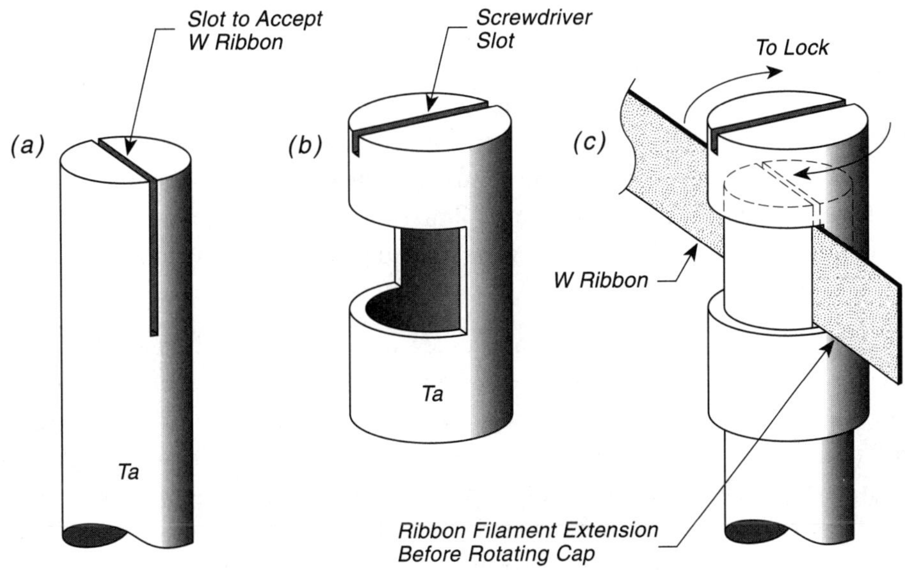


***Fig. 29. Tungsten Ribbon—Attachment to Power Leads***

# 30

# Wrap Connection—Tungsten to Tungsten

Tungsten is a brittle metal that is difficult to bend or wrap. In addition, tungsten is also difficult to weld. The method shown in Figure 30 involves a wrapping technique that produces tungsten coils that fit onto tungsten wires tightly without welding. Such coils can be used as a friction support or locator in high-temperature applications [1].

Splintering or breakage during fabrication of tungsten wire above about 0.25 mm diameter is a common occurrence owing to its inherent brittle character. While this may be avoided by bending tungsten wire while it is being heated to redness in a flame, heavy oxidation will occur and also intricate bends are difficult to achieve before the tungsten cools. The method shown here prevents splintering of 0.5-mm-diameter W wire when it is bent into a tight coil. Such coils may be used to make strong friction contacts to 0.5-mm W wire as will be shown.

A hand drill is mounted in a vise, and a core wire of W, 0.5 mm in diameter, is held in the jaws of the chuck. The second piece of similar W wire to be wrapped is also inserted into the chuck between the edges of the jaw pieces as shown so that it will be prevented from slipping when the wrapping begins. About two revolutions of the chuck are carried out, giving preliminary contact between the core wire and the wrapping wire; the wrapping wire does not start out directed normally from the axis of rotation. A wood-forming clamp is next placed on the wrapping wire coil, and while under pressure, the coil is formed to the desired length. The friction between the wood and the W will prevent fracture. The ends of the wrap wire coil can now be cut to the desired length using tungsten carbide end nippers and dressing on a grinding wheel if necessary.

The coil is removed from the core wire, and a slight bend is placed in the center of the coil, using either two pairs of pliers or a knife blade inserted partially into the coil. This slight bend allows the coil to be forced onto a 0.5 mm W wire and to remain fixed owing to the strain induced in the coil as it is straightened by the inner wire. The coil will not move at high temperature either during ohmic heating or bakeout.

## *References*

[1] Mr. G.L. Fowler, 6816 Kelly Ave., N.E., Albuquerque, NM 87109 (private communication).

***Fig. 30. Wrap Connection—Tungsten to Tungsten***

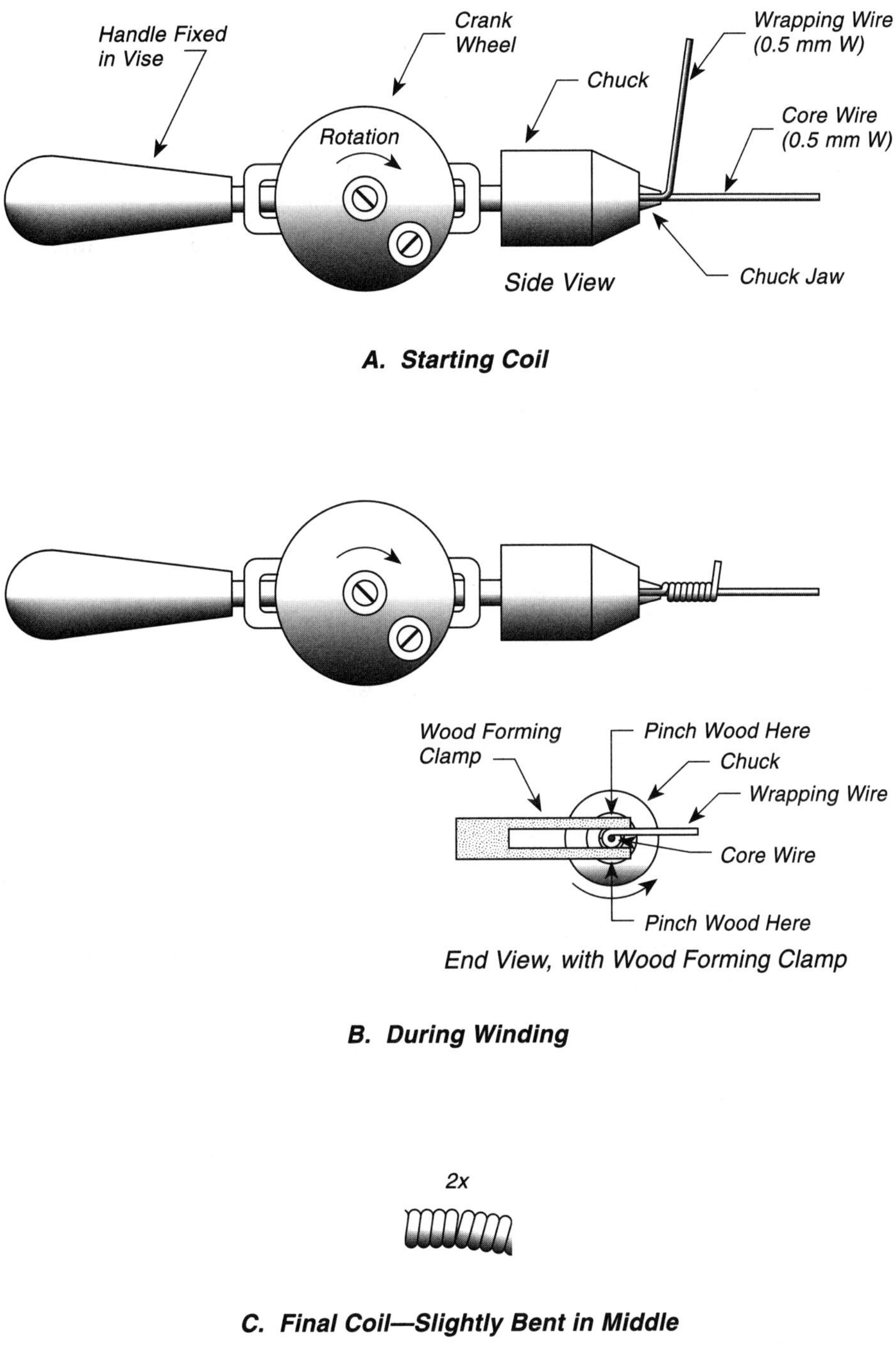

# 31

# Cold-Formed Wire Connector

The fabrication of electrical connections that are conductive for both electricity and heat may be carried out using silver soldering, welding, or screw-type connectors. Another method, involving a press fit of a hard wire into a soft material such as Cu or Al, is shown in Figure 31. Here a vise is used to press the junction, and the various stages of compression are illustrated. The last stage (d) involves deformation of the soft metal and enclosure of the hard wire. The resistance of these junctions between tungsten and copper is typically about 0.05 Ω [1].

## *References*


[1] D. Edwards, Jr. J. Vac. Sci. Tech. **16,** 970 (1979).

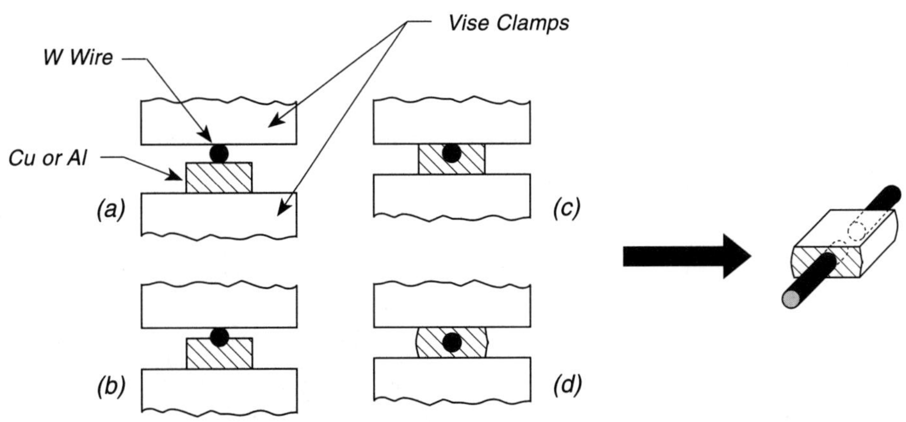


*Fig. 31. Cold-Formed Wire Connector*

# 32

# Ultrahigh Resistance Vacuum Feedthroughs

The measurement of the electrical conductivity of materials with a very high resistance can be plagued by lower-resistance pathways to ground that interfere with the electrical measurements. In particular, $Al_2O_3$ ceramic insulators exhibit resistances of about $10^{12}$ Ω, and these may deteriorate in their resistance with time due to the adsorption of water on the alumina or to the deposition of thin metal films on the vacuum side of the insulator surface. Such problems are termed electrical leakage problems, and prevent measurements into the $10^{16}$ Ω range, which are often needed [1]. In some cases, using an enclosure containing drying agent on the outside of the feedthrough will help, and the more complicated use of forepump evacuation of this region will prevent water adsorption there. For the highest resistivity feedthroughs, other methods must be employed, as shown in Figure 32.

The Figure shows the construction principles of a doubly insulated and guarded design that permits measurements down to the $10^{16}$ Ω range. A special crystalline sapphire-metal feedthrough [2] is employed, and this is brazed into an ordinary Cu tubing feedthrough mounted on a flange by means of an alumina ceramic insulator. The sapphire feedthrough contains a hollow tube. Into this tube is placed either a Ni or Cu conductor wire, which is silver brazed under hydrogen into the hollow tube. At the same time the brazing of the sapphire thimble and the flared Cu tubing is carried out. Great care must be taken to avoid extreme thermal shocks at the sapphire. The assembly on the vacuum side is provided with a second shield as shown, and also shielding of the outer air side of the feedthrough is carried out as shown.

The impedance to ground of the sapphire insulator approached $10^{16}$ Ω when the sapphire was thoroughly cleaned [1]. Spectroscopic grade alcohol, followed by distilled water spraying through a fine glass nozzle, is recommended, followed by blotting the excess water and drying in an oven.

## *References*


[1] A.H. Madjid, J.M. Martinez, and A.A. Vaala, Rev. Sci. Instrum. **43,** 1049 (1972).

[2] Obtained from: Cory Lab, Inc., 823 5th Street, Menominee, MI 49858.

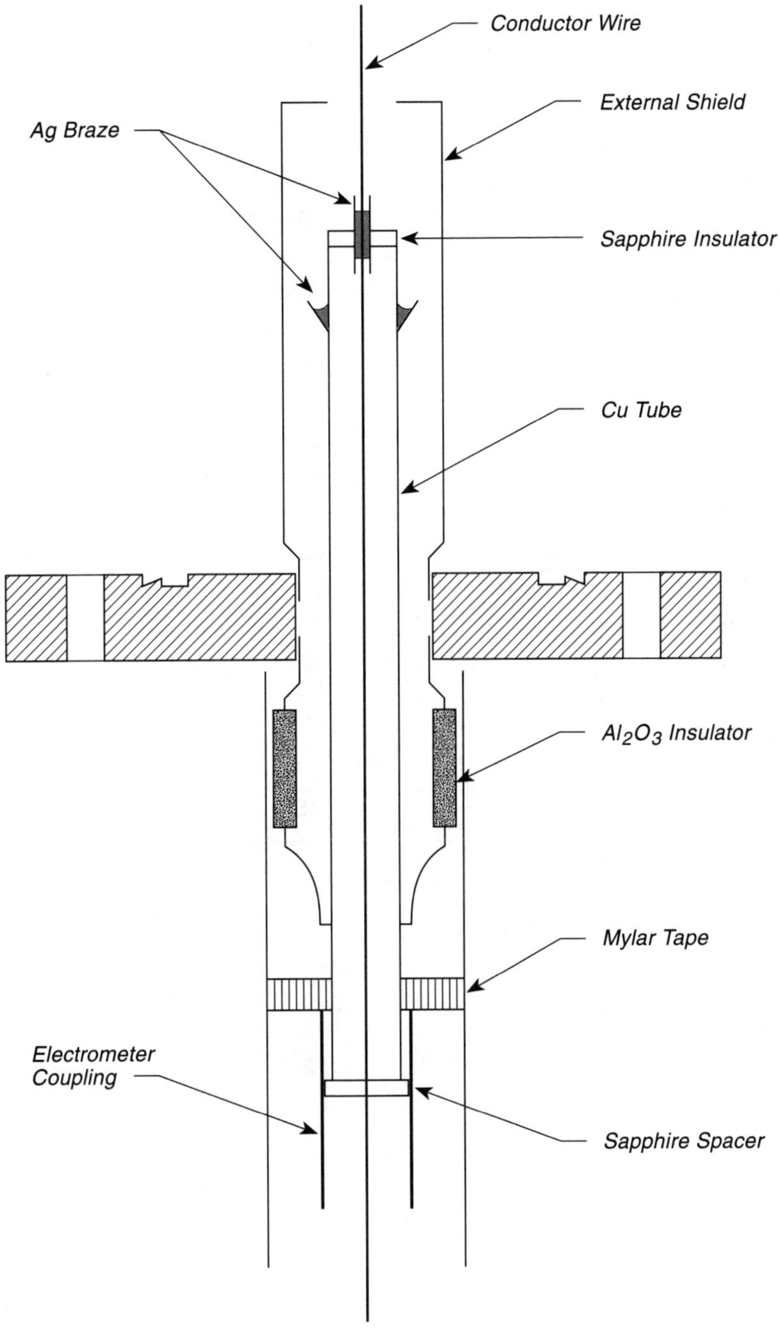


***Fig. 32. Ultrahigh Resistance Vacuum Feedthrough***

# 33

# Sliding Metal Electrical Contact

A need for a sliding electrical contact can occur when flexible coils of ceramic bead-insulated wire are unsatisfactory, perhaps because of the formation of ceramic dust as motion occurs. Figure 33 shows a sliding stainless steel rod that moves through a Teflon bearing that is mounted inside a re-entrant stainless tube connected to a vacuum linear drive of the bellows or magnetic type. Electrical contact to the rod is made by a phosphor bronze spring contact rigidly mounted on a stainless support assembly, and electrically isolated by ceramic insulator cylinders that contain small bolts for holding the spring contact strip. The phosphor bronze spring strip is electrically connected to an electrical feedthrough as shown. The holes drilled in the Teflon-bearing should be about 5% larger than the sliding rod to allow for deformation of the Teflon as well as for its large thermal expansion coefficient. The contact can pass a 2-A current. The stainless steel rod can be moved in and out at 5 cm/s without difficulty, and cold welding in ultrahigh vacuum does not occur [1]. The authors also report that a Ta wire spring may be used successfully against a Mo slider in a rotary device [1].

## *References*


[1] S.M. Newstead, R.A.A. Kubiak, and E.H.C. Parker, J. Vac. Sci. Tech. **A2,** 1603 (1984).

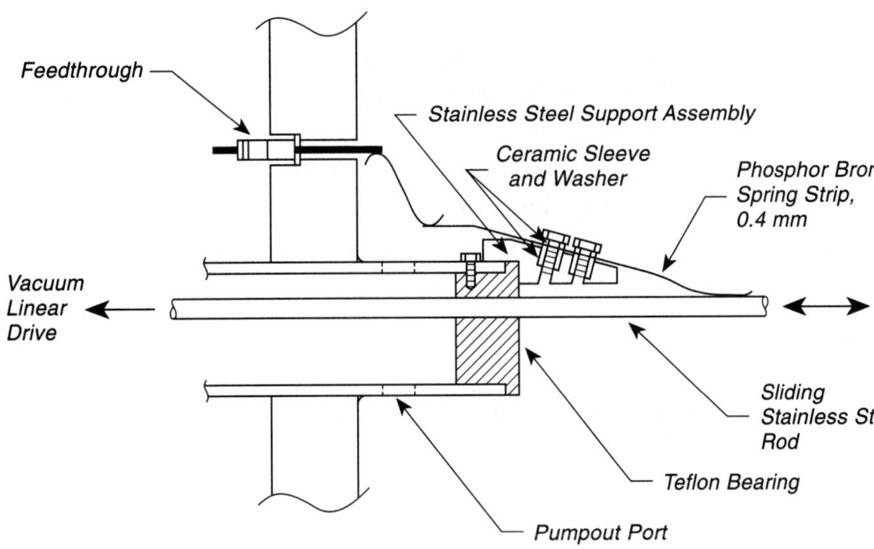


***Fig. 33. Sliding Metal Electrical Contact***

# 34

# Low-Profile Electrical Lead Clamp

There is often a need for a band clamp arrangement to anchor electrical leads to a manipulator rod or other structure. Often, narrow clearance spacing is available around the rod so that the binding on the electrical leads to the rod must be achieved in a limited annular space. Figure 34 illustrates one solution to this problem, using half-hard Cu-Be spring stock. This material is formed as desired by using steel-forming wires of the correct diameter to produce the rounded grooves in the Cu-Be stock by hammering the forming wire into the stock with the forming jig as a guide to fabrication. An extra channel is provided as an additional spring expansion channel to give the band extra elasticity. The tabs at the end of the band are each about 3–4 mm in length. To install, the ceramic insulated wires are placed into position on the rod and temporarily anchored. Overlapping the Cu-Be tabs and folding them into each other will result in a tension holding the band and the ceramic insulated wires in place. The band will not soften or lose its spring quality from -200°C to bakeout temperatures [1].

## *References*


[1] G.L. Fowler, J. Vac. Sci. Tech. **A5**, 2978 (1987).

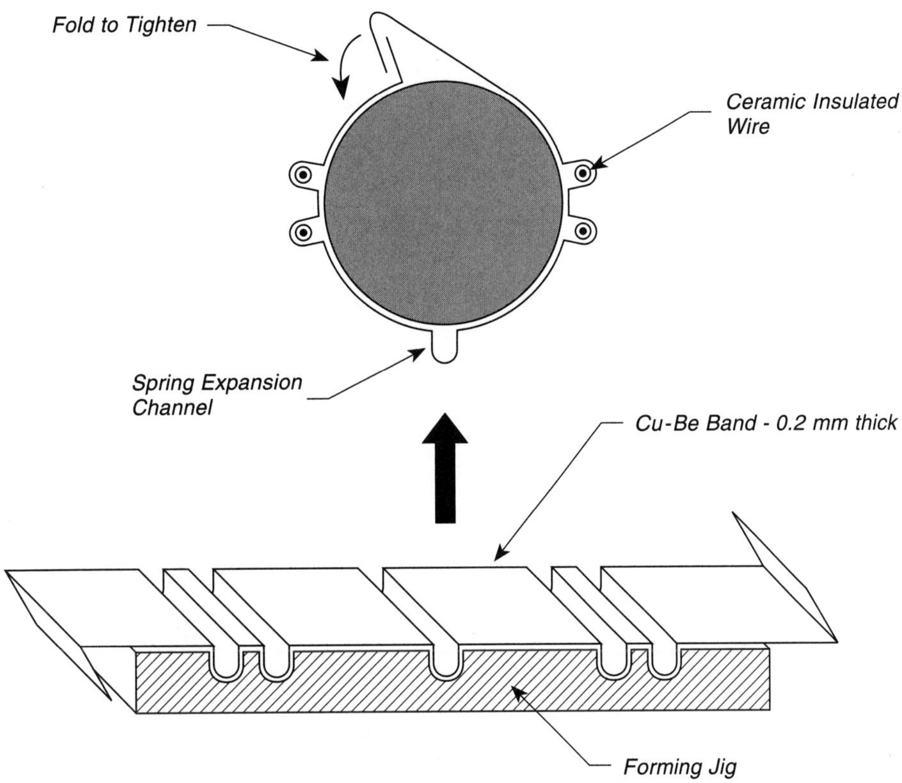


***Fig. 34. Band Clamp for Electrical Leads***

# D. Pumping and Trapping

# 35

# Water Aspirator/Sorption Pump Combination for Efficient UHV System Evacuation

On ion-pumped UHV systems, it is of course necessary to bring the system pressure down from atmospheric to about $10^{-2}$ Torr before starting the ion pumps. This may be done in several ways, including the use of zeolite sorption pumps that are cooled to liquid nitrogen temperatures in order to physically adsorb all of the atmospheric gases. Because of the large quantity of air that must be removed by the zeolite, the process of reaching the limiting pressure sometimes exceeds 1 h. A method to remove the majority of the atmospheric gas has been reported that is free from the possibility of causing oil contamination [1]. The use of a rotary pump does not exclude this possibility.

Figure 35 shows a schematic diagram of the use of a water aspirator for removing the majority of the atmospheric gas. The aspirator is prevented from admitting water to the system (equilibrium vapor pressure of water = 24 Torr at 25°C) by interposing a liquid nitrogen trap. After this aspirator evacuation, which takes about 1–2 min, the aspirator is isolated from the pumping system by closing valves $V_1$ and $V_2$. The sorption pump is then cooled with liquid nitrogen, and valve $V_2$ is opened. It is observed that the pressure in the vacuum system is reduced to about $5 \times 10^{-3}$ Torr in only about 10 min, since the capacity of the zeolite is much greater than the quantity of air that remains to be pumped after the aspirator evacuation. Several hours of pumping with the zeolite trap used alone would be required to go to this pressure range.

Following the initial pumpout of the system, valves $V_4$ and $V_3$ are closed, and the bake-out processing of the system is carried out to reach UHV conditions, as discussed on page 132. This procedure allows the sorption pump to be cycled several times without the need for warmup and vaporization of the sorbed gases.

## *References*


[1] A.M. Russell, Rev. of Sci. Instrum. **36**, 854 (1965).

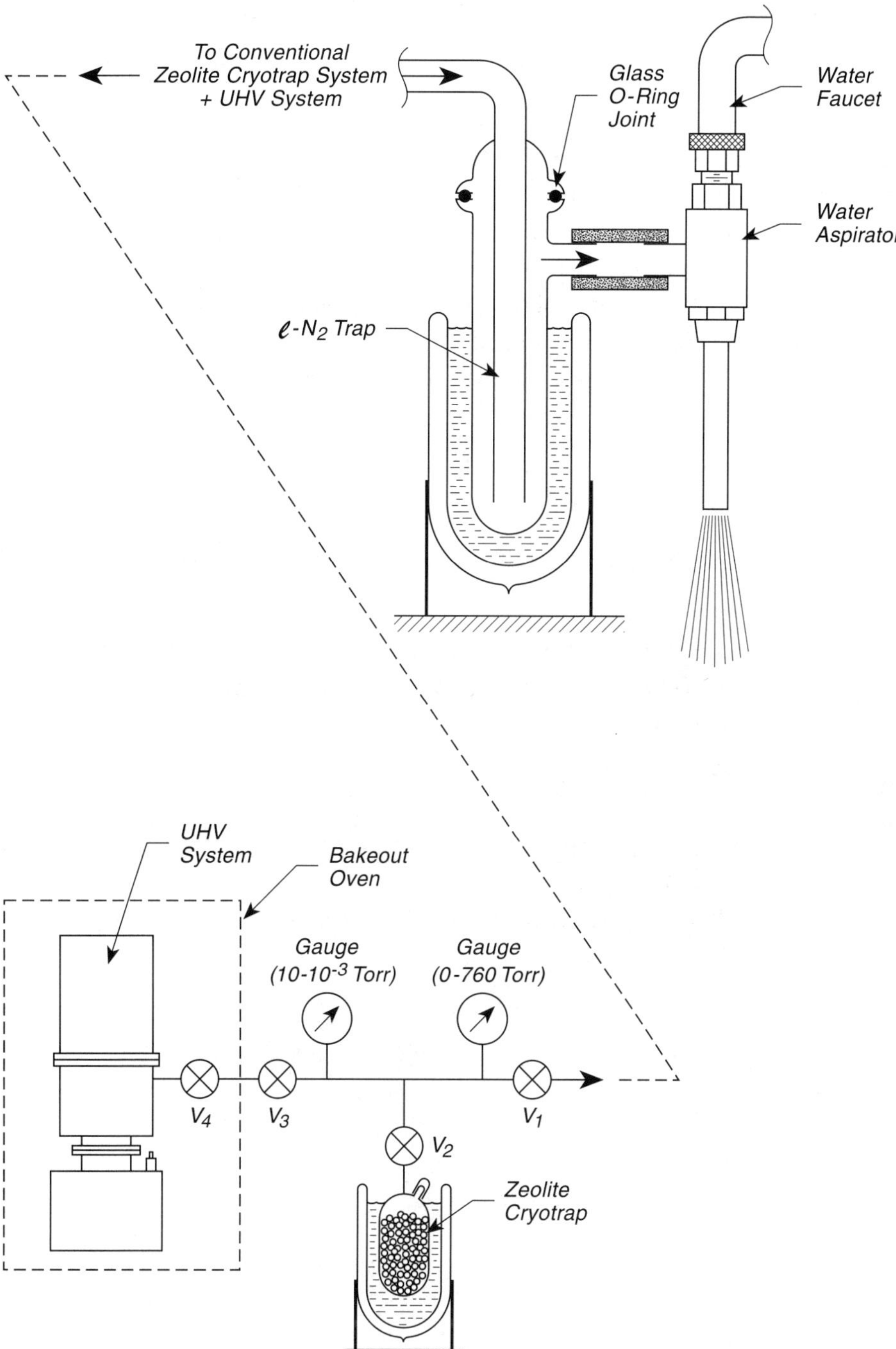


***Fig. 35. Water Aspirator / Sorption Pump Combination***

# 36

# Ballast Pumping—Vibration Free

There are often situations in which the elimination of mechanical vibrations in an ultrahigh vacuum system would require turning off the mechanical forepump. This is feasible using the method to be described [1], since the gas load being pumped from the system is very low if ultrahigh vacuum conditions exist there. The system is forepumped by a large ballast volume of the order of 50 L, which is isolated from the mechanical pump by a valve as shown in Figure 36. The ballast will pump for several days without significant degradation of the turbopump's ability to pump. It can be evacuated in minutes after several days' operation where the forepump has been turned off.

This arrangement is also useful for maintenance of the forepump, since it may be disconnected from the ballast after the isolation valve is closed.

The same idea may be applied to diffusion pumps where large gas loads are not being handled.

A variation on this idea is to back a turbopump with an ion pump over evenings and the weekend. If the power should fail, problems with oil sucking back into the system from the forepump will be avoided [2].

## *References*


[1] Dr. Oliver Schaff, Fritz Haber Institut der Max Planck Gessellschaft, Faradayweg 4–6, D-14195, Berlin (private communication).

[2] Dr. W. F. Egelhoff, National Institute of Standards and Technology, Gaithersburg, MD 20899-0001 (private communication).

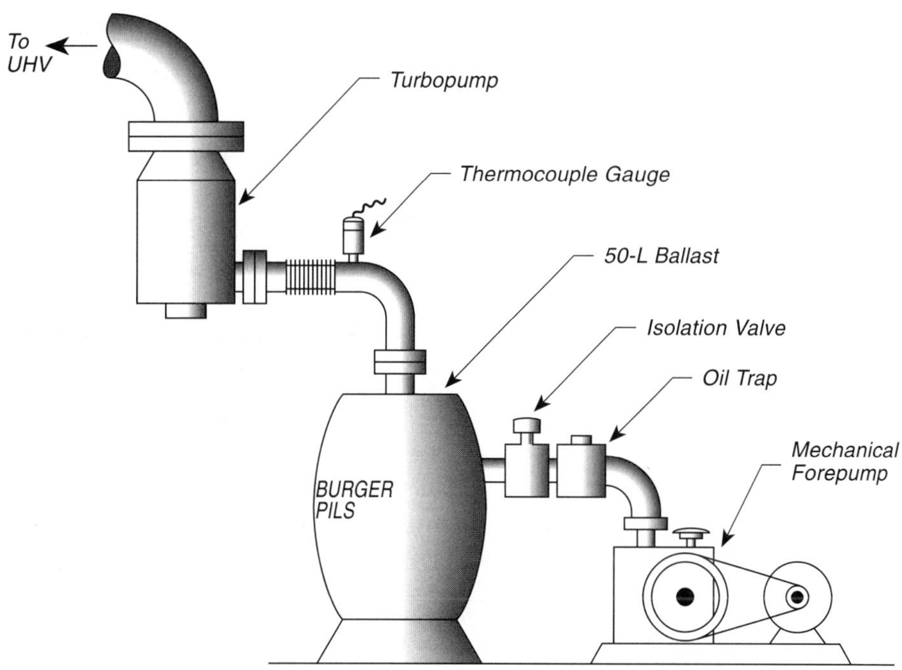


***Fig. 36. Ballast Pumping—Vibration Free***

# 37

# The Use of Appendage Titanium Sublimation Pumps for Pumping Low Gas Loads in Ultrahigh Vacuum

Getter films of active metals have been used for many years in ultrahigh vacuum technology, especially in the days when work was carried out in glass vacuum systems that were sealed off from dynamic pumping [1]. This method of pumping chemically active gases is superior to all other methods insofar as secondary effects are concerned. The principle of getter film pumping is the strong irreversible chemisorption of chemically active gases by clean active metal surfaces. Simple considerations of monolayer capacity indicate that 1 $cm^2$ of a getter surface will pump of the order of 1–3 x $10^{-5}$ Torr L of an active gas. Film roughness factors of 2 or 3 will increase the gettering capacity accordingly.

One example of the use of a Ti sublimation pump (TSP) for pumping low flux gases from a differentially pumped mass spectrometer is shown in Figure 37. Here, a thin film of Ti is deposited on the stainless steel wall of a cylindrical appendage each morning before beginning work. The Ti film is kept at 77 K to maintain its pumping efficiency to the highest level. A single Ti film will remain active all day during use of the mass spectrometer at low gas loads. This particular apparatus also contains a 60-L/s diode ion pump for maintenance of vacuum quality in the mass spectrometer overnight when the titanium pump is not being used [2].

Careful studies, in which the Ti getter film pump has been compared with ion pumping, show that regurgitation effects [3] involving methane liberation from the ion pump are completely eliminated when the Ti pump is employed [2]. This may be seen by looking at the thermal desorption spectra of methyl chloride from an inert surface. The use of the TSP eliminates the $CH_4$ evolution originating in the ion pump. However, wall displacement effects caused by molecules entering the mass spectrometer region from the single crystal when it is heated in thermal desorption experiments can still be detected. These displaced molecules were $H_2$, CO, and $H_2O$ in the particular experiments carried out [2].

The detailed behavior of the residual gases in a liquid nitrogen-cooled titanium sublimation pumping system has been reported [4].

## *References*


[1] P.A. Redhead, E.V. Kornelsen, and J.P. Hobson, Can. J. Phys. **40,** 1814 (1962).

[2] G. Lu, A. L. Linsebigler, and J.T. Yates, Jr., J. Vac. Sci. Tech. **A12,** 384 (1994).

[3] P.A. Redhead, J.P. Hobson, and E.V. Kornelsen, *The Physical Basis of Ultrahigh Vacuum* (AIP, New York, 1993), pp. 403–406.

[4] L.Y.L. Shen, Rev. Sci. Instrum. **43,** 1301 (1972).

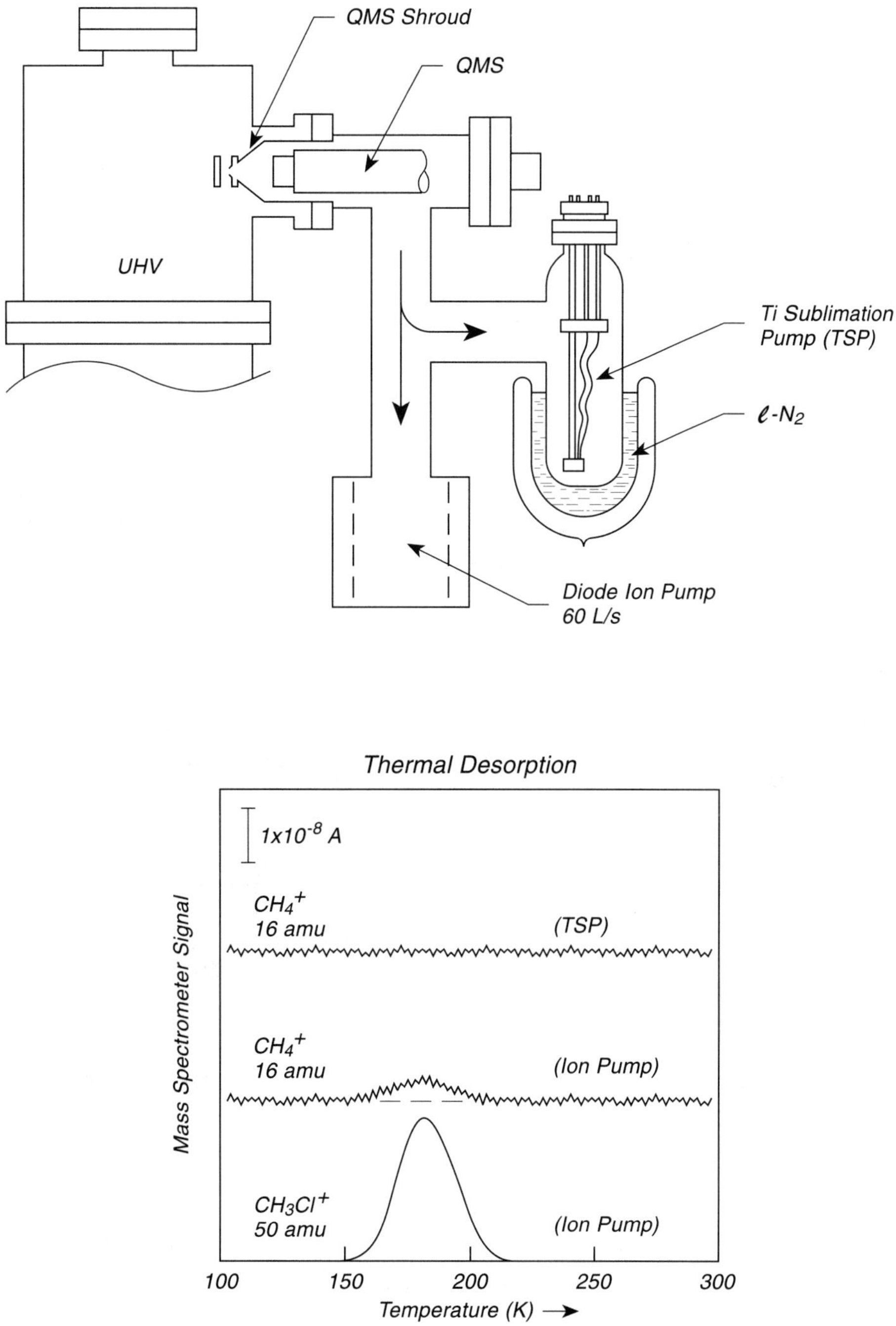


**Fig. 37. Appendage Titanium Sublimation Pump**

# 38

# Pressure-Equalizing Device in Complex Systems

Often in systems such as those used in molecular beam research, it is undesirable for large pressure differences to exist during either pumpdown or during vacuum failure in one section of the apparatus. Large pressure differences, and the associated flow of gas, can damage fragile parts such as thin walls, skimmers, and orifices.

The device shown in Figure 38 is designed to relieve large pressure differences between sections of a vacuum system that are pumped in parallel [1]. Thin metal pressure-release sheets are designed to open in either direction if a large pressure difference occurs. The thin metal sheets are mounted inside flanges that accept Viton O-ring seal rings. The O-ring seal ring is brazed to a disk containing two holes, and the thin metal sheets are mounted on opposite sides of the disk in order to relieve pressure differences in either direction.

## *References*

[1] Dr. R. David, Institut fur Grenzflächenforschung und Vakuum physik, Forschungszentrum Jülich GmbH, Postfach 1913, D-52425 Jülich, Germany (private communication).

***Fig. 38. Pressure-Equalizing Device***

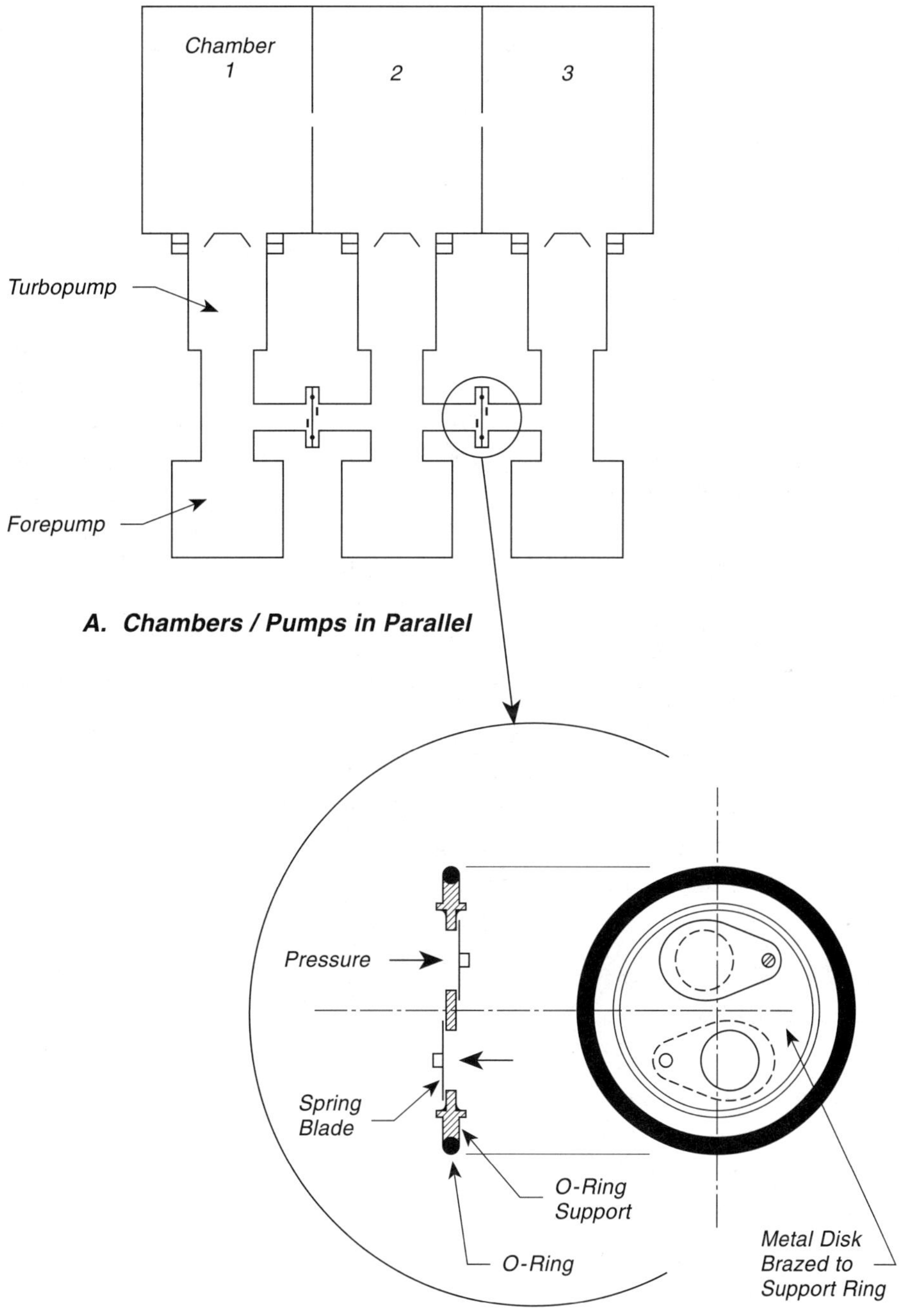


**B. Pressure-Release Device**

# 39

# Diode Ion Pump Performance in He Pumping

There is sometimes a need to fill a vacuum system with a high-pressure inert purge gas, and then to recover ultrahigh vacuum. Diode ion pumps have only a limited pumping speed for inert gases such as He, and their performance is of interest in this application; triode ion pumps are designed to have good pumping speed for inert gases at a small sacrifice in total pumping speed.

In Figure 39, a series of pumping tests were performed on an ultrahigh vacuum system that could be filled with high-purity He at 1 atm pressure. The He was pumped out with a trapped mechanical pump to about $1 \times 10^{-5}$ Torr, and then the diode ion pump was employed. Figure 39 shows selected pumping performance curves after the first He filling and the fifth. The performance degrades in successive tests because of traces of water vapor admitted with the high-purity He and also because of a drop in efficiency of the diode ion pump owing to the regurgitation effect in which previously pumped gas is partially evolved in successive pumping periods. Baking out (BO) the ion pump (IP) only for 4 h at 250°C restores some of the original pumping speed; baking out the ion pump plus the vacuum system restores all of the original performance seen in the first test.

The behavior shown in Figure 39 indicates that He can be pumped well by a diode ion pump from ultrahigh vacuum systems made of stainless steel using highly purified He as a purge gas, and that short bakeouts at 250°C restore system pumping performance. It was suspected that residual water in the He (1.9 ppm) was partially responsible for degradation of the pumping behavior in successive tests [1].

## *References*

[1] P.M. Danielson, J. Vac. Sci. Tech. 5, 89 (1968).

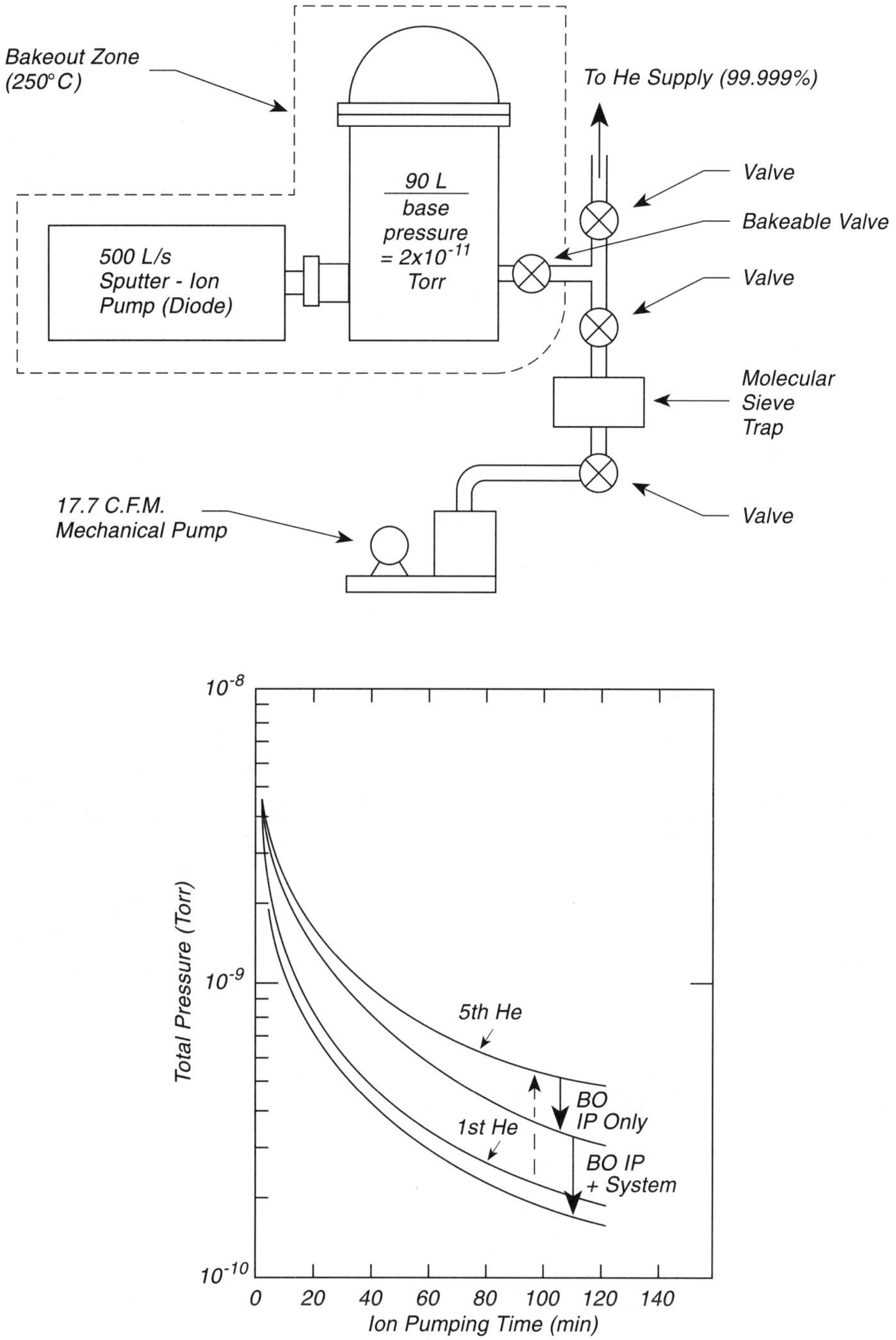


***Fig. 39. Test Chamber and Pumping Behavior after He Purge (1 atm)***

# 40

# Vacuum Applications of Metal Foams

Copper [1], nickel [1] and aluminum [2] foams are commercially available and may be used in a number of vacuum applications [3]. The foams are highly porous, they are easily shaped and machined, and they have some inherent strength. They also can be easily brazed or soldered.

Four vacuum applications of foamed metals are illustrated in Figure 40. In panel A, a plug of foamed Cu is used as a hydrocarbon trap between a forepump and a vacuum system. The high surface area and the torturous pathways in the foam impede hydrocarbon passage through the trap. In panel B, a zeolite trap is retained between porous Cu plugs, and serves also as a hydrocarbon trap. The zeolite must be baked out prior to becoming active and acts as a molecular sieve for hydrocarbon molecules. Panel C shows a plug of Cu foam used as a conductance limiter under molecular flow conditions. To adjust the conductance accurately, additional Cu foam segments can be added as needed. Finally, in panel D, a cylindrical piece of metal foam is used as a diffuser for injection of gases in random directions into a vacuum system.

Metal foam has a cellular internal structure. Low-density foam has cell diameters in the range 0.5–1.5 μm, whereas high density foam has cell diameters from 9 to 40 μm. Typical surface areas range from 0.025 to 0.038 $m^2/g$. For low-density foam the density is typically from 2% to 5.5% of that of the parent metal. For high-density foam, the density ranges from 20% to 60% of the bulk metal. Low-density foam can be compressed by up to a factor of 7.

Measurements of the conductance of 2.4 mm thick disks of 5% density Cu foams made by compression of 1.9% density foam gave a conductance (for air?) of 3.2 L/s $cm^2$.

## *References*


[1] Foametal, Inc., 37645 Vine Street, Willoughby, OH 44904.
[2] Energy Research and Generation, Inc., 900 Stanford Avenue, Oakland, CA 94608.
[3] B.R.F. Kendall, J. Vac. Sci. Technol. 17, 1385 (1980).

## *Fig. 40. Vacuum Applications of Metal Foams*

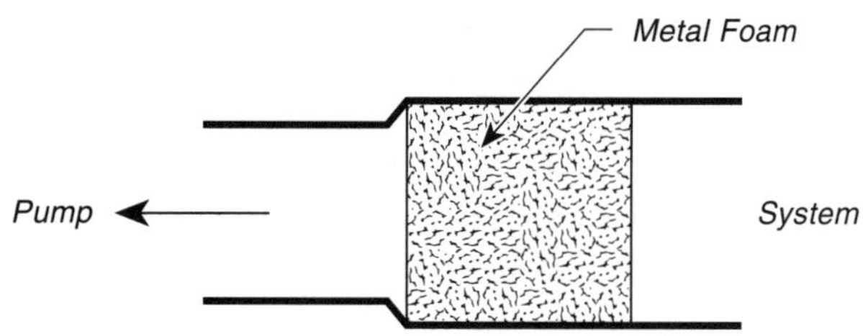


***A. Foreline Hydrocarbon Trap***

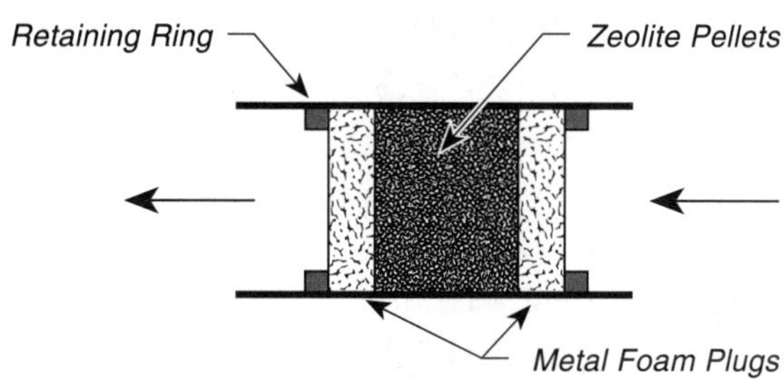


***B. Foreline Zeolite Trap***

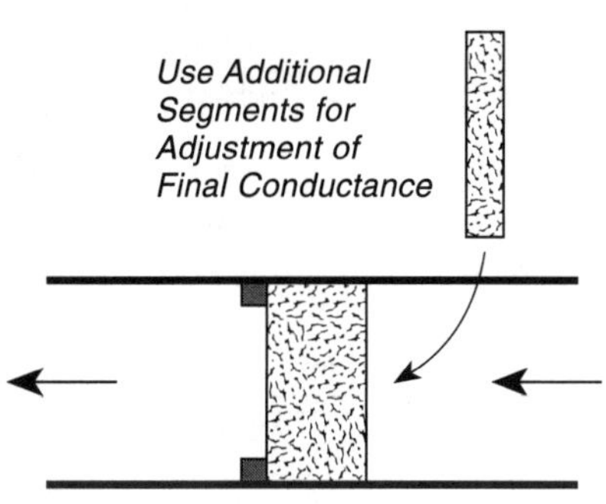


***C. Molecular Flow Conductance Limiter***

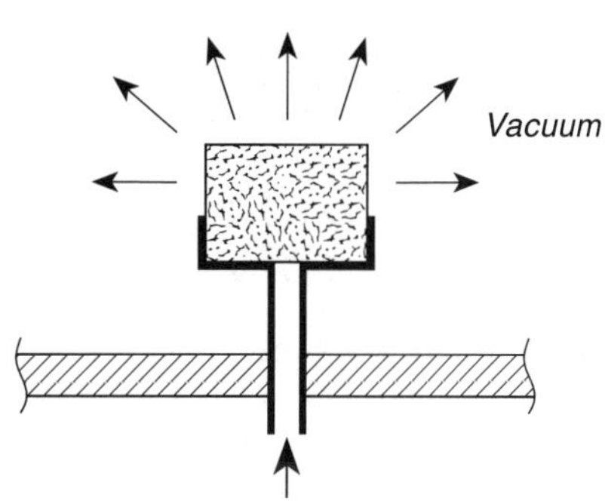


***D. Gas Diffuser***

# 41

# Cleaning of Ion Pumps by Chemical Etching

There are a number of companies that will take in used ion pumps for reconditioning at a fraction of the cost of purchasing a new ion pump [1]. In cases where the pump is a large one, this may be done by disassembly and cleaning or replacement of the individual components. In the case of small ion pumps, it may be necessary to open the pump casing to get at the inner components, and then to reweld the reconditioned components back into the cleaned casing.

The reconditioning of a dirty ion pump can also be done by controlled chemical etching procedures as described for two cases below.

## Case A: Chemical Etching After Disassembly

The pump casing is separated from the Ti cathode assembly as shown in Figure 41. The surface region of the Ti containing implanted impurities is removed by etching the assembly, containing electrode connectors and supports, in a water solution containing 9% $HNO_3$, 12% HF(aq), and 0.4% $H_2O_2$ by weight. A 3-s dip, followed by repeated rinsing in deionized water, should produce shiny Ti surfaces. The pump casing is etched in a standard solution (Varian) containing 30% $HNO_3$(aq), and 3% HF(aq) by volume. The carbon deposit must be removed mechanically by fine abrasive paper; or, alternatively, it may float off with the Ti film during etching. The pump elements are dried in warm air, assembled, and pumped into a second ion pump while baking at 200°C [2].

## Case B: Chemical Etching Without Disassembly

In this situation, the fact that the acid attacks the Ti more readily than the ceramic-metal seal is employed. First the pump is rinsed in deionized water. This is reported [2] to seal the ceramic partially against the acid. A water solution of 30% $HNO_3$(aq) and 3% HF(aq) by volume is added and quickly rinsed away with deionized water. The pump is dried by warm air and baked and pumped as in Case A.

The original report indicates that the hydrogen background pressure from an acid-cleaned pump will be somewhat higher than from one reassembled using new elements [2].

## *References*


[1] One company offering this surface: Duniway Stockroom, Inc., 1600 N. Shoreline Boulevard, Mountain View, CA 94043.

[2] P.H. Schmidt and L.Y.L. Shen, Rev. Sci. Instrum. **44,** 1788 (1973).

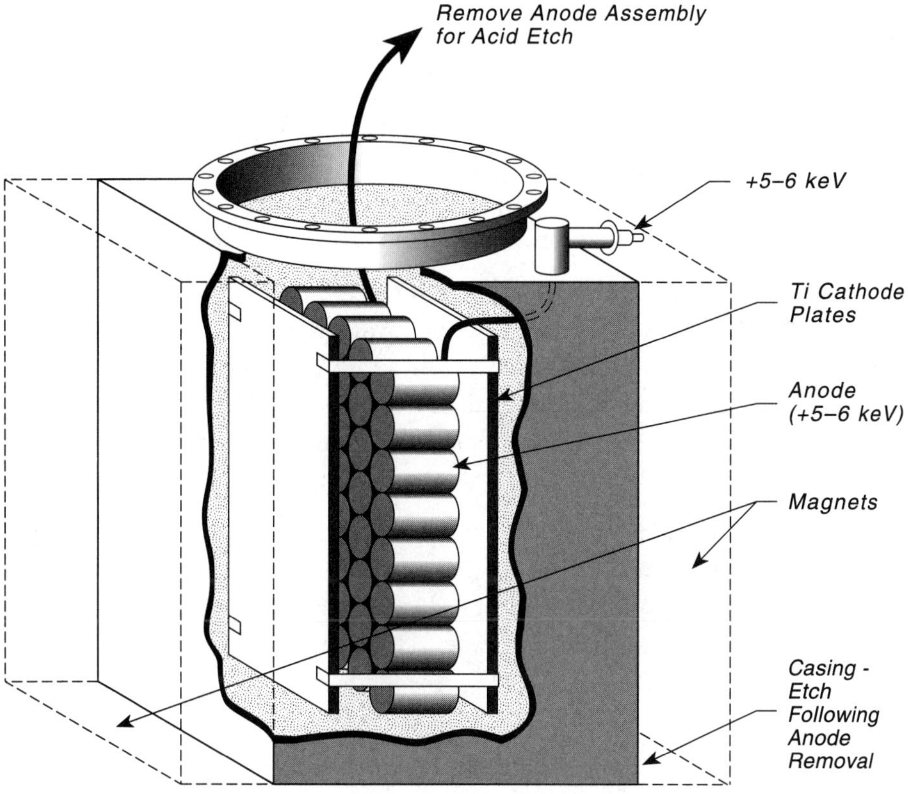


***Fig. 41. Chemical Cleaning—Ion Pump***

# 42

# An Efficient Liquid Nitrogen Trap

Liquid nitrogen traps are highly inefficient because of the discharge of liquid nitrogen out of the trap when the trap is being refilled periodically. A design, shown made of glass in Figure 42, eliminates this problem [1], and reduces liquid nitrogen consumption by severalfold, resulting in great economies over long periods of time. The basic concept embodied in this small glass trap is transferable to large metal traps.

The trap consists of an inner trap of conventional design, labeled the inner chamber, surrounded by a second outer chamber, which is in turn surrounded by a silvered Dewar flask. All of these are sealed together by glassblowing in this design. A resistance temperature sensor is immersed in the second outer chamber.

When the trap is at a condition where refilling is to occur, all liquid nitrogen in the Dewar has been lost, and the level of nitrogen in the outer chamber has dropped to a level causing the sensor to turn on the liquid nitrogen supply, which is driven by compressed nitrogen gas. Liquid nitrogen flows into the Dewar and spills into the outer chamber through the lower hole; it then fills the outer chamber until the sensor delivers a shutoff signal.

The avoidance of liquid nitrogen loss is attributed to the fact that the Dewar does not contain liquid nitrogen at the time of filling, so that the burst of nitrogen gas at the beginning of the fill sequence does not eject liquid nitrogen from the trap. In addition, ice buildup does not occur at the inlet and outlet to the trap [1].

## *References*

[1] R.S. Timsit and A.K.C. Kiang, Rev. Sci. Instrum. **44,** 770 (1973).

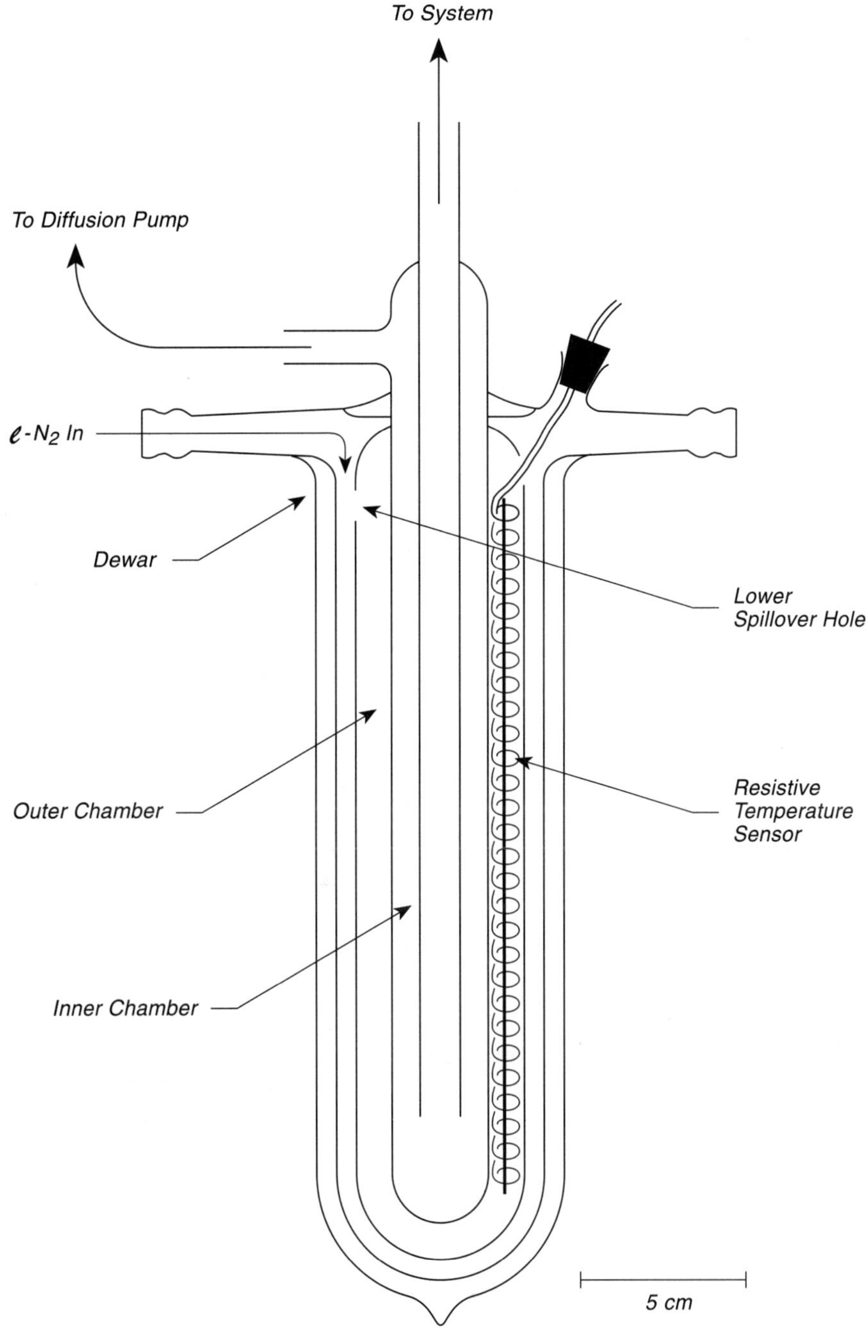


**Fig. 42. Efficient ℓ-$N_2$ Trap**

# E. Bakeout

# 43

# Bakeout of Metal Ultrahigh Vacuum Systems

The removal of adsorbed impurities from the walls of an ultrahigh vacuum system is carried out by heating the walls uniformly while pumping the system. Generally, for stainless steel systems this is done in the temperature range 150–200°C. For preservation of effective ion pumping, bakeout into a turbopump is recommended to avoid introducing a high gas load into the ion pump. Ion pumps have only a finite capacity, and also suffer from regurgitation effects in which previously pumped gas is evolved when another gas is being pumped-as would occur in an experiment in which gas is admitted to the chamber.

Many methods are used to bake out metal systems. Heating tapes, IR radiation heaters on the outside, clamp-on band heaters, glue-on rubberized heaters, and internal light-bulb heaters have all been used. In order to lose as little heat as possible, it is appropriate to surround the system with some sort of reflective bakeout shield [1]. Figure 43 shows a schematic system in which aluminized fiberglass [2] is used to shield the main system. The fiberglass tent is supported internally by temporary metal supports. Radiative heaters are uniformly distributed over the system inside the reflective shield, and for appendages such as the bellows on the top, lightweight reflective metal cylinders may be used for radiation capture. The system should be monitored over many positions with thermocouples to assure uniform heating, and the wattage output of the various radiative heaters may be varied to achieve uniformity. Experience in monitoring the pressure during bakeout will supply information about the bakeout time needed to reach the ultimate low pressure [3]. It is reported that external IR lamps bring the vacuum system to bakeout temperatures much faster than other methods [1].

Quite nice tents made of the aluminized fiberglass may be constructed, and a convenient method for joining the sections is to use a poprivet device to make excellent lap seams. The aluminum surface is placed on the inside of the tent. Even nonaluminized fiber glass curtain material may be used by sandwiching aluminum foil between the layers. A small internal circulating fan inside the oven will lead to improved temperature uniformity.

Experience indicates that clamp-on band heaters applied to metal flanges can thermally shock the flanges to a point where a leak develops. In the event that these heaters are used, a gradual increase in the applied voltage over time is recommended. Internal lamp heaters, working inside the vacuum system, would seem to be ideal for bakeout because of efficient radiation capture by the system walls, but if breakage of the glass envelope occurs, the system will have to be completely cleaned of glass slivers before continued use. Such slivers are very detrimental to welded bellows and to the sealing surfaces of valves.

Turbopumps should be baked at the end of the bakeout period (for about 4 hours). Titanium sublimation pumps should be outgassed until the pressure reading reaches a maximum; long outgassing at high temperature will oxidize the components near the hot Ti. It has been found that cooling the room during measurements will cause a substantial improvement in the limiting pressure under ultrahigh vacuum conditions owing to the reduction in the outgassing rate from the system walls [4].

Often it is appropriate to bake out a new system, with all heat sensitive components removed, to temperatures between 300 and 450°C to remove impurities left from chemical cleaning procedures.

## *References*

[1] B.V. Hess and T.E. Felter, J. Vac. Sci. Tech. **A3**, 2218 (1985).

[2] Fyrepel, P.O. Box 518, Newark, OH 43055.

[3] Dr. R. David, Institut fur Grenzflächenforschung und Vakuum-physik, Forschungszentrum Jülich GmbH, Postfach 1913, D-5170 Jülich, Germany (private communication).

[4] Pittsburgh Surface Science Center.

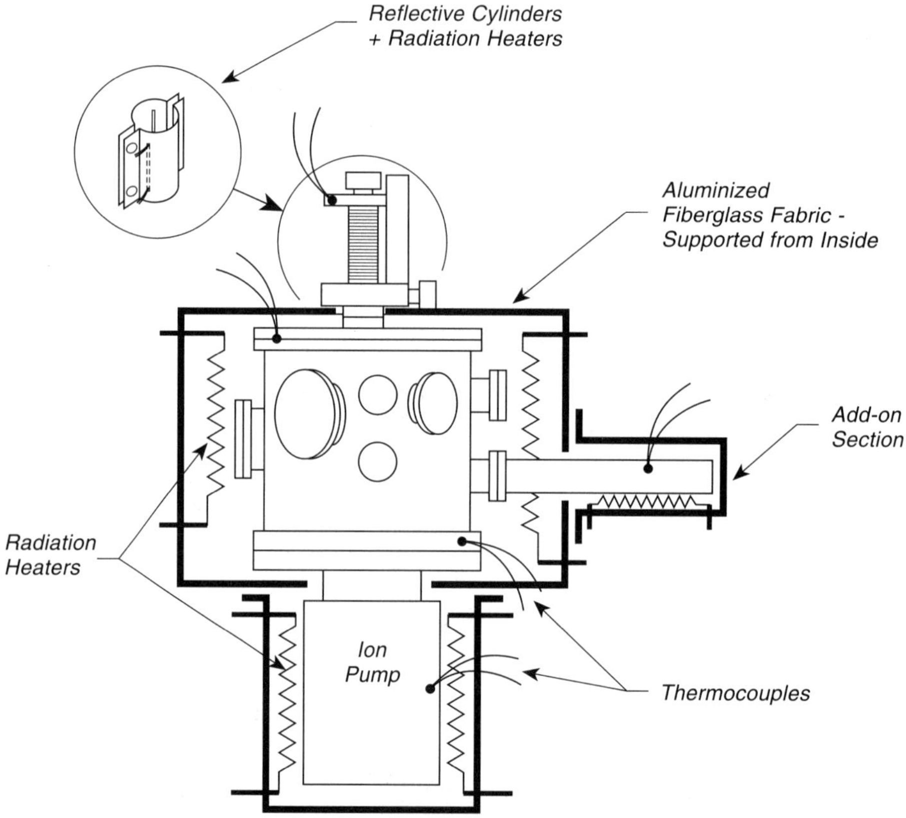


*Fig. 43. Schematic Bakeout Assembly*

# F. Behavior of UHV Systems

# 44

# Wall Passivation in Stainless Steel Ultrahigh Vacuum Systems

To achieve the limiting pressure in an ultrahigh vacuum system, it is necessary to reduce the rate of desorption of molecules from the stainless steel walls to the lowest possible limit. These surface impurities originate from exposure of the walls to impurities during opening and processing of the system. They also result from the diffusion of impurities in the bulk of the stainless steel to the wall surface, especially for the case of dissolved hydrogen [1].

Bakeout in the presence of traces of oxygen or water may result in further oxidation of the walls, leading to some passivation. However, a more effective technique involves the oxidative passivation of the stainless steel walls with a strong gaseous oxidizing agent. The oxidation of the walls and the oxidizable surface impurities is kinetically accelerated using a strong oxidizing molecule. In Figure 44, an energy level diagram (Gibbs free energy) is shown. It may be seen that the nitric oxide molecule is more energetic than a mixture of $O_2$ and $N_2$ to which its energy is referenced. The excess energy is 0.9 eV/NO at 300 K. This excess chemical energy may be sufficient to allow the NO molecule to produce a rapid surface oxidation process that proceeds only slowly over an activation barrier when $O_2$ is the oxidizing agent. The oxidation process on the surface is schematically shown as the formation of a chemisorbed O atom (O(a)), but the formation of $CO_2$ or $H_2O$ from surface carbon or from surface hydrogen could also be the end result of the oxidation process.

These ideas have been used to clean up the walls of an ultrahigh vacuum system during bakeout, achieving lower base pressure [2]. The bakeout at 200°C was carried out in a background pressure of NO of $5 \times 10^{-7}$ mbar achieved by flowing NO through the hot system. Initially, during the first phase of the treatment, high levels of $N_2$, $H_2O$, and $CO_2$ are detected as the NO acts to oxidize the surface and the oxidizable impurities. After 30–180 min, a continuous increase of the partial pressure of NO occurs, and the oxidation of the surface approaches completion. This treatment is particularly good for reducing the displacement of surface impurities from the walls during subsequent adsorption experiments. A simple equilibrium thermodynamics calculation shows that the use of the energetic NO molecule at $5 \times 10^{-7}$ mbar is equivalent to the use of $O_2$ gas for the same purpose at a pressure of about $10^6$ atm!

## *References*


[1] P.A. Redhead, J.P. Hobson, and E.V. Kornelsen, *The Physical Basis of Ultrahigh Vacuum,* (AIP, New York, 1993).

[2] M. Grunze, G. Strasser, and O. Eishazly, J. Vac. Sci. Tech. **A4,** 2396 (1986).

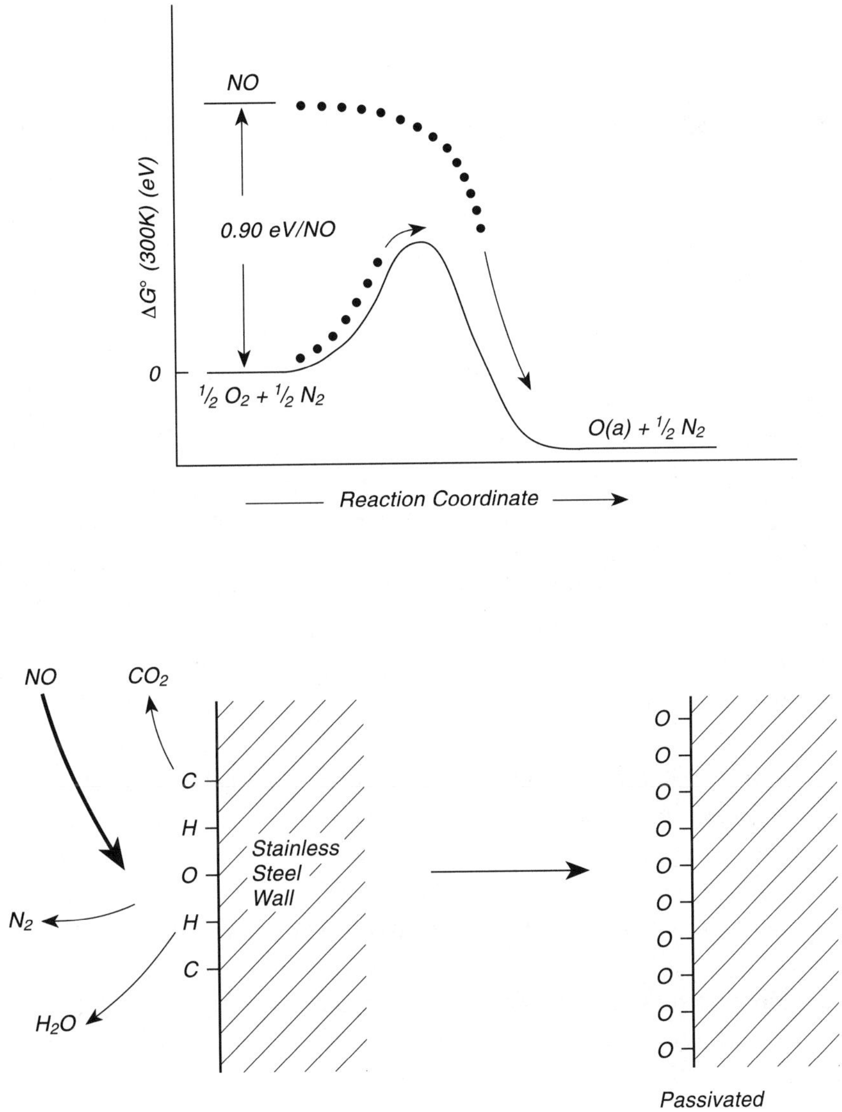


***Fig. 44. Nitric Oxide Surface Reactivity for Passivation of Vacuum System Walls***

# 45

# Minimizing Wall Reactions in Ultrahigh Vacuum Systems—Gas Dosers

The purity of gases used in surface science adsorption and surface reaction studies is of paramount concern just as the production of ultrahigh vacuum is directed toward the same end. Often gas dosers are employed to deliver pure gases to surfaces, and for some gases, the displacement of adsorbed impurities, or the reaction of the dosing gas with the internal surfaces of the gas dosing system itself can be a problem. In the case of a surface reaction, it is sometimes possible to passivate the internal surface of a gas doser to reduce the rate of the surface reaction. Figure 45A shows a typical gas doser made mainly of stainless steel. The construction of this doser and its calibration are discussed elsewhere (page 604); however, it may be seen that the conductance is controlled by a laser-drilled aperture, and collimation of the exiting gas is achieved using a microcapillary array made of glass. Both the internal metal surface and the internal surface of the microcapillary array (which is many tens of $cm^2$) provide reactive surface area that can evolve impurities by displacement or reaction [1]. Displacement processes, such as the exchange of surface OH groups with OD groups in deuterated alcohols, can often best be handled by overnight flushing of the doser with the isotopic gas. In extreme cases of reaction with acid sites on the internal surfaces of the capillaries [1], one must remove the glass microcapillary array from the doser [1,2]. Reaction processes with the stainless steel walls of the doser system can sometimes be reduced in magnitude by passivation of the internal surfaces. This is vividly seen in the behavior of the metal carbonyls, where decomposition of $Fe(CO)_5$ on the inner stainless steel walls has been shown to produce CO(g), leaving iron on the wall [2]. The surface reaction was studied, and it was found that predosing the doser and its stainless steel delivery system with high fluxes of molecular oxygen resulted in significant reduction of the decomposition reaction, as judged by mass spectrometric studies of the gas exiting the doser, line-of-sight, into the mass spectrometer [2]. The internal walls of the ultrahigh vacuum system were found to have little effect on the metal carbonyl, probably because it was exposed to oxygen each day [2]. For the experiments, it was found that the glass microcapillary usually employed with this design of gas doser had to be removed because of deleterious effects on $Fe(CO)_5$.

Another solution to this problem is to make the doser system from materials other than stainless steel [3]. Figure 45B shows such a system, in which the glass tube doser contains a capillary for the control of the flow rate from the connecting tube into the ultrahigh vacuum system through the Pyrex doser tube. The high-pressure (several Torr) section beyond the Teflon-glass valve is made of PFA (polyfluoroalkoxy) plastic and fittings are readily available commercially [4]. The paper describing this system indicates that the capillary tube was produced by hand and was drawn down to a fine diameter. This particular design was used for the delivery of hydrazine ($N_2H_4$). Thermometer tubing, made generally of lead glass, has a ~100–150-μm diameter capillary for sensitive thermometers (1 K/3.8 cm capillary length) and may be used if a lead glass to Pyrex glass graded seal is employed. A Teflon doser for ultrahigh vacuum use also has been described [5], as well as a heatable glass and Teflon doser for admission of organometallics of low vapor pressure [6].

It is unfortunate that glass capillaries often clog up with traces of dust or grease that are present within the vacuum system. A clever method to restore the capillary to a clean, unblocked condition is shown in Figure 45C [7]. Here, the capillary section is surrounded on both ends of the capillary with pointed tungsten electrodes, easily installed by the glassblower. When the capillary becomes blocked, micron pressures are established on the high-pressure side of the capillary, and the vacuum-side electrode is electrically grounded. The application of a Tesla coil to the high-pressure tungsten electrode produces an internal plasma discharge, and the plasma usually cleans out the obstruction in the capillary within seconds. We have found that this harsh internal treatment changes the conductance of the glass capillary slightly, and recalibration after cleaning is suggested.

A third solution that has been used to minimize reactions inside a gas doser is to dilute the reactant gas with an inert diluent such as He or Ar [8]. This may be done in a gas handling line that is dynamically pumped, causing the gas mixture to be replenished rapidly, and removing products of reaction [8]. This method also permits small doses of the reactive gas to be quantitatively monitored since pressure measurements of the mixture may be converted to partial pressure measurements of the active gas. One must remember that the composition of the gas mixture will change under dynamic pumping conditions owing to different rates of pumping for the gas components [8].

Reports of the use of a Teflon coating for passivation of the stainless steel walls of an ultrahigh vacuum system are available [9,10]. Briefly, a weak acid priming treatment was used to remove oxide films to promote adhesion of the Teflon. A Teflon suspension [11] was painted on the walls and cured in the air for several days. The coating was then baked in vacuum at 200°C. Small parts were cured using a heat gun before assembly. This coating was effective in preventing $TiCl_4$ decomposition on the walls. However, the Teflon adsorbs water from the air when the system is opened, and longer bakeout is required to get to a base pressure in the mid $10^{-9}$ Torr range than would be required with an uncoated vacuum chamber [10].

## *References*


[1] P.K. Leavitt and P.A. Thiel, J. Vac. Sci. Tech. **A8,** 148 (1990).

[2] M.A. Henderson, R.D. Ramsier, and J.T. Yates, Jr., J. Vac. Sci. Tech. **A9,** 2785 (1991).

[3] E. Apen, R. Wentz, F. Pompei, and J.L. Gland, J. Vac. Sci. Tech. **A12,** 2946 (1994).

[4] Swagelok fittings are available from Swagelok, 31400 Aurora Road, Solon, OH 44139.

[5] M.A. Mendicino and E.G. Seebauer, J. Vac. Sci. Tech. **A10,** 3590 (1992).
[6] L. A. Jones, F.-X. Wei, and D. F. Thomas, Rev. Sci. Instrum. **66,** 4981 (1995).
[7] D.A. King (private communication) (1970).
[8] F.C. Henn, M.E. Bussell, and C.T. Campbell, J. Vac. Sci. Technol. **A9,** 10 (1991).
[9] W. Muller-Markgraf and M.J. Rossi, Rev. Sci. Instrum. **61,** 1217 (1990).
[10] Professor E.G. Seebauer, Department of Chemical Engineering, University of Illinois, Urbana, IL 61801. From thesis by M.A. Mendicino, Ph.D (1994) (private communication).
[11] Teflon or PTFE suspension is obtainable from: Intek Services, Inc., 7 McMillan Way, Suite 101, Newark, DE 19713.

**Fig. 45. Gas Doser Designs**

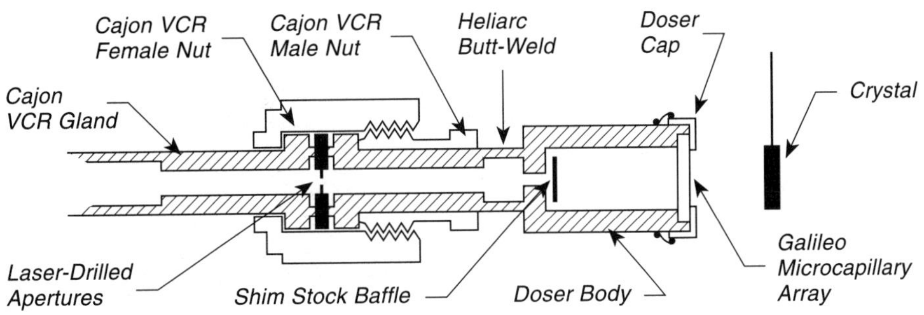


**A. Stainless Doser + Microcapillary Array Collimator**

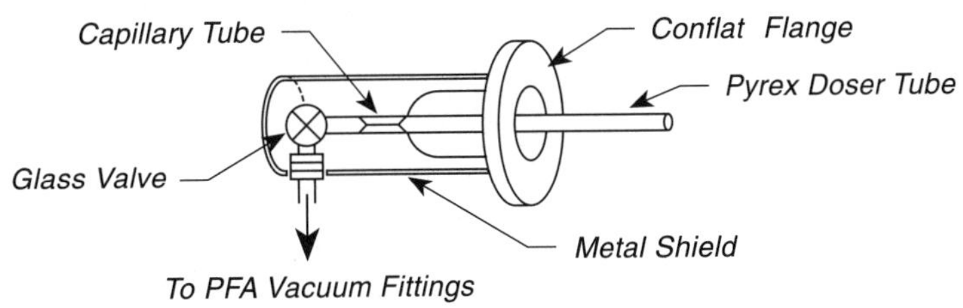


**B. Pyrex - Plastic Doser**

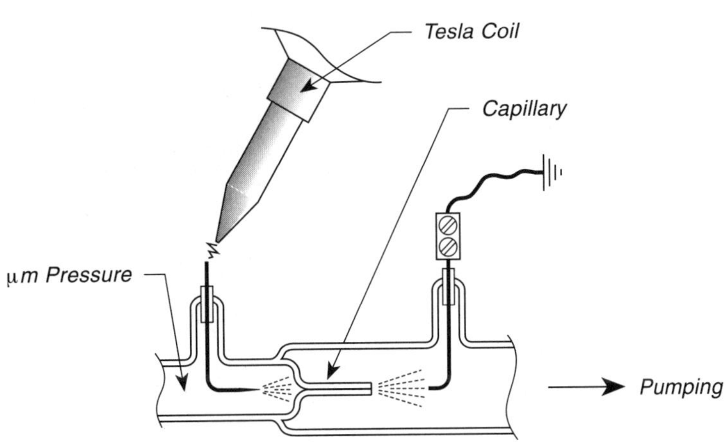


**C. Cleaning Clogged Glass Capillaries**

# G. Mechanical Action on Samples

# 46

# Simple UHV Crystal Cleaver

Often one of the best ways to produce an atomically clean surface is to expose an inner cleavage plane by cleaving the crystal in ultrahigh vacuum. A simple mechanical cleaver design is shown in Figure 46A. Here, a MgO single crystal is mounted on a coolable manipulator inside a small rectangular compartment, held by a screw that presses on an internal plate that distributes the screw pressure over the side of the crystal. A portion of the crystal projects upward, ready for the cleaver blade.

The cleaver is a spring-loaded tungsten carbide blade (25° cutter angle) that slides in a stainless steel channel. Two springs are compressed by turning a pair of screws prior to attaching the cleaver to the manipulator. The manipulator/cleaver assembly is loaded into the vacuum system, and after system processing, the trigger holding the blade back is released by moving the manipulator against the side of the manipulator port. The blade cleaves the crystal, and caution should be exercised not to allow the cleaved-off portion of the crystal to enter fragile apparatus as it bounces around the chamber.

Using a closed cycle He cryocooler, this assembly can reach 44 K. Studies on the freshly cleaved MgO(100) surface, involving Kr adsorption, are shown in Figure 46B, and the results are compared to MgO that has been cleaved outside the vacuum system, transferred, and thermally processed. The sharp vertical step indicating a 2D gas → 2D solid transformation, as studied by LEED, is characteristic of the UHV-cleaved crystal; a less distinct transition is seen on the externally cleaved crystal because of defects produced by water adsorption that cannot be removed by thermal processing. Indeed, after days of storage in UHV, the UHV-cleaved crystal begins to yield an isotherm resembling that of the defective crystal because of slow impurity adsorption.

The authors report that the only disadvantage of this design is that only one cleave can be achieved per loading into the vacuum system because of the necessity to compress the springs outside the vacuum system.

A cleavage device for mica has been described that can be used in ultrahigh vacuum [2].

## *References*


[1] T. Angot, J. Suzanne, and J.Y. Hoarau, Rev. Sci. Instrum. **62**, 1865 (1991).
[2] K. Reichelt, Rev. Sci. Instrum. **44**, 243 (1973).

## *Fig. 46. Simple UHV Cleaver*

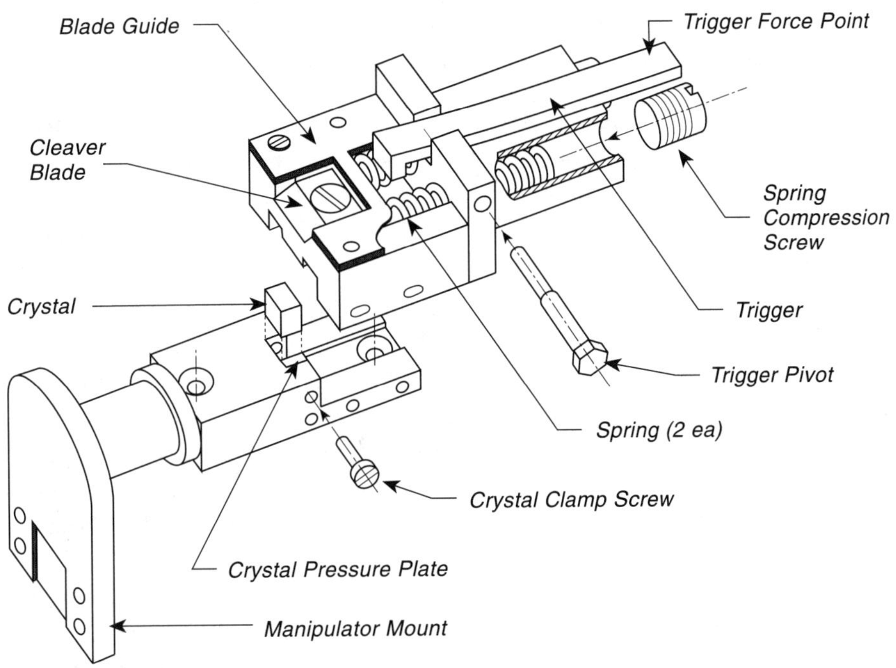


**A. Cleaver Design**

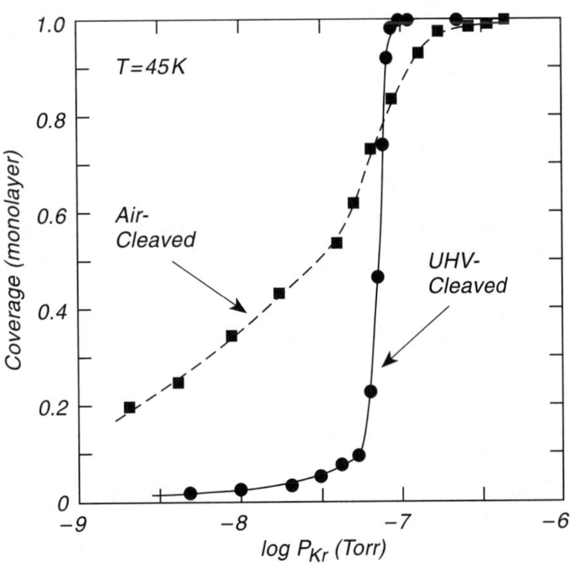


**B. Comparison of Kr Isotherms on MgO(100) Using LEED**

# 47

# Piezoelectric Fatigue Apparatus for UHV Operation

There is much interest in the role of adsorption on the rate of fatigue of metals undergoing periodic stress conditions. Usually these measurements are performed under cycle rates of about 1 Hz. Hydraulic cycling methods often impart noise to the stress-strain measurement system. A piezoelectric drive mechanism, capable of working at 100 Hz frequency in UHV, is described below [1].

Figure 47A shows a schematic drawing of the apparatus. The specimen, coupled to an extensomer, is clamped between two grips that have serrated clamping surfaces to prevent misalignment during testing. A load cell is placed above the moving grip and below the piezoelectric actuator. The translator used has a capacity of 10 kN in compression, and 3.5 kN in tension [2]. It is capable of translation through about 40 μm over an applied voltage range of 0–1500 V. The load is measured with a load cell having a range of 0–2000 N [3]. A strain of about 0.3 % may be measured using an extensometer [4] in a gauge section of 5 mm, and the specimen tested in this case was polycrystalline Ag. The control electronics are described in reference [5]. Figure 47B shows a hysteresis curve measured for Ag in which the stress and strain are measured during compression and expansion. The hysteresis loop changes as the number of cycles increases, and is a measure of the specimen fatigue. The data obtained with the piezodriver exhibits a better signal-to-noise ratio than data taken with hydraulic devices.

The main disadvantage of the device is the small translation range of the driver. The device must account for all of the translations taken up by the components along the axis of the load. This limits the machine to small specimens, and for high-strength materials, smaller gauge lengths and more narrow cross sections would be needed to obtain sufficient strains for accurate measurement [1].

Another design, providing fatigue to a thin metal specimen in ultrahigh vacuum, has been described [6]. It was found that clean Al(111) single crystals exhibited changes in their ultraviolet photoelectron spectrum after 1000 fatigue cycles at 0.2% strain. The involvement of chemisorbed species containing oxygen was eliminated on the basis of the absence of O(2p) emission in the photoelectron spectrum, and no explanation was given for this effect [6].

## *References*


[1] T.S. Sriram, M.E. Fine, and Y.-W. Chung, Rev. Sci. Instrum. **62**, 2008 (1991).

[2] Physik Instrumente, Model 243.37, Option 703.20, sold by PolyTec Optronics, 3001 Redhill Ave., Costa Mesa, CA 92626.

[3] Model 31, SENSOTE Inc., 1202 Chesapeake Ave., Columbus, OH 43212.

[4] Model 632.29C, Option 011, MTS Systems Corporation, Box 24012, Minneapolis, MN 55424.

[5] W.J. Lee, Y.-W. Chung, M.E. Fine, and J.P. Baker, Rev. Sci. Instrum. **57**, 2854 (1986).

[6] T.K.G. Swami and Y.W. Chung, Rev. Sci. Instrum. **51**, 842 (1980).

**Fig. 47. Piezoelectric Fatigue Apparatus—UHV**

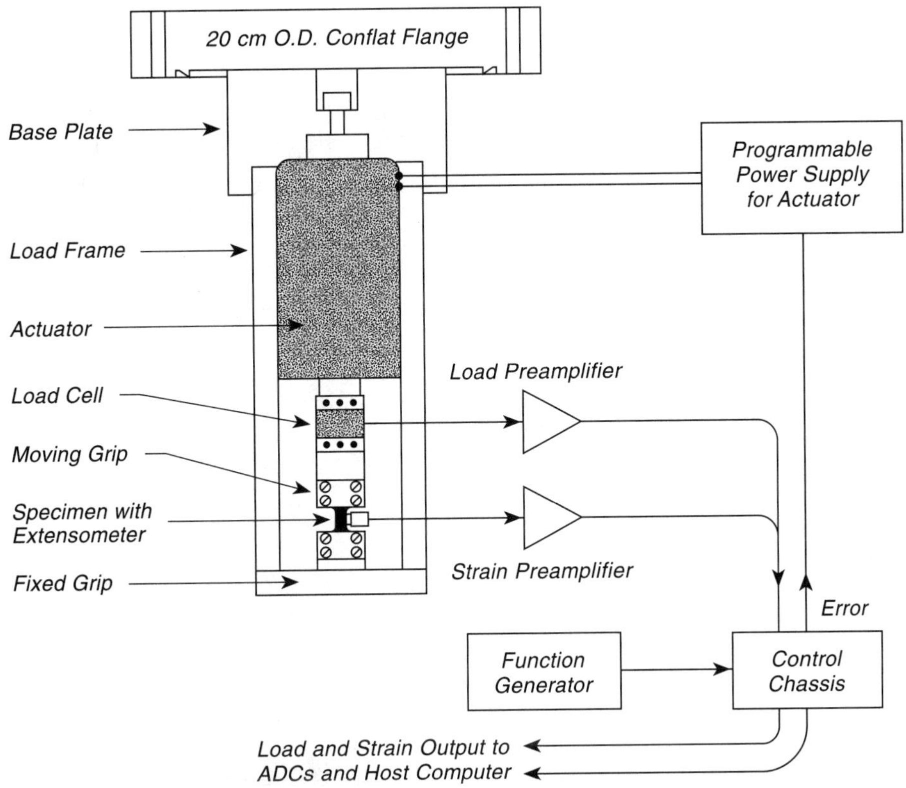


**A. Schematic Design**

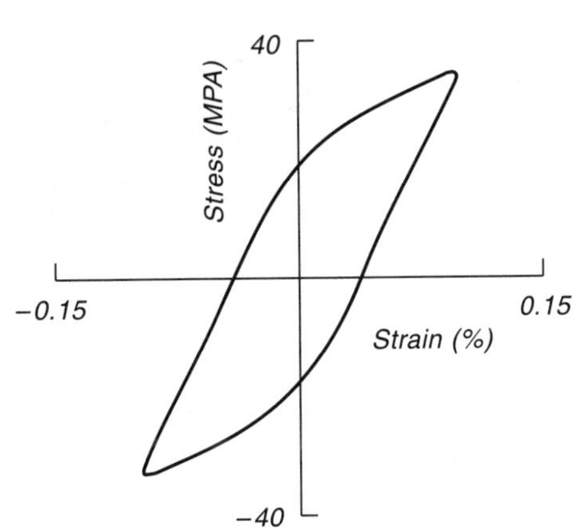


**B. Typical Stress-Strain Curve—Polycrystalline Ag**

# H. Gasket Seals

# 48

# Gasket Seals for Ultrahigh Vacuum Systems

The primary ultrahigh vacuum flange design used universally is the stainless steel Conflat sexless flange, which operates with two opposed knife edges that seal a metallic gasket by deformation when the flange is tightened. A cross-sectional view of the flange and a confined gasket is shown in Figure 48.

In many cases where it may be difficult to seal the gasket by applying the necessary torque to the flange bolts, or where tightening down to the maximum force level is not desired, it is useful to use annealed copper gaskets. These may be purchased from the suppliers, but they can be made quite simply in the laboratory beginning with a nonannealed gasket. The gasket is heated while rotating it around its circumference in an oxygen-gas torch (cool flame) until dull red color is emitted by the hot copper. The gasket is then quenched in water quickly, resulting in softening to almost the softness of lead or tin. This gasket will deform in the seal region with less torque on the bolts and also with less deformation of the flange during tightening. Often, gaskets of this type are needed for spectroscopic windows mounted in stainless steel flanges in order to prevent window cracking during tightening owing to flange flexure.

The annealing process results in considerable oxidation of the copper gasket, and this oxide layer must be removed before use. Two methods are available to do this. In the first, the hot gasket, immediately after annealing, is dropped into methanol instead of water. The methanol, a reducing agent, immediately strips off the oxide from the hot metal, leaving a clean copper surface. Care should be exercised to avoid damage if the methanol should ignite. The second method involves dropping the cool gasket into an aqueous solution of citric acid. Here the citric acid forms a soluble complex with the copper ions in the oxide film, and again clean copper surfaces are produced.

Certain coatings are employed to prevent the oxidation of copper gaskets following relatively high temperature bakeouts. Heavy amounts of copper oxide can fall into the vacuum system when the flange is opened, and also seizure of the gasket in its recess in the stainless steel flanges can occur. Silver plating has been used to prevent this effect [1]. A minimum 5-μm-thick silver film, made by plating, is required. The silver will stand repeated bakeouts at 300–350°C. Contamination of the silver film by sulfur from either the gasket or from sulfur-containing lubricants is to be avoided as this causes discoloration and peeling of the silver film. Thus, bolt lubricants such as molybdenum disulfide cannot be used and graphite or other lubricants must be employed.

Another useful copper gasket coating is titanium nitride, TiN [2]. Using reactive magnetron sputtering from a Ti target in an $N_2$ + Ar mixture (20% $N_2$), films from 50 Å to 150 Å were deposited on copper gaskets. In tests, ten flanges were baked at 550°C in a vacuum of $10^{-5}$ Torr for up to 200 h, and were then tested for leak tightness with a He leak detector, finding no leaks. The flanges could then be disassembled without seizure. Without the TiN film, separation must be done with jacking screws and copper is left on the sealing surfaces, requiring abrasives for removal. A similar coating has also been prepared using diamondlike carbon coatings prepared by RF discharge techniques in methane gas [3].

One of the early methods of sealing stainless steel flanges employed gold wire gaskets [4]. For flat flanges without a knife edge or for a stepped seal used for crushing a metal gasket, gold wire will make an excellent seal. The wire, of about 0.25–0.5 mm diameter, is made into a circular loop on a forming mandrel; then a microtorch is used to butt weld the ends together when the loop is removed from the mandrel. This process is best accomplished by gently clamping together the squared off ends of the gold wire and slowly melting the gold so that surface tension causes liquid gold flow to even out the diameter of the heated region. Too much heating will result in separation of the gold wires with the formation of balls of gold on the ends of each. Practice is, therefore, necessary. The wire once formed is placed between two smooth flange faces, and the bolts are tightened, making an ultrahigh vacuum seal. A stepped seal may also be used with either gold or copper, and here the seal produces a shear seal within the gasket as shown in Figure 48. In the case of gold wire shear flanges, the wire must be butt-welded to fairly exact dimensions using a mandrel of the same size as the male flange for accurately cutting the gold wire to appropriate length. A nice description of the method to make custom copper wire gaskets is given in reference [5].

The reuse of large Cu gaskets for many cycles has been described [6]. Here, a "feeler gauge" such as that used to measure spark plug gaps is employed, and successive seals with the original gasket are made by diminishing the flange-to-flange clearance in a systematic manner.

## *References*

[1] T. Wikberg, J. Vac. Sci. Tech. **5,** 205 (1968).
[2] E.L. Garwin, E.W. Hoyt, and A.R. Nyaiesh, J. Vac. Sci. Tech. **A4,** 2537 (1986).
[3] A.R. Nyaiesh, E.L. Garwin, and E.W. Hoyt, J. Vac. Sci. Tech. **A5,** 2977 (1987).
[4] A. Roth, *Vacuum Sealing Techniques,* Pergamon, Oxford (1966), page 380.
[5] E.M. Beaver, Rev. Sci. Instrum. **51,** 254 (1980).
[6] L.Y.L. Shen, J.E. Rowe, and E.E. Chaban, Rev. Sci. Instrum. **44,** 1747 (1973).

**_Fig. 48. Gasket Designs_**

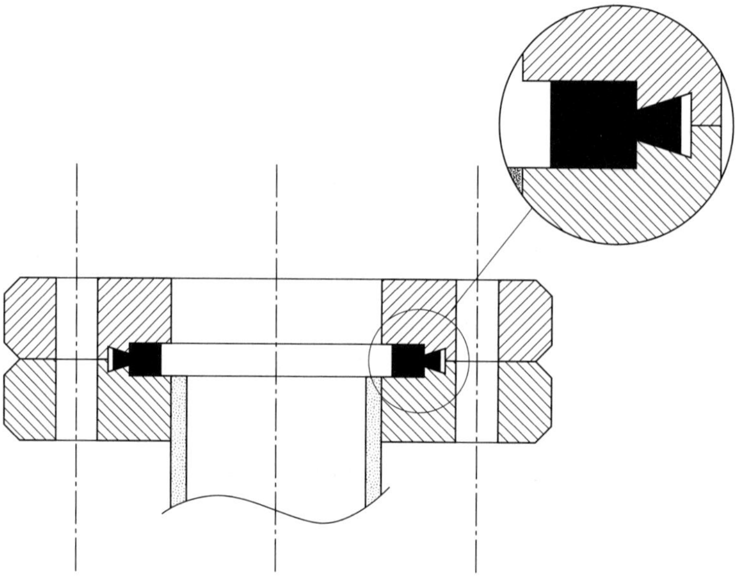

**_A. Conflat Flange + Cu Gasket_**

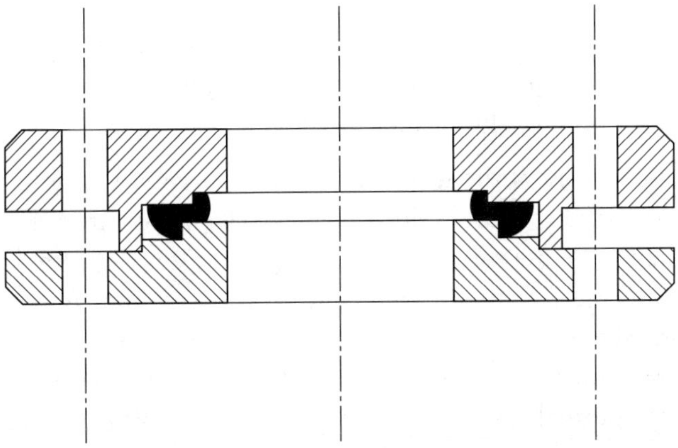

**B. Gold Wire—Shear Seal**

# 49

# Cryogenic Gasket Seals

Stainless steel Conflat flanges are guaranteed to be vacuum tight from 77 K up to their normal bakeout temperature limits. However, there are special situations where vacuum components made of different metals are to be joined by flanges, and in these cases cooling to 77 K may disrupt the vacuum seal because of differences in thermal expansion coefficients.

This problem has been solved for 18-mm-diameter junctions between stainless steel flanges and aluminum Conflat 34 flanges using a gold wire seal (Figure 49). In this design, the mismatch in expansion/contraction occurs in a radial direction along the sealing plane, and this stress can be absorbed by the plastic deformation of the gold wire. The seal is maintained through many cycles as a result of the gold ductility and its good adherence against stainless steel and aluminum surfaces [1].

A cross sectional drawing of the seal is shown in Figure 49. The gold wire ring was 1 mm in diameter, prepared by butt-welding. The screws were 4 mm diameter, made of stainless steel which prevents the need for retorquing after a thermal cycle. The specially designed stainless steel flange has a precision entrance guide to prevent damage to the aluminum knife edge and it is also machined to be very smooth on its sealing surface. In the particular case shown here, the stainless steel, exhibiting poor heat conduction, was needed to reduce the thermal conduction of heat to the aluminum section therefore preserving liquid nitrogen which was housed in the aluminum section.

## *References*


[1] F. Shimoshikiryo, Y. Takakuwa, N. Oyama, and N Miyamoto, J. Vac. Sci. Tech. **A11**, 2874 (1993).

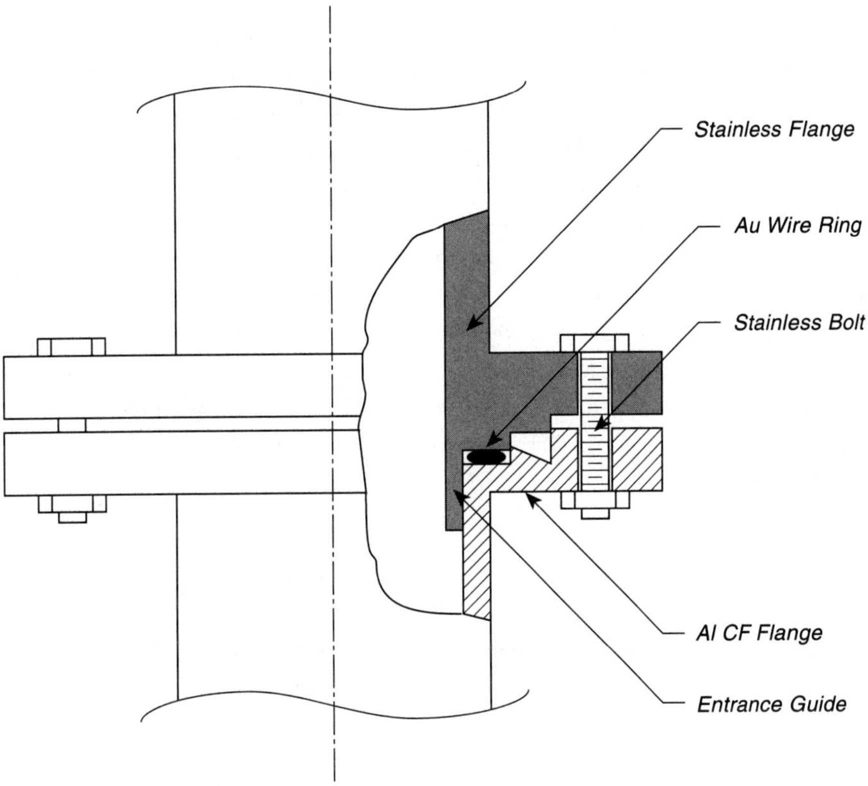


***Fig. 49. Gold Wire Seal—Cryogenic Design***

# 50

# Copper Gasket Removal Devices

There are three published methods that are directed toward easy removal of copper gaskets from Conflat flanges after disassembly [1–3]. These methods are all designed to prevent damage to the sharp sealing knife edges by tools being used to pry up the gaskets. The first of these [1], shown in Figure 50A, employs Conflat flanges that have two opposed slots machined on the inside of the flange. A screwdriver blade can be inserted into the slot and rotated to pop out the gasket and the boundary of the slot prevents the blade from coming into contact with the knife edge. The disadvantage of this design is that one must machine every flange with the two slots [2]. A second design [2] (Fig. 50B), uses the leak-check groove as a slot to control the motion of a specially wedge-shaped and hardened steel blade, shown. The third design (Fig. 50C), shows a modified "parrot-nose" set of ignition pliers that can safely grip the gasket [3]. This design allows only upward force on the gasket and also grips the gasket during removal to prevent it from falling. In certain instances where there is insufficient clearance for the pliers on the outside of the copper gasket, one must resort to a tool of the second design [2,3].

## *References*


[1] J.J. Donelon and J.M. Baker, J. Vac. Sci. Tech. **10**, 894 (1973).

[2] E.E. Chaban, J. Vac. Sci. Tech. **12**, 654 (1975)

[3] B.J. Waclawski, J. Vac. Sci. Tech. **A1**, 99 (1983).

*Fig. 50. Tools for Copper Gasket Removal*

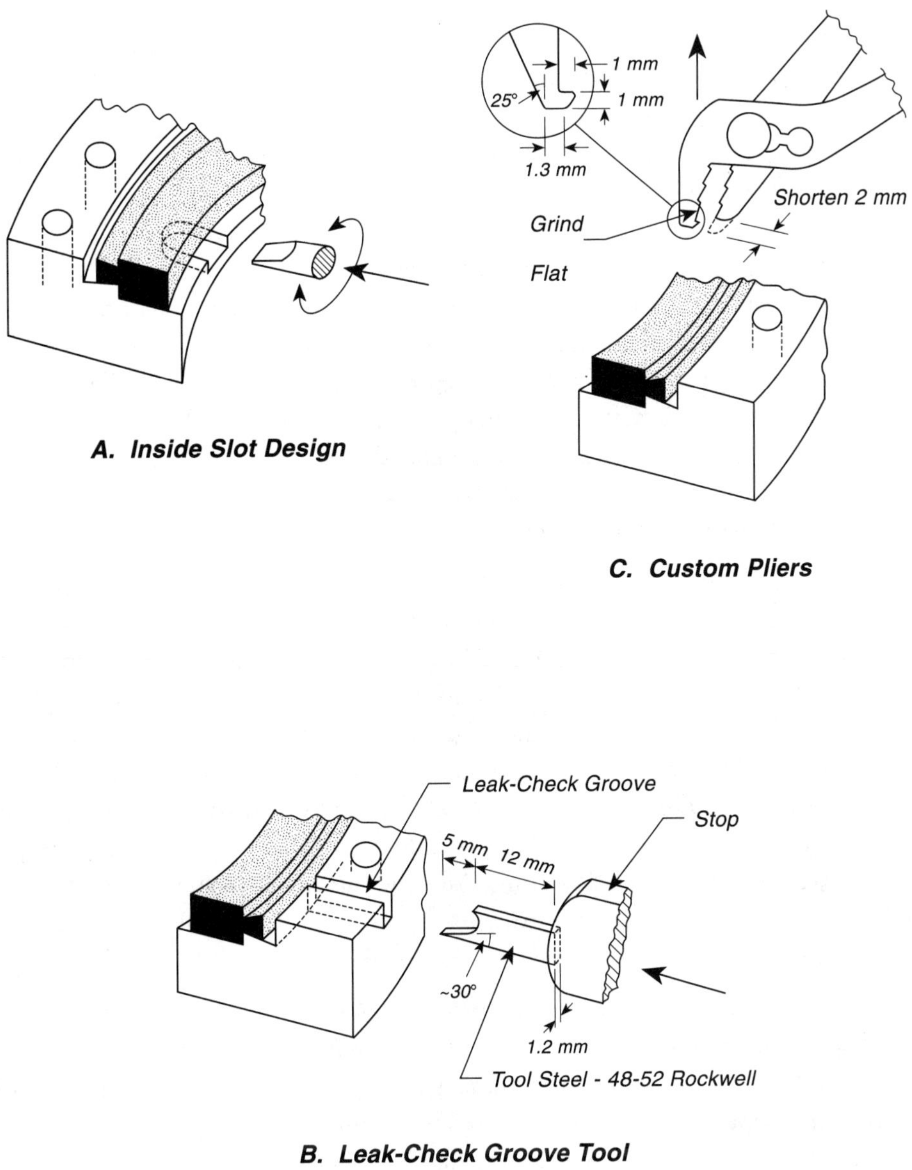

# 51

# Unconventional Compression Seals for Ultrahigh Vacuum

Several unconventional compression seals have been designed for ultrahigh vacuum systems where flange mounting was inappropriate. These seals all have in common the use of a material softer than stainless steel that is deformed under pressure supplied by screwing down the re-entrant seal from the outside of the vacuum system. This forms the vacuum seal, which is always on the inside of the vacuum system. Figure 51A is a seal involving oxygen-free high conductivity (OFHC) copper that is deformed at the edge of a hole drilled into the vacuum system wall. A stainless steel bolt, with a hole drilled axially through its length, is machined to present a hemispherical surface under the head. Copper tubing is spun down on a lathe to closely fit over the bolt. The bolt is inserted from the vacuum side of the hole and tightened with the nut from the outside. Such seals can be cycled between 873 K and 77 K many times, remaining He-leak tight. The seals may be loosened and used repeatedly for five to ten times. Ultimately, the copper becomes thin and hardened as a result of the mechanical deformation. New copper fittings may be used to replace the original used fitting [1].

A second re-entrant bakeable all-metal vacuum seal is shown in Figure 51B. Here a conventional Conflat flange design is used for the sealing surface and the force to produce the seal at the copper gasket is applied externally by means of jacking screws. The assembled seal was designed to apply 2000 lb/in. along the copper gasket seal path. The seal is assembled from inside and initially tightened using the threaded jack ring, which pushes down on a chrome plated thrust washer needed to prevent distortion of the main sealing flange that contains the hole and also to prevent galling of the jackbolts under torque. First torquing of ten bolts to 20 in. lb resulted in a leak tight seal. Further torquing to 40 in. lb reached the plastic-flow limited torque for the No. 8 stainless bolts. Disassembly and reasembly at 40 in. lb torque produced a He-leak-tight seal, which withstood baking to 250°C [2].

A third design, solving a major vacuum problem for sealing thermocouple gauges into ultrahigh vacuum systems, is shown in Figure 51C. For most applications under nonstringent vacuum conditions, thermocouple gauges are sealed at the pipe thread joint by use of Teflon tape, vacuum cements, etc. This is unacceptable for ultrahigh vacuum operation owing to the possibility of gas trapping in the pipe thread or cement. The design shown in Figure 51C uses the pipe thread to compress either a Teflon or aluminum or OFHC copper gasket at the bottom of the tubulation. The end of the stainless steel gauge stem must be smoothed with 600-grit paper, using a removable plug inside the tubulation to prevent entry of debris. Similarly, the sealing surface in the receptor hole must also be polished. This method permits thermocouple gauges to be used on ultrahigh vacuum systems and to withstand normal bakeout temperatures [3].

## *References*

[1] N. Raeuber, K. Stadler, and M.J.O. Strutt, J. Vac. Sci. Tech. **8**, 662 [1971].
[2] P.H. Edmonds and W.D. Shipley, J. Vac. Sci. Tech. **A6**, 165 (1988).
[3] F.N. Rabarchik, J. Vac. Sci. Tech. **16**, 2116 (1979).

**Fig. 51. Unconventional Compression Seals**

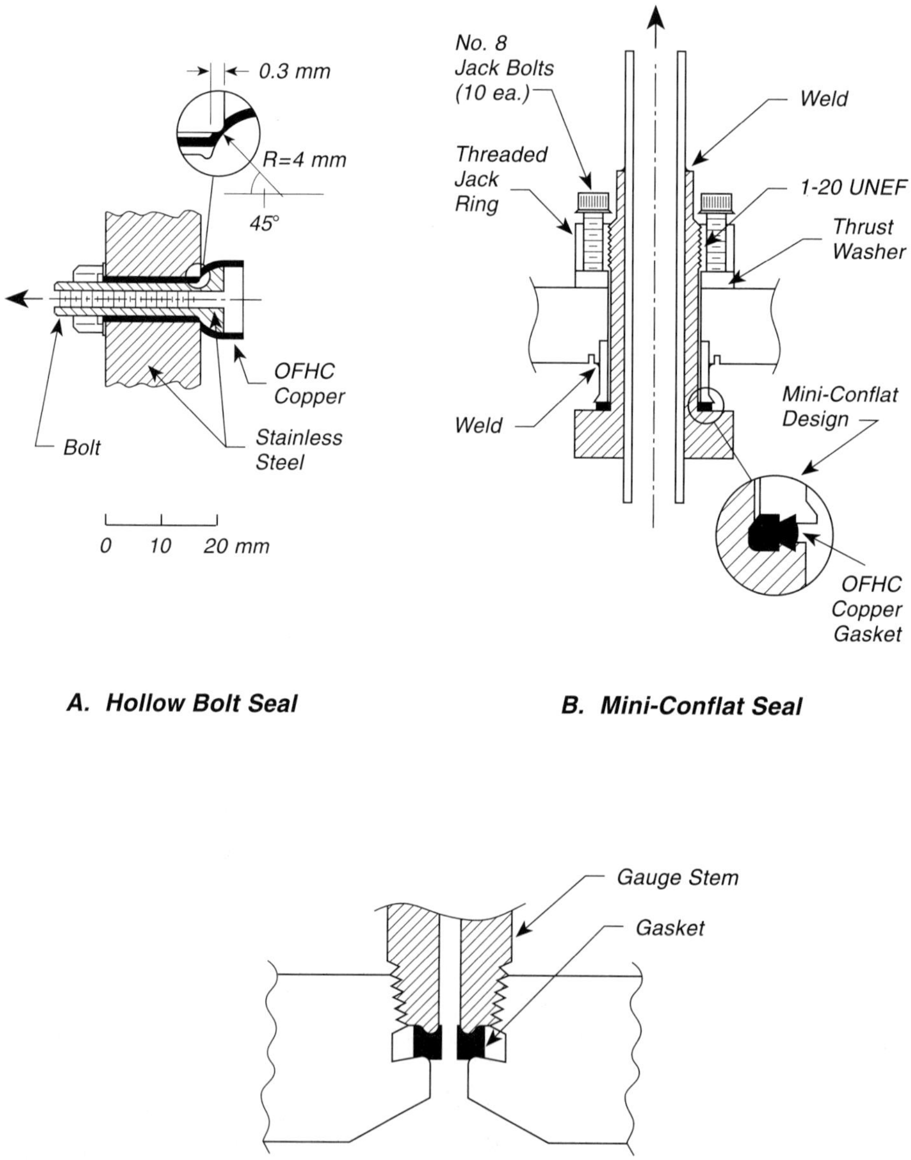


**A. Hollow Bolt Seal**

**B. Mini-Conflat Seal**

**C. Gauge Tube Seal**

# I. Leak Repairs and Detection

# 52

# Coaxial Pumped He Leak Detection Probe

He-leak checking is commonly used, either in the shop with a portable leak detector, or in the laboratory where a mass spectrometer inside the vacuum system is used for detection. A common problem involves the detection of a small leak at some point, where subsequent investigation will show that there is really no leak at that point. This confusing result is caused by the diffusion through the atmosphere of He from the suspect point to the real leak, which is some distance away. The device shown in Figure 52 eliminates this effect by sweeping He away from the test point using a coaxial vacuum sleeve.

Figure 52 shows the device, and the reader is referred to the original publication for construction and assembly instructions [1]. The He flow rate is adjusted to one or two bubbles of He per second as determined by placing the probe under water with the vacuum pump valved turned off. The rate of pumping should be adjusted so that the sensitivity for He is not diminished by too rapid pumping. A pump with a displacement of 5–10 L/min is satisfactory, with a throttle valve for pumping speed adjustment.

Once a leak is noted, a second test should be performed to verify its general location. This consists of placing a plastic bag around the suspected part, sealing the bag with tape, and then filling the bag with He, and observing the leak again. Using this coaxial probe and then using the bag verification test will eliminate carrying non-leaking components to the shop for repair.

In the event that leak sealing compounds are used for temporary repairs, only very small quantities of the sealer should be applied at the leak location, and a tag should be placed on the affected component telling the location of the leak and the date of temporary repair. The leak should be repaired correctly as soon as possible after the temporary repair. The tag is a guide for the memory when the vacuum system is ready to be disassembled.

## *References*


[1] G.L. Fowler, J. Vac. Sci. Technol. **A5**, 390 (1987).

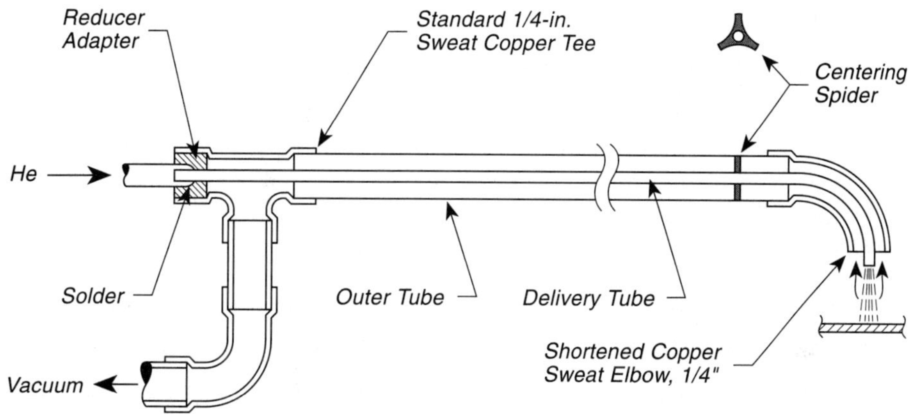


**Fig. 52. Coaxial Pumped He Leak Detection Probe**

# 53

# Temporary Leak Sealing of Welded Bellows

Welded bellows often develop small leaks at times when it is inconvenient to interrupt the work going on in the vacuum system. These leaks are most likely located on the inner welded seal of the bellows and are difficult to locate precisely.

For small leaks a temporary seal may often be established by using a leak sealer diluted with a solvent to reduce its viscosity and then spraying tiny amounts of the liquid onto the leak region. However, continued flexure of the seal after drying can cause the leak to reappear. Therefore, a clever trick [1] was devised in order to extend the lifetime of the bellows while a new one was being ordered. This involves immobilizing the small section affected by the leak. This is done by squeeezing gently several 0.5-mm-thick stainless steel clips around the bellows region of interest immediately after spraying the diluted sealant on the bellows, as shown in Figure 53. Since one leak in a bellows is generally a signal of more to come, this procedure should not be employed as a long-term solution to the leaking bellows problem.

A variety of liquid leak sealers are commercially available, and the reader is referred to the catalogs or to the original reference for a description of a silicone sealer [1].

## *References*

[1] W.F. Egelhoff, Jr., J. Vac. Sci. Tech. **A6,** 2584 (1988).

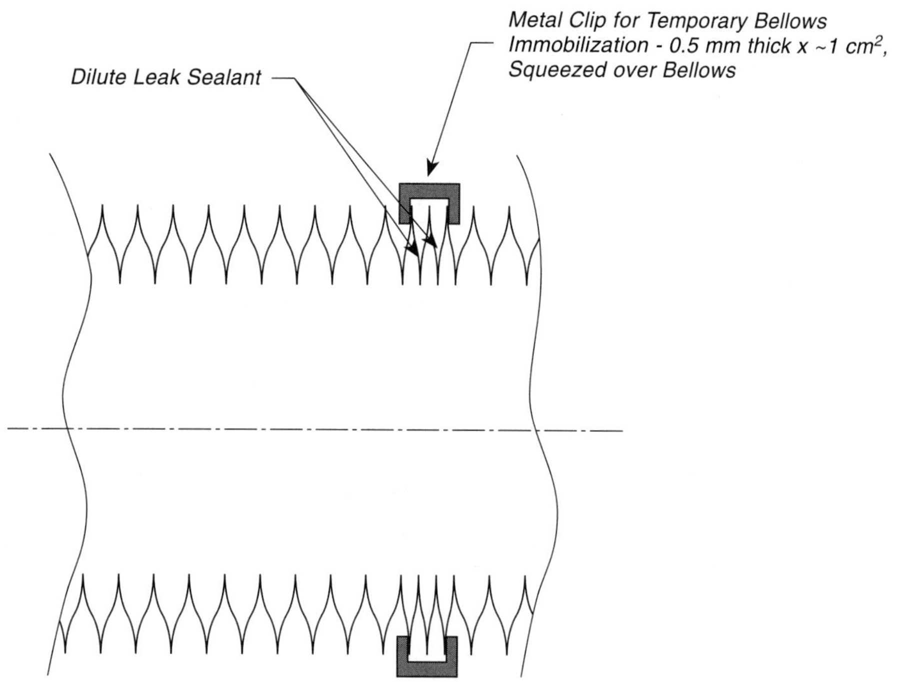


***Fig. 53. Temporary Welded Bellows Repair***

# 54

# Atmospheric Permeation—Viton O-Ring

Viton [1] seals are used in ultrahigh vacuum applications where deformable metal seals cannot be used. Viton seals may be baked out at 200°C and after a 24 h bakeout an outgassing rate of the order of 2 x $10^{-9}$ Torr L/sec $cm^2$ (nitrogen equivalent) has been reported [2]. After bakeout, the main gas evolved was water. Bakeout eliminates hydrocarbons initially detected upon installation of a fresh Viton seal [2].

A significant property of Viton seals is their ability to allow water from the atmosphere to permeate slowly through the outgassed Viton into the ultrahigh vacuum system, causing the partial pressure of water to rise for many days after bakeout. Figure 54A shows an apparatus where atmospheric gas permeation through a Viton O-ring was investigated [3]. A stainless steel vacuum system, equipped with a mass spectrometer and trapped diffusion pumping facilities, contained a test Viton O-ring [3]. The top of the vacuum chamber was baked at about 100°C, while it was estimated that the section containing the O-ring was baked at 50°C. A base pressure of about 5 x $10^{-9}$ Torr was achieved after 3 days' baking. Following 2 days' cooling, the mass spectrum of the residual gases in the chamber was measured for an additional 30 days, as shown in the test schedule (Fig. 54B).

It was found that the major effect was a continuous rise in the partial pressure of water during the 30-day period, with a decreasing rate after about 20 days. Figures 54C and 54D schematically show these tests for conditions of low atmospheric humidity and high atmospheric humidity. The rate of water permeation increases for the humid conditions. There is no detectable amount of permeation of other atmospheric gases such as $N_2$, Ar, or $CO_2$ during the 30-day test, although some $O_2$ permeation was reported [3].

These results suggest that the best application for Viton seals in ultrahigh vacuum systems is in cases where exposure of the seal to the atmosphere does not occur, i.e., where the seal is used internally for isolation of one part of the vacuum system from another. This is the case with Viton sealed gate valves mounted between sections of the vacuum system, where the seal is retained in vacuum during the time when the valve is opened.

A number of studies of outgassing and permeation of vacuum materials have been carried out [4–6].

## *References*

[1] Viton A is a copolymer between vinylidine fluoride and hexafluoropropylene with the formula $[-FC(CF_3)-CF_2-CH_2-CF_2-]_n$. Initial outgassing evolves water, oxygen, hydrogen, nitrogen, and

some hydrocarbons (most prominent mass peak at 43 amu, probably from $C_2F^+$). After bake-out, the main gas observed was water [2].
[2] R.S. Barton and R.P. Govier, J. Vac. Sci. Technol. **2,** 113 (1965).
[3] N. Yoshimura, J. Vac. Sci. Technol. **A7,** 110 (1989).
[4] W.G. Perkins, J. Vac. Sci. Technol. **10,** 543 (1973).
[5] B.B. Dayton, Trans. of the Amer. Vacuum Soc., 6th Vacuum Symposium, 101 (1959).
[6] P.A. Redhead, J.P. Hobson, and E.V. Kornelson, *The Physical Basis of Ultrahigh Vacuum* (American Institute of Physics, New York, NY, 1993), chapter 10.

***Fig. 54. Atmospheric Permeation—Viton O-Ring***

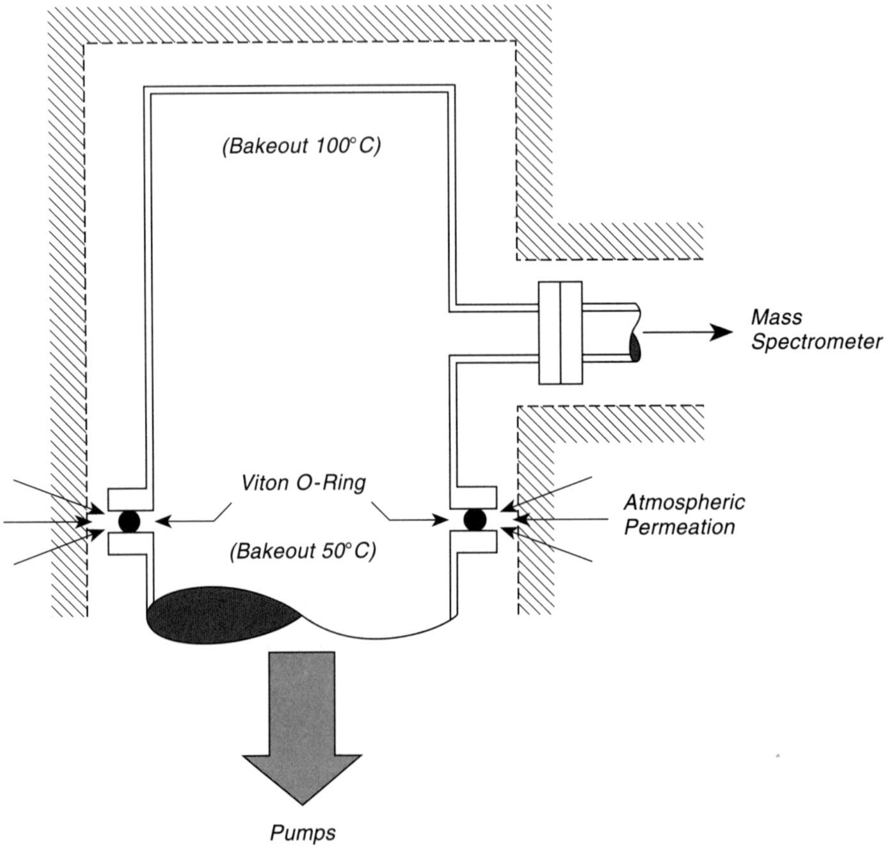


***A. Apparatus for O-Ring Studies***

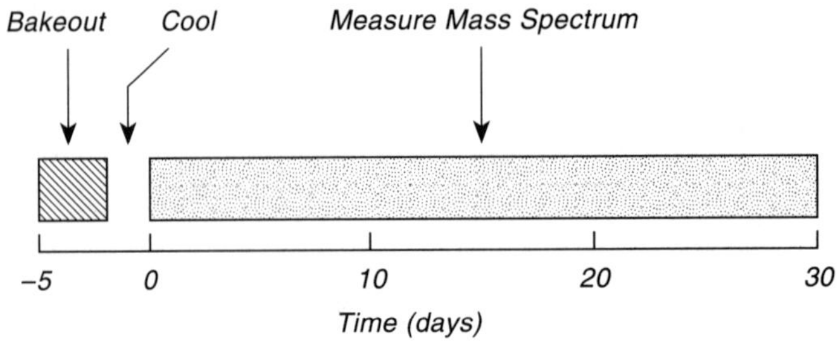


***B. O-Ring Test Schedule***

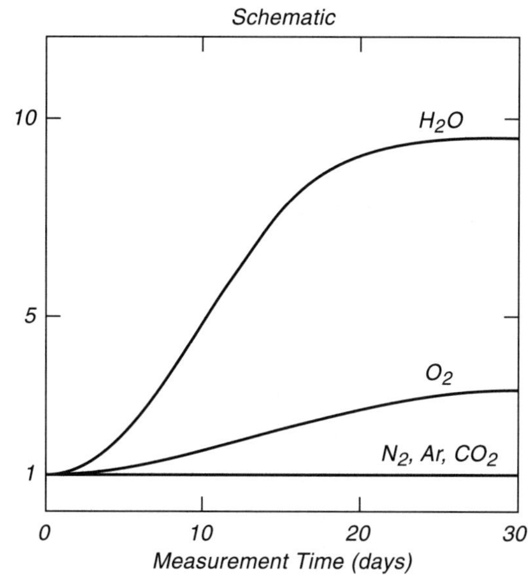


***C. Low Humidity—Permeation***

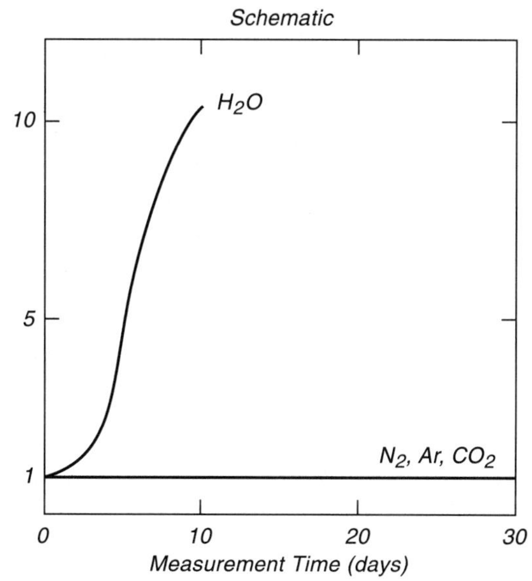


***D. High Humidity—Permeation***

# J. Specialized UHV Systems

# 55

# Aluminum Ultrahigh Vacuum System

There are advantages in using aluminum UHV systems. For proton synchrotrons, the residual radioactivity produced in stainless systems may be reduced by one or two orders of magnitude, 100 hours after machine shutdown, by employing Al instead of stainless steel. For electron synchrotron machines and electron storage rings, aluminum is preferred because of good thermal conductivity, ease of manufacture by extrusion, low residual radioactivity, nonmagnetic character, and low cost. The design here is an all aluminum system made of two types of aluminum alloy; aluminum flanges and gaskets are specially designed [1].

A special aluminum alloy, 2219-T87, is found to have the necessary mechanical properties for making the sealing flanges and nut/bolt sets. As shown in Figure 55, the sealing surface consists of a circular slot on each flange that houses a Helicoflex [2] O-ring containing an elastic spring core inside a pure aluminum sheath. The fixed depth of the O-ring groove was adjusted to correspond to the specified compression radius of the Helicoflex element—a 4.5-mm diameter Helicoflex required two 1.5-mm deep grooves.

The chamber body was made of a 6063-T alloy, which could be welded by an AC TIG welding process. Distortion after welding was negligible and remachining was not necessary. The entire chamber, except the O-ring sealing surfaces, was anodized to a 30-μm thickness. Bellows made of 6063-T6 alloy could be made by a rolling process yielding 0.5 mm thick walls. The bellows are employed for alignment purposes, not for dynamic motion. They also compensate for thermal expansion effects.

The O-rings were sealed by torquing the nut/bolt combinations, bringing the aluminum flange surfaces together. The torque (per unit area of the O-ring) used for tightening was between 140 and 215 kg/cm, with 200 kg/cm being normal.

Following bakeout at 200°C, the leak rate at the Helicoflex seals was measured with a He leak detector. It is reported that the leak rate is less than $10^{-11}$ atm cm/s, where the units on this number suggest that a leak rate per $cm^2$ of sealing surface was calculated. The ultimate pressure in the system was $3 \times 10^{-10}$ Torr after bakeout and the residual gases consisted mainly of $H_2$, CO, and $CO_2$. The 38-L system was pumped by a 220-L/s turbopump and a 160-L/s sputter ion pump. It is indicated that the practical upper bakeout temperature is limited to 170°C by the 6063-T6 aluminum alloy.

A study of the outgassing rate of an aluminum vacuum system during bakeout is given in reference [3].

## *References*


[1] H. Ishimaru, J. Vac. Sci. Technol. **15**, 1853 (1978); see also, H. Ishimaru, J. Vac. Sci. Technol. **A7**, 2439 (1989), which contains references to various papers on this subject.

[2] Helicoflex is a product of Helicoflex Co., P.O. Box 9889, Columbia, SC 29290.

[3] M. Suemitsu, T. Kaneko, and N. Miyamoto, J. Vac. Sci. Technol. **A5**, 37 (1987).

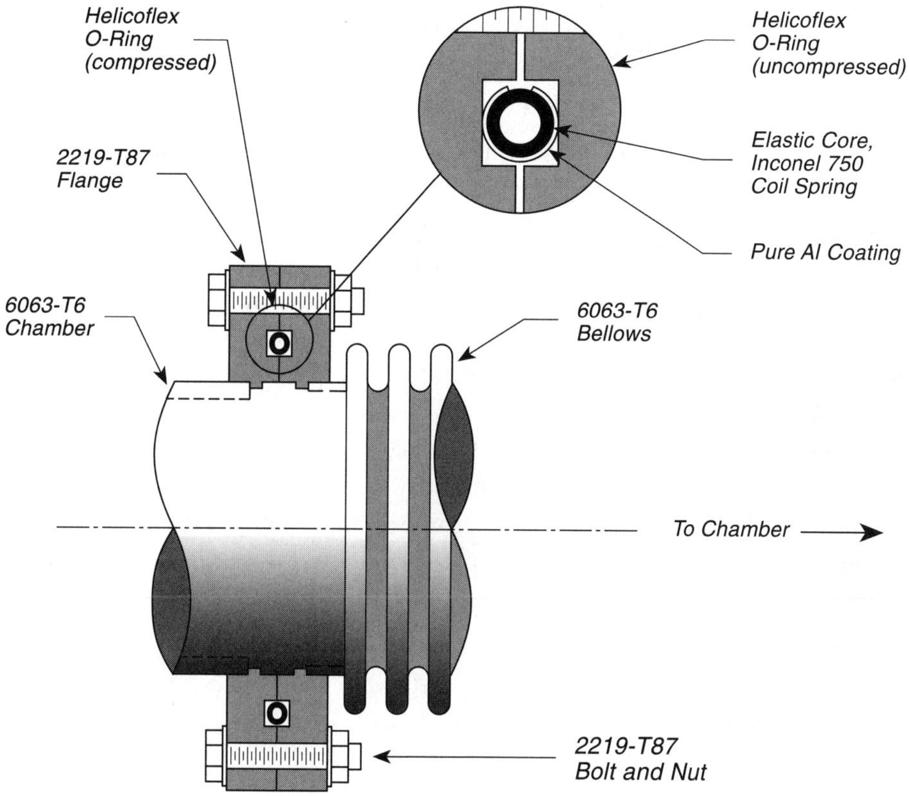


***Fig. 55. Aluminum UHV Chamber, Flange, and Bellows***

# 56

# Surface Electrochemistry Apparatus

The development of the field of electrochemistry from the viewpoint of surface science has depended upon the experimental ability to prepare atomically clean single crystal surfaces and then to transfer these surfaces to an electrochemical environment for measurements of current versus voltage [1]. While a number of designs have been provided for this transfer operation, the design schematically shown in Figure 56 is historically important and remains useful today.

A single crystal is mounted on a sample manipulator in the left-hand UHV system, which contains various facilities for surface cleaning and surface characterization. The sample, once cleaned and characterized, is unplugged in vacuum from the sample manipulator (see pages 90 and 94) and is captured on a bellows type translator coming in from the right through an open gate valve. From here the sample may be immersed into the electrolyte of the electrochemical cell by means of moving the translator vertically and lowering the sample through the open gate valve. Four types of pumps are utilized on the right-hand side of the apparatus: two sorption pumps using zeolite pellets at 77 K; a cryogenic pump containing a charcoal-covered surface cooled to 14 K by a Joule-Thomson refrigerator, which lowers the pressure to about $10^{-7}$ Torr; ionization pumps to bring the pressure down to about $10^{-9}$ Torr; and a titanium sublimation pump for final pumping and for purification of Ar gas which is admitted to high pressure for the sample transfer to the electrochemical cell.

The solution-handling subsystem must exclude dissolved contaminants. Liquids are contained in deaerated ($N_2$ purged) Pyrex bottles. Air is the most difficult impurity to exclude, and Teflon tubing sheathed in Teflon bellows is used under inert gas purging conditions inside the annular space to exclude air diffusion through the Teflon to the electrolyte. Glass vacuum valves rather than stopcocks are used to avoid contact with the air. Teflon compression fittings are used to make tubing interconnections. All parts are soaked in chromic acid cleaning solution until needed, then rinsed with highly purified distilled water.

Potential-dependent interfacial properties of the electrode-sample are measured using a Pt electrode introduced through a 3-mm-diameter section of Teflon tubing. A silver wire, coated with silver chloride, and immersed in a chloride solution is a convenient reference electrode. By applying a potential between the Pt electrode and the sample, and monitoring this potential against the silver chloride/silver electrode, it is possible to plot out the current-voltage curves and to correlate with the nature of the electrode introduced.

A comprehensive listing of other types of electrochemical cells which are coupled to UHV systems is given in references [2] and [3], which also describe X-ray photoelectron spectroscopy (XPS) studies of the corrosive oxidation of a Ni surface.

## *References*

[1] A.T. Hubbard, in *Comprehensive Chemical Kinetics*, edited by C.H. Bamford, D.H.F. Tipper, and R.G. Compton (Elsevier, Amsterdam, 1988), pp. 1–67.

[2] Y. Liang, D.K. Paul, Y. Xie, and P.M.A. Sherwood, Anal. Chem. **65**, 2276 (1993).

[3] P.M.A. Sherwood, Chem. Soc. Revs. **14**, 1 (1985).

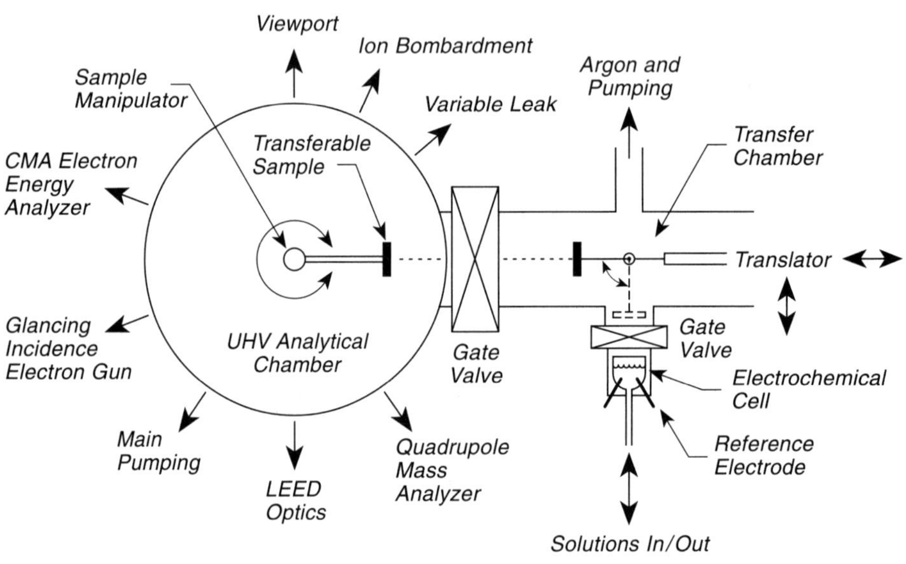


***Fig. 56. Schematic of Surface Electrochemistry Apparatus***

# II. Mechanical Fabrication Techniques


A. Grids . . . . . . . . . . 179
B. Conductive Coatings . . . . . . . . . . 189
C. Phosphor Screens . . . . . . . . . . 193
D. Thermionic Emitters . . . . . . . . . . 201
E. Shielding . . . . . . . . . . 221
F. Single Crystal Fabrication/Orientation . . . . . . . . . . 229
G. Tip Fabrication . . . . . . . . . . 249
H. Spot-Welding . . . . . . . . . . 259
I. Mechanical Procedures . . . . . . . . . . 265

# A. Grids

# 57

# Hemispherical Grids— Formation and Piercing

Formed grids are employed in ion and electron optical devices. Often these grids must be made of highly transparent woven stock, either of stainless steel or tungsten [1].

Woven grid material may be formed into a hemispherical shape using a mold or mandrel, as shown in Figure 57A. Here the grid material is stretched over the mandrel and an annealed aluminum ring is slipped over the stretched grid, capturing it near the base of the mandrel. A steel pressure ring is then seated on the aluminum ring and pressure is applied causing the soft aluminum ring to distort into an L-shaped cross section as shown. This also encases the grid material, fixing it in the aluminum [2].

Another method for attaching a stretched grid to a support ring is shown in Figure 57B. Here a mesh grid is stretched over a brass mandrel, and a stainless steel plate is clamped in place to hold the grid on the mandrel. The plate is tack-welded to the grid in about ten places and the clamps are released and the plate turned over . The grid wires are trimmed and individually spot-welded [3]. A more complex variation on this method is shown in Figure 57C where the grid is captured between two shaped rings, R, which are bolted inside a jig. Spot-welding takes place through the holes, H, after the grid is stretched over the mandrel [4].

Grids must often be pierced in specific locations for the passage of hardware or source beams through the grid. Figure 57D shows a nice way to do this job accurately. The brass mandrel is punched with the location of the center of the hole. A scribe is then used to scratch a circular shape on the mandrel locating the cutting line on the grid. The grid is then stretched and spot welded, and the assembly is immersed into aqueous NaCl solution (50 g/L). An electrode to a spot welder ($S_1$) is attached to the assembly, and a second electrode, $S_2$, made of tungsten wire (0.5 mm) is sharpened to a radius about twice the diameter of the mesh wires to be cut. An electrical arc from the spot welder will cut the wires individually, usually between adjacent wires perpendicularly crossing the wire to be cut. The salt water prevents pitting on the brass mandrel by the electric arc [3].

This stretching method will not make perfect hemispherical grids, since some relaxation will occur upon removal from the mandrel. If more rigid grid material is needed, the stretched grid may be electroplated on the mandrel [5] or it may be fired in an oven. The electroplating stiffens the contact points between the meshes while firing will stiffen the wire itself. In the case of firing, a brass mandrel cannot be used, and release agents on the mandrel will be needed. Either Aquadag (colloidal graphite) or $MoS_2$ will work as release agents.

## *References*


[1] Unique Wire Weaving Co., Inc., 762 Ramsey Avenue, Hillside, NJ 07205-1094; Electronic Space Products International, 1050 Bensen Way, Ashland, OR 97520.

[2] D.A. Shaw and D.A. Bennett, Rev. Sci. Instrum. **43,** 1706 (1972).

[3] P.A. Taylor, J. Vac. Sci.Technol. **A6,** 2583 (1988).

[4] J. Koch, Rev. Sci. Instrum. **45,** 1212 (1974).

[5] P.W.Palmberg, D.C. Johnson, and H.J. Boll, Rev. Sci. Instrum. **35,** 244 (1964).

*Fig. 57. Hemispherical Grids—Formation and Piercing*

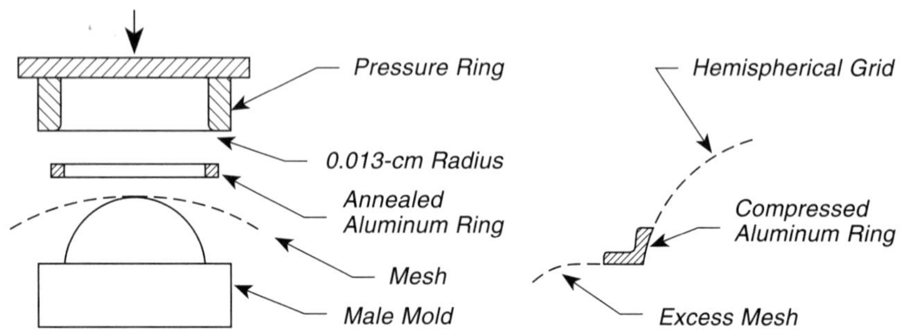


**A. Grid Capture—Al Ring**

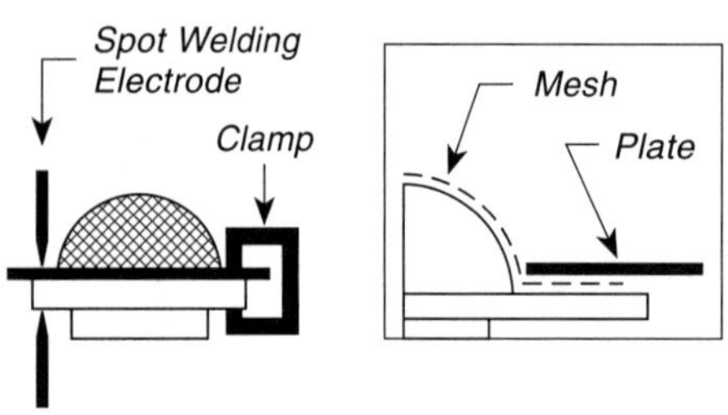


**B. Grid Capture—Spot Welding**

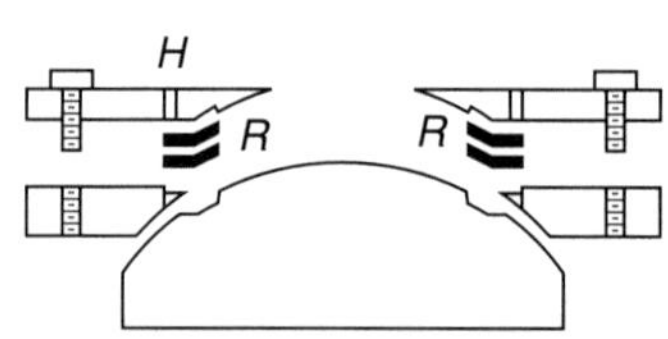


**C. Grid Capture—Spot Welding**

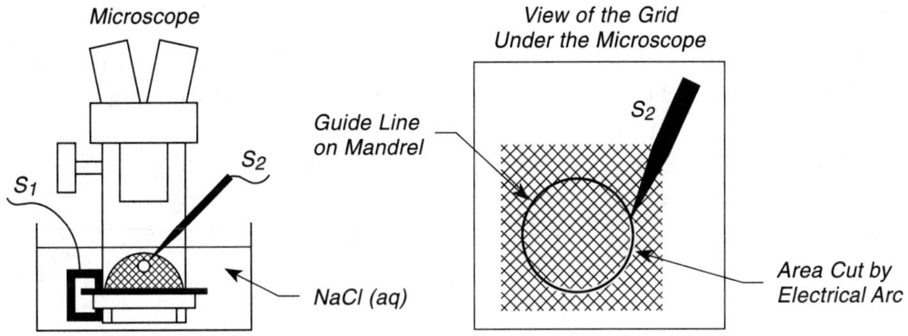


***D. Piercing Grid***

# 58

# Making Flat Mesh Grids of Large Size

Electroformed mesh [1], having up to 85% transparency, may be conveniently used as grid material in electron and ion optical applications. However, it is difficult to construct flat grids of many centimeter dimensions, and irregularities in flatness will influence the electrical field uniformity desired in front or between parallel grids.

Figure 58 shows a convenient method [2] for using a Viton O-ring to produce uniform stresses at the boundaries of electroformed grids, making them flat. The lower plate has a special groove cut in it with the inner wall beveled 30° with respect to the normal. When the sandwich involving the two plates, the grid, and the O-ring is compressed by adjusting bolts around the perimeter, the motion of the O-ring gradually pulls the mesh tight as the O-ring expands outward in the beveled groove under pressure. This produces the desired flat grid.

## *References*


[1] Electroformed mesh is a product of Buckbee Mears, 245 E. Sixth Street, St. Paul, MN 55101, and it may be obtained in various materials, with various thicknesses and optical transmission.

[2] R.J. Noll, L.L. Ochalla, and J.C. Weisshaar, Rev. Sci. Instrum. **62,** 246 (1991).

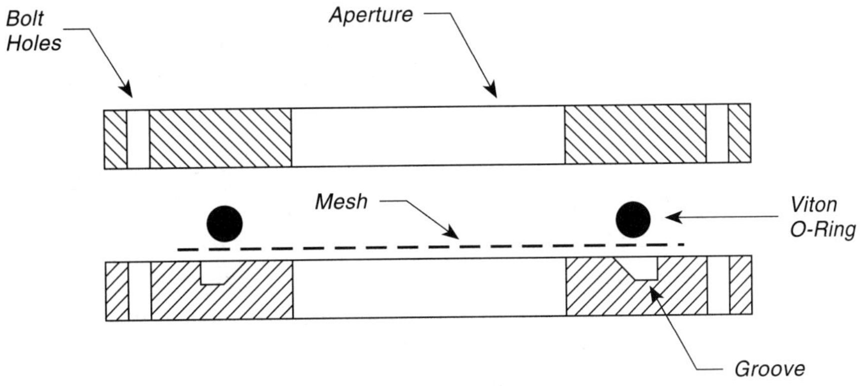


***Fig. 58. Making Large Grids Planar***

# 59

# Grid Fabrication Techniques—Conical Grid

The fabrication of a grid having a precise geometry is accomplished by using a mandrel of the desired shape, and on page 180, methods for making hemispherical grids are shown. Here we describe the fabrication of a conical grid that is used as an ion cage in a high-efficiency molecular beam ionization detector with a short ionization region [1].

The conical cage, which was designed with specific ion optical properties, is shown mounted on its lens plate in Figure 59A. Four angled support wires are welded to the plate and provide rigidity for the grid.

Figure 59B shows the metal mandrel, which has been fabricated to the desired shape so that it can fit the lower electrode mount of a tabletop spot-welder. Ta wire (0.1 mm diameter) is wound on the mandrel on a lathe, operated at slow speed or by hand, and then the ends of the coil are fixed with tape or by a soluble glue dot. After the four support wires are spot-welded to the grid on the mandrel, the grid assembly may be removed by a short immersion in an ultrasonic bath. Following removal of the grid from the mandrel, the angled support wires are welded to the lens plate, using the mandrel for centering the grid assembly on the lens aperture [2].

## *References*


[1] K. Kuhnke, K. Kern, R. David, and G. Comsa, Rev. Sci. Instrum. **65**, 3458 (1994).

[2] Dr. R. David, Institut fur Grenzflächenforschung und Vakuumphysik, Forschungszentrum Jülich GmbH, Postfach 1913, D-52425 Jülich, Germany (private communication).

***Fig. 59. Grid Fabrication Techniques—Conical Grid***

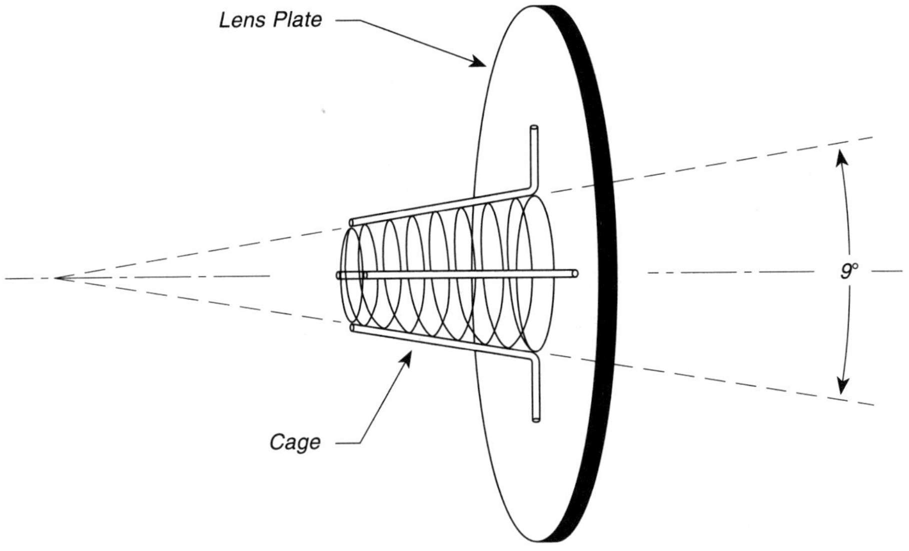


***A. Conical Grid***

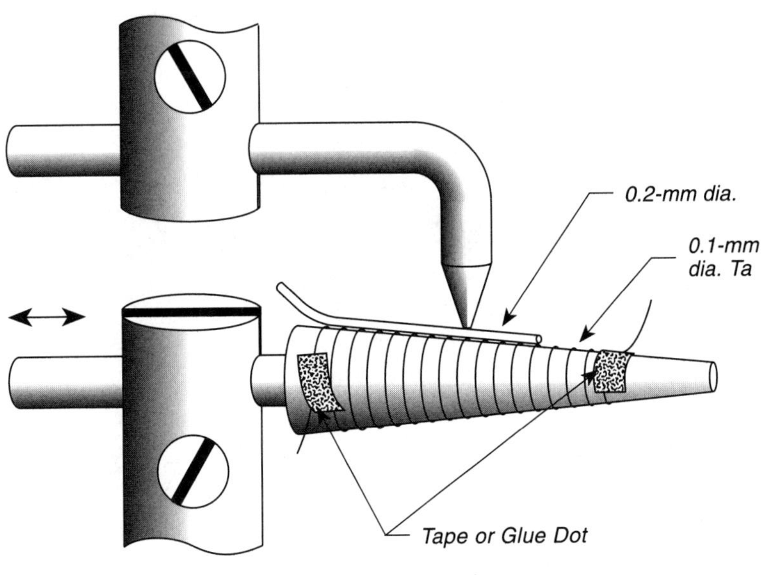


***B. Mandrel + Spot-welding***

# B. Conductive Coatings

# 60

# Deposition of Electrically Conductive $SnO_2$ Films

Many devices used in surface science research require the deposition of electrically conductive films. These films are often used as the backing for phosphor screens in LEED apparatus and other electron beam display techniques.

Figure 60 shows the method used in many laboratories for the deposition of semiconducting $SnO_2$ films on Pyrex glass surfaces. A glass surface, appropriately shielded on the inside and outside with Aquadag (colloidal graphite) [1] to prevent unwanted deposition, is placed into a small furnace or glassblower's annealing oven. The oven is slowly brought to the annealing temperature of Pyrex, 550°C. A saturated solution of $SnCl_2$ in isopropanol is made up and loaded into the reservoir of an atomizer, as shown. Since we wish to deposit $SnO_2$, the atomizer is connected to a regulated oxygen tank, and oxygen gas is used to oxidize the tin from Sn(II) to Sn(IV) upon deposition of the $SnCl_2$ uniformly to the hot surface of the glass. The spraying is carried out in short bursts, and the oven door is then closed between bursts in order to rewarm the glass to the deposition temperature. An experienced operator can judge the conductivity of the film from its color. Blue and barely visible coatings possess approximately kΩ resistances between two points a few centimeters apart, while yellow or brown coatings give less than 1 kΩ resistance.

Caution should be taken not to breath in during the film deposition with the $SnCl_2$, since pungent HCl gas is a major product of the pyrolysis chemistry leading to $SnO_2$ deposition.

Following film deposition, the Aquadag can be removed using an abrasive kitchen cleanser and a cotton swab in water, or by heating in a glass annealing oven for long times, followed by washing.

The electrical contact to the conductive film can easily be made using a glass-metal feedthrough fabricated by the glassblower. To the inner end of the metal feedthrough, a small piece of platinum ribbon or wire is welded in such a manner that it makes good contact with the inner surface of the glass. The contact is made permanent by applying either platinum bright or gold bright paint [2] with a small camel's hair brush, followed by firing in the oven to produce a shiny film of metal which is both adherent and conductive. Alternate methods for making electrical contacts through glass to a conductive $SnO_2$ film are described in reference [3].

The work function of $SnO_2$ films prepared in this way has been measured using a retarding potential method for an incident electron beam. The coatings were found to have a work function of 4.79 ± 0.07 eV. Spatial variations in the work function averaged ± 10 meV, and work function differences between samples were less than 100 meV [4].

The original scientific report of the use of $SnO_2$ films for producing conductive glass may be found in reference [5]. Here also is a study of the resistivity of a number of these films from 1.5 K to 500 K.

Tin oxide films can be removed by "nascent hydrogen" produced near the film surface by dipping a cotton swab first in HCl(aq) and then in powdered zinc dust. This reactive mixture, when rubbed on the $SnO_2$ film, will efficiently remove it by a reduction process. Thorough washing in $H_2O$ should then be carried out.

## *References*


[1] Aquadag colloidal suspensions may be obtained from Acheson Colloids Co., 1607 Washington Avenue, P.O. Drawer 611747, Port Huron, MI 48061-1747.

[2] Organometallic paints containing either Pt or Au may be obtained from the Hanovia Division of Engelhard Industries, 1 West Central Avenue, East Newark, NJ 07029. These paints contain the metal chemically bound with organic ligands which are thermally unstable.

[3] R. Gomer, *Field Emission and Field Ionization,* (AIP, American Institute of Physics, 1993), pp. 169–170.

[4] J. Burns and E. Yelke, Rev. Sci. Instrum. **40,** 1236 (1969).

[5] R. Gomer, Rev. Sci. Instrum. **24,** 993 (1953).

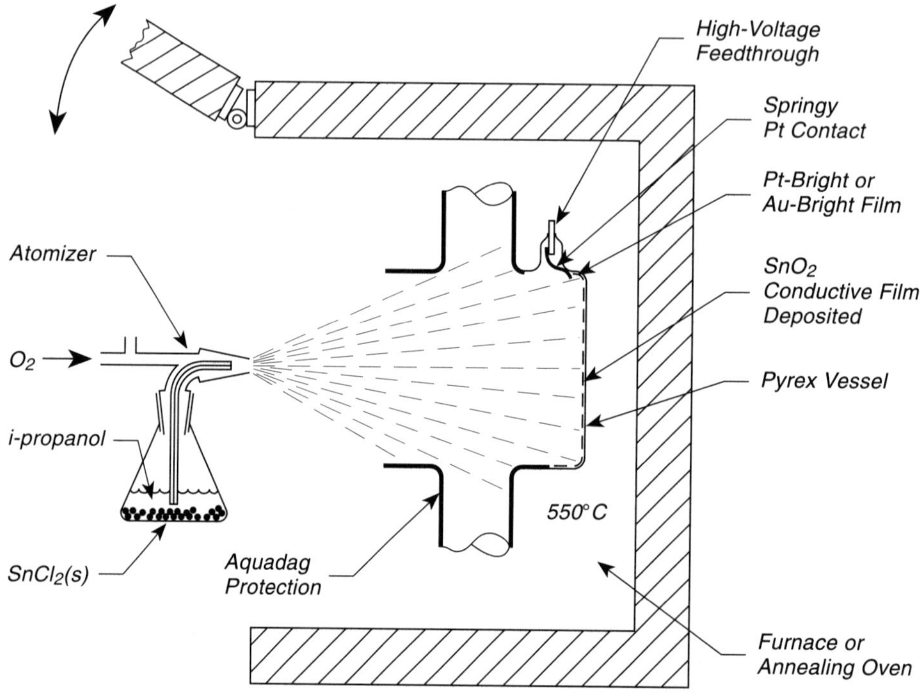


**Fig. 60. Deposition of Conductive $SnO_2$ Films**

# C. Phosphor Screens

# 61

# Sedimentation Method for Depositing Phosphor Screens

A number of methods involving electron beam measurements require the use of phosphor screens that are uniform and adherent. The method described below is a variant of techniques originally described for coating cathode ray tube screens and field emission microscope screens [1–4]. The method may be used for either glass-based phosphor screens, which are viewed by transmission from the back side, or metal-based screens, which are viewed from the front side [5].

The essential requirements are to deposit a phosphor powder uniformly along with a binder substrate that will be adherent during bakeout and use. All such screens are sensitive to mechanical disturbance, including touching with the fingers.

For deposition of metal-based screens the underlying metal substrate must be pickled after chemical cleaning to prepare its surface. Two pickle agents are described for stainless steel substrates: (a) 30 volume % $HNO_3$ + 3 volume % HF in water—warming above room temperature will increase its rate of action and the substrate should be immersed for 30–60 s; (b) 10 volume % $HNO_3$ + 2 volume % HCl + 2 volume % HF in water—immerse screen for 5 min in an ultrasonic bath. One pickle agent is described for aluminum substrates: 12 volume % HF—then rinse with water and dip in 86 volume % HCl for no longer than 60 s. If the aluminum turns black during the HCl dip, repeat the HF step followed by the HCl step [5].

The clean and pickled substrate is then transferred to the 3-L sedimentation chamber shown in Figure 61, and is ready for application of the phosphor screen.

Two sedimentation solutions are prepared as described below:

Solution A: This is an electrolyte solution that facilitates the gelation of the phosphor particles with the silicate. 300–500 ml water + 1.8 g $Ba(NO_3)_2$.

Suspension B: 150–175 ml water + 0.18 g silica gel [6] + 3.2 g phosphor. This suspension should be agitated in an ultrasonic bath or an ultrasonicator (not the same thing).

After mixing thoroughly, transfer suspension B to a large Erlenmeyer flask and dilute to 500 ml with water, using a wash bottle to make a complete transfer. Add solution A and make a final volume of about 1 L. Shake vigorously and transfer to the ultrasonic bath for continued agitation for 10 min. Pour the suspension into a 3–4-L beaker and rinse all material into the beaker. Dilute to about 3 L, and stir rapidly for 5–10 min using a magnetic stirbar of 3-in. size. Turn off the stirrer and allow the suspension to partially settle, during which time the large particles will preferentially settle to the bottom. Decant the liquid into another 3-L beaker, being careful to leave the large particles on the bottom undisturbed. Allow this second suspension to settle for about 7 min and then carefully decant the solution into the sedimentation chamber containing the substrate. This should be done carefully, directing the suspension down the side of the sedimentation chamber using a large funnel and avoiding bubbles. Cover the sedimentation chamber and allow it to sit for 5–12 h without agitation or vibration. Following sedimentation, the most difficult task is to drain off the liquid without mechanically disturbing the screen. The drain stopcock should be opened very slightly so that a very slow drip is achieved, making the flow rate so small that about 6 or more hours are required to drain the liquid.

The drained screen should be air dried in the sedimentation chamber for several hours, keeping it covered to avoid dust. After drying, inspect for homogeneity. A "ring" on the screen indicates a discontinuity owing to a change in the drip rate, possibly due to a particle being trapped in the stopcock. Such discontinuities may not be serious unless photometry is to be employed, where phosphor thickness must be constant. Even without visible evidence of inhomogeneities, the phosphor thickness will vary over the screen surface owing to the variable angle of interception of the settling particles by the curved screen. Following drying, the screen should be baked in an oven at 80°C for 30 min to remove additional water. Screens treated in this manner are satisfactory for use in ultrahigh vacuum and show little or no influence on the base pressure after a bakeout at 150°C for 24 h [5].

For deposition of phosphor screens in glass vacuum envelopes, minor variations of the procedure described above may be used. The sedimentation chamber may be replaced by the envelope itself if the depth of the liquid suspension is sufficient. Alternatively, the glass vacuum envelope can be immersed in a large sedimentation vessel, and in either case the draining can be done by very, very slow tipping of the envelope using a slowly driven motor that turns once in 24 h.

Screens prepared in this manner are approximately 10 μm thick and well suited for electron beam imaging. Thicker screens may be subject to electrostatic charging.

An unusual electrostatic deposition and adhesion method has been described by Cetronio and Cross [7].

## *References*


[1] F. Rosebury, *Handbook of Electron Tube and Vacuum Techniques* (American Institute of Physics, New York, 1993), pp. 251-254.

[2] J. Millman, *Vacuum-tube and Semiconductor Electronics* (McGraw-Hill, New York, 1958).

[3] K.R. Spangenberg, *Vacuum Tubes* (McGraw-Hill, New York, 1948), pp. 429-432 and 821.

[4] R. Gomer, *Field Emission and Field Ionization* (American Institute of Physics, New York, 1993) p. 167.

[5] D.K. Flynn-Sanders, S.-L. Chang, P.A. Thiel, and R. Imbihl, J. Vac. Sci. Technol. **A10**, 413 (1992).

In addition to the phosphor deposition procedure described here, this paper has a description of several phosphors and their light emission properties. One should be especially careful with beryllium-containing phosphors because of their poisonous nature and their possible entry to the human body through cuts in the skin.

[6] Silica gel which works in this application is Davisil, a product of Davidson Chemical which is a division of W.R. Grace, 7221 W. Parkland Court, Milwaukee, WI 53223. The surface of the silica is hydroxylated and the total surface area is about 300 $m^2/g$. In general silica gels have smaller particle sizes than silica powders.

[7] A. Cetronio and J.A. Cross, J. Vac. Sci. Technol. **15**, 1191 (1978).

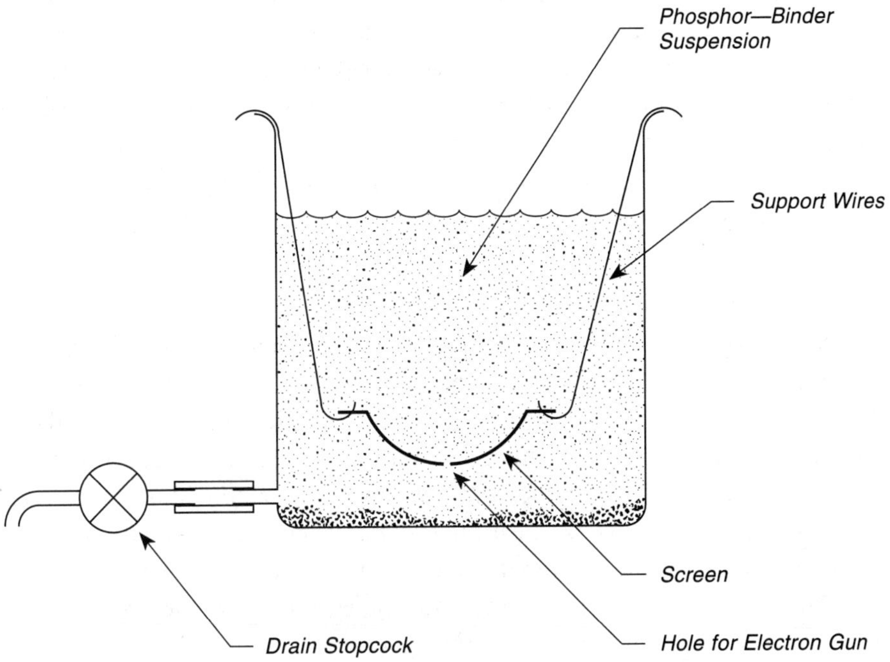


***Fig. 61. Sedimentation Deposition of Phosphor Screen***

# 62

# Dusting Method for Coating Phosphor Screens

Phosphor coating of screens for imaging electron beams may be achieved using a technique developed years ago at A T and T Bell Laboratories [1] and described briefly elsewhere [2]. This method has the advantage over the sedimentation method of producing more uniform phosphor coatings, since it does not sense a curved surface during the deposition procedure.

As shown in Figure 62, a glass or metal substrate, cleaned with a strong hot alkaline detergent, has a stopper made of cork or plastic (but not rubber) that seals the hole for the electron gun or other source. A sticky uniform layer of phosphoric acid, $H_3PO_4$, is deposited by the following procedure:

Clean about 100 glass balls (3-mm diameter) and the screen with ethanol and acetone to degrease. Then from a solution consisting of 3 drops of $H_3PO_4$ in 25 ml of acetone, place 25 drops onto the screen substrate with the stopper in place. Place the glass balls onto the screen substrate and roll them around as the acetone evaporates. After 2–3 min, the acetone will have evaporated, and the phosphoric acid will uniformly wet the substrate. If streaks are observed, repeat the process. Then shake 1–2 $cm^3$ of phosphor powder onto the wet substrate and tap the screen until the phosphor coats the entire surface. Invert the screen and tap to remove the loose powder. Remove the stopper and using a gentle jet of nitrogen, blow off any remaining loose phosphor. The screen should be baked in air at 200°C for 2 h and it is then ready for use in ultrahigh vacuum.

Caution should be exercised to prevent breathing of phosphor or contact with the skin.

A unique method for electrostatically depositing and adhering a phosphor screen is described in reference [3].

Another unique method for depositing a phosphor powder on a flat, tin oxide-coated plate is described in reference [4]. Here, a glass plate which has been chilled in air below the dew point is coated with phosphor powder from an atomizer. The phosphor uniformly adheres to the moist plate. Drying the plate in a dessicator removes the water, and the phosphor adheres to the plate without any adhesive. The adhesion is sufficiently strong to permit the use of the phosphor screen in any position in a vacuum system, yet the coating can easily be removed by wiping under water.

## *References*


[1] R. Carr and E. Chaban, J. Vac. Sci. Technol. **A10,** 3595 (1992).

[2] E.W. Muller and T.T. Tsong, *Field Ion Microscopy* (Elsevier, New York, 1969), p.139

[3] A. Cetronio and J. A. Cross, J. Vac. Sci. Technol. **15,** 1191 (1978).

[4] R.C. Abbott and P.M. Brown, Rev. Sci. Instrum. **38,** 832 (1967).

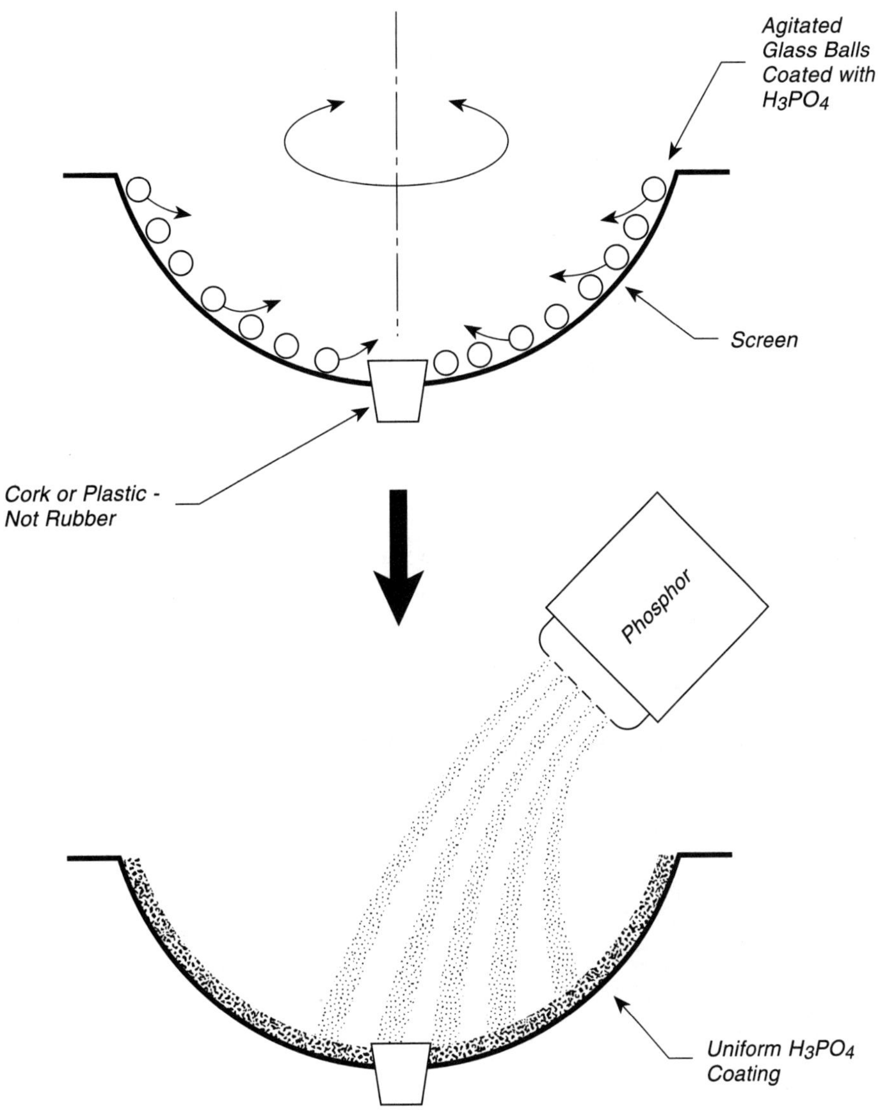


***Fig. 62. Dusting Method for Deposition of Phosphor Screen***

# D. Thermionic Emitters

# 63

# Thoriated Thermionic Emitters

Thermionic emitters are very widely used in surface science experiments as sources of electrons. The emission of electrons in the absence of applied electric fields follows the Richardson equation,

$$I/S = AT^2 e^{-e\phi/kT} \quad (1)$$

where $I/S$ is the electron current density, $\phi$ = the work function of the emitter, $e$ is the electronic charge, $T$ is the temperature in K, and $k$ is the Boltzmann constant. For metals, $A = 120A/\text{cm}^2\text{K}^2$.

Various coatings on thermionic emitters are used to reduce the work function, permitting the thermionic emitter to operate at lower temperatures while delivering the same current density as an unmodified emitter at higher temperatures. Some aspects of the various coatings and their advantages and disadvantages in ultrahigh vacuum systems are described by Redhead et al. [1]. For various reasons, $ThO_2$ coatings on W or Ir surfaces exhibit the most favorable properties for use in ultrahigh vacuum. While long-term operation in an oxygen ambient is not possible with either metal, Ir will withstand short accidental exposure to oxygen without burnout.

Figure 63A shows the simple experimental procedure that may be used for coating a thermionic emitter with a layer of $ThO_2$ by cataphoretic deposition [2]. A suspension of 200 mesh $ThO_2$ powder (5 g) in a solution containing 0.075 g of thorium nitrate in 100 ml of 95% ethanol is used. This suspension is allowed to stand for a few minutes so that the large $ThO_2$ particles settle to the bottom and are removed from further considerations. The suspension is electrically deposited on the filament at a negative potential of a few volts relative to the Pt anode. The required current density depends on the shape of the surface being coated, and is found by trial and error. It is best to electropolish tungsten substrates in NaOH solution (tungsten positive) and to rinse in water before beginning the $ThO_2$ deposition. The white $ThO_2$ powder will form a very uniform deposit on the emitter surface, and when fired in vacuum, this coating will be rather strongly adherent. Coatings between 0.025 and 0.040 mm in thickness are satisfactory [2]. Firing in vacuum to about 1800 K brightness temperature activates the thoria fully [2]. Reference [2] contains many interesting measured properties of $ThO_2$.

The maximum current density that may be drawn from both clean W and $ThO_2$ surfaces is shown in Figure 63B. Since the average work function of W (about 4.5 eV) is reduced to about 3.0 eV by thoriation, a tremendous enhancement of electron emission is achieved, as shown in Figure 63B [1].

One of the reasons for using thoriated thermionic emitters is to reduce the operating temperature for electron emission to levels that involve negligible evaporation of the metal. Figure 63C [1] shows the evaporation rate of W atoms versus electron emission current density for both W and $ThO_2$/W emitters. At the higher emission current densities for pure W, significant evaporation of W will occur over long experimental times, whereas $ThO_2$/W emitters involve negligible W evaporation rates.

It is important not to bombard the $ThO_2$ coating with electrons, which will cause damage. Even using AC accelerating voltages is detrimental in degassing ionization gauge grids, when the hot grid can be a thermionic emitter on the reverse half-cycle [1].

It has been reported that $ThO_2$-coated W filaments often emit unknown impurities that can be contaminants in surface science experiments [3]. This was studied for $ThO_2$/W filaments contaminated by F [4], and significant transfer of F or F-containing species to a clean single crystal of W was observed. Other halogen impurities such as Cl are also likely to be present in $ThO_2$ coatings exposed to new vacuum systems, since both HCl and HF are common pickling agents in cleaning stainless steel [5]. These effects can be eliminated eventually by outgassing new $ThO_2$-coated emitters for long periods of time [4]. Such halogen contamination of the grid surfaces of a Bayard–Alpert gauge renders the gauge useless for low pressure measurements, since electron impact desorption from the halogen-covered grid surface will supply spurious ion currents [6].

In addition to W wire thoriated by surface coatings of $ThO_2$, it is also possible to purchase W wire that contains alloyed Th at the 1%–2% level [7,8]. This small amount of thoria was originally added to decrease crystallite slippage in the polycrystalline wire as it was formed by drawing through a die. These wires also exhibit greatly enhanced thermionic emission, compared to pure W. Despite high vacuum conditions, these emitters are readily poisoned [9]. This poisoning process may be diminished by carburizing the wire in a hydrocarbon gas, such as acetylene at temperatures near 1700°C [8]. An outer layer of $W_2C$, about one-tenth the wire diameter in thickness is produced [8].

A thoriated Ir thermionic emitter has been reported to have high resistance to exposure to air and to exhibit low Th evaporation rates at operating temperatures [10].

Thorium carbide thermionic emitters exhibiting an effective work function of about 3.3 eV have been produced and will emit about 10 A/cm$^2$ at 2150 K [11].

## *References*


[1] P.A. Redhead, J.P. Hobson, and E.V. Kornelsen, *The Physical Basis of Ultrahigh Vacuum* (American Institute of Physics, New York, NY, 1993), Chapter 7, section 7.6.

[2] T.E. Hanley, J. Appl. Phys. **19**, 583 (1948).

[3] P.A. Redhead and J.P. Hobson, Brit. J. Appl. Phys. **16**, 1555 (1965).

[4] J.T. Yates, Jr. and D.A. King, J. Vac. Sci. Technol. **9**, 1256 (1972).

[5] *Stainless Steel Fabrication* (Allegheny Ludlum Steel Corp., 1959) p. 309.

[6] H.F. Winters and J.W. Coburn, Surf. Sci. Reports **14**, 161 (1992).

[7] I. Langmuir, Phys. Rev. **22**, 357 (1923).

[8] R.G. Murray and R.J. Collier, Rev. Sci. Instrum. **48**, 870 (1977).

[9] R.O. Jenkins and W.G. Trodden, J. Electron. Control **12**, 1 (1963).

[10] H.F. Gray, F.H. Harris, T. Pankey, Jr., and G.A. Haas, Rev. Sci. Instrum. **43**, 1113 (1972).

[11] H.H. Glascock, Jr., Rev. Sci. Instrum. **43**, 698 (1972).

***Fig. 63. Thoriated Tungsten Thermionic Emitters***

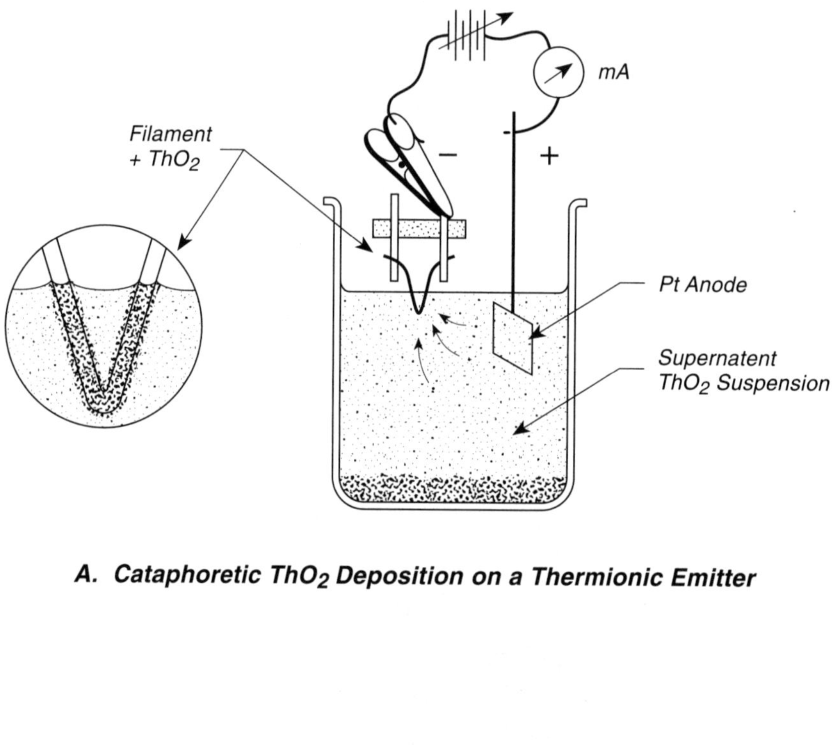


***A. Cataphoretic $ThO_2$ Deposition on a Thermionic Emitter***

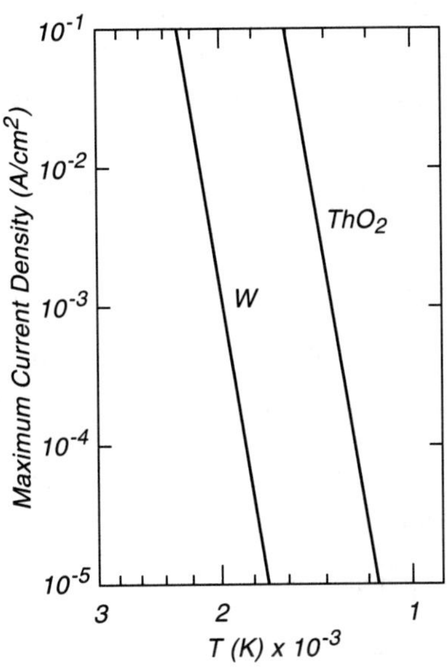


**B. Maximum Current Density for Thermionic Emitters**

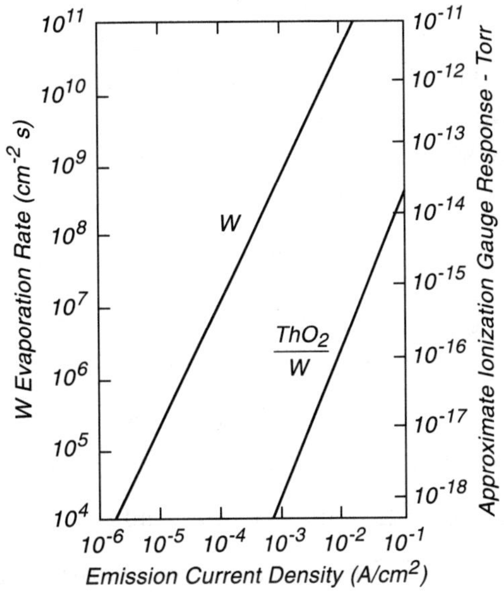


**C. W Evaporation Rate for Thermionic Emitters**

# 64

# Lanthanum Hexaboride Thermionic Emitters—Deposition on Metals

The use of cataphoretic deposition of thorium oxide, $ThO_2$, on thermionic emitter metal surfaces for the purpose of reducing the work function has been described on page 202. A similar technique can be used to produce $LaB_6$-coated metal emitters. Solid $LaB_6$ thermionic emitters have been described on page 210, but these are only useful for well-defined geometries and are costly. Thus for small, disposable $LaB_6$ emitters, cataphoretic deposition on metal ribbons or filaments provides an excellent route to useful thermionic emitters.

Five factors have been identified as being of importance in $LaB_6$ deposition. They are:

1. The particle size and the relationship between particle weight and surface area.
2. The concentration of cations in the suspension holding the $LaB_6$ particles.
3. The ability of cations to adsorb onto the $LaB_6$ surface, providing surface charge to permit particle migration under electrolytic conditions.
4. The electrostatic potential gradient causing the particles to migrate.
5. The adherence of the particles to the metal cathode surface.

The most efficient deposition occurs with particles having a high surface charge and migrating in a strong electrostatic field.

The present study [1] has been built on earlier methods [2,3]. Re ribbons, 0.0025 x 0.76 x 12.7 $mm^3$ were used. These filaments were heated white hot in vacuum to clean them before external cataphoretic deposition. A Ta anode was used, and the current (0.2–1.0 mA) was adjusted to achieve optimum deposition of $LaB_6$ in 30–300 s. For best results, 10 g of very finely divided $LaB_6$ was suspended in 100 ml of methanol and about 0.05 g of $NH_4NO_3$ was added as the electrolyte, dissolving this first in a small volume of methanol. These suspensions have a short lifetime of the order of several weeks owing to hydrolysis of the $LaB_6$, which produces a white coating on the powder that interferes with adhesion to the cathode. A Ta ribbon was used as an anode, and it was noted that the deposition occurs at the Re ribbon edges owing to higher electric field strength there. In order to avoid this, the geometry shown in Figure 64 was used, with the Ta anode parallel to the meniscus of the suspension, and the ribbon pulled almost out of the suspension to reduce the tendency for deposition on the edges. A narrow gap between cathode and anode is preferred to increase the electrostatic field. In addition, deposition against the gravitational field leads to the selection of smaller $LaB_6$ particles which adhere better than large particles. The $LaB_6$ thickness could be adjusted up to 0.5 mm, but little difference in performance was noted for various thicknesses. Deposition on filaments can be done with a large anode and with the electrodes in a vertical position, since the electric field at the cylindrical filament surface is uniform [1,3].

Failure in the adhesion of $LaB_6$ to Re supports has been observed following cataphoretic deposition. It was found that either boron or boric oxide impurity in the $LaB_6$ was responsible for the adhesion loss, and a purification procedure was described which produced $LaB_6$ with excellent adhesion properties on Re surfaces [2].

$LaB_6$ may also be deposited on Ta foil or filament surfaces if the surface is carburized first to produce a barrier layer of TaC that protects the underlying Ta from attack by the $LaB_6$ at high temperatures [4].

## *References*

[1] J.E. Delmore, Rev. Sci. Instrum. **54,** 158 (1983).
[2] L.J. Favreau and D.F. Koenig, Rev. Sci. Instrum. **38,** 841 (1967).
[3] L.J. Favreau, Rev. Sci. Instrum. **36,** 856 (1965).
[4] M. Nasini and G. Redaelli, Rev. Sci. Instrum. **42,** 1765 (1971).

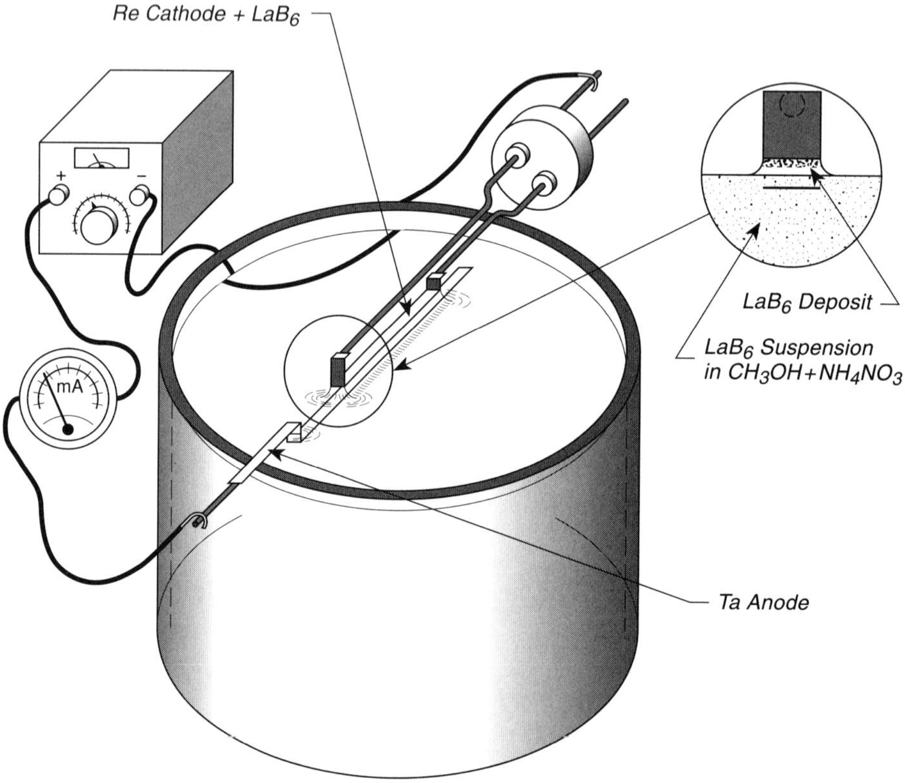


*Fig. 64. $LaB_6(s)$ Cataphoretic Deposition*

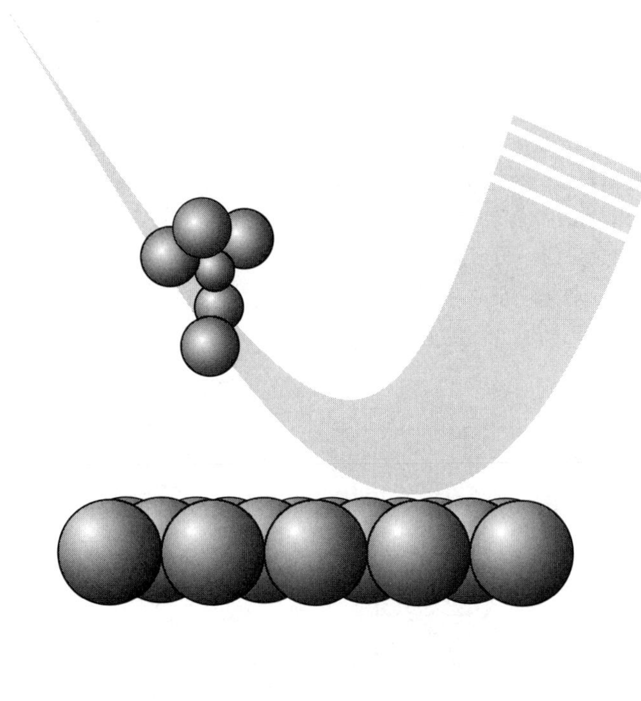

# 65

# Directly Heated Lanthanum Hexaboride Thermionic Emitters

The work function of a tungsten thermionic emitter can be reduced significantly by using a thoria coating (see page 202). In addition, it is well known that $LaB_6$ coatings on refractory metals such as W or Ta serve to produce a low work function thermionic emitter of limited lifetime [1] (see page 206), and $LaB_6$-coated refractory metal emitters are widely used when high electron current densities are desired at low emitter temperatures [1]. The $LaB_6$ structure is also remarkably stable toward damage by positive ion bombardment that may occur in some high-voltage devices [1]. The degradation of $LaB_6$ occurs by means of B diffusion into the W or Ta substrate, accompanied by La vaporization. This effect can be minimized by carburizing the substrate before coating with $LaB_6$. With rhenium as the base metal, this degradation effect is minimized [2]. For bulk $LaB_6$, the surface La atoms vaporize slowly at high temperatures, and are replaced by La atoms from the bulk [1]. Other authors have also investigated the properties of $LaB_6$ as a thermionic emitter [3,4].

Bulk $LaB_6$ may be used also as a thermionic emitter material. Questions about the best mounting procedure to avoid stresses (owing to thermal expansion) and to avoid degradation at the electrical contact points then come into play [2]. Figure 65A shows an unsuccessful mounting procedure for mounting a $LaB_6$ strip between two connecting electrodes made of graphite. While the emitter gave copious emission at 1400°C, it cracked upon cooling as a result of its high coefficient of thermal expansion. To avoid this problem, rectangular and circular filaments were constructed as shown in Figure 65B. These structures survived temperature cycling. The structure in Figure 65B is mounted on graphite holders, while that in Figure 65C is mounted on Mo holders.

Figure 65D gives the temperature versus current for a $LaB_6$ strip, 1.6 x 3 $mm^2$ in cross-sectional dimension. The points are experimental points, and the line is a theoretical line calculated on the basis of the ohmic heating and the radiative losses [2]. Figure 65E shows the thermionic electron emission of $LaB_6$ compared to Ta and W emitters. Approximately four to five orders of magnitude difference in current density is observed compared to the refractory metals.

As mentioned before, one is concerned with La evaporation from $LaB_6$ when it is heated to sufficient temperature. However, one is also concerned about metal evaporation effects when refractory metals are used as thermionic emitters, since they must work at very high temperatures (see page 202). Figure 65F shows a comparison of the evaporation rate from $LaB_6$ compared to W and Ta, obtained from published data, where the ordinate is in g $cm^{-2}s^{-1}$. On this basis, a $LaB_6$ cathode can last for several thousands of hours at current densities of 10–20 A $cm^{-2}$. Above 50 A $cm^{-2}$, the useful life may still be greater than 200 h [3].

A method for welding small $LaB_6$ shapes to Re wire for making compact high-efficiency thermionic emitters has been described [5]. In addition, a method for protecting a W filament from $LaB_6$ attack by interposing a layer of $ThO_2$ has been described [6].

A method for mounting conical tipped $LaB_6$ cathodes for electron beam instruments has been commercially developed. The cathode is mounted between pyrolytic graphite blocks that are clamped between Mo/Re alloy support posts that have a high modulus of elastically at high temperatures. The method is also used for $CeB_6$ cathodes [7].

## *References*

[1] J.M. Lafferty, J. Appl. Phys. **22,** 299 (1951).
[2] K.N. Leung, P.A. Pincosy, and K.W. Ehlers, Rev. Sci. Instrum. **55,** 1064 (1984).
[3] H. Ahmed and A.N. Broers, J. Appl. Phys. **43,** 2185 (1972).
[4] S.P. Gordienko, E.A. Guseva, and V.V. Fesenko, Teplofiz. Vys. Temp. **6,** 821 (1968).
[5] K.N. Ramachandran, Rev. Sci. Instrum. **46,** 1662 (1975).
[6] L. Bakker and Th. J. van Velzen, Rev. Sci. Instrum. **37,** 1404 (1966).
[7] Manufactured by FEI Company, P.O. Box 654, McMinnville, OR 97128.

## Fig. 65. $LaB_6(s)$ Thermionic Emitters

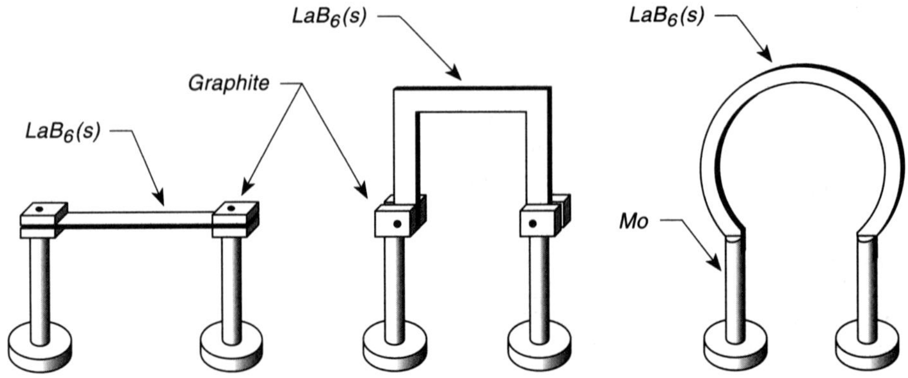


**A. Rigid Mount** **B. Flexible** **C. Flexible**

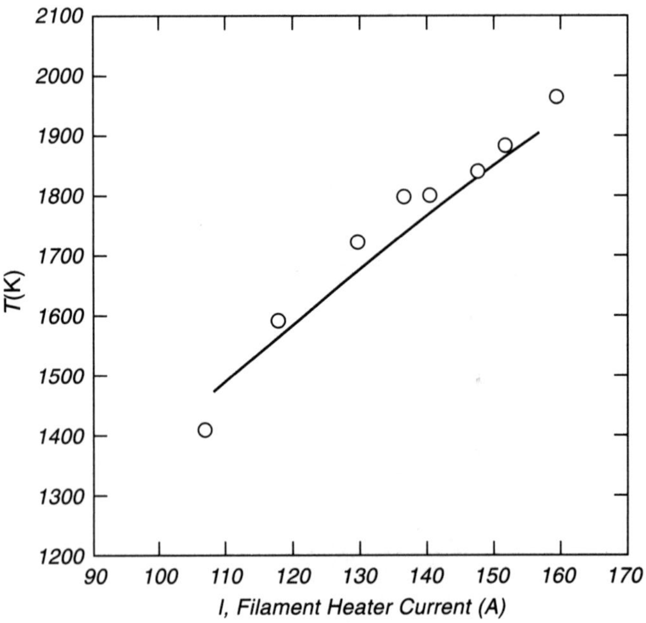


**D. $LaB_6(s)$—T vs. I**

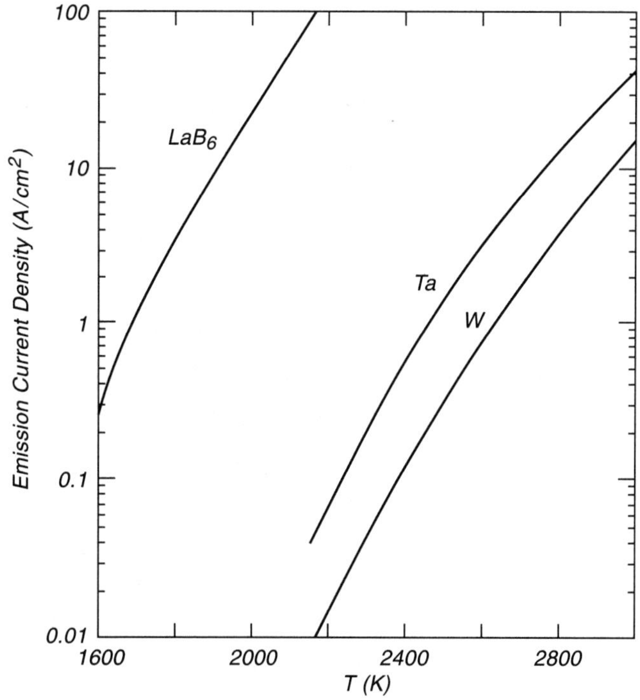


**E. $LaB_6$(s) - $I_e/S$ vs. T**

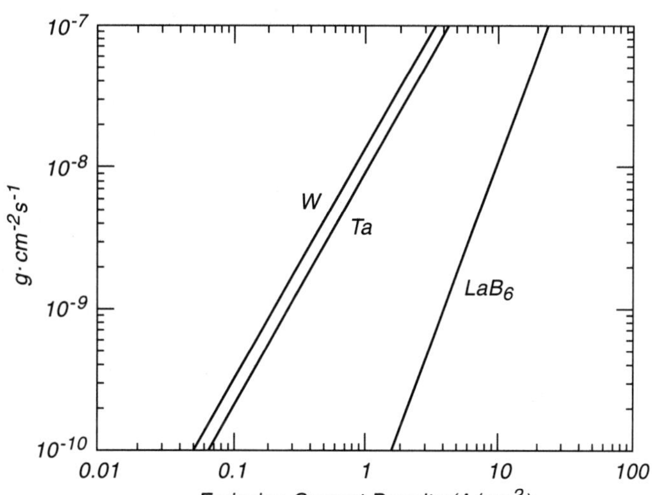


**F. Wt. Loss vs. $I_e/S$**

# 66

# Thermionic Emitter Mounting

The provision of long-lived thermionic emitters for electron bombardment requires special mounting and connection procedures. Figure 66A shows the essentials of mounting a thermionic emitter for stability and also for ease of replacement. A fixed alumina tube with two holes fits a spiral emitter, made from W/3%Re, about 0.2 mm in diameter. The coil is wound uniformly around a mandrel in such a fashion that its ends can be slipped into the holes in the alumina tube without producing any strain on the coil. If this is not done, the coil will distort severely upon first heating. The W/3% Re alloy is preferred over pure W because of its tendency to not become so brittle during heating.

Friction connections are made with heavier Mo conductor wires by slipping in 0.1–0.2-mm-diameter filler pieces from the back side until a tight fit is achieved between the emitter and the conductor wires. This method avoids welding and allows the emitter coil to be disassembled for replacement with ease.

Connection to the Mo conductor wires is conveniently made with Cu braid, and a soft Cu connector tube (annealed in a flame) that is crimped tight several times around the braid and the Mo wire using pliers. The assembly should be arranged such that the braid does not strain the Mo conductor wires by its weight or motion.

Figure 66B shows a similar device that utilizes a Mo electron repeller shield. The alumina tube is flattened on two sides and is placed into a fitted channel at the back of the Mo shield. The Mo shield will charge negatively during operation, but for heavy electron bombardment heating, the crystal must be biased positively.

Macor ceramic supports are inappropriate for these purposes for two reasons. The material degrades at high temperatures, and fluorine compounds are evolved at high working temperatures [2].

## *References*


[1] Dr. R. David, Institut fur Grenzflächenforschung und Vakuumphysik, Forschungszentrum Jülich GmbH, Postfach 1913, D-52425 Jülich, Germany (private communication).

[2] Personal experience.

***Fig. 66. Thermionic Emitter Mounting***

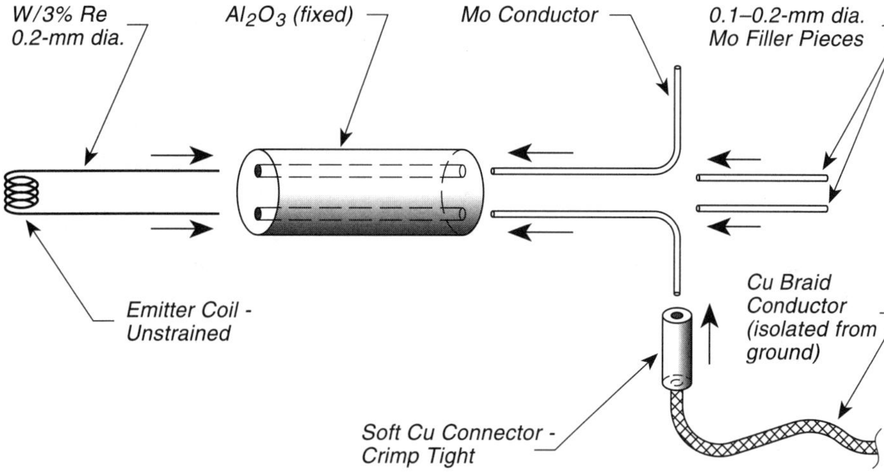


***A. Open Emitter***

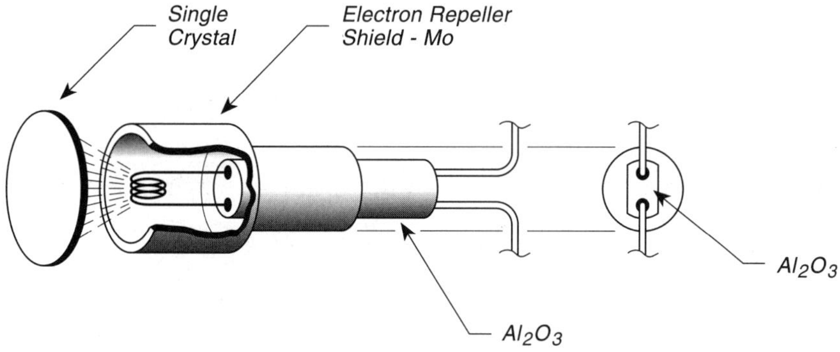


***B. Shielded Emitter***

# 67

# Indirectly Heated Cathodes for High-Temperature Operation

Cathodes for electron guns and other devices are often coated with low-work-function materials that permit higher current densities to be drawn than would be possible with pure refractory metal cathodes such as W or Mo (see page 202). Coated cathodes are often heated to temperatures in excess of 2000 K for activation and operation. Directly heated cathodes may be employed, but the requirement for an emitter to be at equipotential over its surface rules out direct ohmic heating. Thus, indirectly heated cathodes are often used. The design shown in Figure 67A uses an ohmically heated W filament to heat a Mo disk containing a work-function-lowering material on its surface [1].

A self-supporting W filament (0.38-mm diameter), wound in a double helix on a mandrel into a coil 9.5 mm in length, is used to heat a Mo disk cathode mounted on the helix axis. The Mo disk is at the end of a cathode cylinder, 3.8 mm in diameter and 10 mm long. Mounted on the outside of the Mo cathode cylinder are 3 (0.1-mm thick) polished Ta radiation shields. These shields reduce the radiation transmitted to the outer stainless steel support cylinder such that a measurable color temperature for the stainless steel sleeve is not reached.

Figure 67B shows the W heater life behavior versus the heater current for operation in ultrahigh vacuum. The temperatures associated with each point were measured pyrometrically on the Mo cathode surface.

In the example shown here, only radiation heating of the Mo cathode was used. However, this may be assisted by applying a positive potential to the cathode relative to the filament, and drawing electron bombardment current to the cathode. This method of heating is well suited for fast flashing of the cathode which is often necessary during activation of the cathode coating (about 2500 K for $ThO_2$). Electron bombardment also involves lower W heater temperatures, and will not shorten the heater lifetime.

## *References*

[1] J.T. Jensen, Jr., J. Klebanoff, and G.A. Haas, Rev. Sci. Instrum. **38,** 1178 (1967).

***Fig. 67. Indirectly Heated Cathode for High Temperature Operation***

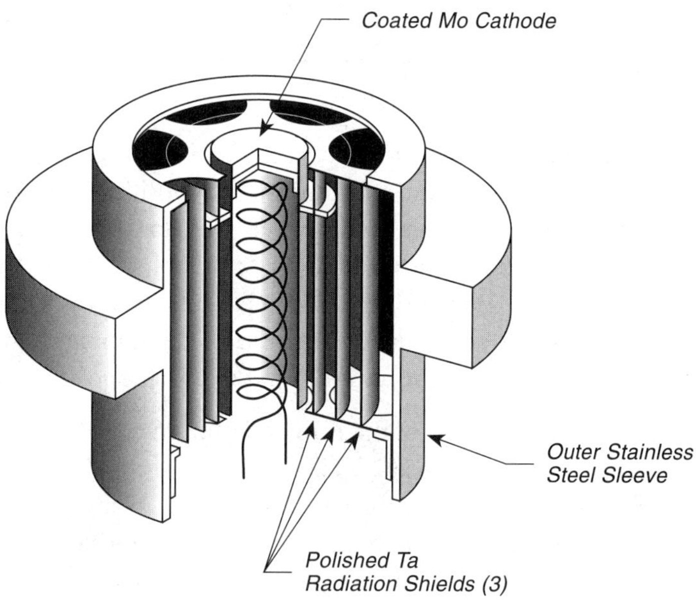


***A. Cathode Design***

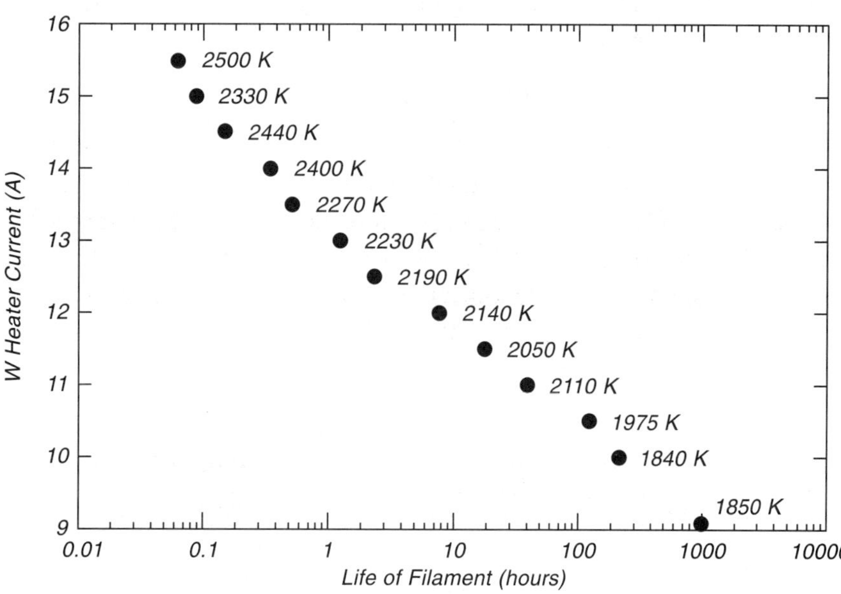


***B. Cathode Lifetime versus Heater Current***

# 68

# Replacing Filaments in Glass Bayard–Alpert Gauges

Filament replacement in a glass Bayard–Alpert ionization gauge is quite simple if a glassblower will move the tubulation from the normal position to a position on the bottom of the gauge where the filament attachment posts are located [1]. Figure 68 shows the location of the new port, and the position of the original tubulation. The new tubulation is sealed to the outer envelope using Corning No. 3320 glass as a grade to Pyrex, and the seal is made with an oxidizing flame, after thorough cleaning of the gauge with Alconox detergent. The gauge should be annealed in the upright position before use.

Replacement filaments can be made of many materials, such as iridium, thoriated iridium, thoriated tungsten, etc. (see page 202). The use of 0.010-in. diameter W is easiest. The bent filament is formed into a tight V-shape, and a hook is fabricated from the filament wire by first twisting the apex region of the wire and then making the hook as shown in Figure 68. The hook is then inserted into a wire loop at the top of the filament support rod. When softer filament materials such as iridium are used, a tungsten hook made of 0.010-in. wire may be used for the upper support. Hemostats are reported to be useful in the filament insertion procedure.

The lower attachment of the two ends of the filament to the filament attachment posts can be made in three different ways:

1. Welding, using a tweezer spot welding assembly, and making small 0.002-in. thick Ta tabs that will weld nicely to W and Ir filaments.
2. The filament post is filed round, and a 1/8-in.-diameter stainless steel barrel is fabricated as shown in Figure 68. The small tungsten locking coil fits onto the filament post, and when the barrel is twisted in one direction, binding occurs. Twisting oppositely results in breaking the junction. A 0.020-in. clearance is needed when the tungsten locking coil is made of 0.010-in. W. The emitter filament is held under an 0-80 screw head placed in a tapped hole on the opposite end of the barrel, using a jam nut as shown.
3. The W filament may just be wrapped several times around the two support posts.

The gauge calibration seems to be well maintained after refabrication [1].

## *References*

[1] D. Bedding, T. Moran, and A. Brown, Rev. Sci. Instrum. **56**, 2170 (1985).

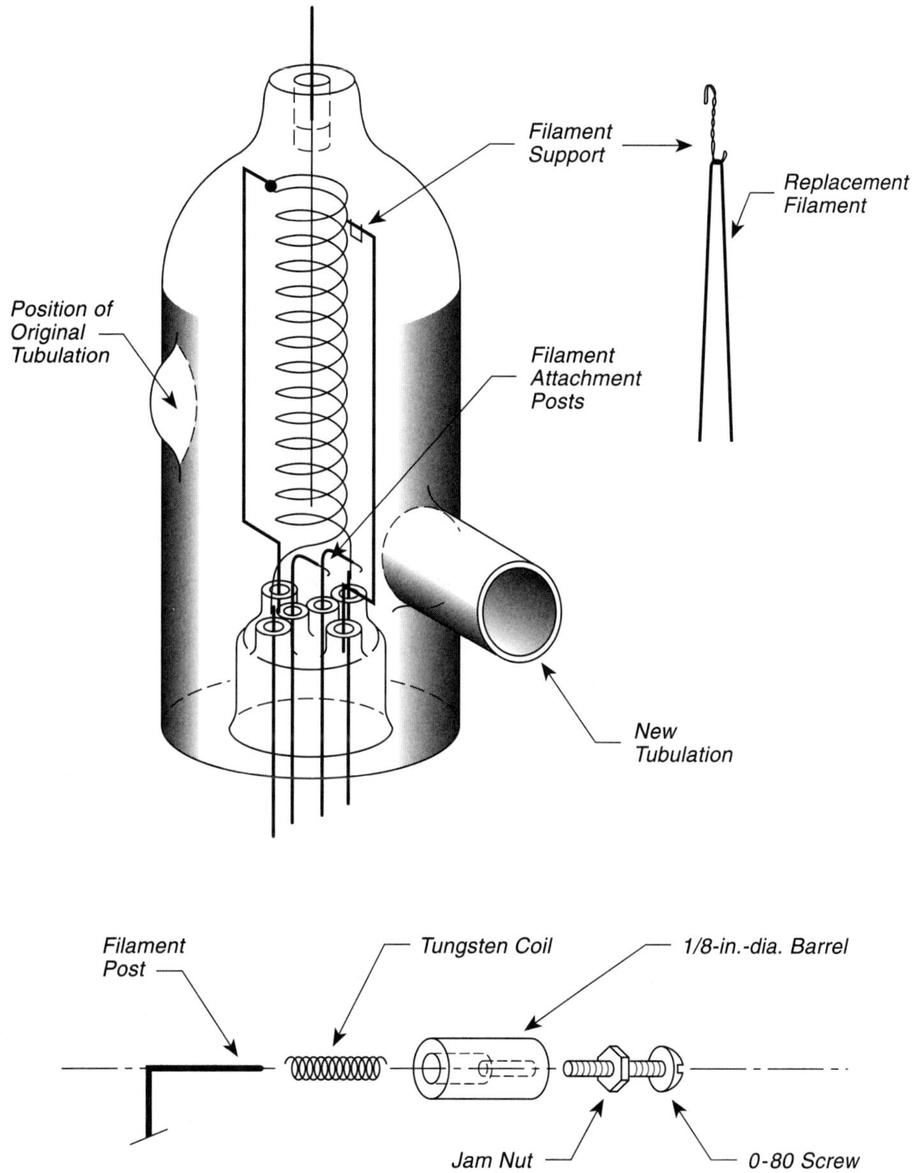


***Fig. 68. Replacing Filaments—Bayard–Alpert Gauges***

# E. Shielding

# 69

# Magnetic Shielding in Ultrahigh Vacuum

It is common to employ magnetic shields inside ultrahigh vacuum systems for the elimination in electron spectrometers of stray magnetic fields caused by the Earth or by magnets associated with ion pumps or other devices near the vacuum system. A number of formulas have been presented for the calculation of the efficiency of a magnetic shield [1–3]. These formulas do not consider deviations from pure cylindrical geometry caused by lips, caps, and pumpout ports. The calculation method studied here is able to approximate the nonideal conditions caused by deviations from perfect cylindrical geometry [4].

A computer code, JASON [5], has been used. This code solves the linear Poisson equation. The Earth's magnetic field is of the order of 500 mG. It is desired in this particular example to reduce the magnetic field to less than 1.0 mG inside the shield shown in Figure 69. The following constraints exist in this design as well as in designs to be made in other ultrahigh vacuum systems: (1) A small number of shield layers should be employed to reduce the additional surface area and outgassing which occurs from these surfaces; (2) Holes must be cut into the shield for pumping by the ultrahigh vacuum system. The pumping speed through a hole for atmospheric gases is about 11 L/cm$^2$ s; (3) The sample must be moved into the magnetically shielded spectrometer along the z axis of the cylindrical shield.

The optimum shield material thickness, *t*, is

$$t \le 3a/2\mu \tag{1}$$

where *a* is the inner radius of the cylinder and μ is the permeability of the magnetic shielding material. If the value of *t* needed, based on Equation (1), is larger than the thickness of available material, two concentric shells may be used in shielding, at the expense of introducing more surface area.

Because the JASON program rotates a two-dimensional array about a cylinder axis, *z*, it is geometrically impossible to specify three-dimensional holes along the shield perimeter. Instead, in a first-approximation method, a gap along the perimeter is used to model the desired hole size where the open area in the gap is equivalent to the hole area. This gap has the permeability of vacuum in the simulations. A second approximation method uses a gap with three times the pumping hole area, containing a magnetic shielding alloy that is two-thirds made of the shielding material, and one-third made of vacuum, i.e., specifying a "hole alloy." The actual hole configuration consists of radial holes of 3-cm-diameter, placed 6 cm apart around the ends of the two outer shields, as shown in Figure 69.

The material chosen is a high permeability material, ad-mu 80 [6]. Its thickness, $t$, is 0.25 mm. The Figure shows a cross section through the optimized shield made of three concentric cylinders that contain pumpout ports as shown, as well as the axial sample insertion region. Approximation (1) gives approximately a 1000-fold attenuation of the Earth's field; approximation (2) gives approximately a 10000-fold attenuation, both measured along the $z$-direction at the center of the spectrometer. To make a rough comparison with actual measurements, the attenuation of an AC magnetic field, naturally present in the spectrometer room, was measured. Attenuation factors in the range $1900(x), 640(y)$, and $>3600(z)$ were measured in general agreement with the calculations.

It should be remembered that mechanically straining magnetic shielding made of mu metal will reduce its effectiveness. The effectiveness of annealing magnetic shielding is dealt with in reference [7].

## *References*

[1] A.J. Mager, IEEE Trans. Mag. **MAG-6,** 67 (1970).

[2] A. Mager, J. Appl. Phys. **39,** 1914 (1968).

[3] D.U. Gubser, S.A. Wolf, and J.E. Cox, Rev. Sci. Instrum. **50,** 751 (1979).

[4] S. Dobscha and W. Siekhaus, Rev. Sci. Instrum. **57,** 3123 (1986).

[5] S. Sackett, JASON, UCRL-18721, Lawrence Livermore National Laboratory, Livermore, CA 94550.

[6] Ad-Vance Magnetics, Inc., 625 Monroe Street, P.O. Box 69, Rochester, IN 46975.

[7] D.L. Martin and R.L. Snowdon, Rev. Sci. Instrum. **46,** 523 (1975).

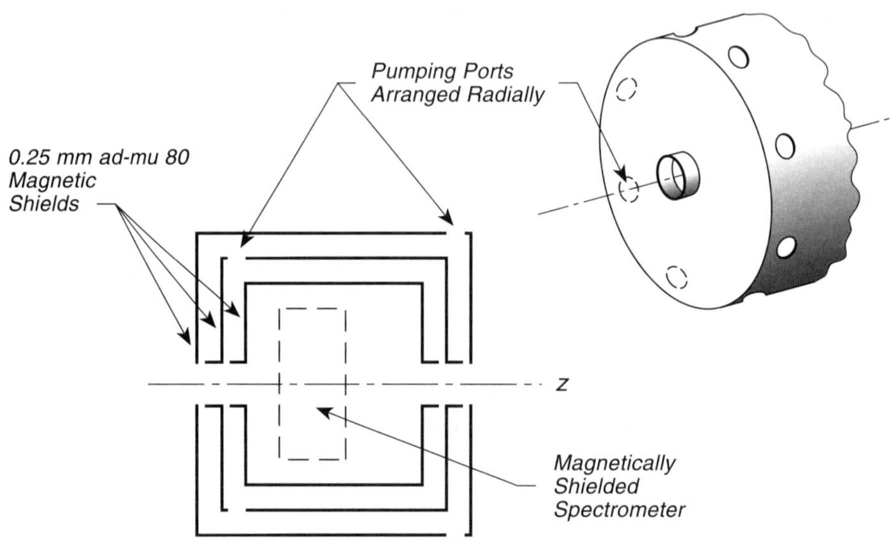


***Fig. 69. Magnetic Shielding in UHV***

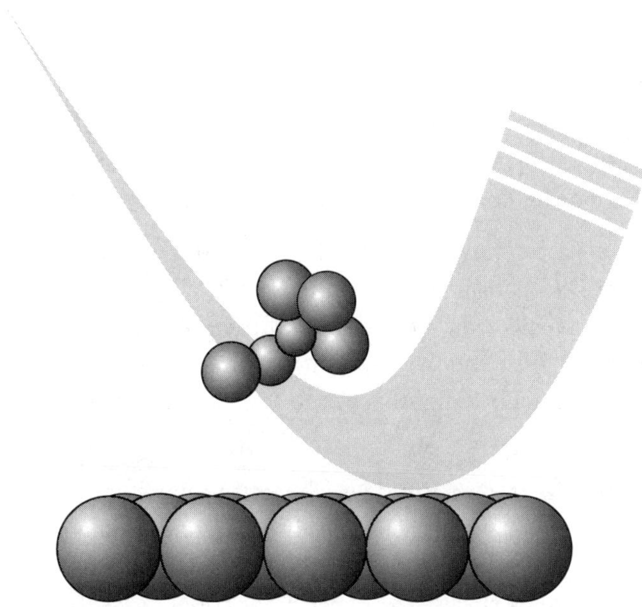

# 70

# Electrical Isolation of UHV Components

It is sometimes necessary to isolate electrically one ultrahigh vacuum component from another. This may be done using a conventional ceramic electrical isolator, connecting the two components with Conflat flanges on either side of the ceramic isolator. This arrangement requires linear space, and also, small lateral forces on the ceramic isolator may cause it to crack. A novel solution to this problem is presented in Figure 70[1].

Polyimide film (Type H), known as Kapton [2], is used. The film thickness is 0.127 mm. Two polyimide washers are cut with punches that have a clearance of 0.0127 mm. Sleeves, made by rolling up the polyimide film, with 2-mm overlap, are prepared and cemented with a fast-setting epoxy cement. This is best done using bolt-clearance holes drilled in a polyethylene block to hold the tubes during glue setting.

Two copper gaskets are sandwiched together around a polyimide spacer ring that is glued between the gaskets using a Dow Corning No. 997 silicone varnish [3]. The varnish is applied, the gaskets are placed on either side of the polyimide spacer, and the assembly is heated at 95°C for 2 h in an oven for curing while being clamped under pressure. The polyimide spacer protrudes slightly beyond the inner and outer edges of the copper gaskets.

As shown in Figure 70, the polyimide washers fit under stainless steel washers. The polyimide washers mechanically and electrically separate the stainless washers and bolt heads from the flange surface. The polyimide sleeves electrically separate the bolts from the bolt holes. The polyimide spacer ring between the copper gaskets separates the flanges from each other electrically.

Thermal shock tests between 260°C and 10°C resulted in no vacuum leak development. The devices do not add gas load to a vacuum system, and the UHV background mass spectrum did not change as a result of 10 of these devices being installed on the same system. The electrical resistance is greater than $10^{11}$ Ω, and electrical breakdown of the assembly did not occur below 17 kV. Breakdown does not result in failure of the vacuum integrity of the seals.

## *References*


[1] J.C. Jones, Rev. Sci. Instrum. **46,** 489 (1975).

[2] Obtained from: Dupont, 1007 Market Street, N-2426-A, Wilmington, DE 19898; see Dupont High Performance Films, Bulletin H-38492 (1983).

[3] Obtained from: Dow Corning, P.O. Box 0994, Midland, MI 48640-0994.

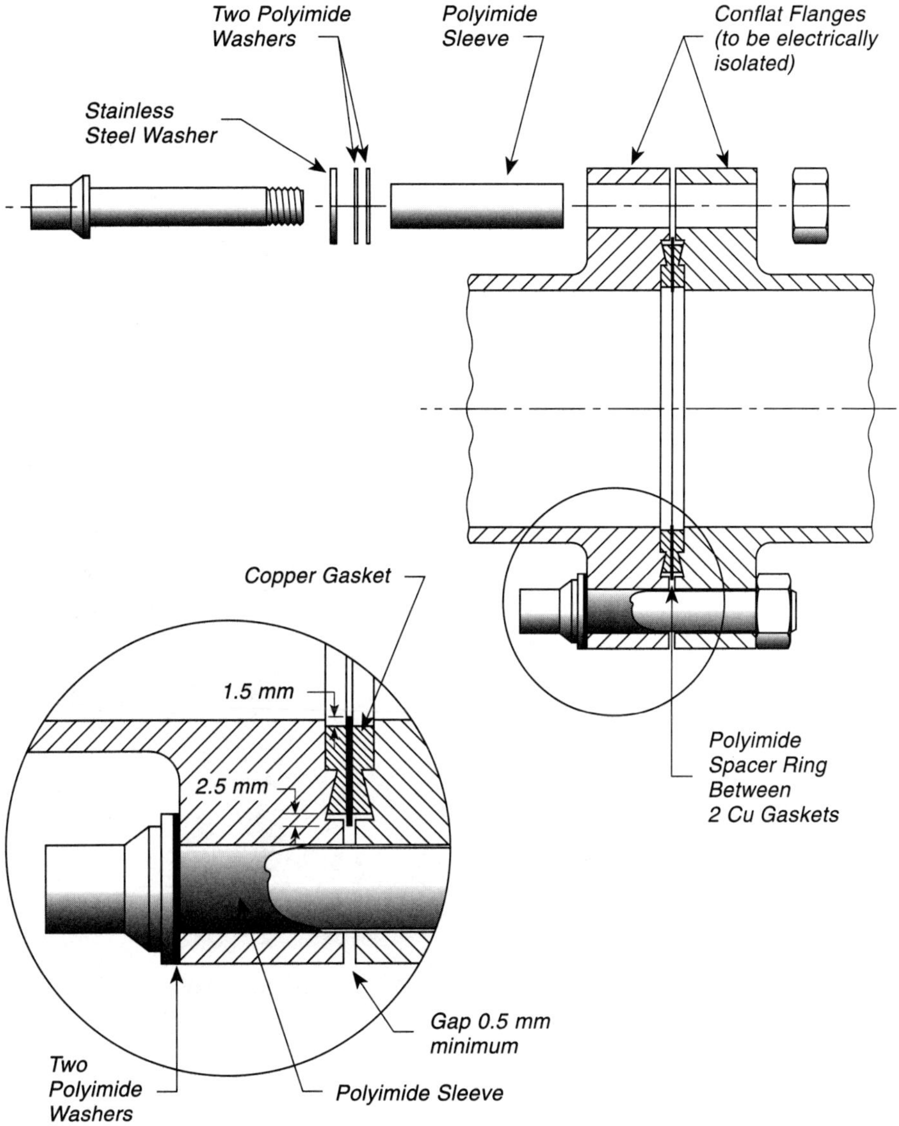


***Fig. 70. Electrical Isolation of UHV Components***

# F. Single Crystal Fabrication/Orientation

# 71

# Single Crystal Orientation, Grinding, and Polishing

The preparation of oriented single crystal surfaces by X-ray diffraction followed by grinding and polishing to a high degree of smoothness and flatness is of central importance in surface science experiments, and there is a long interconnected history of developments in this field [1–9]. At the present time, using the best methods, an orientational accuracy of ±0.05° [5,8] and a flatness of one-twentieth the wavelength of sodium light [5] have been reported.

A brief history of the developments over almost 50 years follows. In 1949, Walker et al. [1] described a cylindrical crystal holder that contained a crystal holder platform that could be tilted so that the crystal could be crystallographically oriented in a mounting attached to an X-ray diffraction apparatus. The holder then could be transferred to cutting, grinding, and polishing apparatus without disturbing the crystal alignment. The cylindrical crystal holder is remarkably similar to a holder now marketed commercially as the Bond Barrell Holder [10]. In 1961, Bond [2] described a jig that held a cylindrical crystal holder for grinding an oriented crystal on a flat grinding plate, called a lapping plate, containing an abrasive powder. Bond's holder was attached to a hardened steel ring that made contact with the lapping plate and provided a reference plane for the mechanical grinding and polishing processes. Bennett and Wilson [3], who used a grinding/polishing jig of different design, were not interested in accurate crystal orientation, but showed that lapping of hard crystals on soft metal plates containing embedded abrasive particles could result in surface flatness of the order of one-eighth the wavelength of sodium light. Butz et al. [5] described a goniometer-jig combination that was used for both orientation and lapping of crystals to an accuracy of 0.25° and to a flatness of one-twentieth the wavelength of sodium light. Berger et al. [7] described a goniometer-jig combination that possessed 2° of crystal tilt freedom, plus translational and azimuthal orientation freedom. They ground and polished a W crystal on a glass lap, with 15 μm $Al_2O_3$, 5 μm $Al_2O_3$, and 6-μm and 3-μm diamond abrasive paste. An interesting feature of this work was the production of a W crystal disk with five desired facet-planes, which were convenient for plane-to-plane comparisons. The accuracy of orienting and polishing single crystals has been brought to a very high level by Linke and Poelsema [8], who showed by He diffraction methods that Pt(111) crystals oriented and polished as described previously [7] exhibited a mean terrace width of 3000 Å and an orientational error of ±0.05°. The reader is referred to the original papers for detailed designs of the goniometer-jig combinations.

Figure 71A shows a Bond Barrel Holder [10] mounted in a simple grinding-polishing jig made of hardened steel [11]. The Pt crystal on the holder platform has been crystallographically oriented using Laue X-ray diffraction, where the Bond Barrel Holder was held in a mount that fitted onto an optical rail that was part of the Laue apparatus. Typically, moving the Bond Barrel Holder from one mount to another can be done with an error of less than 0.1°. The single crystal was ground using a 600 grit abrasive paper mounted on a rotary wheel grinding and polishing machine. The weight of the Bond Barrel Holder forces the crystal onto the abrasive paper. The hardened steel jig is not appreciably affected by this process as should be determined periodically using a micrometer to measure its thickness around its circumference. When the crystal is flat, a series of polishing steps with smaller and smaller abrasive particles is performed as shown in Figure 71B. It is extremely important to remove, by scrupulous washing and rinsing away from all parts of the apparatus, the abrasive from each step, or scratches owing to inadvertent abrasive particles will ruin the next steps in polishing with finer abrasive. The polishing cloth should of course be changed each time as the abrasive particle size is reduced.

Figure 71C shows a measurement made by the observation of interference fringes with an optical flat touching the finished crystal [11]. It was found, in duplicate experiments, that the surface was not flat, but instead had a curved profile with a typical radius from 4 to 2 m near the center, and a slope of from 0.05° to 0.08° at the edges. The limiting slope corresponds to one atomic step for about every 1000 Pt atoms measured horizontally, and for many experiments this is unimportant. Figure 71A shows a hypothesis about the origin of the slightly curved figure of the crystal, where slight buckling of the flexible abrasive paper is postulated to produce a wave that tends to round the crystal slightly. A small misfit between the Bond Barrel Holder and the hardened steel jig can also result in a curved crystal [12].

Grinding and polishing to make highly flat surfaces can be carried out most effectively using rigid soft metal laps that are very flat and contain embedded abrasive particles [3,5,7]. Such an arrangement is shown in Figure 71D. A set of laps, containing different particle size abrasive grains, is required, and cross-contamination with other abrasive grain sizes is to be avoided. Brass and tin are favorite lap materials. Grinding and polishing on laps has produced crystal surfaces flat to one-twentieth the wavelength of sodium light or about 250 Å/cm, which corresponds to only one atomic step for about $4 \times 10^5$ Pt atoms, measured horizontally [5]. At this level of step density resulting from departures from planarity, the step density owing to misalignment will be the dominant factor, being about 2 orders of magnitude greater for a misalignment of 0.05°. A polishing jig giving an accuracy of 0.02° in the orientation of a silicon single crystal has been reported [13]. A method, involving the observation of a single Bragg reflection with a diffractometer, and employing a crystal holder that can be rotated about its axis, has been described, and the reported accuracy is better than 0.003° [14].

Combined mechanical and electrochemical polishing techniques are often used and a description of one method for polishing copper single crystals and alloys is given in references [15] and [16].

## *References*

[1] J.G. Walker, II, J. Williams, and R.M. Bozorth, Rev. Sci. Instrum. **20,** 947 (1949).
[2] W. L. Bond, J. Sci. Instrum. **38,** 63 (1961).
[3] G.A. Bennett and R.B. Wilson, J. Sci. Instrum. **43,** 669 (1966).
[4] D.E. Johnson, Rev. Sci. Instrum. **39,** 238 (1968).
[5] R. Butz, B. Krahl-Urban, and K. Mench, Rev. Sci. Instrum. **44,** 485 (1973).
[6] J.F. Wendelken, S.P. Withrow, and C.A. Foster, Rev. Sci. Instrum. **48,** 1215 (1977).
[7] S. Berger and U. Linke, J. Phys. E: Sci. Instrum. **11,** 1097 (1978).
[8] U. Linke and B. Poelsema, J. Phys. E: Sci. Instrum. **18,** 26 (1985).
[9] Y.K. Chang, Rev. Sci. Instrum. **53,** 515 (1982).
[10] Bond Barrel Holders may be purchased from Blake Industries, Inc., 660 Jerusalem Road, Scotch Plains, NJ 07076.
[11] C.E. Tripa and J.T. Yates, Jr., J. Vac. Sci. and Technol. **A14,** 2544 (1996).
[12] Dr. W.F. Egelhoff, Jr., NIST, Gaithersburg, MD 20899 (private communication).
[13] S. Ipsen Mathiesen, L. Gerward, and O. Pedersen, Rev. Sci. Instrum. **45,** 278 (1974).
[14] M. Deutsch and I. Freund, Rev. Sci. Instrum. **46,** 765 (1975).
[15] J.S. Aherarn, Jr., J.P. Monaghan, Jr., and J.W. Mitchell, Rev. Sci. Instrum. **41,** 1853 (1970).
[16] T.W. Snouse, Rev. Sci. Instrum. **36,** 866 (1965).

## Fig. 71. Single Crystal Orientation, Grinding and Polishing

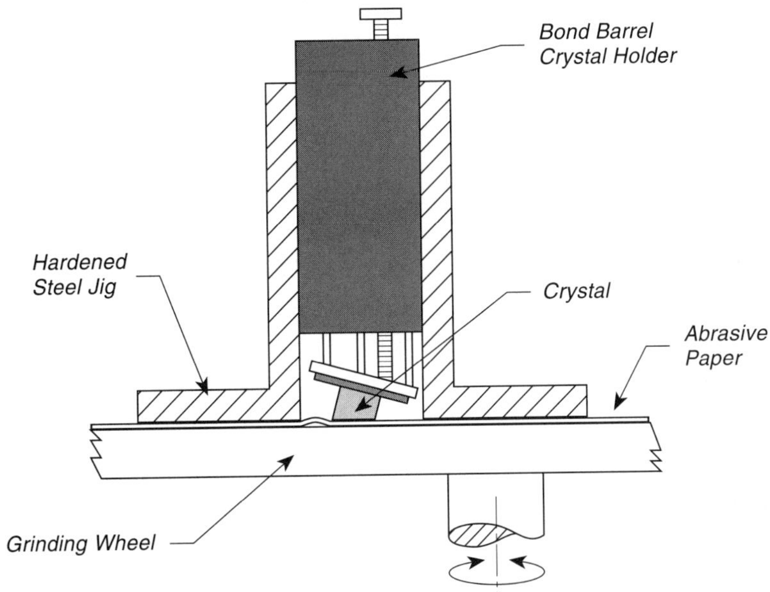


**A. Typical Crystal Grinding**

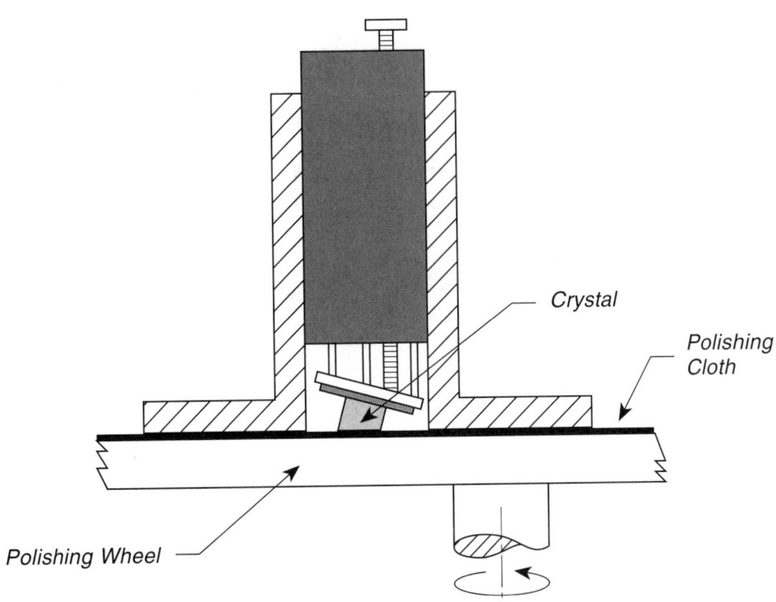


**B. Typical Crystal Polishing**

**Fig. 71. Single Crystal Orientation, Grinding and Polishing**

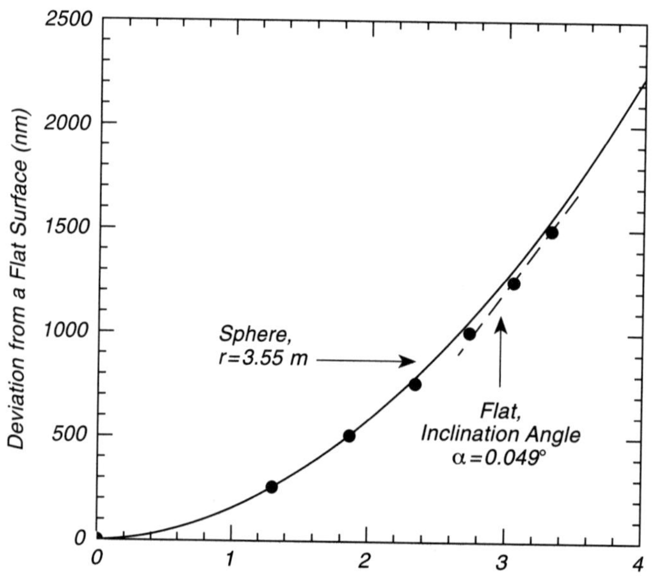


**C. Measured Crystal Curvature**

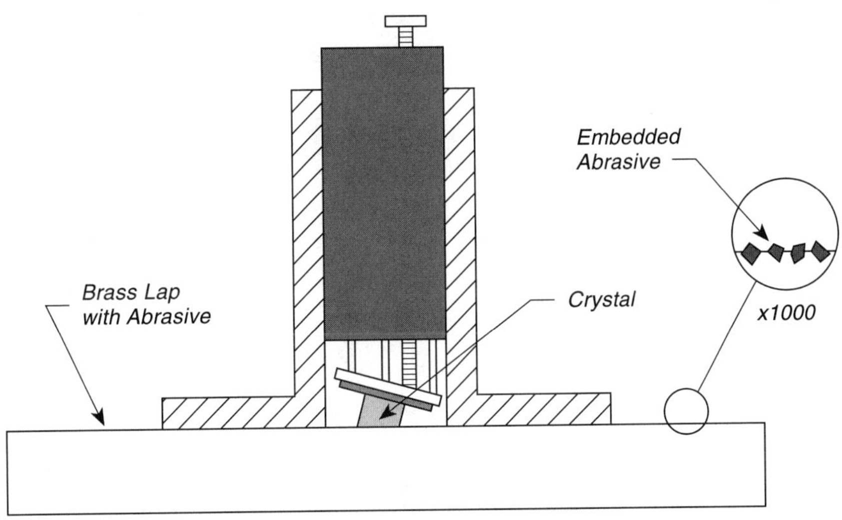


***D. Crystal Lapping for Optimal Flatness***

# 72

# A Simple Goniometer for Cutting Single Crystals

A clever method for cutting single crystal boules along any plane involves a goniometer of very simple design, as shown in Figure 72 [1]. Here, a single crystal boule is fastened into a holder with a crystal bond compound, using a plastic screw that will not deform the crystal when tightened against its outer surface. The holder is placed into a special wedge-shaped device with shoulders cut at angle $\alpha$ with respect to the upper surface. This wedge-shaped device can rotate inside a tilted cylinder with an upper surface inclined at angle $\alpha$ with respect to the base of the device. When the two wedges are rotated relative to each other, the tilt angle of the crystal, $\omega$, will change.

The crystal is aligned by Laue back-reflection X-ray diffraction by manipulating the wedged device. A rotation of 180° will result in swinging the crystal axis by $\alpha$ , as shown in Figure 72. Thus, if the saw cuts in a plane parallel to the baseplate of the goniometer, the crystal may be cut at any angle over the range $\alpha$. An engraved angular scale is provided in two places as shown. The rotations are not simply related to the angles measured on the Greninger chart, used for X-ray diffraction measurements, but the original paper gives a method to achieve the conversion using a small programmable calculator [1].

## *References*

[1] G.L. Price, Rev. Sci. Instrum. **51**, 1000 (1980).

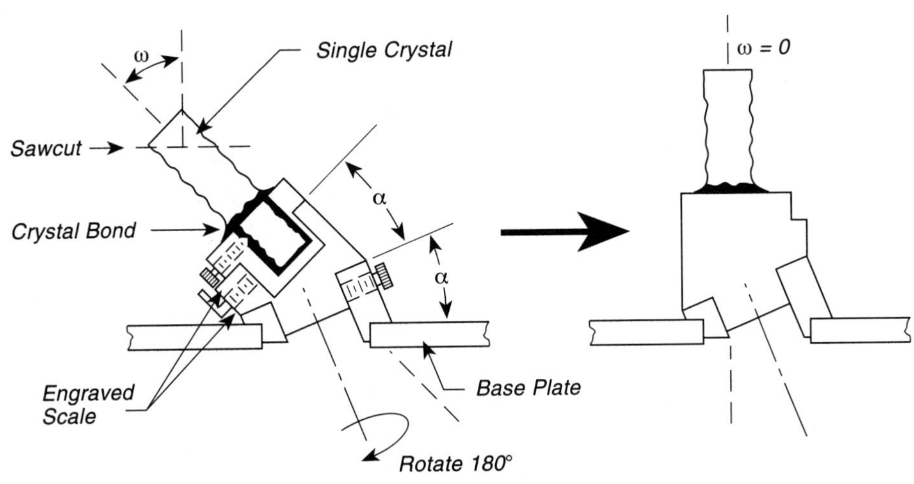


**Fig. 72. Simple Crystal Cutting Goniometer**

# 73

# Ion Beam Polishing of Crystals to Subnanometer Roughness

Crystalline substrates can be polished to near atomic perfection by sputtering under glancing incidence conditions [1–3] as shown in Figure 73. If the glancing incidence angle is chosen such that the surface normal component of the incident beam kinetic energy is sufficiently low to ensure that the beam is reflected with unit probability from atomically flat, defect-free terraces, sputtering only occurs at adatom and step edge sites [2,3]. Even at those defect sites, little damage to the subsurface regions appears to occur. The smoothing technique can be combined with annealing of the substrate, either simultaneously or after bombardment. By monitoring the width and peak position of the angular distribution of the scattered ion beam during sputtering, an indication of the surface quality and the change with time is conveniently obtained. A narrow, near specularly scattered beam indicates a low density of defects, on an atomic scale, over the length of the glancing incidence trajectory at least.

## *References*


[1] Dr. K. J. Snowdon, Department of Physics, University of Newcastle, New Castle upon Tyne, NE 17RU, UK (private communication). The first report of this effect is in U. von Gemmingen and R. Sizmann, Surf. Sci. **114,** 445 (1982).

[2] M. Holzwarth, M. Wissing, D.S. Simeonova, S. Tzanev, K.J. Snowdon, and O.I. Yordanov, Surf. Sci. **331-333,** 1093 (1995). Here polycrystalline Cr films were smoothed with a reduction in the rms roughness by a factor of 5 to 8 from 9.2 nm, as measured by the STM.

[3] M. Wissing, M. Batzill and K. Snowdon, Nanotechnology **8,** 40 (1997). Here, $CaF_2$ single crystal surfaces were smoothed from an rms roughness of 0.6 nm to better than 0.1 nm (the AFM noise level).

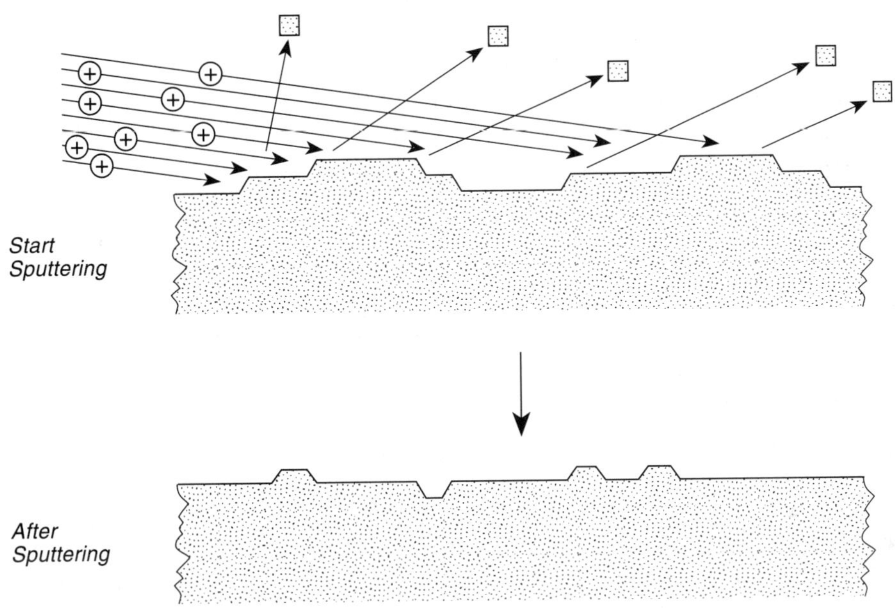


***Fig. 73. Ion-Beam Polishing of Crystals to Subnanometer Roughness***

# 74

# Crystal Optical Alignment

The alignment of highly reflective single crystals by using simple optical methods is commonly employed for the setting of vernier adjustments. In addition to the optical methods, electron beams can also be used for crystal alignment, as shown on pages 244 and 246.

Figure 74A shows a very simple crystal alignment method [1], using only a flashlight in a darkened room. An image of the pupil of the eye is reflected off of the crystal and back into the eye. In addition, the image is also reflected from the window to the eye. When the crystal surface is parallel to the window surface, the two reflected images of the pupil will coincide.

Figure 74B shows the same method used outside the vacuum system to align a crystal in its tilt angle so that the normal of the crystal is parallel to the manipulator mounting flange face [1]. The manipulator is held in a carefully made holder that is square. The image of the eye, illuminated by a flashlight filament, is reflected from the crystal and passes back by a pointer. If the center of the image of the pupil is coincident to the pointer when the pointer is translated along the table, then the crystal normal is parallel to the tabletop and therefore to the manipulator reference flange face.

Figure 74C shows the use of a He/Ne laser for crystal alignment. The laser is moved until its reflection from the window is aligned so that the laser beam returns to the laser as an aligned beam coincident with the incident beam. The crystal is then manipulated in the vacuum system until the reflection of the laser beam from the crystal is also coincident. In this case, then, the normal to the crystal and the normal to the window will be parallel. The accuracy of the laser alignment procedure will increase as the optical path is increased, and the accuracy of coalignment can easily be made to be 0.1°, or better. A partially reflecting glass slide may be used to deflect a fraction of the two reflected laser beams to a card located off the laser axis. The slight divergence of the laser ultimately limits the accuracy of the alignment procedures.

It should be pointed out that the establishment of laboratory coordinates inside an ultrahigh vacuum system, where reference is made to flange alignments, may be in error because of machining errors and welding errors made in the fabrication of the chamber, as well as to distortion of the chamber which occurs both on welding and due to the atmospheric pressure.

## *References*

[1] Dr. Oliver Schaff, Fritz Haber Institut der Max-Planck Gesellschaft, Faradayweg 4-6, D-14195 Berlin (private communication).

***Fig. 74. Crystal Optical Alignment***

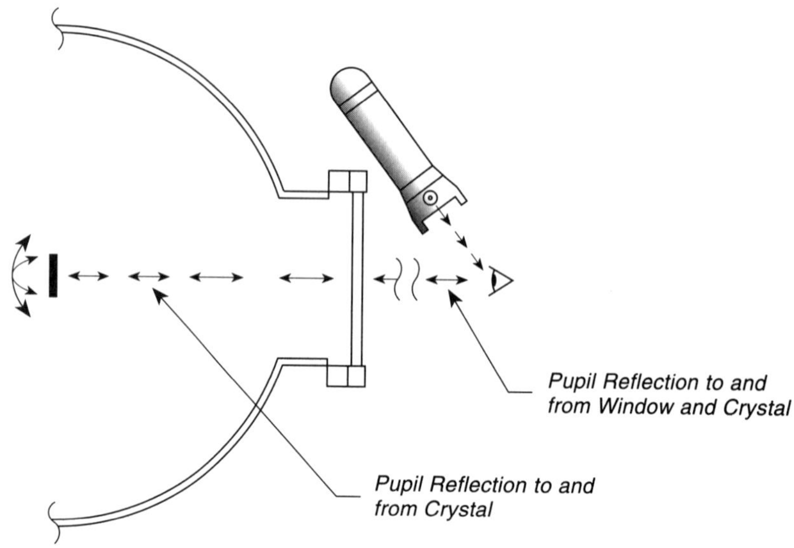


***A. Crystal Alignment by Eye***

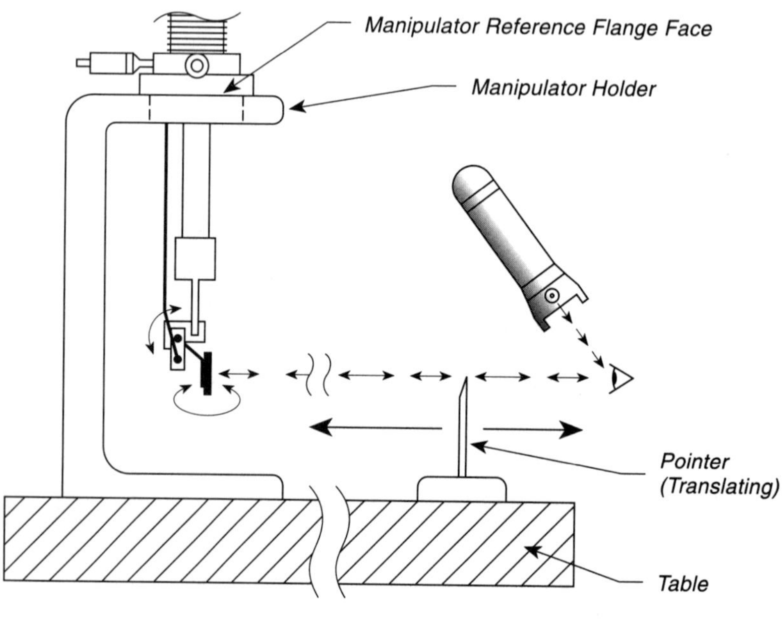


***B. Crystal Alignment by Eye on Table***

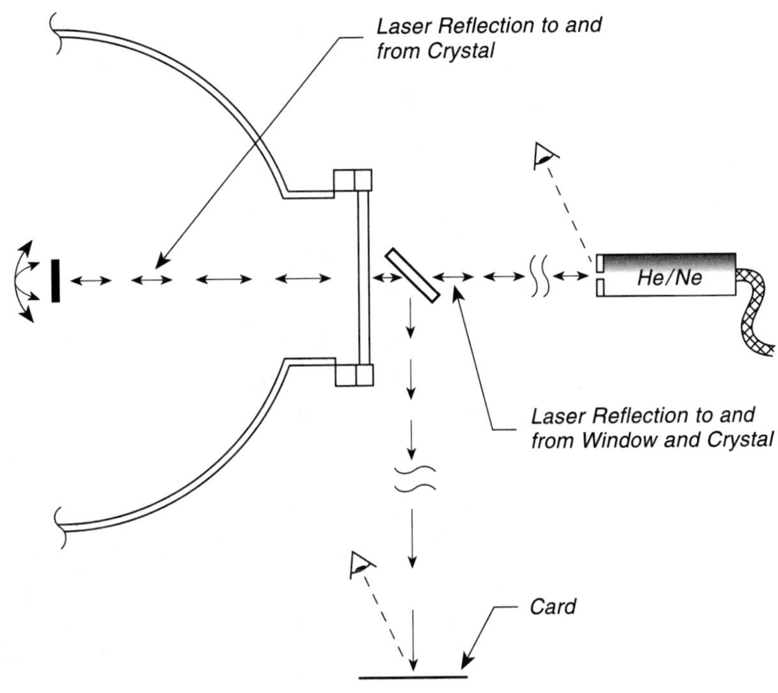


***C. Crystal Alignment by Laser***

# 75

# Measurement of the Angles of Incidence in a LEED Experiment

It is often necessary to determine the angle of incidence of an electron beam in a low energy electron diffraction (LEED) apparatus relative to the crystal being studied. This can be done qualitatively using settings on the crystal manipulator goniometers or by simple measurements of the LEED beam pattern on the screen. However, for accurate work, it should be possible to use all or almost all of the LEED beam positions to do this with high accuracy, exceeding the accuracy of the average goniometer, which has an uncertain relationship to the crystal connected to it. The method to be described [1] outperforms earlier methods involving the use of an internal graduated circle and vernier [2], laser calibration of the manipulator outside of the vacuum system, or measurement of the 0,0 beam position using a Faraday cup on a vernier circle.

Figure 75 shows a LEED screen with its axial electron gun, and the definition of the laboratory directions, $x_l$, $y_l$, and $z_l$. Similar angles relative to a crystallographic direction on the crystal are given by $x_c$, $y_c$, and $z_c$. The azimuthal angle is $\Psi$. The polar angle is given by $\Theta$. The angle $\varepsilon$ is the measured angle between the 0,0 beam and an h,k beam.

By examination of a LEED photograph containing multiple LEED beams and making a series of angular measurements using a transparent angular scale made from circular graph paper, it is possible to map out many angles which permit the accurate determination of $\Theta$ and $\Psi$. The computer program for doing this is available from the authors [1], and accuracies of the order of 0.1° are achieved owing to the use of multiple angular measurements.

## *References*


[1] S.L. Cunningham and W.H. Weinberg, Rev. Sci. Instrum. **49,** 752 (1978).
[2] J.M. Burkstrand, Rev. Sci. Instrum. **44,** 774 (1973).

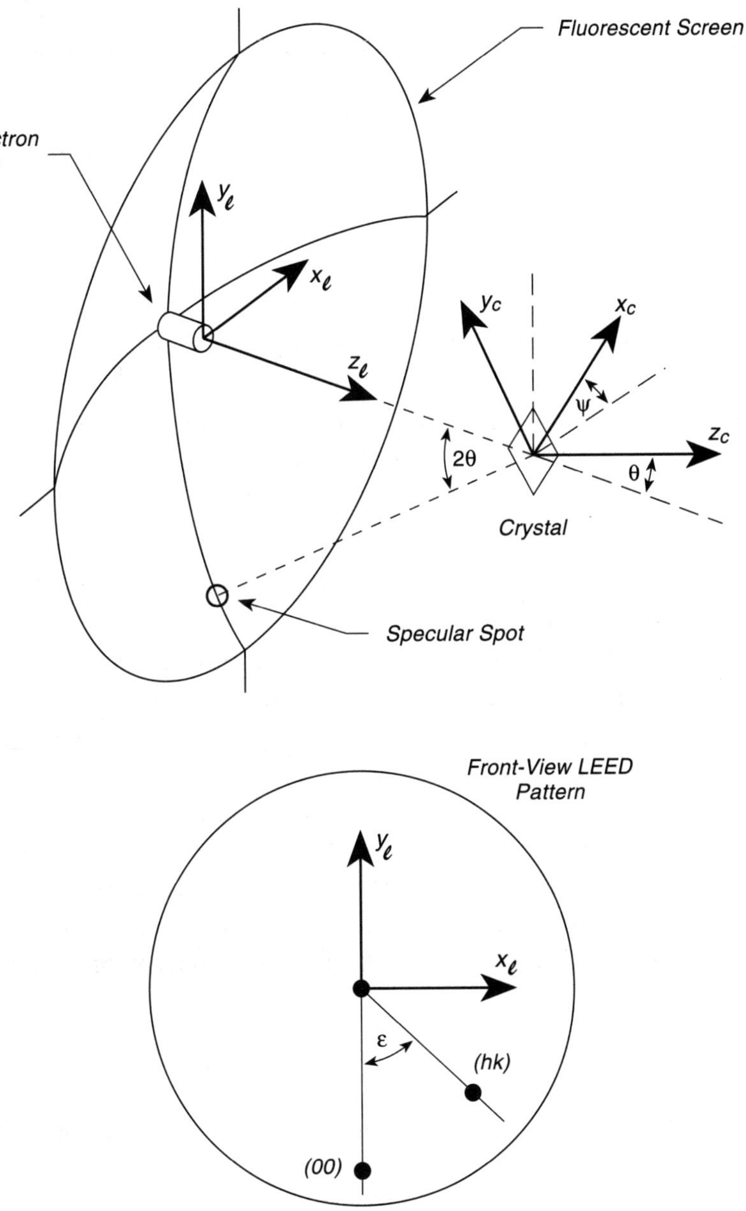


***Fig. 75. Measuring Angle of Incidence in LEED***

# 76

# Measurement of Angle of Incidence of an Electron Beam on a Single Crystal

A simple method for measuring the angle of incidence of an electron beam on a single crystal that involves only the measurement of the collected current is described in Figure 76.

When an electron beam of fixed kinetic energy is received by a single crystal, the reflectivity of the electrons will fall off as the angle of incidence is changed. By differentiating the collected current with angle, plots that are symmetrical about the normal will be obtained, provided the plane of the angular scan is perpendicular to a mirror plane of the crystal. This behavior is a consequence of the band structure of the crystal, and the more free-electron-like a band is, the higher the electron coupling probability is to this band. Inside the band gap region the reflectivity of electrons with low energies can reach values close to unity.

The apparatus shown in Figure 76A shows the electron gun and the electrical circuit used for differentiating the collected current at the crystal. The measurements were taken in the ΓXWK mirror plane of the Cu(100) crystal, perpendicular to another ΓXWK mirror plane. The angle of incidence was varied by rotating the electron gun around the crystal. The data shown in Figure 76B clearly show the symmetry with respect to the normal incidence angle for two electron energies. The symmetry indicates the absence of external magnetic fields and the absence of stray electrical fields. It is estimated that the position of the normal can be estimated to $\pm 0.5°$, although the features in the wings differ in their maxima in the two directions by several degrees. The procedure can be used to determine the azimuthal orientation of the crystal by scanning the azimuthal angle. For off-normal angles of incidence, extrema should be observed when crossing a mirror plane of the crystal [1].

## *References*

[1] R. Schneider, H. Durr, Th. Fauster, and V. Dose, J. Vac. Sci. Technol. **A8**, 3363 (1990).

***Fig. 76. Measuring Angle of Incidence of Electrons***

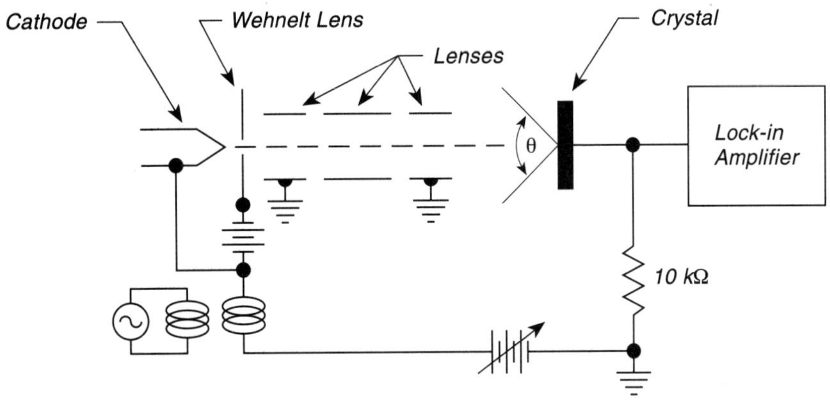


**A. Electron Gun + Circuit**

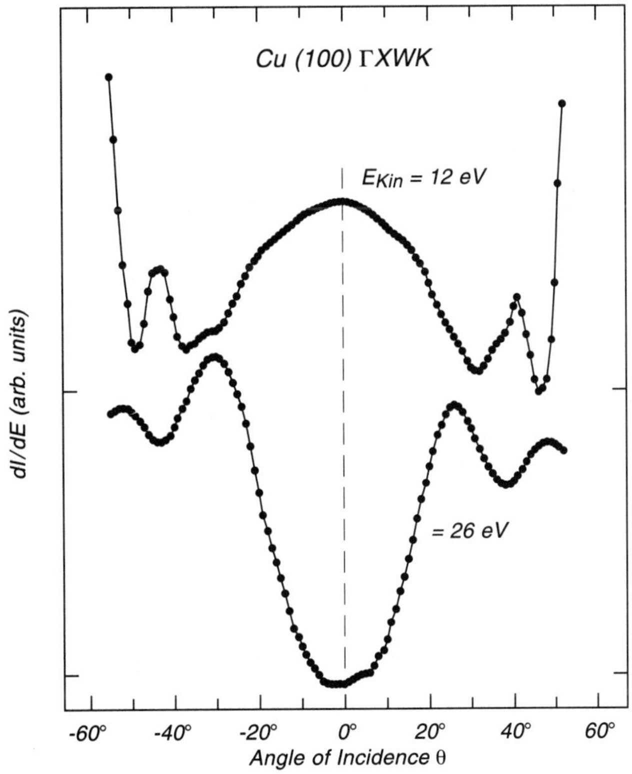


**B. First Derivative—Collected Current**

# G. Tip Fabrication

# 77

# Fabrication of Metal Tips Using Zone Electropolishing

Metal tips are required in field electron and field ion microscopy, as well as in scanning tunneling microscopy. One method for preparing such tips is to use zone electropolishing while viewing the process under an optical microscope [1,2].

Figure 77 shows the setup employed for both wire samples and for irregular specimens such as thin films stripped from their substrates or slivers of metal cut from a sheet. The electrolyte is held on a loop by surface tension, and its volume can be adjusted upward by transfer from a wet glass rod or downward by absorption by a piece of tissue paper. This loop can be moved laterally using an *x,y,z* micromanipulator as shown, and the specimen to be electropolished is attached mechanically and conductively to a specimen holder. For a wire specimen, selected regions of the tip may be electropolished by lateral motion of the loop to the desired region as shown in Figure 77, inset A. Very sharp tips may be produced by first electropolishing a neck behind the tip apex and then reducing the necked-down region to a sharp point in an electrolyte film of about 0.2-mm thickness. Additional sharpening can be achieved by quick thrusts of the electrolyte film onto and away from the tip.

Irregular specimens can be shaped as shown in Figure 77, inset B. Here, working on the sides of the sample, etching is achieved to bring the sample to a tip shape prior to sharpening the tip as shown in Figure 77, inset A.

Either AC or DC electropolishing can be used. In addition, instead of using a micromanipulator, the lateral stage motion of the specimen holder or the stage of the microscope can be used with the sample being held rigidly in place.

The microscope magnification is either x150 or x600 with a long working distance objective. Note the protective coverglass for the objective to prevent etching of the glass lens by droplets of electrolyte produced during electropolishing.

Special considerations in tip construction apply for tips to be used under electrolytes, where Faradaic currents may damage the tip. This damage problem may be eliminated by the use of Apiezon wax to shield the tip shank [3].

An excellent review on methods for sharpening STM tips is given in reference [4].

The use of AC waveform control in shaping tips for field ion microscopy has been described [5,6]. Electrochemical etching of Ag tips has recently been described, using electronic control of the tip movement out of the electrolyte [7,8].

A review comparing the instrumental characteristics of various STM instruments is given in reference [9].

## *References*


[1] A.J. Melmed and J.J. Carroll, J. Vac. Sci. Technol. **A2,** 1388 (1984); see also A.J. Melmed, J. Vac. Sci. Technol. **B9,** 601 (1991).

[2] H. Lemke, T. Goddenhenrich, H.P. Bochem, U. Hartmann, and C. Heiden, Rev. Sci. Instrum. **61,** 2538 (1990).

[3] L.A. Nagahara, T. Thundat, and S.M. Lindsay, Rev. Sci. Instrum. **60,** 3128 (1989).

[4] M. Fotino, Rev. Sci. Instrum. **65,** 159 (1993).

[5] H. Morikawa and K. Goto, Rev. Sci. Instrum. **59,** 2195 (1988).

[6] H. Bourque and R.M. Leblanc, Rev. Sci. Instrum. **66,** 2695 (1995).

[7] K. Dickmann, F. Demming, and J. Jersch, Rev. Sci. Instrum. **67,** 845 (1996).

[8] A.A. Gorbunov and B. Wolf, Rev. Sci. Instrum. **64,** 2393 (1993).

[9] Y. Kuk and P.J. Silverman, Rev. Sci. Instrum. **60,** 165 (1989).

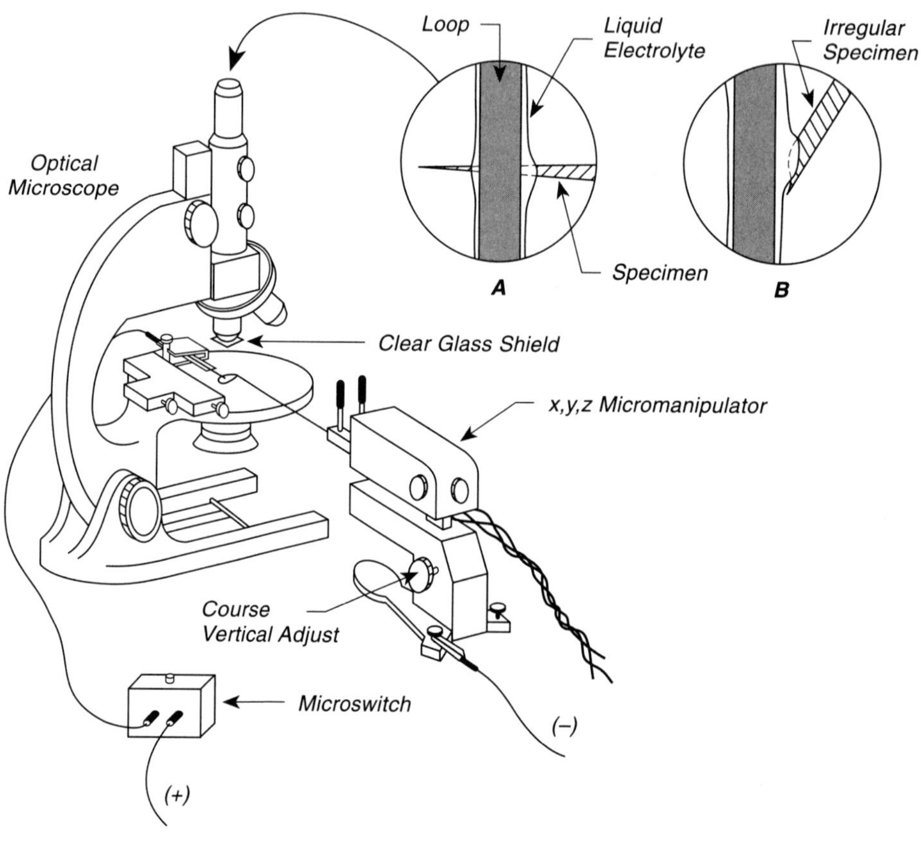


**Fig. 77. Etching Field Ion Microscopy Tips**

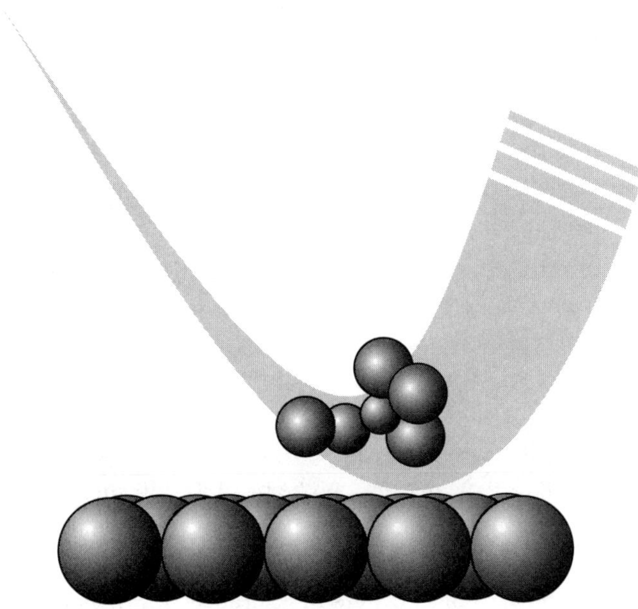

# 78

# Etching STM Tips Reproducibly

A simple apparatus [1] for making a number of STM tips while automatically replenishing the electrolyte is shown in Figure 78. Section (a) shows a three-hole electrolyte support in which thin films of the electrolyte are held by surface tension effects. The central hole of 3-mm diameter is connected to smaller 2-mm diameter supply and withdrawal reservoirs in a 1-mm-thick Cu plate. Section (b) shows the beveled end of a (1-mm outside diameter, 0.45-mm inside diameter) supply tube for fresh electrolyte being fed into the supply reservoir. A similar tube is used to withdraw electrolyte from the other reservoir. The wire to be etched enters the central hole, and projects below the bottom of the electrolyte film. This permits the etched tip to fall away under gravity when the etching is complete, and it is believed that as the lower tip falls away the local strain is likely to produce an atomic asperity [2] for STM imaging. Section (c) shows the application of AC current to the etching apparatus, and the profile of a tip just being started. The thin neck below the desired tip location is useful for making a detachable handle with which to hold the tip, and which can be broken off later.

The flow rate of electrolyte and the configuration of the meniscus can be adjusted by changing the level of electrolyte storage beakers on labjack platforms that connect to the supply and withdrawal reservoirs. Tubing 30 cm in length is approximately correct. For an electrolyte solution of KOH(aq), 4 wt. %, a flow rate of about 1 ml/h is correct for an etching current of 20 mA. A slightly concave upper and lower surface in the etching cell is preferable, achieved by fine height adjustments of the two beakers. Tips have had a wire diameter as small as 0.08 mm. Many tips can be produced in succession by this method. It is helpful to use a wax pencil to outline the three-hole device on both sides to prevent wetting and electrolyte creep across the horizontal surfaces. Another design for tip etching is shown on page 250 as well as others in references [2,3]. The use of Apiezon wax for protection of tip shanks used in electrolytes from Faradaic current effects is discussed in reference [3]. A general review of the performance of various STM instruments is given in reference [4], and a review of tip-making for field ion microscopy tips is given in reference [5]. An Auger spectroscopy study of the surface composition of etched tungsten tips is given in reference [6]. It was found that a 10-nm layer of tungsten oxide is mostly removed by heating to about 1800 K. The tungsten oxide film from an electrochemically etched W tip can also be removed by treatment in concentrated HF(aq) [7].

## *References*


[1] J.O. Fiering and F.M. Ellis, Rev. Sci. Instrum. **61**, 3911 (1990).
[2] P.J. Bryant, H.S. Kim, Y.C. Zheng, and R. Yang, Rev. Sci. Instrum. **58**, 1115 (1987).
[3] L.A. Nagahara, T. Thundat, and S.M. Lindsay, Rev. Sci. Instrum. **60**, 3128 (1989).
[4] Y. Kuk and P.J. Silverman, Rev. Sci. Instrum. **60**, 165 (1989).
[5] E.W. Muller and T.T. Tsong, *Field Ion Microscopy* (Elsevier, New York, NY, 1969).
[6] A. Cricenti, E. Paparazzo, M.A. Scarselli, L. Moretto, and G. Selci, Rev. Sci. Instrum. **65**, 1558 (1994).
[7] L.A. Hockett and S.E. Creager, Rev. Sci. Instrum. **64**, 263 (1993).

(a)

(b)

(c)

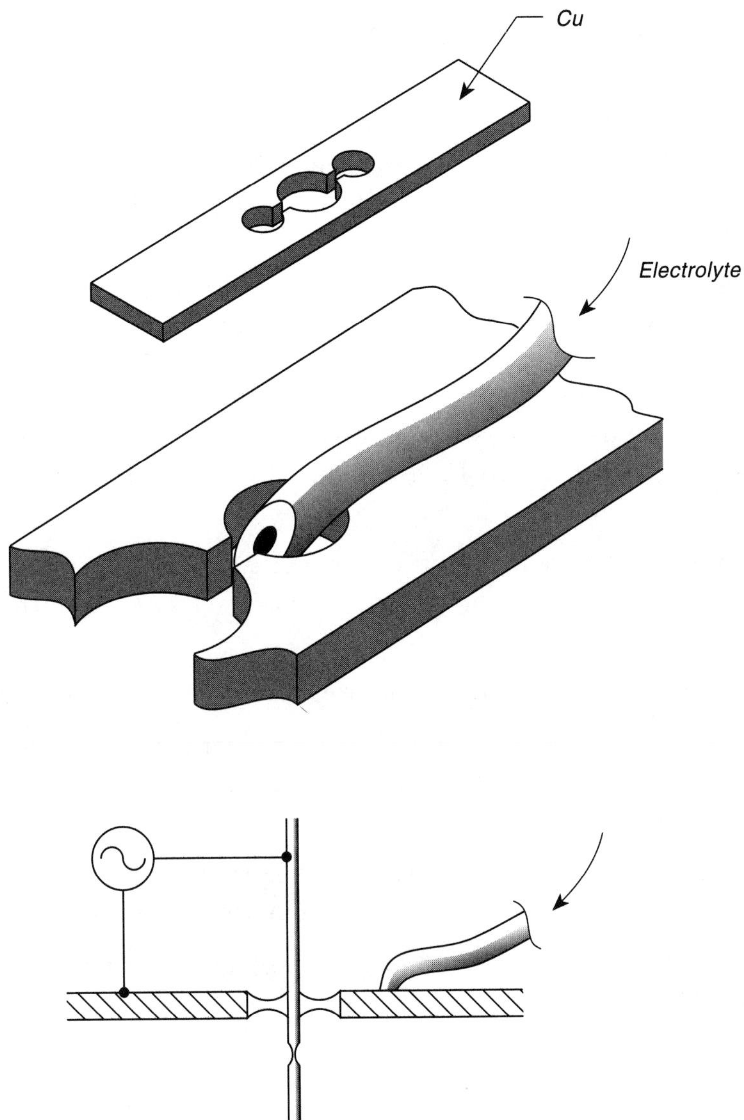


**Fig. 78. Etching STM Tips Reproducibly**

# 79

# Sharpening Single Crystal Metal Tips by Ion Bombardment

Both tungsten and molybdenum single crystal wires can be sharpened into tips by using axial sputtering. The data shown here are for Mo(110) tips [1], but similar results are reported for W(111)-oriented wires [2], except in that work the ion beam direction was from behind the tip, producing very sharp pyramidal shapes.

The Mo(110) crystal wire is mounted such that the axis of the wire is parallel to the ion beam, as shown in Figure 79A. In addition, because of variation in ion current density across the ion beam, the tip should be centered in the beam to achieve symmetrical sputter etching. The ion beam in this case is 350 μm in diameter. It is found that $Xe^+$ ions are more effective in making less roughened tips; $Ar^+$ at low kinetic energies can produce very sharp asperities also. High-energy $Ar^+$ sputtering produces increased surface roughness compared to $Xe^+$.

The profiles of tips produced both by 3 kV $Xe^+$ and 0.5 kV $Ar^+$ sputtering are shown in Figure 79B. It is estimated that the central asperity on the $Ar^+$ sputtered tip has a base diameter of about 10 Å.

Ion sputtering for sharpening field electron and field ion tips has been described [2,3].

Polycrystalline wires cannot be shaped so well by this technique because of the presence of grain boundaries from which preferential sputter etching occurs [3]. However, single crystal tungsten tips have been ion milled to a point that is sharp on the atomic scale [4a,b].

A recent example of the use of a focused ion beam of $Ga^+$ ions for ion milling W and Pt/Ir tips is given in reference [5], which also summarizes much other work in this area.

Copper wires of 30–50 μm diameter have been ion etched into sharp tips, and the dynamics of the process have been studied [6].

## *References*


[1] S. Morishita and F. Okuyama, J. Vac. Sci. Technol. **A9**, 167 (1991).

[2] J.M. Walls, H.N. Southworth, and G.J. Rushton, Vacuum **24**, 475 (1974).

[3] J.A. Kubby and B.M. Siegel, J. Vac. Sci. Technol. **B4**, 120 (1986).

[4] (a) D.K. Biegelsen, F.A. Ponce, and J.C. Tramontana, Appl. Phys. Lett. **54**, 1223 (1989); (b) D.K. Biegelsen, F.A. Ponce, J.C. Tramontna, and S.M. Koch, Appl. Phys. Lett. **50**, 696 (1987).

[5] M.J. Vasile, D.A. Grigg, J.E. Griffith, E.A. Fitzgerald, and P.E. Russell, Rev. Sci. Instrum. **62**, 2167 (1992).

[6] H. Hasuyama, Y. Kanda, T. Soeda, K. Niiya, and M. Kimura, Matls. Sci. and Engr. **90**, 33 (1987).

***Fig. 79. Sharpening Tips by Ion Bombardment***

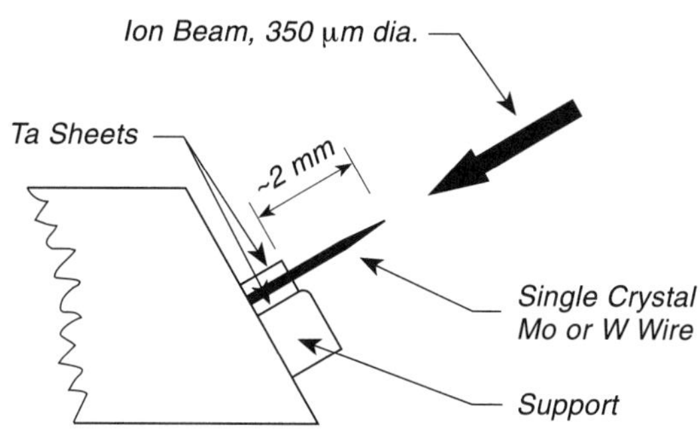


***A. Apparatus***

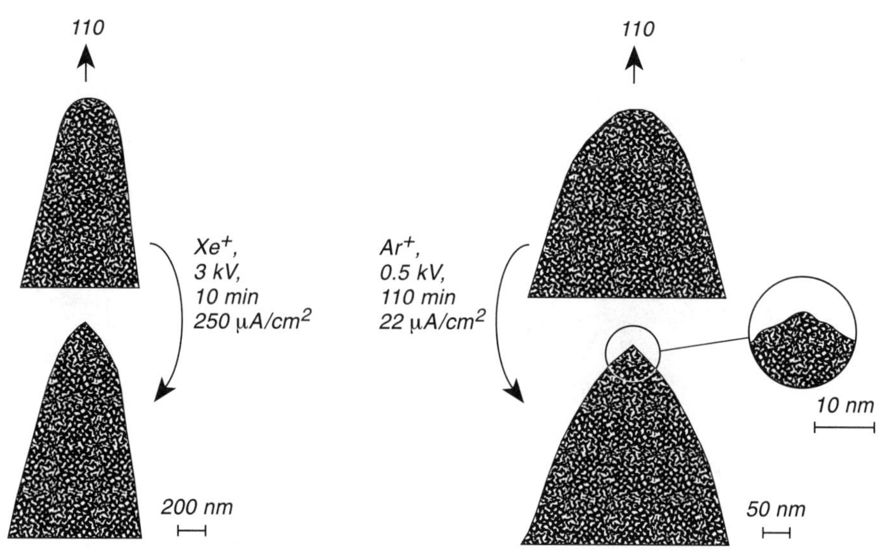


***B. Tip Profiles by TEM***

# H. Spot-Welding

# 80

# Spot-Welding Difficult Junctions

Many metal/metal junctions to be spot-welded are difficult to fabricate as a result of failure to understand the nature of the spot-welding process. A spot-weld is made when an electrical spark causes two surfaces in contact to fuse together as a result of the high local heating. Ideally, the surfaces to be welded should make only a point contact in order to focus the local heating at a single point. Thus, as shown in Figure 80A, two cylindrical wires, placed orthogonally to each other, will theoretically contact each other only at one point, leading to efficient welding. In contrast, two wires oriented parallel to each other will contact along a line, making a broad contact and leading to a weak weld.

Figure 80B shows a way to weld a thermocouple or other wire to a metal single crystal. The crystal is placed on a rigid, clean, smooth surface that is a nonconductor. A single spot welding electrode presses the wire against the back of the crystal, and a second electrode also presses on the back of the crystal to make electrical continuity.

Often it is found that the welding leads stick to the weld when it is completed. This may be due to welding electrodes that have not been smoothed with emery paper before welding, or the application of too much power to make the weld. Sometimes sticking may be prevented by coating the welding electrode with carbon black from a match or candle flame, as shown in Figure 80C.

Oxidation during the spot weld is often a problem owing to local heating. This may be prevented by welding under a droplet of methanol (Fig. 80D), which acts as a shield from the air and is also a reducing agent in the presence of the hot metal at the weld point.

In cases where a good weld is difficult to make, a sandwich weld is sometimes used as an intermediate material between two refractory metals. Tantalum and platinum foil are often good for this, as shown in Figure 80E [1]. It is also reported that changing the welding tip polarity sometimes helps in making a difficult weld, although the reasons for this are not understood [2]

It is very important to only try to weld metals when the surfaces have been freshly cleaned. Thus, W and Mo are cleaned by electropolishing in NaOH(aq) solution, about 1 M in concentration. The surface to be welded is positive during electropolishing, or in some cases AC electropolishing may be used.

Figure 80F shows a clever way to electropolish the metal and then to use the electropolish solution as a protective boundary from air oxidation. Here, a Mo single crystal is to be welded to a Mo wire at the crystal edge. The NaOH (aq) solution in a droplet is placed on the crystal, and electropolishing of the edge is carried out using a Mo wire electrode in series with a light bulb and a transformer. The light bulb supplies a protective load if a short circuit to the crystal occurs. Once the electropolishing is complete, the droplet of solution is left in place, and the spot-weld of a We/Re thermocouple in this case is made under the electrolyte [3]. After spot-welding, the droplet of electrolyte is pipetted away, and the assembly is washed well with distilled water. It was noted that lower power was needed to make effective spot-welds in this manner [3].

It is important to make many practice spot-welds before assembling the desired object. With practice, a careful welder can make ten successful practice welds in series, therefore almost guaranteeing that the eleventh will be good. Adjustments of the power per weld, the pressure used at the weld, and the composition of the welding tips must be made to achieve a good weld. Once the proper conditions are found, a single weld is all that is needed, and in fact multiple welds at the same point can result in metal work hardening and weakening and should be avoided. Metals with very high thermal conductivity (Al, Ag, Cu) are difficult to weld, and slotted crystals of those metals are often used where the slot accommodates a refractory heating/support wire [4]. It is reported that a Pt sandwich will facilitate refractory metal spot-welds to Cu [2].

## *References*

[1] D.C. Carlyle, Rev. Sci. Instrum. **41**, 599 (1970).

[2] Dr. W.F. Egelhoff, Jr., National Institute of Standards and Technology, Gaithersburg, MD 20899-0001 (private communication).

[3] M.E. Jones, B.E. Koel and R.T. Weppner, Rev. Sci. Instrum. **60**, 3067(1989).

[4] J.E. Crowell, J.G. Chen, and J.T. Yates, Jr., Surface Sci. **165**, 37 (1986).

***Fig. 80. Spot-welding Difficult Junctions***

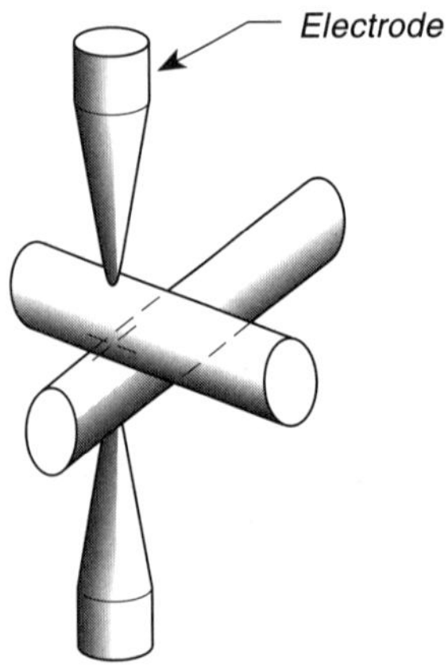


***A. Point Contact Weld***

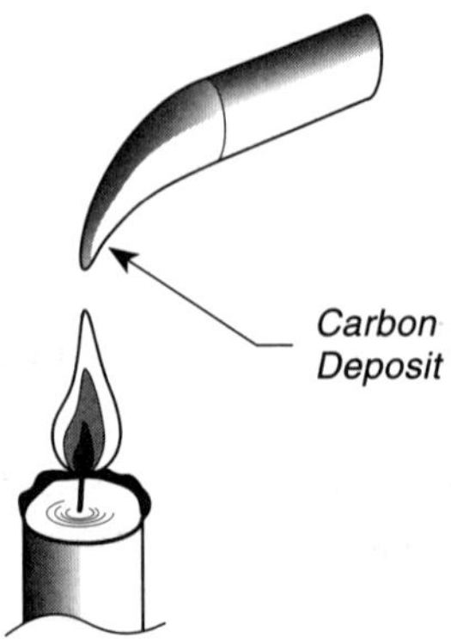


***C. Preventing Electrode Sticking***

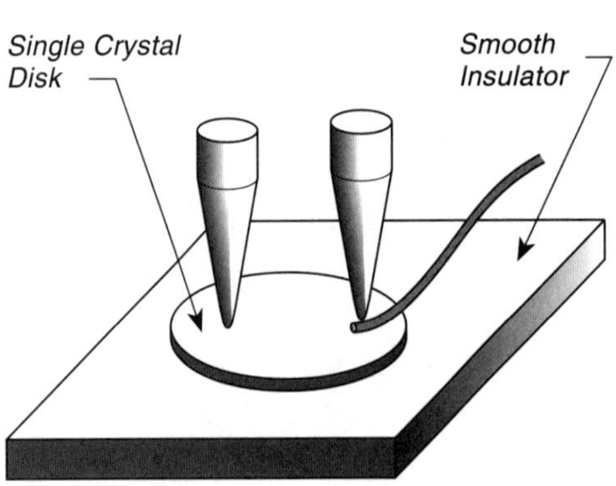


***B. Welding to Single Crystal***

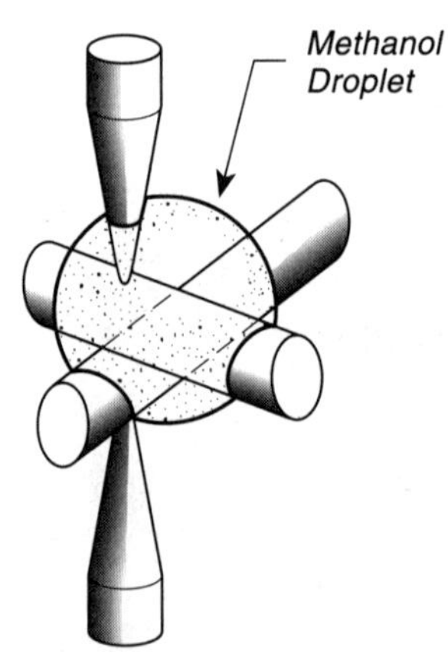


***D. Welding Under Methanol***

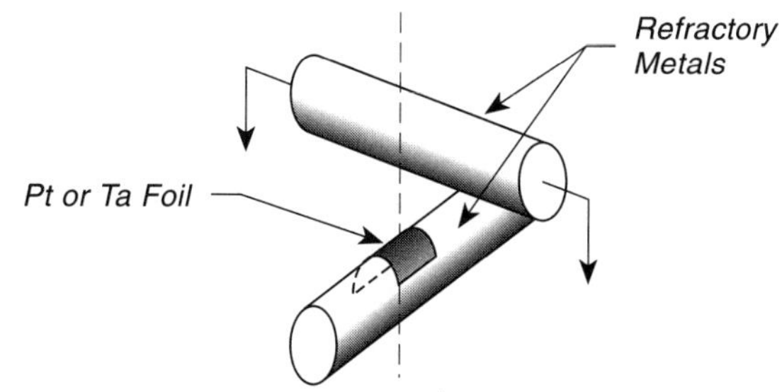


**E. Sandwich Weld Between Refractory Metals**

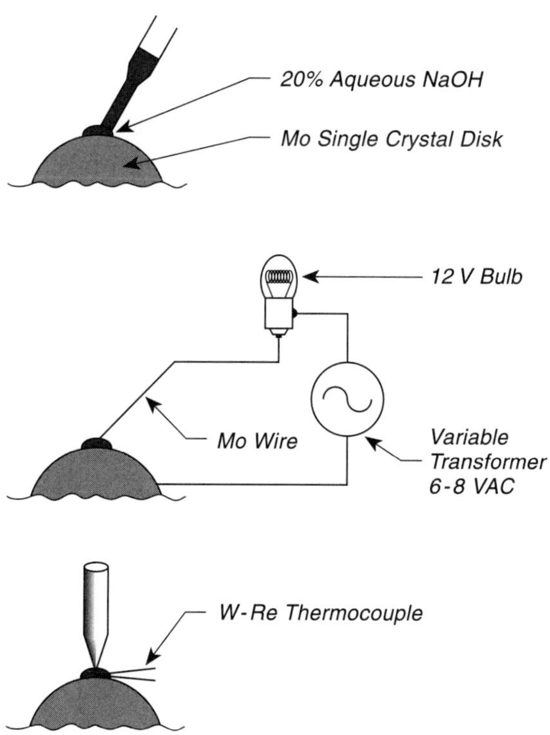


**F. In Situ Electropolishing + Welding**

# I. Mechanical Procedures

# 81

# Stretching Thin Metal Foils to a Wrinkle-Free Condition

In some cases, the provision of wrinkle-free metal foil for windows may be needed. A very simple method, related to the differential thermal expansion of metal and Pyrex glass, may be employed to achieve this objective [1].

Figure 81A shows a short cylinder of fused quartz or Pyrex glass ground flat on its end, mounted on top of the wavy foil. The thin foil is supported on its bottom by a horizontal substrate such as a lapped flat plate of aluminum, coated with a fluorocarbon dry lubricant [2]. The foil is attached to the plate at the edges with an adhesive tape to prevent curling. An epoxy adhesive (Torr seal [3]) is applied to the lower edge of the glass tube and the tube is placed on the wrinkled foil. The assembly is heated in an oven for several hours at about 60°C, and then allowed to slowly cool in the oven overnight. When the assembly is heated, the areal expansion of the metal foil exceeds that of the glass cylinder due to the thermal expansion coefficient differences. The light weight glass cylinder does not impede the foil expansion, and wrinkling does not occur in the center of the glass form. Upon cooling, the foil contracts more than the glass, resulting in a uniformly stretched foil as shown in Figure 81B. This technique can be used for stretching foil over an aperture that cannot be heated; the glass form is larger than the aperture, and the aperture is attached by cement to the foil, previously stretched as shown. A simple consideration of the strain effects in the circumferential and the radial directions indicates that for the circular geometry of the glass cylinder, the strain in both directions is equal, meaning that no shear stress takes place on cooling. This leads to a uniform tension and a flattening of the foil [1]. HAVAR foil (0.005 mm thick) [4], aluminum foil (0.025 mm thick) and Mylar foil (0.025 mm thick) have all been stretched flat in this manner.

## *References*


[1] W.H. Holt and C.M. Blackmon, Rev. Sci. Instrum. **51**, 671 (1980).

[2] Obtained from Miller-Stephenson Chemical Co., Inc., George Washington Highway, Danbury, CT 06810.

[3] Obtained from Varian Vacuum Products, 121 Hartwell Avenue, Lexington, MA 02173.

[4] Obtained from Hamilton Technology, Inc., 101 N. Queen Street, P.O. Box 4787, Lancaster, PA 17604.

***Fig. 81. Stretching Metal Foils***

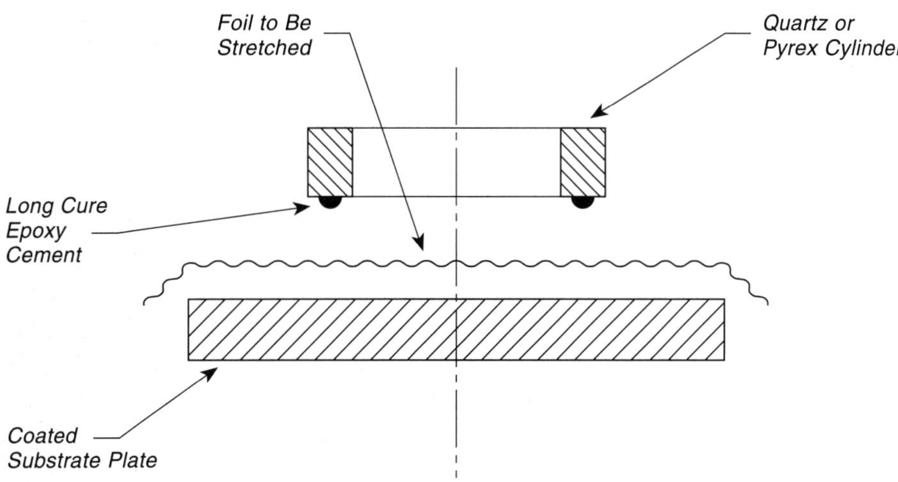


***A. Before Assembly***

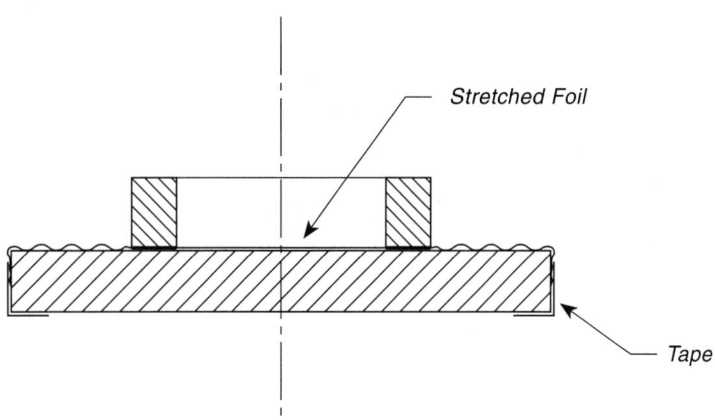


***B. After Cure and Cooling***

# 82

# Making Micron-Size Holes for Supersonic Nozzles

Two methods are outlined for producing micron-sized holes in the ends of tubes to be used as supersonic jet sources in vacuum [1].

## Method I—Figure 82A

A tube of the appropriate length and inner and outer diameter is made with only one end open. An inner rod, which fits into the tube, is made with its length just slightly longer than the inside dimension of the tube. This rod is sharpened on one end, and a cap is machined on its upper end. By varying the cone angle of the sharpened tip, the hole size can be controlled. The sharpened rod is then punched with a hammer until the cap is tightly in contact with the upper tube edge. This should produce a cone-shaped indentation in the end of the tube as shown. The outer surface of the tube is then sanded down carefully until a hole just begins to appear. A microscope is used to judge the hole size using both reflected and transmitted light [2].

## Method II—Figure 82B

A tube of the appropriate length and inner and outer diameter is made with only one end open. The closed end is cut down with a conical surface on a lathe until a tiny hole appears on the end. A punch die with the same cone angle as the tube end is made. The punch die is then hammered on top of the tube end to close down the hole until the desired size is obtained. A microscope is used to monitor the changes in the hole size using both reflected and transmitted light [3].

### *References*


[1] Professor Thomas Engel, Department of Chemistry, Box 351700, University of Washington, Seattle, WA 98195-1700 (private communication). The ideas are not original from the Engel group and are believed to originate from Dr. Adi Scheidemann, from Peter Toennies' group, Göttingen University, Göttingen, Germany.

[2] From Steve Moskovits, T. Engel group, University of Washington.

[3] From Brian Holm, machinist, University of Washington.

***Fig. 82. Making Micron-Size Holes for Supersonic Nozzles***

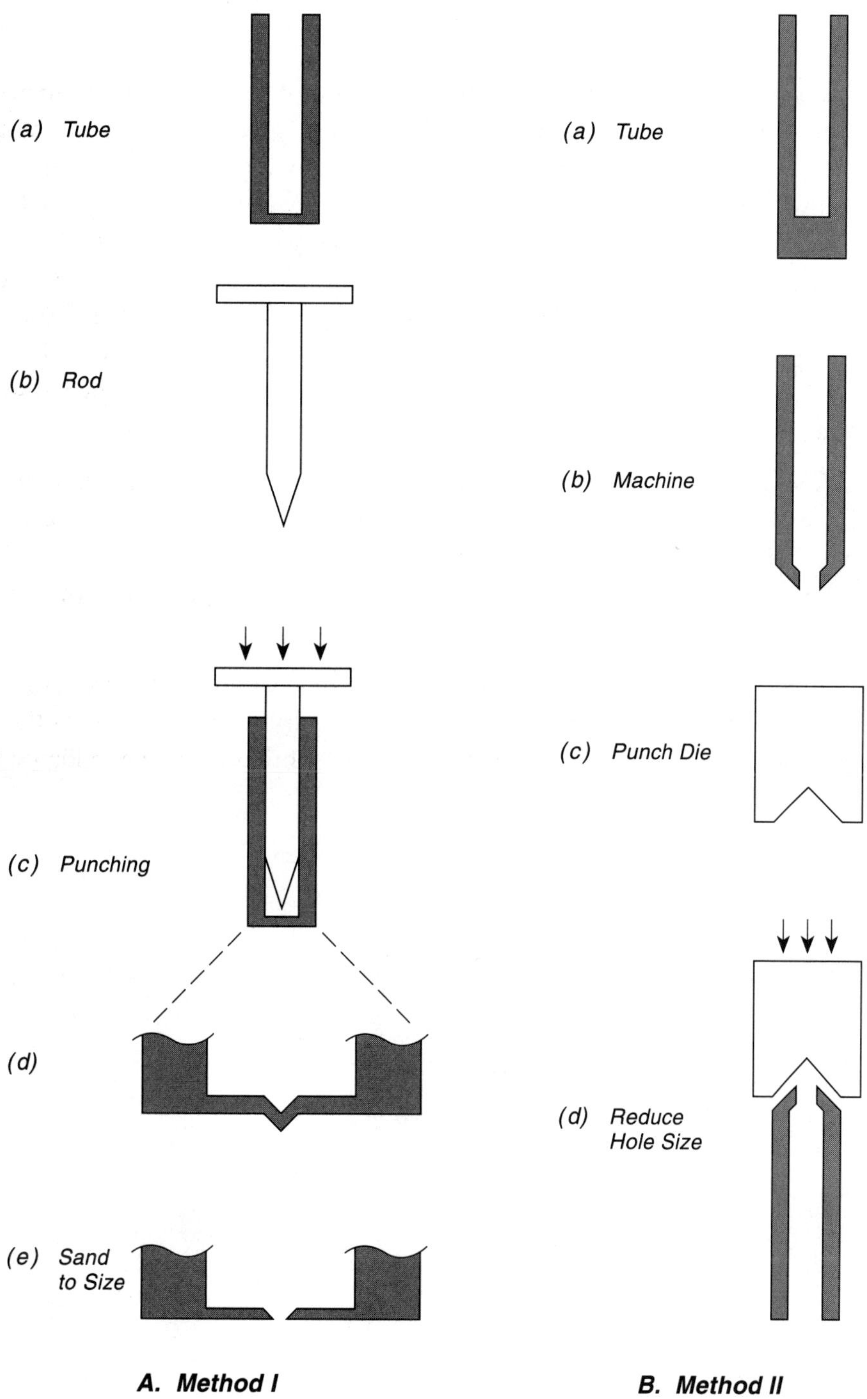

# 83

# Straightening and Braiding Wires

It is often advisable to work with very straight segments of wire, for example, in placing wires along the outside of a manipulator shaft (see page 110), where bends might interfere with the linear construction or where a tight fit through tubulations is needed. Figure 83A shows a method which works for nonrefractory metal wires such as Cu, Ni, Ag, and stainless steel [1]. The end of the wire is rigidly clamped in a vise, and with a pair of pliers, the wire is stretched only a fraction of a percent. The elastic limit should not be exceeded, and one can feel the give in the wire as stretching occurs below the elastic limit. It is wise to have prepared a place to store the straightened wire and a storage tube is recommended for this. After straightening, many wires become work hardened so that further straightening by this method will not work unless the wire is annealed.

Thin refractory wire will not straighten by this method, and another method must be used [1]. Lay the wire on a paper tablet and stroke the wire with a round tool at right angles to the wire axis, keeping the arch upward by rotating the wire during straightening. A pair of smooth jaw pliers will hold the wire in the desired position. Work on only a few inches at a time.

Braided wire may be conveniently produced in a uniform bundle by placing the wires together and holding one end in a vise. A hand-cranked drill is chucked to the other end of the wire bundle and turned slowly to make a uniform braid, as shown in Figure 83B.

## *References*

[1] Mr. Gerald L. Fowler, 6816 Kelly Ave., N.E., Albuquerque, NM 87109 (private communication).

**Fig. 83. Straightening and Braiding Wires**

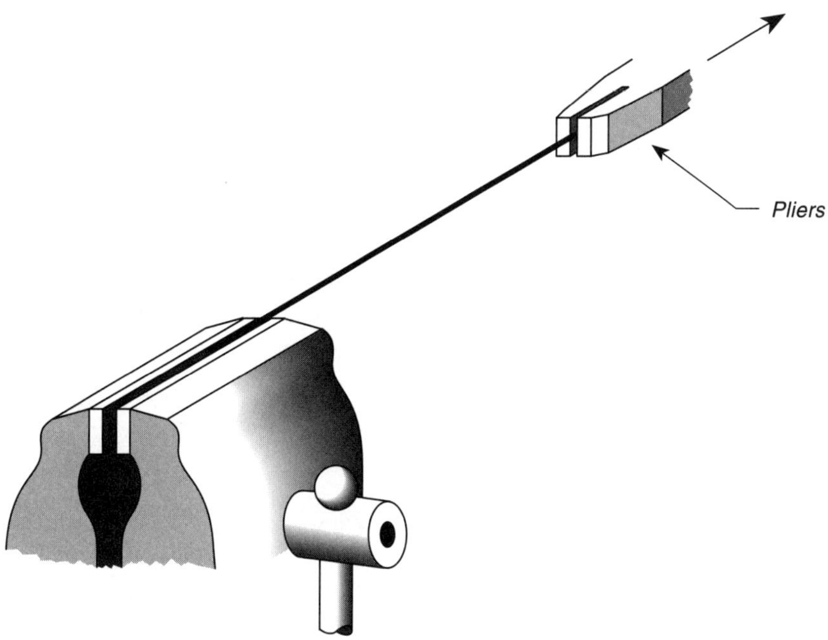


**A. Stretching a Wire to Straighten**

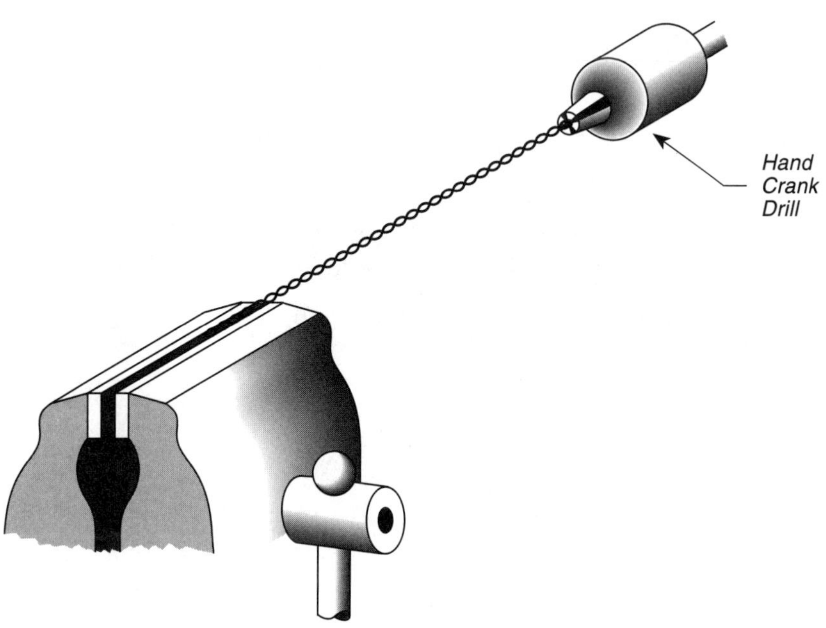


**B. Making Uniform Braided Wire**

# III. Measurement Methods

A. Electrons . . . . . . . . . . 273
B. Ions. . . . . . . . . . 307
C. Photons (UV) . . . . . . . . . . 337
D. Kinetics of Adsorption/Adsorbate Coverages . . . . . . . . . . 357
E. Mass Spectrometry . . . . . . . . . . 365
F. Temperature Programmed Desorption/Reaction. . . . . . . . 385
G. Gas Chromatography—Enhanced Sensitivity. . . . . . . . . . 407
H. Work Function . . . . . . . . . . 411
I. Radioactive Adsorbates . . . . . . . . . . 435
J. Thin Film Deposition . . . . . . . . . . 439
K. Infrared Spectroscopy . . . . . . . . . . 455
L. Calorimetric Heats of Adsorption—Single Crystals . . . . . 475
M. Surface Debye Temperature . . . . . . . . . . 479
N. Friction . . . . . . . . . . 485

# A. Electrons

# 84

# Electron Gun Design and Behavior

Many descriptions of the design and properties of electron guns are presented in the literature [1–4] and it is not the purpose of this treatment to summarize this practically important field. Instead, an ordinary electron gun [5] is described to give the reader a beginning picture of this technology [6].

Figure 84A shows a drawing of the arrangement of elements in an electron gun like those used in LEED and other surface science probes. It consists of a thermionic emitter filament or ribbon, usually coated with a material that lowers the work function, yielding higher operating currents at lower filament temperatures. For this directly heated tungsten cathode, either thoriated tungsten, or $ThO_2$, or $LaB_6$ coatings can be used, as discussed on pages 202 and 206. Some electron guns, such as the widely used Varian Model 981-2125, have the thermionic emitter mounted off axis to avoid line-of-sight exposure to the crystal reducing the possibility of contamination as well as reducing light output. The thermionic cathode is housed in a Wehnelt cylinder that surrounds it except for the beam extraction aperture. The Wehnelt cylinder can be biased either positively or negatively, and is usually kept near the potential of the cathode. This may be done by placing a resistor between the cathode and the Wehnelt, which results in self-biasing of the Wehnelt; a typical value of 570 kΩ will produce a small negative bias of the cylinder relative to the cathode and will maintain beam focus over a several hundred volt range. The electron beam is extracted by the first lens, $G_1$, which then focuses the initially divergent beam. This lens forms an image of either the emitter or the aperture of the Wehnelt cylinder, depending on geometrical factors and the potentials applied. The focal length is of the order of 10–30 cm. The drift tube, the lens $G_3$, and the exit aperture form an Einzel lens that focuses the beam. The beam diameter is typically about 1 mm, and this can be reduced by decreasing the first two aperture sizes (0.4-mm apertures in guns working at 100 eV energy typically deliver 1 μA current). For 0.1-μA beam currents, a minimum beam size of 0.2 mm and a divergence of 0.2° are possible. Changing the bias on the Wehnelt cylinder with respect to the cathode drastically affects the beam current and its stability.

Figure 84B shows an experimental plot of the electron beam width for the electron gun in Figure 84A, and Figure 84C shows the beam divergence versus electron energy. These properties were measured with the multiple Faraday cup assembly and procedure shown on page 294, and on this book's cover.

Modeling of the behavior of an electron gun with cylindrical symmetry such as this one can be carried out using an electrostatic computer program that is readily available [7].

High-quality electron guns for high-resolution LEED measurements have been described and often involve field emission electron sources [8].

A high-resolution low-energy electron source involving a pair of 127° cylindrical selectors and delivering $3 \times 10^{-8}$ A with 30-meV resolution has been described [9].

A simple technique for precise regulation of the emission current from an electron gun has been published. It uses the returning ground current which is matched to a reference current. The difference is optically coupled to control the filament to Wehnelt cylinder potential allowing beam currents of 0.1 μA to be regulated to 1% [10].

A pulsed electron gun with picosecond pulses has recently been described. The pulse is generated by a pulsed laser that produces photoemitted electrons [11].

## *References*


[1] K.R. Spangenberg, in *Vacuum Tubes* (McGraw-Hill, New York, 1948); see also, J.A. Simpson and C.E. Kuyatt, J. Res. Nat. Bur. Stand. **67C,** 279 (1963).

[2] J.R. Pierce, *Theory and Design of Electron Beams* (D. Van Nostrand Co., Inc., New York, 1949).

[3] F. Rosebury, *Handbook of Electron Tubes and Vacuum Techniques* (American Institute of Physics, New York, 1993).

[4] A particularly excellent account of practical aspects of electron optics is given in J.H. Moore, C.C. Davis and M.A. Coplan, *Building Scientific Apparatus* (Addison-Wesley, Reading, MA, 1983), chapter 5.

[5] Cliftronics, Inc., 15 South Street, Clifton Springs, NY 14432. Model 406-S; Another supplier of electron guns is Kimball Physics, Inc., 311 Kimball Hill Road, Wilton, NH 03086-9742.

[6] M.G. Lagally and J. A. Martin, Rev. Sci. Instrum. **54,** 1273 (1983).

[7] This program, which is periodically updated, is available from Idaho National Engineering Laboratory, EG&G Idaho, P.O. Box 1625, Idaho Falls, ID 83415.

[8] R.Q. Hwang, E.D. Williams, and R.L. Park, Rev. Sci. Instrum. **60,** 2945 (1989). See also, R.N. Lee, Rev. Sci. Instrum. 39, 1306 (1968).

[9] M. Proulx, P. Marmet, and R. Dutil, Rev. Sci. Instrum. **53,** 778 (1982).

[10] M.K. Debe and D.C. Johnson, Rev. Sci. Instrum. **47,** 37 (1976).

[11] M. Aeschlimann, E. Hull, J. Cao, C.A. Schmuttenmaer, L.G. Jahn, Y. Gao, H.E. Elsayed-Ali, D.A. Mantell, and M.R. Scheinfein, Rev. Sci. Instrum. **66,** 1000 (1995).

### Fig. 84. Electron Gun Properties

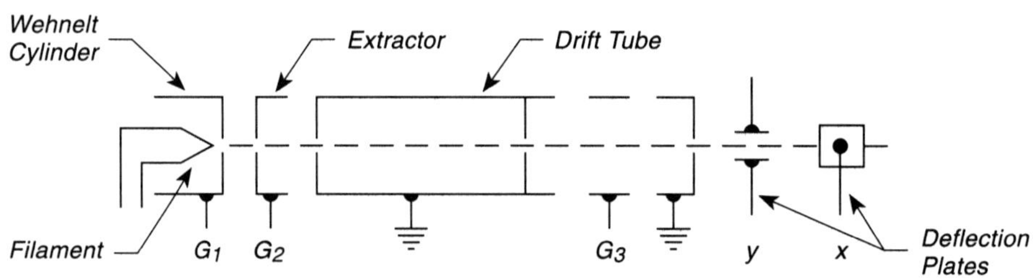


**A. Ordinary Electron Gun**

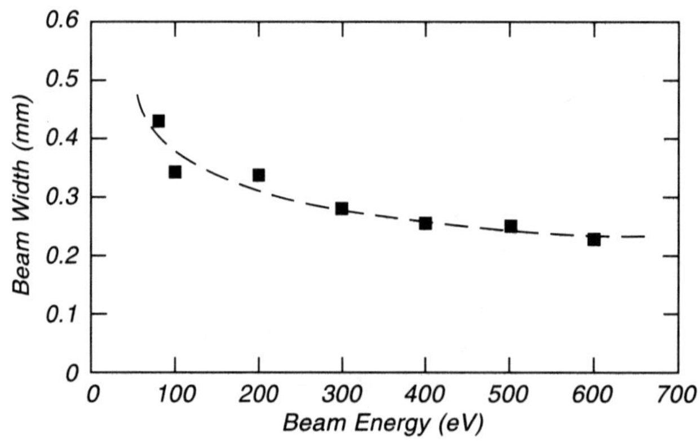


**B. Beam Width**

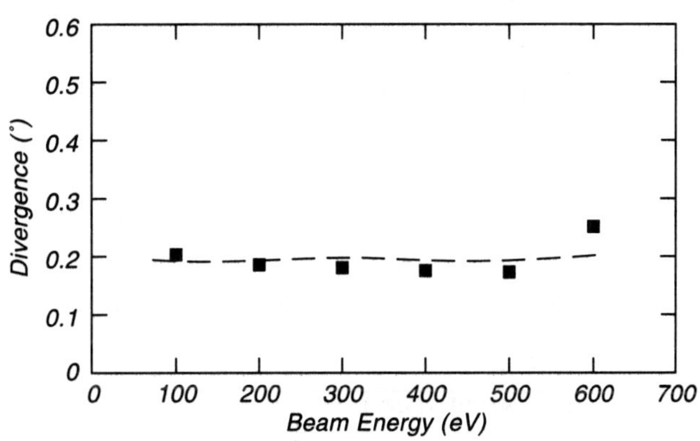


***C. Beam Divergence***

# 85

# Low-Energy Electron Gun for Broad-Beam Irradiation

Electron sources for irradiation of $cm^2$ area surfaces often employ stray electrons that originate from thermionic emitters in mass spectrometers and ionization gauges, or from bare filaments some distance from the target. While this is often satisfactory, without adequate modeling of the distribution of electrons originating from these sources, one may be dealing with non-uniform current densities on the surface being irradiated. Extended electron beams are employed in cases where surface analysis over a broad target area is to be done following damage to the surface by electron bombardment.

A simple electron gun [1] designed to give a uniform current density across an extended surface is shown in Figure 85A. The gun contains a thoriated tungsten filament (0.019-cm diameter) (see page 202) which is hidden from line-of-sight to the target by an L-shaped shield. Electrons are accelerated through a mesh grid (80% transparency) and then decelerated on the way to the target. The SIMION electrostatic lens and analysis program [2] was used to model the behavior of the electron gun as a function of various potentials, and Figure 85B shows some electron trajectories calculated for various electron energies. A fairly uniform flux in the vertical plane is observed for electron energies above 2 eV.

Because of end effects on the thoriated emitter [3], to achieve a uniform current density laterally, it would be wise to employ a gun width that is perhaps two times larger than the target width, as was done here.

The current collected by the mesh grid and the target is shown in Figure 85C. Currents at the crystal in the 2-20 μA range are measured for electron energies in the 2 eV to 31 eV range. Even at the lowest electron energies, about 1 in 5 electrons emitted are collected by the target for mesh potentials above 10 V.

## *References*


[1] J-L. Liang and J.T. Yates, Jr., J. Vac. Sci. Technol. **A12,** 2795 (1994).

[2] This program, which is periodically updated, is available from Idaho National Engineering Laboratory, EG&G Idaho, P.O. Box 1625, Idaho Falls, ID 83415.

[3] J.C. Lin and R. Gomer, Surf. Sci. **172,** 183 (1986).

Fig. 85. Low-Energy/Broad-Beam Electron Gun

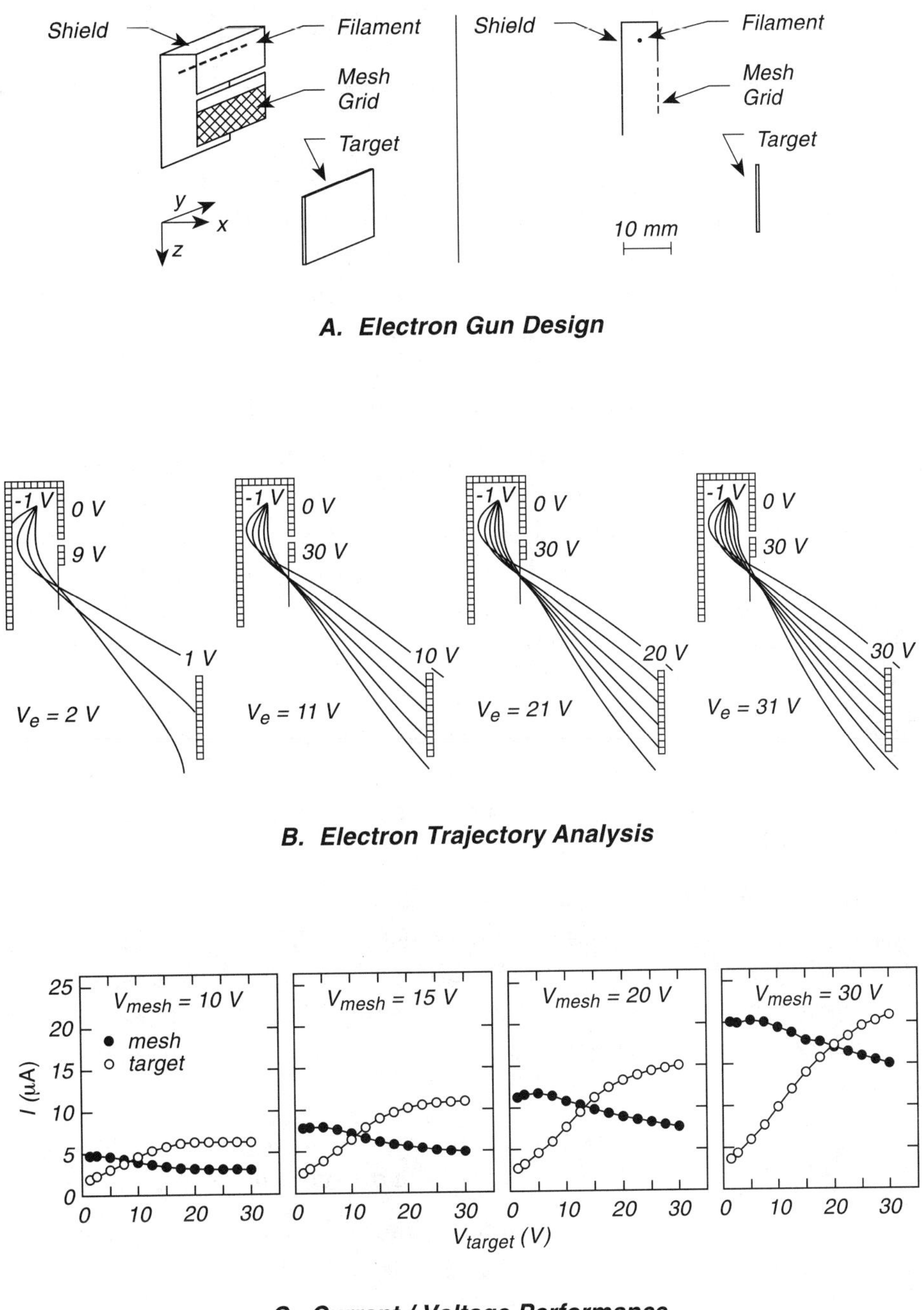


C. Current / Voltage Performance

# 86

# Low-Energy Electron Gun for Broad-Beam Irradiation—Cylindrical Symmetry

The electron gun described here provides a broad, cylindrically symmetric electron beam, as opposed to a simpler gun described on page 206, which has only planar symmetry. Low-energy electron beams produced by guns of these designs are used to cause surface chemical reactions by controlled electron beam damage, where the entire surface of a macroscopic sample is to be irradiated with uniform current density. Since electron-induced surface reaction is expected to happen at low electron energies, special design provisions must be made for delivery of high current densities at low energies.

Figure 86A shows a cylindrical gun design [1]. A BaO-coated disk cathode [2] is mounted inside a cylindrical repeller electrode. An anode plate with a 2-mm-diameter, 4-mm-long cylindrical hole is located close to the emitter disk. Four cylindrical lenses are used to focus the electron beam, and these are stacked between ceramic spacers [3]. A single positioning ring at the end of the stack holds the assembly together, and the assembly is housed inside a metal tube housing. To minimize work function differences along the gun, the inside of the lens is gold plated.

Figure 86B shows calculated electron trajectories for different electrode potentials. At the top, an anode at high positive potential withdraws electrons from the cathode, and the trajectories are shaped by the four lenses set for potentials from +30 V to 0 V as shown. This focuses the beam on the crystal at ground potential

Figure 86C shows the electrostatic potential in the cathode region in a three-dimensional plot for electrons. The geometry and potential of the repeller and anode create a steep potential drop for electrons. The shape of the potential is designed to increase the lateral momentum of the electrons in order to achieve a broad electron beam. The broad beam exhibits very uniform current density as may been seen from the trajectory spacings in Figure 86B.

In the simulations, it was found that the emission is very sensitive to the exact position of the cathode disk relative to the repeller cap and the anode opening.

Figure 86D shows the performance of the gun for various electron beam energies. At an electron energy of 10 eV, a current of 7 uA is obtained at the crystal, comparable to the performance of the gun shown on page 280. The optimized anode and repeller potential are shown for various electron energies in Figure 86E.

## *References*


[1] T.P. Barefoot, H.D. Ebinger, and J.T. Yates, Jr., J. Vac. Sci. Technol., **A15**, 2740 (1997).

[2] BaO disk cathode, ES-015 on AEI base, Kimball Physics, Inc., 311 Kimball Hill Road, Wilton, NH 03086-9642.

[3] $Al_2O_3$ ceramic insulator—Finnigan # 40005-20070, Scientific Instrumentation Services, Inc., 1027 Old York Road, Ringoes, NJ.

## Fig. 86. Low-Energy Electron Gun—Broad Beam

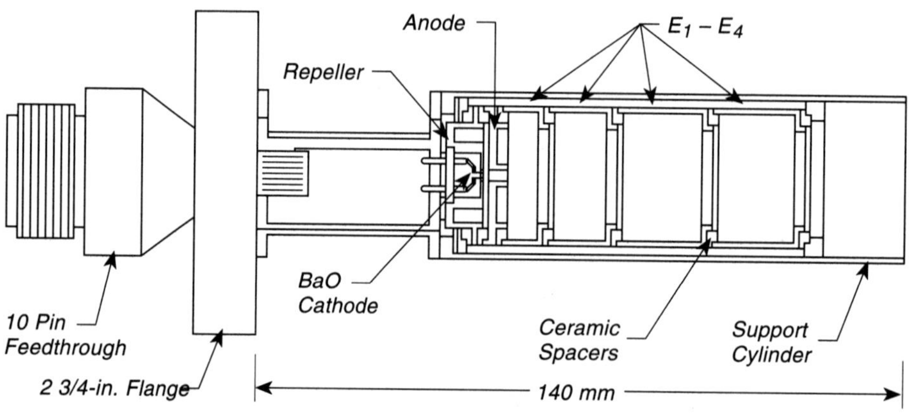


A. Design of the Low-Energy Electron Gun

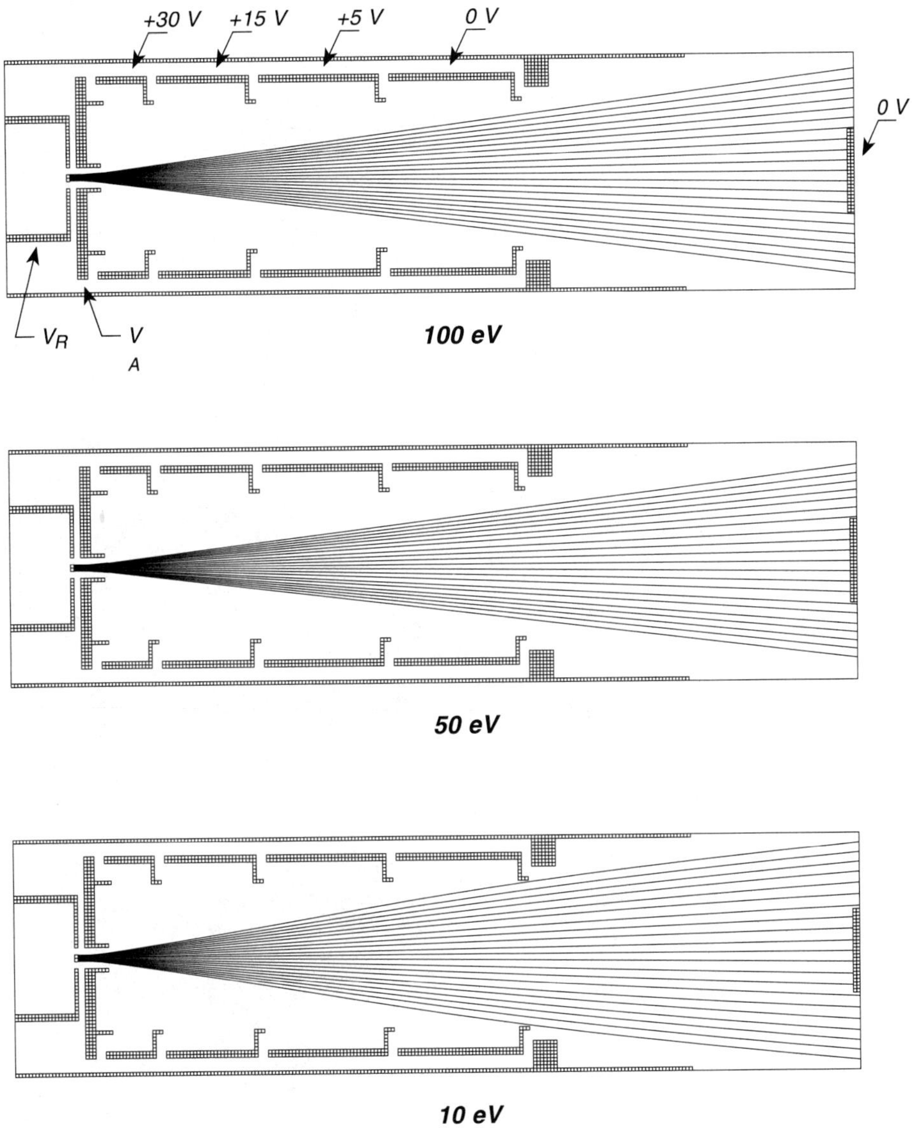


***B. Beam Profiles for Different Electron Energies***

**Fig. 86. Low-Energy Electron Gun—Broad Beam**

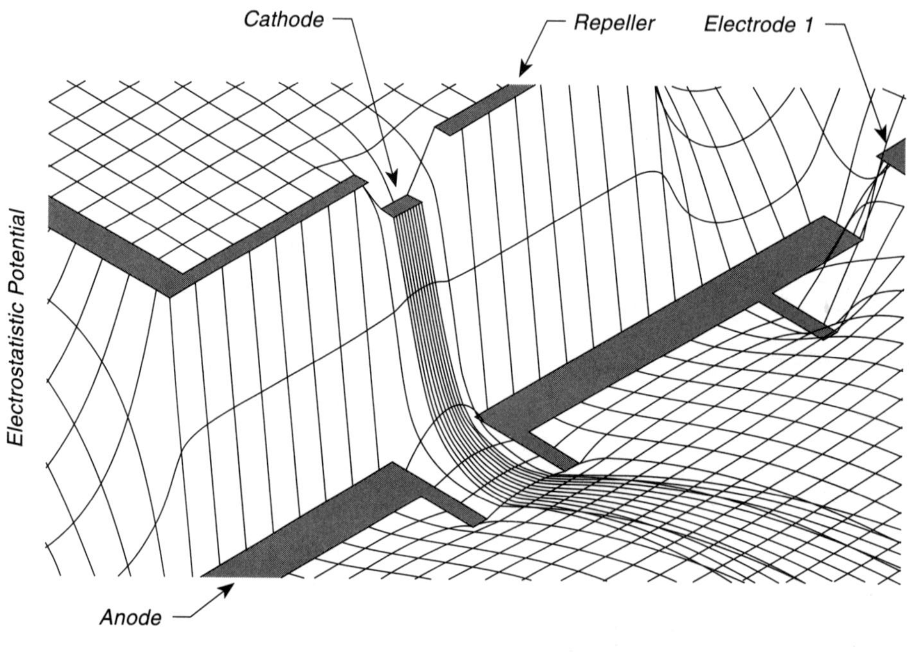


**C. Shape of the Potential Near the Cathode**

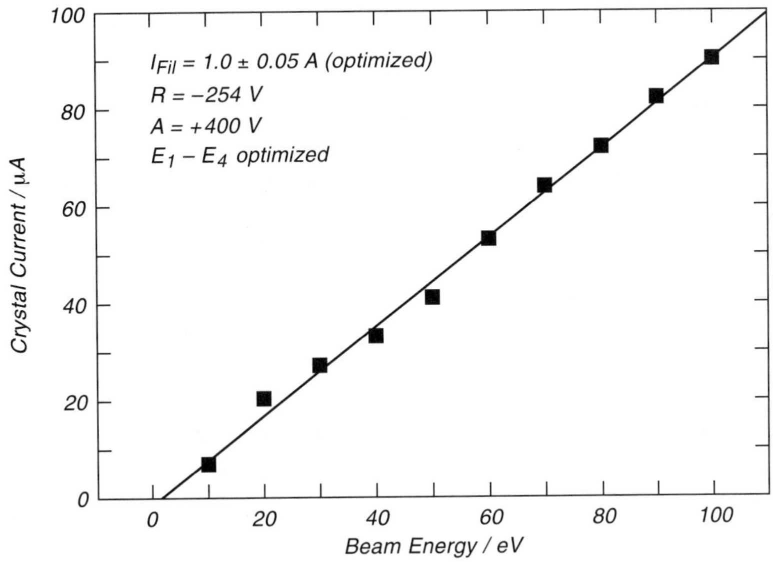


**D. Emission Current from Low-Energy Electron Gun**

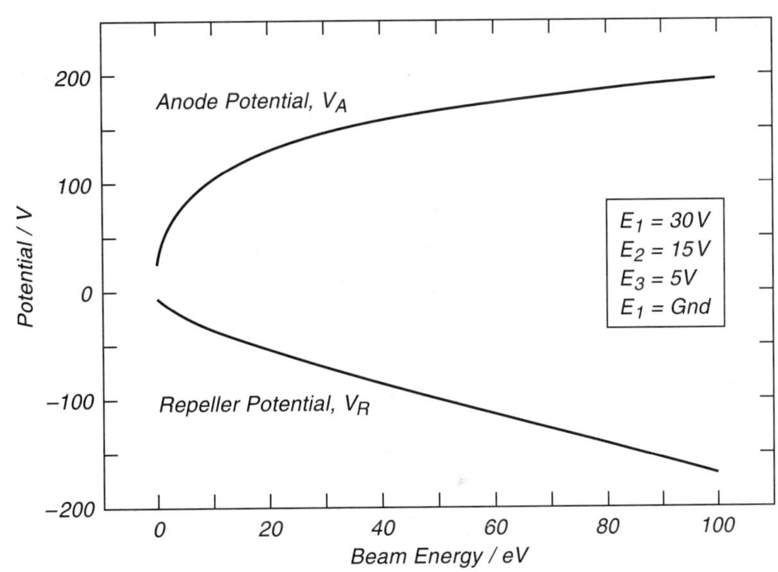


**E. Potential Settings for Broad Beam Operation and Constant Electrode Potentials**

# 87

# Electron Energy Analyzers

It is not the purpose of this book to discuss the details of the various forms of electron energy analyzers that are used for the analysis of surface phenomena. Instead, a practical guide to the selection, construction, and use of electron energy analyzers is available [1]. In addition, an extensive bibliography dealing with the various design analyzers is also presented in reference [1].

Six types of electron energy analyzers are described in reference [1], and these constitute the most widely used instruments. The analyzers are briefly categorized below.

## Figure 87A—Time-of-Flight Analyzers

These analyzers are used primarily for ion detection owing to the extremely high velocity of electrons. However, time-of-flight methods have also been used for rapid pulsed sources of slow electrons. Electrons are allowed to drift in a field-free region, and their arrival at a microchannel plate detector is recorded by a suitable waveform digitizer. Microchannel plate systems can provide gains of $10^6$–$10^7$ and pulse widths of 1 ns. Drift-tube lengths of a few meters are employed, and in such long tubes magnetic field effects may be significant.

## Figure 87B—Retarding Grid Analyzers

Here, the source is located at the center of two hemispherical grids and a spherical collector. The source and the inner grid are grounded to make a field-free intervening region. The outer grid, biased at $-V_R$, retards the electrons, and the collector is biased positive relative to ground to collect all electrons that pass the retarding grid. The energy resolution is estimated from the assumption that the source is spherical, of radius $r = a$, and that the radius of the collector is $r = b$. A schematic of the transmission of the device versus kinetic energy is shown when the retarding grid is set at $V_R$. The energy resolution, $\Delta E$, at $V_R$ is given by the equation in Figure 87B.

## Figure 87C—Tandem Reflection-Transmission Analyzer

This type of analyzer allows bandpass filtering of the energy distribution of electrons emitted from the source. The first grid system is a low-pass filter. Electrons having energies greater than $eV_{REF}$ can overcome the retarding field and are collected on the ellipsoidal collector. Lower-energy electrons are reflected and focused through a grounded aperture, and thence to a retarding grid analyzer something like that shown in Figure B. The two-dimensional image detector makes possible the simultaneous detection of the energy and the emission angle of the electrons received in the band pass of the two analyzers, as shown in the transmission diagram for the tandem analyzers. Analyzers of 100-meV resolution and large angular acceptance have been constructed.

## Figure 87D—Plane Mirror Deflection Analyzer

This represents the most elementary type of the deflection analyzers. Electrons, constrained to a certain angular range by the apertures, enter from the left into a region of width = $h$ in which field-free conditions exist. From there, they enter a second region of width = $d$, where a constant electric field exists, which field is produced by making the upper plate negative. Here, the electrons travel in parabolic paths in the uniform field. Electrons having a specific energy arrive at the exit slits on the right and are measured below the lower grounded plate.

## Figure 87E—Cylindrical Mirror Deflection Analyzer

Here principles similar to those used in the plane mirror analyzer apply. However, because of the cylindrical symmetry, a wide solid angle of acceptance exists, leading to high throughput and sensitivity. The cylindrical geometry is also convenient for the placement of the electron gun on axis, and has made this geometry popular for Auger electron spectroscopy (AES).

## Figure 87F—Cylindrical Deflector Analyzer

This type of analyzer is widely used in high-resolution electron energy loss spectroscopy (HREELS). The optimum angle between the entrance and exit slits is $\theta = 127°$. Analyzers of this type are particularly useful for measuring the angular distribution of electrons emitted from a surface.

## Figure 87G—Spherical Deflection Analyzer

This type of analyzer is widely used in photoelectron spectroscopy. The optimum angle between the entrance and exit slits of the analyzer is $\theta = 180°$.

Improvements in the resolution and throughput of these analyzers can be achieved by preretarding the incident electrons prior to entry into the analyzer [1].

### *References*

[1] J.L. Erskine, in *Experimental Methods in the Physical Sciences,* Vol. 29A (Academic Press, Inc., 1995), p. 209.

**Fig. 87. Electron Energy Analyzers**

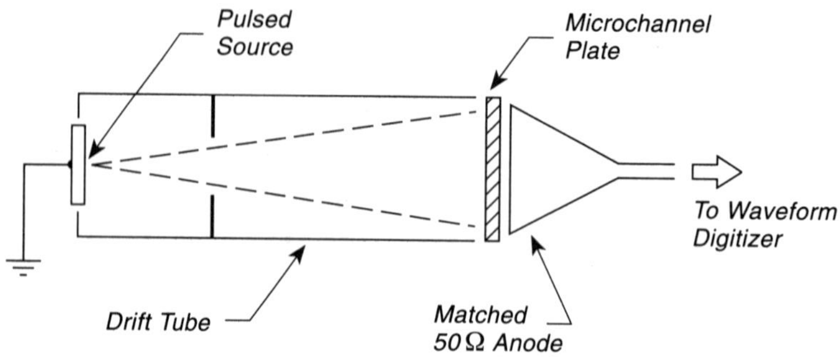


**A. Time-of-Flight Analyzer**

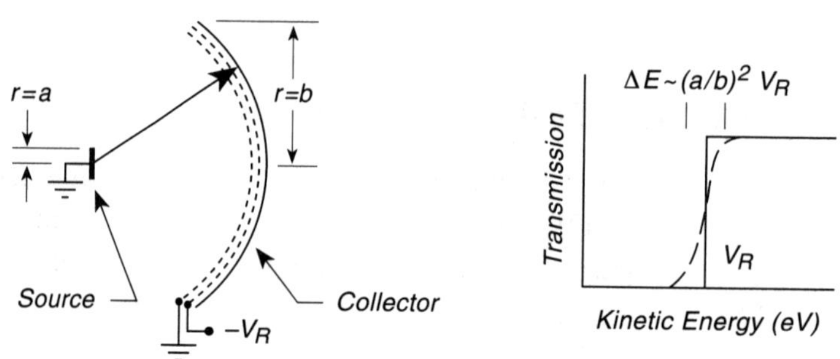


**B. Retarding Grid Analyzer**

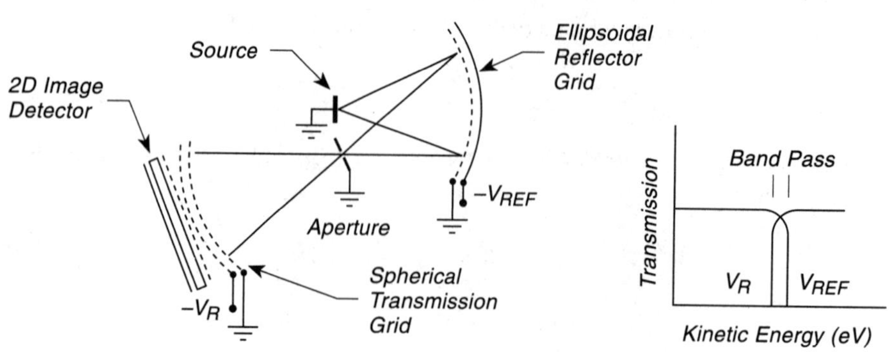


**C. Tandem Reflection-Transmission Analyzer**

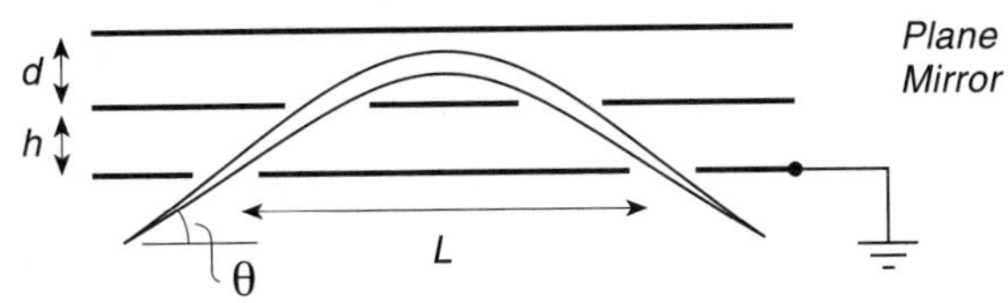


***D. Plane Mirror Deflection Analyzer***

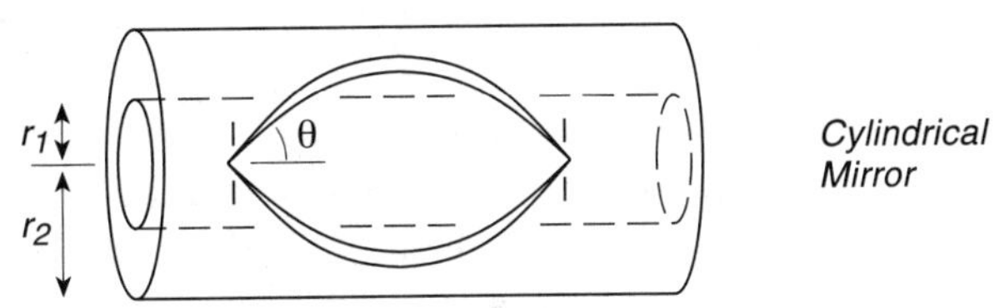


***E. Cylindrical Mirror Deflection Analyzer***

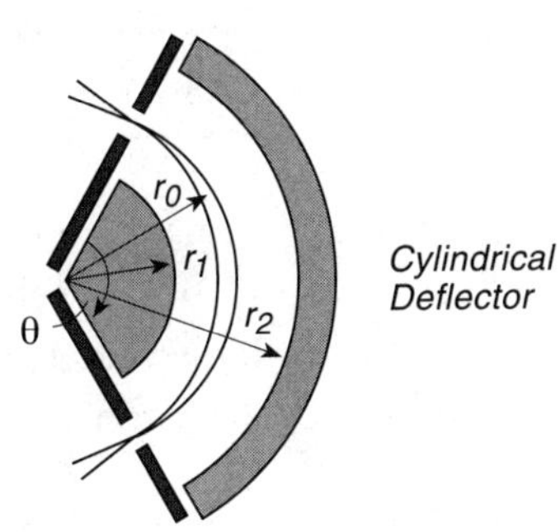


***F. Cylindrical Deflector Analyzer***

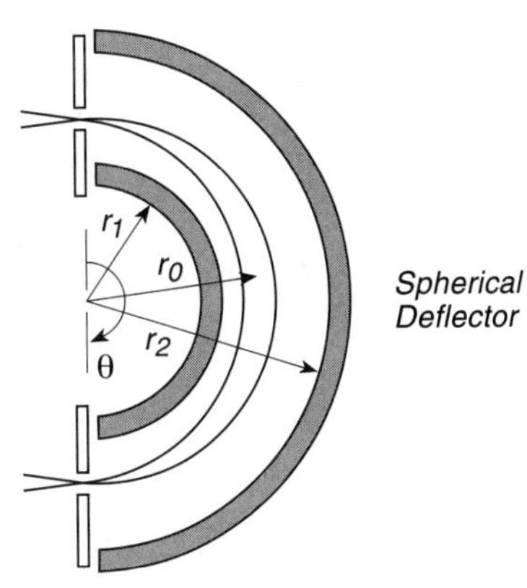


***G. Spherical Deflector Analyzer***

# 88

# Electron Spectrometer Calibration Using Graphite

The accurate location of the Fermi level ($E_F$, sample) is often an important requirement for the calibration of electron spectrometers. Often in ultraviolet photoelectron spectroscopy (UPS) the Fermi level can be determined by measuring the Fermi edge feature of a clean metal surface. However, if UPS is not available, the precise calibration of the Fermi edge position often cannot be carried out. A method to do this, using the secondary electron emission from electron impact excitation, X-ray photons, or UV photons on a graphite reference sample is described below [1].

Figure 88 shows the secondary electron emission region of a sample of graphite (highly oriented pyrolytic graphite, or HOPG) and a sample of Aquadag (colloidal graphite) [2] that have been excited by He I (21.22 eV) radiation. In addition, the upper energy photoelectrons in the vicinity of the Fermi edge are also shown from a cleaned Ta sample in electrical contact with the graphite samples. The fact that the upper energy measured is at 21.22 eV indicates that the spectrometer energy scale is properly calibrated in this measurement. It can be seen in the secondary electron emission region that there is a distinct emission feature centered above the Fermi level at $7.57 \pm 0.03$ eV and superimposed on the broad secondary emission tail. This feature is related to electronic excitation of the $\sigma$ state present in the conduction band of the graphite. This state arises from the $\sigma$ electrons in the C—C bonds in the graphite lattice and has been well studied previously [3,4]. The energy of the $\sigma$ peak from graphite is independent of the energy of the exciting radiation or electrons and hence can be used to locate the Fermi level accurately when it is otherwise not accessible.

The behavior of the $\sigma$ state is independent of the annealing temperature of the Aquadag over the range 230–750°C and is independent of air exposure. Also, the energy is invariant with the angle of emission. From the low-energy cutoff in the secondary electron spectrum, the sample or electron analyzer work function can be determined [1].

## *References*


[1] P. Oelhafen and J. L. Freeouf, J. Vac. Sci. Technol. **A1,** 96 (1983).

[2] Aquadag may be obtained from Acheson Colloids, 1607 Washington Avenue, P.O. Drawer 611747, Port Huron, MI 48061-1747. In applications of this type, it may be painted onto the sample to yield the desired reference energy above the Fermi level.

[3] J. Krieg, P. Oelhafen and H.-J. Guntherodt, Solid State Comm. **42,** 831 (1982).

[4] R. F. Willis, B. Feuerbacher, and B. Fitton, Phys. Rev. **B4,** 2441 (1971).

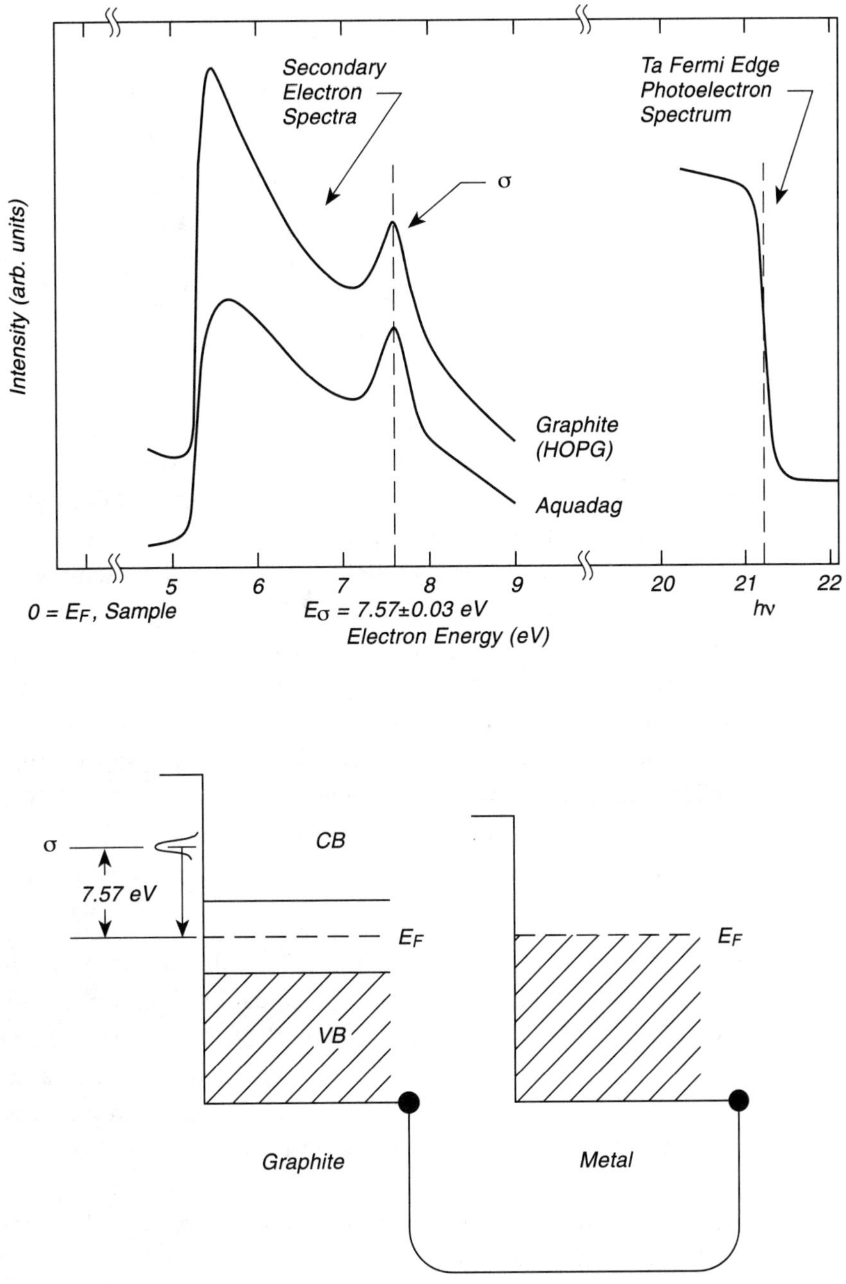


**Fig. 88. Electron Spectrometer Calibration**

# 89

# Measurement of Electron Beam Angular Divergence

The response of an instrument measuring an electron diffraction beam is determined by many factors. Included are the source extension (or the angular divergence of the electron beam), the beam width, and the electron energy distribution. The most difficult factor to measure is the angular divergence of the beam, and this factor has generally been measured by the elimination of other parameters.

The method to be described can measure both the beam width and the angular divergence of the beam. Figure 89 shows a schematic view of the apparatus [1]. A stationary electron gun is directed toward a set of three Faraday cups that are translated by an electrically driven bellows. The Faraday cup detectors are located at different distances along the electron beam, and each 2-mm-diameter hole is half-covered with a razor blade acting as a knife edge. The current to each Faraday cup is measured as the assembly is translated across the beam. Thus, the shape of the electron beam is measured at three *z*-positions, and this permits the divergence to be obtained by comparing the full width at half maximum (FWHM) at each of the three cup positions. Placing one of the Faraday cups at the distance of the sample in normal use allows the beam character to be measured at the sample position. The loss of current through secondary electron emission is minimized by using Faraday cups that are at least three times as deep as their 2 mm apertures, and losses of a few percent are found in the 50–600-eV region [2]. The cups are not biased, and coaxial cables are used inside and outside the vacuum system to minimize noise.

An AC method may be used to make measurements of the first derivative of the collected current versus translation position. A small sinusoidal modulation is applied to one pair of deflection plates to provide a slight motion of the beam perpendicular to the knife edge on the collector. A lock-in amplifier measures the derivative signal as a function of translation position (measured electrically) and an accurate beam profile is plotted out with improved stability and noise compared to using the DC method. No mechanical instability was observed in measuring 50–600-eV beams having a typical FWHM of 0.1 mm. A beam divergence of less than 0.1° may be measured, and typical electron guns exhibit divergences of about 0.2°.

The application of information obtained by the above method to actual LEED instruments is described in an excellent review article [3].

For the approximate measurement of electron and ion beam (>400 eV) dimensions, a phosphor screen may be used. A nice method to measure the beam dimensions involves the use of a small stainless steel plate with a precise grid overlay attached by welding. A grid that has 0.2 x 0.2 $mm^2$ openings, photolithographically etched, is ideal. A grid thickness of 0.025 mm is appropriate [4]. The phosphor, mixed with ethanol, is applied to the screen, and excess phosphor is removed from the outer surface of the grid overlay. When dry, the phosphor will adhere only inside the grid spacings, and the known dimensions of the grid provide a direct calibration scale for the electron beam [5].

## *References*

[1] J.A. Martin and M.G. Lagally, J. Vac. Sci. and Technol. **18,** 58 (1981).

[2] M.G. Lagally, Ph.D. Dissertation, University of Wisconsin, 1968.

[3] M.G. Lagally and J.A. Martin, Rev. Sci. Instrum. **54,** 1273 (1983).

[4] Photolithograph W grid is available from Buckbee Mears Corp., 245 East 6th Street, St. Paul, MN 55101-1959.

[5] Dr. R. David, Institut fur Grenzflächenforschung und Vakuum-physik, Forschungszentrum Jülich GmbH, Postfach 1913, D-5170 Jülich, Germany (private communication).

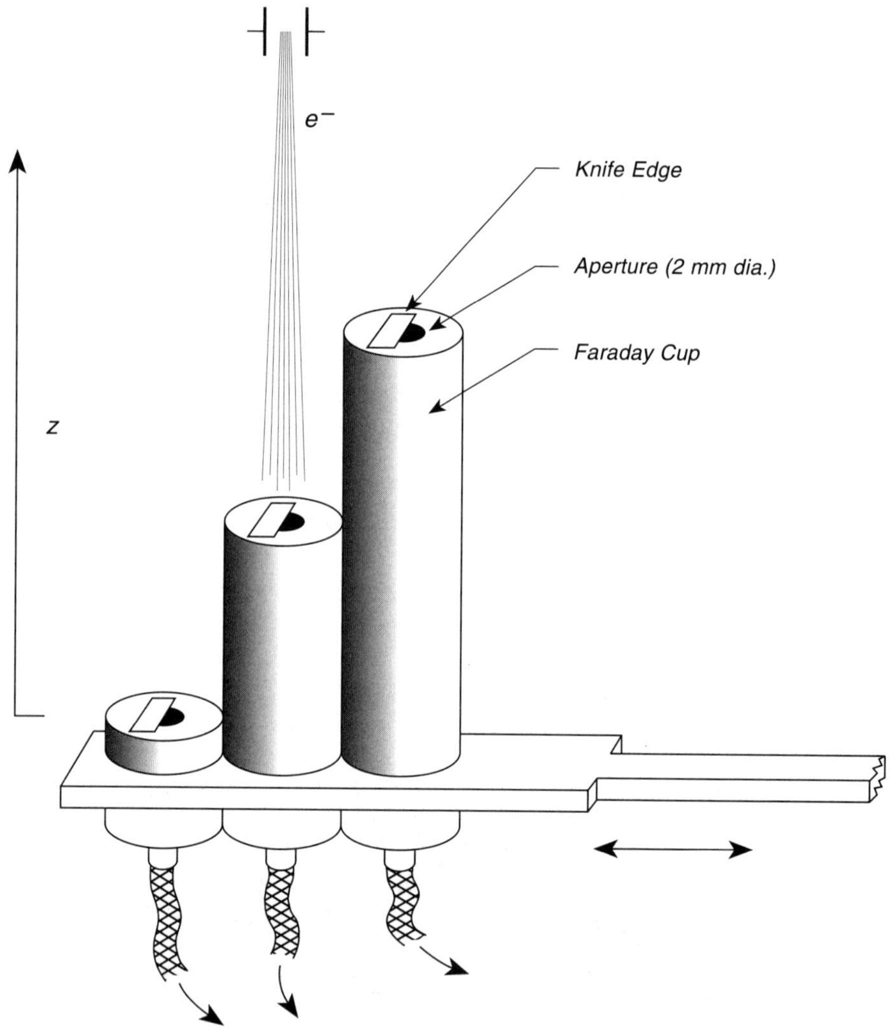


**Fig. 89. Measurement of Electron Beam Divergence**

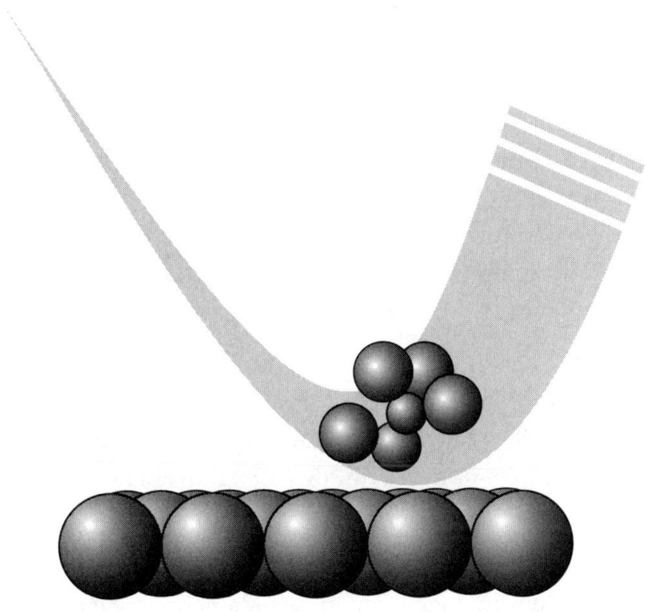

# 90

# Spot Photometer for LEED Intensity Measurements

The measurement of LEED beam intensities using a spot photometer may be carried out in the absence of a Faraday cup or electronic image-enhancing display method. The use of a uniform phosphor screen makes these measurements satisfactory for many purposes.

The construction of a spot photometer is shown in Figure 90. The detector is equipped for synchronous detection, eliminating interference from background light. A lens focuses the light from a LEED beam on a photodiode that is mounted behind an aperture placed in a thin reflective metal screen. The LEED pattern can be observed visually through a port using the reflective screen and the spot photometer is moved laterally for measurement of individual beams. A photodiode mounted behind the aperture integrates the light received from a selected beam plus the surrounding background. The diffraction pattern is modulated at 100 Hz by applying a chopping voltage to the suppressor grid of the order of $V_e + \delta$, which is sufficient to suppress the LEED pattern [1].

At a spacing of 30 cm from the LEED screen, the photodiode sees a circular area of about 5.4-mm diameter, about five times the average diameter of a diffraction spot in the apparatus. The photodiode is a low-noise planar-diffused silicon sensor photodiode [2]. The reader may see examples of the use of this apparatus in references [3,4].

Newer methods involving vidicon detectors are widely employed for the study of LEED patterns and intensities on phosphor screens. It is possible to employ a bandpass optical filter to remove stray light from high temperature filaments and emission from a hot sample. For a P11 phosphor, with a narrow band of emission at 459 nm, an interference filter with a central wavelength of 460 nm and a bandwidth of 50 nm is excellent for removal of the stray light, permitting more accurate measurement of LEED intensities on a background that is closely similar to the dark current of the vidicon [5].

## *References*


[1] A.G. Schrott, M.D. Chinn, C.G. Shaw, and S.C. Fain, Jr., J. Vac. Sci. Technol. **21,** 101 (1982).
[2] PIN-5DP, United Detector, 12525 Chadron Avenue, Hawthorne, CA 90250.
[3] C.G. Shaw, S.C. Fain, Jr., M.D. Chinn, and M.F. Toney, Surf. Sci. **97,** 128 (1980).
[4] R.D. Diehl, M.F. Toney, and S.C, Fain, Jr., Phys. Rev. Lett. **48,** 177 (1982).
[5] R. J. Phaneuf and H.-C. Kan, Rev. Sci. Instrum. **65,** 3871 (1994).

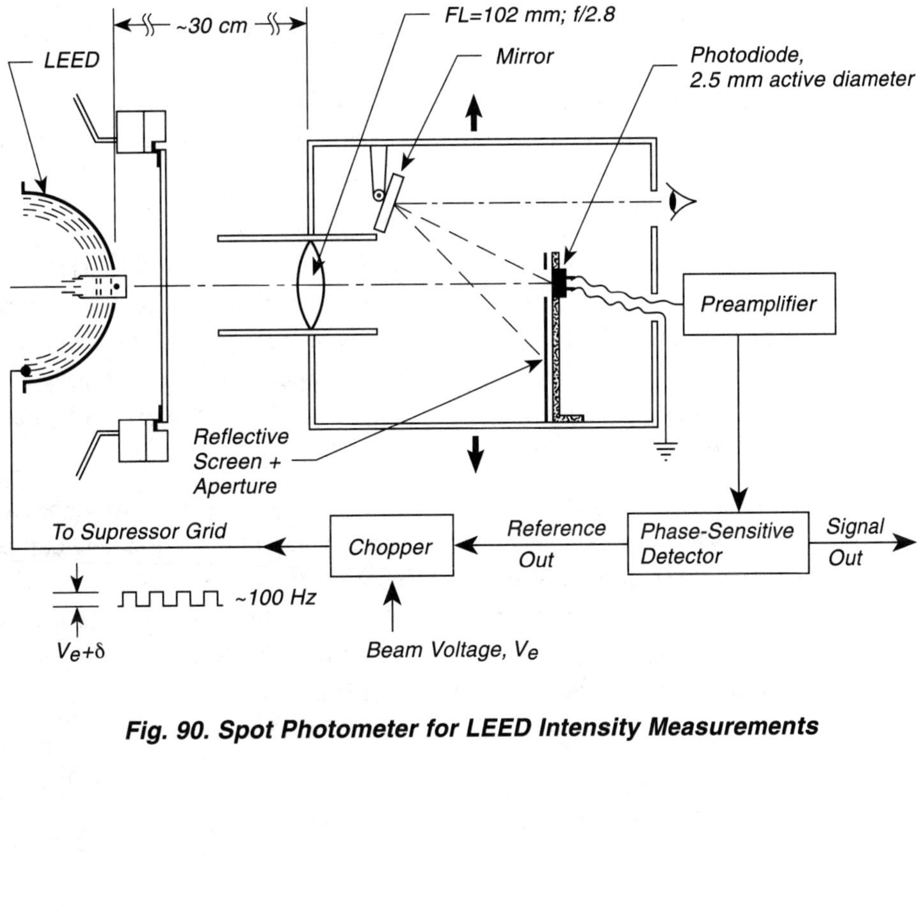


***Fig. 90. Spot Photometer for LEED Intensity Measurements***

# 91

# Modified Faraday Cup for Electron Current Measurement

Faraday cups are employed for the collection of electrons in collimated beams. One problem is that specularly reflected electrons from the bottom of the Faraday cup may be reemitted, causing the measured collected current to be diminished. Various methods for the suppression of specularly reflected and secondary electron emission are employed, and one principle employed is to make the Faraday cup depth a number of times the aperture width. But even this design criterion does not prevent the reflection of specular electrons from the bottom of the cup. Surfaces are known to have high reflectivities for low-energy electrons of a few eV energy.

To avoid this problem, a stainless steel cone was inserted in the cup [1], as shown in Figure 91A. The specularly reflected electrons from this conical surface maintain their forward component of velocity along the optical axis of the cup. These electrons have no velocity component along the optical axis toward the aperture. Thus this design is the electron optical equivalent of a light trap. The design invalidates the principle that the depth of the cup should be large compared to the aperture.

The energy distribution of the collected electrons ($10^{-11}$ A) versus retarding potential is shown in Figures 91B,C for the unmodified Faraday cup and the Faraday cup containing the inner conical surface. It is seen that the ratio $I/I'$ can be increased by a factor of 2 from 2.5 to 5 when the cup contains the conical insert. At the same time, $\Delta E$ increased from 1.2 eV to 1.6 eV. No explanation for the existence of the plateau extending into the loss region by about 8 eV was presented in the paper [1].

## *References*

[1] W.R. Miller, Jr. and N.N. Axelrod, Rev. Sci. Instrum. 37, 1096 (1966).

## Fig. 91. Modified Faraday Cup

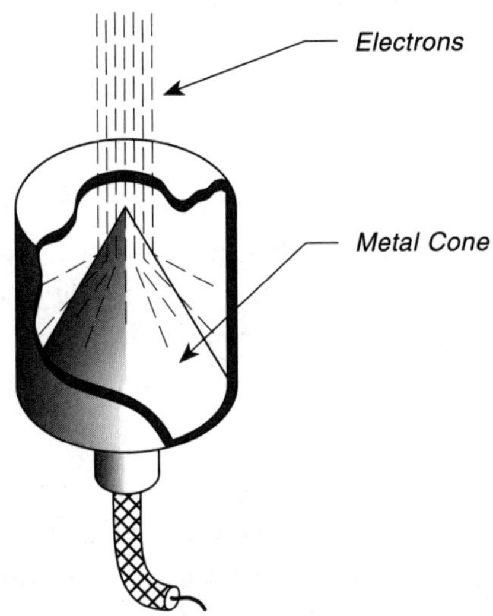


A. Faraday Cup with Internal Cone

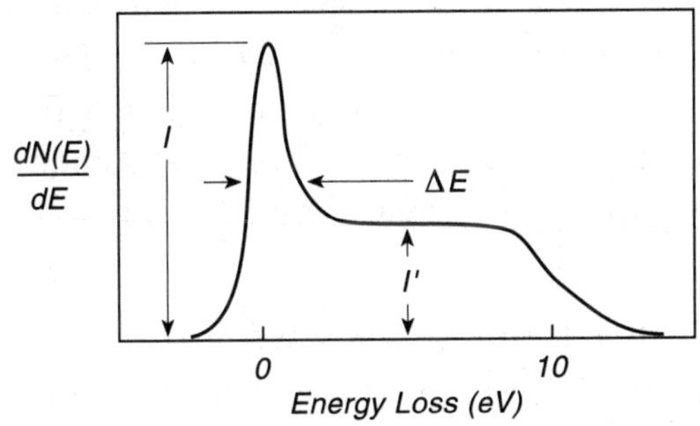


B. Electron Energy Distribution—No Cone

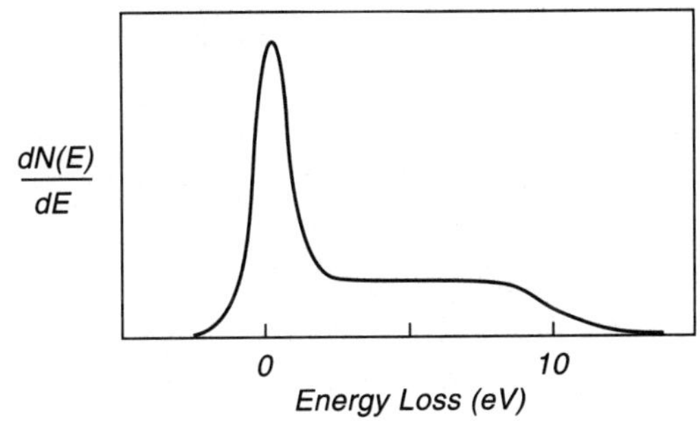


C. Electron Energy Distribution—With Cone

# 92

# Rejuvenation of Cu/Be Electron Multipliers

The deposition of thin films onto electron multipler surfaces occurs after long use, particularly in vacuum systems having a background of oil vapors or in systems that have been used for reactive gases, such as fluorine [1]. A simple rejuvention method has been reported for Cu/Be dynodes exposed to fluorine to the extent that the multiplier gain had dropped to 1/300 of its original gain. Complete rejuvenation back to the original gain was observed [1].

For the particular case here, it was decided to expose the dynodes to a plasma discharge in hydrogen to remove the fluorine, and then to expose them to a plasma discharge in oxygen to reoxidize the dynode surfaces. This was done, of course, by removing the multiplier and treating it in a separate metal vacuum system. In Figure 92, a sequence of rejuvenation steps is carried out, in which neighboring dynodes are used as anode/cathode combinations, working up the multiplier chain. For hydrogen plasma generation, a pressure of about 1 Torr of $H_2$ was used, under flow conditions, with a 350 V DC power supply connected through 5000 Ω between adjacent dynodes. A Tesla coil was used to initiate the plasma, and the pressure was increased in the range 1–5 Torr until about 30 mA of current passed. Hydrogen plasma treatment was carried out for 2 min. After this, for $O_2$ plasma generation, about 10 mA of current was used for 1 min. The multiplier anode was not treated.

A cleaning and oxidation treatment, designed to avoid copper oxide formation on the dynode structures and to place beryllium oxide films on the dynodes, has been described [2]. Here, oxidation in water vapor at pressures up to $5 \times 10^{-2}$ Torr was carried out in cycles in an RF-heated furnace at 625°C. Beryllium oxide exhibits a higher secondary electron yield than copper oxide [2].

## *References*


[1] G.E. Salser, Rev. Sci. Instrum. **37,** 674 (1966).

[2] M.B. Pongratz, Rev. Sci. Instrum. **43,** 1714 (1972).

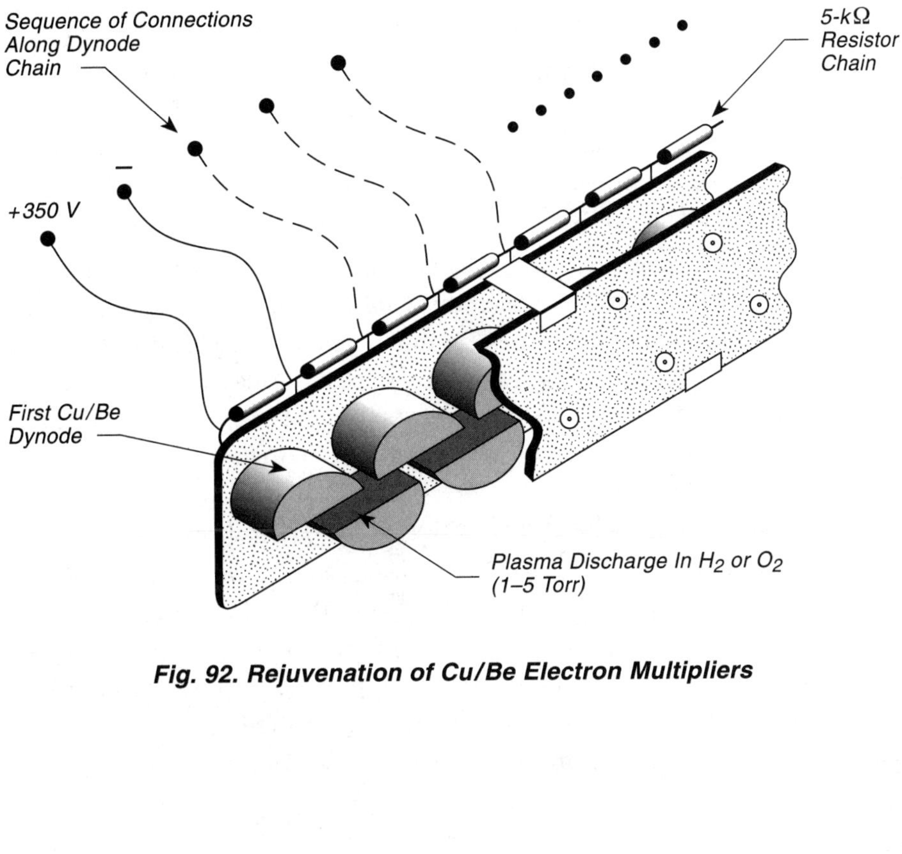


***Fig. 92. Rejuvenation of Cu/Be Electron Multipliers***

# 93

# Geiger-Müller Counter for Inverse Photoemission

Inverse photoemission involves the emission of ultraviolet photons from a surface when the surface is excited by electron impact. This method samples the empty density of states of the surface and near surface regions of a solid. The intensity of photon emission is measured as a function of incident electron energy. A Geiger-Müller counter, with windows passing ultraviolet light, is used for the measurements [1].

Figure 93 shows the counter, which consists of two parts, the housing unit and the insertion unit. The housing unit is mounted on a 2 3/4-inch Conflat flange. A conical ring presses either $CaF_2$ or $SrF_2$ windows [2] onto the housing containing a Viton O-ring. The inner surface of the ring is beveled at 45° to minimize photon emission from this surface by stray electrons. The window has a diameter of 20 mm and a thickness of 2 mm.

The insertion unit is mounted inside the housing unit, and can be removed for bakeout. It too is sealed with a Viton O-ring at the 2 3/4-inch flange. Both the inner wall of the insertion unit and the counting wire (1.5-mm diameter) are polished. The counting wire is electrically isolated from the apparatus by a Macor insulator, which is glued to the outer flange and to the counting wire with vacuum cement [3]. The active length of the counting wire is 40 mm and its end is located 10 mm from the window. The glue joint is made with the counter wire accurately centered in the insertion unit. The port serves to evacuate the counter as well as being an inlet for Ar background gas.

The counter is assembled with the help of a He leak detector to assure tight sealing between the housing unit and the entrance window with minimum force on the front-end bolts. After installation, the assembled unit is pumped overnight, then opened and a few crystals of iodine are dropped in. After pumping again for 30 min, Ar is introduced to a pressure of 100–200 mbar, and the counter is valved off.

Typical operating voltages are 400–500 volts. All pulses should be of identical height (about 0.2 V). Occasionally the counter is pumped again and refilled with Ar. Repeated fillings with Ar may also help if the counter does not work initially. Deterioration of the counter performance (requiring a steady increase in voltage) usually indicates insufficient vacuum conditions.

## *References*


[1] V. Dose, Rev. Sci. Instrum. **39,** 1055 (1968); V. Dose, Th. Fauster, and R. Schneider, Appl. Phys. **A40,** 203 (1986).

[2] $CaF_2$ and $SrF_2$ windows are available from Harshaw Optical/Bicron, 6801 Cochran Road, Solon, OH 44139.

[3] The vacuum cement used is Araldit AT1, Ciba-Geigy GmbH, Oeflinger Str. D-79664 Wehr (Baden), Germany.

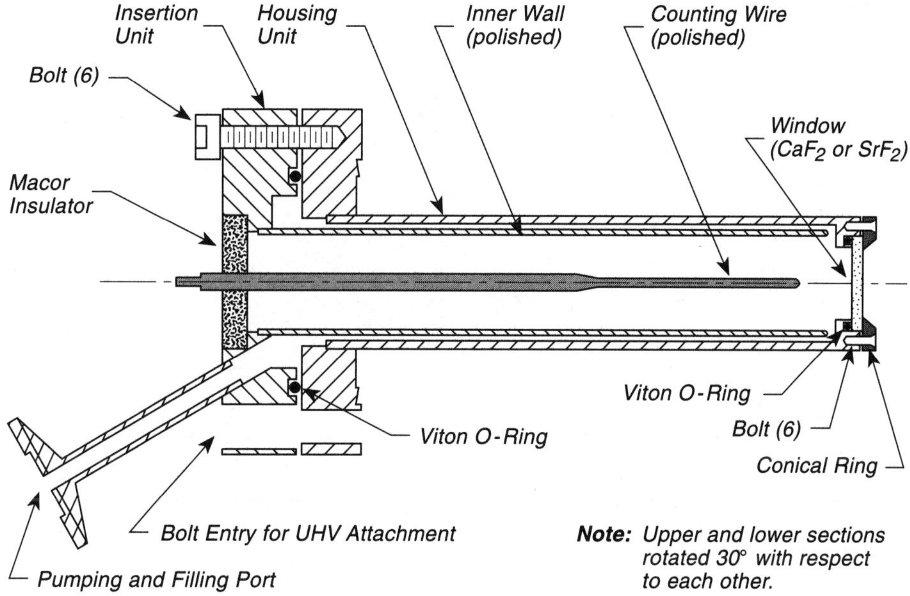


**Fig. 93. Geiger-Müller Counter for Inverse Photoemission**

# B. Ions

# 94

# Ion Sputter Gun with Low Gas Emission

Normal ion guns operate in an ambient pressure of the gas to be used for sputtering. This pressure needed to generate a satisfactory ion bombardment current is normally in the $10^{-6}$ to $10^{-5}$ Torr range [1]. Improved ion gun designs involve a direct feed of the inert gas into the ionization region, and these designs normally involve an ambient pressure in the $10^{-8}$ Torr range. However, any rise in the pressure in the main chamber during the cleaning process is detrimental, since it is known that gas release from the chamber walls, from the ionization pump (if left on) and from the sublimation pump is caused by an increase in the pressure of even an inert gas. This undesirable gas evolution depends upon the previous history of the vacuum system and influences the time required to pump down to the base pressure after ion sputtering.

The design shown in Figure 94A involves direct injection of the sputtering gas into the ion gun and direct pumping from the ion gun into a turbopump. The gas beam inside the ion gun is created by a multicapillary array collimator (see page 604 for gas doser designs employing this collimation principle). The angular distribution of sputter gas molecules exiting from the array may be calculated [2,3]. The sputtering gas is injected into the ion gun through transparent Ta grids that are welded over windows cut into the walls of the outer shield of the ionization gun. A multicapillary array of 13-mm diameter and 3-mm thickness is used, with 10-μm diameter capillaries. Control of the gas pressure is achieved with a leak valve, as shown in Figure 94B. Two gate valves are used to isolate the main chamber from the turbopump after sputtering when an ion pump is used for maintenance of UHV. Normally the 360-L/s turbopump is used to pump the main chamber in parallel to an ion pump and a Ti sublimation pump (not shown).

While an IQ 10/35 ion gun is shown in the example here, many ion gun designs will work with this modification.

## *References*


[1] Data sheet for Ion source IQ 10/35, SPECS, GmbH, 1996.

[2] C.T. Campbell and S.M. Valone, J. Vac. Sci. Technol. **A3**, 408 (1985).

[3] A. Winkler and J.T. Yates, Jr., J. Vac. Sci. Technol. **A6**, 2929 (1989).

[4] Design provided by Dr. J. Ahner and D. Mocuta, Surface Science Center, University of Pittsburgh.

***Fig. 94. Ion Sputter Gun with Low Gas Emission***

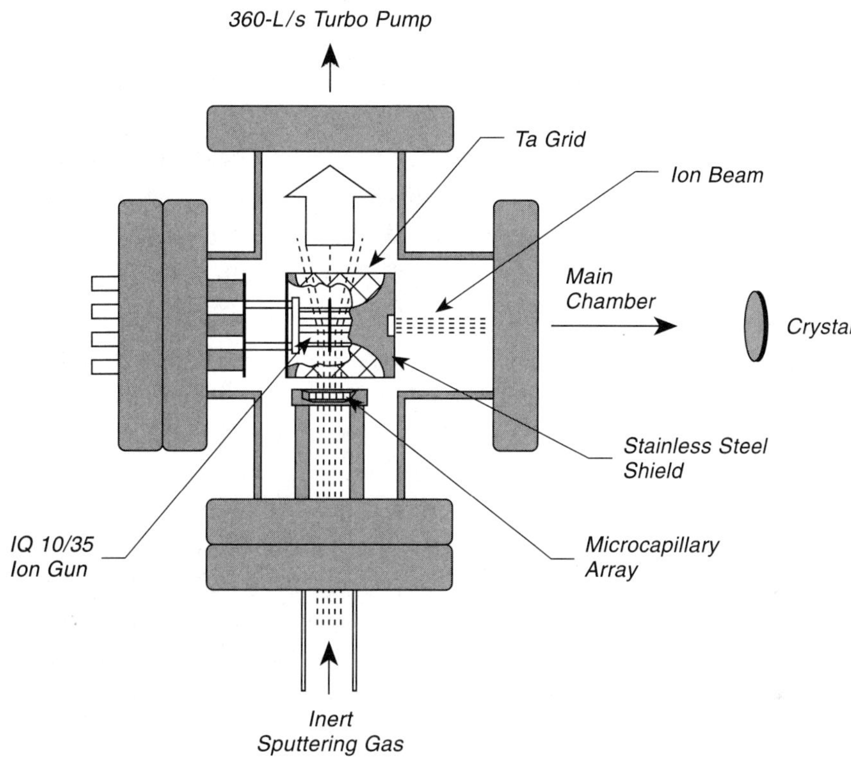


***A. Modified Ion Gun***

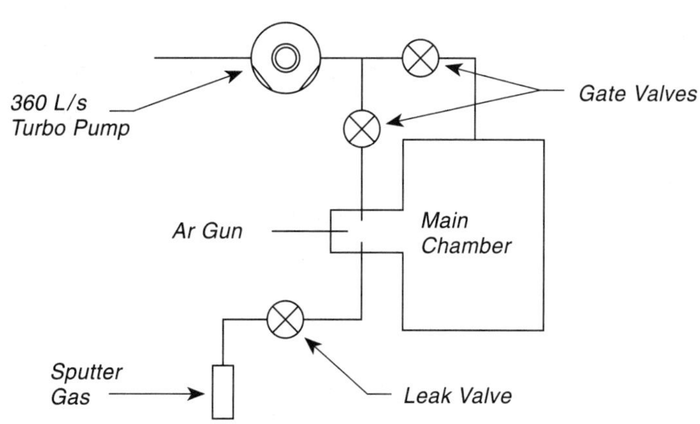


***B. Schematic of Vacuum System***

# 95

# Alternate Ion Bombardment Sources

Ion bombardment of elemental crystal surfaces is widely used for removing both surface and bulk impurities. In the case of bulk impurities, cleaning by sputtering is often accomplished at elevated temperatures, permitting bulk impurities to diffuse to the surface for removal. Normally an ion gun is employed for this and guns giving both focused beams and more diffuse beams are available. Admission of Ar gas directly to the ionization region of the ion gun will reduce the gas load in the vacuum system and hence reduce contamination effects during and after sputtering (see pages 132–140).

Alternately, very simple ion bombardment sources have been reported, and two examples of these are illustrated here. In Figure 95A, an open ion gauge is used as a source of $Ar^+$ ions [1]. The vacuum system is filled with Ar gas to about $10^{-3}$ Torr, permitting ample gas density within the ionization region of the gauge for producing an ion current which is collected at the sample biased at -600 V. It is reported that for an electron emission current of 10 mA, ion currents of 10–15 μA are collected at the crystal, depending on the negative bias. The gauge becomes inaccurate for pressure readings under these conditions. In addition, sputtered metal from the elements of the gauge will soon coat the insulator, which isolates the central collector electrode, causing leakage currents that disable the gauge as a pressure-measuring device. A shield below the grid structure preventing metal deposition on the collector insulator has been used to prevent these effects. The gauge is normally cleaned extensively by electron bombardment of the grid before use as an ion source, keeping the sample positioned elsewhere without its electrical bias.

Ion bombardment sources made without the use of carefully designed ion extraction optics have the capability of delivering impurity metal to the sample from sputtering processes inside and around the gauge components. Therefore, one should be very certain in controlled, and high-sensitivity Auger studies, that these contamination effects do not occur. For the design shown here, with the dimensions and mounting geometry indicated in Figure 95A, Pt and Cu crystals were successfully cleaned and comparisons with a conventional ion gun showed that no differences in the surface preparation could be detected by Auger spectroscopy and other surface characterization methods [1].

Figure 95B shows an even simpler ion bombardment source. A tungsten filament is mounted on two tungsten feedthrough wires between a curved anode and the sample. Electron emission current of the order of 18 μA is collected by the curved anode at a bias potential of +500 V, and at an Ar pressure of about 5 x $10^{-5}$ Torr, an $Ar^+$ current density of about 20 $\mu A/cm^2$ is collected by the sample at ground potential. At 3 x $10^{-4}$ Torr Ar pressure, ion current densities of 50–60 $\mu A/cm^2$ are reported. The ion source should be scrupulously cleaned by heating the emitter for a long time to remove desorbing gases, and also by electron bombarding the curved anode extensively to remove adsorbed gases that might undergo electron stimulated desorption during use of the source for sputtering. Again, with this device, special care should be taken to eliminate the possibility of metal deposition from extraneous sources, although the report [2] indicates that this effect is not observed on a GaAs wafer sample.

A more complex ion gun source made from a Bayard–Alpert gauge is described on page 314.

## *References*

[1] T. McCabe, J. Somers, M.E. Bridge, and D.R. Lloyd, J. Vac. Sci. Technol. **A5**, 119 (1987).

[2] V.M. Bermudez and R.E. Thomas, J. Vac. Sci. Technol. **A1**, 1557 (1983).

Fig. 95. Two Alternate Ion Bombardment Sources

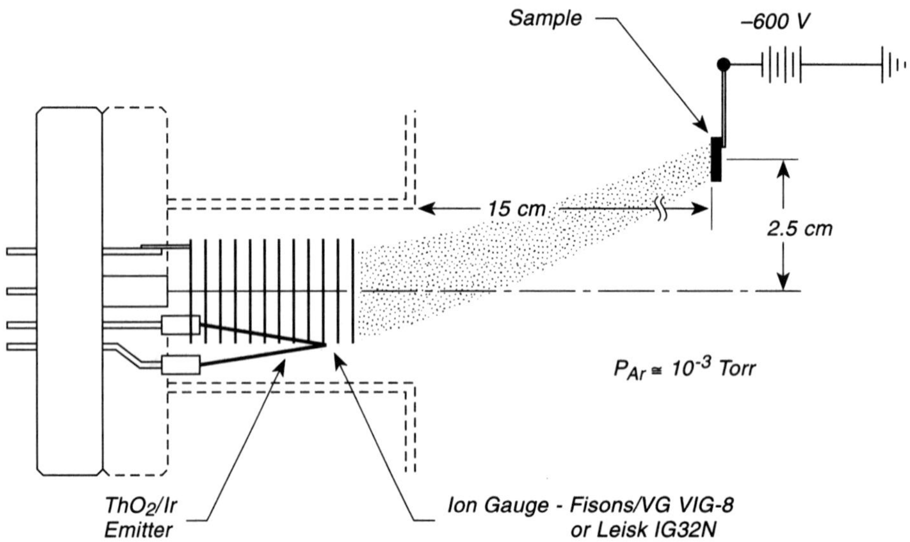


A. Ion Gauge Sputter Source

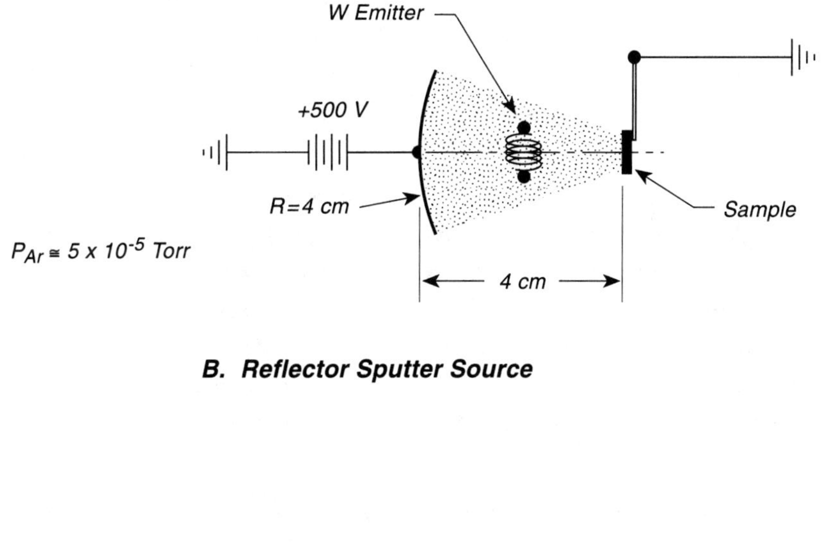


**B. Reflector Sputter Source**

# 96

# Ion Gun Based on Bayard–Alpert Gauge

A more complex ion gun than that described on pages 308 and 310 is presented here [1]. A Bayard-Alpert gauge is modified in two ways: first, by removal of the top part of the cylindrical grid cage for addition of the ion extraction electrodes; and second, by the addition of a metal shield around the gauge body and the thermionic emitter.

Figure 96A shows a schematic of the gun design and the electrical components. A tantalum cylindrical shield to repel electrons is mounted around the gauge body and its electrical potential is maintained at the same potential as the positive end of the thermionic emitter filament. An extractor lens shaped with a waist withdraws ions from the region enclosed by the grid and these pass into an accelerator region, and thence into a cylindrical lens and a quadrupole lens, where the ions are ejected into vacuum and to the target. The flange holding the ion gun is equipped with radial feedthroughs about its outer edge for supplying potentials to the ion gun elements.

The extractor lens accelerates mainly off-axis ions to an on-axis focal point, and its potential can be as high as -200 V. The divergence of the ion beam is controlled by the bore of the accelerator electrode, and the final focusing is done by the lens. Since all potentials in the gun, except for the lenses, are referenced to the primary energy, $U_p$, the extraction of ions is independent of the final beam energy. At low energies, the ions are of high energy when they arrive at the lens, and are decelerated there, while being focused. This leads to high currents at low ion energies. The quadrupole lens consists of four plates at 90° to each other symmetrically placed around the gun axis, but its use is not described in the original article [1].

The spatial distribution of the ions may be adjusted as shown in Figure 96B. Either a sharply peaked ion beam with high current density at the center may be produced, or alternatively, a flat-topped beam profile may be produced by adjusting the acceleration and lens voltages.

In Figure 96C, the current and current density are plotted for a pressure of 1 x $10^{-4}$ Torr Ar pressure. Current densities between 0.1 and 120 μA/cm$^2$ are obtained at various ion energies. The current collected near 1 keV energy is limited by space charge effects in the extractor region. The collector in these cases was located 4 cm from the end of the gun.

Figure 96D shows the energy distribution measured with a retarding field analyzer and showing that at 1.2 keV, an ion beam halfwidth of 1.6 eV may be obtained using an extractor voltage of about -30 V. The ion beam was also mass analyzed, and at an electron energy of 70 eV, a 1:10 ratio of $Ar^{++}$ to $Ar^{+}$ was measured.

An earlier version of this design for an ion gun has been published [2].

## *References*

[1] J. Kirschner, Rev. Sci. Instrum. **57,** 2640 (1986).
[2] M. Khan and J. M. Schroeer, Rev. Sci. Instrum. **42,** 1348 (1971).

## *Fig. 96. Ion Gun Based on Bayard–Alpert Gauge*

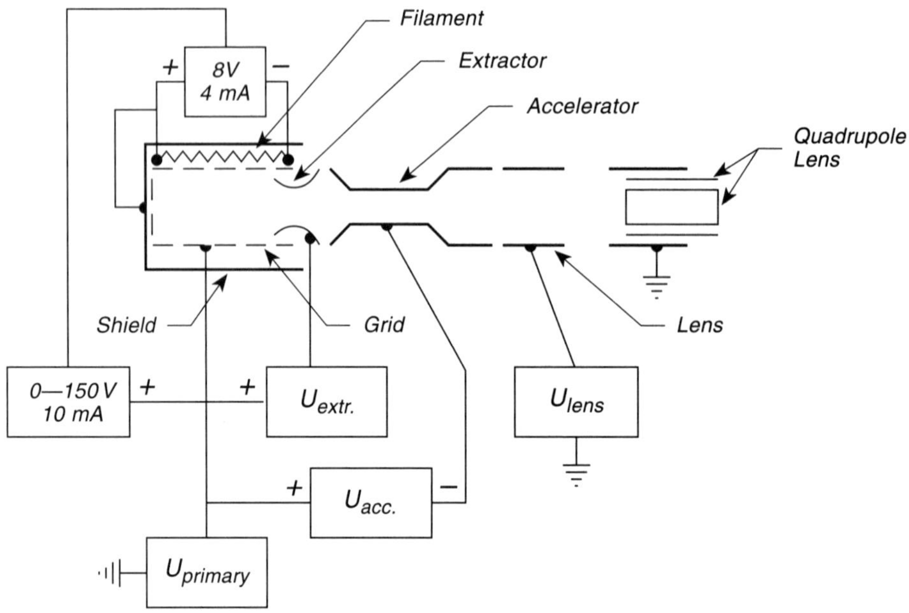


***A. Ion Gun and Circuit***

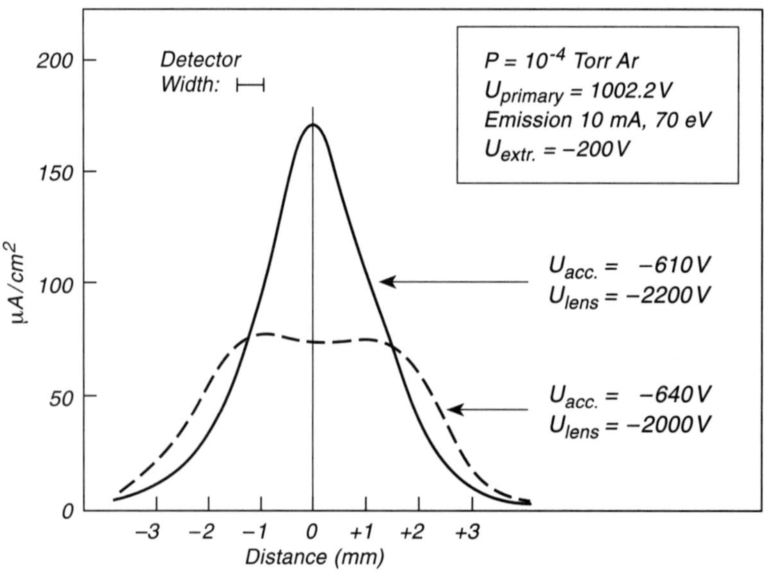


***B. Ion Spatial Distribution***

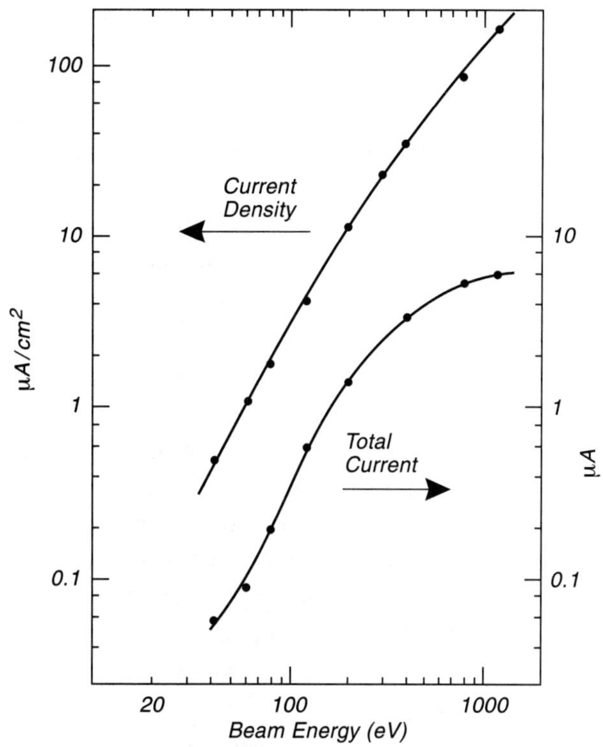


**C. Current and Current Density**

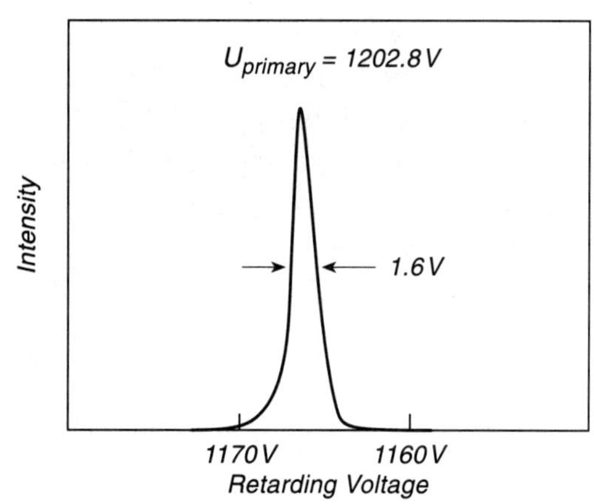


**D. Ion Beam Energy Distribution**

# 97

# Sputter Ion Gun—Cold Cathode

A cold cathode ion gun, useful for sputter deposition, can be made, as shown in Figure 97 [1]. The gun is useful for sputter deposition from any conducting target material. The ion gun was first described in reference [2], and this is a modification of that design. The gun is capable of producing an ion beam of 1-mA current at a kinetic energy of 2–6 keV, impinging on an area 5–10 mm in diameter. There is no filament, and the gun can be operated for hundreds of hours without maintenance. Currents in excess of 1 mA cause tip heating which shortens the tip life. Sputter deposition rates of 1–4 Å/s can be obtained.

The beam gas is admitted from a pumped gas handling system through two ceramic insulator nipples and through a metal frit particle filter into a Conflat flange-mounted electrical feedthrough passing a 1/4-in. tube that is connected via Swagelok connectors to a hollow tube made of 1/16-in.-diameter tungsten or tungsten carbide alloy. This needle is 0.625-in. long and has an orifice of 0.05–0.075-mm diameter. It is called a wire bonding tool [3]. The high electrical field at the tip ionizes the gas as it passes through the tip, resulting in a plasma cone that extends to the target. The diameter of the cone depends on the orifice diameter, supply gas pressure, and target distance. A 2.5 Torr beam of gas in the gun, combined with a tip-target distance of 6 mm and a 0.050-mm-diameter orifice produces a beam crater diameter of 7–10 mm.

A principal difficulty is keeping particles out of the tip. The tip may be unplugged by passing a 0.003-in.-diameter W wire from the back end. This wire may be made from larger stock by electrolytic etching in NaOH (aq) solution. Final cleaning of the tip is done in an ultrasonic bath with acetone.

The beam current and stability depend on the pressure of gas entering the gun. Stable operation is obtained when the chamber pressure is in the $5 \times 10^{-4}$ to $3 \times 10^{-3}$ Torr range. If the supply pressure is too low, arcing will occur in the high-voltage nipples on the gas supply line. If the pressure is too high, or the orifice diameter is too large, a glow discharge will occur in the chamber with little apparent localization. The beam current increases with supply gas pressure up to the point of glow discharge initiation. The high-voltage power supply should be current limited to prevent more than 5–10 mA of current from flowing during such an event.

Another design of a cold cathode sputtering source for metal deposition may be found in reference [4].

## *References*

[1] Dr. A.R. Krauss, Argonne National Laboratory, 9700 South Cass Avenue, Argonne, IL 60439 (private communication).

[2] J.F. Mahoney, J. Perel, and A.T. Forrester, Appl. Phys. Lett. **38,** 320 (1981).

[3] The wire bonding tool is available from Technical Devices, Inc., 5721 Prescott Road, Prescott, AZ 86301.

[4] C. Kim, D.F. Ogeltree, M.B. Salmeron, X. Godechot, G.A. Somorjai, and I.G. Brown, Appl. Surf. Sci. **59,** 261 (1992).

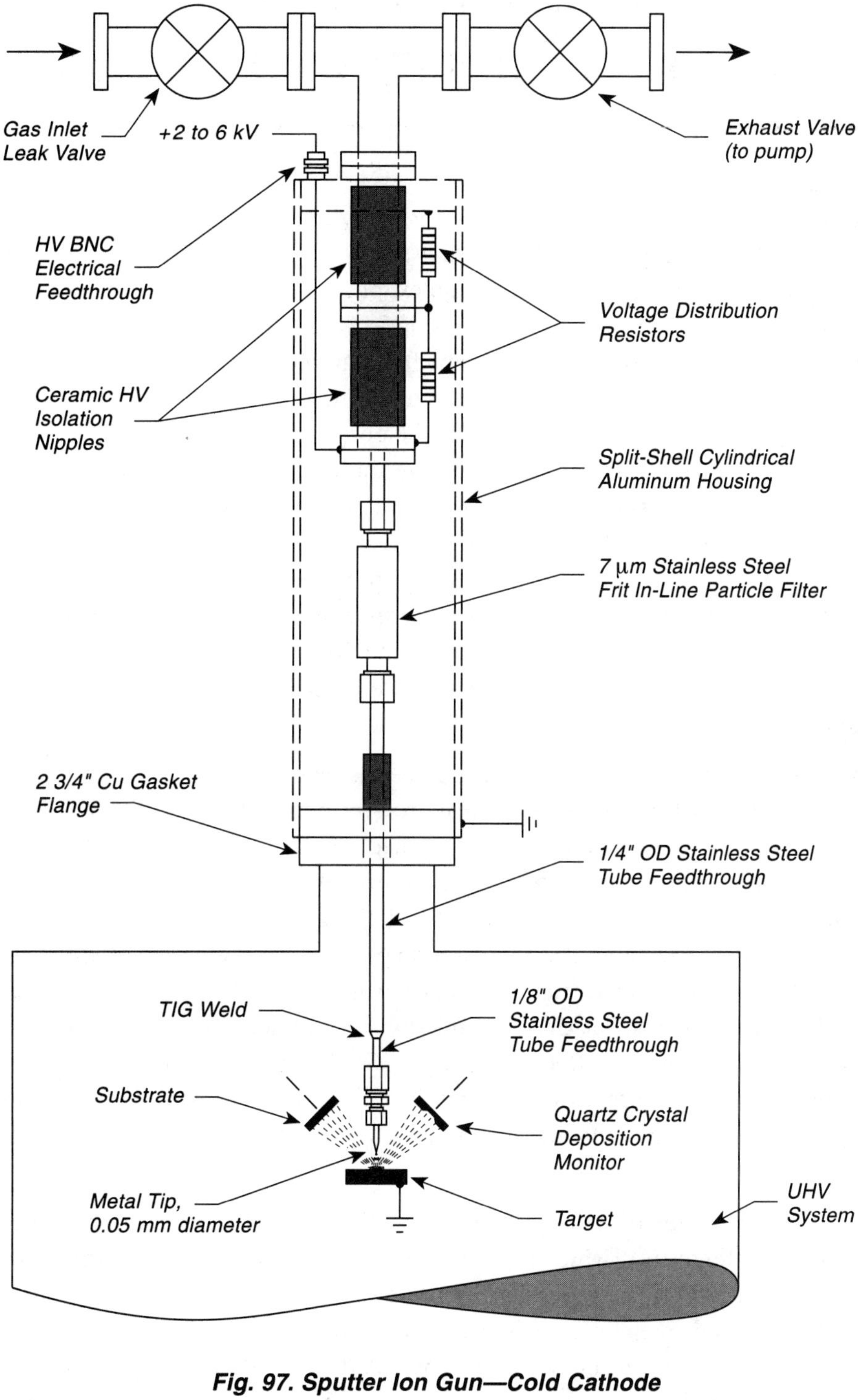


*Fig. 97. Sputter Ion Gun—Cold Cathode*

# 98

# Broad Ion Sources—Improved Stability with Oxygen

Broad-beam ion sources are employed for many purposes including sputter deposition, etching, etc. Figure 98 schematically shows a typical conventional source, which illustrates the basic principles of more complex sources [1]. The cylindrical chamber contains a cylindrical anode and an axially positioned cathode. Gas for ionization is admitted from the rear of the source. A magnetic field, supplied by either an electromagnetic winding or a permanent magnet, is aligned axially, and the magnetic field diverges slightly as one approaches the front of the source. The purpose of the magnetic field is to cause the electrons emitted from the cathode to have a spiral path length sufficiently long such that each will cause a gas phase ionization event before reaching the anode. The source has two grids on its front side—a screen grid and an accelerator grid. The discharge chamber is normally operated under low-pressure conditions where the mean free path of an electron for ionization is much longer than the apparatus dimensions. The plasma potential is usually within a few volts of the anode potential, and the primary electrons emitted from the cathode therefore have energies close to the potential difference between cathode and anode [1].

When Kaufman sources [1] are used with inert gases or reactive gases at low power levels, they operate for long times without failure. However, at high pressures with reactive gases like $O_2$, the ion source lifetime is shortened. This is due to failure of the refractory cathode (such as W or Ta), or to buildup of an insulating oxide layer on the anode. Such an insulating layer on the anode reduces the discharge current until it is no longer possible to support a discharge [2]. This lifetime problem can be reduced significantly by using thorium oxide coated Ir cathodes (see page 202) instead of the more standard bare refractory filaments. For electron emission current densities of 0.2 A/cm$^2$, the thoriated filament lifetime is 1000 h or longer [3]. As the emission current density is increased, the lifetime decreases as the square of the current density. Multiple filaments, used in series, can extend the lifetime by reducing the current density at each filament [2].

To avoid the buildup of an insulating layer on the anode, it may be sputter-coated with a 1-μm-thick film of $IrO_2$, which is a conductive metallic oxide. Because this oxide cannot be oxidized further, operation in an intense plasma will not form an insulating layer, as with many anode metal materials. The source, containing both a thoria-coated Ir filament and an $IrO_2$-coated anode has been operated at 100-W discharge power for about 50 h without degradation of source performance. With an uncoated stainless steel anode, failure occurred in 30 min [2].

Other possibilities exist for using this idea. The anode may be coated with other conductive metal oxides such as those from Ru or Re [4,5]. Alternatively, the anode could be made from the oxide by chemical vapor deposition. Finally, the anode could be made from solid Ir, since an insulating oxide layer will result from its oxidation by the atomic O and other active oxygen species produced in the source [2].

Steinbruchel [6,7] has studied the emission of both neutral and ionized species produced from a Kaufman-type ion source operating with fluoro- and chlorocarbon species, as well as Ar, Ne, $O_2$, $N_2$, and $BCl_3$. This work involved a line-of-sight quadrupole mass spectrometer as the detector.

Ensinger [8] has written an excellent review of a broad range of ion source designs for many purposes.

## *References*


[1] H.R. Kaufman, J.J. Cuomo, and J.M.E. Harper, J. Vac. Sci. Technol. **21,** 725 (1982); J.M.E Harper, J.J. Cuomo, and H.R. Kaufman, J. Vac. Sci. Technol. **21,** 737 (1982).

[2] C.R. Guarnieri, K.V. Ramanathan, D.S. Yee, and J.J. Cuomo, J. Vac. Sci. Technol., **A6,** 2582 (1988).

[3] G.A. Haas, *Thermionic Emission Sources* (AD266039 Electron Tubes Branch, Electronic Division, U.S. Naval Research Laboratory, Washington D.C., 1961), p. 13. (For sale by Superintendent of Documents, U.S. Government).

[4] D.B. Rogers, R.D. Shannon, A.W. Sleight, and J.L. Gillson, Inorg. Chem. **8,** 841 (1969).

[5] L.F. Mattheiss, Phys. Rev. **181,** 987 (1969).

[6] Ch. Steinbruchel, J. Vac. Sci. Technol. **B2,** 38 (1984).

[7] Ch. Steinbruchel, J. Appl. Phys. **59,** 4151 (1986).

[8] W. Ensinger, Rev. Sci. Instrum. **63,** 5217 (1992).

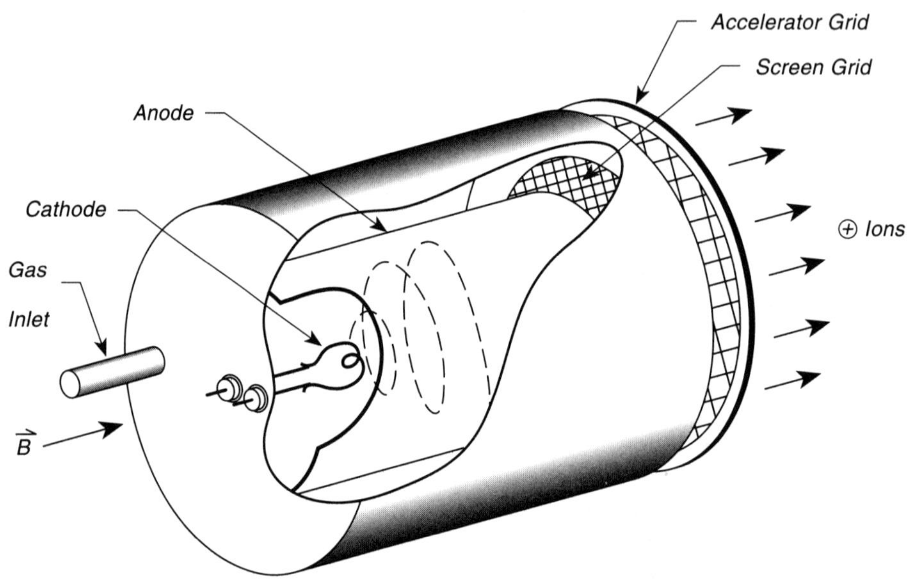


***Fig. 98. Ion Source—Axial Magnetic Field***

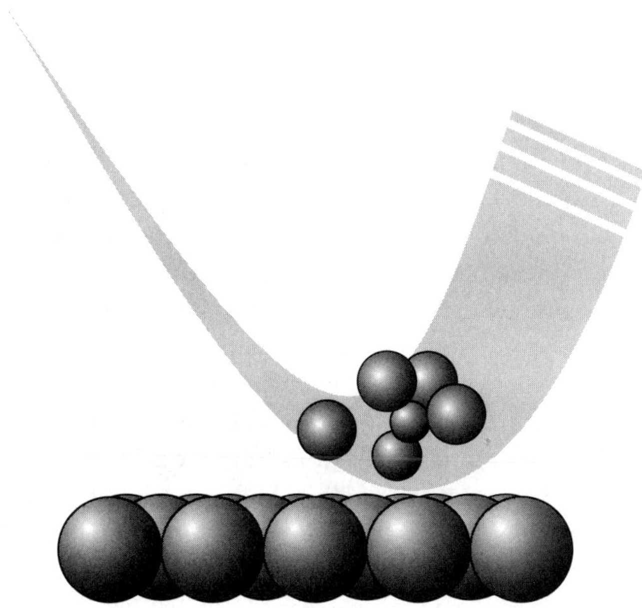

# 99

# Focusing of Low-Energy Ions by Magnetic Field

There is some interest in producing surface chemical reactions with low-energy ions impinging on the surface. In order to do this effectively, ion current densities at very low energies must be maintained at sufficient levels to carry out surface reactions in a reasonable time. A method is described in which the efficiency of collection of a beam of ions with a diameter of about 7 mm is markedly enhanced with an axial magnetic field [1].

The ions are generated by a Colutron GD-2 ion beam source, and ions are produced with an energy spread of as low as ±0.1 eV. The ions are extracted, focused by an Einzel lens, and mass-filtered with a Wien filter. These ions are directed at an aperture plate of 12-mm thickness containing a 7-mm-diameter hole, as shown in Figure 99A. A tungsten grid (90% transparency) is placed across the aperture plate. The gap between the grid and the sample to be bombarded is kept between 2 and 5 cm and the ions are decelerated here. The difference between the beam potential and the positively biased sample defines the ion energy at the sample surface. A 2000 G Co-Sm permanent magnet (4-cm diameter) is placed behind the sample at a distance of about 0.5 cm. This magnet can be electrically biased, and sometimes a magnet potential of +200 V above that of the sample has been found to increase the measured ion current collected by the sample. This effect may be due to the collection of secondary electrons from the grid. The 0.5-cm gap is varied during ion collection to maximize the collected ion current at the sample. The magnet can also be removed when surface analytical methods influenced by external magnetic fields are employed. Obviously the sample must be nonmagnetic.

Figure 99B shows the effect of magnetic focusing on an $Ar^+$ ion beam incident on a Pt sample. A dramatic increase in the collected current below an ion energy of 10 eV is noted, where the initial ion current at 1000 eV is 400 nA. The ion beam energy spread at the sample is about ±0.5 eV. The paper [1] describing this work contains an analysis of the effect of the cyclotron motion of the ions in the magnetic field on their divergence. This method is simple and cost effective compared to electrostatic focusing methods when a small ion beam spot diameter is not a critical concern.

## *References*

[1] D.R. Strongin, J.K. Mowlem, K.G. Lynn, and Y. Kong, Rev. Sci. Instrum. **63,** 175 (1992).

**Fig. 99. Focusing of Low-Energy Ions by Magnetic Field**

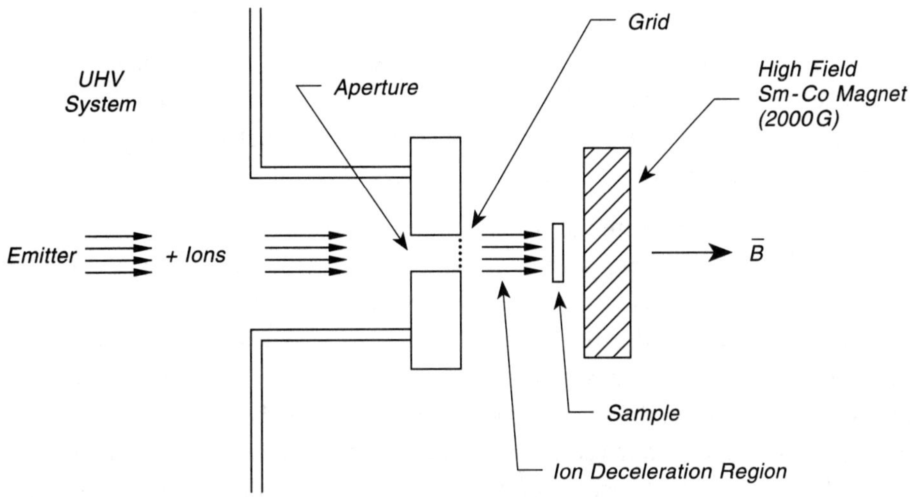


**A. Ion Deceleration in a Magnetic Field**

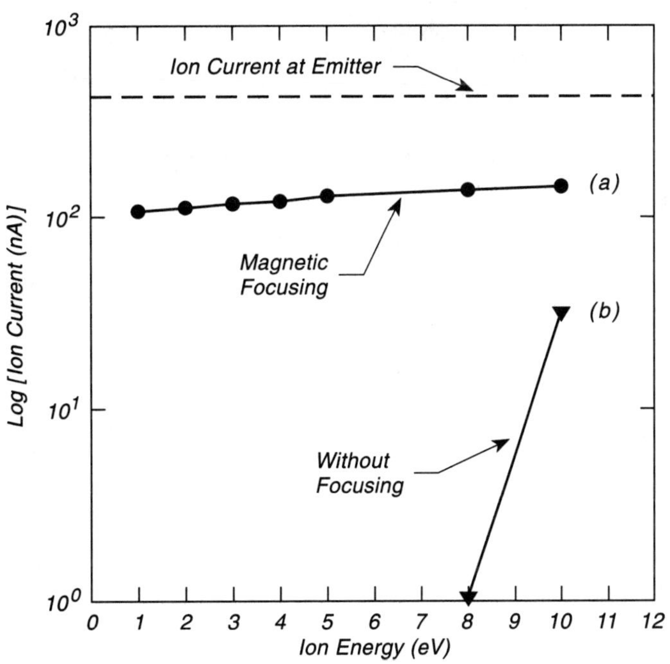


**B. Ion Collection Efficiency**

# 100

# Measuring Ion Beam Dimensions

The size of ion beams used in research can be measured easily [1] by using the fact that ion-irradiated $Gd_2(MoO_4)_3$, called GMO, becomes passivated toward attack by dilute hydrofluoric acid, HF(aq). This passivation occurs for the *c*-surfaces (001 plane) of GMO crystals, when irradiated with 1-2 keV $He^+$, $Ne^+$, and $Ar^+$.

The GMO crystal is polished down to a surface finish achieved with 1-μm alumina abrasive and then may be irradiated with the ion beam as shown in Figure 100. Following irradiation, the crystal is removed and etched with 1% HF(aq). The area exposed to the ion beam is almost unaffected by the HF, whereas the surrounding areas are etched away. The imaging of the ion beam is reported to have 1-μm lateral resolution, and to be much more sensitive than the competitive technique in which a $Ta_2O_5$ anodic film on Ta is bombarded until metallic Ta is exposed. The authors warn that the technique is an integrating technique, and the pattern will therefore reflect any wandering of the ion beam during the irradiation period. The technique is quite useful for grazing incidence ion beams.

Three more accurate methods for measuring ion beam sizes and the divergence of an ion beam are given in reference [2]. Here, a HgCdTe substrate covered with a photoresist is used in one method, or movable collectors are used in the latter two methods.

## *References*

[1] A.S. Bhalla and L.L. Tongson, J. Vac. Sci. Technol. **21,** 1039 (1982).
[2] S. Maniv, Rev. Sci. Instrum. **62,** 1179 (1991).

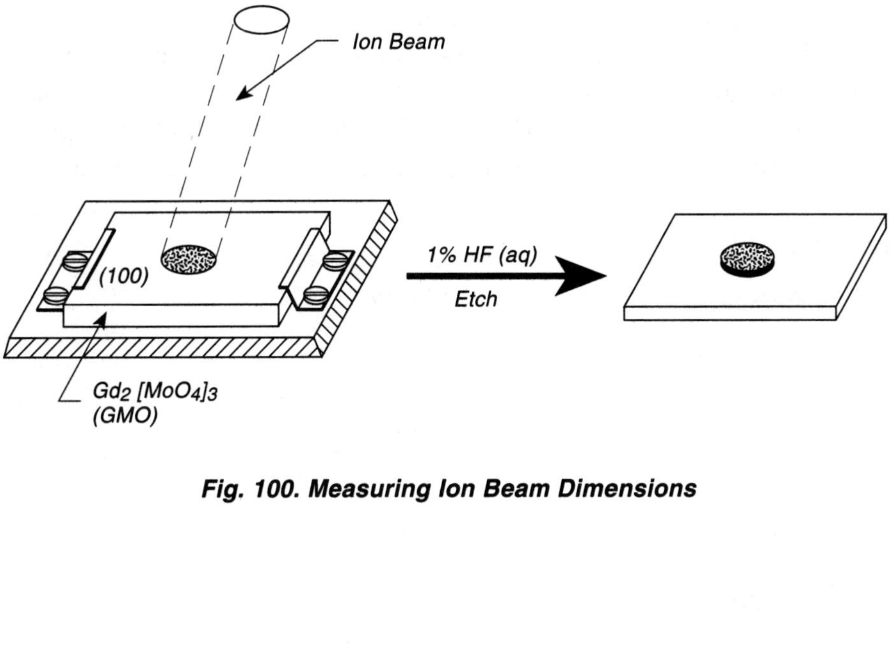


***Fig. 100. Measuring Ion Beam Dimensions***

# 101

# Beam Position Monitor—Charged Particles

The alignment of a beam of ions or electrons may be facilitated by the use of a segmented aperture device [1] shown in Figure 101. A simple bridge circuit is used, with a current measuring device to monitor the imbalance in collected current between opposing segments of the aperture. Only one of the two bridges is shown in the Figure. Several of these monitors may be placed along the beam axis, and any ions or electrons scattered by the edges of the apertures are removed by other apertures further down the beam axis. The variable shunt around the monitor allows the fraction of the collected current going to the current monitor to be adjusted. Varying the value of the resistance, *R*, over a wide range has little effect on the meter deflection, but *R* should be kept as small as possible to avoid charge buildup on the segments that could cause beam aberrations.

## *References*


[1] K.J. Snowdon and C.H. Barber, J. Phys. **E12**, 923 (1979).

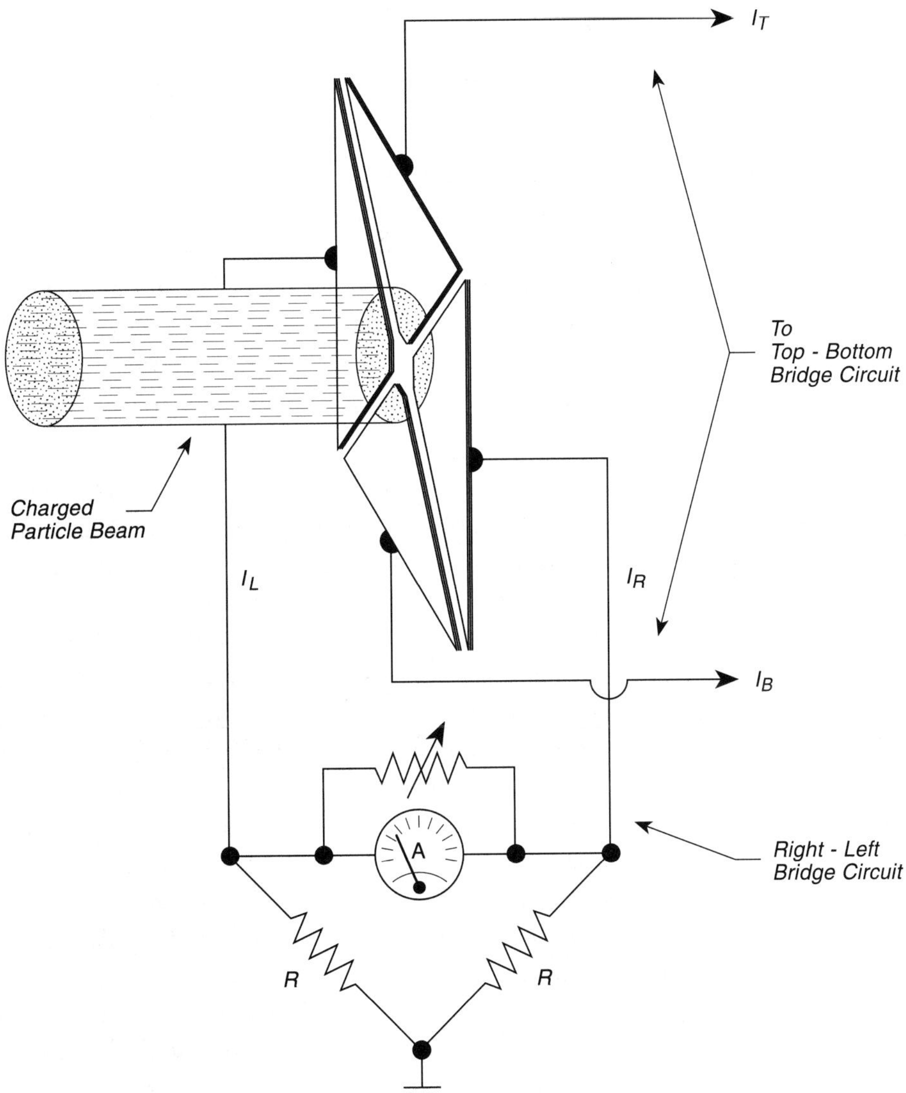


*Fig. 101. Beam Position Monitor—Charged Particles*

# 102

# Pulse-Counting LEED/ESDIAD Analyzer—Using Microchannel Plate Detection

There are many advantages in pulse counting during the measurement of either electron diffraction or ion angular distribution patterns produced by electron beam excitation of surfaces (ESDIAD). Pulse counting affords the maximum sensitivity to particle measurement, assuming each pulse is counted. The digitized information arising from pulse counting allows easy handling in computer analysis and error analysis. But most importantly, the high sensitivity of pulse counting techniques allows one to work with electron beams having low intensities, thereby minimizing damage to the surface layer during measurement [1,2].

Figure 102A shows a LEED/ESDIAD apparatus [3] that uses four grids in front of a pair of microchannel plate (MCP) amplifiers [4]. Charged particles (and, in some cases, excited neutral particles) are transmitted through the grid system to the microchannel plate amplifiers, where each particle exciting the microchannel plate produces a cascade of electrons that exit the back of the second microchannel plate and are accelerated into a two-dimensional resistive anode detector [5]. This pulse of about $10^5$ or $10^6$ electrons expands radially from the point of impact and is collected at the four corners of the resistive anode. Since the local intensity of expanding charge decreases as $1/r$, the pulse height arriving at the four corners can be used to calculate the coordinates of the point of impact (the centroid). The pulses at the four corners are capacitively coupled into four preamplifiers, A-D. These amplified pulses then travel into the position computer [6]. The position computer then generates a pair of pulses which are proportional to the $x$- and $y$-coordinates of the impact point on the resistive anode. The $x$- and $y$-pulses are then sent to a pair of analog-to-digital converters that form the input to either a multichannel analyzer [7] or to a computer. A three-dimensional display of the collected angular distribution of the electron or ion or excited neutral pattern is then displayed. It is important to point out that the excitation of the crystal surface by the electron beam will produce a background due to soft X-rays, and this must be subtracted from the measured patterns [8]. The ultimate spatial resolution of the apparatus is about 0.3°. Typical electron beam currents of the order of $1 \times 10^{-9}$ A are employed over time periods of the order of 100 s. Damage to even the most electron-beam-sensitive surface species is generally less than 1%. The positive ion patterns viewed in this apparatus may be compressed by applying a positive potential, $V_c$, to the crystal.

Examples of both typical LEED patterns and ESDIAD patterns are shown in Figures

102B and C for the case Cu(110) and CO chemisorbed on Cu(110) [9]. An earlier example of this technology applied to LEED measurements alone can be found in reference [10].

Earlier examples of the use of microchannel plate amplification for ESDIAD or LEED measurements involved the production of images on phosphor screens and either photographic [11–13] or video camera [14–17] data acquisition. The video camera method is efficient in the sense that a single video camera can be used on multiple systems.

An alternate high-resolution position detector is the wedge-and-strip anode [12] that captures the electron pulses on a conductive sheet containing accurately shaped wedge and strip conductive regions. The spatial resolution of these devices exceeds the inherent resolution limitations of the microchannel plate amplifiers [12].

A LEED analyzer built along the principle of the MCP/resistive anode combination has been reported by others [16,18,19]. In addition, a gated movable channeltron detector, coupled with a chopped electron beam, has been used to reduce background effects [20].

The counting efficiency for ions on a MCP detector has been carefully measured using a field evaporation apparatus [21]. It was found that the application of a positive bias on the front surface of the channelplate did not increase the capture/counting efficiency much above that expected from the fractional area of open channels present on the plate. It was, however, argued by others that a small positive potential (20 V) facilitated capture of secondary electrons and hence enhanced the counting efficiency [22], and this effect was verified in a careful paper in which the absolute detection efficiencies for $He^+$, $H^+$, and $O^+$ ions was directly measured using a Faraday cup to compare with the count rate. Ion detection efficiencies ranged from 18% to 63% in the energy range 250 eV to 5000 eV [23]. These workers also measured the detection efficiency as a function of incident angle, finding that an incident angle of about 10° yielded maximum efficiency, with the efficiency falling 24% in the range 20–40° off the channel axis direction [23].

The basic physical properties of straight through MCP arrays, chevron MCP arrays, and curved channel MCP arrays are discussed in some detail in reference [24].

Field penetration effects from one region of the LEED apparatus to another can be prevented in most cases by using neighboring grids at the same bias potential to screen elements at high potential (such as the front of the MCP) from the interior of the grid system [25].

The hemispherical grid analyzer may be used as a retarding field energy analyzer for either electrons or ions. For these measurements, the retarding potential ramp is applied to a central grid along with a small sinusoidal modulating voltage, and lock-in detection is applied to the transmitted current. This has been discussed in detail in reference [26].

A planar grid ESDIAD apparatus has been used to monitor the angular distribution of ions emitted from an insulator surface as a result of excimer laser pulses [27].

## *References*


[1] C.G. Pantano and T.E. Madey, Appl. Surf. Sci. **7,** 115 (1981).

[2] R.D. Ramsier and J.T. Yates, Jr., Surf. Sci. Reports **12,** 243 (1991).

[3] J.T. Yates, Jr., M.D. Alvey, K.W. Kolasinski, and M.J. Dresser, Nucl. Inst. and Methods in Phys. Res. **B27,** 147 (1987). The first apparatus of this sort for low incident current LEED studies was reported in P.C. Stair, Rev. Sci. Inst. 51, 132 (1980).

[4] Galileo Electro-Optics Corp., Galileo Park, P.O. Box 550, Sturbridge, MA 01566.

[5] No. 3394D Quantar Technology, Inc., 3004 Mission Street, Santa Cruz, CA 95060-5700.

[6] No. 2401 Quantar Technology, Inc., 3004 Mission Street, Santa Cruz, CA 95060-5700.

[7] No 76B Canberra Industries, Inc., Nuclear Products Group, 800 Research Parkway, Meriden, CT 06450.

[8] M.J. Dresser, M.D. Alvey, and J.T. Yates, Jr., Surf. Sci., **169,** 91 (1986); J. Vac. Sci. Technol. **A4,** 1446 (1986).

[9] J. Ahner, D. Mocuta, R.D. Ramsier and J. T. Yates, Jr., J. Vac. Sci. Technol. **A14,** 1583 (1996).

[10] P.C. Stair, Rev. Sci. Instrum. **51,** 132 (1980).

[11] J.J. Czyzewski, T.E. Madey, and J.T. Yates, Jr., Phys. Rev. Lett. **32,** 777 (1974).

[12] M.G. Lagally and J.A. Martin, Rev. Sci. Instrum. **54,** 1273 (1983); D.F. Ogletree, G.S. Blackman, R.Q. Hwang, U. Starke, G.A. Somorjai, and J.E. Katz, Rev. Sci. Instrum. **63,** 104 (1992).

[13] V.M. Bermudez, J. Vac. Sci. Technol. **A5,** 2975 (1987).

[14] T.E. Madey , J. Vac. Sci. Technol. **A4,** 257 (1986); Science **234,** 316 (1986).

[15] D.F. Ogletree, G.A. Somorjai, and J.E. Katz, Rev. Sci. Instrum. **57,** 3012 (1986).

[16] R.A. Malic, Rev. Sci. Instrum. **59,** 1951 (1988).

[17] G. Held, S. Uremovic, C. Stellwag, and D. Menzel, Rev. Sci. Instrum. **67,** 378 (1996).

[18] E.G. McRae, R.A. Malic, and D.A. Kapilow, Rev. Sci. Instrum. **56,** 2077 (1985).

[19] F.-K. Men, B.L. Clothier, and J. L. Erskine, Rev. Sci. Instrum. **64,** 1883 (1993).

[20] H. Niehus and B. Krahl-Urban, Rev. Sci. Instrum. **52,** 56 (1981).

[21] T. Sakurai and T. Hashizume, Rev. Sci. Instrum. **57,** 236 (1986).

[22] J.A. Panitz and J.A. Foesch, Rev. Sci. Instrum. **47,** 44 (1976).

[23] R.S. Gao, P.S. Gibner, J.H. Newman, K.A. Smith, and R.F. Stebbings, Rev. Sci. Instrum. **55,** 1756 (1984).

[24] J.G. Timothy, Rev. Sci. Instrum. **52,** 1131 (1981).

[25] C.W. Caldwell, Jr., Rev. Sci. Instrum. **36,** 1500 (1965).

[26] N.J. Taylor, Rev. Sci. Instrum. **40,** 792 (1969).

[27] J.T. Dickinson, S.C. Langford, J.J. Shin, , and D.L. Doering, Phys. Rev. Lett. **73,** 2630 (1994).

*Fig. 102. Pulse Counting LEED/ESDIAD*

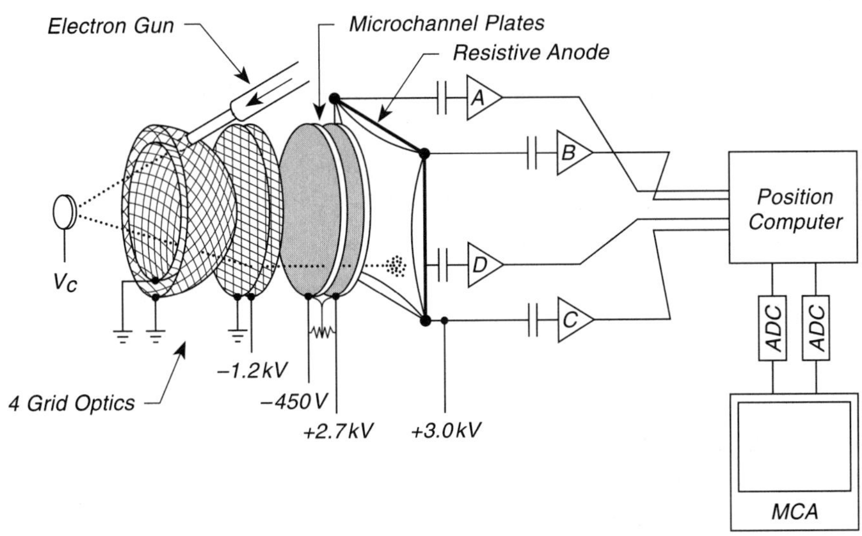


***A. Pulse Counting LEED / ESDIAD Analyzer***

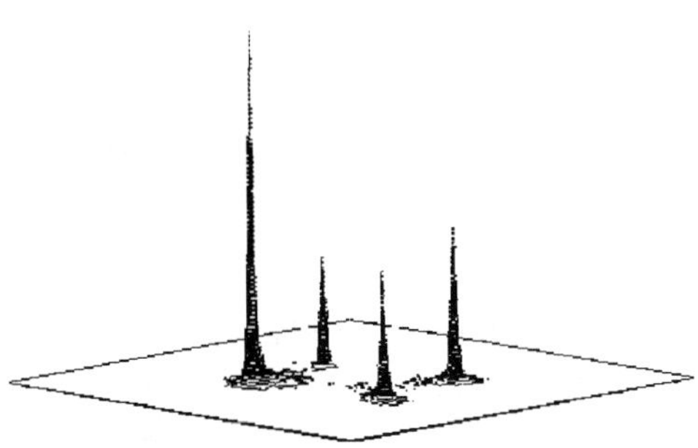

***B. Typical LEED Pattern - Cu(110)***

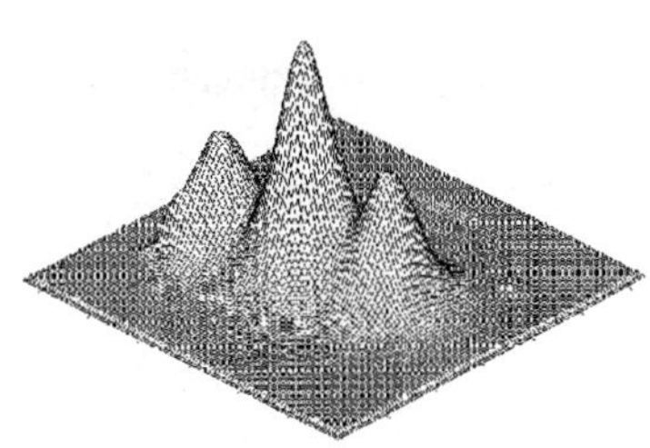

***C. Typical ESDIAD Pattern - CO / Cu(110)***

# 103

# Time-of-Flight Detection for Laser-Ionized Neutrals

The detection of neutral atoms and molecules that have been generated from surfaces using scattering, electron- and photon-stimulated desorption, and ion-induced desorption is facilitated by tunable nanosecond pulsed lasers. Laser excitation permits particle detection with quantum state specificity using laser-induced fluorescence (LIF) or resonance-enhanced multiphoton ionization (REMPI) methods. The REMPI technique has proven to be more popular because of the high sensitivity of the ion detection technique and the ability to image ion angular distributions analogous to the ESDIAD technique (see page 330).

The apparatus shown in Figure 103 was designed to detect laser-ionized neutral atoms and molecules that have undergone electron stimulated desorption from a single crystal surface [1,2]. It is based on the design developed by N. Winograd et al. [3] for the detection of neutral atoms desorbed by ion bombardment. The pulsed electron gun shown in the Figure therefore can be replaced by a pulsed ion gun or a pulsed laser, and the detection scheme is the same. The principle is as follows. Neutral products that leave the surface as a result of excitation by the pulsed source must travel a specified distance before they enter the ionizing laser beam, which is directed parallel to the crystal surface. The laser generated ions are then accelerated into the chevron electron multichannel plate array (CEMA) detector [4]. The output of the detector is coupled to a phosphor screen. Provided that the laser is focused in a sheet-like fashion, the visual display on the phosphor screen provides angular resolved patterns of the desorbed neutrals, as in the ESDIAD method, but now with full internal quantum state resolution [2,3]. The integrated signal can be detected with a gated integrator or the two-dimensional image can be digitized using a charge-coupled device (CCD) camera.

By pulsing the electron beam rapidly (1 μs), the velocities of the neutrals can be measured by the time delay between the pulses of the electron gun and the laser. Desorbed ions can be a problem, but they can be swept away prior to the laser pulse by an applied potential to the first grid and/or the crystal. It is also possible to discriminate against unwanted signals by pulsing the applied voltage on one of the multichannel plates. Recently several groups have oriented the detection assembly 90° to the surface normal, so that the spread in velocities is obtained simultaneously with the angular distribution [5].

## *References*

[1] A.R. Burns, E.B. Stechel, and D.R. Jennison, in *Desorption Induced by Electronic Transitions,* DIET III, edited by R.H. Stuhlen and M.L. Knotek (Springer-Verlag, Heidelberg, 1988), p. 67.

[2] A.R. Burns, Surface Sci. **280,** 349 (1993).

[3] P.H. Kobrin, G.A. Schick, J.P. Baxter, and N. Winograd, Rev. Sci. Instrum. 57, 1354 (1986).

[4] CEMA assemblies are available from Galileo Electro-Optics Corp., Galileo Park, P.O. Box 550, Sturbridge, MA 01566.

[5] M. Menges, B. Baumeister, K. Al-Shamery, B. Adam, T. Mull, H.-J. Freund, C. Fischer, D. Weide, and P. Andresen, in *Desorption Induced by Electronic Transitions,* DIET V, edited by A.R. Burns, E.B. Stechel, and D.R. Jennison (Springer-Verlag, New York), 1993, p. 275.

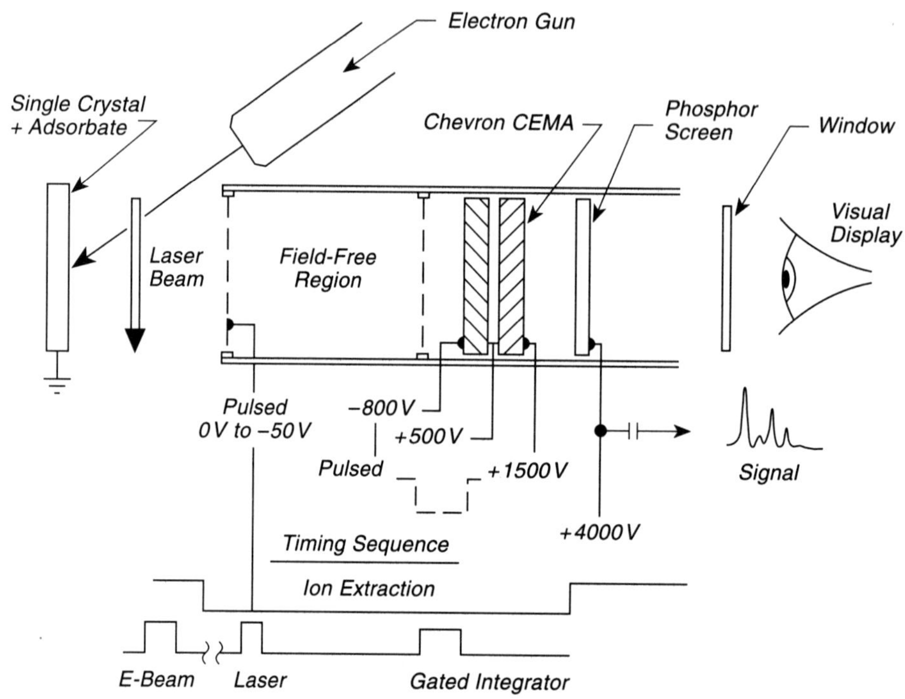


**Fig. 103. Time-of-Flight Detection for Laser-Ionized Neutrals**

# C. Photons (UV)

# 104

# Hollow Cathode Resonance Lamp for He(II) (40.8 eV) Photoemission

The use of vacuum ultraviolet sources for valence band photoemission studies of the surfaces of solids is widespread, and many such studies are made on synchrotron sources. Small laboratory vacuum UV sources predated synchrotron sources, and are still widely used. A high-intensity source delivering both He(I) and He(II) radiation, depending upon the discharge conditions, is shown schematically in Figure 104A [1]. He gas is admitted to a differentially pumped region containing a hollow cathode, and is pumped away continuously by a trapped forepump. A second stage of differential pumping occurs closer to the ultrahigh vacuum chamber using a diffusion pump that is backed by the same forepump. The forepump (rated at 500 L/min) speed was measured to be 100–200 L/min at the lamp and 30–70 L/min at the back of the diffusion pump for typical operation [2]. The forepump was trapped with a water-cooled Cu trap and a zeolite trap. An 8-cm-diameter port at the lamp leading to the liquid nitrogen trap before the diffusion pump was used [1], as an aid to improve pumping speed over that originally used [2]. This large port was made from a straight-through valve. This permits the lamp to be isolated from the UHV system.

Figure 104B shows the details of the He lamp. A high-voltage UHV feedthrough holds a hollow cathode cylinder made of 0.5 mm Ta sheet. This cathode, 2 cm in diameter and 4–6 cm in length, is held inside a glass tube in a location in which a stainless steel capillary (2.3-mm internal diameter) picks up radiation at a distance of 1–1.5 cm from the end of the cathode. The glass discharge region can be removed for cleaning. A needle valve admits He into the side of a Conflat flange that mates to the flange holding the first stainless capillary tube. Below the first capillary is the region pumped by the forepump, region A–B. A second capillary, mounted between B and C, directs the radiation to a third capillary which enters the UHV chamber. The He pressure in the system with the lamp operating is approximately $1 \times 10^{-8}$ Torr. Contaminant gases are measured to be below $10^{-10}$ Torr, although such gases will be highly directed as a beam to the sample by the capillaries, and one should be alert to the possible build up of surface contamination.

The performance of the lamp for both He(I) (21.2 eV) and He(II) (40.8 eV), and for 48.4 eV radiation, is shown in Figure 104C, where an evaporated Au film was studied, using a double-pass CMA electron spectrometer for measurement of the photoelectron spectrum at a constant resolution of 0.2 eV. The pressure of He in the discharge was about 1 Torr for optimum He(I) intensity and 0.1–0.2 Torr for 40.8 eV and 48.4 eV photons. This lamp has improved He(II) intensity compared to earlier lamps utilizing a microwave discharge [3].

## *References*


[1] J.E. Rowe, S.B. Christman, and E.E. Chaban, Rev. Sci. Instrum. **56,** 759 (1985).

[2] J.E. Rowe, S.B. Christman, and E.E. Chaban, Rev. Sci. Instrum. **44,** 1675 (1973).

[3] T.V. Vorburger, B.J. Waclawski, and D.R. Sandstrom, Rev. Sci. Instrum. **47,** 501 (1976).

**Fig. 104. Hollow Cathode Resonance Lamp—He (II) Source**

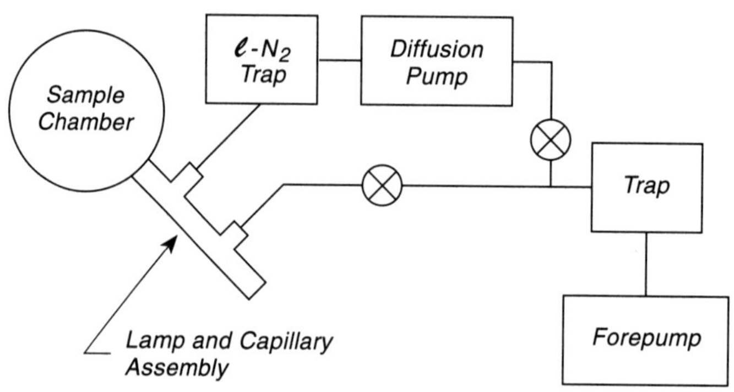


**A. Pumping System**

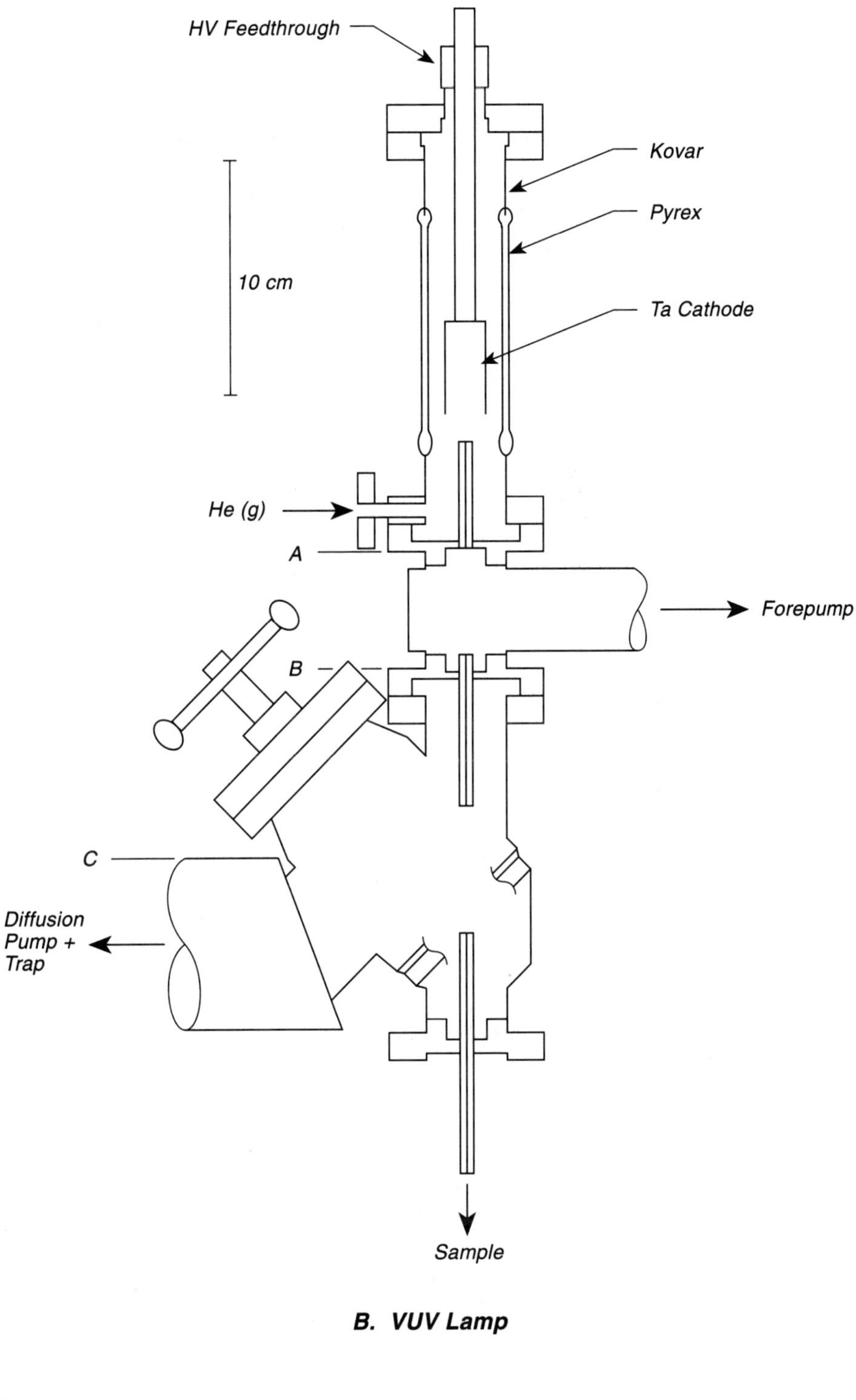
HV Feedthrough
Kovar
Pyrex
10 cm
Ta Cathode
He (g)
A
Forepump
B
C
Diffusion
Pump +
Trap
Sample

***B. VUV Lamp***

**Fig. 104. Hollow Cathode Resonance Lamp—He (II) Source**

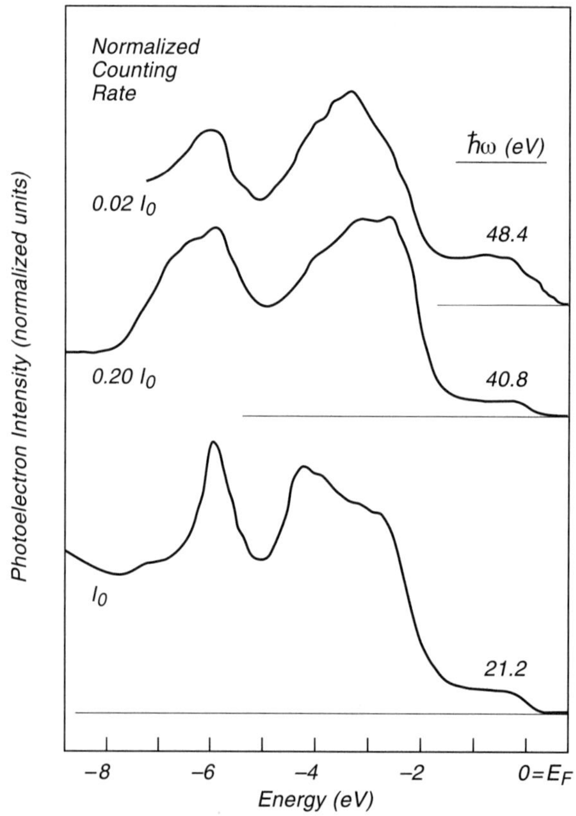


**C. Au UPS Spectrum**

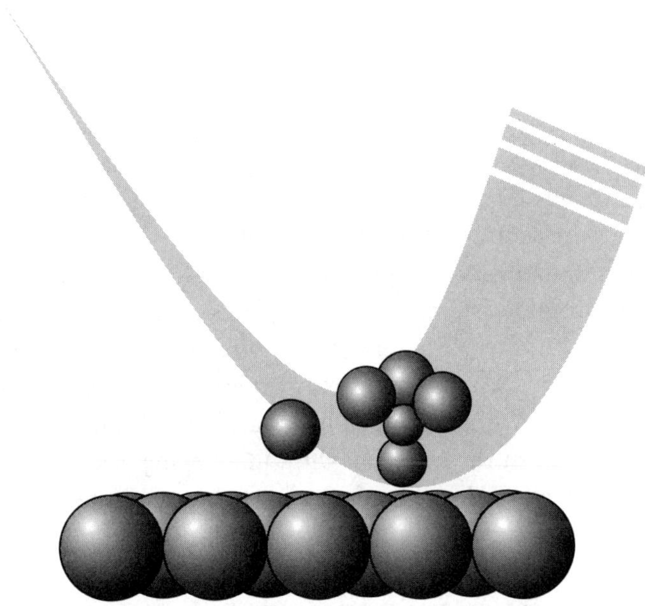

# 105

# He(I) and He(II) Ultraviolet Resonance Source

A laboratory source of vacuum ultraviolet radiation useful for photoelectron spectroscopy in an ultrahigh vacuum chamber is described [1]. The source is based on the principle of the charged particle oscillator [2–5]. The production of maximum He(I) (21.22 eV) and He(II) (40.81 eV) radiation occurs at a discharge pressure of He of about $8 \times 10^{-2}$ Torr. By means of two stages of differential pumping in the source, the pressure in the main UHV chamber can be maintained at $2 \times 10^{-10}$ Torr during operation.

Figure 105A shows the UV source. It consists of a grounded outer cathode cylinder made of 304 stainless steel. Two 3/16-in.-diameter tungsten rods serve as the anode, and they are supported and separated from the outer cathode by machinable ceramic fittings. The W rods can expand and slide in loosely fitting holes in the ceramic as the lamp heats during operation. The ceramic also contains larger holes for conductance of He. The anode potential of 0 to +5 kV is supplied through a high-voltage feedthrough at the rear of the source. Grounded stainless steel end plates cover the inner side of the ceramic fittings to reflect electrons traveling in the axial direction back into the source volume. Photons emerge through a 0.1-in.-diameter exit aperture between the anodes at the front of the lamp. An aperture with a diameter less than one-third the inner diameter of the cathode allows sampling of the radiation without allowing electrons to escape [6]. The gas used here was 99.9995% He, which flows through an activated charcoal trap at 77 K and a variable leak valve. A 0 to +5 kV DC power supply (0.500 mA) is connected through a variable resistance ballast set at 16 kΩ. This prevents current surges during initiation of the discharge whose resistance varies from infinity (before activation) to several kΩ during operation.

The pumping system in the lamp consists of two stages of differential pumping, followed by a third stage in the UHV chamber. The first stage, across a 1.5-mm-i.d. Pyrex capillary, 3 cm long, is pumped by a chilled water-trapped 325-L/s oil diffusion pump. This stage can be eliminated by opening a bypass valve between entrance region and the first stage of diffusion pumping. The second stage is also across a similar capillary and is pumped by a liquid-nitrogen-trapped 325-L/s oil diffusion pump. The third stage of pumping across the exit capillary (2.5 mm i.d, 22.86 cm long) is pumped by the UHV system pumps, which consist of a 450-L/s turbopump and a 500-L/s ion pump. A gate valve is used to isolate the source from the UHV system. The original paper shows plots of photoelectron intensity from a Au surface versus pressure [1].

Under optimal operating conditions, the He(I)/He(II) intensity ratio produces a 10/1 ratio of photoelectrons from the two spectral lines from the Au surface.

## *References*

[1] G.M. Lancaster, J.A. Taylor, A. Ignatiev, and J.W. Rabalais, J. Electron Spectrosc. Relat. Phenom. **14,** 143 (1978).
[2] A.H. McIlraith, Nature **212,** 1422 (1966).
[3] A.H. McIlraith, J. Vac. Sci. Technol. **9,** 209 (1972).
[4] R.K. Fitch, T. Mulvey, W.J. Thatcher, and A.H. McIlraith, J. Phys. **D3,** 1399 (1970).
[5] R.K. Fitch, T. Mulvey, W.J. Thatcher, and A.H. McIlraith, J. Phys. **E4.** 533 (1971).
[6] A.H. McIlraith (private communication to J.W. Rabalais).

***Fig. 105. He (I) and He (II) UV Source***

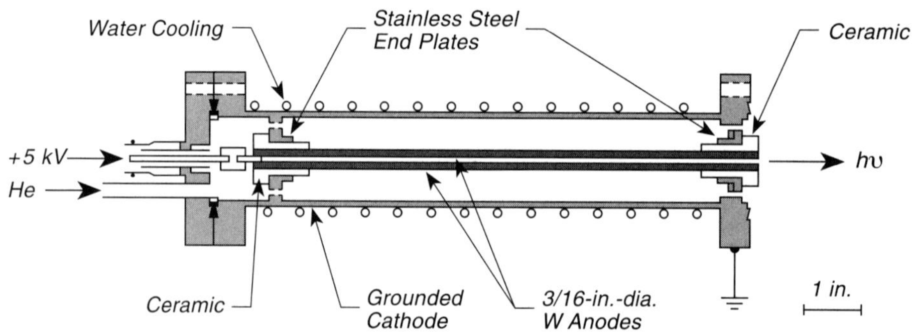


***A. Cold Cathode Discharge UV Source***

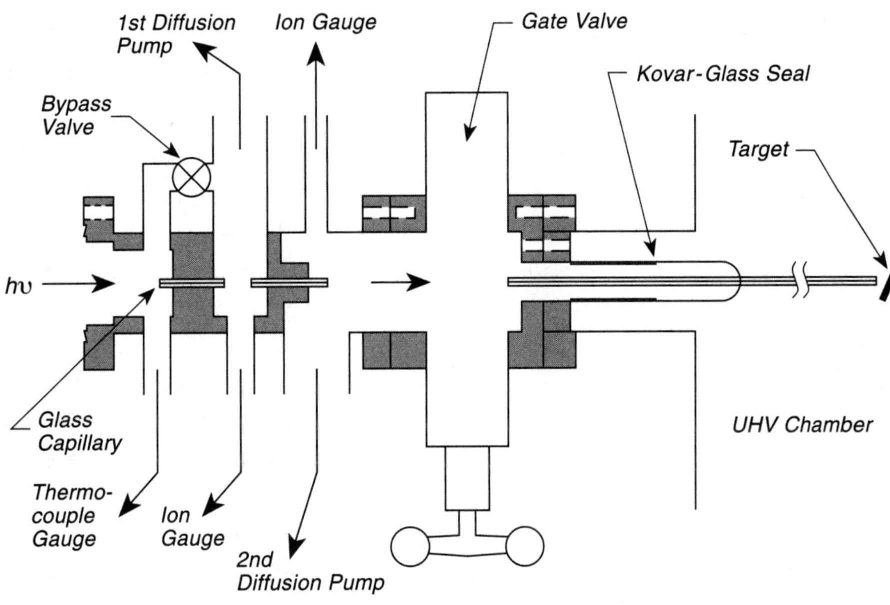


***B. Differential Pumping System***

# 106

# Photochemistry on Surfaces Using an Ultraviolet Lamp

Although ultraviolet induced photochemistry research has been carried out for more than 60 years in homogeneous phase, detailed studies of photochemical processes on well-defined single crystal surfaces are much more recent. Because ultraviolet lamps emit more photons over an extended exposure time than most laboratory lasers, these lamps are often used for inducing photochemical processes on surfaces. Both intensity measurements and photon energy range selection using filters are conveniently carried out, permitting good measurements of cross sections versus photon energy for surface process induced photolytically.

Figure 106A shows an ultrahigh vacuum chamber [1] set up so that a single crystal may be irradiated with ultraviolet light from a high-pressure Hg arc source [2]. The single crystal is located in front of an aperture that leads to a quadrupole mass spectrometer in a differentially pumped section, permitting the direct measurement of photodesorbed species. In addition, the mass spectrometer may be used also for thermal desorption studies after ultraviolet irradiation to access the changes that have occurred due to irradiation of an adsorbed layer. The ultraviolet light passes through a 10-cm distilled water filter that filters out infrared radiation. A 200-mm f/1 quartz condensing lens focuses the light onto the crystal through a UV-grade sapphire window. Interference filters (Oriel Corporation) centered on the intense lines of the mercury arc are often used to produce narrow band radiation and have a transmission of from 15%- 50% at the center of the band with a full width at half maximum (FWHM) of 10 nm, except for the 237-nm filter, which has a FWHM of 20 nm. One must be careful in using these filters for long irradiation times, since the transmission band center can shift up to 2 nm when extensive irradiation occurs.

The apparatus contains a quartz flat partial reflector used to measure continuously the light intensity. The quartz flat reflects roughly 10% of the light, and this is directed to a neutral density filter (1% transmission) and then into a blackened housing which holds a calibrated photodiode [3]. The calibration of the photodiode relative to an absolute thermopile detector can be done externally on an optical bench, with the thermopile being placed in the same position as would be occupied by the crystal. Such measurements have found that the intensity of UV radiation from the arc source is not uniform, and it is interesting to map out the intensity pattern which gives a view of the arc location in the source in the most intense region of emission.

A commercially available UV quartz prism polarizer may be placed into the beam if either *s*- or *p*-polarized radiation is desired (see page 352).

By combining the photoirradiation experiments at various photon energies with analysis of the surface before and after irradiation (using thermal desorption in this case) it was possible to observe two major separate photochemical processes for molecular oxygen chemisorbed on Pd(111)-desorption of $O_2$ and dissociation to produce chemisorbed O. The cross sections for these two processes are plotted versus photon energy in Figure 106B, giving an idea of the random errors inherent in this type of measurement [1]. These cross sections were also measured in another laboratory using laser-induced photochemistry and the results were in agreement to within a factor of 2–3 [4].

A photoreactor for studies of powders and single crystals has been described in reference [5]. This device can operate at high or low gas pressures.

## *References*

[1] L. Hanley, X.-C. Guo, and J.T. Yates, Jr., J. Chem. Phys. **91**, 7220 (1989).
[2] (Model 66014, 350 watt). Oriel, 250 Long Beach Boulevard, P.O. Box 872, Stratford, CT 06497-0872.
[3] The 1-cm$^2$-area circular photodiode-preamplifier package is sensitive from 185 to 1150 nm, and is calibrated by the manufacturer with standards traceable back to the National Institute of Science and Technology. (EG and G HUV-4000B).
[4] F. Weik, A. de Meijere, and E. Hasselbrink, J. Chem. Phys. **99**, 682 (1993).
[5] A.Z. Moshfegh and A. Ignatiev, Rev. Sci. Instrum. **59**, 2202 (1988).

***Fig. 106. Photochemistry Using Ultraviolet Lamp***

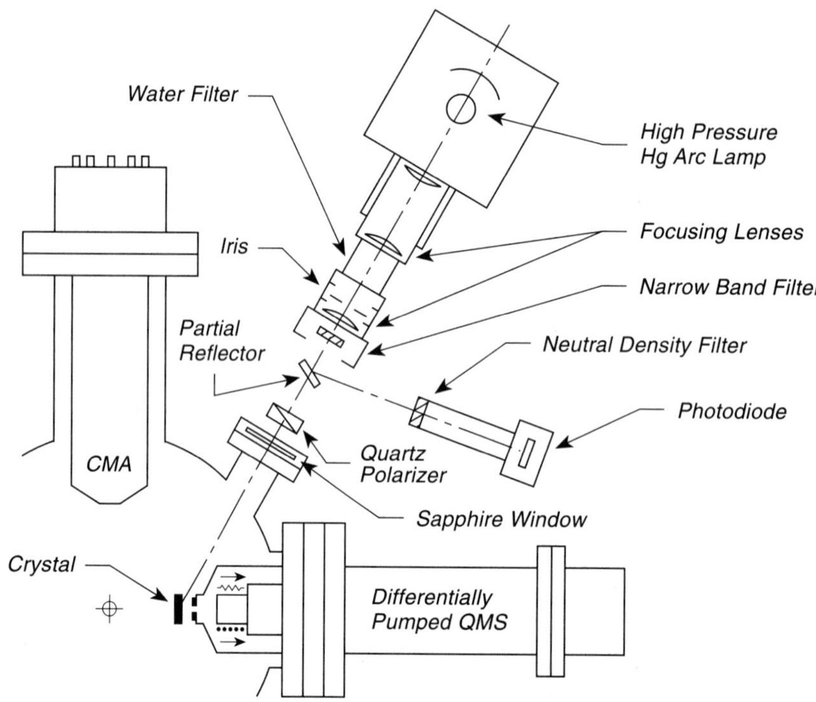


***A. Ultraviolet Lamp for Photochemistry Using Filters and Polarizer***

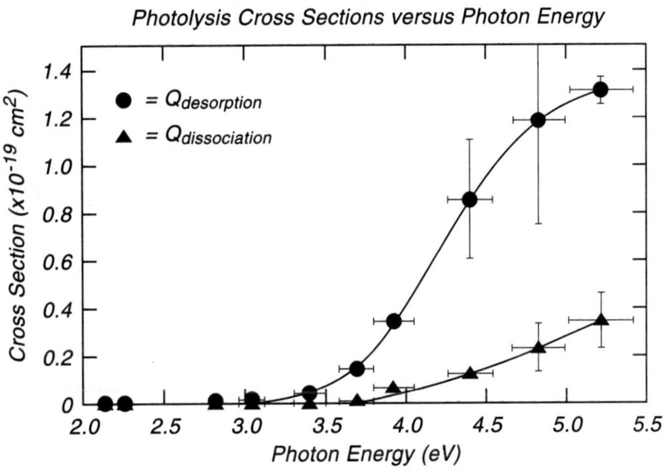


**B. Typical Cross Section Measurements From TPD Analysis**

# 107

# Use of Polarized Ultraviolet Light for Photochemistry

The use of polarized UV radiation to induce photochemical processes on surfaces is of importance because of the ability to use this information to obtain information about the mechanism of the electronic processes responsible for photoactivation of the surface [1,2]. The photochemical activation of chemisorbed molecular $O_2$ is a model system, discussed briefly on page 348, and here we show how polarized radiation may be used for studies of the same system.

Figure 107A shows the configuration of the photochemical system [3]. Ultraviolet light, originating from a lamp source, may be polarized by using a UV-grade quartz prism polarizer, directing either *p*- or *s*-polarized radiation to the single crystal surface. The single crystal may be used as a "calorimeter" to measure the thermal energy deposited from the radiation incident on its surface. This absorbed energy will vary in accordance with Fresnel's laws and the optical constants of the crystal absorber. Figure 107B shows a calorimetric experiment in which polarized broad band radiation having 0.23 W/cm$^2$ power density is allowed to warm the crystal from 79 K. For an incidence angle of 60°, one can see that the *p*-polarized radiation is more efficiently absorbed than the *s*-polarized radiation, as indicated by the relative increases in temperature. The measured difference in absorptivity is in fair agreement with the optical constants considering the errors inherent in this type of experiment [3].

Figure 107C shows the comparative effect of *s*- and *p*-polarized radiation on a molecular $O_2$ layer on Pd(111) as studied by thermal desorption following various periods of irradiation. The $\alpha_1$- and $\alpha_2$-molecular oxygen states desorb by photostimulation, and also photochemically dissociate to produce a thermal desorption feature labeled $\beta_2$. It may be seen that *p*-polarized radiation is more efficient in causing these desorption and dissociation processes than is *s*-polarized radiation [3]. This is consistent with a process in which electronic excitation of the metal by ultraviolet light is responsible for the surface photochemical processes, but more detailed studies would be needed to prove this.

## *References*


[1] S.A. Buntin, L.J. Richter, D.S. King, and R.R. Cavanagh, J. Chem. Phys. **91**, 6429 (1989).

[2] E. Hasselbrink, S. Nettesheim, M. Wolf, A. Cassuto and G. Ertl, Springer Series in Surface Science, Diet IV **19**, 75 (1990).

[3] J. Yoshinobu, X.-C. Guo, and J.T. Yates, Jr., Chem. Phys. Lett. **169**, 209 (1990).

***Fig. 107. Measurements with Polarized UV Light***

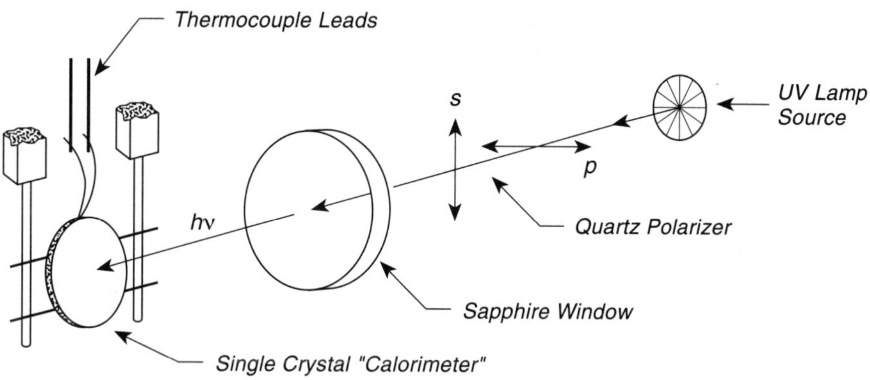


***A. Configuration for Irradiation with Polarized UV Light***

## Fig. 107. Measurements with Polarized UV Light

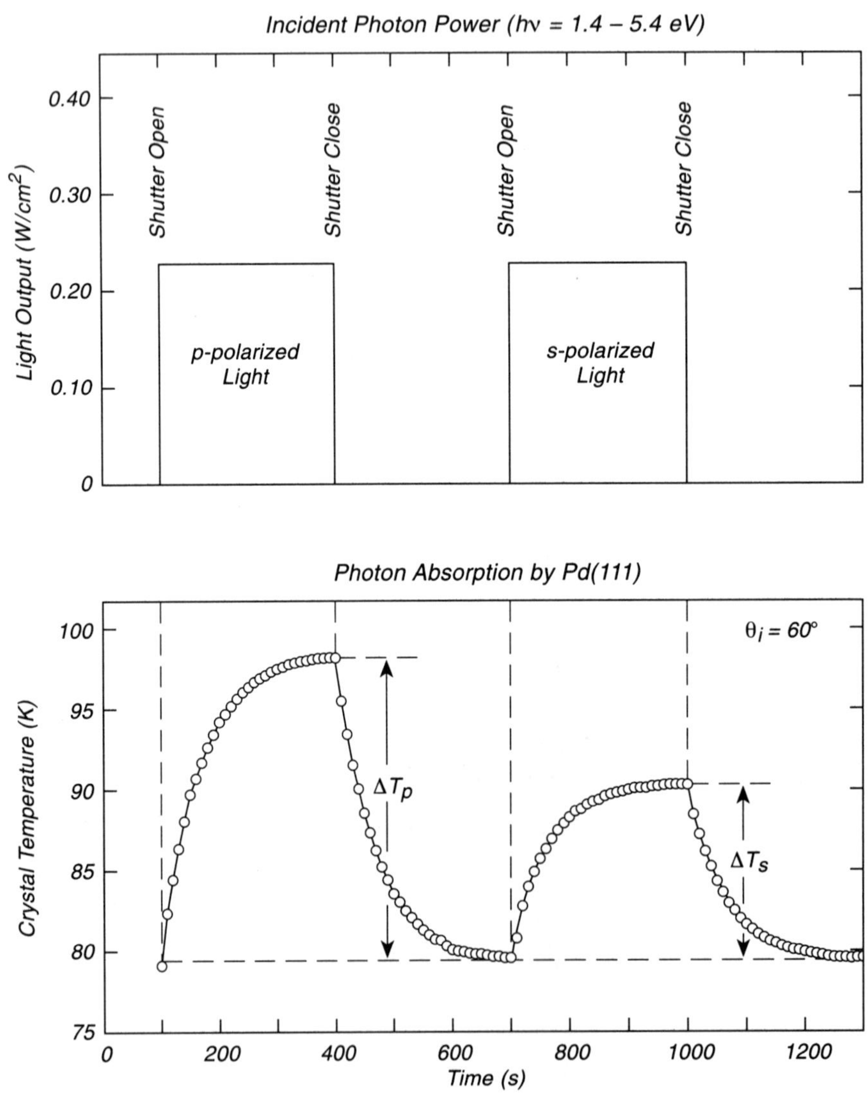


**B. "Calorimetric" Measurement for p- and s-Polarization**

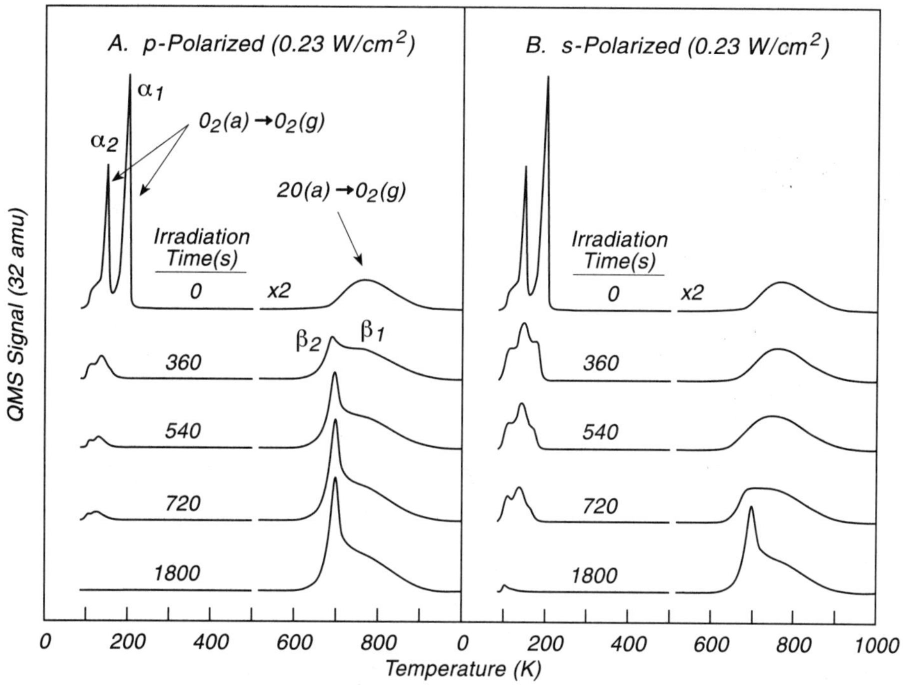


**C. Effect of Polarization on Oxygen Desorption State Distribution**

# D. Kinetics of Adsorption/ Adsorbate Coverages

# 108

# Absolute Adsorption Uptake Measurements from a Calibrated Effusive Molecular Beam

The measurement of the kinetics of adsorption and the absolute saturation coverage of an adsorbate on a single crystal is basic to our understanding of surface chemistry. While relative surface coverages, compared to the saturation coverage, are fairly easy to measure using various surface spectroscopies, the measurement of the absolute coverage and the absolute sticking coefficient are more difficult. One of the absolute methods employed in some cases involves the use of nuclear activation of the adsorbed layer, where absolute counting of nuclear disintegration events after activation may be used with high accuracy [1]. Unfortunately, not all adsorbates are amenable to this method.

Figure 108A shows an apparatus that is capable of measuring both the absolute sticking coefficient and the absolute surface coverage during an adsorption experiment. The method is a version of the King and Wells method, developed in 1972 [2]. A multicapillary doser, with an absolutely known flux of effusing molecules in a known angular distribution (see page 604) is incident on a single crystal of known size, located at a known distance from the doser. A quadrupole mass spectrometer, shielded from direct line of sight from the crystal and doser, measures the random flux of gas in the vacuum system as a function of the exposure of the crystal to the incident beam. A shutter can be rotated to intercept the beam, shielding the crystal from the direct flux, and converting all of the incident beam into a random flux by reflection from its surface. In order to make an uptake measurement, the crystal is cleaned, brought to adsorption temperature, and positioned behind the shutter. Gas is admitted, as shown in Figure 108B, and the system pressure rises quickly to a sharp point where the rate of rise decreases significantly. At time $t_1$, the shutter is opened, and gas begins to be adsorbed on the crystal, causing the random flux of gas to decrease. Region A of the uptake curve represents the first phase of adsorption, and the horizontal trace indicates that the sticking coefficient is constant (and probably unity) in this region. Following adsorption to point $t_a$, it is observed that the random flux begins to rise in region B as the sticking coefficient decreases below its value in region A. At the end of the experiment, the random flux has asymptotically risen to the dashed line, which indicates that, at the sensitivity level of this experiment, the uptake is complete. Placing the shutter into the beam causes little effect at point $t_b$, since the complete reflection of incident molecules by the inert shutter is identical in magnitude to the reflection of incident molecules by the saturated crystal. The area of region A and region B may be used to compute the final absolute coverage at $t_b$ [3].

There are three issues that contribute to the uncertainty of this measurement:

1. The nonhorizontal character of the dashed line against which the integrals are measured. This error is probably due to slow pumping by the vacuum system walls, which decreases as the exposure is increased. The empirical solution to this problem is to treat the dashed limiting line as a baseline; it is designated $P_{eq}$.
2. The adsorption of gas from the random flux. A small correction can be made for this effect, which typically amounts to 1% in the region from $t_0$ to $t_1$. This initial coverage is designated $N(t)_{initial}$.
3. The low accuracy of determining the saturation point, $t_b$. Here, Auger spectroscopy was employed to show that for long exposures beyond $t_b$, the rise in adsorbate coverage was 6% as shown in Figure 108C for the example chosen here.

The equation that governs the adsorption is

$$N(t)_{total} = N(t)_{initial} + (F + F_r) \cdot \int S(t)dt, \qquad (1)$$

where $N(t)_{total}$ = the total coverage at saturation; $F$ = the beam flux; $F_r$ = the random flux; and $S(t)$ = the sticking coefficient. $S(t) = [P_{eq} - P(t)]/fP_{eq}$, and $f$ = the fraction of the beam intercepted by the crystal. The value $N(t)_{total}$ is then increased to account for the small correction mentioned above in section (3).

A variation on this technique uses a diffuse source of adsorbate from an uncalibrated leak valve and a collimated detector used to measure the rate of adsorbate reflection from the surface during adsorption [4].

## *References*


[1] Z.H. Lu, K. Griffiths, P.R. Norton, and T.K. Sham, Phys. Rev. Lett. **68,** 1343 (1992).

[2] D.A. King and M.G. Wells, Surface Sci. **29,** 454 (1972).

[3] C.C. Cheng, R.M. Wallace, P.A. Taylor, W.J. Choyke, and J.T. Yates, Jr., J. Appl. Physics **67,** 3693 (1990).

[4] A.M. Glines, M.A. DeAngelis, and A.B. Anton, J. Vac. Sci. Technol. **A10,** 3179 (1992).

***Fig. 108. Absolute Adsorption Uptake Measurement from Calibrated Beam***

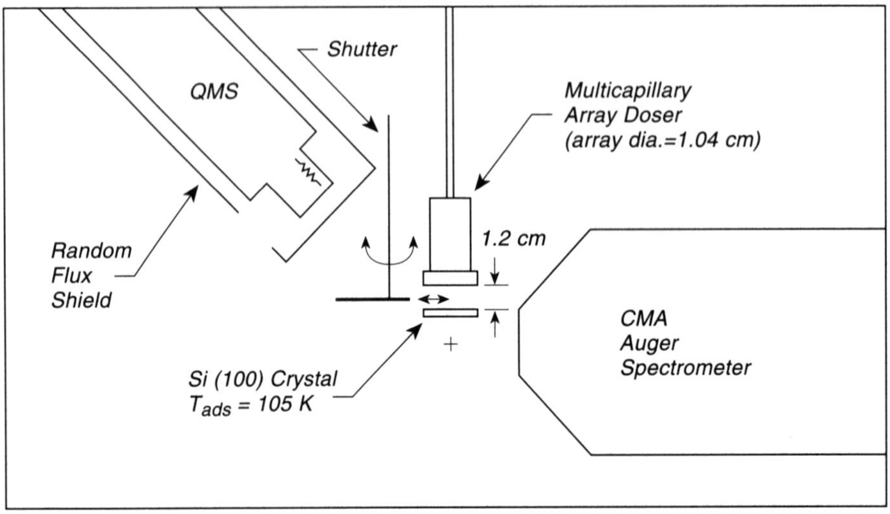


***A. Apparatus for Absolute Adsorption Uptake Measurements***

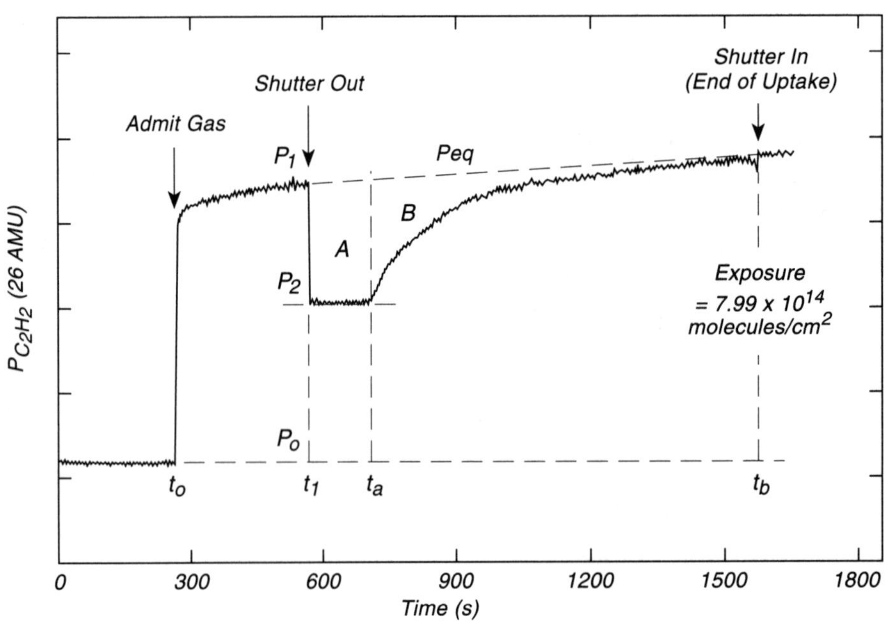


***B. Typical Uptake Curve $C_2H_2$ / Si (100)***

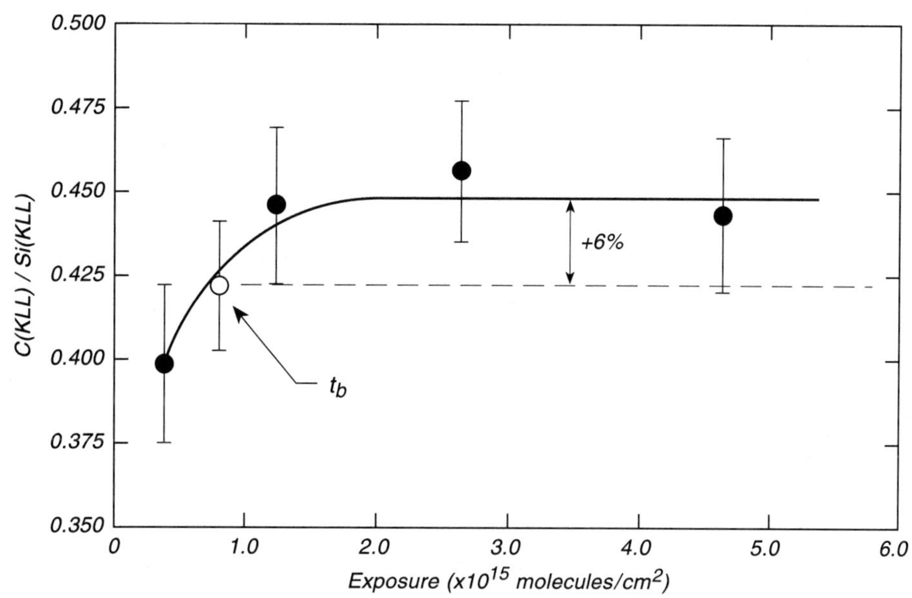


**C. Increase in $C_2H_2$ Coverage Beyond $t_b$ Owing to Continued Slow Adsorption**

# 109

# Measurement of Absolute Sticking Coefficient at Desorption Temperature

Conventional measurement techniques measure the net sticking probability, which is related to the difference between the absolute rate of adsorption, $r_a$, and the absolute rate of desorption, $r_d$, according to Equation (1).

$$s_{net} = (r_a - r_d)/z \quad (1)$$

Here $z$ = the impingement rate for the incident gas.

It is often desired to measure the absolute sticking coefficient, $s_{abs}$, which is given by equation (2).

$$s_{abs} = r_a/z \quad (2)$$

The absolute sticking coefficient is equal to the net sticking coefficient only when the rate of desorption is zero, so in order to achieve this condition, conventional sticking coefficient measurements are made at temperatures below the desorption temperature.

Substituting Equation (1) into Equation (2) gives Equation (3).

$$s_{abs} = s_{net} + r_d/z \quad (3)$$

Equation (3) yields

$$s_{abs} = \lim_{1/z \to 0} s_{net}(z) \quad (4)$$

Normally, a plot of $s_{net}$ versus $1/z$ at constant temperature will give an intercept of $s_{abs}$ and slope $r_d$. Figure 109 shows a three-dimensional plot with the linear extrapolation of the family of curves ( black points) to the limit of $1/z = 0$, yielding the white points which form a trace of the absolute sticking probability in the limit of very high impingement rate, z. Here the symbol ML = monolayer.

The data shown in the Figure are taken at 400 K, whereas the desorption temperature for CO in this case is in the range 350 K to 550 K [1]. The method works well when $r_d < z$, or correspondingly, when $1/r_d > 1/z$.

An alternate method, particularly useful for irreversible dissociative chemisorption, has been described by Engstrom et al. [2].

## *References*


[1] X-C. Guo and D.A. King, Surf. Sci. **302,** L251 (1994).

[2] L.-Q. Xia, M.E. Jones, N. Maity, and J.R. Engstrom, J. Vac. Sci. Technol. **A13,** 2651 (1995).

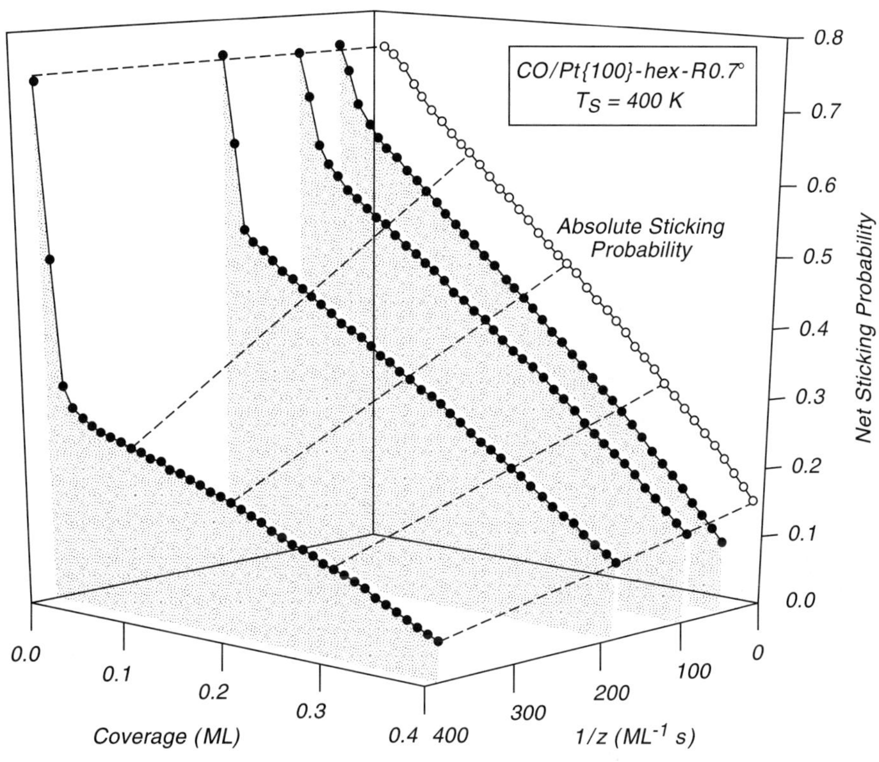


***Fig. 109. Absolute Sticking Coefficient Method at Desorption Temperature***

# E. Mass Spectrometry

# 110

# Shielded Quadrupole Mass Spectrometers—Temperature-Programmed Desorption

One of the most widely used and versatile techniques for the study of adsorbed species on single crystal surfaces is temperature-programmed desorption (TPD), often called thermal desorption spectroscopy (TDS). The method extends back to the early work of Taylor and Langmuir, 60 years ago, who studied the desorption of Cs from polycrystalline tungsten filaments [1]. The method was extended to other adsorbates on polycrystalline surfaces by Becker and Hartman [2], Ehrlich [3], and by Redhead [4]. A review of the early work can be found in reference [5]. Temperature-programmed desorption is now used extensively for studies on single crystal surfaces of metals, semiconductors, and insulators. A comprehensive review of ten methods for analyzing TPD data, and an evaluation of each was recently written by de Jong and Niemantsverdriet [6].

The collection of TPD data has progressed from the early use of ionization gauges to the almost universal employment of quadrupole mass spectrometers (QMS) for the detection of desorbing gases. Using the mass spectrometer in the multiplex mode, one can simultaneously measure the evolution of various gas products as the temperature of the crystal is raised. This can also be extended to the study of isotopic mixing between labeled adsorbates, as was first demonstrated for adsorbed CO desorbing from polycrystalline W [7]. Large differences in the desorption kinetics from different single crystal faces was first demonstrated for $H_2$ desorbing from W(100), W(111), and W(110) by the combined works of various authors [8–11].

The early work with single crystals employed unshielded mass spectrometers that sampled the random flux of desorbing gases originating from the front of the crystal, from the electrical support leads to the crystal, and from the back of the crystal disk. These effects could be eliminated to various degrees by using a differentially pumped and shielded mass spectrometer that preferentially collected desorbing gas only from the front of the crystal. The enhanced sensitivity to desorbing gas as a result of the shield is related to the fact that the shield retains gas molecules for an average time period of $\tau = V/S$, where $V$ = volume and $S$ = pumping speed, both relating to the shield region itself. Values of $\tau$ = 50–100 ms are common, permitting each trapped molecule to pass through the ion source many times before being pumped away. The apertured shield is selective for desorption from the front of the crystal, but it is not measuring significant desorption for molecules making a single pass, compared to the contribution from trapped molecules within the shield which make multiple passes through the ionization source. The value of $\tau$ must be kept below the characteristic time for the desorption process measured, which is usually of the order of seconds for temperature scan rates of a few K/s.

Figure 110A shows a glass shield that was just slipped over the ionization source of a mass spectrometer. Pumping occurs from the annular space at the rear of the shield [12]. Variations on this design (Figure 110B) involve a metal shield with pumping holes present on the side of the shield behind the ionization source. For $N_2(g)$ at 300 K, a 1-$cm^2$ aperture will provide a pumping speed, $S = 11$ L/s, where the total pumping aperture area considered should also include the area of the entrance aperture. Figure 110C shows a quadrupole mass spectrometer mounted in an apertured shield and differentially pumped by a dedicated pump [13], which may be either an ion pump, a titanium sublimation pump (TSP) [14], or a turbopump. The reader is referred to page 118 for a discussion of the relative merits of an ion pump and a TSP in this application.

The ionization source of the quadrupole mass spectrometer contains a thermionic emitter producing electron emission at mA currents. A fraction of these electrons escape from the axial aperture and may be collected by the crystal, causing electron beam damage of the adsorbate. Stray currents of the order of 10 μA are often observed. These effects may be eliminated by either biasing the crystal at a negative potential to retard the collection of electrons, or by using a separately negatively biased aperture, as shown in Figure 110D [15]. The effect of electron beam damage from stray electrons in the absence of these precautions can be profound in alterating the chemistry on the crystal surface.

Figure 110D also shows a helpful device useful for reproducibly locating the crystal at a fixed distance from the aperture [15]. A Ta contact with a gently rounded end is connected to the electrically isolated aperture, and the connection to the bias potential is opened. An ohmmeter, working on the MΩ scale, is connected between the Ta contact and ground. The crystal is slowly advanced to the Ta contact until MΩ resistance is measured as a result of very light contact. This allows one to calibrate the horizontal micrometer drive for the crystal at a fixed point in the apparatus. The micrometer drive is then retracted by a measured amount, and knowing the position of the Ta contact allows one to calculate the position of the crystal relative to the aperture.

The fractional interception of desorbing gas from the crystal will depend on the distance between the crystal and the aperture and also on the angular distribution of the desorbing gas, expressed as $\cos^n\theta$. For a crystal of 3.5 mm radius, and an aperture of 0.8 mm radius, the relative collected signal for various values of $n$ is shown in Figure 110E. It is noted that only small losses are observed in the first 1-mm crystal-aperture distance, and also that an error in distance will make a smaller effect in the first 1-mm range compared to errors at larger ranges [16].

A critical review of the advantages and disadvantages of using a shielded mass spectrometer for temperature programmed desorption studies has been given by Schlichting and Menzel [17]. A shielded mass spectrometer such as that shown in Figure 110A will have a constant pumping speed determined by the geometry of the aperture and the opening at the rear, whereas the pumping speed of a vacuum system used with an unshielded mass spectrometer will be variable. Also effects due to non-Boltzmann energy distributions and noncosine angular distributions in the desorbing flux will be reduced by the shield, which is primarily sensing molecules scattered from its inner walls. These effects can be totally eliminated by internally blocking the ionization source from the sample by means of a shutter [17]. The disadvantages are the dependence of the detected signal on the distance from the crystal to the aperture (Fig. 110E), and the contribution of readsorption on the crystal during the TPD measurement because of gas readmission from the shielded mass spectrometer through the sampling aperture. By considering these factors, the dynamic range of the TPD coverage and kinetic measurement technique has been extended from monolayer detection down four or five orders of magnitude [17].

For the desorption of complex molecules, or mixtures of species, it is important to employ multiplex methods in which various mass peaks are rapidly monitored in sequence many times during the desorption process. A recent development involves monitoring the complete mass spectrum consecutively for many cycles during the evolution of thermal desorption products [18,19].

A comparison between the sensitivity of a mass spectrometer in measuring gas pulses and also in measuring steady-state pressures, compared with a nude ionization gauge, has been published [20]. This paper shows that for eight different gases, the ionization gauge signal and the mass spectrometer signal are linear with the gas pulse size over six orders of magnitude for pulses ranging from $10^{-11}$ atm cm$^3$ to $10^{-5}$ atm cm$^3$. The authors were surprised at the high sensitivity of the Bayard–Alpert gauge in these measurements [20].

## *References*

[1] J.B. Taylor and I. Langmuir, Phys. Rev. **44,** 423 (1933).
[2] J.A. Becker and C.D. Hartman, J. Phys. Chem. **57,** 153 (1953).
[3] An excellent review may be found in: G. Ehrlich, Adv. Catal. **14,** 255 (1963).
[4] P.A. Redhead, Trans. Faraday Soc. 57, 641 (1961).
[5] J.T. Yates, Jr., in *Methods of Experimental Physics,* Vol. 22. (Academic Press, Orlando, FL, 1985) p. 425.
[6] A.M. de Jong and J.W. Niemantsverdriet, Surf. Sci., **233,** 355 (1990).
[7] T.E. Madey, J.T. Yates, Jr., and R.C. Stern, J. Chem. Phys. **42,** 1372 (1965).
[8] P.W. Tamm and L.D. Schmidt, J. Chem. Phys. **51,** 5352 (1969).
[9] T.E. Madey and J.T. Yates, Jr., in *Structure et Properties des Surfaces des Solides No. 187* (CNRS, Paris, 1970), p. 155.
[10] P.W. Tamm and L.D. Schmidt, J. Chem. Phys. **54,** 4775 (1971).
[11] T.E. Madey, Surface Sci. 29, 571 (1972); T.E. Madey and J.T. Yates, Jr., Surface Sci. **63,** 203 (1977).
[12] P. Feulner and D. Menzel, J. Vac. Sci. Technol. **17,** 662 (1980).
[13] S.M. Gates, J.N. Russell, Jr., and J.T. Yates, Jr., Surface Sci. **146,** 199 (1984).
[14] G. Lu, A. L. Linsebigler, and J.T. Yates, Jr., J. Vac. Sci. and Technol. **A12,** 384 (1994).
[15] V.S. Smentkowski and J.T. Yates, Jr., J. Vac. Sci. and Technol. **A7,** 3325 (1989).

[16] A. Winkler and J.T. Yates, Jr., J. Vac. Sci. Technol. **A6,** 2929 (1988).
[17] H. Schlichting and D. Menzel, Surface Sci. **285,** 209 (1993).
[18] L.H. Dubois, Rev. Sci. Instrum. **60,** 410 (1989).
[19] A.C. Liu and C.M. Friend, Rev. Sci. Instrum. **57,** 1519 (1986).
[20] W.M. Graven and D.E. Gilmartin, Rev. Sci. Instrum. **55,** 1827 (1984).

***Fig. 110. Shielded Quadruple Mass Spectrometers—Temperature-Programmed Desorption***

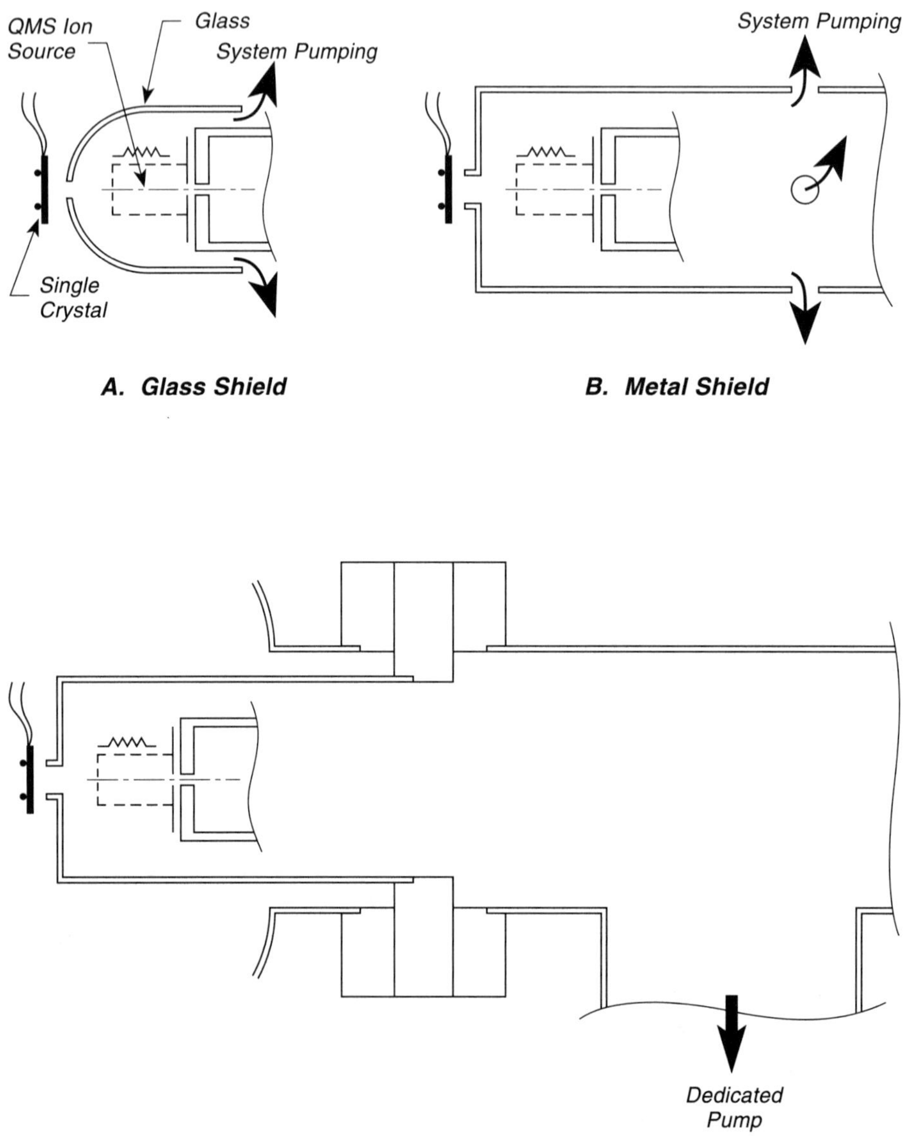


***A. Glass Shield***

***B. Metal Shield***

***C. Differentially Pumped Shield***

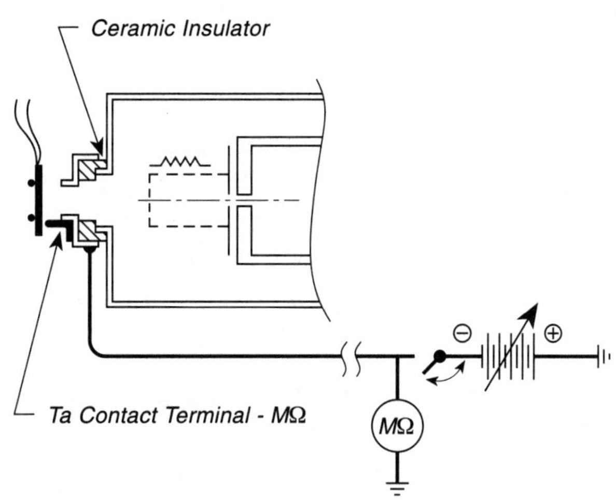


**D. Stray Electron Suppression + Crystal Positioning Contact**

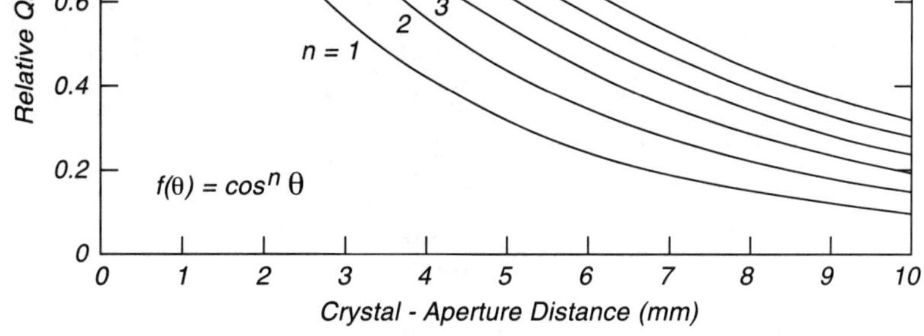


**E. Relative Signal for Desorption into Aperture**

# 111

# Enhanced Signal-to-Noise in Quadrupole Mass Spectrometers

The quadrupole mass spectrometer is widely used in surface science for residual gas analysis as well as for the study of desorbing species from surfaces. Thermal desorption, and electron- and photon-induced desorption (including laser-induced desorption) have been widely studied. For most quadrupole mass spectrometers, a channeltron multiplier is employed for detection of the mass-separated ions. A method is described here for improvement of the signal-to-noise ratio of a channeltron detector; and in addition, the behavior of a channeltron detector is also described [1].

An electron multiplier has a low inherent noise, leading to a dark current. In the case of a channeltron multiplier mounted behind quadrupole rods with line of sight to the ionization source, there are additional background effects that increase as the electron emission current increases and also as the pressure in the mass spectrometer is increased. While these effects are not fully understood [1], they are thought to be due to soft X-rays and UV photons produced in the ion source, and also to excited neutral species, both of which pass down through the quadrupole rods without deflection. The use of an off-axis ion multiplier enables one to eliminate these effects to a large degree, and the modification described here accomplishes this while not causing unacceptable reduction of the ion collection efficiency.

Figure 111A shows the off-axis channeltron detector located beneath the quadrupole rods. Between the exit region of the quadrupole rods and the channeltron entrance are two ion deflection plates, designed to steer ions into the channeltron. The channeltron is located 10 mm off axis. A multiplier shield plate, 8 mm in diameter, is mounted 2 mm in front of the channeltron entrance cone. A deflector plate, with a 45° angle to the shield plate, is mounted on the axis of the quadrupole, and a second deflector plate is mounted in a direction perpendicular to the shield plate, and 15 mm from the quadrupole axis. The distance between these deflector plates and the exit aperture of the quadrupole lens is 2 mm. The 45° deflector plate may be biased by an isolated lead, which is separately shielded. The entire structure is housed in a shield can to prevent both photons and charged particles from entering the channeltron.

Figure 111B shows the performance of the channeltron for various applied multiplier voltages, and also for various positive deflector plate voltages. The large enhancement of output current for a slight increase of multiplier voltage is evident from the three curves; in addition, a large effect of the positive deflector plate is also seen, with maximum transmission of ions being observed at +10 V plate bias.

Figure 111C shows the behavior of the dark current versus electron emission current in the ionization source of the mass spectrometer. A slope of 1.3 in this log-log plot indicates that slight nonlinearities exist as the filament temperature is raised in the ionization source to produce a larger electron emission current. A 100-fold improvement in the dark current is observed for the off-axis detector.

Figure 111D shows the behavior of the dark current as a function of Ar pressure in the mass spectrometer, with data taken at 1 mA emission current. The dark current rises as the 1.1 power of the pressure in both of these measurements. The off axis detector reduced the dark current here by a factor of 140.

The performance of the mass spectrometer with the two configurations of the channeltron detector can be summarized in Equations (1) and (2) below:

$$\text{For on-axis detector:} \quad I_i = 3.5 \cdot P^{1.1} \cdot I_e^{1.3} \tag{1}$$

$$\text{For off-axis detector:} \quad I_i = 0.0025 \cdot P^{1.1} \cdot I_e^{1.3} \tag{2}$$

where $I_i$ = measured ion current; $P$ = Ar pressure; and $I_e$ = electron emission current.

An ion current of the order of $10^{-18}$ A can be detected with the off-axis channeltron.

Earlier improvements in the detection efficiency of quadrupole mass spectrometers are described in reference [2]. The regeneration of Ag-Mg and Cu-Be dynode multipliers by heating to 300°C in oxygen gas at 1 atm has been described [3].

## *References*

[1] F. Nakao, Rev. Sci. Instrum. **46,** 1489 (1975).

[2] P. Marchand, C. Paquet, and P. Marmet, Rev. Sci. Instrum. **37,** 1702 (1966).

[3] J.R. Young, Rev. Sci. Instrum. **37,** 1414 (1966).

*Fig. 111. Enhanced Signal-to-Noise—QMS*

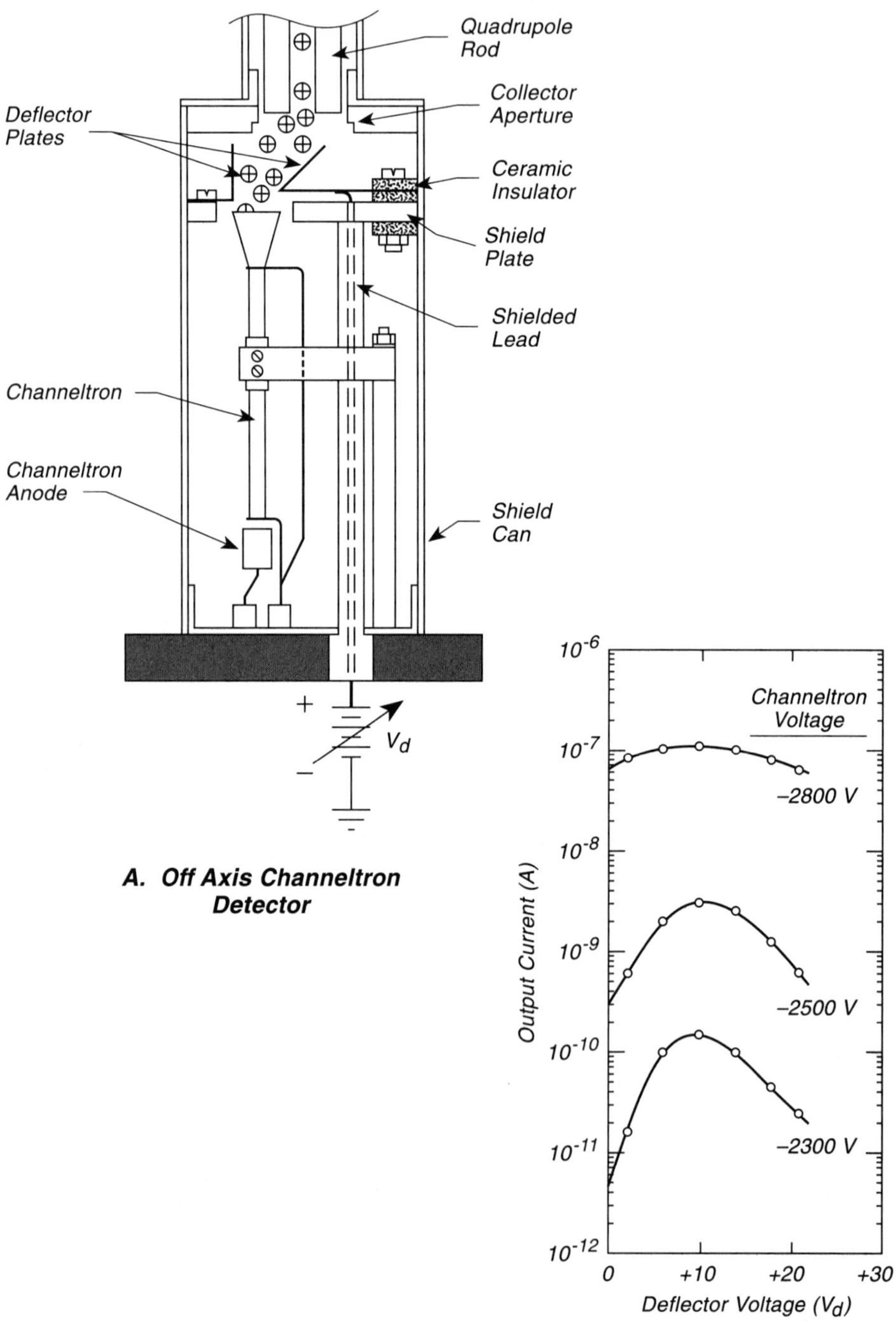


**A. Off Axis Channeltron Detector**

**B. Detector Performance**

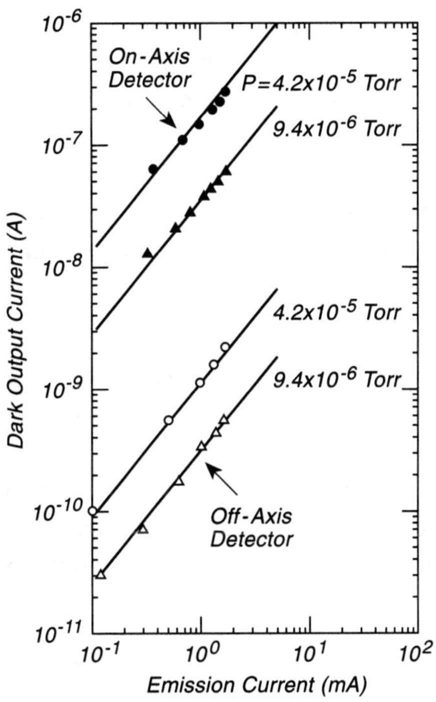


**C. Dark Current Behavior**

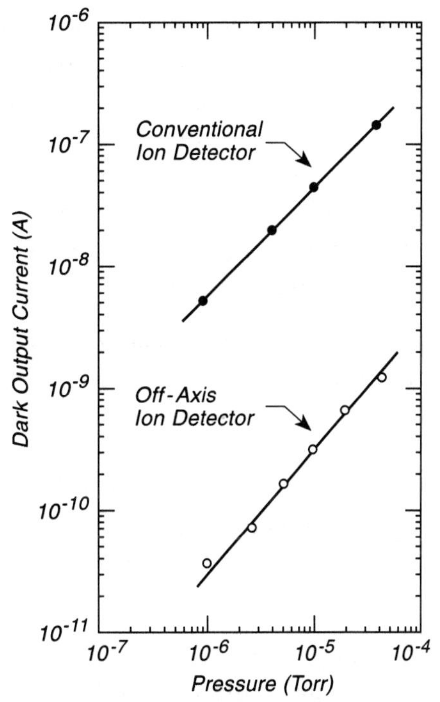


**D. Dark Current Behavior**

# 112

# Elimination of Spurious Electron Emission from a Quadrupole Mass Spectrometer

Electrons emitted from the ion source of a mass spectrometer can be a source of error in measurements when the electrons strike and damage a sample in front of the mass spectrometer. For an enclosed mass spectrometer, with an aperture, placing a negative bias on the aperture will repel electrons and prevent this effect (see page 366). The effect of spurious electron emission may also be eliminated by biasing the sample negative to repel electrons. A third method involves the use of a floating grid surrounding the mass spectrometer ion source, which is automatically brought to the potential of the thermionic emitter (usually -70 V in a mass spectrometer). This occurs as a result of the charging of the isolated grid by the emitted electrons [1].

Figures 112 A–C show the steps in rebuilding a mass spectrometer ionizer to provide the floating grid. In Figure 112A, an inner grid cap is placed over the existing inner grid cage. In Figure 112B, the existing electron repeller grid cage is placed on the ion source. Inside the repeller grid is the thermionic emitter assembly, not shown in the drawing. Finally, in Figure 112C, the floating grid, consisting of a cylindrical grid cage capped with a planar grid, is installed. The floating grid is electrically isolated by means of a ceramic ring that can be bolted to the mass spectrometer housing by means of three standoff bolts as shown.

The floating grid at negative potential is capable of extracting positive ions from the ionizing region inside the inner grid. This is the reason for the addition of the inner grid cap on the ionizing region, which acts as a shield to prevent positive ion extraction.

The floating grid collects electron current so that the electron emission current to the inner grid cage is somewhat reduced. Thus, in the UTI mass spectrometer, the filament protect circuit must be deactivated, since it will trip when it senses too little current at the inner grid. With the floating grid design, spurious emitted electron currents are reduced from about 10 μA to less than 1 nA [1].

A large $O^+$ signal and a large $F^+$ signal may be observed in the mass spectrometer operating with the floating grid (the latter especially if fluorinated molecules are introduced to the chamber deliberately). These signals probably originate from electron stimulated desorption effects from the ceramic ring and from the grid material itself [1]. Fluorine, a well-known surface contaminant in UHV systems, with a very high cross section for electron-stimulated desorption as a positive ion, is often seen when contaminated surfaces are electron bombarded [2].

## *References*


[1] Professor Patricia Thiel, Department of Chemistry, Iowa State University, Ames, Iowa 50011 (private communication); taken from the Ph. D. Thesis of M.M. Walczak, Iowa State University (1989); the original idea comes from A.L. Johnson, Ph.D. Thesis, University of California, Berkeley (1986).

[2] J.T. Yates, Jr. and D.A. King, J. Vac. Sci. Technol. **9**, 1256 (1972).

**Fig. 112. Elimination of Spurious Electron Emission from a Quadrupole Mass Spectrometer**

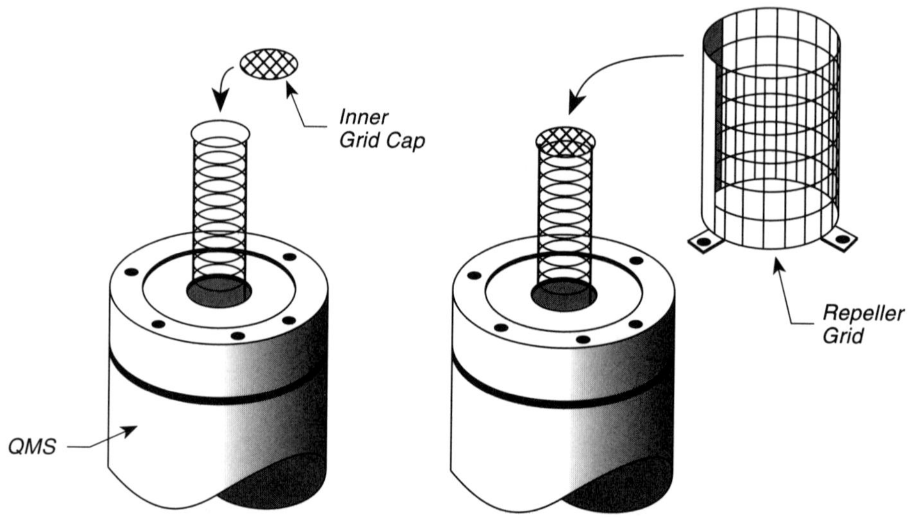


**A. Placing Inner Grid Cap**

**B. Placing Repeller Grid + Emitter (not shown)**

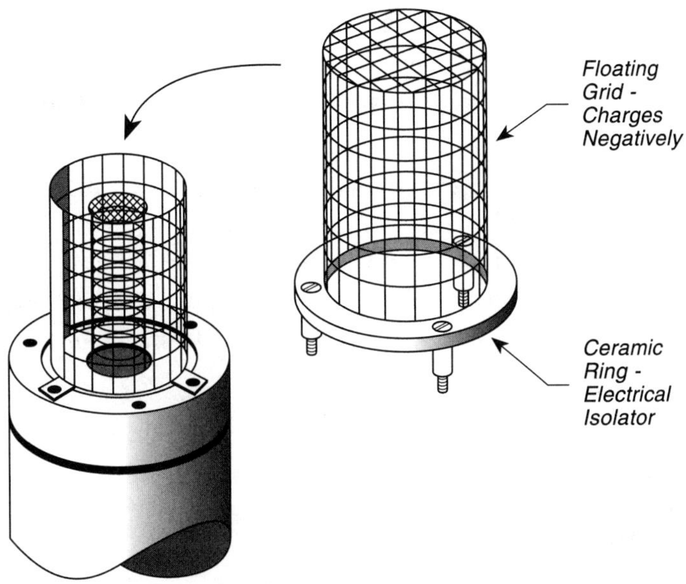


***C. Placing Floating Grid***

# 113

# Ta Mass Spectrometer Filament

Many mass spectrometers use either W, thoriated W, or thoriated Ir thermionic emitters as described on page 202. The use of Ta as an emitter offers some advantages [1] as described below.

Ta is much more ductile than W at room temperature and can be wound and welded easily compared to W. Ta wire (0.006-in. diameter) is wound around a rod (typically 0.040-in. diameter) to form a coil before being installed into the mass spectrometer as shown in Figure 113.

A Ta emitter has a built-in property that is sometimes useful. Ta has a sufficiently high vapor pressure that at filament operating temperature there is a detectable signal from $^{181}Ta$, the most abundant isotope. This signal for the $^{181}Ta^{+2}$ ion, observed at M/q = 90.5, may be used as a mass spectrometer calibration point, and interference from other ions in this region will be absent. The lifetime of the Ta filament will be years despite its slow evaporation during use.

## *References*


[1] Professor J.B. Hudson, Materials Engineering Department, Rensselaer Polytechnic Institute, Troy, NY 12180 (private communication).

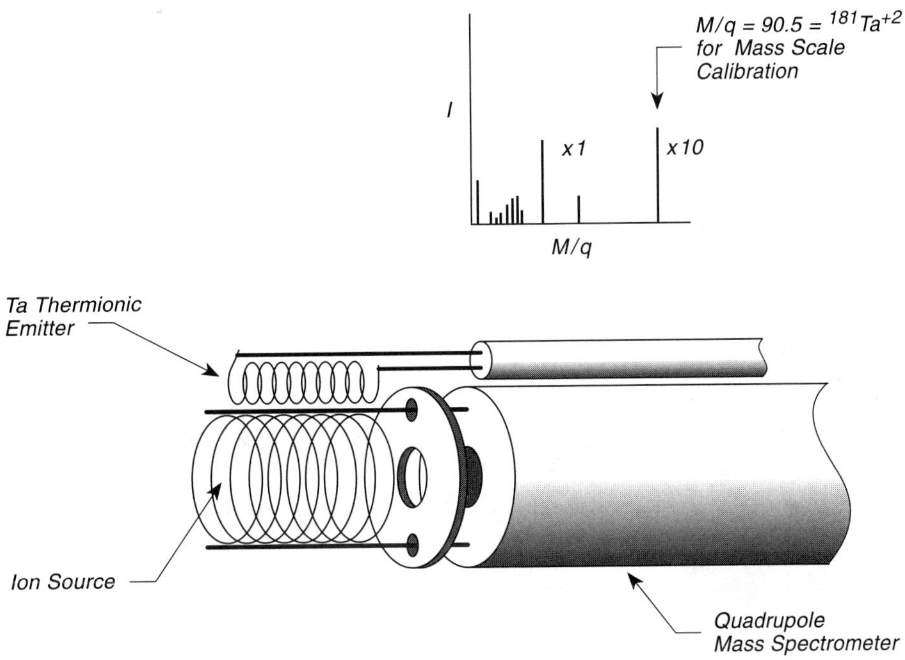


**Fig. 113. Ta Mass Spectrometer Filament**

# 114

# Mass Spectrometer Calibration by Molecular Decomposition

Many factors make it difficult to control the partial pressures of two or more gases in a pumped vacuum system and to know the absolute ratios of partial pressures, much less the absolute partial pressures themselves. The factors that prevent such calibrations have to do with fundamental issues such as the pumping speed, $S_x$, for individual gases, the relative differences in sticking coefficients on the chamber walls for different molecules, and the absence of absolute knowledge of gas fluxes from calibrated dosers, etc. One way to overcome some of these difficulties is to utilize the thermal decomposition of molecules, containing known stoichiometric ratios of the elements, and to measure the relative signals from the gaseous decomposition products under dynamic pumping conditions [1].

Figure 114 shows a general arrangement that may be utilized for these measurements. A molecular beam consisting of the pure compound is directed at a heated surface that is known to decompose the compound, producing only gaseous products. A mass spectrometer samples the desorbing gas products and yields signals of the mass spectrometer cracking products and parent ions of the product molecules from the decomposition.

The absolute mass spectrometer sensitivity, $\beta_x$, is defined as

$$\beta_x = kT_g / S_x \eta_x \tag{1}$$

Here we define $\eta_x$ as a proportionality constant such that $P_x = \eta_x I_x$ and is often called the mass spectrometer sensitivity, converting an ion signal into a partial pressure. The quantity $\eta_x$ contains factors such as the ionization probability of the molecule, the transmission probability of the ions through the mass spectrometer, and the electron multiplier gain, which is generally considered to have a $M^{-1/2}$ dependence, where $M$ = molecular mass.

On this basis, the ratio $\beta_x/\beta_y$ will be the relative mass spectrometer sensitivity for the two species, $x$ and $y$.

If we can generate the two species in a known stoichiometric ratio by decomposition of a molecule on a hot surface, and if we know the pumping speeds for the two molecules from their molecular weights (scaling as $M^{-1/2}$), then the relative sensitivities $\beta_x/\beta_y$ can be determined. It is of course important to measure changes in $I_x$ above the background and to consider that a particular mass spectrometric ion may contain contributions from the cracking of more than one type of molecule. It is also important to remember that this method assumes that the angular distribution of all of the product molecules from the molecular decomposition is the same, and probably of cosine form if surface equilibrium is achieved before release of the product molecules.

The method has been used for two molecules to determine the relative sensitivies for four gases as diagrammed below:

$$H_2CO \rightarrow H_2 + CO \tag{2}$$

$$4DCOOH \rightarrow D_2 + 2HD + H_2 + \text{other products} \tag{3}$$

In the case of the isotopically labeled formic acid, an equilibrium mixture of the three hydrogen isotopes will be produced, since possible kinetic isotope effects changing this ratio are likely to be small at the high surface temperature of the decomposition.

For the vacuum system used in reference [1], in which a UTI-100C mass spectrometer was employed, the following relative sensitivities were determined:

$$\beta_{H_2} : \beta_{HD} : \beta_{D_2} : \beta_{CO} = 1.00 : 0.80 : 0.63 : 0.98 \tag{4}$$

## *References*

[1] Y.-K. Sun and W.H. Weinberg, J. Vac. Sci. Technol. **A7**, 3121 (1989).

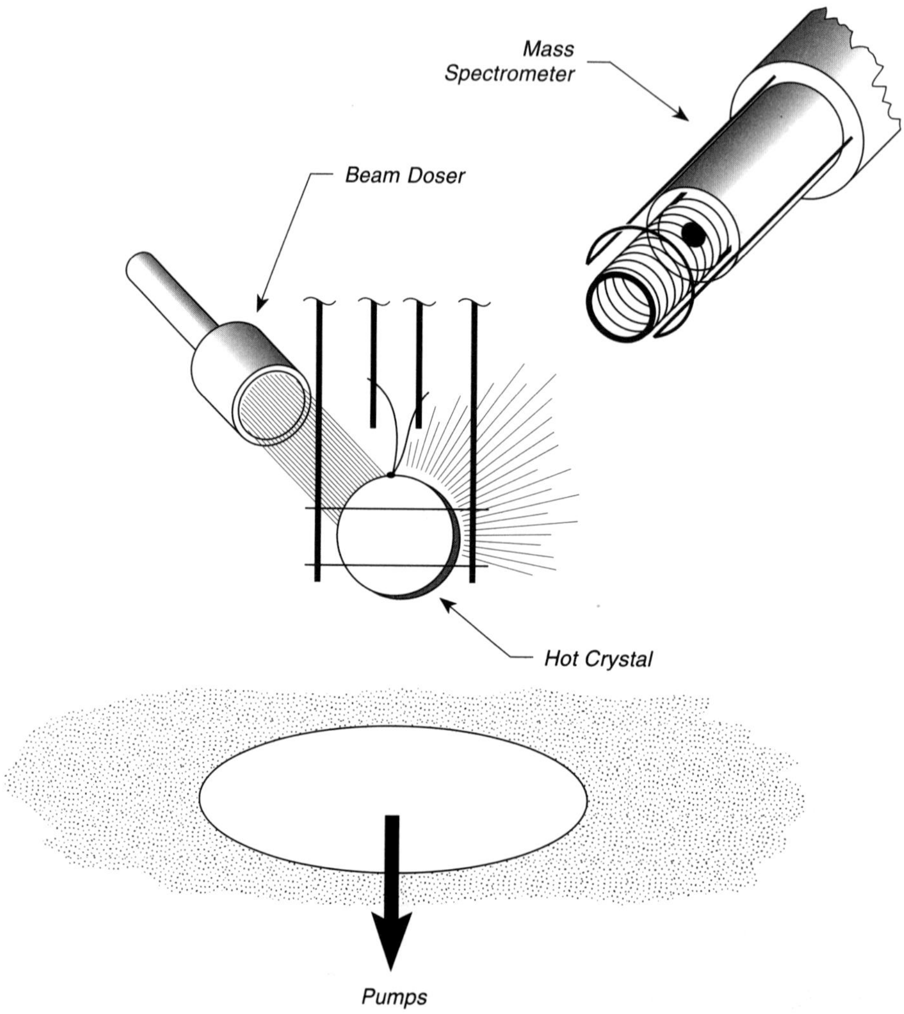


**Fig. 114. Mass Spectrometer Calibration by Molecular Decomposition**

# F. Temperature-Programmed Desorption/Reaction

# 115

# Thermal Desorption—Angular Resolved

The measurement of the angular distribution of molecules desorbing from single crystal surfaces provides information on the dynamics of the desorption process. For desorption from completely equilibrated surface species, a cosine distribution is expected. However, often a noncosine distribution is observed, and this may be related to processes such as atom recombination at a point on a repulsive adsorbate-surface potential energy curve in the region of the transition state. A simple apparatus to measure the angular distribution is shown in Figure 115A [1].

A collimator is used to select particles desorbing at a certain angle, $\Theta$ , with respect to the surface normal. The crystal is of rectangular geometry (10 mm x 10 mm x 1 mm), and the apertures have dimensions $a_1$ = 1.5 mm x 6 mm; $a_2 = a_3$ = 30-mm diameter; $a_4$ = 6 mm x 10 mm). The length of the collimator is 100 mm and the distance to the crystal is 23 mm in these measurements. The acceptance angle outside and inside the collimator is 2.1°, and a region on the crystal measuring 3 x 9 mm$^2$ will be sampled in the normal direction; at 70°, a 9 x 9 mm$^2$ region will be sampled. An angular distribution of form $\cos^n\Theta$ will give a signal proportional to $\cos^{(n-1)}\Theta$.

The collimator is internally pumped by means of a Ti getter film which is deposited by evaporation from a Ti wire source, and this will act as a very fast internal pump for the collimator (see page 118).

The ionization gauge is located off-axis so that only a random flux of gas, thermally equilibrated with the walls of the chamber containing the gauge and ion pump, will be measured. The ionization gauge was thermostatted by means of a water bath, and the ion pump was operated with a stabilized DC power supply to reduce fluctuations in pumping speed. It was found that a pulse of $H_2$ into the ionization gauge chamber resulted in a short period decrease in the background signal, so for measurements of $H_2$, a continuous minute constant flux of $H_2$ was added. No difficulties of this kind were observed with CO.

Figure 115B shows typical desorption spectra for $H_2$ from Ni(111) that has been saturated with $H_2$. The angular distributions of both $H_2$ and CO from Ni(111) are shown in Figure 115C. It can be seen that CO desorbs in the two azimuths measured with a $\cos\Theta$ distribution, whereas $H_2$ has a sharper distribution, described as $\cos^n\Theta$, where $n$ = 4.6 or 4.8.

## *References*

[1] H.P. Steinruck, A. Winkler, and K.D. Rendulic, Surf. Sci. 152/153, 323 (1985).

## Fig. 115. Thermal Desorption—Angular Resolved

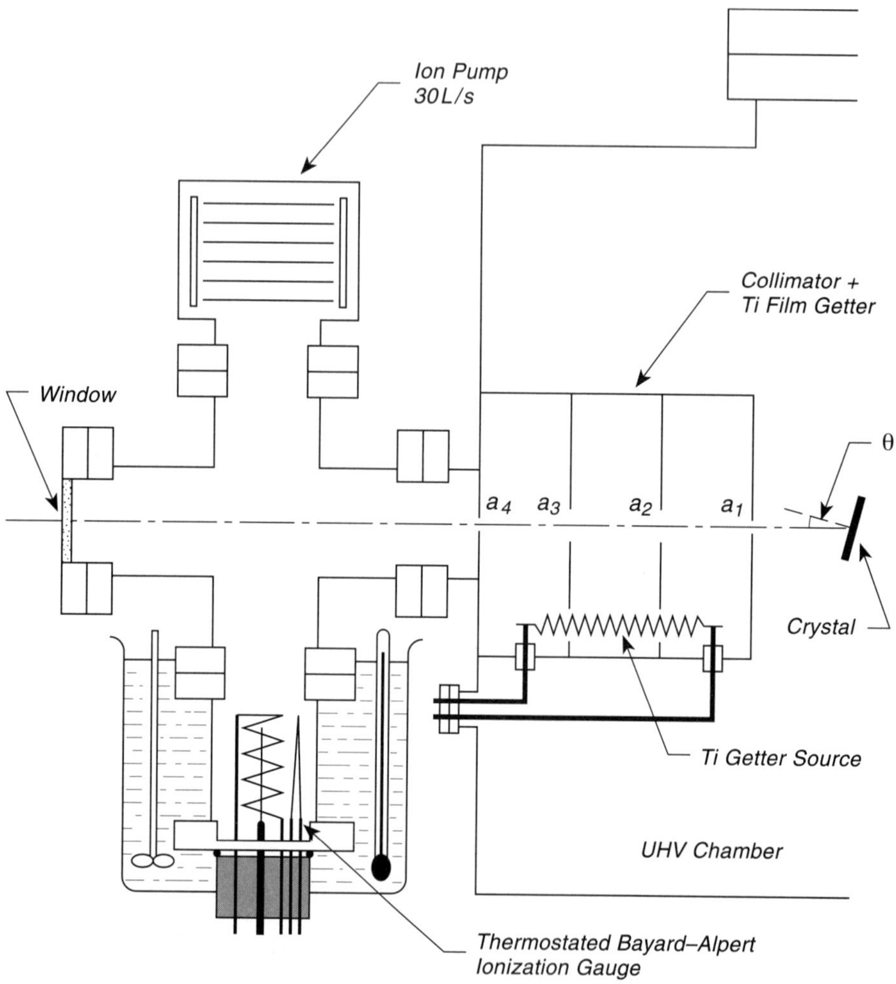


**A. Desorption Apparatus**

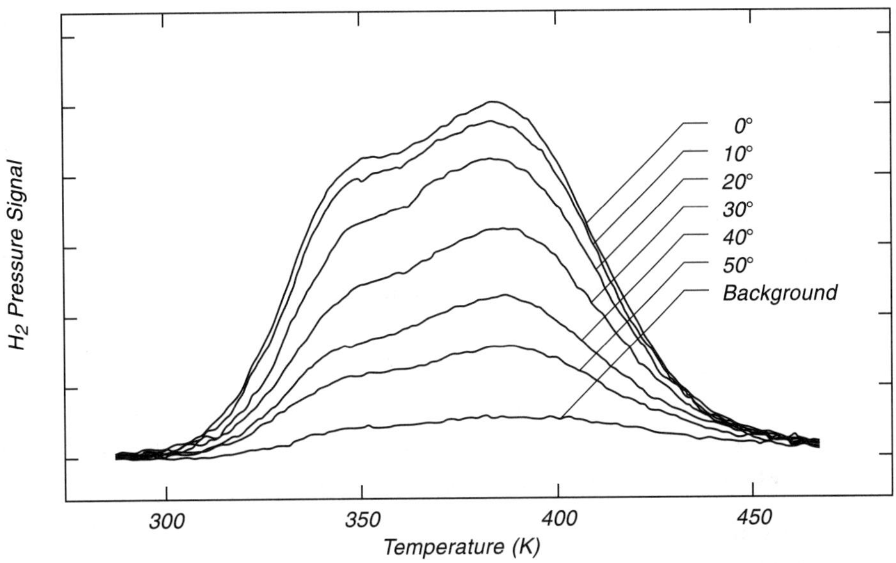


**B. Desorption Spectra - $H_2$ / Ni (111)**

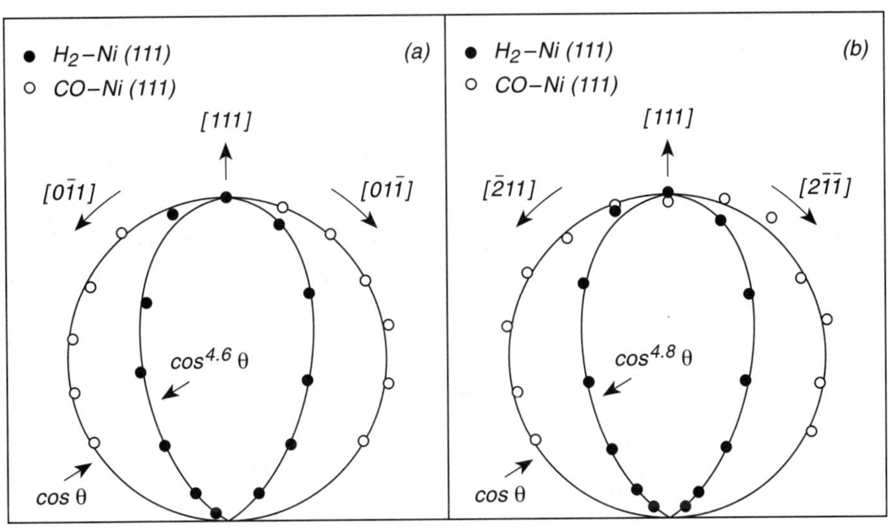


**C. Measured Angular Distribution for $H_2$ and CO**

# 116

# Magic-Angle Thermal Desorption Mass Spectroscopy

The thermal desorption of species from surfaces is widely used throughout the field of surface science for studies of adsorbates on metal foils and single crystal surfaces (see page 386). Often an unshielded mass spectrometer is used for detection of the desorbing particles and it will generally measure the density of gas produced in a random flux within the vacuum chamber. More modern methods involve the use of apertured mass spectrometers that are differentially pumped within the surrounding shield (see page 118). The aperture is often close-coupled to the crystal surface along the surface normal in order to collect the majority of the desorbing gas evolved from the region of the crystal viewed by the aperture. Under these conditions of efficient collection, considerations of the angular distribution of the desorbing gas are of second-order importance.

However, when the apertured mass spectrometer is located far from the crystal surface, considerations of the angular distribution of the desorbing gases become very important, especially if measurements are made of the relative signal from different gases desorbing simultaneously (and sampled by multiplex methods). Thus, gases are known to desorb with angular distributions approximated by $\cos^n\Phi$ where $\Phi$ is the angle from the normal. Here $n$ can have a value of unity, or values running to about 20 in extreme cases. If the value of $n$ differs significantly for two species desorbing simultaneously, then errors in estimating the relative signal intensities will occur, and these errors can approach 400% for collection along the normal direction when the mass spectrometer is distant from the crystal. Pauls and Campbell [1] have considered this problem and have found that there is a magic angle for the mass spectrometer location that will minimize the error in comparing desorption signals with $n$ varying over the range 1 to 9. Figure 116A shows the geometry of the mass spectrometer collection optics, where $r$ is the crystal radius, and $d$ is the distance to the aperture. The polar angle $\Phi$ is shown. Figure 116B shows the magic angle, which minimizes the error in the relative magnitude of the desorption signals having varying angular distribution, where an integration over the emitting surface of the crystal has been carried out. The function shown varies as the value of the ratio $d/r$ varies. For typical values of $d/r$ from 2 to 5, the magic angle varies from 42° to 34°, and at large values of $d/r$, the optimal angle of $\Phi_m = 34°$ is found. The variations in the signal for the range of $n$ considered is only about 26% at $\Phi_m = 42°$.

It should be mentioned that the mass spectrometer detection behavior may vary between two limits. For very fast pumping within the shield, only a single pass of the desorbing species occurs through the ionizing optics, and considerations of the molecular velocities of the gas molecules would be necessary in measuring relative fluxes [2]. For very slow pumping in the shield, which is the normal method of operating the shielded mass spectrometer at high sensitivity (see page 366), the desorbing particles will probably acquire the temperature of the ionizing source and then these considerations of molecular velocity will be unneeded. The detector signal is then truly proportional to the flux entering the aperture [1].

These considerations also apply to the measurement of any particle desorption technique, for example sputtering of neutral species from surfaces [3].

## *References*

[1] S.W. Pauls and C.T. Campbell, Surf. Sci. **226,** 250 (1990).

[2] G. Comsa and R. David, Surf. Sci. Rep. **5,** 145 (1985).

[3] A. Wucher, F. Novak, and W. Reuter, J. Vac. Sci. Technol. **A6,** 2265 (1988).

Fig. 116. Magic-Angle Thermal Desorption Spectroscopy

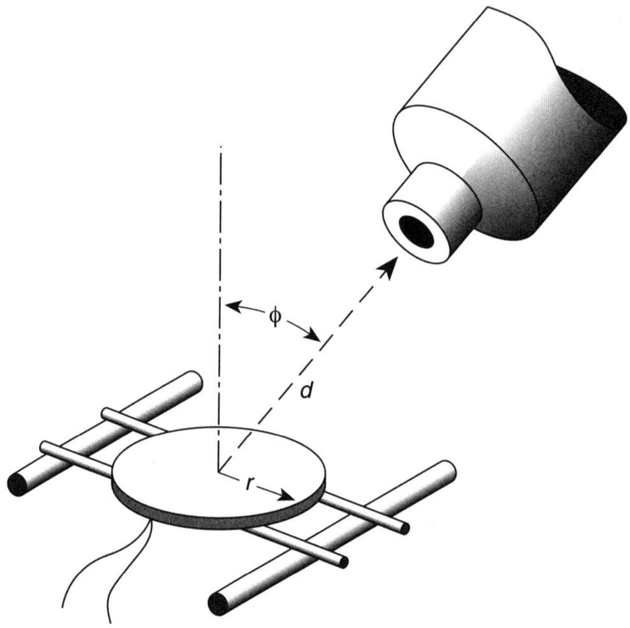


A. Mass Spectrometer Aperture—Collection Angle

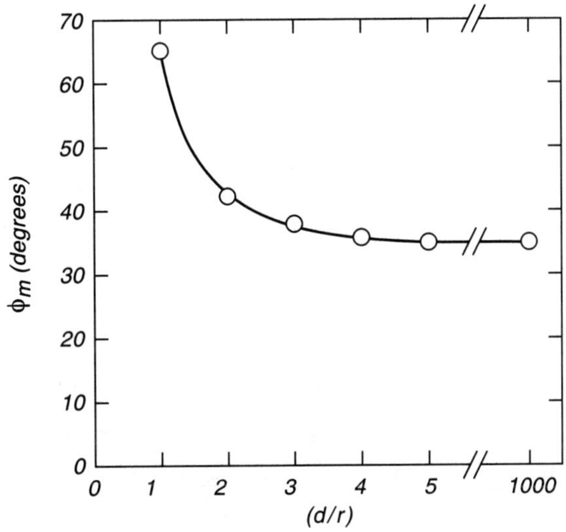


**B. Magic-Angle, $\phi_m$, for Aperture Collection**

# 117

# High-Temperature Thermal Desorption Spectroscopy

The extension of thermal desorption spectroscopy into temperature regions well over 2000 K involves the use of electron bombardment heating rather than ohmic heating. Electron bombardment heating introduces experimental complications into the method, which have been overcome as shown below [1].

The experimental difficulties introduced by electron bombardment heating are listed below:

1. The accelerating potential for electrons must be in the kV region. If the crystal is grounded, and the emitter is biased negatively, then spurious desorption may occur from regions other than the crystal that are also at ground potential. Therefore, biasing the crystal positive is the preferred mode. This requires that the thermocouple be measured using a high-voltage isolation amplifier, and the amplifier must be able to handle short high-voltage common mode signals if shorting occurs between the sample/thermocouple assembly and grounded regions of the support assembly. A suitable high-voltage precision isolation amplifier is commercially available [2].

2. Because of radiative heating of the emitter filament by the incandescent crystal, causing enhanced thermionic emission, the heating current to the emitter must be controlled by a computer working in feedback style. The LabVIEW computer (see page 498) and program are ideal for this, but another program was used in the work described here.

3. The mechanical mounting of the spiral emitter filament is critical. It must be rather close to the back of the crystal, and it should be mounted in a strain-free condition prior to first heating. Even then, distortion of the filament may occur, and because of its brittle character after heating, readjustment of its position by bending the W is impossible. However, if the filament is mounted on ductile Ta posts, they may be slightly adjusted by bending after the filament is embrittled, moving the filament to the correct position. Spiral W emitters, wound to specified dimensions, are commercially available [3].

4. The thermocouple wires, because of their small radius of curvature, will concentrate the electric field lines to themselves, and will undergo electron bombardment with a higher current density than the flat crystal, causing excess local heating. This may be avoided by mounting the thermocouple close to the top of the crystal, and placing the emitter closer to the bottom as shown in Figure 117A.

The desorption of Pt from a W(211) crystal is shown in Figure 117B, where a quadrupole mass spectrometer, working in the line-of-sight mode was employed for detection [1].

## *References*

[1] K. Pelhos and T.E. Madey, Department of Physics and Astronomy, Laboratory for Surface Modification, Rutgers, The State University, Piscataway, NJ 08855 (private communication).

[2] Model AD210AN. Analog Devices, One Technology Way, Norwood, MA 02062-9106.

[3] C.M. Furnaces, Inc., 103 Dewey Street, Bloomfield, NJ 07003.

***Fig. 117. High-Temperature Thermal Desorption Spectroscopy***

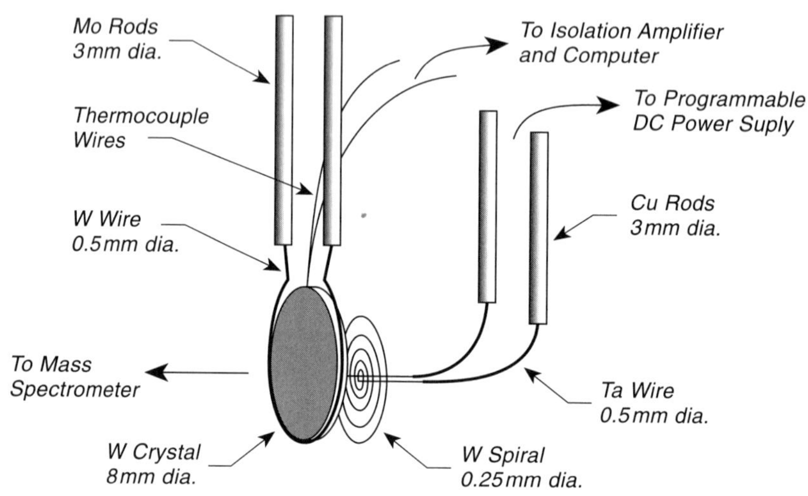


***A. Electron Bombardment Design***

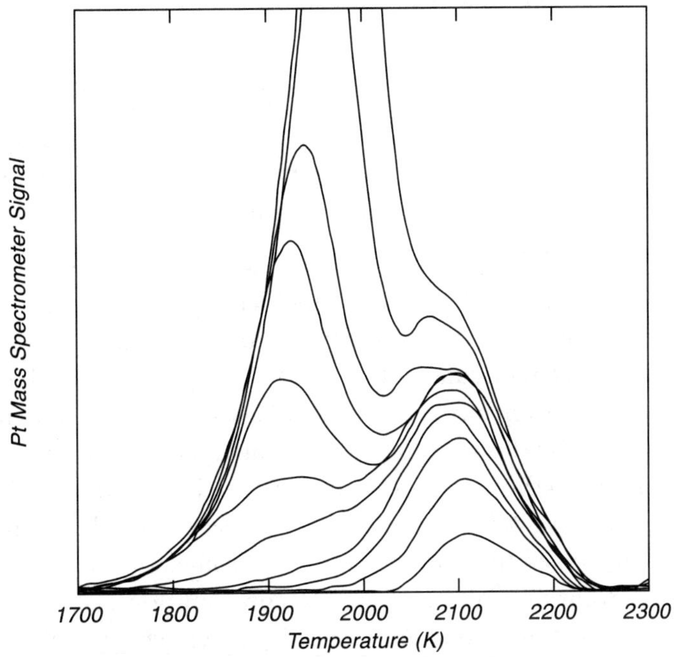


***B. Pt Thermal Desorption from W(211)***

# 118

# Scanning Kinetic Spectroscopy—A Survey Method for Investigation of Surface Reaction Processes

Multiple reaction pathways are available to a polyatomic molecule interacting with a solid surface. It is often useful, in a survey experiment, to delineate the temperature regions where various reaction steps are either active or inactive. The scanning kinetic spectroscopy (SKS) method [1] permits this to be done efficiently in many cases [2,3].

Figure 118A shows an SKS apparatus for the study of a molecule's decomposition routes on a single crystal surface. The cleaned single crystal is first covered with a monolayer of the adsorbate to be studied at a temperature below the dissociation temperature of any of the bonds in the molecule. At this point, the crystal is rotated into a beam of the molecules originating from a collimated beam doser, as described on page 604. Initially, the incident beam is reflected off of the saturated monolayer on the crystal, providing a reference signal in the mass spectrometer which is located behind an apertured shield (see page 366) and is differentially pumped (see page 118). The crystal is then programmed upward in temperature, and the mass spectrometer records, in the multiplex mode, all of the relevant mass peaks related to consumption of the parent reactant molecule and to liberation of product molecules, as shown schematically in the bottom section of Figure 118A.

Figure 118B shows a typical set of SKS measurements, made in multiplex mode, which characterizes the temperatures where the various elementary steps occur. The steps themselves are shown in Figure 118C. Here, the reaction of the methanol molecule on Ni(111) is investigated.

Let us follow the processes observed in Figure 118B and outlined in Figure 118C. In this example, the isotopically labeled molecule, $CD_3OH$, is being employed as the reactant at a known flux onto the crystal. The covered crystal is rotated into the $CD_3OH$ beam at the double-headed arrow on the left. In the top panel of Figure 118B, at temperatures below 187 K, $CD_3OH$ merely reflects off of the monolayer of $CD_3OH$ previously adsorbed, and a flat line for mass 33 is observed in this region. Upon beginning the temperature program, α-$CD_3OH$ is first observed to desorb, originating from undissociated methanol molecules. At the dashed line, 290 K, a β-$CD_3OH$ feature is sharply evolved. The β-process is due to an elementary step involving the recombination of adsorbed H (from $CD_3O$-H bond scission) with adsorbed methoxy, $CD_3O$. It is informative to note that when the isotopic molecule, $CD_3OD$, is monitored at mass 36, the β-process is also observed, and the D atom involved here can only come from the breaking of a C—D bond in adsorbed $CD_3O$. This probably occurs by the production of adsorbed formaldehyde, $CH_2O$ (a), shown (Fig. 118C) in process A, where the isotopic composition is not written for any of the molecules in Figure 118C. When mass 4 is monitored in Figure 118B, $D_2$ is observed to desorb at 330 K, opening up sites on the surface. When mass 28 is monitored in Figure 118B, CO is observed to desorb at 420 K, also opening up sites.

Three processes labeled $\gamma$ are also indicated on the top SKS spectrum. The process $\gamma_1$ for adsorbed $CH_3O$ formation by O—H bond scission is not defined, since a volatile product is not liberated from this process, but O—H bond scission must begin below 290 K since the $\beta$-recombination process begins at 290 K. The process $\gamma_2$ and $\gamma_3$ involve the consumption of the reactant molecule since the reflected flux of reactant falls below the reference line in this temperature region. These consumption processes are closely correlated with the $D_2$ and CO desorption processes, shown near the bottom of Figure 118B, which open Ni sites for adsorption and reaction processes.

The SKS method is a fast survey method, giving one a partial view of the characteristic temperatures for various adsorption, desorption, and surface dissociation processes. It has been effectively used with intentional poisoning of the surface by added species, and with intentional addition of known reactants to the surface [1]. It has been successfully used to observe deuterium kinetic isotope effects that give insight into the rate-controlling steps in surface reactions involving the hydrogen moiety [1,3]. The original concept of temperature scanning while making reaction kinetic measurements on single crystal surfaces was first presented in 1981 by Goschnick and Grunze [4]. The SKS method also resembles the familiar temperature-programmed reaction spectroscopy (TPRS) [5,6] (see page 404), but SKS has added advantages of working like a flow reactor where fresh reactant is continuously supplied during the measurements.

The ability to rapidly scan many mass peaks and to record the time behavior of selected mass peaks has been a familiar subject for a number of years. Recently, a fast multimass measurement method has been introduced that permits up to 15 mass spectra of 50 amu width to be recorded per second, using a quadrupole mass spectrometer [7]. Another variation on this method scans up to 100 complete mass spectra per thermal desorption experiment. Enhanced signal-to-noise is achieved by integrating each of the mass peaks rather than by measurement of the peak height only [8].

There are many other mass-spectrometer based methods used for the study of surface reactions in which transient surface species may be detected. Among these is laser induced desorption, in which a Fourier transform mass spectrometer is often used [9]. Here, the laser may be used to desorb surface species without extensive thermally induced fragmentation.

## *References*

[1] S.M. Gates, J.N. Russell, Jr., and J.T. Yates, Jr., Surface Sci. **159**, 233 (1985).

[2] I. Chorkendorff, J.N. Russell, Jr., and J.T. Yates, Jr., J. Chem. Phys. **86**, 4692 (1987).

[3] P.L. Hagans, I. Chorkendorff, and J.T. Yates, Jr., J. Phys. Chem. **92,** 471 (1988).

[4] J. Goschnick and M. Grunze, J. Vacuum Sci. Technol. **18,** 561 (1981).

[5] R.J. Madix and J. Benziger, Ann. Rev. Phys. Chem. **29,** 285 (1978).

[6] R.J. Madix, in *Chemistry and Physics of Solid Surfaces,* Vol. II, edited by R. Vanselow (CRC Press, Boca Raton, FL, 1979).

[7] G. Zagatta, H. Muller, N. Bowering, and U. Heinzmann, Rev. Sci. Instrum. **65,** 359 (1994).

[8] L.H. Dubois, Rev. Sci. Instrum. **60,** 410 (1989).

[9] D.P. Land, C.L. Pettiette-Hall, D. Sander, R.T. McIver, Jr., and J.C. Hemminger, Rev. Sci. Instrum. **61,** 167 4 (1990).

***Fig. 118. Scanning Kinetic Spectroscopy (SKS)***

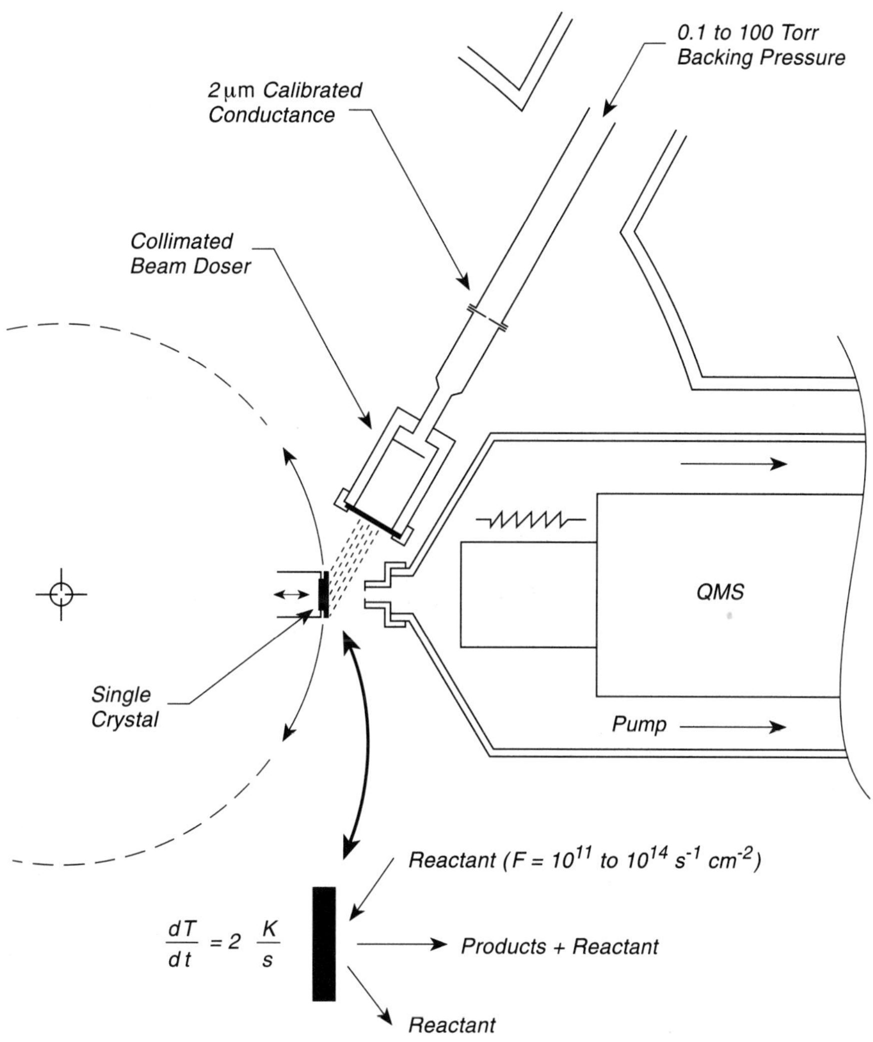


***A. SKS Apparatus***

***Fig. 118. Scanning Kinetic Spectroscopy (SKS)***

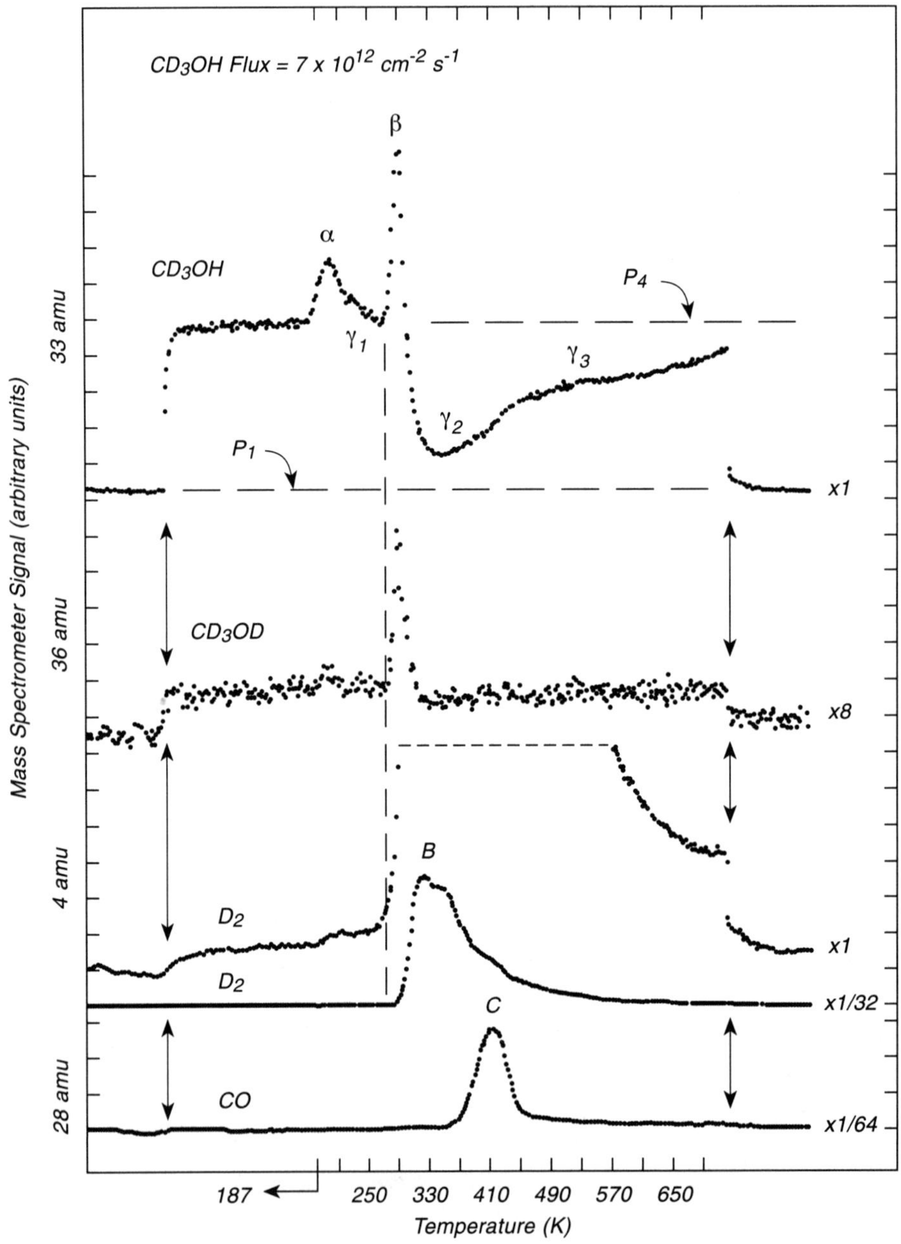


**B. Typical SKS Data**

| | |
|---|---|
| *Adsorption* → / ← *Desorption* | $CH_3OH\ (g) \underset{\alpha}{\rightleftarrows} CH_3OH\ (a)$ |
| $CH_3O$ *Formation* | $CH_3OH\ (a) \xrightarrow{\gamma_1} CH_3O\ (a) + H\ (a)$ |
| *Recombination* | $CH_3O\ (a) + H\ (a) \xrightarrow{\beta} CH_3OH\ (g)$ |
| $CH_3O$ *Decomposition* | $CH_3O\ (a) \longrightarrow CO\ (a) + 3\,H\ (a)$ |
| *Selective Decomposition* | $CH_3O\ (a) \xrightarrow{A} H\ (a) + CH_2O\ (a)$ |
| $H_2$ *Desorption* | $2\,H\ (a) \xrightarrow{B} H_2\ (g)$ |
| *CO Desorption* | $CO\ (a) \xrightarrow{C} CO\ (g)$ |

***C. Elementary Reaction Steps as Inferred from B***

# 119

# Temperature-Programmed Reaction Spectroscopy (TPRS)

This method extends the basic concepts of temperature-programmed desorption (TPD) (see page 366) to examine complex surface reactions. Multiple mass channels are monitored simultaneously during programmed heating. The method has been widely used to unravel the kinetic steps in both decomposition and synthesis reactions on single crystal surfaces. The method also resembles scanning kinetic spectroscopy (SKS) described on page 398.

An example of the use of TPRS is given here [1]. The interaction of ethylene glycol, $(CH_2{}^{16}OH)_2$, with fractional monolayers of chemisorbed $^{18}O$ on Ag(110) has been studied. First, up to 200 mass peaks are scanned to survey the products. Then, using methods which scan 10 mass peaks in a multiplex mode in 20 ms, the temperature dependence of evolution of a variety of products was measured as shown in Figure 119A. The two cross-hatched desorption processes occur in surface reaction limited processes and contain no $^{18}O$, and are the end result of the oxidative dehydrogenation of the ethylene glycol. The five remaining product peaks are indicative of reactions involving the chemisorbed $^{18}O$.

The desorption of $H_2{}^{18}O$ at 255 K occurs as a result of the direct reaction of O—H bonds of the two O—H groups of the ethylene glycol with adsorbed $^{18}O$, as illustrated in Figure 119B. The desorption together of $H_2{}^{18}O$ and $CH_2{}^{16}O$ with a peak temperature of 300 K is indicative of a surface reaction controlled by C—H bond scission in the $CH_2$ groups of the 1,2 ethanedioxy species as shown in Figure 119C. The simultaneous evolution of $H_2$ and $C^{16}O^{18}O$ at 415 K is indicative of the decomposition of adsorbed formate, $HC^{18}OO$, on the surface. Similar reactions were observed in separate experiments with adsorbed formate, and are illustrated in Figure 119D.

There are other possible mechanisms that might participate in the surface reactions, and these were eliminated by using D substitution for H in the $CH_2$ groups of the ethylene diol and looking to see whether a primary kinetic isotope effect was seen, in which the scission of the C—D bond occurs at a significantly higher temperature than the C—H bond in the unlabeled molecule. This isotope effect was observed, pointing to C—H bond scission as being a key elementary step in the surface chemistry.

Two reviews on TPRS have been written [2,3].

## *References*


[1] A.J. Capote and R.J. Madix, Surf. Sci. 214, 276 (1989).

[2] R.J. Madix and J. Benziger, Ann. Rev. Phys. Chem. 29, 285 (1978).

[3] R.J. Madix, in *Chemistry and Physics of Solid Surfaces*, Vol. II, edited by R. Vanselow (CRC Press, Boca Raton, FL, 1979).

***Fig. 119. Temperature-Programmed Reaction Spectroscopy (TPRS)***

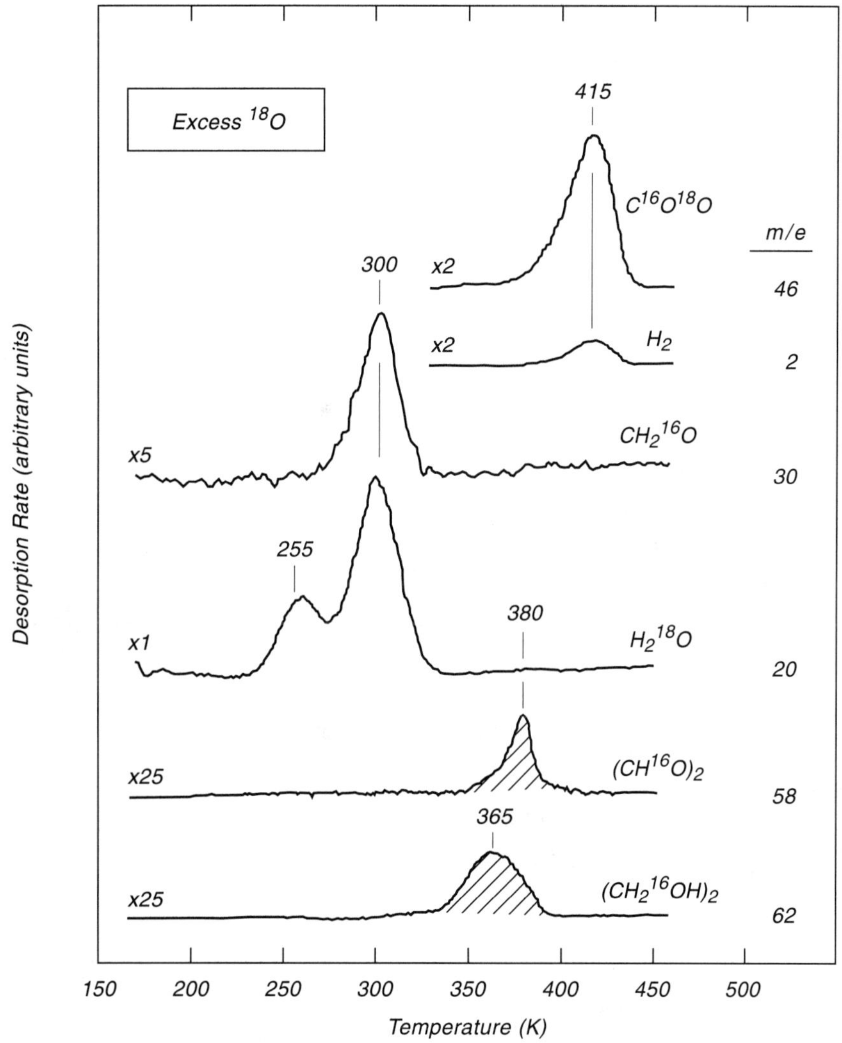


***A. TPRS Data—$(CH_2{}^{16}OH)_2$ + $^{18}O$/Ag(110)***

***Fig. 119. Temperature-Programmed Reaction Spectroscopy (TPRS)***

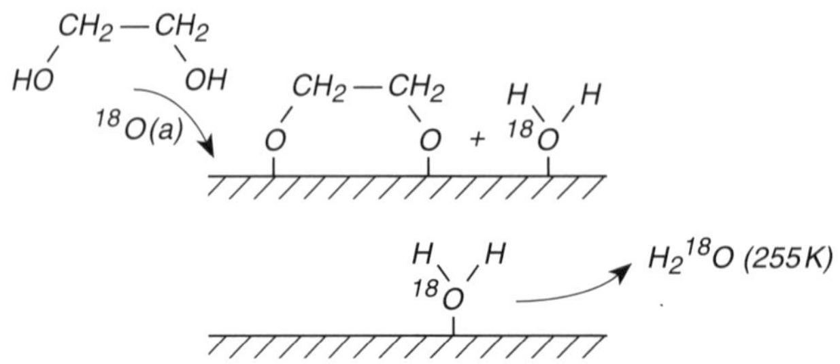


***B. Dissociative Adsorption Followed by $H_2{}^{18}O$ Evolution***

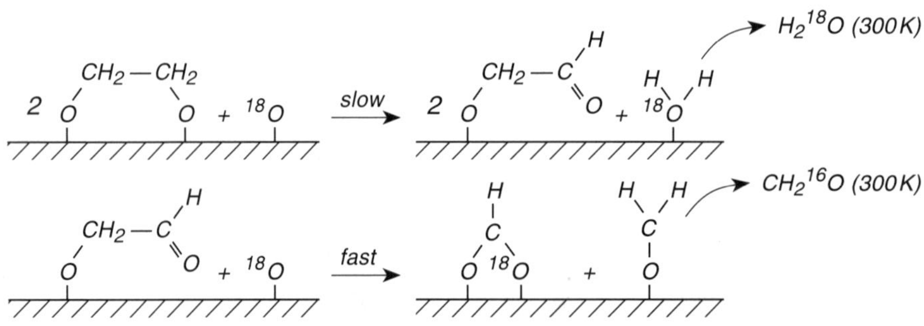


***C. $H_2{}^{18}O$ and $CH_2{}^{16}O$ Evolution***

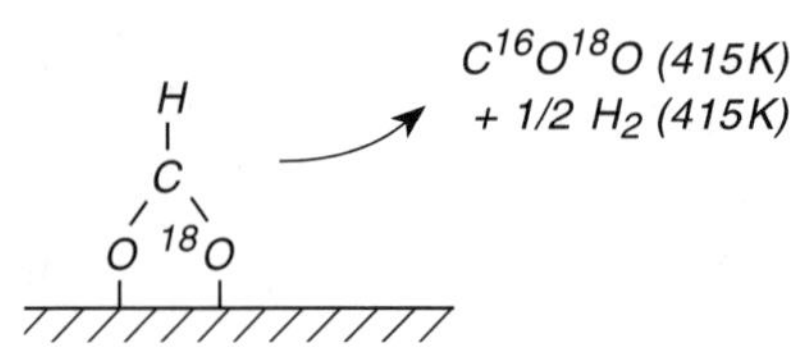


***D. $HCO^{18}O$ Decomposition***

# G. Gas Chromatography—Enhanced Sensitivity

# 120

# Gas Chromatography—Sensitivity Enhancement

The sampling of gases for gas chromatography analysis suffers from the fact that gas expansion in the sampling lines invariably causes loss of analytical sensitivity. This behavior may be reversed by using a high-pressure pulse of the carrier gas to compress the low-pressure gas sample into the sample loop, where it is then swept into the gas chromatograph. A factor of 20 or more sensitivity enhancement may be achieved. The application shown here involves sampling a high-pressure reactor being used to study the kinetics of a catalytic reaction over a single crystal [1].

As shown in Figure 120A, the reactor volume to be sampled is connected by valve $V_1$ to a length of capillary tubing containing an aliquot volume. The tube is then connected via a tee to a gas chromatography sampling valve. Gas from the reactor is expanded into the aliquot volume and the thin tubing ($V_5$ open) and its pressure is read with a low-volume pressure gauge. The valve $V_2$ to the pump is closed and valve $V_3$ is opened to the high pressure of the carrier gas regulator (about 800 Torr). Valve $V_4$ is then opened, and the carrier gas acts as a piston, compressing the sample gas into the sample loop located on the sampling valve. The contents of the sample loop are then injected into the gas chromatograph, as shown in Figure 120B. In the case shown here, the sampling volume is the aliquot volume plus the volume of the thin tubing, valve, and sampling loop. By keeping valve $V_5$ closed during sampling, the sampling volume is just the aliquot volume. It was shown that two consecutive analyses using this method differed by less than 4%.

## *References*

[1] Personally contributed by Professor C.T. Campbell, Department of Chemistry, University of Washington, Seattle, WA 98195. See also C.T. Campbell and K.A. Daube, J. Catalysis **104,** 109 (1987).

**Fig. 120. Gas Chromatography—Sensitivity Enhancement**

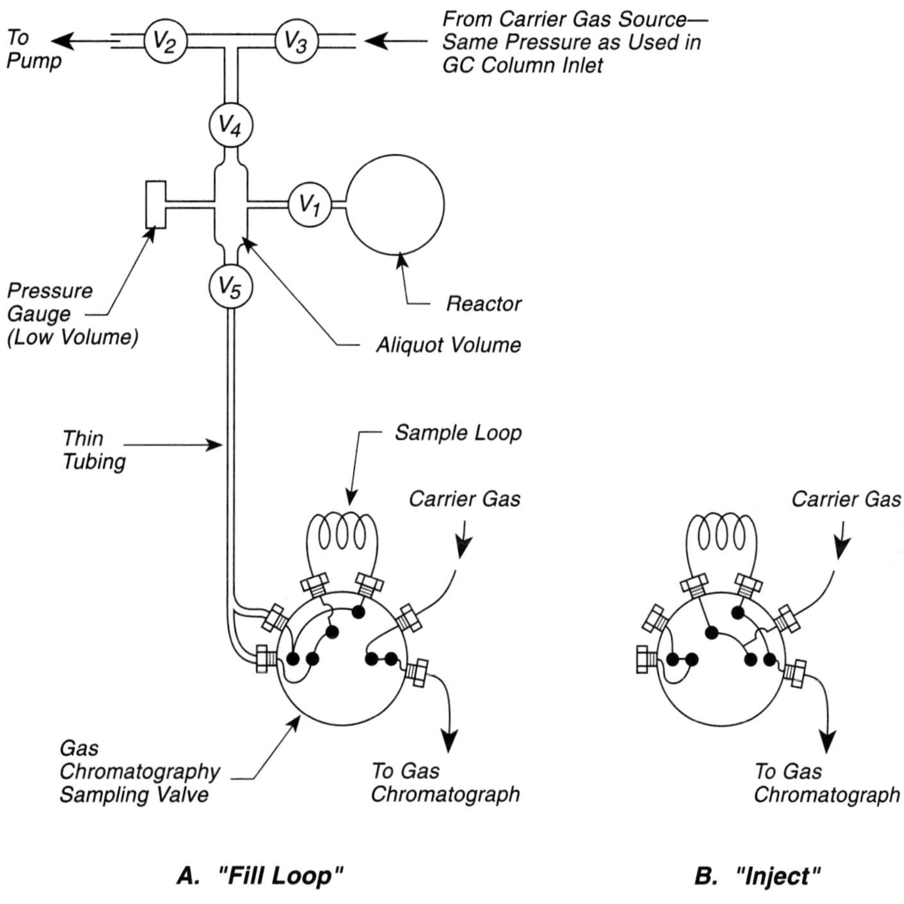

# H. Work Function

# 121

# Work Function Changes Using Retarding Diode Method

The use of a retarding diode method for measuring work function changes upon adsorption is one of the oldest and simplest methods for measurement in surface science. The literature related to this method is mostly from the 1960s or 1970s [1–6]. Other methods for measuring work function changes are described elsewhere in Chapters 122–126.

In Figure 121A is shown a thermionic emitter housed in an ultrahigh vacuum system and facing a sample whose work function change, $\Delta\phi$, is to be measured as a result of adsorption or desorption. The emission current collected by the sample as a function of the applied voltage, $U$, is given by

$$I(U) = AT^2 \exp[-e(U + \phi)/kT] \tag{1}$$

where $A$ is a constant (proportional to the cathode area), $T$ is the absolute emitter temperature, $k$ is the Boltzmann constant, and $\phi$ is the sample work function, and $e$ is the electronic charge. Equation (1) applies only for the initial section of the retarding potential curve where space charge effects are absent [4].

In Figure 121B, a plot of the collected current versus the retarding potential, $U$, is shown for a sample which has undergone a work function change, $\Delta\phi$. The difference in the retarding potential measured at constant current on the linear sections of the curves is equivalent to the work function change, assuming that the emitter work function has not changed and that the emitter temperature remains constant. This difference measurement is made where the two "linear" portions of the curve are exactly parallel.

In practice, the resistance, $R_e$, of the emitter is kept constant using a Wheatstone bridge, and if the dimensions of the emitter filament have not changed due to oxidation of the material of which it is made or to sublimation of the emitter, the temperature will also be constant. The temperature of the cathode emitter will be sensitive to the heat loss through the leads to the emitter, and thermostating these at constant temperature with a water cooling jacket will remove this variable.

There are several factors that can affect the measurements using this simple device. One possible problem is the emission of impurity gases from the hot emitter filament, which will influence the work function of the sample. Second is the effect of the hot filament on the residual gases in the vacuum system (for example, $H_2$ dissociation) making them reactive with the sample. This effect can be eliminated by turning off the emitter during gas admission. The method will not work well at pressures above about $10^{-4}$ Torr, since otherwise the cathode might be cooled by heat conduction through the gas. Finally, the effect of electron bombardment and radiation on the surface layer on the sample must be understood when using this method.

The retarding diode method has been employed in a conventional LEED apparatus with continuous derivative measurement of the work function change during adsorption. A sensitivity of about 5 meV was obtained, using the lock-in amplifier method for making continuous measurements of the changing work function [7]. Reference [7] also summarizes background papers in this area.

## *References*

[1] Z. Knor and V. Ponec, Coll. Czech. Chem. Commun. **31,** 1172 (1966).

[2] T. Jirsak, V. Nikolajenko, and Z. Knor, Czech. Chem. Commun. **59,** 1709 (1994).

[3] See literature quoted in V. Ponec, Z. Knor, and S. Cerny, *Adsorption on Solids* (Butterworth, London, 1974).

[4] A. Eberhagen, Fortschritte d. Physik **8,** 245 (1960).

[5] F.C. Tompkins, *Chemisorption of Gases on Metals* (Academic Press, London, 1978).

[6] G. Ertl and J. Küppers, *Low Energy Electrons and Surface Chemistry* (Verlag Chemie, Weinheim, 1974), Chapter 8.

[7]. J. L. Taylor and W.H. Weinberg, J. Vac. Sci. Technol. **15,** 1811 (1978).

***Fig. 121. Retarding Diode for Measuring Work-Function Changes***

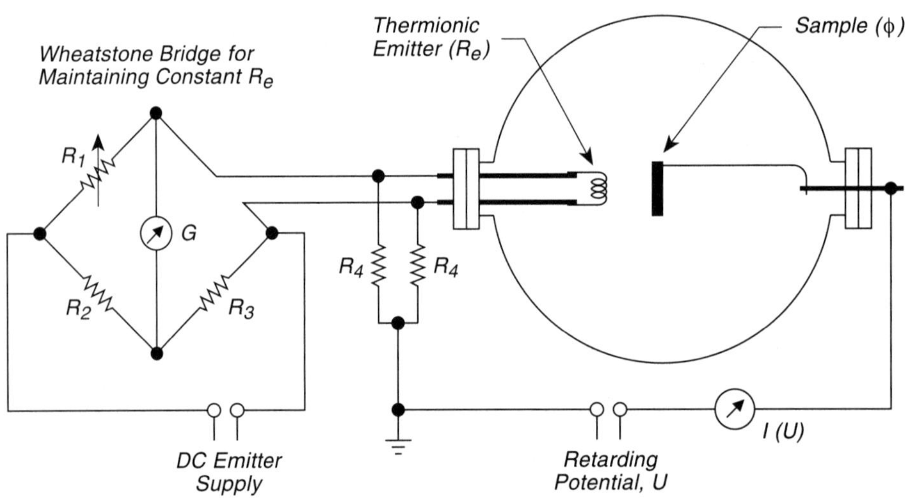


***A. Schematic of Circuit***

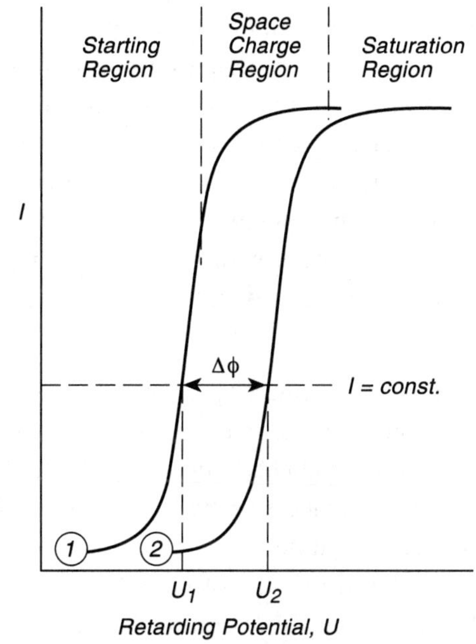


***B. Collected Current vs. Retarding Potential***

# 122

# Work Function Measurements Using an Electron Gun

One of the simplest methods to measure the work function change during adsorption on a metal single crystal is to employ a retarding potential method, using an ordinary electron gun and an interposed grid or aperture. Figure 122A shows a simple device first used for the study of hydrogen, nitrogen and xenon adsorption on tungsten surfaces [1–3]. An electron gun of ordinary design supplies current to the crystal through an aperture made of coiled tungsten in this case. The coiled tungsten is also used as an electron emitter for thermally cleaning the tungsten crystal at 2500 K. The crystal is welded to a thin wire support which may be cooled with liquid nitrogen. By using 25 eV electrons, and retarding between the aperture and the crystal surface, I-$V_R$ plots are obtained, and from the shift in electrical potential of the knee region, the work function change may be deduced as gases are adsorbed on the crystal. The reproducibility of duplicate measurements is about ±20 meV.

Figure 122B shows the linear work function change versus coverage for $H_2$ adsorption on W(100). The linearity versus measured coverage (using thermal desorption spectroscopy) suggests that depolarization effects between neighbor H atoms are small in this case [1]. Figure 122C shows measurements of the adsorption of $N_2$ on W(110) at 123 K. Here, the work function change is nonlinear with coverage, and this indicates that depolarization effects exist between the adsorbed molecular $N_2$ molecules [2]. The plot shown in Figure 122C is a fit to the immobile Topping model, and the $N_2$ polarizability may be obtained from this fit [2,4].

The use of an electron gun in a display LEED apparatus for retarding potential measurements of work function changes has been described [5].

## *References*


[1] T.E. Madey and J.T. Yates, Jr., Colloques Internationaux du Centre National de la Recherche Scientifique, **No. 187,** 155 (1970).

[2] J.T. Yates, Jr., R. Klein, and T.E. Madey, Surf. Sci. **58,** 469 (1976).

[3] M.J. Dresser, T.E. Madey and J.T. Yates, Jr., Surf. Sci. **42,** 533 (1974).

[4] J.R. MacDonald and C.A. Barlow, Jr., J. Chem. Phys. **39,** 412 (1963).

[5] J.H. Fritz and C.A. Haque, Rev. Sci. Instrum. **44,** 394 (1973).

***Fig. 122. Work-Function Changes Using Electron Gun***

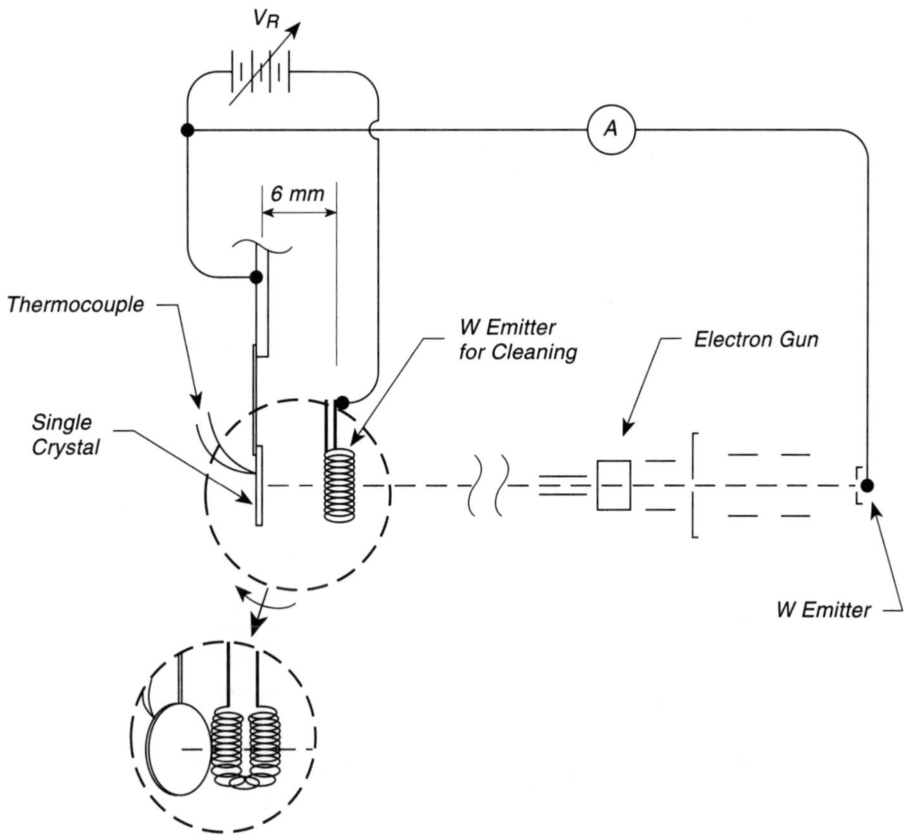


***A. Retarding Potential Measurement System Using an Electron Gun***

## *Fig. 122. Work-Function Changes Using Electron Gun*

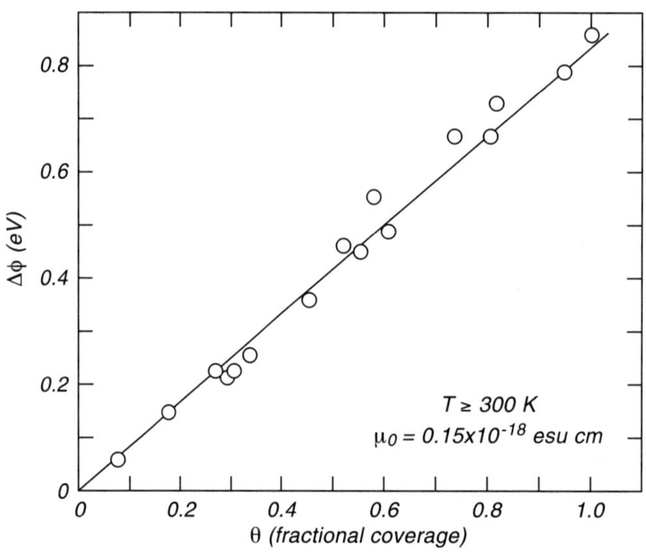


***B. Work Function Change - $H_2$/W (100)***

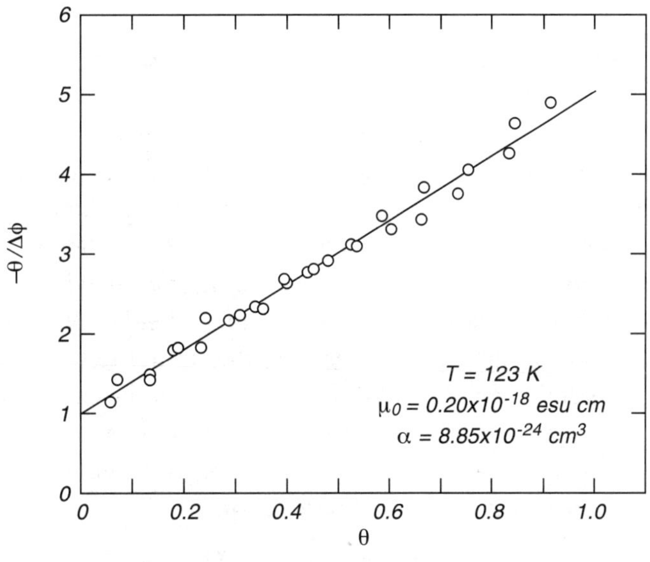


**C. Topping Model Plot - $N_2$/W (110)**

# 123

# Work Function and Electron Reflection Coefficient Measurements with the Shelton Diode

The use of the Shelton diode [1] to measure the work-function difference between two surfaces, an emitter and a collector, has been employed to determine the difference in work function between two crystal planes of tungsten. In addition, the work-function change during adsorption of a gas on the collector, and the energy-dependent reflection coefficient for electrons incident normally on the collector surface, can be measured with this device [2].

Figure 123A shows the schematic arrangement used [2]. Two single crystals of different crystallographic orientation are mounted by wires and a tungsten spring in support rings. These crystals face a movable septum containing a small aperture. Either crystal may be heated by electron bombardment to a temperature where thermionic emission occurs, and a feedback stabilized controller is used to maintain emitter and crystal temperature constant, using the current collected from the crystal emitter by the septum for regulation of the temperature of the spiral tungsten emitter located behind the emitter crystal. The entire assembly is placed in an axial magnetic field of 3600 G, causing the emitted electrons to travel in tight spiral paths from the emitter to the collector crystal. The movable septum permits sampling of different regions of the emitter and collector. The collector crystal current is plotted against the retarding potential applied between emitter and collector, and a typical retarding potential plot, as shown in Figure 123B, may be measured. The collected current follows the equation

$$I = I_0 \exp - [e(V_R + \Delta\phi)/kT] \tag{1}$$

where $I_0$ is the emitted current from the emitter, $V_R$ is the retarding potential applied, and $\Delta\phi$ is the difference in work function between the collector and the emitter, and $e$ is the electronic charge. Thus as the work function of the collector changes, the retarding potential plot (Fig. 123B) will shift. The shift of the "knee" in the plot is a measure of the work-function change of the collector, and may be used as a measure of the adsorption of a gas on the collector, the emitter remaining clean at its high operating temperature. The device shown in Figure 123A measures collected currents of the order of 0.01 μA and to avoid problems with leakage currents, the terminals to each crystal are shielded from evaporating metal films by the hat-shaped shields shown on each feedthrough. The device is symmetric, so that it may be operated with either crystal acting as the emitter, but such a symmetrical arrangement is of course unnecessary if one is interested in only one crystal. In contrast to similar devices [1] that used ohmic heating of the emitter, this device does not have a potential drop across the emitter region sampled by the aperture and should therefore perform better by having less uncertainty of the electrical potential distribution in the sampled emitter region.

By varying the electron energy and measuring the collected current, curves like those shown in Figure 123C may be generated. These curves are related to the effective electron reflection coefficient in the plateau region and have to do with band structure and other possible effects at the collector.

The accuracy of work-function measurements made with this device is of the order of ±20 meV over many months of operation with tungsten crystals.

## *References*


[1] H. Shelton, Phys. Rev. **107,** 1553 (1957); P. Kisliuk, Phys. Rev. **122,** 405 (1961).

[2] T.E. Madey and J.T. Yates, Jr., Supplemento al Nuovo Cimento, No. 2, Series I, **5,** 483, (1967).

***Fig. 123. Work Function Measurements—Shelton Thermionic Diode***

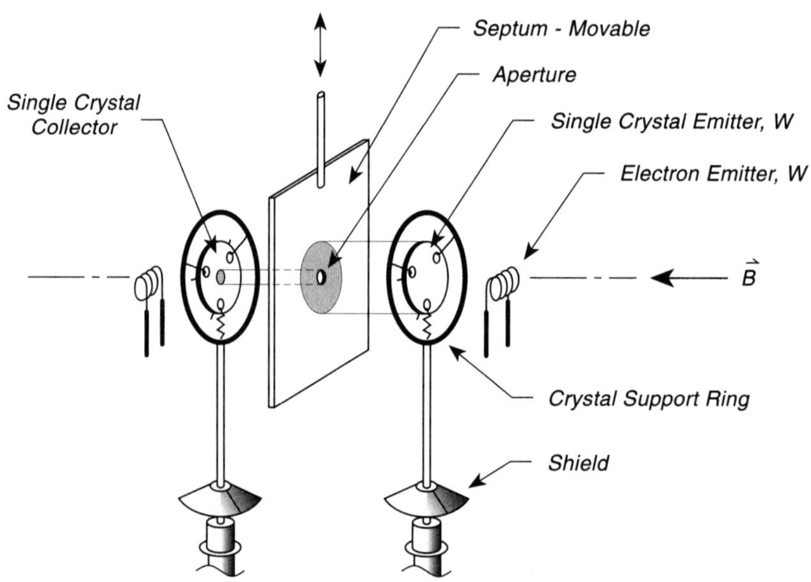


***A. Shelton Diode***

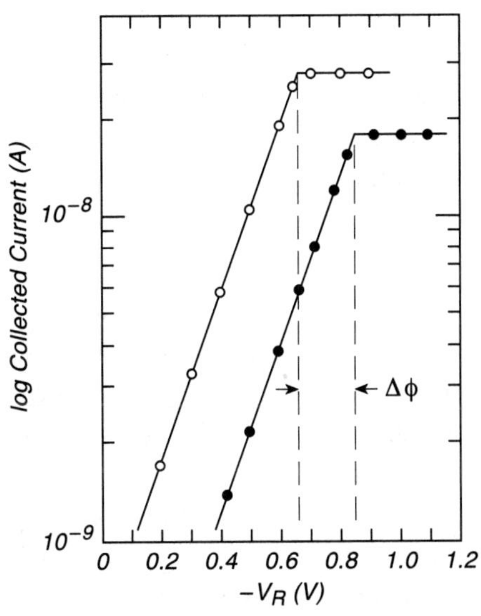


**B. Retarding Potential Plot**

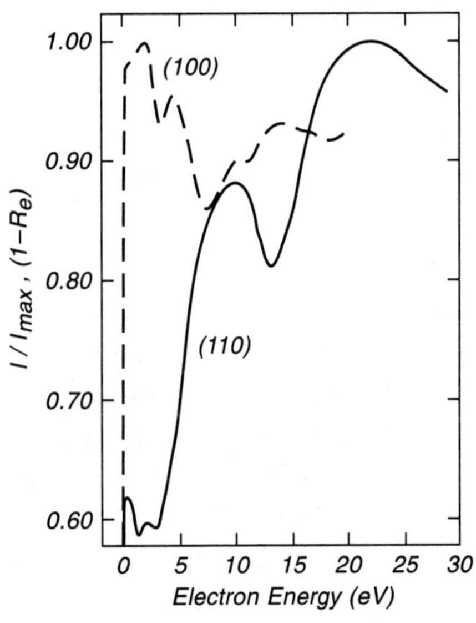


**C. Reflection Coefficient Plot**

# 124

# Improved Piezoelectric Drive Kelvin Probe

The Kelvin-Zisman vibrating capacitor method for measuring changes in the work function of a surface is widely used in surface science studies. The principles of this method are well described in the literature [1,2] and will not be discussed in detail here. Basically, a reference electrode is caused to oscillate at audio frequency and the electrical potential of the reference electrode is adjusted to yield a null signal in the capacitance circuit. As the work function of the sample changes due to adsorption or to changes in temperature, the potential applied to the reference electrode is adjusted to maintain a null signal in the AC circuit. The change in the applied potential is equal to the change in the work function or surface potential of the sample. It is of course important to know that the reference electrode is not changing its work function during the measurements, and inert electrodes such as Au or $SnO_2$ films are often used. The vibrational motion of the reference electrode is commonly produced by mechanical excitation of the resonant frequency of the device, by an electromagnetic driving mechanism, or by the use of a piezocrystal oscillator.

The design shown in Figure 124A involves the use of a piezoceramic foil actuator [3]. This avoids the interaction of stray magnetic fields from an electromagnet driver with the sample, which could cause problems if the sample is magnetic. In addition, the measured signal is taken from the reference electrode, not the sample, avoiding complications owing to the connection of the sample to both the heating circuit and the thermocouple circuit. The device described in Figure 124A is a modification of the piezoelectric Kelvin probe built by Besocke and Berger [4,5]. The device was used to study magnetic materials and yielded an accuracy of ±1 meV [6]. This compares to a reported sensitivity of better than 0.1 meV in the original device built by Besocke and Berger [4]. The reader is referred to the original paper [3] for dimensional details of the Kelvin probe shown in Figure 124A. The reference electrode is an L-shaped piece of gold, connected by a tungsten wire to a quartz rod and thence to a shield plate. This assembly is actuated by a piezoceramic foil [7], containing vapor deposited Ag contacts on both surfaces, and glued using UHV leak sealer to a tungsten reed, 20 mm x 4 mm x 0.1 mm thick. The signal from the Au reference electrode is measured through a very thin wire connected to a support ceramic before exiting the apparatus. The stable resonance frequency of the reed is about 130 Hz, the amplitude of vibration is about 0.1 mm, and the AC driving voltage is about 1.0 $V_{rms}$. The spacing between the reference electrode and the sample is about 0.5 mm.

Before operating, the piezoceramic foil is polarized by a DC voltage pulse above 300 V. The polarization remains as long as the piezoceramic is not heated significantly.

Figure 124B shows a block diagram of the self-compensating feedback circuit which allows a continuous measurement of the work function change. It is estimated that the drift is about 1 meV/h and the noise is about ±0.5 meV. The response time constant of the system is less than 10 s. The variable DC power supply supplies a variable compensating voltage, $V_b$. The authors note that electron emission from thermionic sources can disturb the electrical isolation of the sample and the reference electrode [3].

Figure 124C shows a record of the increase in the work function of a polycrystalline Ni sample as it cools through the Curie temperature, $T_C$.

Excellent reviews and analysis of the effects of stray capacitance [8] and noise [9] on the performance of the Kelvin probe have been published. Kelvin probes, driven by an electromagnetic voice coil and possessing superior noise levels, have been described [10].

The Kelvin probe has been used in dynamic measurements of the desorption of one desorbing surface species from a single crystal surface [11]. The method was shown to give superior resolution of overlapping desorption processes compared to temperature programmed desorption into a large chamber with gas monitoring by a quadrupole mass spectrometer [11]. Also, the dynamic work function change method can be used over a wider range of $dT/dt$ than the mass spectrometer method [11].

For more complicated surface phenomena in which surface decomposition of a molecule may occur without product desorption, the use of the Kelvin probe for observing work- function changes associated with decomposition has proven helpful [12]. The Kelvin probe may also be used to measure work-function changes on semiconductor surfaces when adsorption occurs [13].

## *References*


[1] G. Ertl and J. Küppers, *Low Energy Electrons and Surface Chemistry* (Verlag Chemie, GmbH, D-694 Weinheim, 1974), p. 120.

[2] J. Holzl and F.K. Schulte, in *Solid Surface Physics*, edited by G. Hohler (Springer, Berlin, 1979), Vol. 85, p. 1.

[3] S. Saito, T. Soumura, and T. Maeda, J. Vac. Sci. Technol. **A2,** 1389 (1984).

[4] K. Besocke and S. Berger, Rev. Sci. Instrum. **47,** 840 (1976).

[5] G. Wedler and R. Ruhmann, Appl. Surf. Sci. **14,** 137 (1982-1983).

[6] S. Saito and T. Maeda, J. Vac. Soc. Jpn. **25,** 190 (1981).

[7] Piezofoil can be obtained from Kyoto Ceramic Ltd., Kyoto, Japan.

[8] I.D. Baikie, E. Venderbosch, J.A. Meyer, and P.J.Z. Estrup, Rev. Sci. Instrum. **62,** 725 (1991); see also J.S.W. de Boer, H.J. Krusemeyer, and N.C. Burhoven Jaspers, Rev. Sci. Instrum. **44,** 1003 (1973).

[9] I.D. Baikie, S. Mackenzie, P.J.Z. Estrup, and J.A. Meyer, Rev. Sci. Instrum. **62,** 1326 (1991).

[10] (a) I.D. Baikie, K.O. van der Werf, H. Oerbekke, J. Broeze, and A. van Silfhout, Rev. Sci. Instrum. **60,** 930 (1989); (b) C.S. Kumar, A. Subrahmanyam, and J. Majhi, Rev. Sci. Instrum. 67, 805 (1996).

[11] P. Feulner and D. Menzel, J. Vac. Sci. Technol. **17,** 662 (1980).

[12] L. Hanley, X. Guo, and J. T. Yates, Jr., J. Phys. Chem. **93,** 6754 (1989).

[13] J. Abelson and G. de Rosny, J. Physique **44,** 993 (1983).

## Fig. 124. Improved Kelvin Probe—Piezodrive

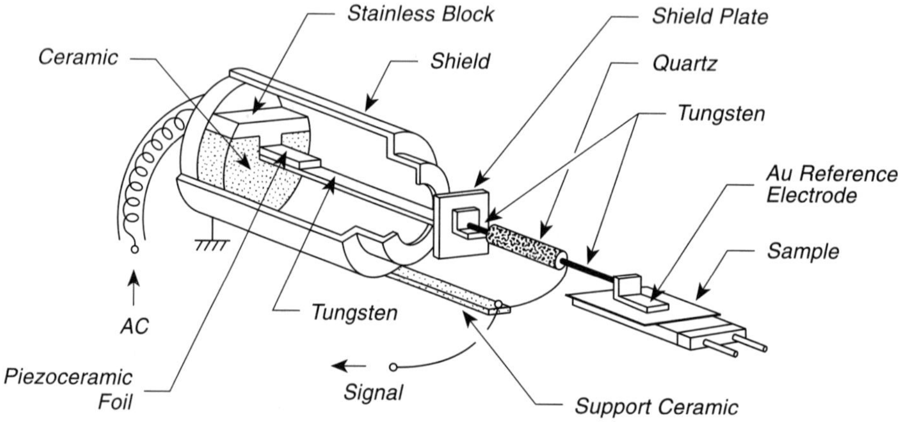


**A. Piezodriven Kelvin Probe**

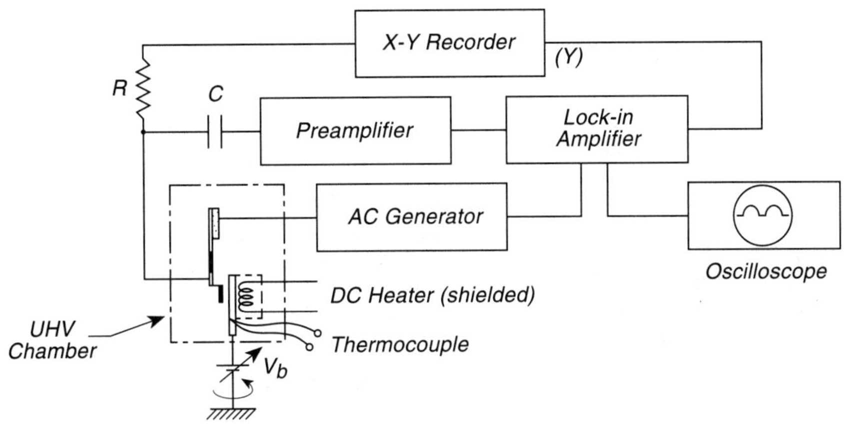


**B. Circuit for $\Delta\phi$ Measurements**

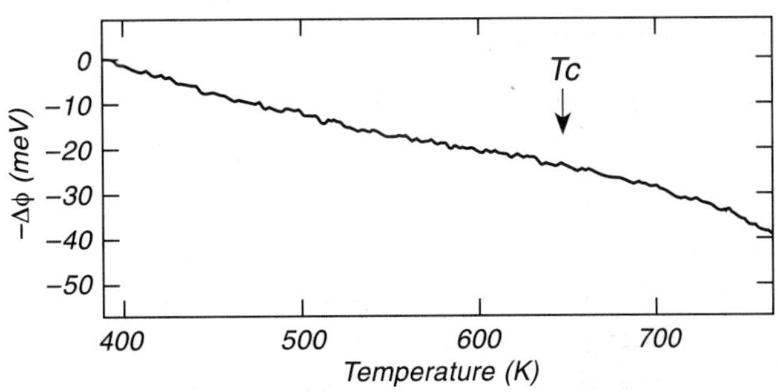


**C. Work Function Change on Cooling Ni (poly.)**

# 125

# Improved Electromagnetic Drive Kelvin Probe

In comparison to the piezodriven Kelvin probe shown on page 424, an electromagnet-driven device is described here with somewhat different features.

In the design shown in Figure 125A, great attention has been paid to the elimination of stray capacitances [1]. This is accomplished by electrical adjustments of a potential applied to the electrical shielding as described in the original paper [1]. The reference electrode is a grid of 80% transparent stainless steel mesh, stretched across a 3-4-mm-diameter Mo ring, and coated with $SnO_2$ (see page 190) to produce an inert surface of constant work function during gas adsorption. The spacing between the reference electrode and the sample is 0.3–0.5 mm. The transparent grid electrode is used in order to allow the sample to have almost full access to gas during adsorption rather than having partial blocking by a solid electrode. The frequency of the device is about 160 Hz, but it is advantageous to operate at higher frequencies, so that an overtone frequency of 1150 Hz was used. Working at such high frequencies has the effect of removing disturbances owing to low frequency noise, caused by effects such as bumping the apparatus. Because the resonant overtone frequency is sharp (<0.5 Hz) a quartz oscillator was used for stability. The amplitude of the vibration is about 0.1 mm. The response time constant was measured by applying a sudden potential to the sample, and was 20 ms.

The Kelvin probe is constructed using a light stainless steel tube to connect the reference electrode to the holder. A lightweight mu metal platelet was welded onto the tube in the position of the electromagnet, and the assembly was clamped in a stainless steel holder, electrically isolated by ceramic insulators from the electromagnet mounting. The electromagnet is made of soft iron and is wrapped with a few turns of aluminum winding wire, which has an anodized oxide insulating surface film for isolation. All electrical measurements are made from the isolated reference electrode assembly, using rigid partially screened conductor wires that are free of microphonics, and that are connected to vacuum feedthroughs. Figure 125B shows a measure of the noise and long term stability of the device.

The quality of the measurements made with this apparatus permit one to study the time derivative of the work-function change on the sample as adsorbate surface coverage changes. Assuming that the work function is linearly proportional to coverage, the thermal desorption spectrum obtained with the mass spectrometer should agree with that obtained from the derivative of the work function. It was found that an improvement in the resolution of the thermal desorption spectroscopy could be achieved by using the Kelvin probe, compared to the mass spectrometer detector of desorbing gas. However, the two kinetic curves do not agree because of the nonlinear behavior of work function with coverage. Figure 125C shows data for the thermal desorption of CO from Ru(001) measured both ways [1].

Reference [1] gives an excellent review of the development of the vibrating capacitor method for work-function measurements.

An excellent analysis of the effect of stray capacitance on Kelvin probe measurement is given in reference [2], and the effect of noise on Kelvin probe measurements is given in reference [3]. Reference [4] gives a new design for a Kelvin probe involving a voice coil driver. Reference [5] gives design and performance features for a Kelvin probe with a reproducibility of 1 meV over long periods of time. In addition, methods for measuring the work-function change during ohmic heating of a metal crystal are described [5].

## *References*

[1] H.A. Engelhardt, P. Feulner, H. Pfnur and D. Menzel, J. Phys. **E10**, 1133 (1977).

[2] I.D. Baikie, E. Venderbosch, J.A. Meyer, and P.J.Z. Estrup, Rev. Sci. Instrum. **62**, 725 (1991).

[3] I.D. Baikie, S. Mackenzie, P.J.Z. Estrup, and J.A. Meyer, Rev. Sci. Instrum. **62**, 1326 (1991).

[4] I.D. Baikie, K.O. van der Werf, H. Oerbekke, J. Broeze, and A. van Silfhout, Rev. Sci. Instrum. **60**, 930 (1989).

[5] M. Chelvayohan and R. Gomer, Surf. Sci. **172**, 337 (1986).

*Fig. 125. Improved Kelvin Probe—Electromagnet Drive*

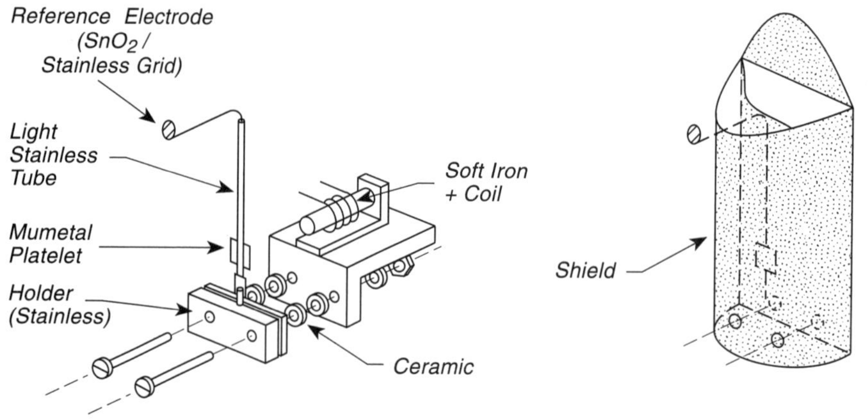


**A. Electromagnet-Driven Kelvin Probe**

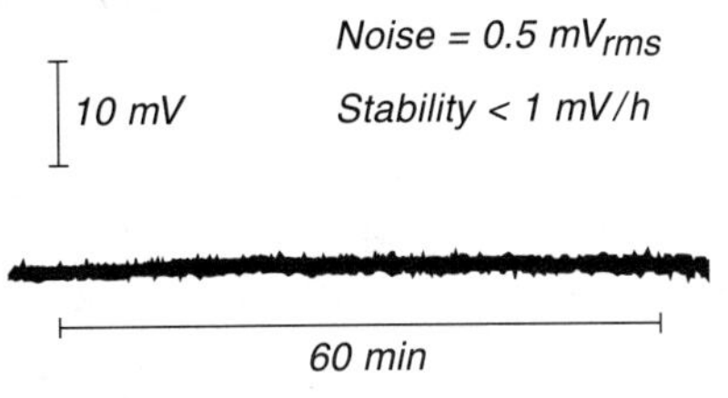


B. Stability and Noise

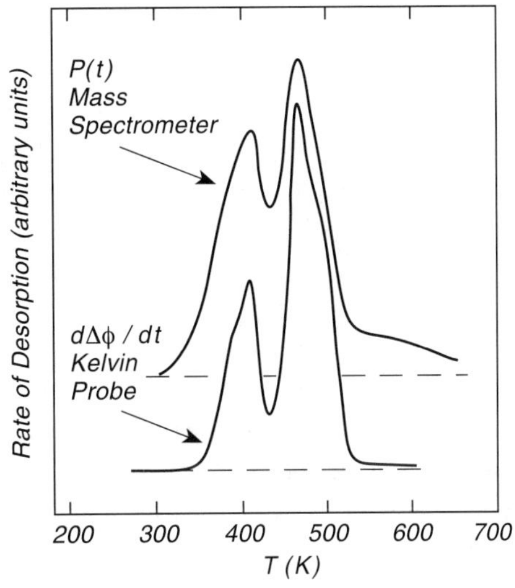


C. Thermal Desorption of CO/Ru(001) Measured Two Ways

# 126

# Kelvin Probe—High Temperature Flow Reactor

The vibrating Kelvin probe has been described on pages 424–431. Both mechanical and electromechanical activation of the motion of the reference electrode have been employed. For electromechanical activation, a piezoceramic actuator is often used, but because the upper temperature limitation is near 415 K for these actuators, they are not generally useful for high-temperature studies.

Figure 126 shows a method for using a piezoceramic actuator [1] to drive a Kelvin probe at some distance from a high-temperature sample [2]. The device, with a balanced quartz tube lever of 180-mm length, has the actuator on one end and the Au Kelvin probe electrode on the other. The quartz tube (2-mm diameter) lever arm is balanced on sapphire rods, as shown in the drawing. The actuator is coupled to the quartz tube lever with a light weight stainless steel spring. This refinement avoids the addition of a large mass to the actuator, and the resonant frequency of the device is close to that of the free actuator (in this case, 325 Hz).

The apparatus is used to study chemical reactions on a single crystal sample placed under the Kelvin probe. This sample may be heated in flowing gas while the work function is being measured. The large distance between the hot sample and the actuator avoids heating of the actuator. In addition, the large separation distance prevents electromagnetic pickup in the work function measurement from the piezodrive unit. A sampling tube permits the gas composition to be measured in the vicinity of the sample, and this provides a measurement of the local rather than the incident gas composition.

The resolution of the probe was estimated to be better than 5 meV, using known applied potentials on the sample for calibration. It is believed that improvements in the electrical measurement methods will lead to substantially lower noise and higher resolution [2].

## References


[1] A Phillips PXE 5 piezoactuator with length of 27 mm was used.

[2] S. Lundgren and B. Kasemo, Rev. Sci. Instrum. **66,** 3976 (1995).

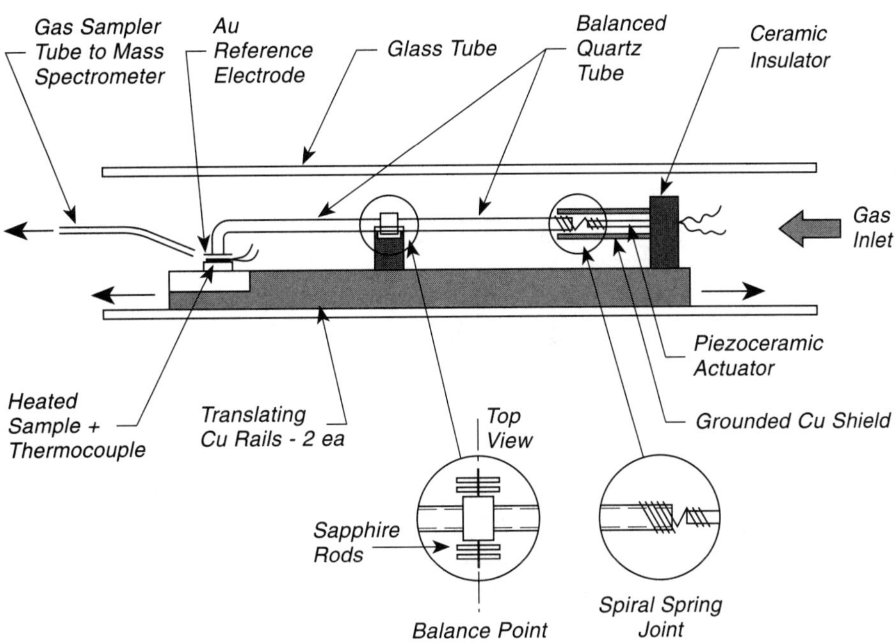


**Fig. 126. Kelvin Probe—High-Temperature Flow Reactor**

# I. Radioactive Adsorbates

# 127

# Radiotracer Techniques for Surface Studies

The high sensitivity of radiochemical methods suggests that these methods would have utility in studies of chemisorption and reaction on single crystal surfaces. This description of the use of a silicon surface barrier detector for the measurement of $^{14}C$ in chemisorbed layers on Pt(111) represents one of the few attempts to make measurements of this type [1]. Earlier studies using a Geiger-Müller counter were also successful [2,3].

The surface barrier detector [4] is a large area diode consisting of a depleted slice of *n*-type Si mounted in an insulator ring as shown in Figure 127A. The front and back surfaces are metallized. The thickness of the Au front surface coating is optimized for low energy β counting (158 keV) from $^{14}C$. Also, the depth of the depletion region (300 μm) is chosen to optimize the detection of the β-electrons from $^{14}C$, which have a range of about 150 μm in Si. Free charge carriers produced by the β-emission are separated by an applied electric bias (50 V). Integration of the current induced between the front and back electrodes yields an output pulse that is proportional to the ionizing energy. Either β-spectra may be collected or a ratemeter may be used to count the level of radioactive material, using the appropriate discriminator settings to eliminate noise.

The detector, ~ 20-mm in diameter, is mounted on a rotary feedthrough that allows it to be positioned in front of the Pt crystal at a distance of 5–10 mm, as shown in Figure 127B. In the opposite orientation, background counting may be done. When a bias is applied to the detector, the temperature must be maintained between -30°C and +30°C. This is done by using a flexible Cu braid (see page 564) that is anchored on a copper block maintained at the desired temperature. To avoid contributions from detector noise, the lower discriminator was set at 35 keV, and this resulted in rejection of about 40% of the $^{14}C$ β-signal as determined in separate calibrations. The counting efficiency was determined as 2.9 ± 0.3% for the geometry and discriminator settings used here.

Figure 127C shows the correlation between the measured count rate and the Auger intensity from carbon deposited as $^{14}C—C_6H_6$ on the Pt (111) crystal. The linear correlation between the two measurements indicates that both are measuring the carbon coverage as expected.

The sensitivity of the $^{14}C$ counting method depends upon many factors such as geometry, detector parameters, and the specific activity of the carbon-containing compounds used. While many hydrocarbons and other molecules may be obtained with activities up to 64 mCi/mmol, a nominal value of 10 mCi/mmol is more common. Assuming a detection efficiency of 3%, and a background signal of 3 counts/min, the detection

Table 127. Partial Listing of Radioisotopes Useful for Surface Studies Using a Barrier Detector [1]

| Isotope | $E_{max}$ (keV) | Half-life (years) |
|---|---|---|
| $^{14}C$ | 158 | 5,720 |
| $^{22}Na$ | 544 | 2.6 |
| $^{26}Al$ | 1160 | 740,000 |
| $^{32}Si$ | 210 | 710 |
| $^{35}S$ | 167 | 0.24 |
| $^{36}Cl$ | 714 | 300,000 |
| $^{137}Cs$ | 510 | 30 |

sensitivities for 20-min counting, at nominal labeling levels, are $1 \times 10^{13}$ (23% standard deviation), $5 \times 10^{13}$ (10% standard deviation), and $1 \times 10^{14}$ (4% standard deviation) molecules/cm$^2$. Thus, coverages of 1% of a monolayer may be determined accurately. It is found that repeated adsorption experiments lead to increasing background radioactivity during measurements over a day; pumping overnight restores the background to a low level.

A second example of the use of this method for the study of hydrocarbon chemistry on Pt(111) can be found in reference [5].

The use of radioactive adsorbates is not limited to those labeled with $^{14}C$ as may be seen in Table 127.

## *References*

[1] S.M. Davis, B.E. Gordon, M. Press, and G.A. Somorjai, J. Vac. Sci. Technol. A19, 231 (1981).
[2] K. Klier, Rev. Sci. Instrum. 40, 372 (1969).
[3] K. Klier, A.C. Zettlemoyer, and H. Leidheiser, Jr., J. Chem. Phys. 52, 589 (1970).
[4] Ortec TA-23-25-300. Obtained from Ortec, P.O. Box 1397, Easley, SC 29640.
[5] S.M. Davis, F. Zaera, B.E. Gordon, and G.A. Somorjai, J. Catalysis 92, 240 (1985).

**Fig. 127. Radiotracer Techniques for Surface Studies**

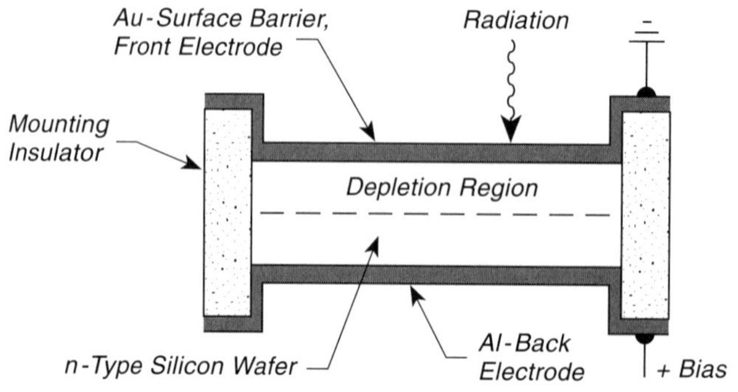


**A. Si Surface Barrier Detector**

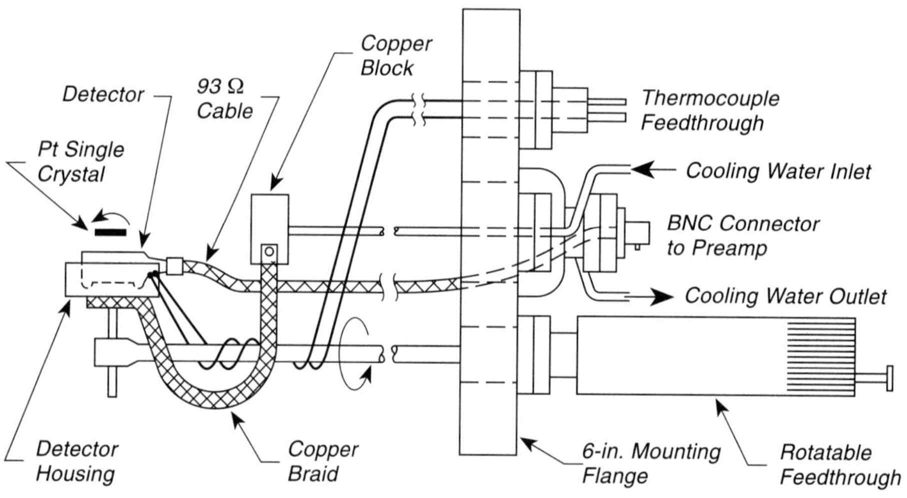


**B. Detector Geometry**

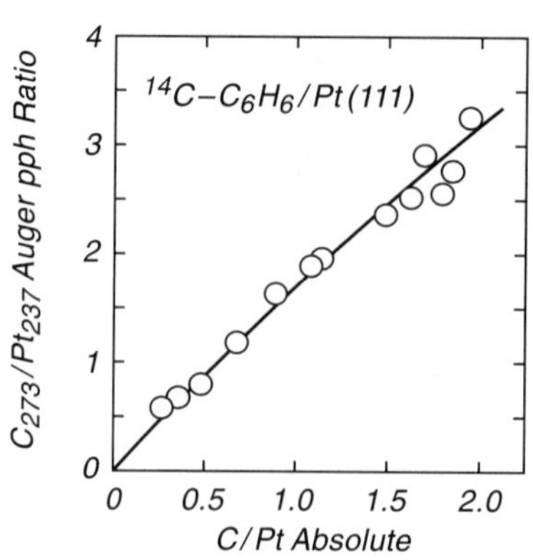


**C. C Auger Signal vs. Radioactive $^{14}C$ Coverage**

# J. Thin Film Deposition

# 128

# Thin Film Deposition to Absolute Surface Coverages

The ability to evaporate clean thin films and to achieve desired surface coverages that are absolutely known is of fundamental importance in many areas of surface science research. Fortunately, quartz thickness monitors (QTM) are commercially available that have sensitivities of the order of a nominal thickness of 1 Å of deposited material. With appropriate use, these monitors can be used to monitor, absolutely, the deposition of thin films on substrates of interest.

Figure 128A shows a successful arrangement for producing absolute coverages of metal on a single crystal [1]. In the case of this work, the metal deposited was Pt and the substrate was atomically clean Mo(110). The metal deposition source consists of a tungsten hairpin filament, mounted on degassing loops (see pages 584 and 628), which is shielded by a metal shield containing a 15-mm-diameter hole near the source. A thermostated QTM is able to be moved quickly between three positions—1, 2, and 3. When the QTM is in position 1, the Pt source is completely shielded, so that metal does not escape into the vacuum chamber. Position 1 is used for through outgassing of the source. When the QTM is moved to position 2, the QTM collects evaporated metal and radiation from the hot source. Thermostatting with circulating water from a temperature-regulated source maintains the QTM temperature constant to ±0.5 K, and this condition is necessary for absolute accuracy of the QTM. When the QTM is rotated to position 3, the crystal receives the metal vapor and the QTM is out of line-of-sight with the metal source. Stable operation of the metal source can be achieved by monitoring the current and voltage drop across the source and making the product of these constant. In practice, the source is stable over hours of operation without sophisticated wattage regulation. The relative flux at the crystal and the QTM is determined from the distances by the ratio, $d^2_{QTM}/d^2_{crystal}$.

Figure 128B shows the duty cycle described above as monitored by the QTM. In position 1, no Pt is received either by the QTM or the crystal. In position 2, the mass of the deposited platinum on the QTM rises linearly with time. In position 3, Pt is deposited on the Mo(110) substrate and the QTM again does not show any change in mass. Finally, in position 2 for the second time, the QTM again receives Pt, to check that the rate of evaporation has not changed significantly during deposition. The two slopes in position 2 agree to within 0.4%.

Following each deposition cycle, the Mo(110) crystal was rotated to the CMA Auger spectrometer position for measurement of the Pt(NVV) intensity, which is a secondary measure of the extent of Pt deposition. It can be seen in Figure 128C that the Pt Auger signal grows linearly in at least three branches of different slope. The break points, located analytically, correspond to the completion of a monolayer of Pt and the start of the formation of a second layer of Pt as growth proceeds by a layer-by-layer mechanism. It is interesting to note that the three labeled break points correspond well, in absolute Pt coverage as measured with the QTM, with the number of Pt atoms in a close-packed Pt(111) layer which has a coverage of 1.504 x $10^{15}$ Pt/cm$^2$. The location of the break points were used to calculate the average attenuation length for 64 eV Pt(NVV) electrons, using the bulk (111) interplane spacing for Pt of 2.47 Å. These experiments gave an average attenuation length of 4.1 ± 0.3 Å, in close agreement with calculated values for this quantity [1,2].

Even with the care taken in this experiment, a propagation of systematic and random errors results in a ±16 % error estimate in the absolute coverages measured.

The use of quartz microbalance measurements stretch back to the pioneering work of Sauerbrey [3], Slutsky et al. [4], and to Kasemo et al. [5], who showed the usefulness and sensitivity of these devices for adsorption and thin film studies. A good review of the history of the development of the quartz microbalance may be found in ref. [6]. A paper concerned with improvement in the accuracy of the quartz microbalance to allow the measurement of the acoustic impedence ratio of the film to the crystal, using two non-harmonically related resonant modes is given in reference [7].

A method to compensate for the effects of temperature changes owing to radiation heating is described in reference [8]. Here, a thermocouple pair measures the difference between the temperature of the quartz crystal detector and a matched reference detector in the vacuum system. A heater, driven by a feedback circuit working off the thermocouple pair adjusts the reference crystal temperature to within 0.1 K. An accuracy of thickness measurement of at least 0.01 nm is estimated [8].

A description of the calibration of a quartz microbalance against an electrostatic balance is given in reference [9]. The probable error of the microbalance was found to be 0.6%, compared to the theoretical expectation.

A description of the use of a quartz microbalance at cryogenic temperatures is given in references [10–12].

A description of the use of a quartz microbalance for the calibration of an Auger spectrometer used to measure oxygen chemisorption on Ni is found in references [13,14].

## *References*

[1] A. Linsebigler, G. Lu, and J.T. Yates, Jr., Surf. Sci. **294,** 284 (1993).

[2] S. Tanuma, C. J. Powell. and D.R. Penn and W.F. Egelhoff, Jr., J. Vac. Sci. Technol., **A8,** 2213 (1990).

[3] G. Sauerbrey, Zeit. fur Physik **155,** 206 (1959).

[4] L.J. Slutsky and W.H. Wade, J. Chem. Phys. **36,** 2688 (1962).

[5] B. Kasemo and E. Tornqvist, Surf. Sci. 77, 209 (1978).

[6] C. Lu and A.W. Czanderna, in *Applications of Piezoelectric Quartz Crystal Microbalances* (Elsevier, New York, 1983), Chapter 2.

[7] A. Wajid, Rev. Sci. Instrum. **62,** 2026 (1991).

[8] G. Kremer and L.A. Moraga, Rev. Sci. Instrum. **56,** 1467 (1985); see also B. Ramadan, K. Piyakis, and J.F. Kos, Rev. Sci. Instrum. **50,** 867 (1979).

[9] R.M. Mueller and W. White, Rev. Sci. Instrum. **39,** 291 (1968).

[10] L.L. Levenson, Supple. al Nuovo Cimento, Serie I, **5,** (No. 2), 321 (1967).

[11] L.L. Levenson, J. Vac. Sci. Technol. **8,** 629 (1971).

[12] C. Wang and R. Gomer, Surf. Sci. **74,** 389 (1978).

[13] A, Winkler, K.D. Rendulic, and K. Wendl, Appl. Surf. Sci. **14,** 209 (1982-83).

[14] A. Winkler, Fresenius Z. Anal. Chem. **319,** 635 (1984).

***Fig. 128. Thin Film Deposition to Absolute Surface Coverage***

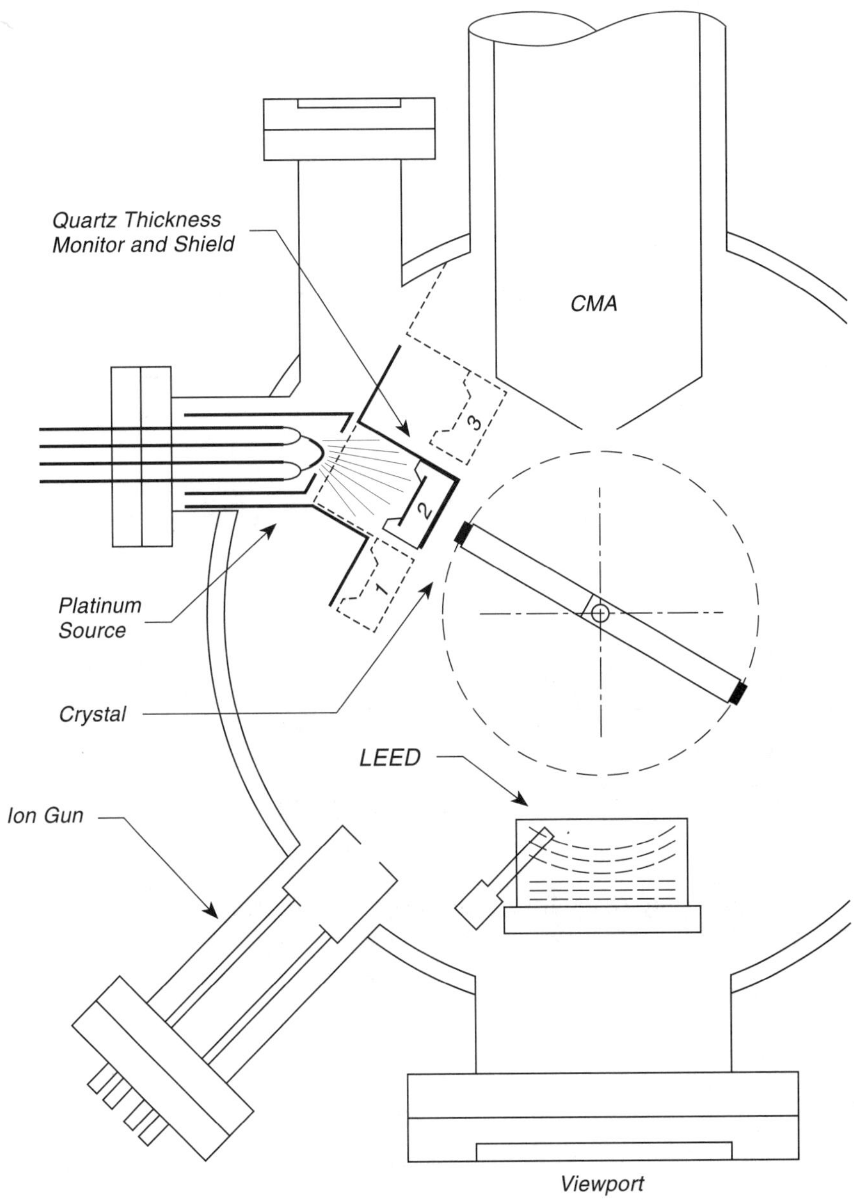


***A. Metal Evaporation / Quartz Thickness Monitor***

Fig. 128. Thin Film Deposition to Absolute Surface Coverage

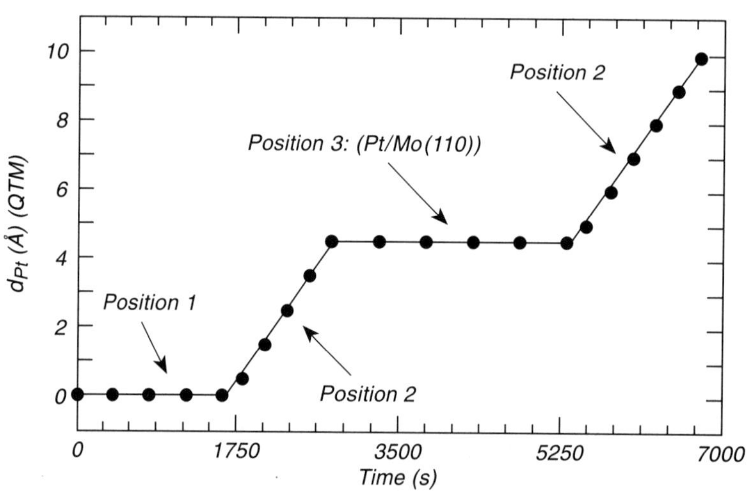


B. Duty Cycle for Evaporation / Measurement

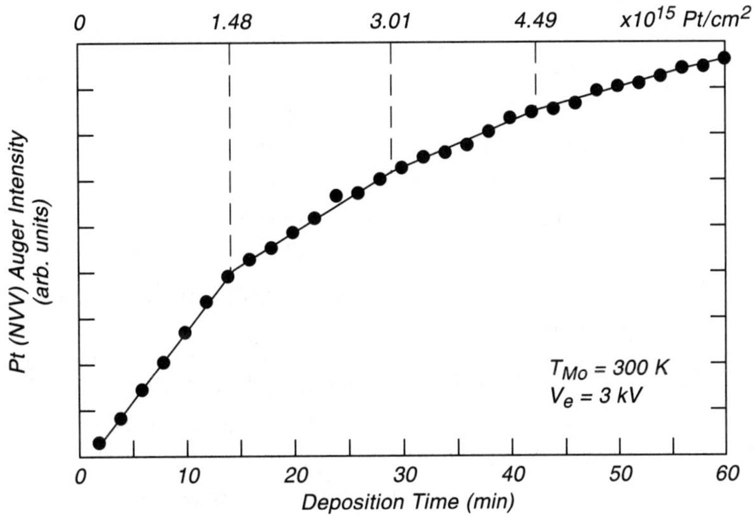


**C. Layer-by-Layer Growth Kinetics**

# 129

# Ionization Gauge Measurement of Film Evaporation Rate

A convenient and inexpensive secondary standard for the measurement of the rate of metal evaporation may be established using a pair of ionization gauges with matched geometry. The flux of metal atoms through one gauge is measured by an increase in ion current owing to the metal atom steady state gas density during evaporation. A second gauge is used to make a small correction for the presence of residual gases produced during evaporation. The calibration of the ion gauge is made against a quartz thickness monitor (QTM) [1] (see page 440).

Figure 129A shows the design of the apparatus, where a translatable QTM is located line-of-sight to a metal evaporation source that contains two separate sources housed in a shield. It is desired to evaporate known coverages of metal onto the crystal, and the QTM is located very near the crystal location. After outgassing of the source, the crystal is moved to a line-of-sight position and measurements of the mass received by the QTM are made. Simultaneously, measurements of the difference between the two ionization gauge readings are integrated versus time, and this difference is related to the metal vapor steady state gas density inside the ionization gauge assembly, which is located line-of-sight between the source and the crystal and QTM.

Figure 129B shows a cross calibration obtained for Sn evaporation. The linear correlation over a wide Sn thickness range indicates that both measurement methods are behaving as expected. However, the sensitivity of the ion gauge is high compared to the commercial QTM, so that small film thicknesses can be measured with the ionization gauge conveniently.

The dual-ionization gauge arrangement could also be calibrated against other thickness standards such as RHEED intensity measurements made during layer-by-layer growth [2].

## *References*


[1] D.S. Campbell, in *Physics of Nonmetallic Thin Films*, edited by C.H. Dupuy and A. Cachard (Plenum, NY, 1976), p. 179. I first learned of this technique from Dr. W.F. Egelhoff, Jr. of the National Institute of Standards and Technology (NIST). The work shown here is from P.J. Chen, M.L. Colaianni, J. T. Yates, Jr., and M. Arbab, J. of Non-Crystalline Solids **155,** 131 (1993).

[2] D.A. Steigerwald, I. Jacob and W.F. Egelhoff, Jr., Surf. Sci. **202,** 472 (1988).

## *Fig. 129. Ionization Gauge Measurement of Film Evaporation Rate*

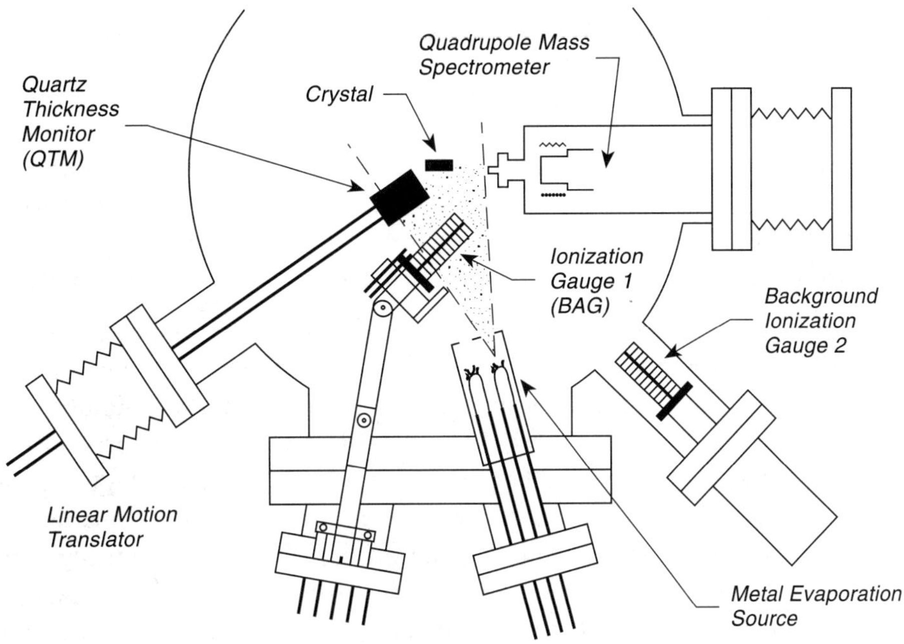


***A. Ionization Gauge Geometry***

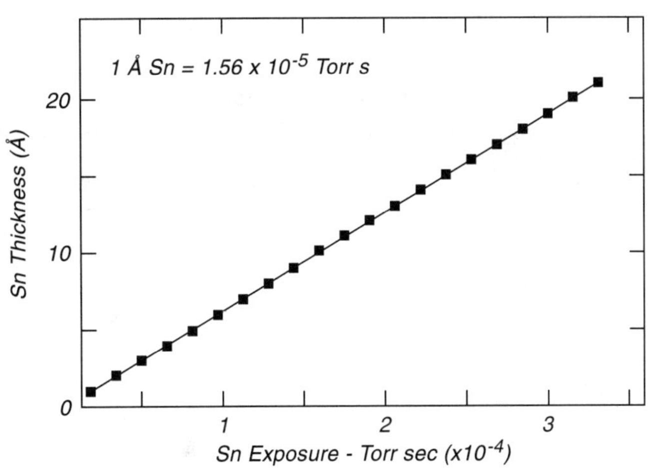


***B. Correlation of QTM and BAG***

# 130

# Reusable Quartz Crystals for Film Thickness Measurement

Quartz crystals used in thin film measurement are able to hold only so much deposited material before becoming unusable. Typical thickness limitations are approximately $10^5$ Å of metal deposit. The thin films can sometimes be etched away to recover the clean quartz oscillator, but strong chemical etchants will often destroy the metallic electrodes that activate the crystal, requiring redeposition of the electrodes [1].

A simple solution has been found to allow thin film overlayers on quartz crystals to be removed without harming the electrodes [2]. A photoresist coating is deposited on the clean quartz crystal before use. Later, using acetone as a solvent for the photoresist, the thin film to be removed is stripped away by dissolving the underlying photoresist layer as shown in Figure 130. It was found that a photoresist layer of at least 3000 Å thickness is needed to permit stripping of all overlayer deposits cleanly. This loads the quartz crystal only to about 1% of its mass capacity. The photoresist [3] is deposited using a spin-coating method, baked at 120°C for 20 min, and then patterned to cover the area on the crystal used for deposition. The photoresist coatings made in this manner can be stored in air but should not be baked at temperatures approaching 200°C because this makes the photoresist insoluble in acetone.

## *References*


[1] Recoating services are provided by Lebow Company, 5960 Mandarin Avenue, Goleta, CA 93117.

[2] W.K. Chao and H.K. Wong, J. Vac. Sci. Technol. **A8,** 150 (1990).

[3] Photoresist AZ-1300-31 purchased from Shipley Company, Inc., 2300 Washington Street, Newton, MA 02162.

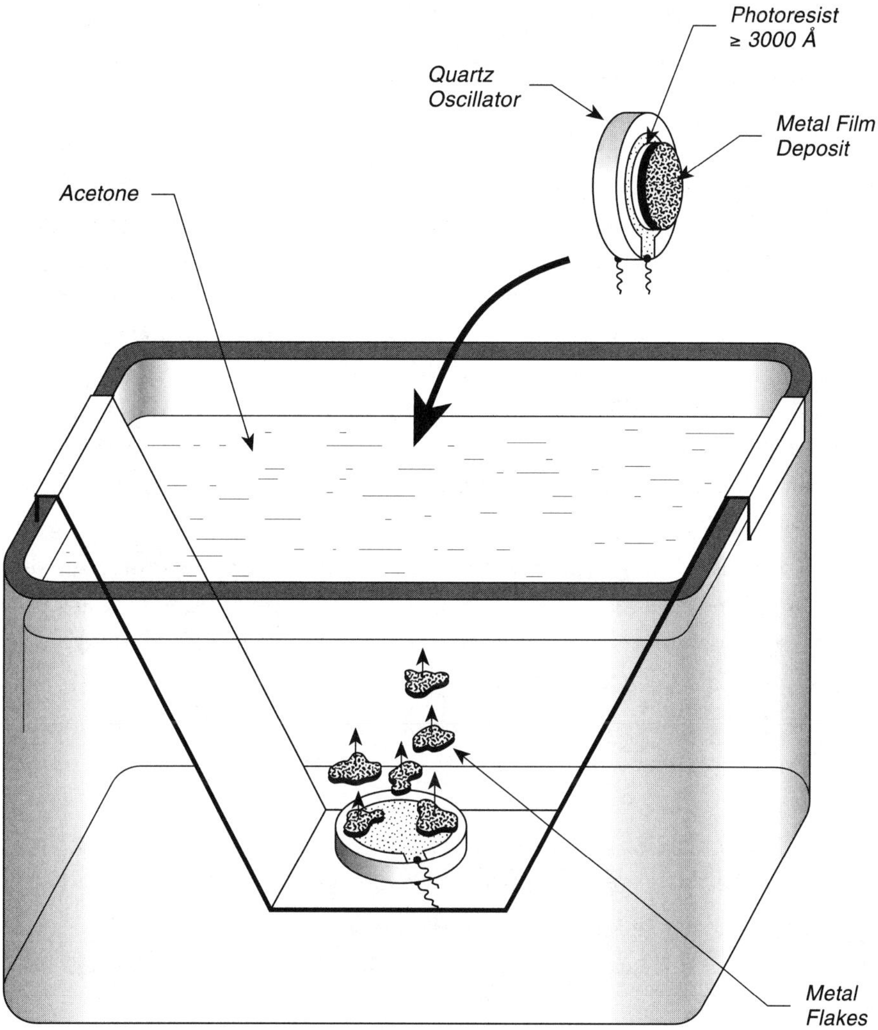


***Fig. 130. Reusable Quartz Crystal Oscillator***

# 131

# Hybrid Thin Film Deposition Processes—Stable and Metastable Films

The deposition of thin films with desirable chemical and physical properties is widely used for the production of electronic materials, tribological coatings, and optical coatings. The processes used may be divided into those that produce thermodynamically stable compound films and those that produce metastable materials. It is beyond the scope of this book to describe the wide range of film deposition methods and materials that have been deposited in thin film form. Books and reviews are widely available. Below, a short description of the some of the hybrid film deposition methods is given [1], and figures schematically describing each method are shown. Figure 131A–D illustrates ion bombardment from an external plasma. Figure 131E illustrates the use of ions extracted from a plasma near the substrate, and this method is now used widely in plasma assisted vapor deposition processes.

Figure 131A—Ion Beam Deposition: Ions are extracted from the source where they are produced by plasma excitation of the appropriate gas. The method has the advantage that the energy of deposition of the ions can be controlled rather than working with a mixture of ions, neutrals, and excited neutrals.

Figure 131B—Ion Beam Sputter Deposition: Ions are produced in an ion source by a plasma. These are used as a means for sputttering a target, and collection occurs on the substrate.

Figure 131C—Dual Beam Sputtering: Two ion beams are used. One sputters a target; the second bombards the growing film on the substrate.

Figure 131D—Ion Bombardment Evaporation: Material is evaporated thermally using electron beam or ohmic heating. The growing film is bombarded by ions extracted from an ion source.

Figure 131E—Reactive Sputtering with Dissociable Reactive Gas: Metals are sputter-deposited in a reactive gas such as silane to produce metal silicide films.

Many variations on the schematic methods shown in Figure 131A–E are employed and a wide variety of preparation methods are described in the literature [2,3].

## *References*


[1] R.F. Bunshah, Mineral Processing and Extractive Metallurgy Review. **9,** 147 (1992).

[2] R.F. Bunshah, in *Evaporation in Deposition Technologies for Films and Coatings,* Second Edition, edited by R.F. Bunshah (Noyes Publications, 1994) p. 131.

[3] C.V. Deshpandey and R.F. Bunshah, Thin Solid Films **163,** 131 (1988).

## *Fig. 131. Hybrid Thin Film Deposition Processes*

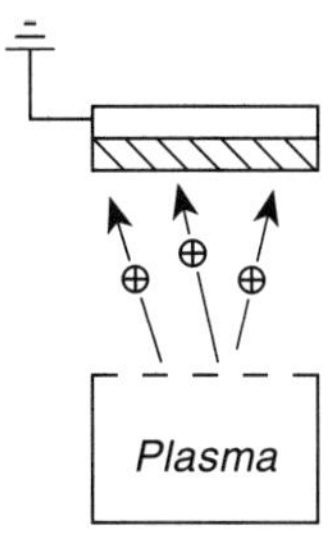


***A. Ion Beam Deposition***

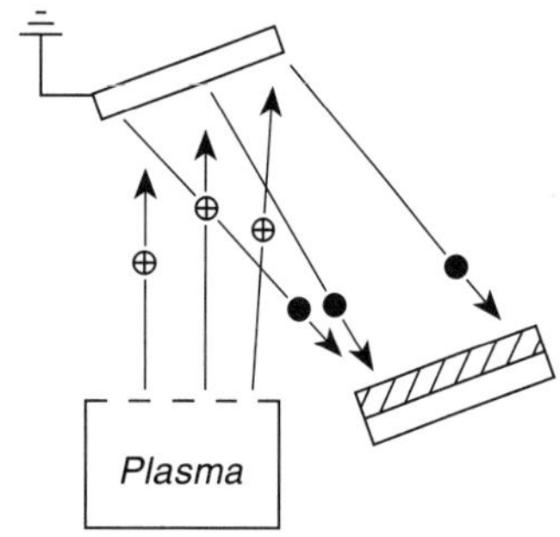


***B. Ion Beam Sputtering***

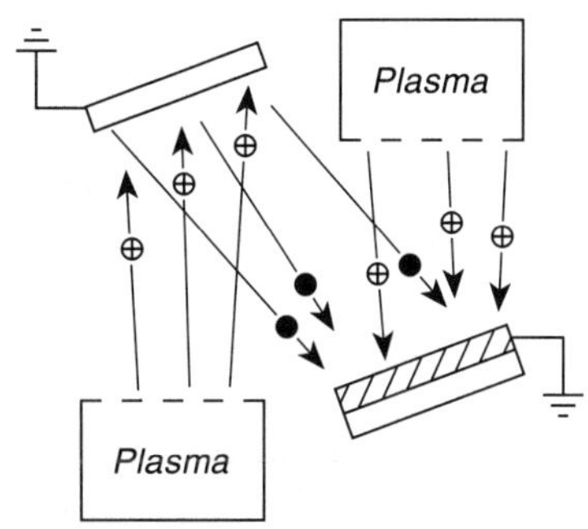


***C. Dual Beam Sputtering***

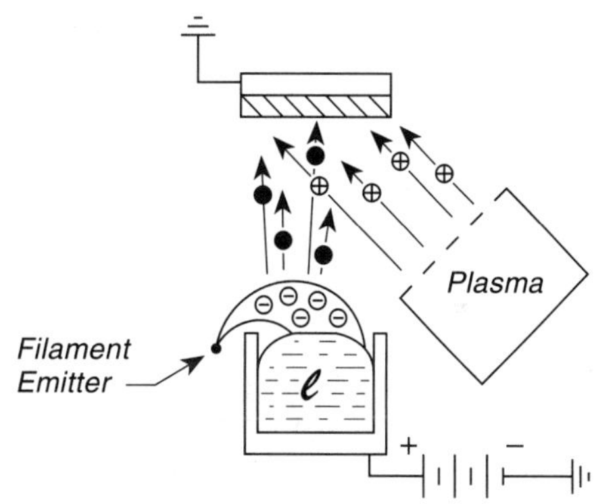


***D. Ion Bombardment - Evaporative Deposition***

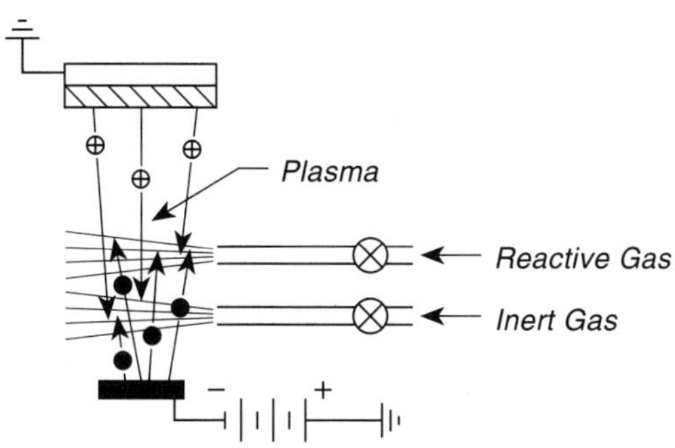


***E. Reactive Sputtering - Dissociable Reactive Gas***

# 132

# Sputter-Coating Inner Walls of Tubes

The need to coat the inside of a metal or insulator tube with a metal film can be met by both electrochemical and chemical vapor deposition methods. It is also possible to deposit pure metal films inside a tube using a special sputtering procedure described here [1].

As shown in Figure 132, a conical metal target, placed on the axis of a polished low alloy steel tube to be coated, is exposed to a beam of ions that are directed down the tube. The conical metal target is moved along the tube toward the ion source by an electrical drive mechanism. In the particular case studied here, the inner diameter of the tube was 6 mm, and a conical Ag target of 5 mm diameter was employed. The target was pushed through the tube at a rate of 0.05 mm/s, while a 2 keV $Ar^+$ beam was incident. A sputter yield of about 7 $Ag/Ar^+$ was measured for a 60° incidence angle.

The conical target was chosen so that the maximum sputter yield could be achieved using a 60° angle of incidence for the ions on the target surface. This optimum $Ar^+$ incidence angle for Ag sputter deposition was determined experimentally using a separate apparatus that permitted the collection of the Ag metal sputter deposit and its measurement by chemical analysis.

To determine that the Ag film deposited was uniform in thickness, the tube was cut along its axis in four sections. Visual inspection showed that a complete Ag film was deposited along the tube length, and sputter-depth profile measurements showed that depth of the film was uniform to within about 10%.

Experiments with Pt deposition on Ta tubes were also carried out and subsequent mechanical embrittlement tests showed that the Pt deposit was very effective in preventing hydrogen-induced embrittlement of the Ta tubes.

## *References*

[1] W. Ensinger, Rev. Sci. Instrum. **67**, 318 (1996).

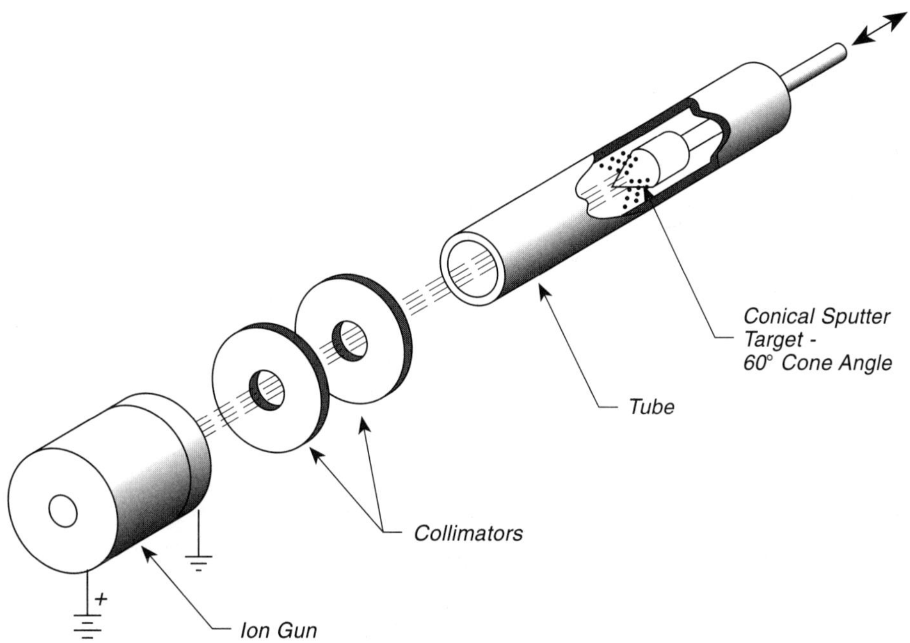


*Fig. 132. Sputter-Coating Inner Walls of Tubes*

# K. Infrared Spectroscopy

# 133

# Infrared Reflection Absorption Spectrometer

The use of infrared reflection-absorption spectroscopy (IRAS) has grown in popularity since the publication of the classic papers by Greenler on this subject. While the first IRAS measurements in ultrahigh vacuum were made with grating spectrometers [1–3] the measurement technique has been revolutionized by Fourier transform (FTIR) spectrometers, which possess greatly enhanced signal-to-noise capabilities at high spectral resolution [4]. Various technical methods have been used to improve the performance of FTIR spectrometers operating as IRAS instruments [5–11]. The field has been well reviewed [12,13].

The instrument shown in Figure 133A,B has special characteristics which are designed to improve the performance beyond that of conventional single beam IRAS instruments. A collimated IR beam from the FTIR spectrometer is reflected into an external optical bench by a flat mirror. The beam is split into two parts by a right-angled aluminized beam splitter (made from a 90° glass prism) in the external optical bench. One beam enters the IRAS cell containing the crystal, and the second beam enters the reference cell, which is at ultrahigh vacuum. The beam splitter is able to be moved laterally by a micrometer drive, a novel feature which allows the ratio of the two beam intensities to be accurately and reproducibly adjusted. Both beams are focused with KBr lenses. A beam collector, which is identical to the beam splitter, collects both infrared beams from the IRAS cell and the reference cell and reflects them onto the focussing mirror to the mercury cadmium telluride (MCT) detector. Two electrically driven shutters in the infrared beams enable the detection of these beams independently. For study of adsorbates on metal single crystals, the use of a *p*-polarizer in the incident beam, as shown, will improve the spectroscopic signal-to-noise ratio, since the polarization selection rule is strongly obeyed in this case, and *s*-polarized light is ineffective in detecting the adsorbate.

The collection of infrared spectra is interfered with by the presence of $H_2O$ and $CO_2$ in the atmosphere within the optical path outside the ultrahigh vacuum system. The "double-beam" arrangement shown here reduces this contribution significantly. As an example, if a $H_2O$ band is 5% of the total transmission, a 1% drift of the $H_2O$ partial pressure will yield, in a direct ratio measurement, a 0.05% $H_2O$ absorption band which would be greater in intensity than many weak surface species' absorption bands. Therefore, the optical path is purged extensively with very dry nitrogen gas from boil off of a liquid nitrogen dewar. With the double-beam method, small variations in the purged gas background spectrum can be reduced as shown next.

In single-beam measurements, if the background spectrum signal is $S_0$ and the adsorbate spectrum signal is $S_1$, then the desired adsorbate spectrum is $\log(S_1/S_0)$, which includes any possible changes during the time period between the measurements of $S_0$ and $S_1$. By the double beam method, we record the background absorbance spectrum as $\log(S_0/R_0)$ and the adsorbate absorbance spectrum as $\log(S_1/R_1)$. Then the difference absorbance spectrum is obtained by subtracting the background absorbance spectrum from the adsorbate absorbance spectrum,

$$\log(S_1/R_1) - \log(S_0/R_0) = \log(S_1/S_0)(R_0/R_1) \quad (1)$$

In Equation (1), the term $R_0/R_1$ produces a cancellation of all changes in environment during the period of taking the two spectra. This could include drifts of composition of the purging gas, drifts in the source and detector, etc. In doing measurements on the crystal in the optical cell with a high pressure of gas also present, the spectral contribution from the gas phase can be canceled almost completely.

As shown in Figure 133B, the single crystal is moved by a long throw manipulator from the surface analytical positions in the ultrahigh vacuum chamber (Level 2 and Level 3) up into the IRAS cell (Level 1). The crystal can be rotated through 360° by means of a doubly differentially pumped Teflon rotary seal (see page 8) for precise and reproducible angular orientation to the IR beam, which typically has an angle of incidence of about 86°. The windows of the IR cell can be made of single crystal $CaF_2$ mounted in Conflat flanges [4] or can be made of KBr single crystals mounted by means of a differentially pumped double Viton seal assembly (see page 742). The small optical chamber is sealed by a metal-seal gate valve that does not have the problem of Viton O-ring outgassing under compression nor the problem of permeation. For a sealed-off static cell, this is important. For high pressure experiments, the composition and pressure of the gas phase may be determined by spectroscopic measurements through the reference cell for infrared active gases.

Figure 133C shows a comparison of operation of the IRAS apparatus in single beam mode and in double-beam mode. The optical path is purged with dry nitrogen. After 6 h, a severe intensity drift in atmospheric $H_2O$ (about 0.4% T) and atmospheric $CO_2$ (about 0.2% T) can be seen in the spectrum obtained by the single beam method. In the spectrum obtained by the double-beam method, this effect is eliminated. The comparison on Figure 133C shows the ability of the double-beam method to eliminate background effects owing to slow drifts during long periods. For 900 scans for both $S$ and $R$ infrared measurements, a noise level of about $1 \times 10^{-4}$ absorbance unit may be routinely obtained. The amount of light delivered to the detector is significant in determining the noise level. Therefore, large single crystals of the order of 15 mm diameter are favored in this work.

A FTIR spectrometer mounted on a precision optical mount has been employed for IRAS measurements in which the path length between the spectrometer and the external detector has been minimized by using only KBr lenses for focusing the radiation [14].

Both the spectrometer and the detector are coupled to the UHV chamber by small bellows, permitting each unit to be independently adjusted with precision XYZ optical mountings.

Recently, an instrument designed to measure the mass of adsorbate on a thin film in addition to the infrared spectrum has been described [15]. The working temperature range of the quartz crystal oscillator (see page 440) used for adsorbate mass measurements is 80–350 K [15].

## *References*

[1] J. Pritchard and M.L. Sims, Trans. Far. Soc. **66,** 427 (1970).
[2] M.L. Kottke, R.G. Greenler, and H.G.Tompkins, Surf. Sci. **32,** 231 (1972).
[3] J.T. Yates, Jr., R.G. Greenler, I. Ratajczykowa, and D.A. King, Surf. Sci. **36,** 739 (1973).
[4] B.E. Hayden, in *Vibrational Spectroscopy of Molecules on Surfaces,* edited by J.T. Yates, Jr. and T.E. Madey (Plenum, New York, 1987).
[5] E. Schweizer, B.N.J. Persson, M. Tushaus, D. Hoge, and A.M. Bradshaw, Surf. Sci. **213,** 49 (1989).
[6] W. Erley, J. Electron Spectrosc. and Related Phenom. **44,** 65 (1987).
[7] W. Erley, Surf. Sci. **205,** L771 (1988).
[8] D. Hoge, M. Tushaus, and A. M. Bradshaw, Surf. Sci. **207,** L935 (1988).
[9] T. Buffeteau, B. Desbat, and J.-M. Turlet, Mikrochim. Acta (Wien) **II,** 23 (1988).
[10] W.G. Golden, in *Fourier Transform Infrared Spectroscopy,* edited by J.R. Ferraro and L.J. Basile (Academic, New York, 1985), Vol. 4, p. 315.
[11] A. E. Dowrey and C. Marcott, Appl. Spectrosc. **36,** 414 (1982).
[12] F. M. Hoffmann, Surf. Sci. Rep. **3,** 107 (1983).
[13] Y.J. Chabal, Surf. Sci. Rep. **8,** 211 (1988).
[14] Q. Wu and L. Hanley, J. Phys. Chem. **97,** 2677 (1993).
[15] G. Zhuang, G. Chottiner, I. T. Bae, E. Hwang, and D.A. Scherson, Rev. Sci. Instrum. **65,** 2494 (1994).

***Fig. 133. Infrared Reflection—Absorption Spectroscopy—IRAS***

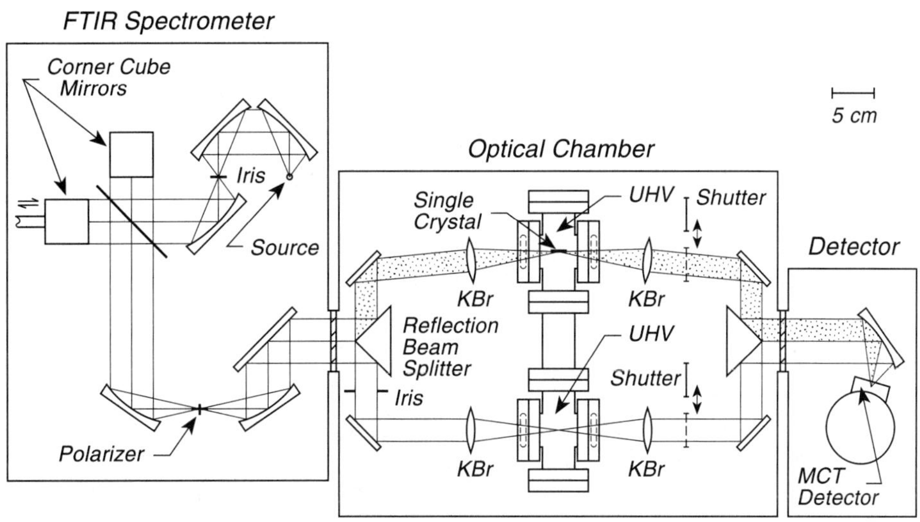


***A. IRAS-FTIR Spectrometer for Surface Studies ($N_2$-Purged) - Top View***

***Fig. 133. Infrared Reflection—Absorption Spectroscopy—IRAS***

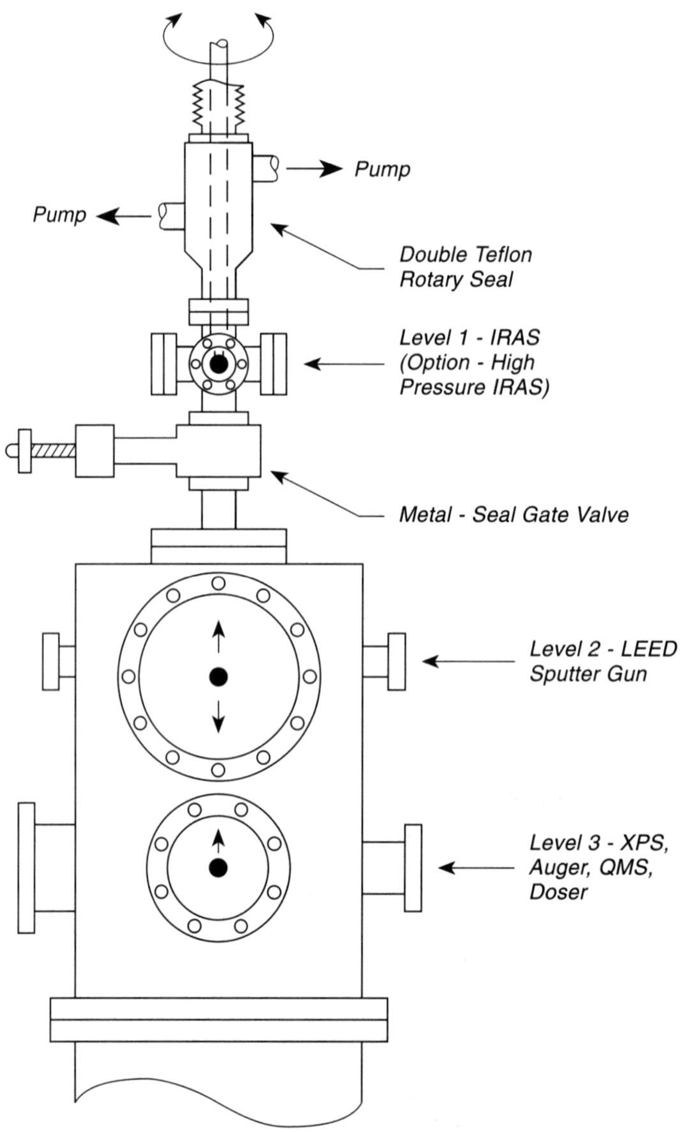


***B. Vacuum System - Side View***

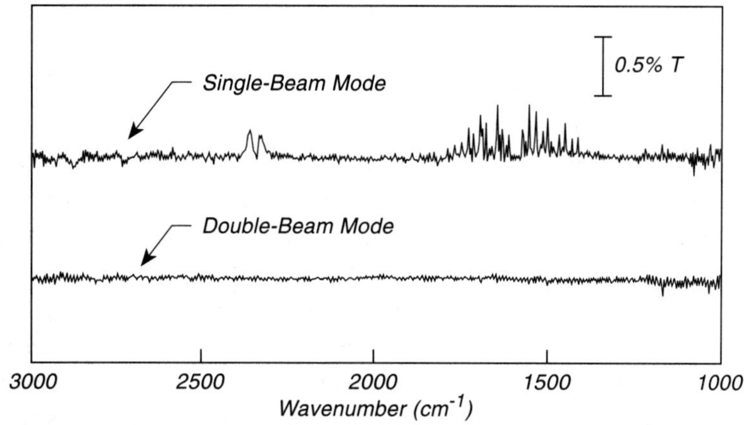


**C. Comparison of Two Data Acquisition Methods in IRAS (6 h after base line)**

# 134

# IRAS Measurements at High Pressure

The IRAS apparatus described on page 456 has the special capability of working at elevated pressures. At these pressures, electron spectroscopies fail because of the reduction of the mean free path of electrons to lengths comparable to or less than the electron traveling distances inside the apparatus. Fortunately, no such limitations exist for optical techniques, assuming the gas phase at high density is optically transparent in the required spectral regions.

An example is shown of the study of $N_2$ adsorption on the Ni(111) surface. Figure 134A shows that as the $N_2$ pressure is raised over 4 orders of magnitude from $1 \times 10^{-8}$ mbar to $2.9 \times 10^{-4}$ mbar, the equilibrium coverage of chemisorbed $N_2$ at a crystal temperature of 89 K increases as indicated by the increase in the absorbance of the chemisorbed $N_2$. Data of this sort were collected as a function of crystal temperature in the range 89–115 K, and the results are shown in Figure 134B, where the vertical scale is the integrated IRAS absorbance. It is clear that as the temperature is increased, higher pressures of $N_2$ must be employed to produce a given integrated absorbance for the chemisorbed $N_2$. If one assumes that the integrated absorbance is proportional to surface coverage (an approximation of unknown accuracy) then the "isoabsorbance" plots versus reciprocal absolute temperature can be made, as shown in Figure 134C, and from the slope of these plots the enthalpy of adsorption may be deduced at particular integrated absorbances. The inset to Figure 134C shows this plot, indicating that the enthalpy of adsorption is almost constant, or rises very slightly with increasing absorbance (or coverage within the approximation used above).

Work of this type, in which an initially clean single crystal is exposed to high pressures of a gas, must be carefully done with highly purified gas in order to avoid contamination. For the example of about $10^{-4}$ mbar of gas, impurity levels of only 10 ppm will give sufficient impurity partial pressure that in about 1000 s of exposure, one monolayer of impurity could be adsorbed if the impurity has a sticking coefficient of unity over its coverage range in the presence of the dominant gas. Effects of this type can often be discovered by examining the reversibility of the spectral changes, or by desorption of the dominant adsorbate, followed by readsorption. The purification of gases is discussed on pages 718–737.

The slight changes in the vibrational spectrum of the chemisorbed $N_2$ are thought to be due to various island structures that develop at various equilibrium coverages. The fact that so little shift occurs in the N—N stretching mode (a 24 $cm^{-1}$ range) suggests that the electronic character of the chemisorbed $N_2$ is not strongly affected by changes in coverage, and suggests that the approximation that integrated absorbance is proportional to coverage may be valid [1].

## *References*

[1] J. Yoshinobu, R. Zenobi, J. Xu, Z. Xu, and J. T. Yates, Jr., J. Chem. Phys. **95**, 9393 (1991).

## Fig. 134. IRAS Measurements at High Pressure

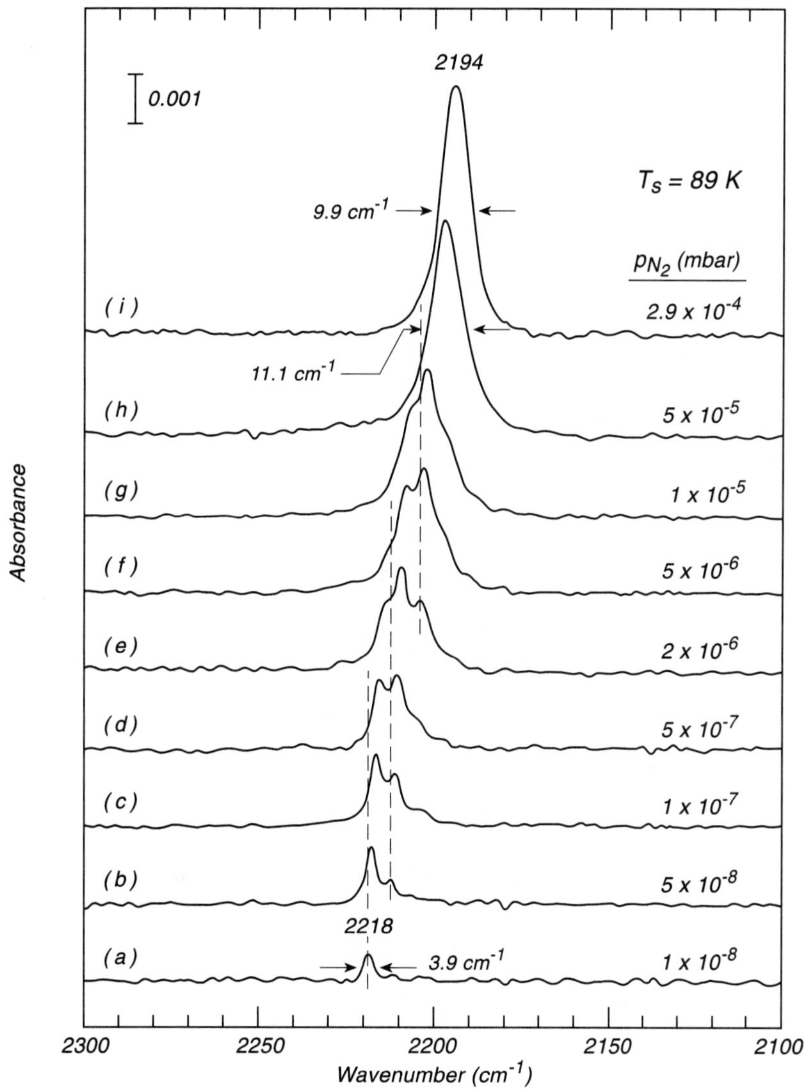


**A. Development of IRAS Spectra of $N_2$ / Ni(111)**

## Fig. 134. IRAS Measurements at High Pressure

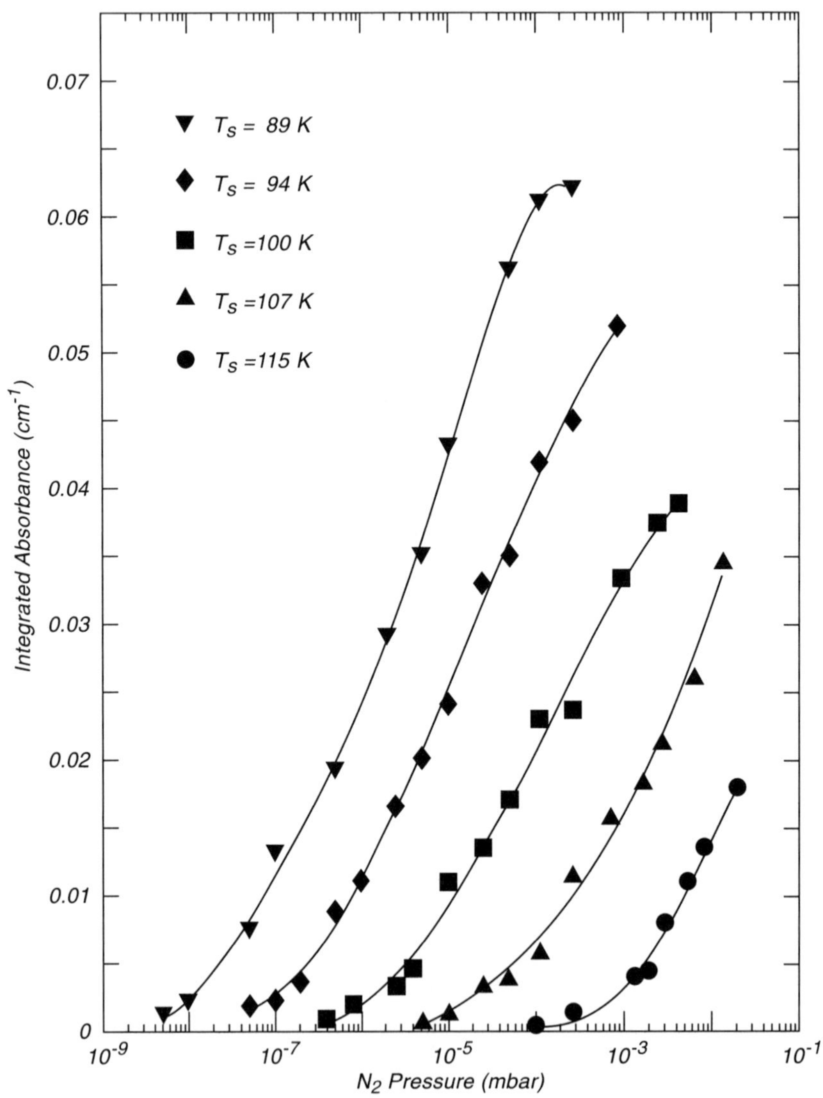


**B. $N_2$ Adsorption Isotherms on Ni(111) by IRAS**

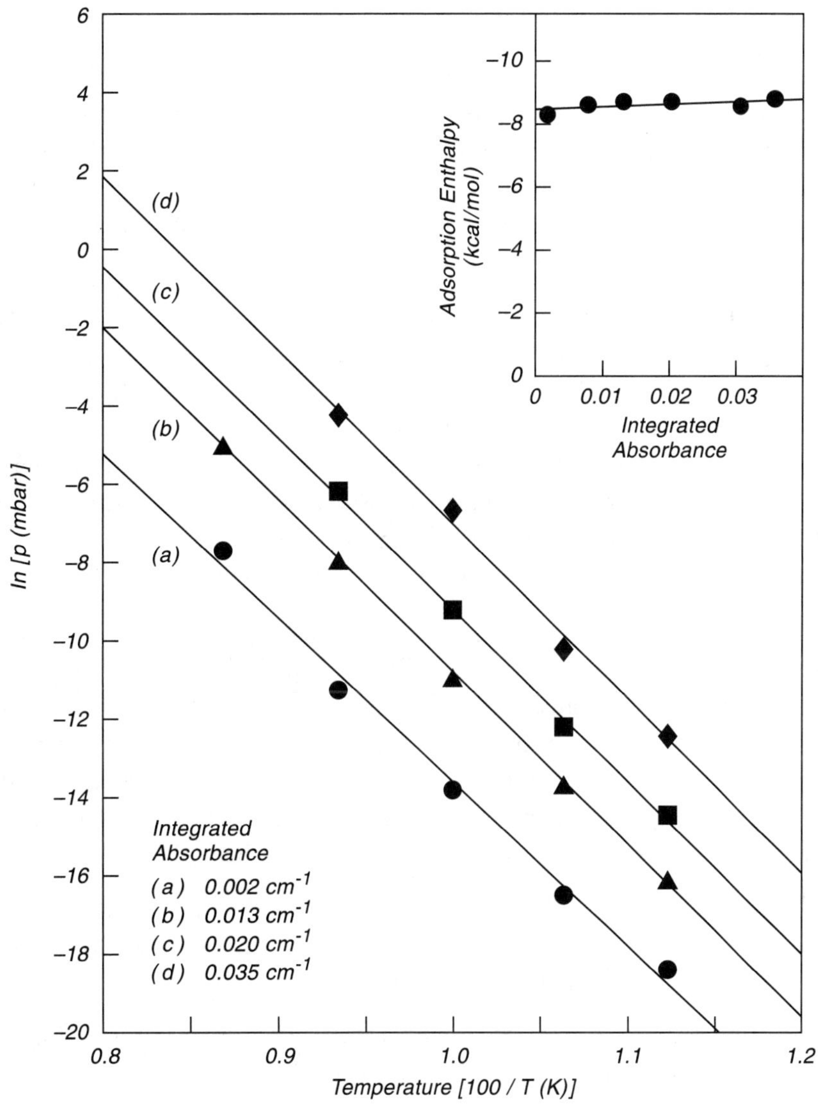


**C. Isosteric Heat of Adsorption**
**$N_2$ / Ni(111)**

# 135

# Internal Reflection IR Spectroscopy—Silicon

The use of infrared spectroscopy to probe the vibrational spectrum of adsorbates on semiconductor surfaces has opened up a wide range of new physical understanding of surface species structure and dynamics. In contrast to the reflection IR method used on metallic surfaces (see page 456), it is most appropriate to use internal reflection IR spectroscopy for semiconductor surfaces. This permits the IR beam to make many reflections at the interface, increasing the signal-to-noise ratio significantly.

Figure 135A shows the geometrical arrangement used for holding a silicon single crystal on a manipulator and for making connections for electrical heating [1]. Here, Ta clips are attached to the top of the U-shaped crystal and the crystal may be heated by the passage of current. Conductive cooling to the clips causes a temperature gradient across the crystal (100 K at 1000 K), but in the portion probed by the IR beam, only a small gradient is observed. The edges of the crystal are beveled and polished for admittance of the focused IR beam. The vector diagram for *p*- and *s*-polarization is also shown in Figure 135A. Oxidation of the top surface deactivates this surface for adsorption.

Figure 135B shows the UHV system and the associated optical system for use in the FTIR internal reflection measurements [2]. The U-shaped silicon crystal is mounted on a manipulator and may be rotated to an Auger spectrometer or to a LEED apparatus. In its position for internal reflection spectroscopy, the focused IR beam from the FTIR spectrometer passes through a CsI IR window (see page 742) and enters the silicon sample at the proper incident angle. After many internal reflections, the exiting radiation passes through another CsI window and is focused by a mirror through a polarizer and onto the detector.

For silicon, the absorption of infrared light below about 1450 $cm^{-1}$ prevents transmission with typical sample lengths of 3–5 cm.

Excellent reviews of this field, giving many other optical arrangements for doing IR surface spectroscopy on semiconductors, are available [3–8].

## *References*

[1] Y.J. Chabal, Bell Laboratories, Lucent Technologies (personal communication). See Y.J. Chabal, Phys. Rev. Lett. **50,** 1850 (1983).

[2] Y.J. Chabal, Lucent, AT&T (personal communication). See Y.J. Chabal, E.E. Chaban, and S.B. Christman, J. Electron Spect. Rel. Phenom. **29,** 35 (1983).

[3] N.J. Harrick, *Internal Reflection Spectroscopy* (Wiley, New York, 1967).

[4] F.M. Mirabella and N.J. Harrick, *Internal Reflection Spectroscopy*: Review and Supplement (Harrick Scientific Corporation, Ossining, NY, 1985).
[5] Y.J. Chabal, in *Semiconductor Interfaces: Formation and Properties,* edited by G. LeLay, J. Derien, and N. Boccara, Springer Proc. Phys. **22,** 301 (1987).
[6] Y.J. Chabal, Surface Sci. Reports **8,** 211 (1988).
[7] Y.J. Chabal, in *Internal Reflection Spectroscopy: Theory and Applications,* edited by F.M. Mirabella, Jr. (Marcel Dekker, New York, 1993), printed in Practical Spectroscopy, **15,** 191 (1993).
[8] M. Katiyar and J.R. Abelson, J. Vac. Sci. Technol. **A13,** 2005 (1995).

***Fig. 135. Internal Reflection IR Spectroscopy—Semiconductors***

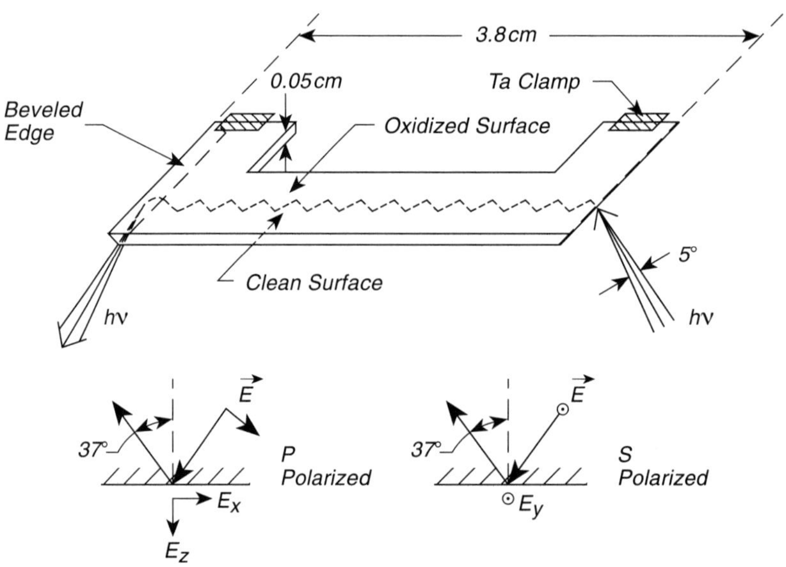


***A. U-Shaped Si Single Crystal Plus Vector Diagram for p- and s-Polarization***

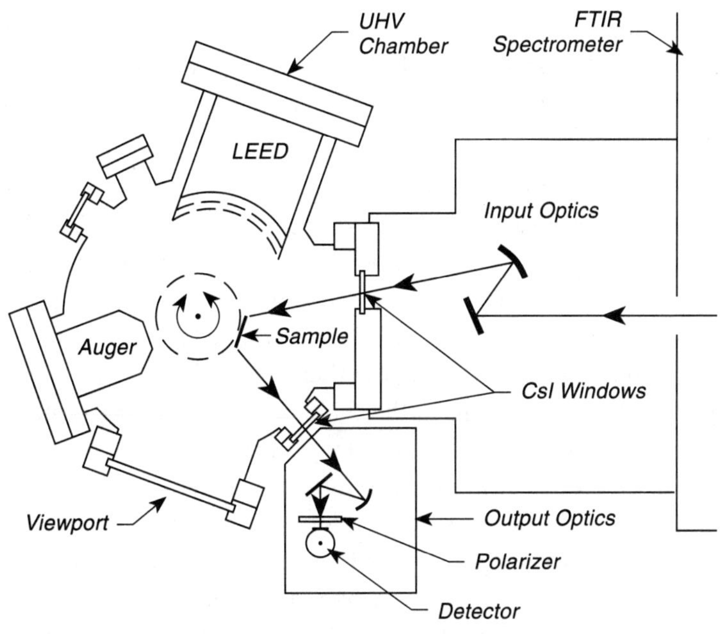


**B. UHV Optical System**

# 136

# Internal Reflection IR Spectroscopy—Gallium Arsenide

The use of internal reflection IR spectroscopy for silicon single crystals has been discussed on page 466. Different problems exist when the same measurement technique is applied to gallium arsenide. In order to permit IR transmission through GaAs, the carrier concentration needs to be kept below about $10^{16}/cm^3$ [1]. GaAs with this level of carriers is commercially obtainable from one supplier, which produces 1 Ω-cm material [2]. Normally, wafer GaAs is sold with conductivities of about 106 Ω cm (semi-insulating), or about $10^{-2}$ Ω cm. While the semi-insulating material would be satisfactory for IR measurements, it cannot be conveniently heated by resistive heating; alternatively, the highly doped material can be heated electrically, but its resistance is so low that large currents are needed and temperature nonuniformities are produced [1]. This material is also opaque in the infrared [1].

The mounting of 1 Ω-cm material for internal reflection infrared spectroscopy is shown in Figure 136. The infrared beam enters and exits through polished beveled edges. Heating current flows perpendicular to the IR beam direction, using Ta mounting clips on the top and bottom of the crystal. A thermocouple is conveniently glued to the crystal using a refractory cement, used also for thermocouple attachment to other semiconductor crystals such as SiC [3]. Welding the thermocouple first to a Ta foil square (1.2 mm x 1.2 mm) and then gluing this to the GaAs provides good agreement between the thermocouple and pyrometry [1]. The advantages of this mounting arrangement are that either surface can be studied, or if both are studied together a two-fold improvement in the signal strength in the FTIR measurement is obtained. The two sides of the crystal, if terminated by the same element, exhibit orthogonal orientation of the dangling bonds. Thus, using *s*- and *p*-polarized IR radiation, one can probe the adsorbate orientation [1]. A disadvantage of working with the 1 Ω-cm material is that the resistance becomes too great for electrical heating when the sample is cooled to 120 K. This problem can be alleviated by warming the sample slightly with a projector lamp (see page 532) [1].

## *References*

[1] Dr. J. R. Creighton, Dept. 1126, MS 0601, P.O. Box 5800, Sandia National Laboratories, Albuquerque, NM 87185-0601 (personal communication). See also J.R. Creighton, Appl. Surf. Sci. **82/83,** 171 (1994).

[2] GaAs with no intentional dopants (HB material, about 1 Ω-cm) is available from Sumitomo Electric, 551 Madison Avenue, New York, NY 10022.

[3] F. Bozso, L. Muehlhoff, M. Trenary, W.J. Choyke, and J.T. Yates, Jr., J. Vac. Sci. Technol. **A2,** 1271 (1984). Aremco 516 ceramic cement is used. Obtainable from Aremco Products, Inc., P.O. Box 429, 23 Snowden Avenue, Ossining, NY 10562-0429; the cement should be preserved in the refrigerator and has a limited shelf life.

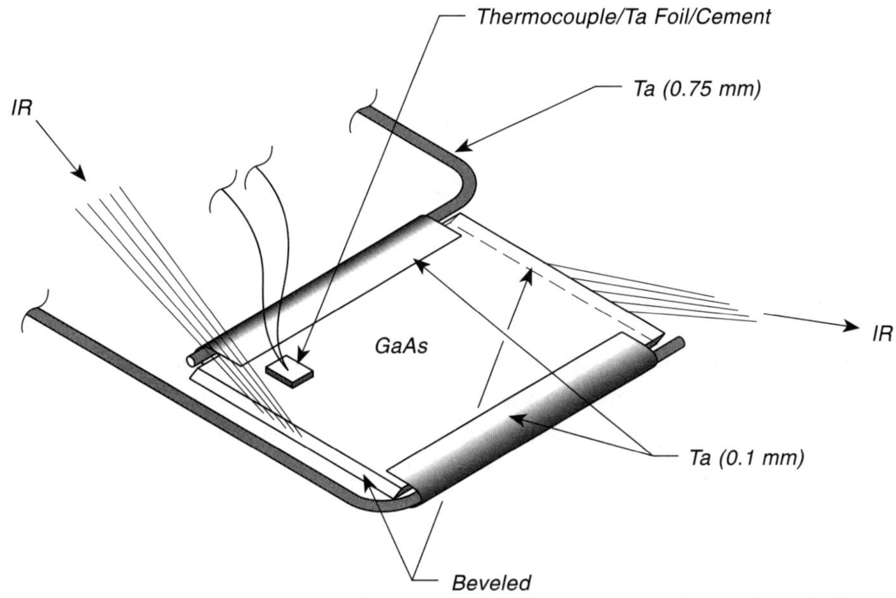


***Fig. 136. GaAs—Mounting for Internal Reflection IR Spectroscopy***

# 137

# Infrared Spectroscopy—Continuous Pumping of Detector Dewar

InSb and HgCdTe infrared detectors are used in the mid-IR range (4000 $cm^{-1}$ to 500 $cm^{-1}$) for the study of adsorbed species on surfaces. The detector element is mounted on the end of a small metal Dewar which is kept filled with liquid nitrogen during operation, as shown in Figure 137. The Dewar is often pumped on a separate vacuum system and then sealed with a Viton-sealed valve. In addition, the detector window is often sealed with a Viton O-ring. Thus, over time, slow permeation, outgassing, and leakage can be expected (see page 166), and the condensation of water and other materials may occur on the detector surface, causing interference with infrared measurements. These problems can be prevented by using a small appendage ion pump to continuously pump the detector [1].

The IR detector is connected to a metal bellows-sealed valve and thence to a tee as shown in Figure 137. One branch of the tee is connected to the ion pump via a 1.5-cm-diameter bellows that extends up to about a 30-cm length as it makes the transition from the detector to the ion pump. The second branch is temporarily connected to an oil diffusion pump (liquid nitrogen trapped) for preliminary pumping of the Dewar and the appendage pump.

Initially the entire system is pumped out by the diffusion pump. The valve on the detector is then closed and the ion pump (not operating) and the metal tubing are baked out with heating tape. When the pressure is low enough that the ion pump can be operated, the valve to the diffusion pump is closed and the pump is removed from the vacuum system. The heating is discontinued and when the line is cooled, the valve to the detector is opened. Under these conditions, the IR detector will be continuously pumped and condensation effects will be eliminated on the detector. It is wise to place a conical shield just below the top of the Dewar entrance to prevent liquid nitrogen from accidentally pouring over the detector window and the vacuum seal for the amplifier wires, since these vacuum seals may be broken by such accidental cooling [2].

## *References*

[1] Contributed by Dr. Victor M. Bermudez, Naval Research Laboratory, Code 6863, Washington, DC 20375-5347.

[2] Developed in author's laboratory.

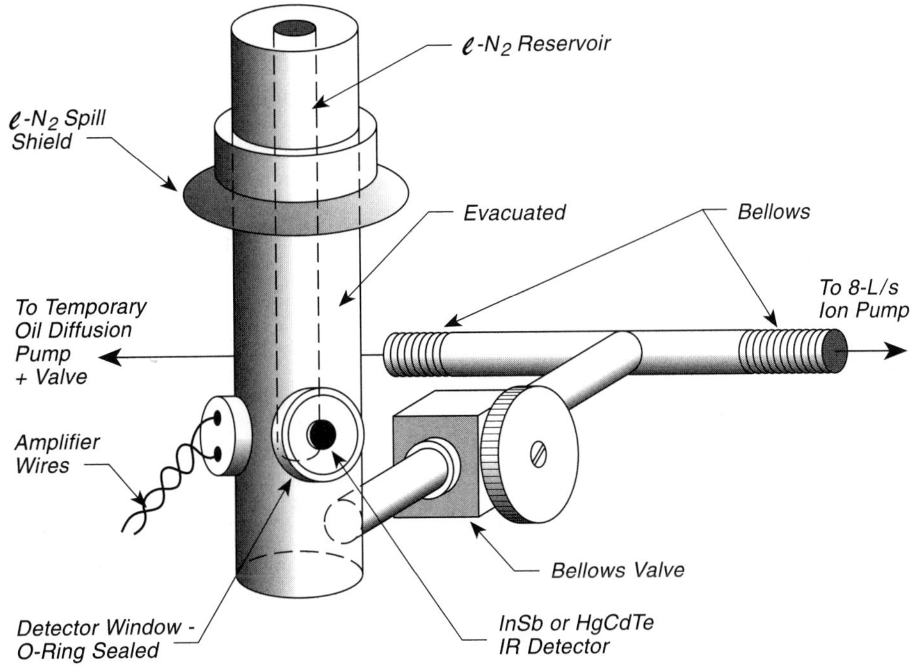


*Fig. 137. Ion-Pumped IR Detector*

# L. Calorimetric Heats of Adsorption—Single Crystals

# 138

# Measurement of the Calorimetric Heat of Adsorption on Ultrathin Metal Single Crystals

There are many methods for measuring the energy required for desorption of a molecule from a single crystal surface using either isosteric heat measurements under equilibrium conditions, or using desorption kinetic methods. The interpretation of the isosteric measurements requires perfect reversibility in adsorption and desorption, a condition not met when compound formation or dissolution of the adsorbate in the crystal occurs. The interpretation of the kinetic measurements are beset with difficulties having to do with the inability to write the desorption rate laws that apply with accuracy. The calorimetric method avoids these difficulties by directly measuring the heat evolved upon adsorption and by using only the first law of thermodynamics.

In order to employ a calorimetric device, the heat capacity of the calorimeter must be made very small in order for a measurable temperature rise to be produced. This has been cleverly done, using single crystal films of metals which are only 2000 Å in thickness [1], using apparatus as shown in Figure 138. These films are grown epitaxially on an alkali halide crystal surface, and the continuous metal film is then lifted off of the salt crystal as it dissolves in water. The metal crystal film is transferred and attached to a ring, made of the same metal, and may be placed at the focus of a pulsed molecular beam of the adsorbate. The thin crystal may be cleaned in UHV by ion bombardment and by heating in vacuum or in reactive gases used for cleaning. Standard surface analysis and LEED studies are routinely performed on the thin crystal after cleaning. As the gas pulses arrive and are adsorbed on a 2-mm-diameter spot, the temperature of the thin crystal rises, and the pulsed emission of infrared radiation is measured from the rear of the crystal by focusing the radiation on a mercury cadmium telluride detector held at liquid nitrogen temperature. Temperature rises on the crystal per pulse are below 0.1 K for chemisorbing gases, where each pulse delivers about 0.01 ML of chemisorbed species to the surface. The emissivity of the backside of the crystal is significantly enhanced by initially coating it with amorphous carbon from a candle flame. For calibration, a pulsed laser beam of known energy per pulse is used, and its reflectivity at near normal incidence is measured. The temporal character of the laser beam is matched to that of the pulsed molecular beam in the calibration measurements. The energy losses after each pulse are found to be due to radiation losses only.

In addition to the errors in measuring the small pulses of heat, errors in the coverage measurement (and the sticking coefficient) currently limit this method [1]. The sticking coefficient and absolute coverage are measured using the method of King and Wells [2], which measures the intensity of the reflected gas from the crystal during each pulse. A gold flag is employed in these measurements. A stagnation detector (see page 616), which is a Bayard–Alpert gauge with a small aperture, is employed for measurements of the absolute beam flux, as discussed on page 446.

## *References*

[1] C.E. Borroni-Bird and D.A. King, Rev. Sci. Instrum. **62,** 2177 (1991).
[2] D.A. King and M.G. Wells, Surf. Sci. **29,** 454 (1972).

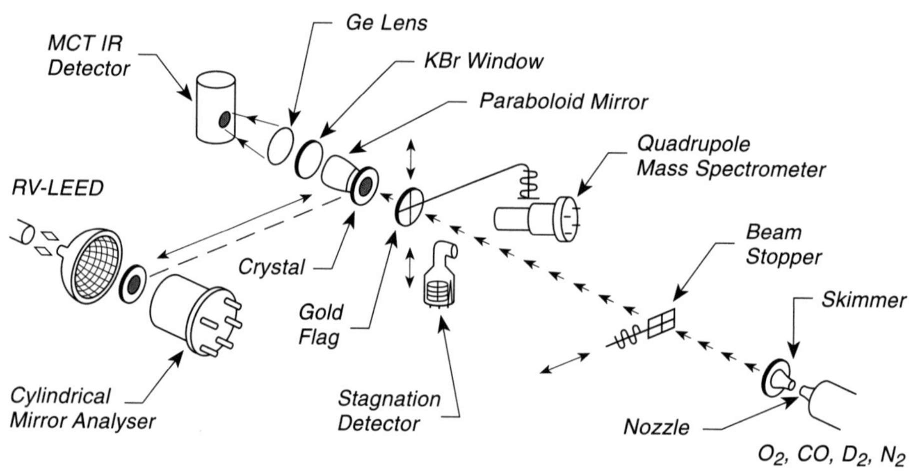


**Fig. 138. Calorimetric Adsorption Heat—Ultrathin Metal Films**

# M. Surface Debye Temperature

# 139

# Measuring the Surface Debye Temperature of a Single Crystal

The Debye temperature of a bulk solid is a measure of the stiffness of the bonds within the crystal and is related to the Debye characteristic frequency, $\nu_D$. This frequency corresponds to the highest allowed vibrational mode in the crystal in the distribution of frequencies following a mode frequency distribution function with a magnitude which is proportional to the square of the mode frequency. The Debye temperature is $\Theta_D = h\nu_D / k$, where $k$ is Boltzmann's constant.

The Debye-Waller factor, $2M$, is connected to the Debye temperature through Equation (1), and determines the intensity of scattering of an electron beam incident on the crystal through Equation (2).

$$2M = 24m_e(E\cos^2\alpha + V_o)T/m_a k(\Theta_D)^2 \tag{1}$$

where $m_a$ = mass of the atoms in the crystal, $m_e$ = mass of the electron, $\alpha$ = the angle of incidence, $V_o$ = the inner potential, and $E$ is electron energy.

For an electron beam incident on a crystal, in the kinematic approximation,

$$I = I_o e^{-2M} \tag{2}$$

where $I_o$ is the incident intensity and $I$ is the diffracted intensity.

Figure 139A shows a four-grid LEED apparatus with a movable Faraday cup to intercept a selected diffraction beam [1]. The crystal surface in inclined by angle $\alpha$ from the incident beam, and the diffracted 0,0 beam is being collected in the drawing. As the crystal cools, the intensity of the diffracted beam increases as a result of the decrease in the vibrational motion of the atoms in the region of the solid surface being sampled by the incident electrons. As the energy of the incident electron beam is decreased in successive experiments, a lower sampling depth is involved in the diffraction process, and the measured Debye temperature decreases as a result of the looser binding of the crystal atoms near to the surface. Figure 139B shows a Debye-Waller plot of $\ln(I - I_{bk})/I_o$ versus temperature for various values of $\Delta K$, the perpendicular component of the momentum transfer between the electron and the metal atoms in the crystal. Here, $I_{bk}$ is a background scattering contribution which can either be measured in the vicinity of the 0,0 beam, or treated parametrically as a variable. In this case, for low angles of incidence, one can only measure an effective perpendicular Debye temperature as $\Delta K$ is varied [2].

Figure 139C shows typical behavior of the Debye temperature measurements for three angles of incidence on a Rh(111) crystal as a function of $\Delta K$. It may be seen that all of the measurements show that as $\Delta K$ increases, and the depth of sampling of the crystal by the electron beam increases, the measured value of $\Theta_D$ increases to a plateau at $\Theta_D = 370$ K, which is characteristic of the bulk of the crystal. The lower limit of $\Theta_D$ is near 200 K, and is characteristic of the surface metal atoms.

These measurements were made with a LEED apparatus mounted between two pairs of Helmholtz coils to cancel the earth's magnetic field. The values of the diffracted beam intensity were measured as the crystal slowly cooled to avoid the serious influence of the magnetic field associated with the electrical heating current. The Faraday cup aperture subtended an angle of $4.6 \times 10^{-3}$ steradians, and was large enough to collect the entire beam, $I$.

## *References*

[1] C.M. Chan, P.A. Thiel, J.T. Yates, Jr., and W.H. Weinberg, Surf. Sci. **76**, 296 (1978).

[2] G.A. Somorjai and H.H. Farrell, Adv. Chem. Phys. **20**, 215 (1971).

**Fig. 139. Measuring the Surface Debye Temperature of a Single Crystal**

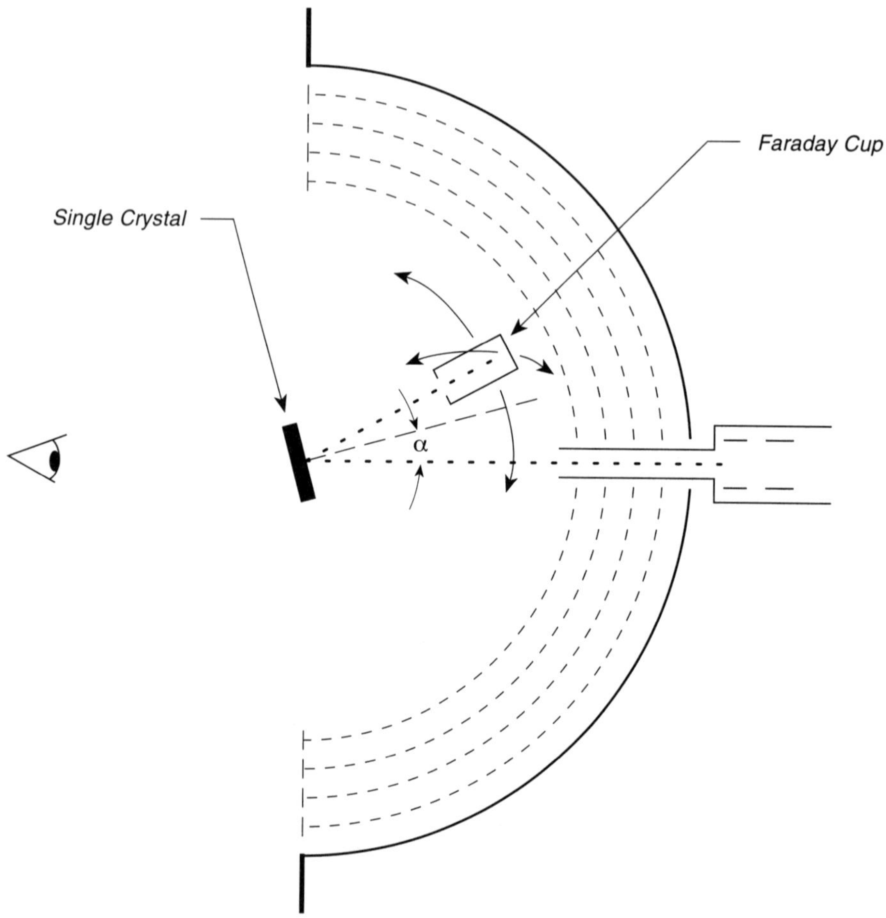


**A. LEED Apparatus with Faraday Cup**

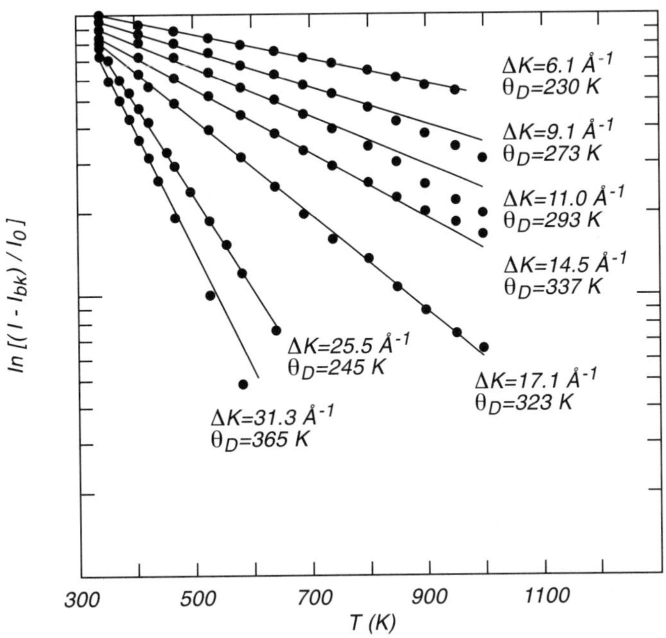


**B. 0,0 Beam Intensity, Rh (111)**

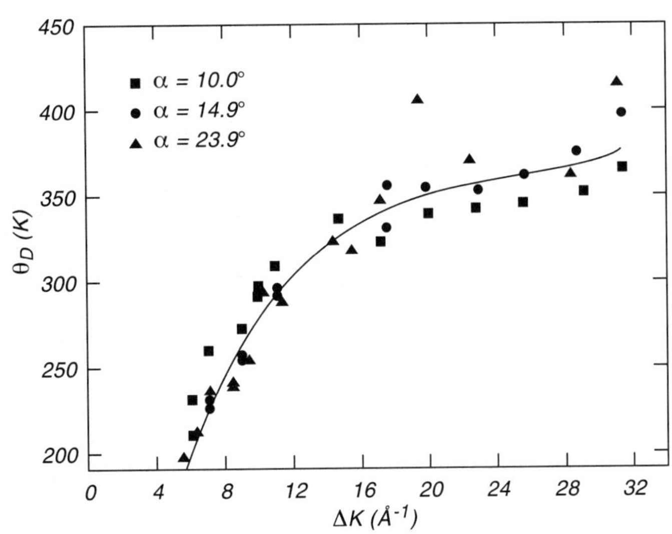


**C. Debye Temperature, θD (K)**

# N. Friction

# 140

# UHV Tribometer for Measuring the Coefficient of Friction

Many technologies depend upon knowledge of the coefficient of friction between a surface and a probe that slides across the surface. Since the coefficient of friction $\mu = F_s/F_n$, where $F_s$ is the shear force, and $F_n$ is the normal force, a device that simultaneously measures both components of force during sliding is needed. In addition, such a device should be able to be applied to surfaces that are initially atomically clean, followed by studies on the surface which has a known coverage of lubricant. These research demands have resulted in the development of the UHV tribometer [1].

Figure 140A shows a schematic picture of the tribometer. A single crystal is mounted on two support wires that connect to a rigid frame. Behind the crystal is a coiled thermionic emitter used for heating the crystal to achieve a condition of atomic cleanliness.

The frame surrounding the sample is connected by means of two L-shaped Cu/Be alloy springs (0.006-in. thick by 0.25-in. wide) and each arm of the L is 0.5 in. in length. Glued to each of these springs are strain gauges (see page 488) thermally matched to the Cu/Be alloy. A total of eight strain gauges are employed, connected to two Wheatstone bridges (normal; shear).

Measurements are made by bringing the sample in contact with the slider (also a single crystal) which is mounted on a manipulator which can be translated at constant speed, under an applied normal load. Figure 140B shows the displacement components, $\delta$, produced by applying a normal load to the sample, and it can be seen that both normal displacements and shear displacements are sensed by the strain gauges. However, for a normal force applied, the shear displacements counterbalance, and are not measured by the Wheatstone bridges. In the same way, for shear forces the normal displacement is counterbalanced and not measured by the bridges. This arrangement therefore permits the coefficient of friction to be continuously measured during the sliding experiment.

It has been found on single crystal metal surfaces that a single monolayer of a low molecular weight molecule is effective in reducing the coefficient of friction. Three frictional regimes of increasing coverage of ethanol on an interface, a S-covered Ni(100) crystal surface, have been observed. Below a monolayer, sticking is observed; near a monolayer, stick-slip motion is observed; and above a monolayer, slip motion is observed [1].

Another tribometer design [2] is shown on page 488.

## *References*


[1] A.J. Gellman, J. Vac. Sci. Technol. **A10**, 180 (1992).

[2] T.G. McDonald, D.E. Peebles, L.E. Pope, and H.C. Peebles, Rev. Sci. Instrum. **58**, 593 (1987).

## *Fig. 140. UHV Tribometer for Measuring the Coefficient of Friction*

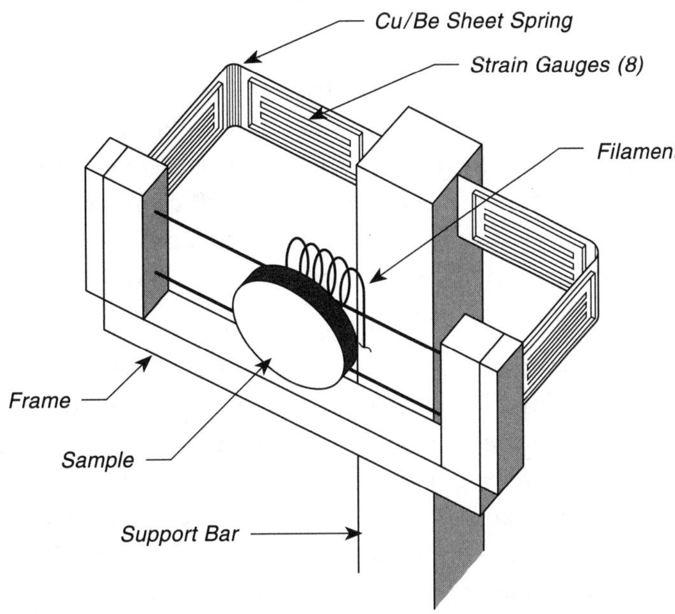


***A. Strain Gauge Arrangement***

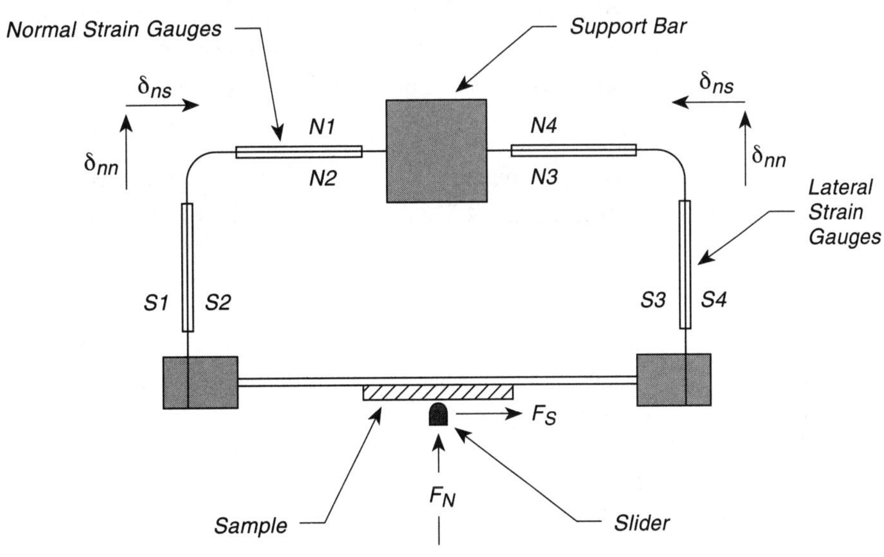


***B. Forces at Slider/Sample Interface and Displacement Components***

# 141

# Pin-On-Disk Measurement—UHV

The measurement of the coefficient of friction (see page 486) is of importance in understanding lubrication. To do this without the influence of adsorption processes from the atmosphere, tribometers operating in ultrahigh vacuum have been devised. The device shown here has the ability to make friction coefficient measurements, as well as measuring the electrical resistivity of the interface between the pin and the disk. In addition, Auger spectroscopy measurements of the surface composition of the wear track can be made, using a scanning instrument [1].

Figure 141A shows an exploded view of the tribometer. The disk sample is mounted on a rotary platform that may be turned in UHV by an external motor. A pin sample bent in the form of a loop may be pressed onto the disk sample by means of a spring tensioner that attaches to a cantilever beam. The cantilever beam is free to move in the $z$-direction on bearings in a trunnion that can be moved in the $z$-direction. Electrical contact is made to the pin for current and voltage measurements. Two brushes (Cu/Be alloy) make electrical contact with a commutator electrically connected to the sample; one of these is used for current, $I$, measurements, and the second for voltage, $V$, measurements. The lateral force, $F$, is measured by two semiconductor strain gauges [2] that are accurately and squarely mounted on a thinned section of the cantilever beam, using strain gauge cement [3]. The normal force, $N$, is adjusted by means of a screw adjuster on the spring tensioner. Figure 141B shows the vector forces and the electrical measurement method employed as the disk sample rotates. The entire pin-on-disk apparatus may be moved inside the vacuum system by means of internal $x$-,$y$-, and $z$-bearings; and rotary motion is provided by a rotary manipulator shaft.

The apparatus has been used to measure the effect of atmospheric gases on contact electrical alloys under wear conditions [4,5].

## *References*


[1] T.G. McDonald, D.E. Peebles, L.E. Pope, and H.C. Peebles, Rev. Sci. Instrum. **58**, 593 (1987).

[2] P-doped U configuration, 350 Ω, 130 gauge factor, obtained from Kulite Semiconductor Products, Inc., 1 Willow Tree Road, Leonia, NJ 07605.

[3] M-Bond 610 Epoxy Adhesive.

[4] L.E. Pope and D.E. Peebles, J. Vac. Sci. Technol. **A4**, 1659 (1986).

[5] L.E. Pope and D.E. Peebles, *Electrical Contacts-1986* (IEEE, New York, 1986), p. 65.

***Fig. 141. Pin-On-Disk Measurement—UHV***

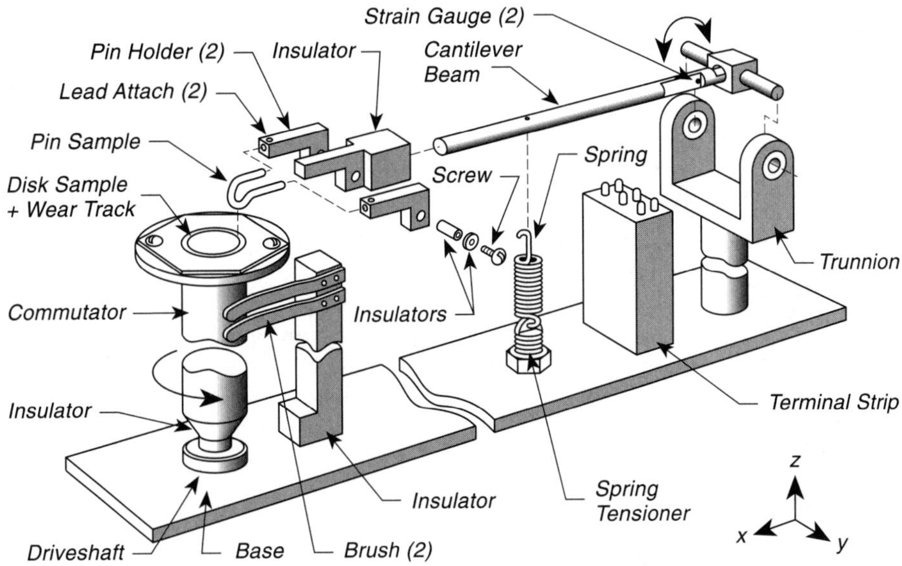


***A. Exploded View***

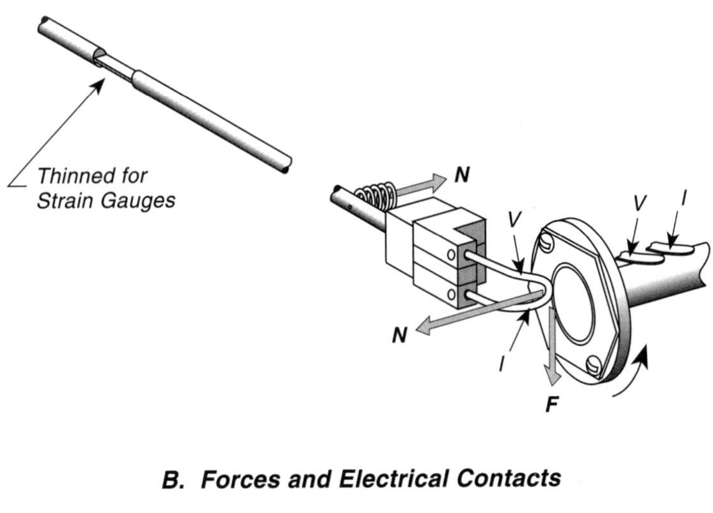


***B. Forces and Electrical Contacts***

# IV. Thermal Control

A. Heating Samples . . . . . . . . . . . . . . . . . . . . . . . . . . . . . . . . . . 491
B. Cooling Samples . . . . . . . . . . . . . . . . . . . . . . . . . . . . . . . . . . 543
C. Temperature Measurements. . . . . . . . . . . . . . . . . . . . . . . . . 579

# A. Heating Samples

# 142

# Electronic Temperature Programmer—Metal Crystals

The accurate and reproducible programming of the temperature of metal single crystals is used for studies of temperature-programmed desorption (see Section III.F, page 385) as well as in many other applications. A review of the literature in this area to 1984 may be found in reference [1].

Figure 142A shows a block diagram of a temperature programmer design that has been thoroughly tested in many applications [1]. An input signal from the thermocouple, welded to a metal crystal, is fed into a thermocouple amplifier [2]. The input amplifier adds gain and offset to this signal, and delivers a 0 to +10 V signal to the process controller. The thermocouple amplifier protects the single crystal from being destroyed in the case of an open thermocouple circuit, which drives the output of the amplifier to a high positive level. An electronic thermocouple reference junction eliminates the need for an external reference and is routinely checked against a reference junction maintained at 77.3 K or 273.2 K and connected temporarily to the input for calibration.

The ramp generator is programmed by the laboratory worker, as shown schematically in Figure 142A. A more detailed block diagram of the ramp generator is shown in Figure 142B. A fixed frequency oscillator drives the divider circuit that feeds a 12-bit binary counter. The 12-bit binary word is the input to a digital-to-analog converter, which with scaling and offset, delivers a 1–5-V signal to the process controller set point. The start point input presets the counter to a binary count specified by the operator. The end point input determines the temperature at which the ramp stops and holds constant temperature.

The power driver delivers up to 50 A with a burden voltage of 4 V. In addition, the crystal may be isolated and a DC voltage may be applied. A commercial 5-V DC supply was used, with a current regulator added to the output with appropriate scaling and isolation to permit the process controller to control the power output. An input of 1–5-V DC produces a 0–50-A DC output.

The process controller [3] (Fig. 142A), was selected as the primary control device for the following reasons: (1) the ability to program the readouts in arbitrary units; (2) accuracy (±0.05 %, full scale); (3) speed (updates 3 times/s); (4) digital input of tuning parameters.

An illustration of the performance of the programmer is shown in Figure 142C. The programmer tuning constants were adjusted to produce the linear temperature vs. time performance from about 95 K to 520 K as shown. The temperature stability at 272 K is shown, where fluctuations of only ±0.35 K were observed. Scan rates from about 2 K/s to 8 K/s may be programmed into the controller.

In order to achieve optimum performance, appropriate values of the process controller constants (gain, rate, and integral time) need to be established. These three control constants are dependent on sample properties such as mass and resistivity and on the sample-mounting configuration. Initially the rate and integral time can be set to zero while an optimum gain is being established. An optimum gain value is one which minimizes oscillations in the sample temperature, both during the heating ramp and while the crystal is being maintained at constant temperature, and the optimum gain may differ slightly for the two operating conditions. It is also necessary to optimize the gain such that the difference between the ramp voltage and the measured thermocouple voltage is minimized during the heating ramp. The other control constants can then be adjusted to fine tune the operation of the controller. Tuning of the integral time is essential, particularly when trying to avoid an overshoot at the end of temperature ramp.

Another electronic controller, designed to correspond to the system's thermal response and capable of programming at 20 K/s without overshoot, has been described [4].

## *References*

[1] R.J. Muha, S.M. Gates, P. Basu, and J.T. Yates, Jr., Rev. Sci. Instrum. **56,** 613 (1985).

[2] Model 2B50B, obtainable from Analog Devices, One Technology Way, Norwood, MA 02062-9106.

[3] Model UDC 500, obtainable from Honeywell, Micro Switch Division, 11 W. Spring Street, Freeport, IL 61032-4353. More advanced controllers are now available from Honeywell, such as the UDC 5000 ultrapro.

[4] M.A. DeAngelis and A.B. Anton, J. Vac. Sci. Technol. **A10,** 3507 (1992).

## FIG. 142. Electronic Temperature Programmer—Metal Crystals

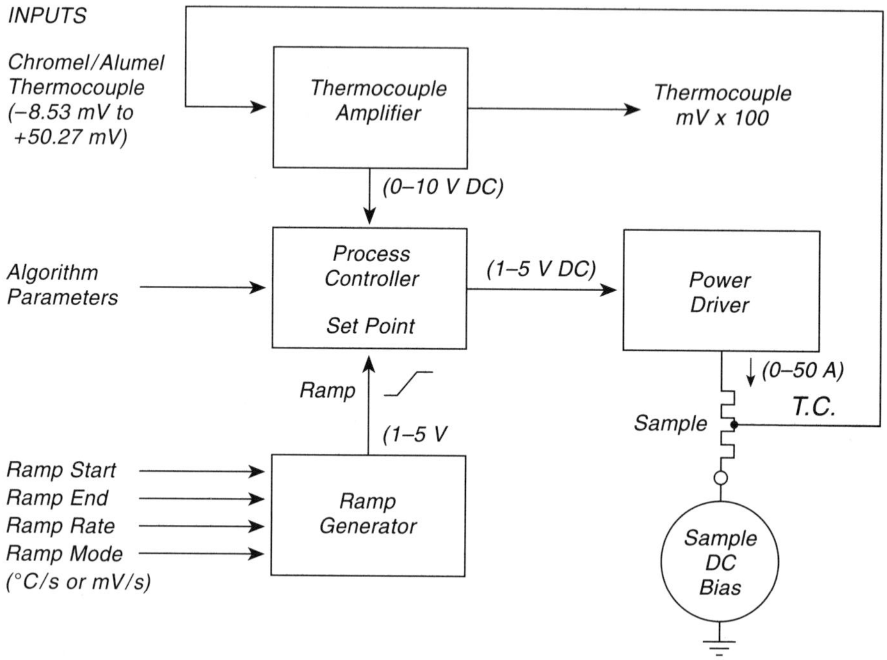


**A. Block Diagram—Temperature Programmer**

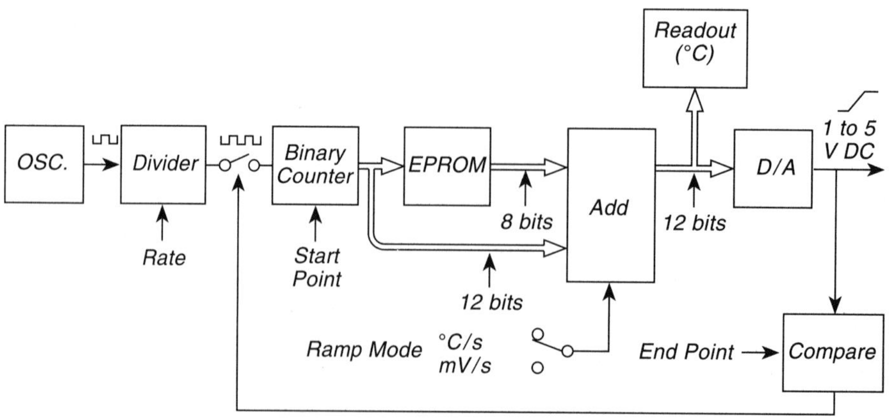


**B. Block Diagram of Ramp Generator**

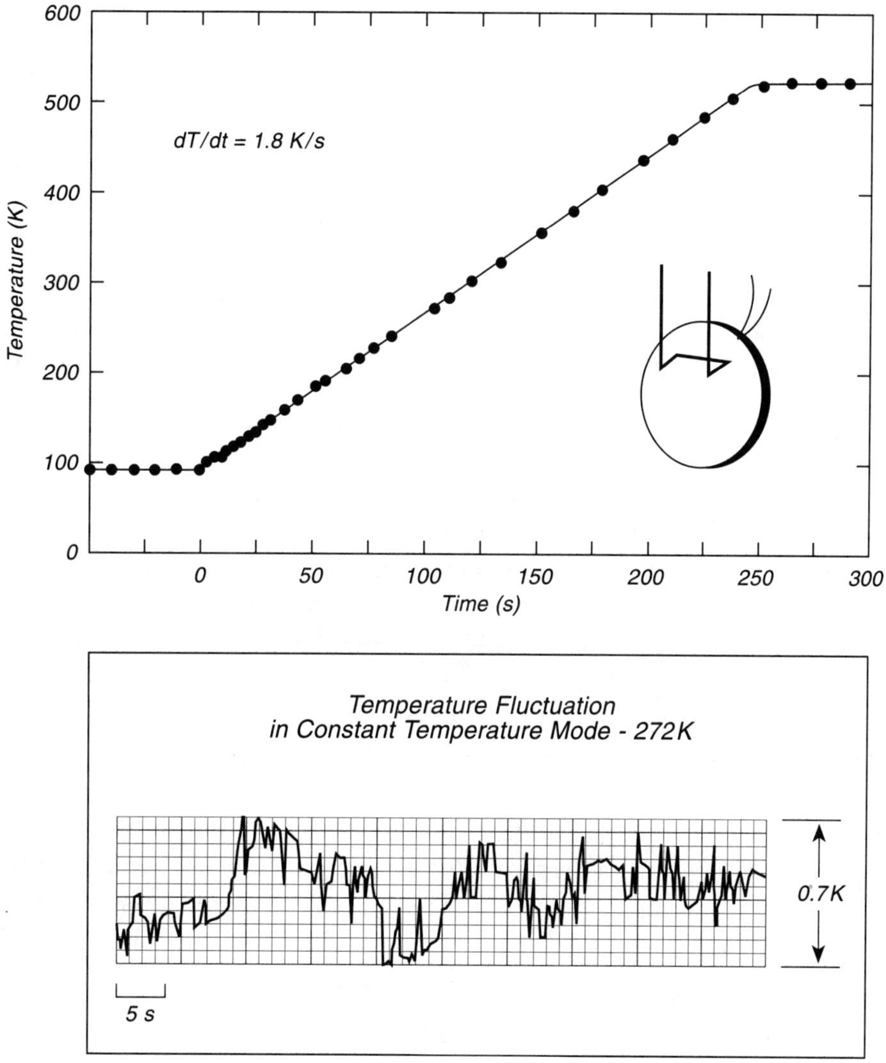


**C. Performance of Programmer**

# 143

# Temperature Programming of Metal and Semiconductor Crystals

Almost all of the studies of adsorption, desorption, and reactivity at surfaces involve control of the temperature of the surface as well as control of temperature programming. A traditional solution involves specialized controlling units, one of which is mentioned on page 494. These controllers provide a technically simple but rigid solution for the temperature control. An alternative solution employs an acquisition board in a personal computer controlled by a program written in any programming language. This method provides great flexibility and can be used for any type of temperature sensor and any desired temperature program. A graphical programming language, LabVIEW [1], is especially appropriate for these tasks. It has a powerful graphical interface, a large library of subroutines for data acquisition, and a variety of mathematical operations. An example of a virtual instrument created by LabVIEW for temperature measurement and control of a silicon crystal is described below.

Figure 143A shows a block diagram of the measurement setup for a silicon crystal having a room temperature resistivity up to 100 Ω cm. The crystal temperature is measured by a thermocouple that has a known relationship to the true temperature (measured by pyrometer) as a function of temperature. A cold junction compensator [2] is used to establish, electronically, a reference voltage corresponding to the ice temperature, 273.2 K. The signal is fed to a separating amplifier and amplified by 50–100 times, which minimizes the noise level received at the acquisition board. A second purpose of the separating amplifier is the rejection of the DC offset caused by the potential applied to the crystal for heating. The acquisition board can only handle 10–20 V offsets, which would be normal for metal crystals because of their higher conductivity compared to a semiconductor crystal. The signal from the separating amplifier is fed to the acquisition board [3], which has an analog input/output with a maximum input and output voltage of 10 V. Two DC power supplies are needed, one to power the separating amplifier (±5V) and a heating supply (60 W, 120 V, and a maximum current of 50 A).

Figure 143B shows a block diagram of the LabVIEW controlling program that controls the heating power supply according to a predetermined program. A sample of a complex heating/cooling program and the result is shown in Figure 143C. This program provides the desired heating ramp from -200°C up to 850°C, and then cooling by various ramp functions, as shown on the front panel readout in Figure 143C. The lower right-hand side of the front panel readout shows the actual program achieved compared to that desired.

Figure 143D shows a detailed view of the linearity of the heating ramp and the error that was measured. Here, $\beta = 3$ K/s, the linear ramp function. The lower difference plot shows the accuracy of this particular ramp and all of the data points fell in a 1.5 K noise range up to about 1000 K, where the mechanical contact (see page 508) between the thermocouple and the Si crystal began to fail.

The ability to control the isothermal temperature of the crystal for long periods of time is also excellent, with better than 1 K temperature constancy and error spread. There is often an initial overshoot of the order of several K as one ramps up to a desired temperature, and this can be minimized by adjusting the proportional integrating differentiating (PID) parameters in the program, Figure 143B.

The advantages of the LabVIEW approach to temperature programming are listed below:

1. Flexibility—the temperature profile can be changed as desired and nonlinear programs are possible. Any type of temperature sensor with a non-linear output can be used (see page 580). Any type of correction for errors, such as the error in a thermocouple because of inefficient thermal contact (see page 508), can be made.
2. High speed.
3. Universal interface—the acquisition board with multiple inputs and outputs coupled to the LabVIEW program, can be used for computer control of other instruments (e.g., a mass spectrometer). However, the interface cannot be used at the same time with another acquisition board driven by a program working in DOS.
4. Convenient graphical programming with a large library of subroutines.

## *References*

[1] LabVIEW is available from National Instruments, 6504 Bridge Point Parkway, Austin, TX 78730-5039, USA.

[2] Type MJC-C, obtainable from Omega Engineering, Inc., P.O. Box 4047, Stamford, CT 06970, USA.

[3] Type Lab-PC+, obtainable from National Instruments, see reference [1].

***Fig. 143. Temperature Programming of Metal and Semiconductor Crystals***

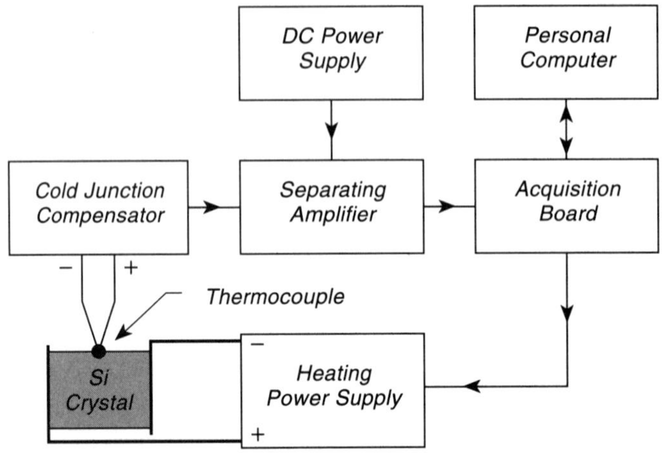


***A. Block Diagram—Resistive Heating Control***

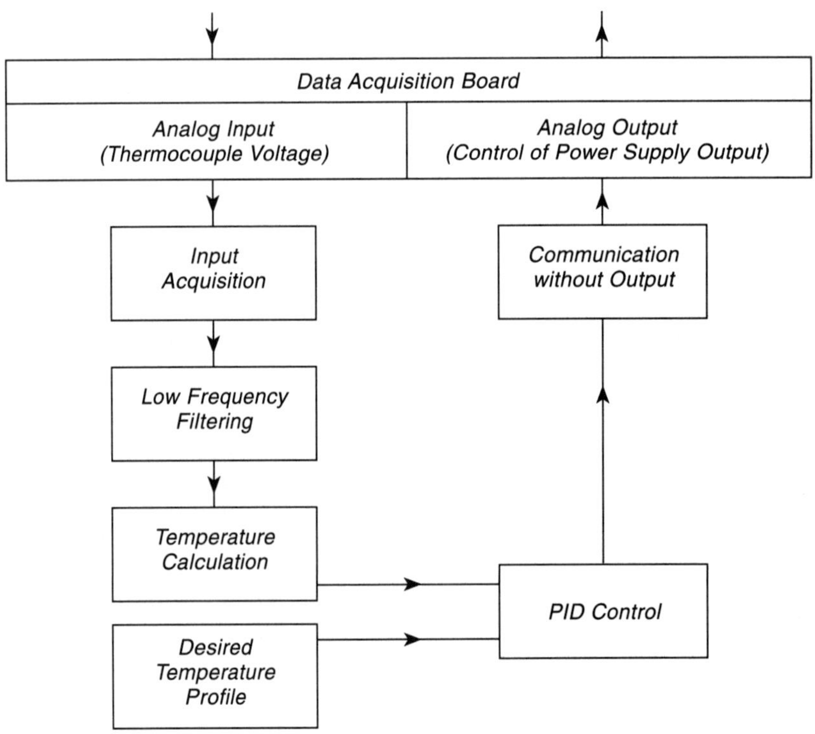


***B. Block Diagram—LabVIEW Controller Logic***

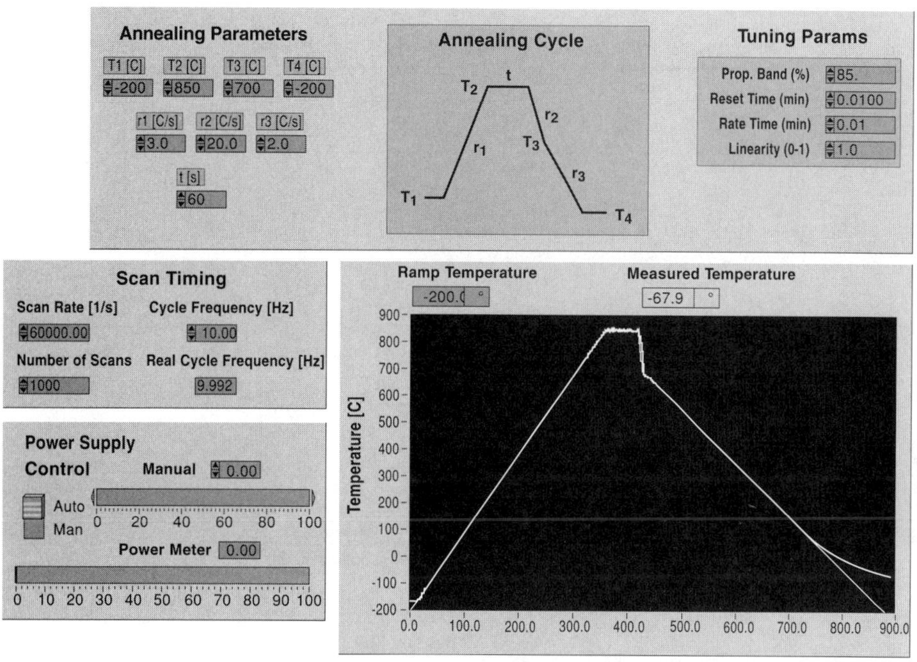


*C. Control Parameters for Temperature Program*

***Fig. 143. Temperature Programming of Metal and Semiconductor Crystals***

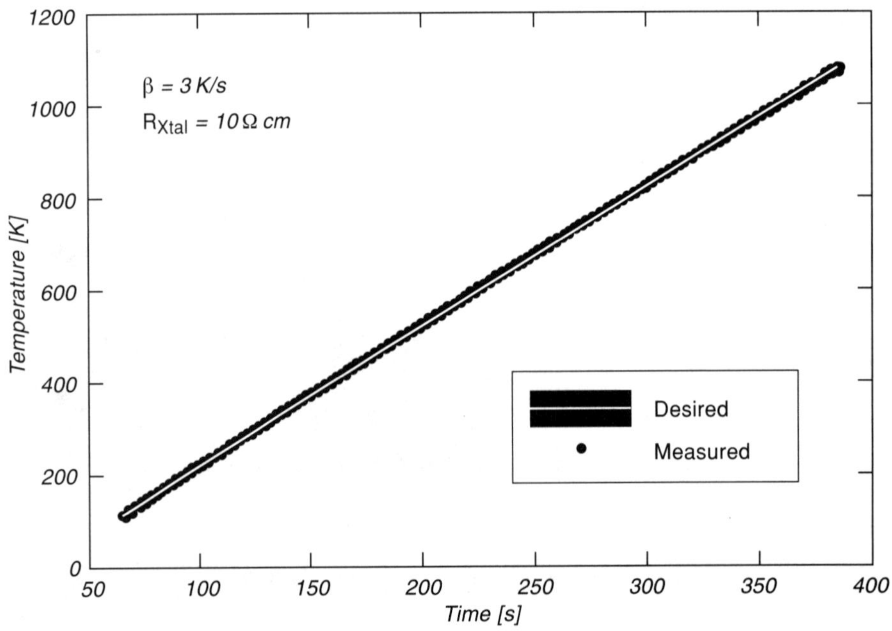

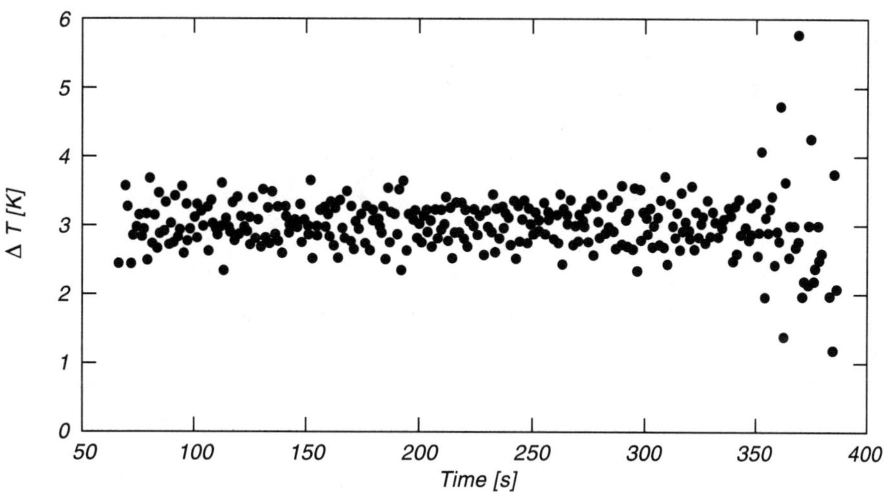


***D. Linearity and Error in Heating Ramp***

# 144

# Mounting Nonweldable Crystals for Resistive Heating and Cooling

Crystals such as Ag, Al, Au, and Cu possess very high thermal conductivity and cannot easily be spot-welded to other metals because of the lack of melting at the low temperatures achieved at the junction during welding. The mounting of such crystals to heating/cooling wires requires a special method that will provide good mechanical, thermal, and electrical contact over many cycles of heating and cooling, and a design for this is shown in Figure 144.

The crystal holder is fabricated out of OFHC copper, sapphire, and tungsten. These materials all possess optimum thermal conductivity properties below 100 K.

Figure 144 shows a scheme to mount such crystals. A slot is carefully cut into two opposite edges of the crystal. The slot makes a tight sliding fit for the tungsten wire heating lead, and the crystal should not be distorted to make this fit.

The tungsten heating/cooling leads are fabricated with a pair of U-loops that remove the effects of linear expansion that otherwise would push the wire from the slot. Thus, the crystal, incorrectly mounted as shown in the right-hand insert, is subject to variation in the cooling and electrical contact as the wire is heated and cooled owing to bowing of the wire in the slot. The looped wires on the other hand undergo thermal expansion effects which are almost counterbalanced on either arm of the U-loops.

Make the assembly as follows using gloved fingers:

1. Mount the preshaped W wire (0.38-mm diameter) on the lower set of Ta screws and tighten.
2. Fix the preshaped upper W wire on the right hand screw and do not tighten the screw.
3. Fix the crystal with its thermocouple (hole-mounted—details not shown) on the lower wire.
4. Rotate the upper wire counter clockwise until it fits into the slot and then tighten both screws carefully.

This mounting procedure produces a stable junction over many cycles that permits cooling of the crystal to about 32 K when liquid He is used in the reservoir intended for liquid nitrogen [1].

## *References*


[1] J. Ahner, D. Mocuta, R.D. Ramsier, and J.T. Yates, Jr., J. Chem. Phys. **105**, 6553 (1996).

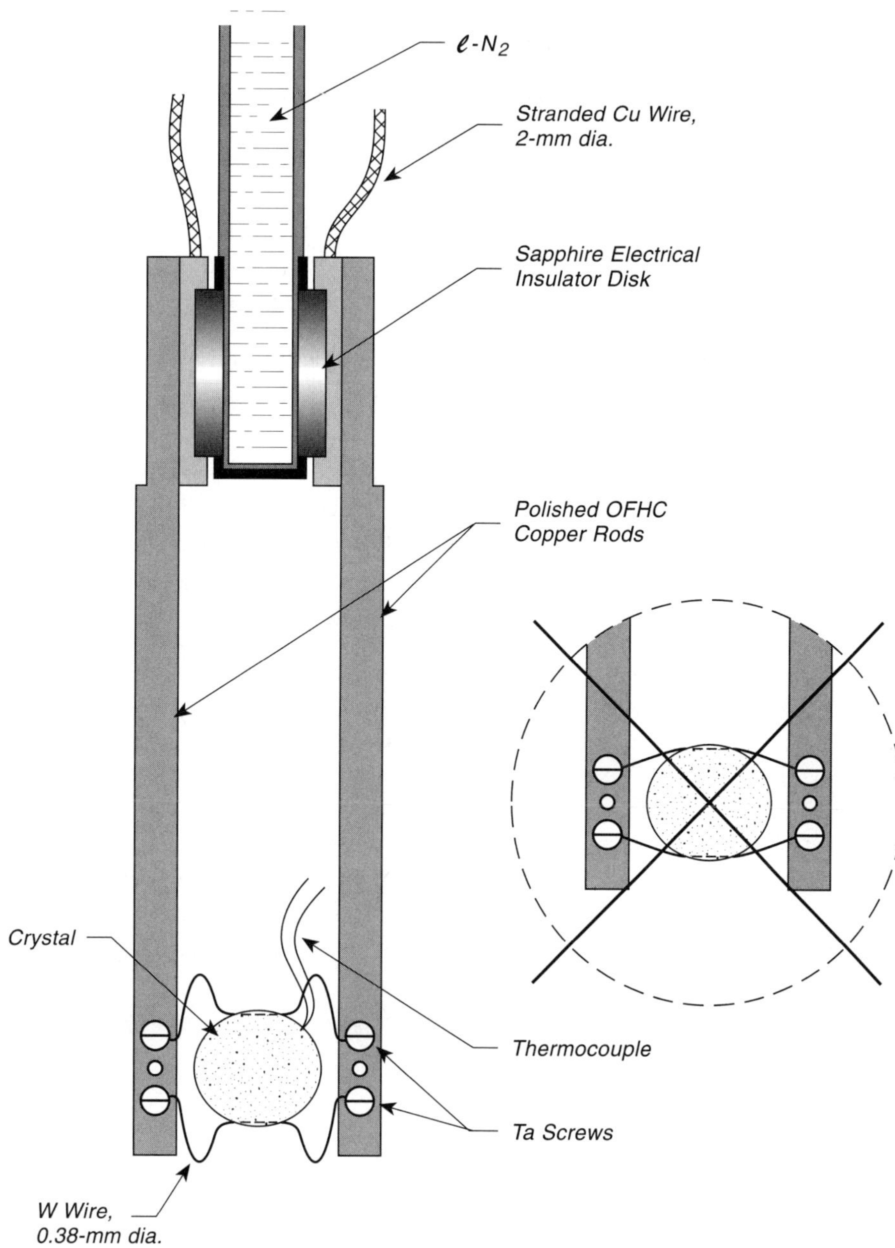


***Fig. 144. Mounting Nonweldable Crystals for Resistive Heating and Cooling***

# 145

# Strain-Free Single Crystal Mounting—Azimuthal Motion

In certain applications, single crystals must be mounted in such a manner that during heating and cooling they are free from strain owing to expansion and contraction effects in the metal to which they are welded. This is particularly true in atom scattering experiments, where tilting of the crystal can cause significant problems.

The design shown in Figure 145 eliminates this problem by allowing the conduction leads welded to the crystal to freely expand and contract [1]. This is facilitated by having one end of each lead free to slide. The square cross-section Ir rods (0.7 x 0.7 mm$^2$) are welded to Ta pads on the right-hand side, but are free to slide inside of sapphire bearing supports on the other end. The sapphire support bearings are held in place by Ta foils welded to the Ta support plates. Ta conductor leads (about 30-A capacity) are connected in pairs to the Ta support plates on the left, and current passes through these and the support plates to the Ir rods and thence through the crystal.

With a Ru crystal, temperatures above 1600 K may be reached, and cooling to 95 K or 50 K was achieved using liquid nitrogen or liquid helium, respectively. The Ru crystal was spot welded to the square Ir rods through a Ta sandwich foil.

## *References*


[1] Dr. K.L. Kostov, Institute of General and Inorganic Chemistry, Bulgarian Academy of Sciences, 1113 Sofia, Bulgaria ] (private communication). This design is used in the laboratory of Professor Peter Toennies, Max Planck Institut, University of Göttingen, Stroemungsf, Bunsenstr. 20, D-37037 Göttingen, Germany.

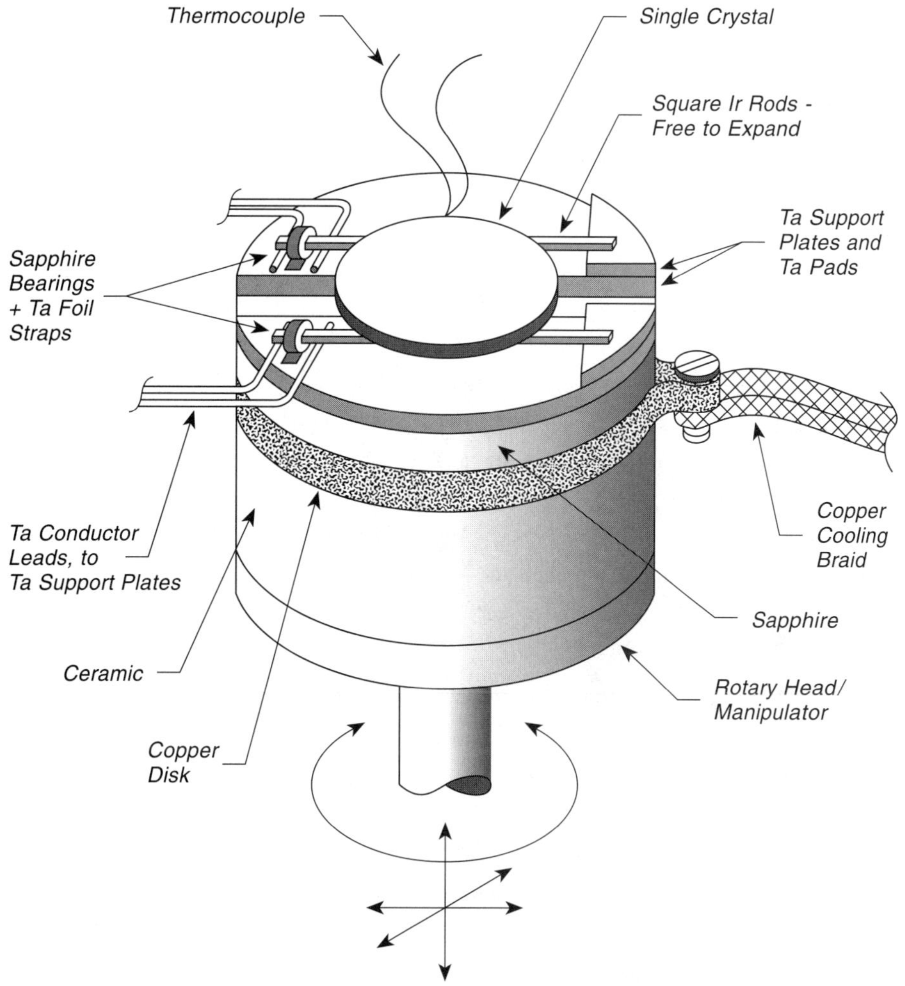


***Fig. 145. Strain Free Single Crystal Mounting—Azimuthal Motion***

# 146

# Heating Silicon Crystals and Temperature Measurements

One of the most difficult tasks in the laboratory is uniformly heating silicon single crystals by resistive heating and measuring the temperature of the crystal. A variety of methods of making electrical contact to silicon crystals have been devised that involve attachment to metal springs [1,2] or elastic metal sheets [3,4]. These contact methods often result in the formation of hot spots at certain regions of contact and can lead to nonuniform heating. Ta and Mo are used in these applications, since other metals can lead to metal contamination of the silicon crystal, either through metal diffusion through the bulk (Ni; Fe) or by transport by vaporization.

The attachment of thermocouples is also a difficult task owing to metal diffusion or metal vaporization. Here, too, Ta-protective envelopes must be used, as in the design of Bozack et al. [4], which employed Ta-protected W26%Re/W5%Re thermocouples inserted into slots cut into the crystal edge [4]. We have found that chromel-alumel (type K) thermocouples are unsatisfactory owing to the high vapor pressure of Ni ($10^{-5}$ to $10^{-4}$ Torr at 1450–1500 K) [5]. A few atomic percent of Ni produces, on Si(100), the well-known split-off dimer defects [6]. Thus, W26%Re/W5%Re thermocouples, shielded by Ta, have been used successfully in our own work.

Adhesive thermocouple contacts, such as zirconia-based Ultratem (Aremco), have been found to diffuse through Si above 1100 K. A Ta thermocouple tab contact, joined by a layer of Ge that alloys with both Ta and Si was proposed [2], but adhesion was lost on heating to high temperatures [7].

Figure 146A shows a design [8] that addresses the clamping of Si crystals to provide rather uniform electrical contact and temperature distribution. A Ta sheet (20 x 9 x 0.125 $mm^3$) is bent as shown in the diagram to form a bracket shape, which is then welded to a W rod (2-mm dia.), using a copper plate as an alignment jig, producing the structure shown in Figure 146A(d). Four slips of Si wafer material are prepared (4.5 x 8.5 x 0.4 $mm^3$) and used to sandwich a Si wafer slice (23 x 8 x 0.4 $mm^3$) as shown in Figure 146A(e). The entire sandwich structure is compressed by a flat U-shaped Ta spring that is wedged into the slot between the top Si slip and the top of the bracket, Figure 146A(f). The fully assembled mounting is shown in Figure 146A(g).

Figure 146B shows a method for attaching a W26%Re/W5%Re thermocouple (0.08-mm dia.) to the crystal by a U-shaped clip design [8]. To prevent direct contact of the thermocouple material to the crystal, it is shielded with a Ta tube (0.19-mm inner dia., 0.06-mm wall thickness, 5-mm length). The prewelded thermocouple is inserted into the Ta tube at the end, which has been previously cut off at an angle (Figure 146B[a,b]), and the tube is flattened (Figure 146B[b]), and formed into a bracket shape that tightly clips over the crystal edge (Figure 146B[c]),. The thermocouple exits from the angled end and is bent upward as shown in Figure 146B(c). The thermocouple voltage is converted to temperature using a calibration method similar to that of Smentkowski et al. [9], as discussed on page 580. A cold junction compensator is used for the reference junction at 273 K.

Two designs for thermocouple attachment to Si crystals have been compared using a calibrated optical pyrometer. Figure 146C shows the two designs where the slotted crystal design from Bozack et al. [4] and the design shown here [8] are compared. The deviation for the slotted design is from 50 to 150 K, whereas the deviation of the U-shaped clip design is from 50 K to 90 K over the same temperature range.

The uniformity of heating of the Si crystal with this mounting design is found to be ±15 K at 1400 K, as measured by an optical pyrometer. At temperatures above about 1100 K, temperature oscillations are sometimes observed with the U-shaped clip thermocouple.

Great care must be exercised in making the mount and the thermocouple clip. All Ta parts must be boiled in concentrated HCl solution to remove Ni contaminants [6]. All tools should be made of plastic to avoid metal contamination, and fabrication should be done under clean conditions without contact to metals other than cleaned Ta. In addition, Mo parts, which are more springy, may be substituted for Ta parts, although the HCl cleaning procedure followed by STM studies has not been done for crystals with Mo/Si contacts, as has been done for W/Ta contacts [6].

The use of interferometry for measuring the temperature changes in semiconductor crystals has been reported [10]. Here, both temperature dependent changes in the refractive index and in the thickness of the crystal wafer cause the optical path in the wafer to change in comparison to $\lambda/2n$, where $\lambda$ = the laser radiation wavelength, and $n$ = the refractive index. Temperature changes as small as 0.2 K have been detected, but one must start at some known temperature in order to make temperature measurements at other temperatures.

## *References*


[1] R.C. Henderson and R.F. Helm, Surface Sci. **30,** 310 (1972).

[2] J.A. Stroscio and W. Ho, Rev. Sci. Instrum. **55,** 1672 (1984).

[3] K. Jacobi and W. Ranke, J. Electron Spectrosc. Relat Phenom. **8,** 225 (1976).

[4] M.J. Bozack, L. Muehlhoff, J.N. Russell, Jr., W.J. Choyke, J.T. Yates, Jr., J. Vac. Sci. Technol. **A5,** 1 (1987).

[5] Metal vapor pressures are listed and graphed in F. Rosebury, *Handbook of Electron Tube and Vacuum Techniques* (American Institute of Physics, New York, NY, 1993), pp. 142-148.

[6] V.A. Ukraintsev and J.T. Yates, Jr., Surface Sci. **346,** 31 (1996).

[7] W. Ho (private communication).

[8] H. Nishino, W. Yang, Z. Dohnalek, V. Ukraintsev, and J.T. Yates, Jr., J. Vac. Sci. Technol. **A15,** 182 (1997).

[9] V. Smentkowski and J.T. Yates, Jr., J. Vac. Sci. and Technol. **A14,** 260 (1996).

[10] V.M. Donnelly and J.A. McCaulley, J. Vac. Sci. Technol. **A8,** 84 (1990).

***Fig. 146. Heating Silicon Crystals and Temperature Measurements***

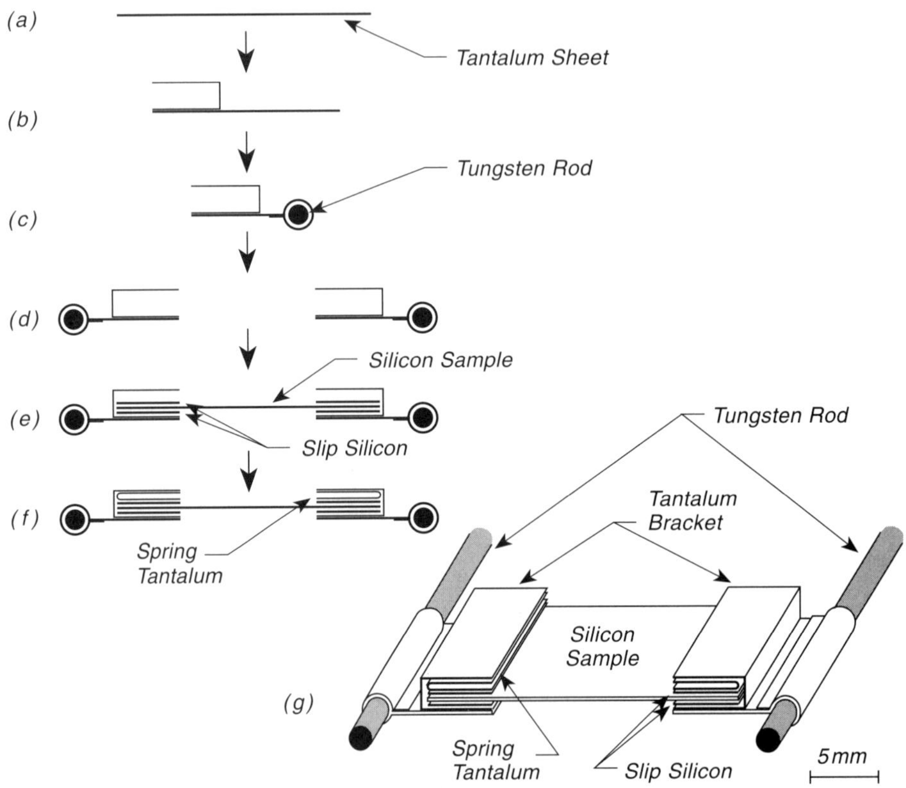


***A. Mounting Design - Si Sample***

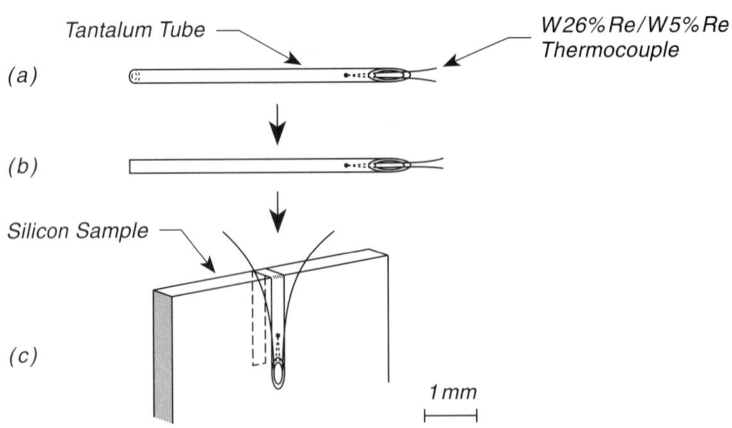


***B. Thermocouple Attachment***

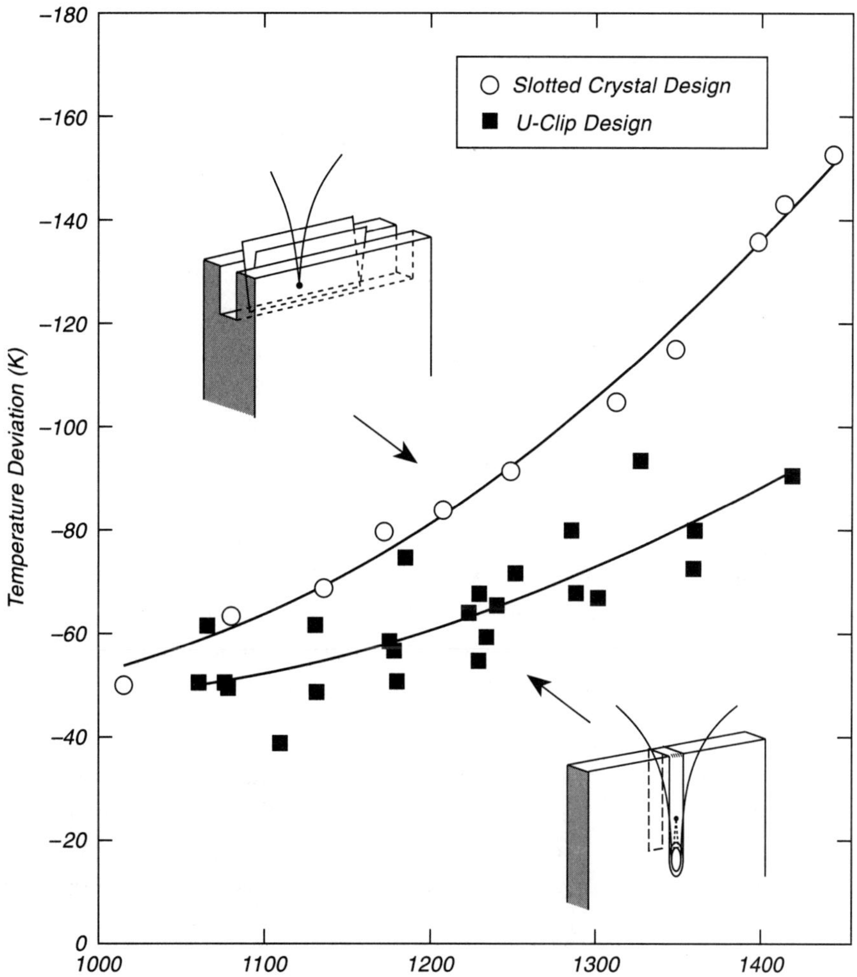


***C. Temperature Deviations—Two Thermocouple Attachment Methods***

# 147

# Avoiding Extraneous Electrical Effects Owing to Heating Crystals

Single crystals of metals may be heated in several ways for studies at constant temperatures. Unfortunately the presence of large AC or DC currents used for heating and the concomitant presence of stray electric and magnetic fields will be a severe disturbance to measurements involving electron currents to or from the crystal. A general method used to avoid this problem involves pulsing the heating current and making current measurements from the crystal during the time when the heating pulse is off.

Figure 147A shows a typical mounting for a metal single crystal on a manipulator using ohmic heating for the crystal. Large-diameter (1.0–1.5 mm) tungsten support leads are connected to current and cooling leads on the manipulator and conduct current to a pair of tungsten wires (0.3-mm diameter) welded to the back of the crystal. Tungsten is preferred to tantalum because of its higher coefficient of thermal conductivity at low temperatures. In this configuration the crystal is heated by thermal conduction from the hot wires rather than by ohmic heating of the crystal itself, and at incandescence, it will be seen that the parallel wires are comparatively bright in the region between the crystal and the support leads. The uniformity of welding to the crystal and the support leads can be judged by the relative brightness of the four intermediate regions. Poor welds will yield higher brightness and will possibly cause premature burnout. The thermocouple welded to the crystal edge is used to measure the crystal temperature.

Figure 147B shows a typical arrangement for heating by electron bombardment. The tungsten coil emitter located about 3–5 mm behind the crystal is kept at ground potential, and the crystal is biased positively (several hundred volts to 2 kV) (see page 516). Thermionically emitted electrons are used for heating the crystal and any crystal can easily be melted accidentally with this method. In addition to electron bombardment heating, it is also possible to use radiation heating to achieve crystal temperatures approaching 1000 K. The upper temperature limits for radiation heating depend upon the magnitude of the heat leakage to the support wires through the thinner wires welded to the crystal.

Three circuits for interrupted heating of single crystals have been reported in the literature [1–3]. For ohmic heating, at 1000 K, the temperature controller [1] required a duty cycle of 2/3. The switching rate was approximately 0.3 Hz, and the temperature regulation was ±15 K. Better control would be possible at higher switching rates [1]. For electron bombardment heating, a second circuit has been published in which a mechanical relay is used [2]. Here, bounce within the relay limits the time period for data acquisition to about 40 ms at 10 Hz. It is necessary to switch both the filament heater current and the bias voltage and since the power supplies are slow to recover from the switching, dummy loads are used in the off cycle. A difficulty reported for switching large currents and high voltages is that transients are produced that disturb electronic measuring equipment being used. Proper shielding of the elements of the circuit and the use of shielded cables eliminates these effects. A third circuit, which has shown favorable utilization in conjunction with LEED, Auger, and HREELS measurements is reported in reference [3].

The circuits shown in the two references [1,2] employ older electronic components, and the user should consult with an electronic specialist for designing similar circuits using modern components.

## *References*


[1] G.E. Thomas and W.H. Weinberg, J. Vac. Sci. Technol. **16,** 87 (1979).

[2] G.B. Hoflund and R.E. Gilbert, J. Vac. Sci. Technol. **18,** 56 (1981).

[3] J.A. Stroscio, L.J. Richter, and W. Ho, Rev. Sci. Instrum. **55,** 732 (1984)

***Fig. 147. Avoiding Extraneous Electrical Effects Owing to Heating Crystals***

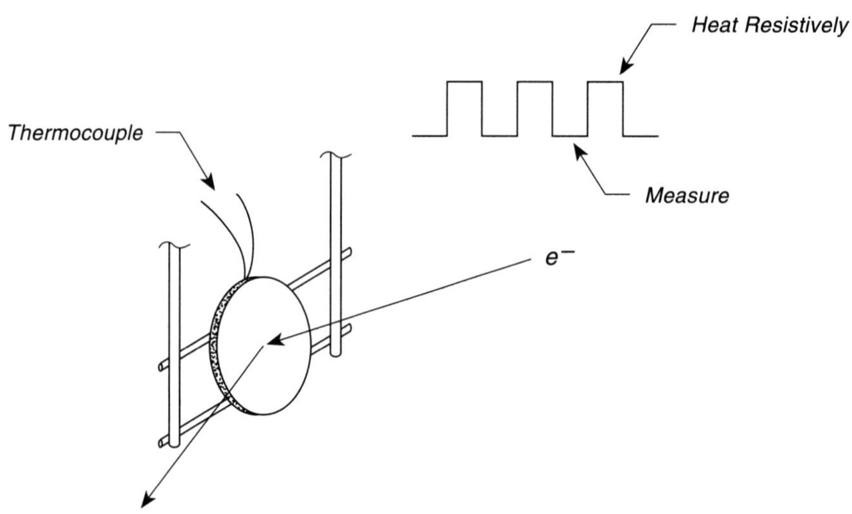


***A. Interrupted Ohmic Heating***

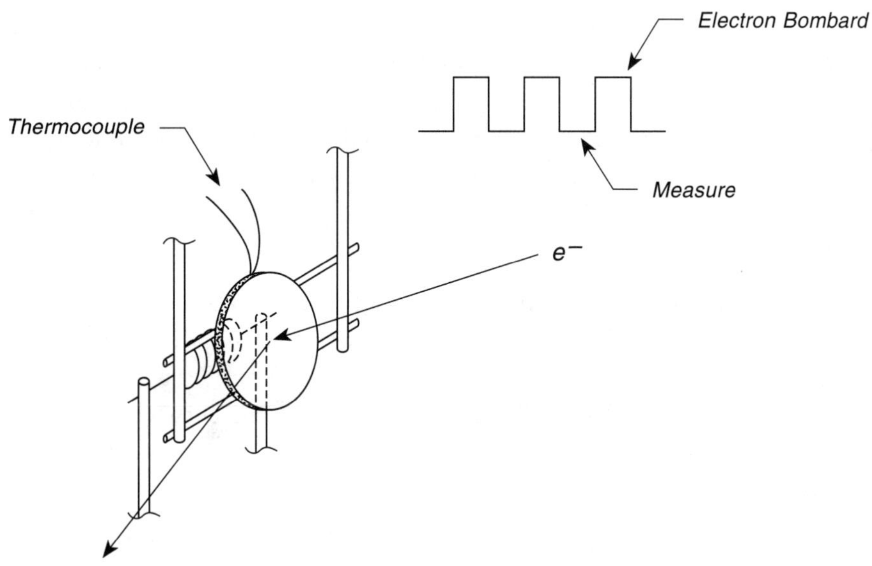


**B. Interrupted Electron Bombardment Heating**

# 148

# Electron Bombardment Crystal Heating

Electron bombardment heating of a metal crystal can often be used to bring the crystal to higher temperature than can be done conveniently by ohmic heating. Thus, for the highly refractory metals such as W, Ta, Mo, and Re, all of which require very high temperatures to remove impurities, electron bombardment is often used. (See also page 520.)

A typical design for electron bombardment heating, adapted from reference [1], is shown in Figure 148. The crystal is supported on a manipulator by one or more welded wires that connect to a stub on the end of the manipulator. This stub is electrically isolated from the manipulator shaft by a high-voltage ceramic spacer that will work at positive bias potentials of several kilovolts. The high-voltage connection to the stub is provided by a well-insulated wire that passes through a ceramic sheath, and this sheathed wire is firmly attached to the manipulator shaft by means of a ceramic support designed to hold it tightly, and in isolation from the rotary manipulator shaft or any other grounded parts.

The very simple electron bombardment gun is made of a stainless steel tube with an aperture cap placed over the open end, and a custom made ceramic base to hold the heavy filament leads. Filament current is supplied through copper braid that is connected to the ceramic support, and thence to flexible power leads that connect to feedthroughs on the manipulator flange. The electron gun is mechanically connected to the ceramic support by a rod, as shown.

The filament for the electron gun can be made of pure W wire, or thoriated W wire (see page 202). Emission currents of 10–50 mA with a positive bias of 1 kV on the crystal will supply 10–50 W, usually enough for heating to the desired temperature, depending on the thermal conductivity of the welded leads to the crystal and the radiation loss at the desired crystal temperature. For gentle heating, only radiation heating from the filament can be used, and in this case a non-thoriated W filament would be preferable. While the winding of W filaments is easily done, professionally wound filaments of uniform shape may be purchased [2]. A very compact and tightly wound W filament can be obtained from a slide projector quartz-halogen lamp, and the quartz envelope can be cut off by a glassblower, leaving an accurately mounted W filament attached to the quartz base. The filament should not be removed from the lamp base as it will be difficult to reweld it to the leads without strain and subsequent distortion upon heating.

It is emphasized that the coiled W filament (typically wire diameter = 0.2 or 0.3 mm) should be welded to its supporting leads under conditions where there is no strain left in the coil after the second weld. Strain will result in extreme filament distortion upon heating to incandescence.

The temperature of the crystal can be measured with an optical pyrometer for adjustment (see page 588). However, usually a thermocouple is employed for this, and one must remember that *the thermocouple will also be biased at kilovolts when the crystal bias is on.* Since these voltages are lethal, proper care must be taken in the insulation of the thermocouple leads, and in the isolation of the readout device connected to these leads.

One might consider reversing the bias and making the electron gun operate at a negative bias of a kilovolt or so, and grounding the crystal. This is not advised, since then electron bombardment will occur broadly to all regions at ground potential in the vicinity of the crystal, resulting in excessive outgassing and damage to devices and screens.

Even simplier electron guns may be constructed, the simplest being just an emitter filament supported a few mm from the back of the crystal.

## *References*

[1] J. Lecante, R. Riwan, and C. Guillot, Surface Sci. **35**, 271 (1973).

[2] Standard and custom wound W filaments may be obtained from C.M. Furnaces, Inc., 103 Dewey Street, Bloomfield, NJ 07003.

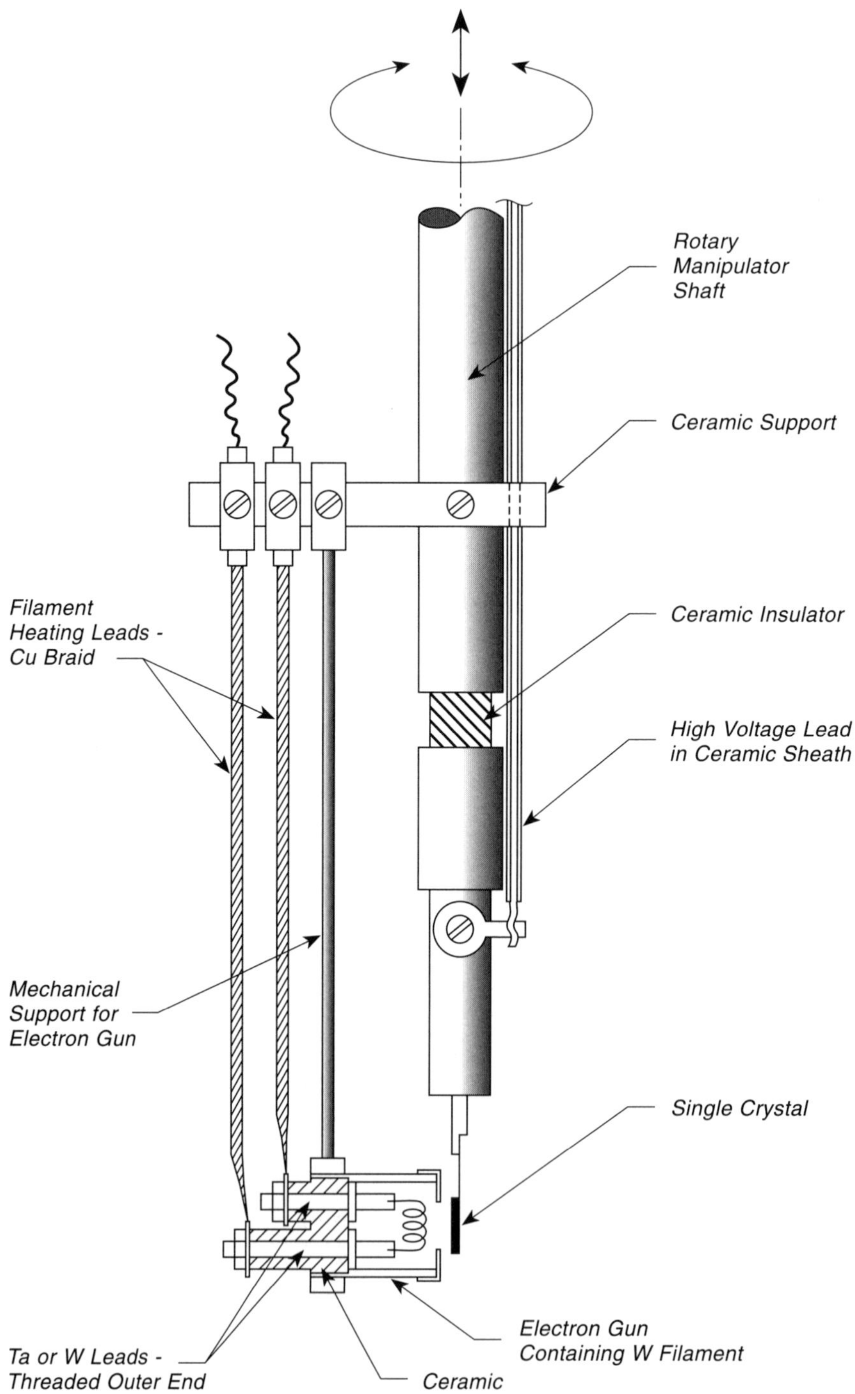


***Fig. 148. Electron Bombardment Crystal Heating***

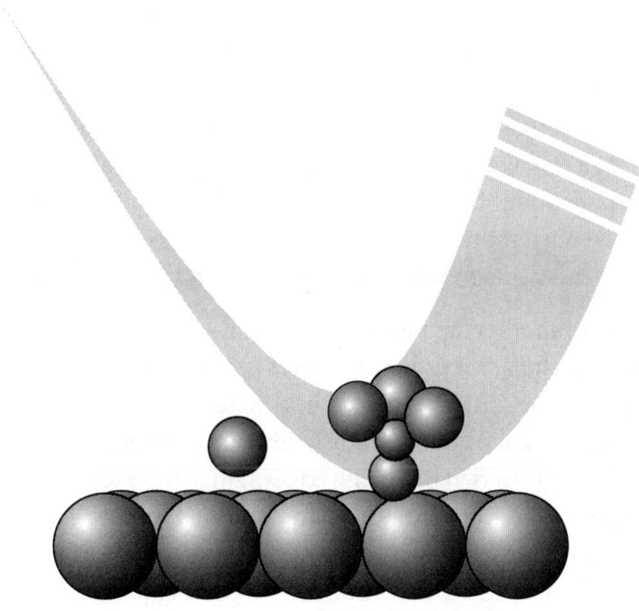

# 149

# Indirect Sample Heating

The use of indirect heating of samples is especially important for materials that are nonconductive and therefore cannot be heated by ohmic heating. However, indirect heating of conductor materials is also a favored method of heating, since excellent control of temperature can be achieved by indirect heating methods. Both heating by radiation and heating by electron bombardment are termed indirect heating, since neither method depends on the electrical resistance of the sample, as does direct ohmic heating.

Figure 149A shows a typical method for indirectly heating a sample using a button heater [1] that is supplied commercially [2]. The sample is held onto the top surface of the Mo cylindrical shell that surrounds the W heater wire. The W heater is immersed (potted) in an aluminum oxide powder. In the case shown here, the sample is held in place by means of a Mo sleeve that fits snugly over the Mo cylinder and has a lip that grips the front surface of the sample. This arrangement may be used up to about 1500 K without damage to the heater assembly. Since the heater wire is noninductively wound, electron spectroscopic measurements may be made with heating power on [1].

Figure 149B shows another arrangement that may be used at temperatures above 1500 K, but which also is useful at lower temperatures [3]. A Mo cylindrical can is polished flat on its top surface, and the sample is placed in contact with the upper surface. In the case shown here, a single crystal of diamond is held by two tungsten or tungsten alloy wires by means of parallel slots in two opposing edges. The wires are spot-welded to the edges of the can and tightly bind the diamond to the polished top of the can. Inside the can is a flat W pancake spiral made by winding a W wire onto a mandrel into a watch spring configuration as a tight coil. No aluminum oxide potting powder is present, so that the Mo can is heated only by radiation from the W spiral, or by radiation plus electron bombardment if more power is needed. A reflective radiation shield behind the Mo can assists in increasing the efficiency of heating the Mo support by the incandescent filament. A more complex electron bombardment heating device involving a shielded thermionic emitter has been reported [4].

The performance of the heating assembly in Figure 149B has been measured for a diamond single crystal [3], and the results are shown in Figure 149C,D. In Figure 149C the temperatures of two thermocouples, buried in laser-drilled holes in the edge of the diamond crystal, are shown for static heating experiments, where the assembly is brought to steady temperature in ultrahigh vacuum. The two thermocouples differ by only 8 K in these experiments, indicating that the true temperature of the diamond is being measured. However, at a diamond temperature of 1400 K, the difference between the Mo support temperature and the diamond temperature amounts to 140 K. This difference decreases at lower temperatures. Since diamond is an excellent thermal conductor, it is likely that heat transfer from the Mo surface to the diamond back surface is the limiting factor in these experiments. Similar temperature differences between the temperature of a heated supporting surface and a contacting sample have been measured by others at about $10^{-3}$ Torr pressure. The sample was contacted to the supporting surface using a silver paint that worked up to 1073 K, where a 22 K temperature difference was measured [5].

Figure 149D shows the difference between the heating behavior of the diamond crystal and the Mo support can under conditions of temperature programming at a rate of 1.3 K/s. At a diamond temperature of 1150 K, a difference of temperature between the Mo support and the diamond of 225 K was noted (compared to only 90 K under static conditions as shown in Figure 149C). The Mo can cools faster than the diamond when the heating is interrupted, as would be expected for cooling controlled by thermal conduction from the assembly.

The maximum filament heating power used to heat the diamond may be reduced by employing a mixture of radiation heating combined with electron bombardment heating. Thus, for example, a positive bias of 250 V produces thermionic heating of 12 W at a diamond temperature of 1273 K, whereas 47 watts of filament heating power was needed under conditions of zero bias to achieve the same temperature.

A successful variation of this method, involving ohmic heating of the support assembly [5] is shown in Figure 149E. Here, a polished Ta block of the same size as a $TiO_2$ single crystal is heated by means of W wires which pass through the body of the Ta. The temperature of the crystal is measured with an embedded thermocouple cemented [6–8] into a diagonal slot in the crystal corner, and programming of the heating power is controlled by this thermocouple which acts to supply a feedback signal. Somewhat similar methods for heating and cooling alkali halide crystals, and using an In metal film as a heat transfer medium have been reported [9].

## *References*

[1] J. Klebanoff, V.H. Ritz, and R.E. Thomas, J. Vac. Sci. Technol. **A2**, 1396 (1984).
[2] Spectra-Mat Inc., 100 Westgate Drive, Watsonville, CA 95076.
[3] V.S. Smentkowski and J.T. Yates, Jr., J. Vac. Sci. Technol. **A11**, 3002 (1993).
[4] P.H. Holloway and J.B. Hudson, Rev. Sci. Instrum. **43**, 1045 (1972).
[5] T.E. Jones, W.C. McGinnis, and J.S. Briggs, Rev. Sci. Instrum. **65**, 977 (1994).
[6] J.C.S. Wong, A. Linsebigler, G. Lu, J-F. Fan, and J.T. Yates, Jr., J. Phys. Chem. **99**, 335 (1995).
[7] F. Bozso, L. Muehlhoff, M. Trenary, W.J. Choyke, and J.T. Yates, Jr., J. Vac. Sci. Technol. A2, 1271 (1984). Here the cement used was mixed with graphite in order to permit electrical conductivity for electron spectroscopy of a SiC single crystal.
[8] A zirconia-based refractory cement was employed which matched the thermal expansion coefficient of $TiO_2$. This particular cement is Aremco 571. The address is Aremco, P.O. Box 429, Ossining, NY 10562-0429.
[9] E.P. Marsh, F.L. Tabares, G.A. Bach, and J.P. Cowin, Rev. Sci. Instrum. **57**, 3134 (1986).

Fig. 149. Indirect Sample Heating

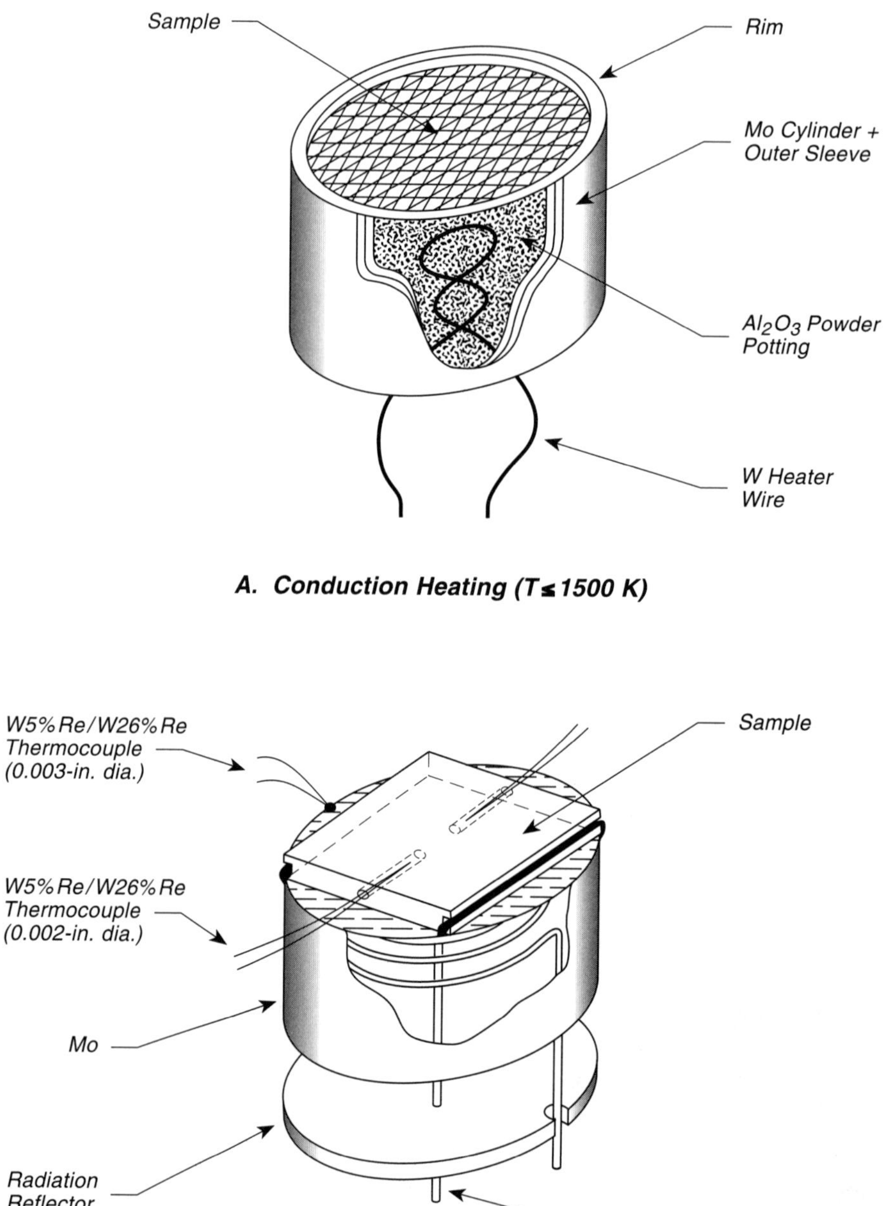


A. Conduction Heating (T ≤ 1500 K)

B. Radiation and E-Bombardment Heating (T ≥ 1500 K)

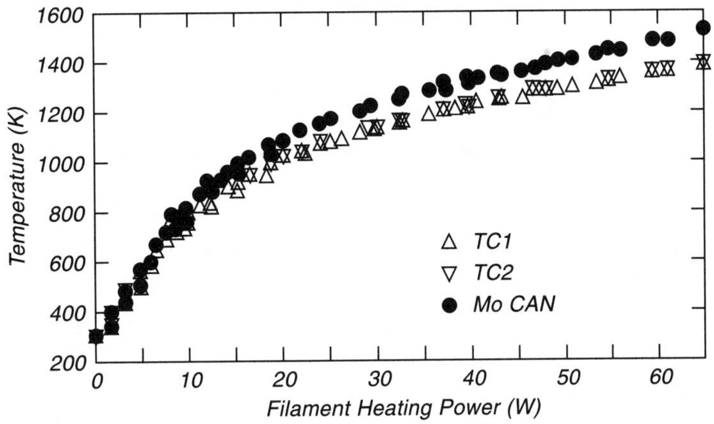


**C. Static Heating - Config. B**

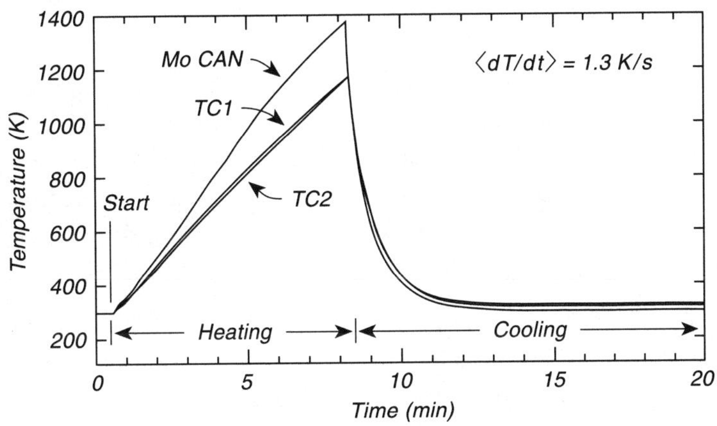


**D. Programmed Heating and Cooling - Config. B**

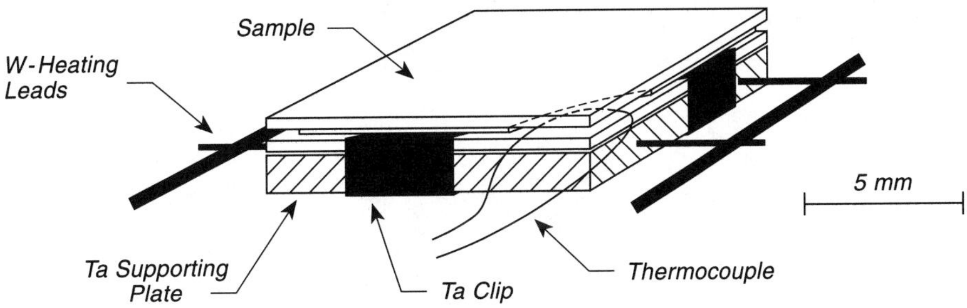


**E. Ohmic Support Heating**

# 150

# Indirect Heating of Compound Semiconductors

A variety of indirect heating methods for semiconductors which cannot be heated by electrical conduction have been discussed on page 520. GaAs is a special material because of its fragile mechanical properties and because many samples are of the incorrect conductivity for direct resistive heating (see page 470). The method shown here is especially good for this type of material [1].

As shown in Figure 150 the GaAs crystal is bonded to a sapphire plate using an indium film bond. The sapphire plate is coated on its back with a thin Ta film that is used for resistive heating. A thermocouple is attached to the Ta film for the measurement of the temperature.

The indium film is applied conservatively so as to not cause a short circuit between the Ta clip leads that support the sapphire plate on 0.75-mm Ta support leads. The indium film application may be easily done by using the Ta film resistor to raise the sapphire temperature above the melting point of In (157° C). The Ta resistor film requires 1.5–2 A at 8–10 V to heat the unit to about 900 K. The GaAs sample is grounded by using a Ta spring contact that is welded to a Ta clip.

The disadvantages of this technique are that the In reacts slowly with GaAs, causing failure after long heating. This effect can be minimized by using only small amount of In. In addition, at high temperatures, the In has a significant vapor pressure ($10^{-7}$ Torr at 833K; $10^{-6}$ Torr at 895 K).

## *References*


[1] Dr. J.R. Creighton, Dept 1126, MS 0601, P.O. Box 5800, Sandia National Laboratories, Albuquerque, NM 87185-0601 (personal communication).

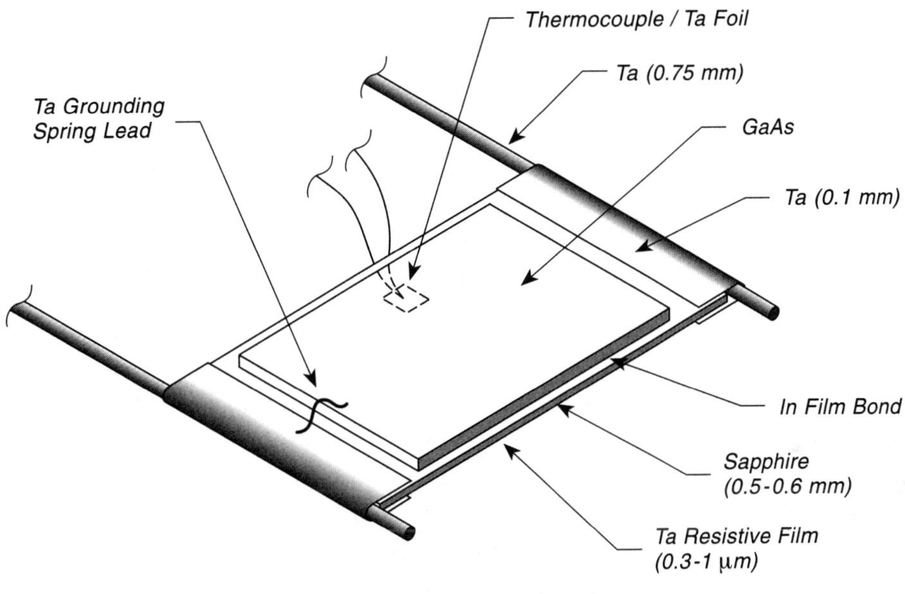


***Fig. 150. GaAs—Mounting for Indirect Heating***

# 151

# Heating Design—Insulator Crystals

A mounting design for electrically nonconducting crystals must provide a method for uniformly heating the crystal. When ohmic heating is to be used, various methods of placing the crystal on a heatable metal substrate have been described in other sections.

A method for making intimate contact between an insulator crystal and a metal [1] is shown in Figure 151. The method involves first making a sputtered film of the metal on the back of the crystal, as shown in Figure 151A. In the case of a MgO crystal described here, about 1 μm of Ni was sputter deposited in a period of 30 min. This film provides an intimate contact to the back of the crystal while permitting the electrodeposition of a thicker metal film. Figure 151B shows the crystal with the sputter-deposited Ni film and with a Ni heating wire lying across the back. This assembly is placed into a vessel (Fig. 151C), and a thick coating of Ni is electrodeposited, making good contact with the underlying sputter-deposited Ni film as well as holding the heating wire in place. Between steps A and B, the cement used is removed and the crystal is placed on a new glass slide (No. 2) to remove the extraneous Ni film deposited on slide No. 1, which would also be electroplated.

The cement used in this procedure to mask the crystal edges is a commercial UHV leak sealer, Vacseal [2], and is acetone soluble. It is important to electroplate the thick Ni film without stress, and special procedures for doing this must be followed. A nickel sulfamate plating solution (1 L volume) was employed containing nickel sulfamate (600 g/L), nickel bromide (45 g/L), boric acid (2 g/L), saccharin (0.5 g/L), and a surfactant X-100 (2 drops). The bath temperature is 40°C, and its pH is 4.0 ± 0.1, which can be adjusted using sodium hydroxide solution. The approximate current density is 3.4 A/dm$^2$. The use of six to eight electrolytic grade Ni anodes [3] is important, only one of which is shown in Figure 151C. Before use the electrolytic bath should be treated with 30 g of activated charcoal, filtered, and then plated onto a dummy anode for 2 h. Deposition rates on the back of the insulator crystal are typically 50 μm/h. The bath should be stirred with a magnetic stirring bar and final attachment of the Ni heating wire takes place as shown in Figure 151D. To check on the stress in the Ni film being deposited, a thin strip of Cu (1 x 10 x 0.05 cm) is used to receive the Ni film as a test. After plating for 30 min, the test strip should bend 1.5–2 mm toward the copper backside after the protective coating of lacquer is removed, indicating that the Ni film adheres well and is under slight compressive stress. The thermal expansion coefficient of the Ni, 13.3 x $10^{-6}$ $K^{-1}$, should match that of the crystal, as in the case of MgO (13.5 x $10^{-6}$ $K^{-1}$). For a MgO crystal the Ni film could be cycled in vacuum many times to 1073 K without failure.

## *References*


[1] J.C. Burns, P.C. Stair, and R. Haidle, J. Vac. Sci. Technol. **A8**, 3886 (1990).

[2] Vacseal is made by Space Environment Laboratories, P.O. Box 1061, Boulder, CO 80306.

[3] *Metal Finishing Guidebook and Directory* (Metals and Plastics, Hackensack, NJ, 1989), pp. 244-265.

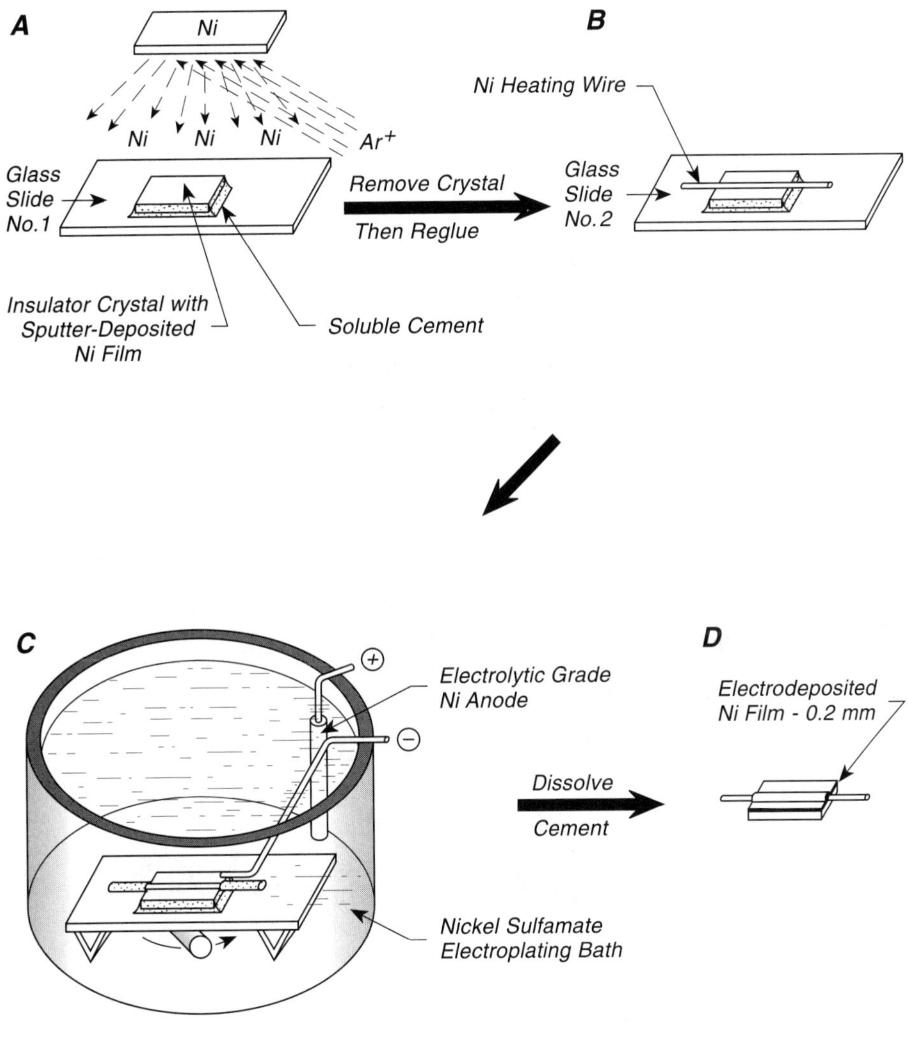


**Fig. 151. Heating Design—Insulator Crystals**

# 152

# Cement Mounting—Semiconductor Crystals

A number of methods for mounting semiconductor crystals in ways where they can be electrically heated and cooled are discussed throughout Section IV.A beginning on page 491. In the case of crystals which cannot be heated by electrical conduction, either because of unfavorable conductivity or because of unfavorable shape, it is possible to use a refractory cement that holds the crystal to a tungsten disk that is heated through welded contacts to refractory metal heating wires [1,2].

In Figure 152, an irregularly shaped SiC crystal has been cemented to a W disk using Aremco Ultratem 516 cement, which has a $ZrO_2$ base [3]. The W disk has previously been welded to a pair of Ta heating wires that pass current from the heavy W (1.5-mm-diameter) support leads. The cement is made electrically conductive by mixing with the appropriate amount of powdered graphite. This permits the electron spectroscopies to be used without sample charging.

A chromel-alumel thermocouple is connected to the back side of the SiC crystal by means of an alloy solder (90% Au/10% Ta) that anchors a tiny Ta tab to the crystal. The connection to the SiC crystal must be done in a preparative vacuum chamber to avoid gross oxidation of the crystal during induction heating.

Measurements with AES and XPS indicate that no migration of the ceramic cement occurs to the front of the SiC crystal up to 1400 K. The cement is adequate for mounting light crystals, but it is by no means robust.

At 1000 K, the temperature indicated by the thermocouple may be about 100 K too high compared to the true SiC temperature. The temperature range from 85 K to 1400 K was employed, and minimal outgassing occurs at 1400 K [1].

## *References*


[1] F. Bozso, L. Muehlhoff, M. Trenary, W.J. Choyke, and J.T. Yates, Jr. J. Vac. Sci. Technol. **A2**, 1271 (1984).

[2] M.J. Bozack, L. Muehlhoff, J.N. Russell, Jr., W.J. Choyke, and J.T. Yates, Jr., J. Vac. Sci. Technol. **A5**, 1 (1987).

[3] Obtain from Aremco Products, Inc., P.O. Box 429, 23 Snowden Avenue, Ossining, NY 10562-0429; the cement should be preserved in the refrigerator and has a limited shelf life.

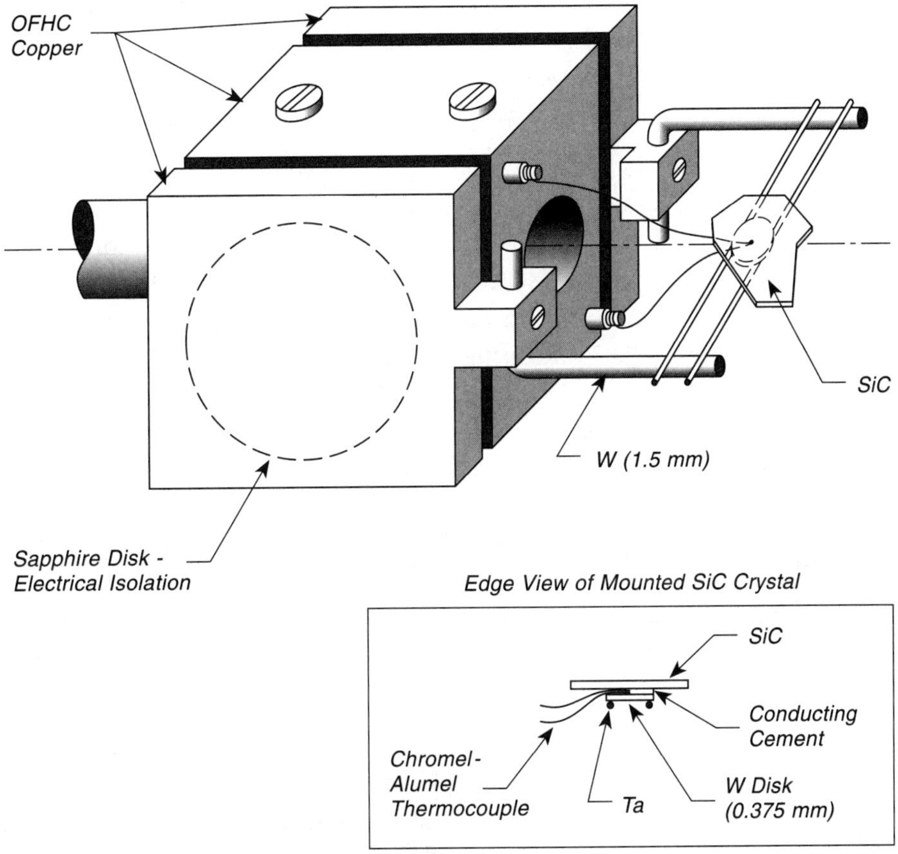


***Fig. 152. Cement Mounting—Semiconductor Crystals***

# 153

# Low-Stress Mount for Fragile Semiconductor Crystals

A system for mounting thin semiconductor crystals in an ultrahigh vacuum system for heating and cooling is described [1]. The method avoids stresses on the heated crystals, allowing them to remain flat at high temperatures. The method also allows a thermocouple to be mounted between two semiconductor samples, as shown in Figure 153.

Two L-shaped copper electrical conductor bars are attached to coolable electrical feedthroughs that are in contact with a refrigerated Dewar tube. The Cu conductors are separated and aligned by a Macor spacer. On the front of the L-shaped conductor bars are tapped holes accommodating 0–80 stainless steel machine screws that have had their threads removed under their heads to avoid snagging on the semiconductor wafer samples. Chromel wire springs (0.016-in.-diameter wire, formed around an 0.0625-in.-diameter rod with 38 turns/in.) provide compression force in the sandwich structure to be fabricated. A sandwich structure, consisting of two identical semiconductor samples, is placed between Ta clips (which have elongated holes allowing expansion to occur). Thermocouple wires are welded to the 0.005-in.-thick Ta foil that is sandwiched between the semiconductor samples. A hole drilled in the back semiconductor sample allows the thermocouple to extend back to its terminals on the manipulator feedthrough. The entire sandwich structure is compressed by turning and locking the nuts on the stainless steel machine screws. For two 0.014-in.-thick Ge samples, the compression in the chromel springs is typically 0.1 in.

This arrangement heats samples of Ge up to their melting point (1211 K) with virtually no temperature gradient across the samples. The thermocouple readings are within 2 K of pyrometer readings. By cooling the Dewar with liquid nitrogen, temperatures near 200 K are reached. The chromel springs lose their ability to compress the sample when long annealing near 1273 K is carried out.

The reader is warned that the stainless steel and type K thermocouple material used in this configuration is likely not to work for Si, which must be heated to much higher temperatures for cleaning. Both stainless steel and type K alloys will transfer Ni impurities to Si, giving unsatisfactory modifications of the Si surface structure and its surface properties (see page 790).

## *References*

[1] K.A. Schultz and E.G. Seebauer, Rev. Sci. Instrum. **63**, 218 (1992).

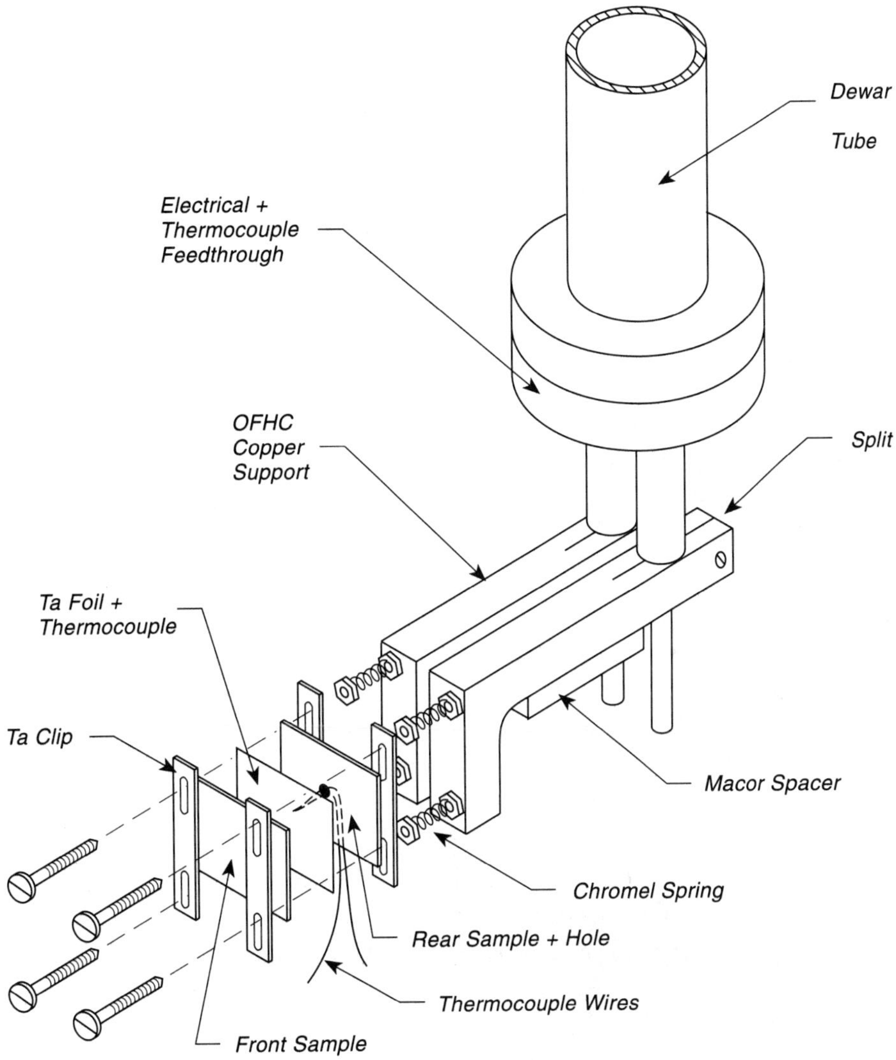


**Fig. 152. Thin Semiconductor Crystal Mounting—Heating and Cooling**

# 154

# Radiation Heating of Crystals Through Glass Windows

Single crystals may be heated by radiation from hot filaments mounted internally in the ultrahigh vacuum system, and this method possesses the best efficiency of coupling the radiant source (usually a hot W wire) to the crystal. For special cases, however, it may be preferable to employ radiation heating from external tungsten halogen lamp sources, which typically have maximum power limits in the 250–1000-W range. Usually the maximum temperature that can be reached with an external lamp is insufficient for cleaning the crystal and electron bombardment or ohmic heating methods must be employed for initial cleaning.

A very early use of external radiation heating for programming the temperature of semiconductor crystals can be found in the work of Palmberg and Peria [1]. In Figure 154A, their glass ultrahigh vacuum apparatus is shown. A track allows the translation of the crystal, using a magnetic device, from location to location in the linear vacuum system. Shown is a 200-W halogen lamp inside a spherical reflector (silvered glass flask section). Figure 154B shows a variation on the design of Figure 154A, where a commercial 750-W lamp (operating at reduced power) is housed in an ellipsoidal reflector, which has reflective aluminum focusing surfaces behind and in front of the lamp [2]. This arrangement permitted crystal temperatures near 1000 K to be achieved using the lamp alone, and cooling of the hemispherical glass surface was necessary to prevent damage to it by the lamp. A nonlinear heating rate was observed. The crystal was mounted on a 0.25-mm-diameter W support wire, which minimized cooling by conduction to the liquid-nitrogen-cooled support lead. A tungsten thermionic emitter is shown behind the crystal and this is used for electron bombardment cleaning at 2500 K, assisted by magnetic collimation. Figure 154C shows a lens arrangement for focusing light from a 900-W lamp [3]. The single-glass double-convex lens, f/1.32, has a magnification of unity. Lenses of this type are used as condenser lenses in photographic enlargers. The window is Pyrex glass with 95% transmission in the near infrared, where the spectral radiancy is greatest for tungsten. A silicon crystal could be raised to 900 K with this arrangement. The temperature program using the lamp in Figure 154C is shown in Figure 154D, and the reproducibility over a period of several days is within ±5 K.

## *References*


[1] P.W. Palmberg and W.T. Peria, Surface Sci. **6,** 57 (1967).

[2] T.E. Madey and J.T. Yates, Jr., *Structure et Proprietes des Surfaces des Solides,* No. 187, (Editions de Centre National de la Recherche Scientifique, Paris, 7-11 July 1969), p. 155.

[3] M.J. Bozack, L. Muehlhoff, J.N. Russell, Jr., W.J. Choyke, and J.T. Yates, Jr., J. Vac. Sci. Technol. **A5,** 1 (1987).

***Fig. 154. Radiation Heating of Crystals Through Glass Windows***

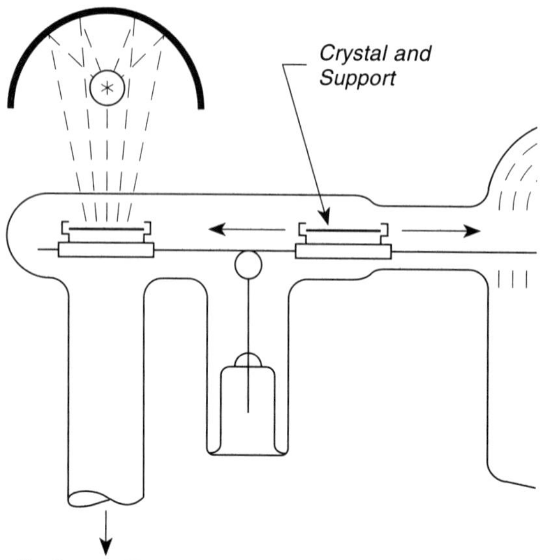


***A. Hemispherical Reflector***

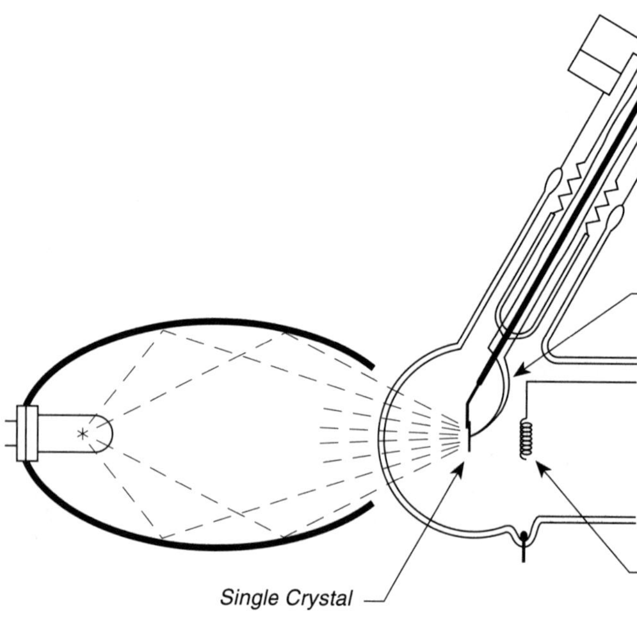


***B. Ellipsoidal Reflector***

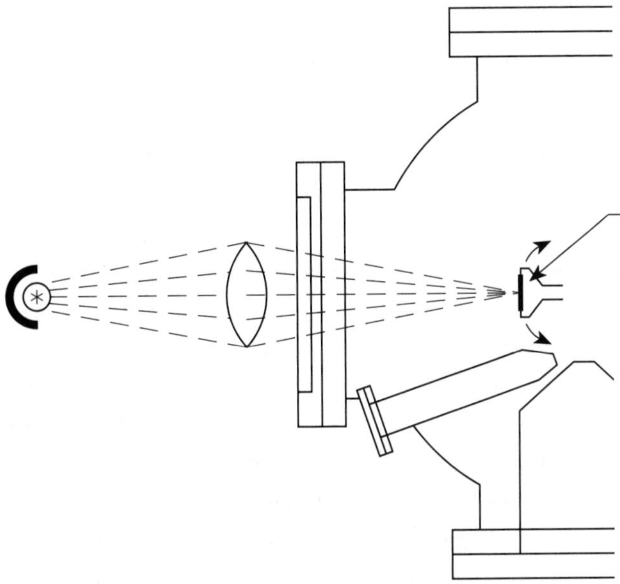


**C. Lens + Hemispherical Reflector**

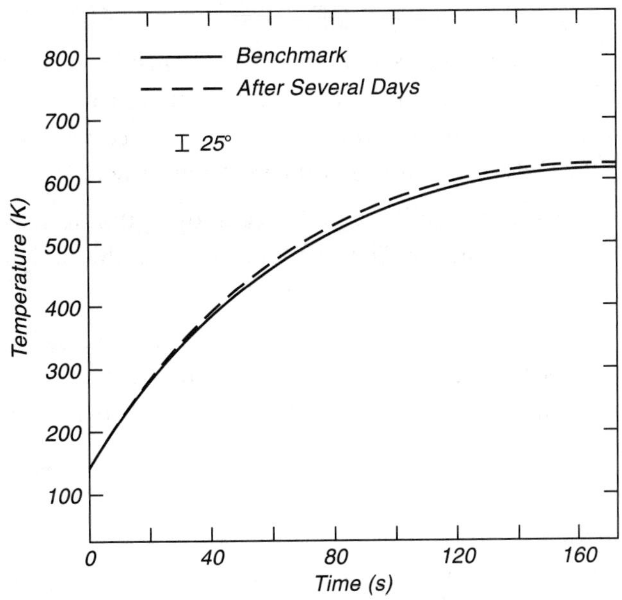


**D. Temperature Program for C**

# 155

# Pyrolytic Graphite Heating Element for UHV

Heavy power deposition in an ultrahigh vacuum system can be achieved using electron bombardment, with the inherent possibility of burning out the emitter filament. The resistive heater described below avoids this problem.

The heater shown in Figure 155A was constructed of chemical-vapor-deposited (CVD) pyrolytic graphite, which is highly anisotropic in its electrical conductivity [1]. The heater is fabricated from a slab of the graphite so that the current flow is in the C direction, which has a resistivity about $10^3$ times higher than in the A or B directions. The width of the device in the C direction is governed by the thickness of the pyrolytic graphite supplied by the manufacturer [2], which is about 25 mm. The slab is cleaved along the A/B direction into slices 4–5 mm in thickness. Resistive sections are separated from each other using a band saw or preferably a diamond saw wheel. Care must be taken not to fracture the graphite during sawing.

The heater is mounted on threaded Mo connection rods, as shown in Figure 155A. These rods are not heated too much since the thermal conductivity along the *C*-direction is small compared to the other directions. When the temperature in the central region is in the 2000 K range ±30 K, typical power dissipation is about 2.0 kW for 155 V applied with 13.5-A current.

Figure 155B shows the heater used as in insertion device inside of a block to be heated to about 1200 K. The pyrolytic graphite heater was mounted in a rectangular hole and was insulated from the metal by an alumina spacer. Electrical contact was made with a tungsten foil spring attached to the bottom of the heater as shown.

Although the heater is mechanically fragile, once in operation it may be repeatedly cycled in temperature without damage. The authors report that this design is compatible with ultrahigh vacuum.

The resistivity of the pyrolytic graphite heater was measured versus temperature, as shown in Figure 155C. Relatively poor agreement was obtained compared to the manufacturer's specifications. This level of variation is typical for CVD pyrolytic graphite measured in the *C*-direction.

## *References*


[1] C.S. Pitcher, P.C. Stangeby, G.E. Austin, and G.M. McCracken, J. Vac. Sci. Technol. **A5**, 383 (1987).

[2] Pfizer, Inc., 235 E. 42nd Street, New York, NY 10017.

## *Fig. 155. Pyrolytic Graphite Heating Element*

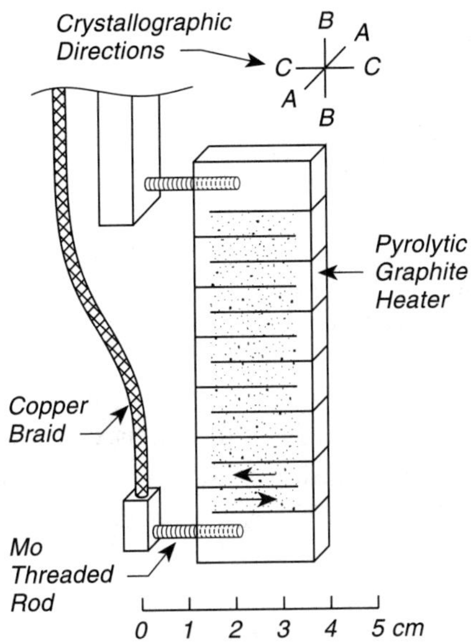


**A. Heater Design - Current Flow in C Direction**

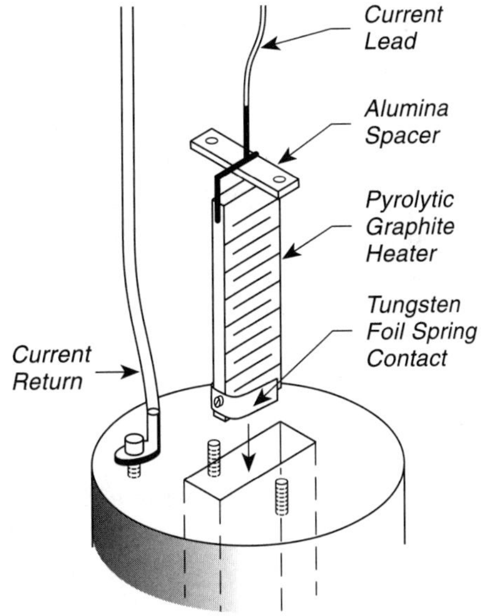


**B. Design for Internal Heating Element - Metal**

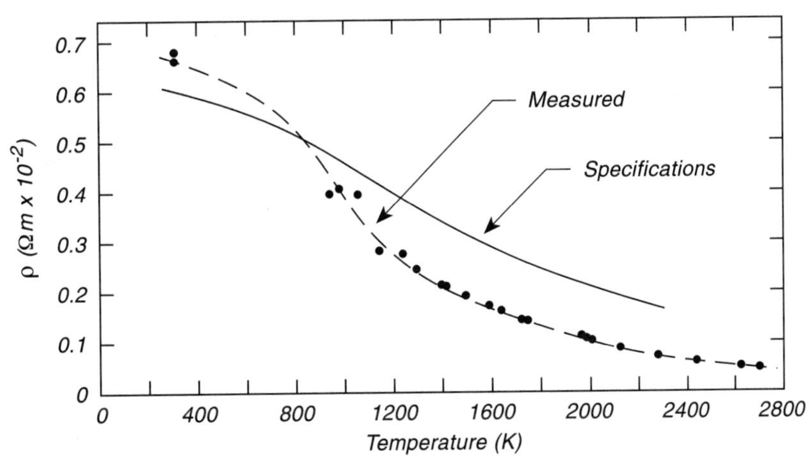


**C. Resistivity in C Direction**

# 156

# Heater—Paint-on Type

A simple resistance heater which can be painted on an insulating surface [1] may be useful in indirectly heating components in an ultrahigh vacuum system [as shown in Figure 156]. A thin layer of Aquadag [2], which is a colloidal graphite suspension, is painted onto a roughened quartz surface and outgassed for 0.5 h by baking in an oven at 300°C, which is reported to be satisfactory for complete outgassing [2]. The graphite layer can then be smoothed with fine abrasive paper. Contact to the graphite layer is made using Ni or Pt strips, 0.1 mm thick, which are sandwiched between the graphite layer and a second quartz plate that is clamped to the first to assure good contact.

For a 27-mm-diameter plate, typical room temperature resistances are about 15 ohms. With a variable transformer, up to 20-V maximum, the heater could be used continuously for long periods at temperatures of 500°C, and operation at higher temperatures may be possible. The application of these ideas to heating irregularly shaped surfaces is possible.

## *References*


[1] P. Winsemius and H.P. Lengkeek, Rev. Sci. Instrum. **44,** 229 (1973).

[2] Obtain "Dag 660" from Acheson Colloids, 1607 Washington Avenue, P.O. Drawer 611747, Port Huron, MI 48061-1747.

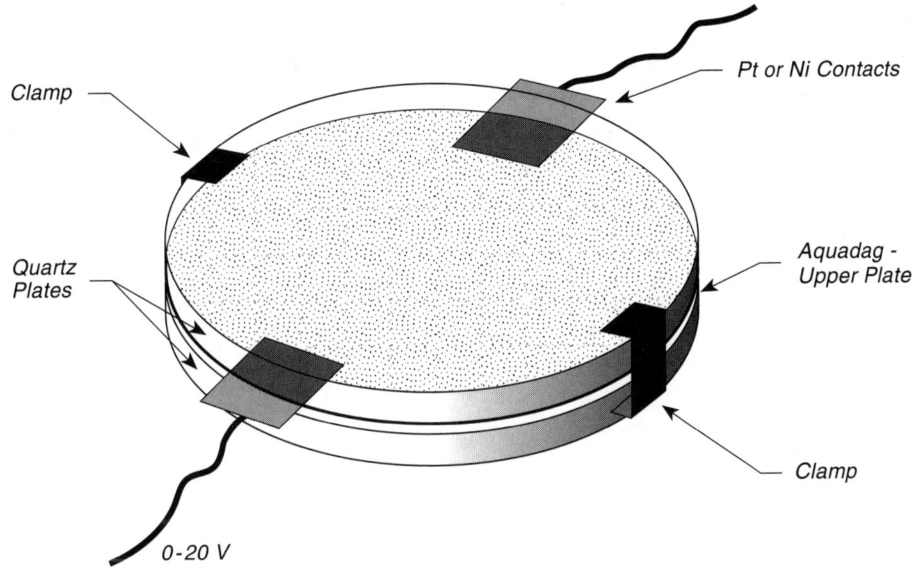


*Fig. 156. Heater—Paint-on Type*

# 157

# Mounting High-Temperature Tungsten Filaments

The proper mounting of tungsten filaments so that operation at high temperatures does not melt or vaporize the filament mount is sometimes a problem in instrument design. A useful method of doing this without using complex screwing or clamping arrangements is shown in Figure 157. A Mo rod, threaded to fit into an electrical mount, is drilled with an axial hole somewhat larger in diameter than the tungsten wire. The drilling process should be done slowly (1000 rpm) to avoid splitting the Mo rod [1]. The ends of the tungsten wire are then coiled by winding around a mandrel, followed by stretching, such that the helix formed makes a tight sliding fit into the hole in the Mo rod.

## *References*

[1] J.R. Conrad, S.F. Horne, D.R. Pirkle, and R.P. Torti, Rev. Sci. Instrum. **51**, 144 (1980).

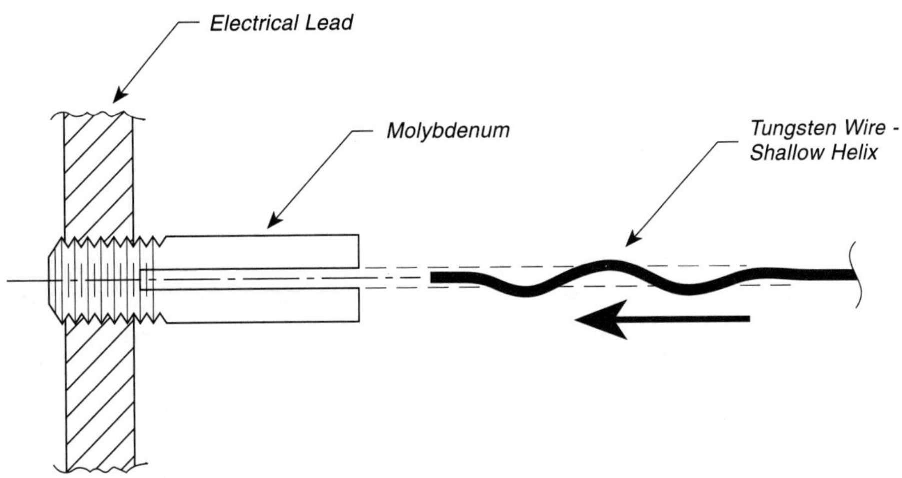


***Fig. 157. Mounting of High-Temperature Tungsten Filaments***

# B. Cooling Samples

# 158

# 6 K—Cryogenic Crystal Holder

Cooling single crystal samples in ultrahigh vacuum below 77 K was at one time considered to be technically difficult because of the perceived need for cooled radiation shields for the structure holding the crystal. Recently, excellent cooling properties have been achieved without a radiation shield using a number of refinements that have permitted rapid cooling to 6 K, control of temperature to 30–70 mK at 10 K, and measurements with absolute temperature errors of about 0.7 K at 10 K and 1.5 K at 100 K [1].

Figure 158 shows the design for efficient cooling [1]. A liquid He cryostat [2] containing concentric ducts for liquid admission and gas transport away is mounted on a standard UHV manipulator. While the publication [1] shows the use of OFHC Cu cooling members, recent refinements have substituted highly polished Ag members for Cu primarily because the surface of Cu oxidizes slowly in use, giving a nonreflective surface that is not as efficient as the shiny Ag surface in reducing warming from the radiation from the surroundings [3]. All heat-conducting interfaces are made by pressing smooth surfaces together and not by welding, which produces poor thermal contact. A 10 x 10 x 55 $mm^3$ block of Ag is bolted to the bottom of the cryostat and holds two 1-mm-thick sapphire plates that are separated from each other by a gap. Three 2-mm-diameter screws, with ceramic isolation washers, apply force to hold two Ag electrical conduction leads in good contact with the sapphire plates. The screws pass through the lower Ag bar in large clearance holes that assure that no short circuits can occur. Each Ag bar is covered with a stainless steel force plate to avoid distortion of the Ag owing to the force of the screws. The two Ag conduction leads have 4-mm-diameter stainless steel power leads exiting from their back and passing up the cooled cryostat surface. These power leads are heat-stationed twice each, in series, to the cooled cryostat wall to reduce heat flow downward to the crystal assembly. The heat stations involve sapphire electrical isolation junctions, mounted 50 mm apart on the cryostat shield.

The crystal is mounted on the Ag power leads using tungsten rods (5 mm long x 2 mm in diameter). These snugly fit into holes in the end of the Ag leads and are tightly connected, but not welded, to Ta stubs of about 2-mm length at right angles, as shown in Figure 158. The crystal is mounted on two different contact leads ( 6 mm long) which consist of pairs of W/25% Re wires of 0.5-mm diameter and two Ir wires of square cross section, 0.45 mm x 0.45 mm. The logic for this mounting structure is that at low temperatures pure metals do not have the needed resistivity for ohmic heating, and an alloy must be used for this purpose. Conversely, the Ir rod has good thermal conductivity at low temperatures. Unfortunately, the thermal expansion coefficient is different for the two wires, so that heating the crystal puts torque on the welds. To reduce this stress, the flexible Ta leads are incorporated, and the spacing between the W/Re and the Ir leads is kept small.

The thermocouple (in this case, type K) is welded to the back of the crystal and the leads pass upward after also being heat-stationed on the cryostat with isolation by sapphire. Heating rates from 0.01 to 100 K/s are possible. A heating current of 70 A DC at about 8 V is sufficient to bring the crystal to 1600 K. The cooling time between 650 K and 6 K is 3 min.; from 1600 K to 6.5 K the time is 5 min. At steady state, the consumption of liquid He is 4 L/hr. With the cryostat cool, the base pressure is below$10^{-11}$ mbar.

Reference [1] is highly instructive about the pitfalls of using cryogenic methods in ultrahigh vacuum applications.

Others have employed closed-loop He cryostats for cooling of crystals in UHV that do not achieve the very low temperatures described here [4–6].

## *References*

[1] H. Schlichting and D. Menzel, Rev. Sci. Instrum. **64,** 2013 (1993).
[2] Constant flow cryostat, Air Products, APD-Cryogenics, P.O. Box 2802, Allentown, PA 18105.
[3] H. Schlichting (private communication).
[4] K. Wandelt, S. Daiser, R. Miranda, and H.-J. Forth, J. Phys. **E17,** 22 (1984).
[5] A. Winkler, M. Hausenblas, M. Leisch, and K. Rendulic, Vacuum **40,** 39 (1990).
[6] M. Milun, P. Pervan, T. Valla, and P. Dukic, Rev. Sci. Instrum. **62,** 3116 (1991).

***Fig. 158. 6 K—Cryogenic Crystal Holder***

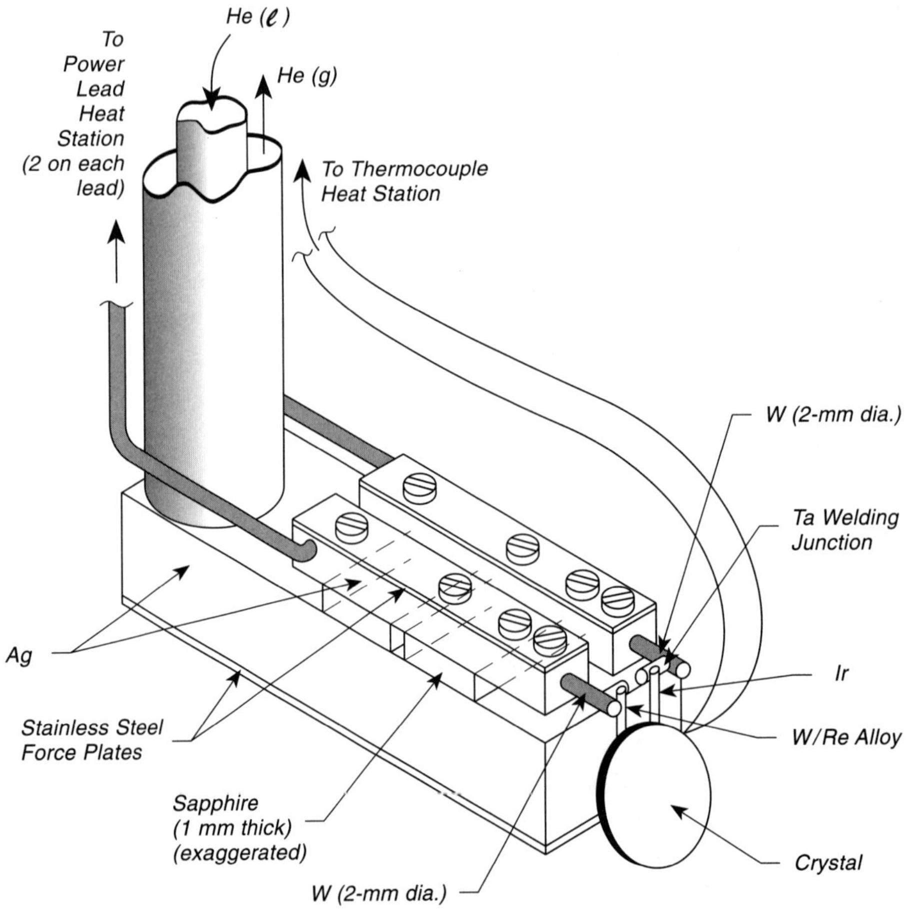


***Schematic Design***

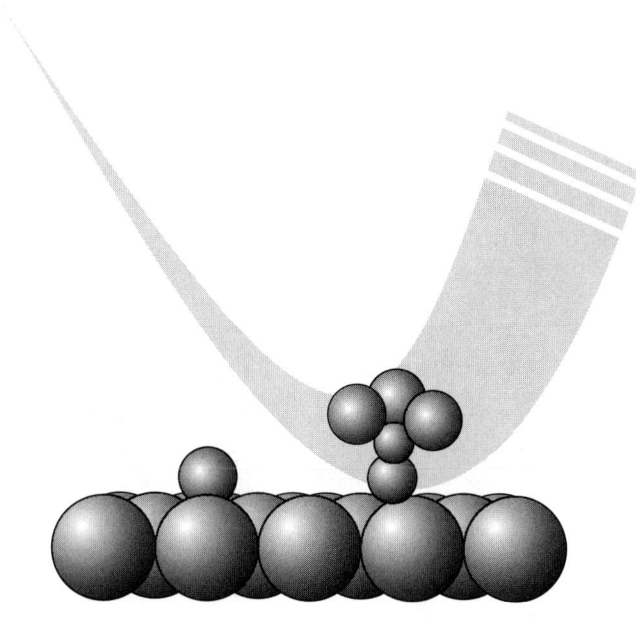

# 159

# Cooling Crystals by Contact with a Thermal Reservoir

One of the best ways to employ a low-temperature, closed-cycle cryogenic source with limited cooling capacity is to use a heat absorption reservoir that is continuously cooled, but only used periodically for cooling the sample. A design using this principle in shown in Figure 159 [1], based on an earlier design [2].

The crystal and its mounting assembly, shown in Figure 159A, is held on the end of a conventional manipulator, and it may be heated either by radiation from the W filament or by electron bombardment. The entire crystal and the heating assembly is electrically isolated from the manipulator by means of a machinable ceramic support. The sample may be contacted to the thermal reservoir by means of a conical thermal switch in order to initiate cooling. The stainless steel spring is placed under compression when this contact is made to yield the best thermal contact possible between the two OFHC Cu components.

The cooling rate using this device is very impressive, and cooling from 1023 K to about 20 K occurs in 300 s, as may be seen in Figure 159B.

A second design in which contact between the sample and the closed cycle refrigerator is maintained continuously is given in reference [3].

## *References*


[1] J. Yoshinobu, H. Ogasawara, and M. Kawai, Surf. Sci. **360,** 234 (1996).

[2] Professor S. Masuda, Department of Chemistry, College of Arts and Sciences, The University of Tokyo, Komaba, Meguro-ku, Tokyo 153, Japan.

[3] A. Winkler, M. Hausenblas, M. Leisch, and K.D. Rendulic, Vacuum **40,** 39 (1990).

## Fig. 159. Spring-Loaded Cooling by Contact

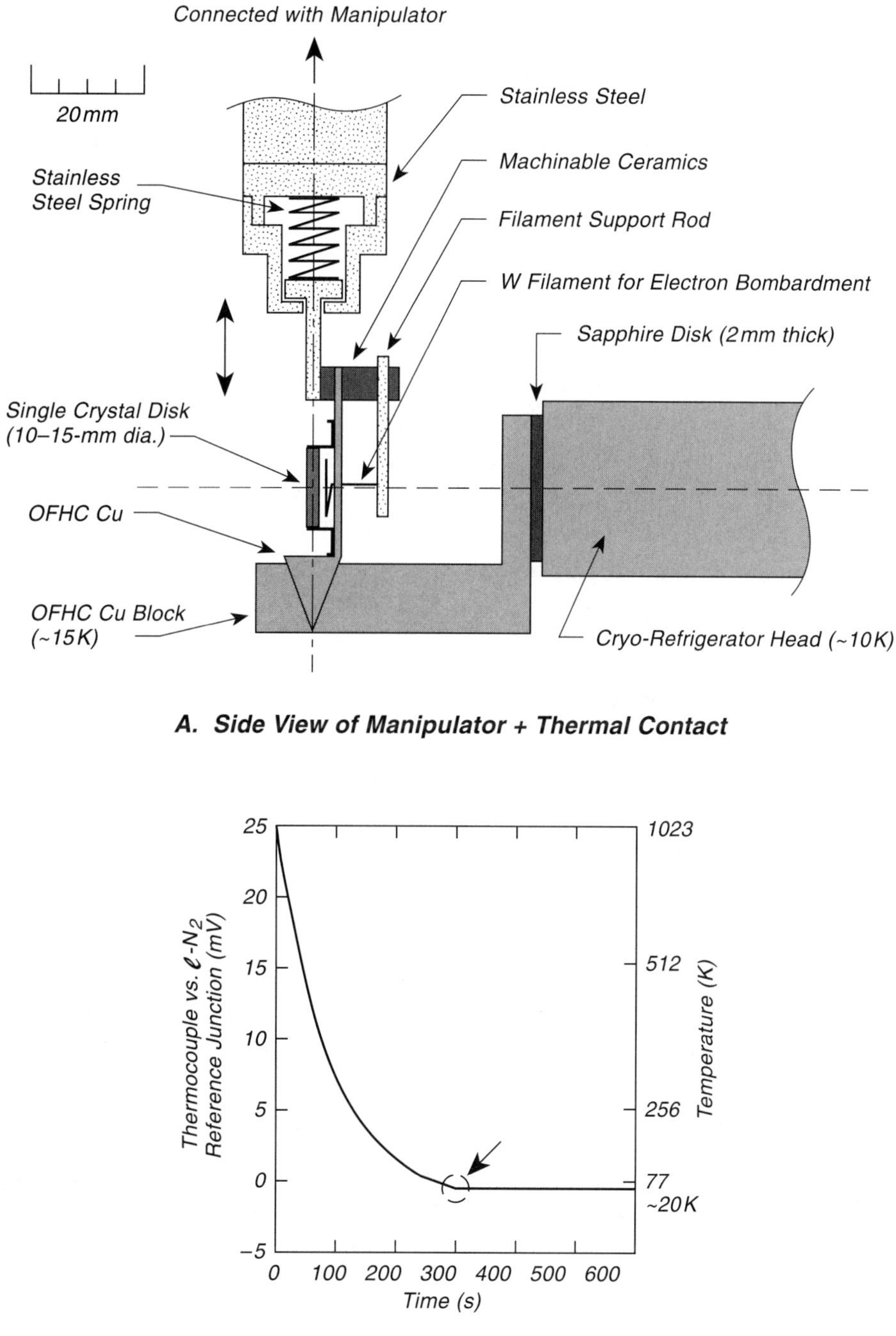


B. Cooling Curve

# 160

# Cooling Samples on Manipulators—Thermal Conductivity Issues

A variety of designs for holding samples on manipulators is available in the literature, and most of these have a requirement that the crystal be isolated electrically from the manipulator, while being in contact thermally with a cryogenically cooled reservoir, usually made of oxygen-free high conductivity copper (OFHC Cu). Comparisons of cooling rates for different methods of electrical isolation have been made and are presented below [1].

Figure 160A shows the experimental apparatus used in this work. A Cu reservoir with polished planar faces holds two Cu support bars that are clamped with equal force to the reservoir. The total contact area is 3.13 cm$^2$ for both support bars. The temperature of each bar ($T_1$ and $T_2$) can be measured during cooling, using type K thermocouples that are wedged into small holes in the support bars. In addition, the crystal temperature can also be measured as cooling occurs to determine the influence of the 0.025-mm-diameter tungsten wire (and the associated spot welds) on the crystal cooling rate, compared to the support bar cooling rate.

Figure 160B shows cooling experiments comparing three different insulator materials. The slowest cooling rate of the support bars is achieved using Kapton films, 0.037 mm in thickness. Faster cooling of the support bars is achieved using 0.500-mm sapphire single crystal insulators. The fastest cooling is achieved when an indium film is used to join the sapphire to the cooling reservoir and the support bars. For cooling times up to 90 s, cooling of the central reservoir to 77 K is taking place, and the data in this time period are disregarded in the measurements.

Figure 160C shows semilogarithmic plots of the cooling rate after the reservoir has reached constant temperature. The slope of these plots is a measure of the rate of cooling through each insulator junction. It may be seen that the Kapton exhibits cooling at 81% of the rate of the sapphire, and that the indium coating on the sapphire enhances its cooling rate by about 6%.

Figure 160D shows a comparison of the cooling rate of the sapphire-insulated support, compared to the cooling rate of a metal single crystal. This comparison was carried out in order to test the effect of the thermal conductivity of the tungsten wire (and its spot-welds and screw pressure contact). The crystal was found to cool at a rate of 93% of the cooling rate of the copper bar supports, indicating that this crystal support design is certainly satisfactory.

The differences in the sapphire- and Kapton-controlled cooling rates are in accordance with the measured thermal conductivity of these materials [2]. Thus, sapphire, or indium-coated sapphire, exhibit superior qualities compared to Kapton of the thickness employed here. Thinner Kapton is likely to be mechanically damaged when it is placed under pressure or baked out and may therefore exhibit electrical leakage.

## *References*

[1] C. Rusu and J.T. Yates, Jr., J. Vac. Sci. and Technol **A15**, 436 (1997).

[2] Tabulated values for thermal conductivity are sapphire, 41.90 J/sKm at 273 K; 100 HN Kapton, 0.10–0.35 J/sKm at 276 K. *Goodfellow Catalog* (Goodfellow Metals, 1995), p. 433; p. 501. In addition reference [1] gives information and references on thermal conductivities as a function of temperature. Sapphire exhibits a greatly enhanced thermal conductivity at 80 K of 960 J/sKm.

## Fig. 160. Cooling Samples on Manipulators

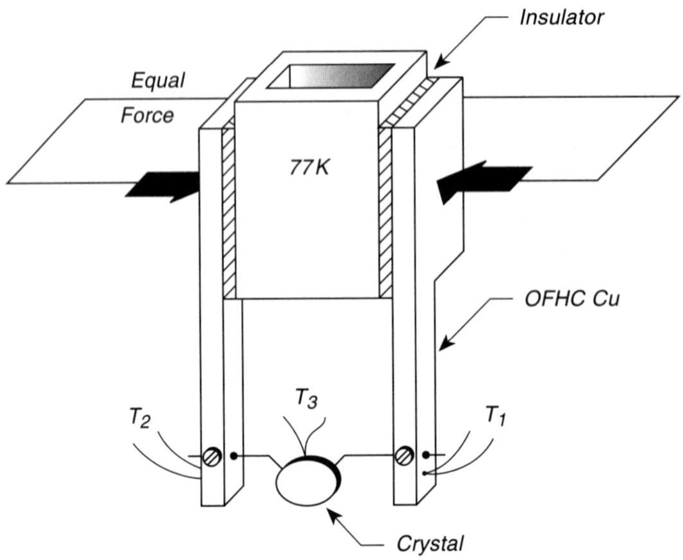


**A. Manipulator Cooling Section**

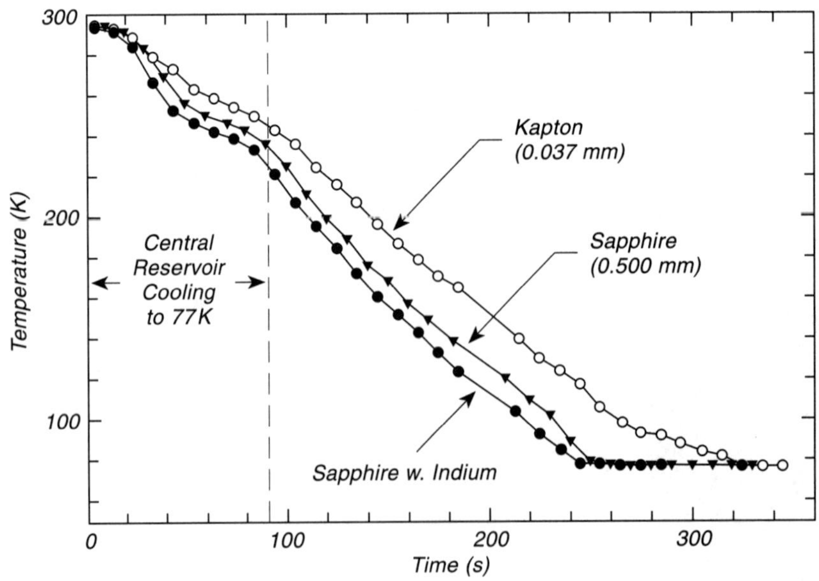


**B. Comparative Cooling Rates**

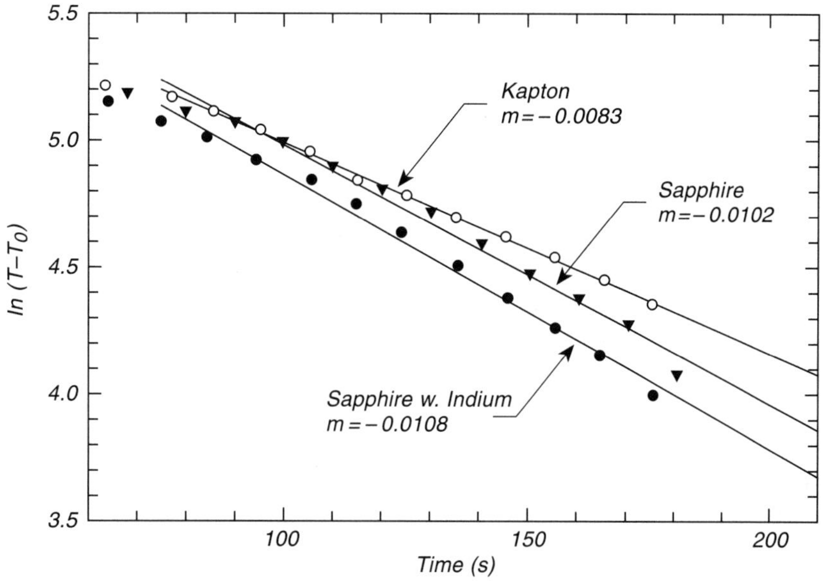


**C. Comparison of Cooling Rates - Various Insulators**

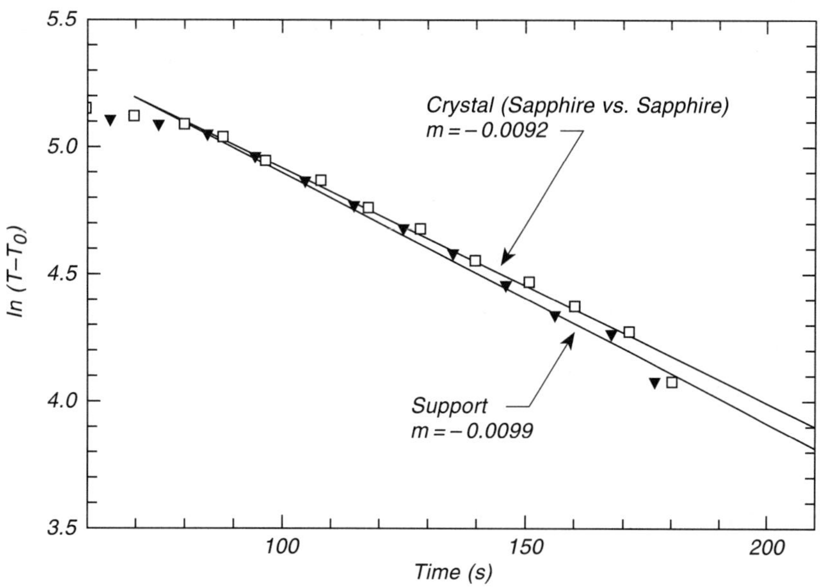


**D. Rate of Crystal Cooling vs. Support Cooling**

# 161

# Vacuum-Jacketed Cryogenic Manipulator Rod

Long manipulator rods filled with liquid nitrogen can act as cryogenic surfaces that can physisorb or condense gases used in surface science studies. As the liquid nitrogen level drops, reevolution of these gases can cause problems in experiments conducted under ultrahigh vacuum. The design shown in the Figure 161A eliminates this problem by using a vacuum jacket to isolate the outer wall of the manipulator rod from the inner tube, which is cooled with liquid nitrogen [1].

A copper reservoir is supplied with liquid nitrogen through a 1/8-in. diameter stainless steel tube. The excess liquid and vapor exits through a 5/16-in.-diameter outer tube, which is concentric with a 1/2-in.-diameter stainless tube welded to a Conflat flange. The outer tube is pumped and contains an adsorbent material for maintenance of the vacuum, as is sometimes used in metal Dewar flask vacuum jackets. The liquid nitrogen inlet and outlet tubes are able to rotate inside a compression fitting that contains a spring-loaded Teflon seal that allows the sample to be rotated while the nitrogen delivery tube and the outlet tube remain fixed. Liquid nitrogen is drawn through a latex rubber tube into the assembly from an unpressurized storage Dewar by means of a small carbon vane vacuum pump [2] attached to the outlet tube. A valve at the inlet of the pump is used to adjust the flow of liquid nitrogen. This pump is isolated from the liquid nitrogen reservoir by a length of copper tubing that permits the liquid nitrogen to convert to gaseous nitrogen before entering the pump. The length of the manipulator is 1.3 m below the Conflat flange.

Owing to thermal contraction, the inner tube will decrease in length upon cooling. Since this is fixed at only one end, it may contract and expand freely. The differential expansion of the outer two tubes is taken up by a formed bellows that is welded to the compression fitting. Some contraction and expansion of the outer tube still occurs as the inner tube changes temperature.

The sample holder shown in Figure 161B is made of OFHC copper and is held on the base of the copper reservoir by four screws that pass through a 0.5-mm-thick sapphire plate that permits efficient cooling (see page 550) as well as electrical isolation. A Ta support structure consisting of two arms for the sample is bolted to the copper support arms. Heating leads and thermocouple leads to the sample pass on the outside of the vacuum-jacketed manipulator rod and are heat-stationed on the lower cooled sections of the assembly.

In operation, the sample is cooled to 90 K when liquid nitrogen is the refrigerant. Measurements indicate that the outer tube slowly cools to about 200 K owing to cooling from its lower end, where it contacts the copper reservoir. This could be prevented by the installation of a small electrical heater on the outer tube.

Because of the length of the manipulator rod, a lower bearing is supplied in which the manipulator rod rotates. This prevents vibration of the sample inside the apparatus. Rotation and translation of the sample are supplied by a standard differentially pumped rotary seal (see page 8) and a standard *x*-*y*-manipulator stage.

## *References*

[1] D.J. Frankel, B. Fruhberger, R.H. Jackson, and D.J. Dwyer, Rev. Sci. Instrum. **64,** 2368 (1993).
[2] Model 0522, Gast Manufacturing Corp., P.O. Box 97, Benton Harbor, MI 49023-0097.

## Fig. 166. Vacuum-Jacketed Cryogenic Manipulator Rod

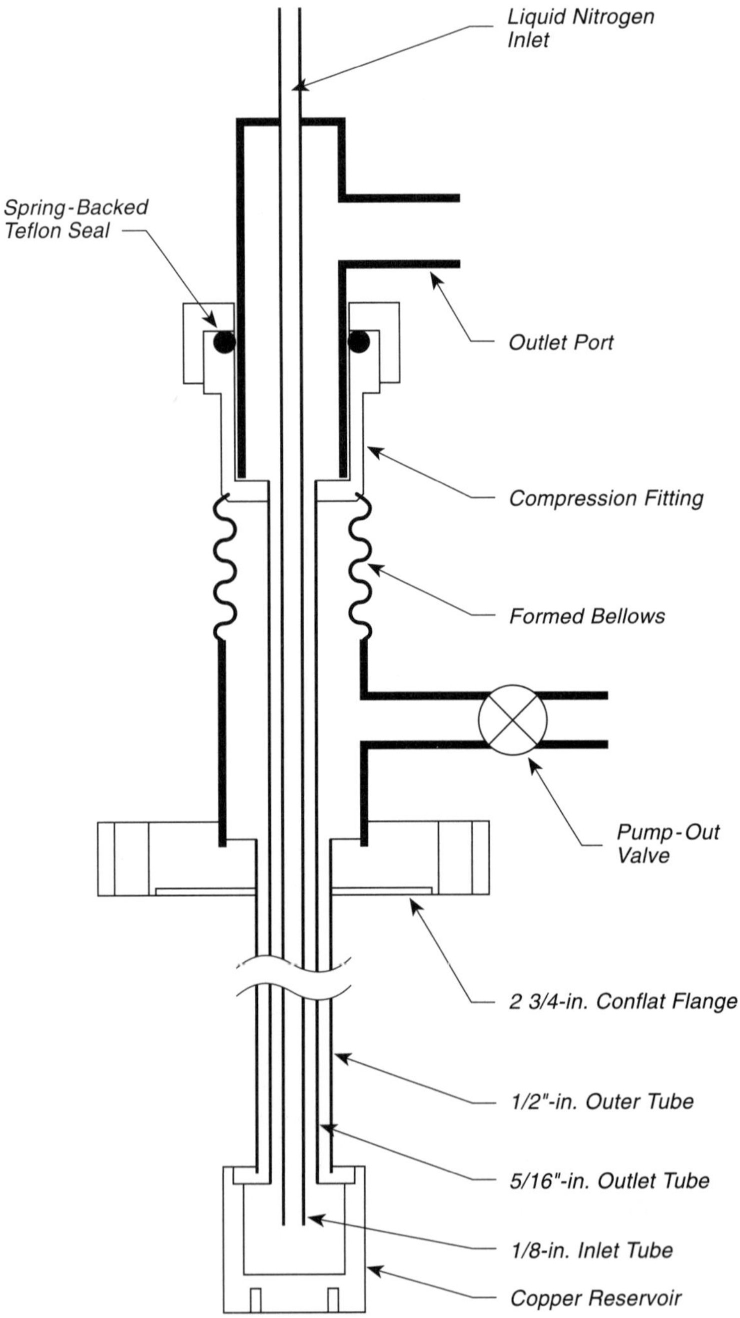


**A. Vacuum-Jacketed Sample Support Tube**

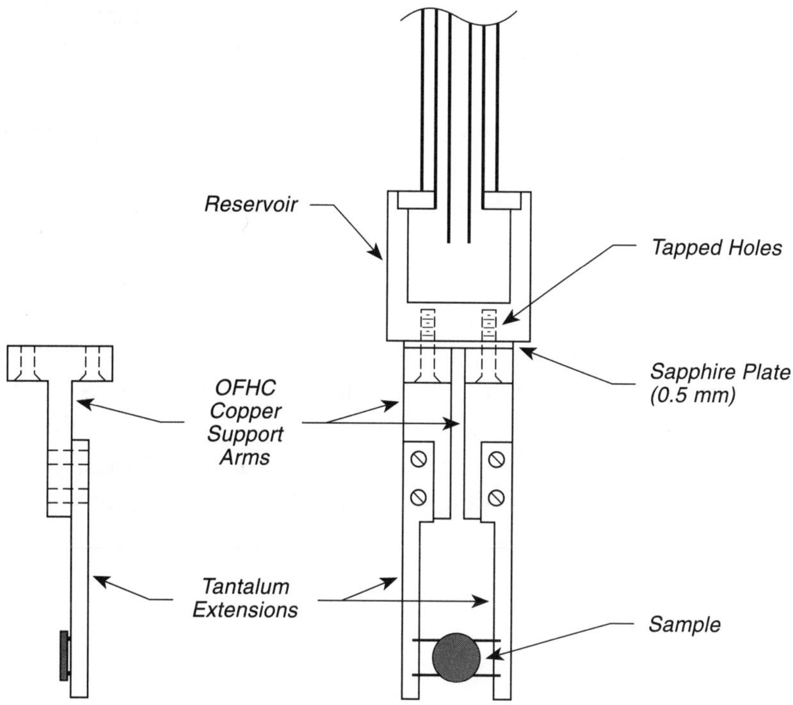


***B. Sample Holder Cryogenically Coupled to Copper Reservoir***

# 162

# Manipulator Component with Cryogenic Azimuthal Rotation

A simple design [1], involving indirect heating of a sample that can be cooled to cryogenic temperatures and undergo limited azimuthal rotation, is shown in Figure 162. The sample is mounted on a metal cap fixed to a sapphire cylinder that makes thermal contact with a Cu-Be cup through a novel ring of Cu-Be finger stock [2]. The Cu-Be cup is held by a hollow shaft that rotates in the Cu-Be block through stainless steel ball bearings that permit azimuthal rotation of the cup and the associated parts connected to the sample. The sapphire is a good thermal conductor but is an electrical insulator. In order to provide more cooling of the sample, four Cu braids are connected between the sample mount and the Cu-Be block, which contains internal pathways for flowing liquid nitrogen refrigerant. Azimuthal rotation is achieved using a rack and pinion connection allowing linear translation of the rack to produce rotation of the Cu-Be cup and the sample. An internal heater may be used for radiation heating or electron bombardment of the sample and its mount. It is electrically isolated and consists of a two-hole ceramic tube fixed in position with a ceramic collar so it does not rotate as the Cu-Be cup is rotated. The azimuthal freedom is ±45°, limited by the length of the copper cooling braids. The temperature range of the sample is from 130 K to 1000 K, with cooling times of about 45 min.

## *References*


[1] P.G. Strupp and P.C. Stair, J. Vac. Sci. Technol. **A9,** 2410 (1991).

[2] Catalog No. 97-216. The springs are obtainable from Instrument Specialties Co., P.O. Box A, Delaware Water Gap, PA 18327. These springs make good mechanical and thermal contact and permit dimensional changes due to differing thermal expansion coefficients of the sapphire and the Cu-Be alloy.

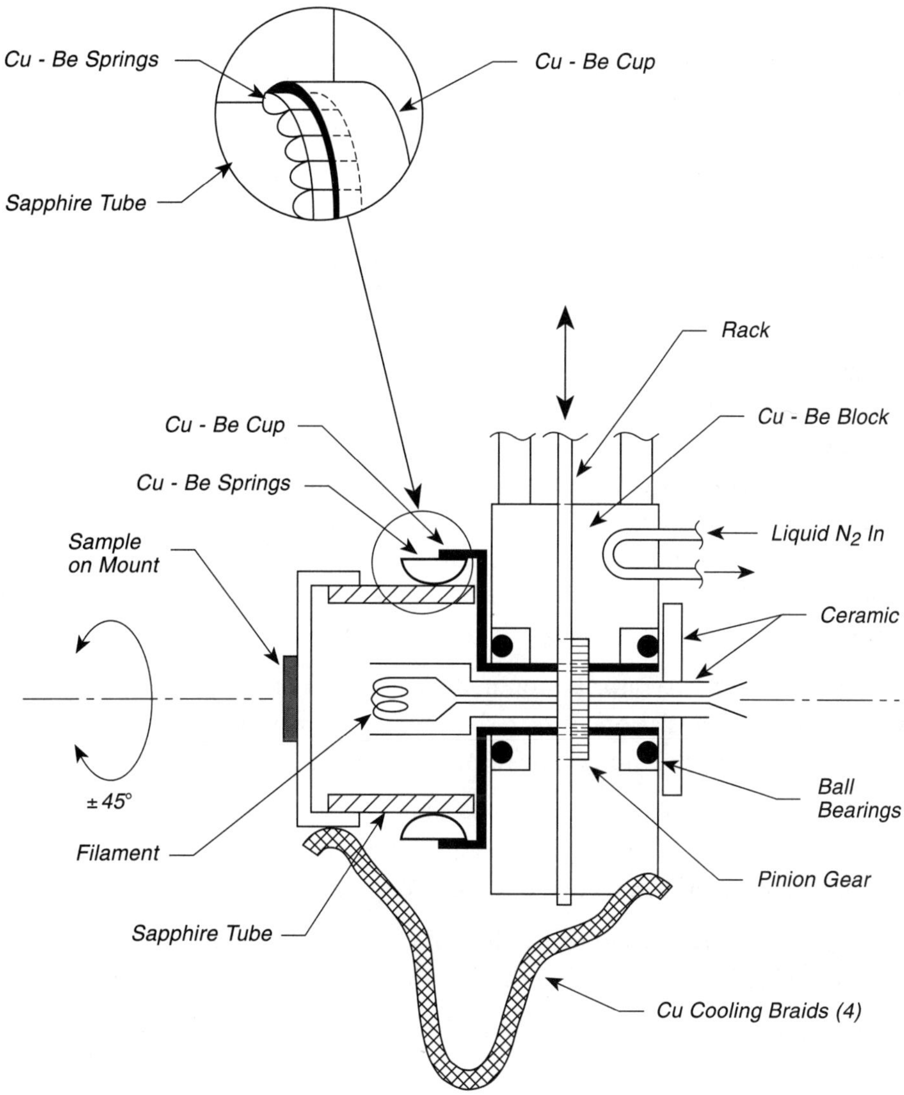


***Fig. 162. Azimuthal Rotation—Cryogenic***

# 163

# Cooling Using Direct Contact

The use of cooling braids to supply flexibility in cooling of samples also causes a limitation in cooling rate owing to the low thermal conductivity associated with braids having a desired flexibility. Therefore, methods to cool movable samples by direct contact with a cold reservoir have been developed.

The first of the methods [1] to be described is shown in Figure 163. Here a 200-g Cu tank is connected directly to a closed-cycle He refrigerator. The cold second stage is shielded from external radiation by a first stage shield as shown. The Cu tank contains a slot for the insertion of the sample holder containing a freshly cleaved semiconductor sample in this case. This sample holder makes contact with a droplet of molten Ga for the production of a metallic joint of high thermal conductivity. The Ga can be melted by radiation heating from an attached heater as shown in Figure 163. The Cu tank can then be cooled to 20 K over a 140 minute cooling period after starting the refrigerator. The long cooling time is probably a result of the large heat capacity of the massive Cu tank in this design. Temperature adjustment can be achieved by using the radiative heater in opposition to the He refrigeration unit. During heating, outgassing of the assembly with bursts of pressure is minimal since cooling is being maintained and slow heating occurs.

The closed cycle refrigerator is mounted on a bottom-entry manipulator permitting rotation and $x,y,z$-translation. Translation in the $z$-direction is especially important because of thermal contraction of the device on cooling.

A thermal switch arrangement involving a high area Cu-Cu contact between a liquid-He cooled reservoir and a crystal support assembly has produced a fast cooling performance and a crystal temperature of 20 K after long equilibration times [2].

A design of a highly efficient attachment of a removable closed-cycle refrigerator to a crystal mount has been described, with a limiting low temperature of 11 K [3].

Other arrangements for direct contact cooling are shown on pages 548 and 566.

## *References*


[1] I.M. Vitomirov, C.M. Aldao, G.D. Waddill, and J.H. Weaver, J. Vac. Sci. Technol. **A8,** 3368 (1990).

[2] E.E. Chaban and Y.J. Chabal, Rev. Sci. Instrum. **54,** 1031 (1983).

[3] A. Winkler, M. Hausenblas, M. Leisch, and K.D. Rendulic, Vacuum **40,** 39 (1990).

***Fig. 163. Cooling Without Thermal Braid***

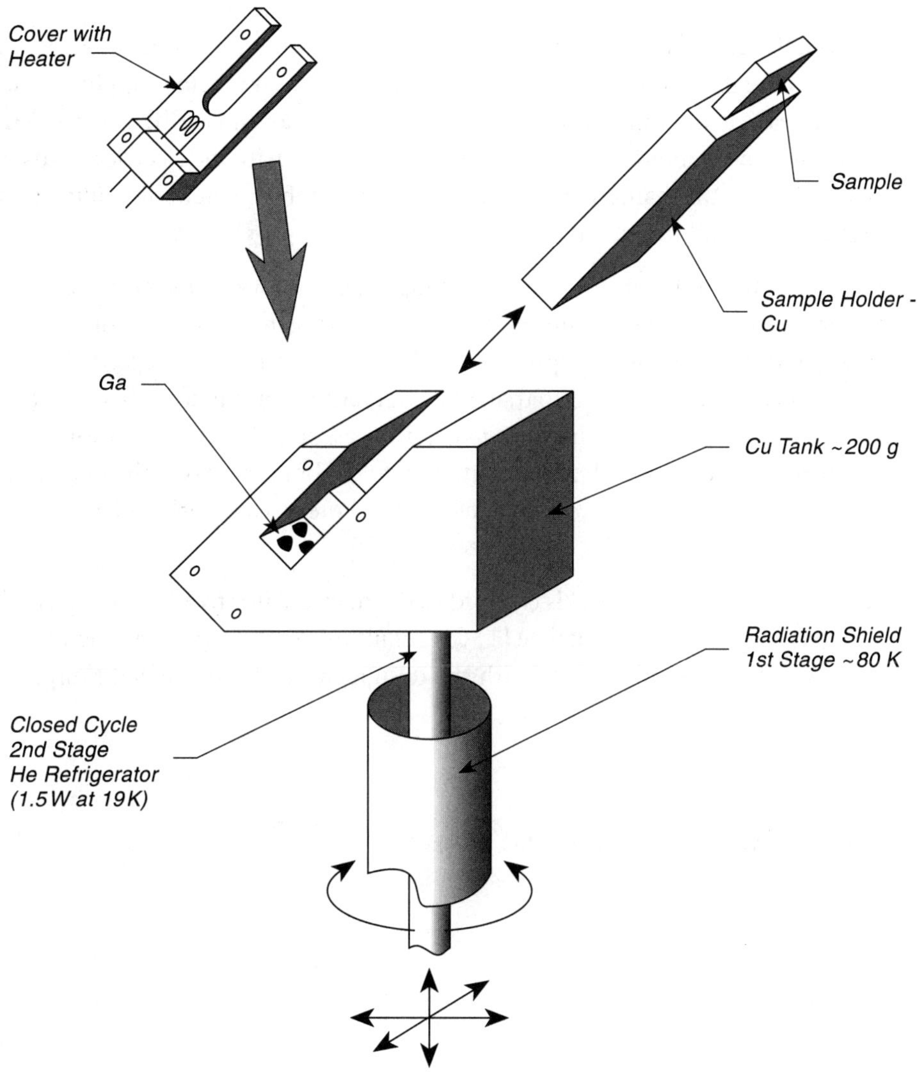


***Cooling Using Direct Contact***

# 164

# Cooling Reservoir—Horizontal Manipulator

A horizontal manipulator is inherently difficult to cool with liquid nitrogen since gravity may cause the refrigerant to drain from the tubing as it is being supplied. In addition, rubber seals at the inlet end of the manipulator may cause difficulty when the manipulator is rotated about its axis because of interference with the electrical leads passing into the center of the manipulator as well as because of the inherent difficulty of retention of refrigerant at all rotational angles.

The design shown in Figure 164 uses two copper disks as heat transfer devices, and at all rotational angles the disks are immersed in liquid nitrogen which is confined by gravity in the expanded end of the manipulator [1]. The two current carrying leads are brazed into holes in alternate copper disks so that efficient thermal conduction will occur to each lead at all rotational angles. The use of two Cu disks assures electrical isolation between the current carrying leads, which are insulated up to the braze points with Teflon sleeves. The thermocouple leads pass through an elongated hole in each disk and are insulated from the disks and each other by Teflon sleeves.

The design shown in Figure 164 is coupled to the manipulator by VCR fittings, and the reader is referred to the original article [1] for details of this arrangement and the alignment procedures for assembly. Tests with the design shown have reached sample temperatures in the range 103–123 K.

## *References*

[1] G.L. Fowler, J. Vac. Sci. Technol. **A9**, 360 (1991).

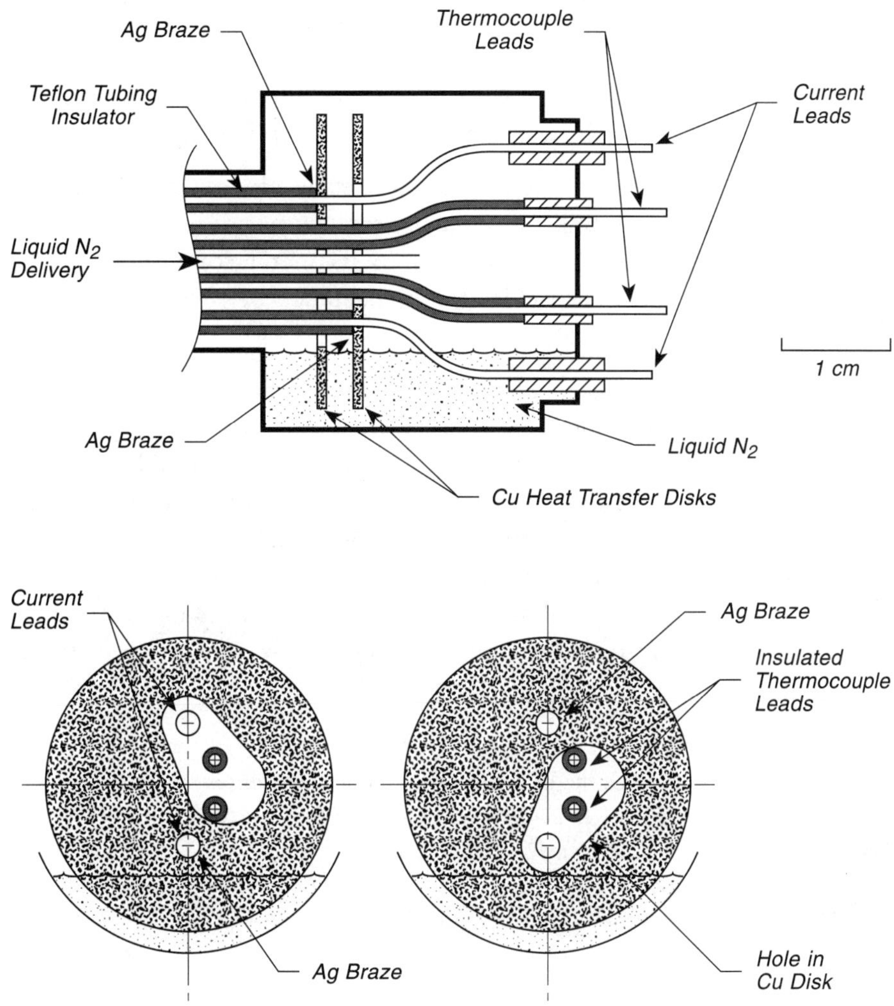


**Fig. 164. Cooling Reservoir—Horizontal Manipulator**

# 165

# Cryogenic Connections

The design of efficient heat exchange methods that also permit motion of a cooled sample is made difficult by two factors: (1) the provision of sufficient thermal conductivity in the materials making the thermal connection, and (2) the provision of excellent thermal contacts between interfaces along the conduction pathway.

The best method for making a thermal contact between interfaces along the conduction path is by welding [1,2]. Brazing, while better than most pressed mechanical connections, provides an alloy that is inferior to its metal components in thermal conductivity. Wherever possible, welded thermal contacts between the same materials (such as Cu or Ag) should be made.

Figure 165 shows a braid assembly made of OFHC Cu that is connected to a cryostat and to a Cu sample holder by means of welds at two of the three contact points. Because the Cu flexible braid work hardens upon extended usage, it should be considered an expendable component. The Cu heat conductivity can be improved by hydrogen firing at 1 bar pressure at a temperature of 600–800°C for 3–5 min. Longer hydrogen firing may cause hydrogen embrittlement of the braid. The effect of limited hydrogen treatment is probably to remove copper oxide films that impede heat transfer at interfaces in the braid and on its connectors. Hydrogen firing should be done only by experienced persons!

## *References*


[1] Dr. R. David, Institut fur Grenzflächenforschung und Vakuumphysik, Forschungszentrum Jülich GmbH, Postfach 1913, D-52425 Jülich, Germany (private communication). See also R. David et al., Rev. of Sci. Instrum. **57**, 2771 (1986).

[2] Detailed procedures for the preparation and welding of Cu braid are presented in the literature. See G.L. Fowler, J. Vac. Sci. Technol. **A5**, 2976 (1987).

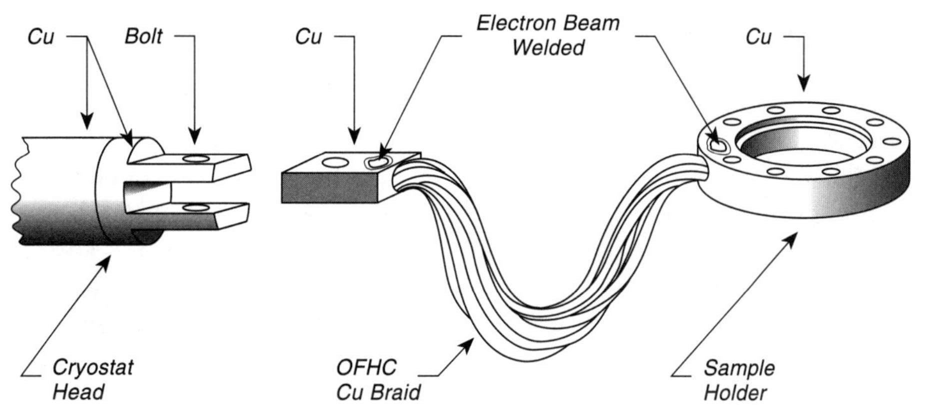


***Fig. 165. Cryogenic Connections***

# 166

# Sample Cooling Using Thermal Braid

Many experiments require sample cooling below room temperature and below the temperature of liquid nitrogen. The use of copper thermal braid for this purpose provides flexibility in sample motion, and with proper design copper braid can be employed for efficient cooling.

In Figure 166A, a copper braid is employed with a detachable thermal switch to provide intermittent cooling of a sample mounted on a movable manipulator [1]. The braid is firmly connected to the sample mount by means of clamps made of Cu. The other end is connected to a solid Cu cylinder, weighing about 100 g which makes contact with a Cu cup. The Cu cup is cooled by flowing liquid nitrogen through a coil wrapped and brazed in place. The use of brazing is highly recommended, since the thermal conduction through a brazed joint is far superior to any other mechanical contact except for a weld between identical metals. A reversible metal contact between the Cu cylinder and the Cu cup is made using a liquid gallium droplet (mp = 32°C at a vapor pressure of $10^{-14}$ Torr). Thus, to melt the Ga, hot air is passed through the coil on the cup; the weight is lowered into the cup by moving the sample holder to which it is attached, and subsequent cooling freezes the Ga, making a good thermal and mechanical contact. Sample cooling to $93 \pm 10$ K then occurs within about 200 s.

In Figure 166B, a thermal braid is permanently attached to the sample holder [2]. The manipulator swings about a radius of 15 cm, and various degrees of freedom are present. The sample holder is thermally isolated from the manipulator shaft and most of the arm. The thermal braid is attached to a closed-cycle He refrigerator [3], which has two stages of cooling to 80 K (15 W) and to about 10 K (1 W). The 80-K stage provides cooling for a radiation shield that surrounds the low- temperature stage, as shown schematically in the Figure. Modeling of the energy transfer was done in order to compute the optimal dimensions for the sample holder and the braid. The minimization of thermal conduction to the sample holder makes thermal radiation the primary source of unwanted heat. Since the ratio of the surface area of the sample holder to that of the thermal braid is about 1:12, a cooled radiation shield around the sample is unnecessary in this design. It was calculated that the cross-sectional area of the Cu braid was 0.075 cm$^2$ on the basis of thermal conduction and thermal radiation considerations. It is of interest to note that this design achieved sample temperatures of 14–15 K routinely with only 1 W of cooling in the second stage of the closed-cycle refrigerator.

Copper braid should be made of oxygen-free, high-conductivity (OFHC) Cu for maximum efficiency. UHV-compatible closed-cycle refrigerators are commercially available [3]. It should be noted that efficient thermal connections to copper braid can be made by brazing the ends with silver solder, which can then be deformed by clamping. Procedures for achieving the same objective without brazing have been described [4].

## *References*

[1] E.E. Chaban, P.H. Citrin, and F. Sette, J. Vac. Sci. Technol. **A6,** 3018 (1988).
[2] V.K.F. Chia, D.K. Veirs, and G.M. Rosenblatt, J. Vac. Sci. Technol. **A7,** 108 (1989).
[3] Cryomech GB-220: Cryomech, Inc., 1630 Erie Boulevard E., Syracuse, NY 13210.
[4] G.L. Fowler, J. Vac. Sci. Technol. **A5,** 2976 (1987)

***Fig. 166. Cooling Using Thermal Braid***

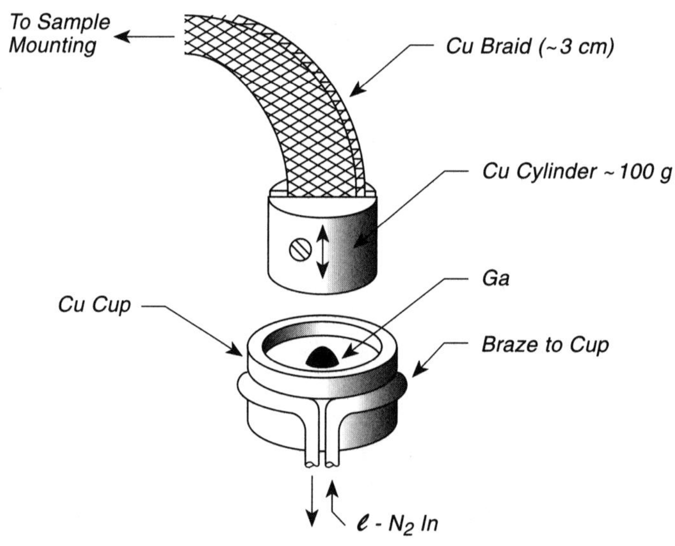


***A. Detachable Cooling Assembly***

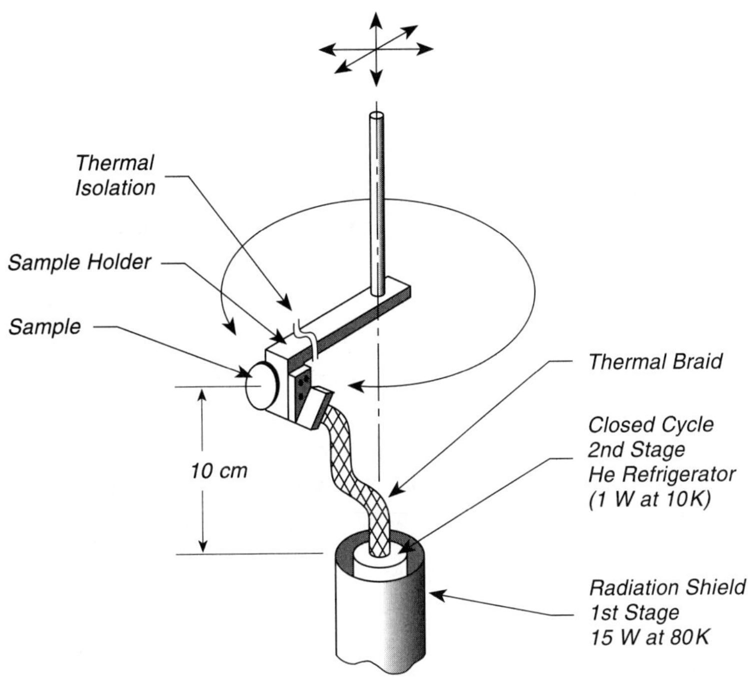


***B. Continuously Coupled Cooling Assembly***

# 167

# Flexible Cooling Reservoir

The use of copper braid for cooling samples on a manipulator is widespread. This method of cooling avoids complex crystal motion devices that must also conduct heat efficiently, and mountings with six degrees of motional freedom are often built with braid cooling.

As the crystal and the manipulator shaft move, there will be the need for relief of the strain in the copper cooling braid. While a long braid may be used to give a sufficient degree of motion, this decreases the cooling efficiency of the system. It is also possible, however, to mount the cooling reservoir itself on a flexible mount as shown in Figure 167 [1], thereby permitting a shorter braid to be used.

A mini-Conflat flange, mounted on the manipulator flange, is welded to a standard Swagelock Tee. A stainless tube enters the vacuum system through this flange and is connected to a very flexible bellows and then to a double kovar/glass seal, which acts as an insulating break, electrically isolating the cooled reservoir from the vacuum system and ground. An OFHC Cu reservoir is brazed to the end of the tube assembly, and contains a polished flat to which the copper braid and its copper lug are mounted (see page 564). An indium foil gasket makes good thermal contact when compressed by a bolt that passes through the flat and is drawn down by a nut on the back side of the flat. The alternate use of threads in the soft copper flat is not recommended because of the possibility of stripping threads.

Liquid nitrogen is supplied to the copper reservoir by suction from a styrofoam cup or small Dewar through 1/8-in.-diameter Teflon tubing that can be sealed inside the Swagelock fittings. Sample temperatures below 100 K can easily be achieved.

## *References*

[1] Dr. Lawrence H. Dubois, DARPA/DSO, 3701 N. Fairfax Dr., Arlington, VA 22203-1714 (private communication).

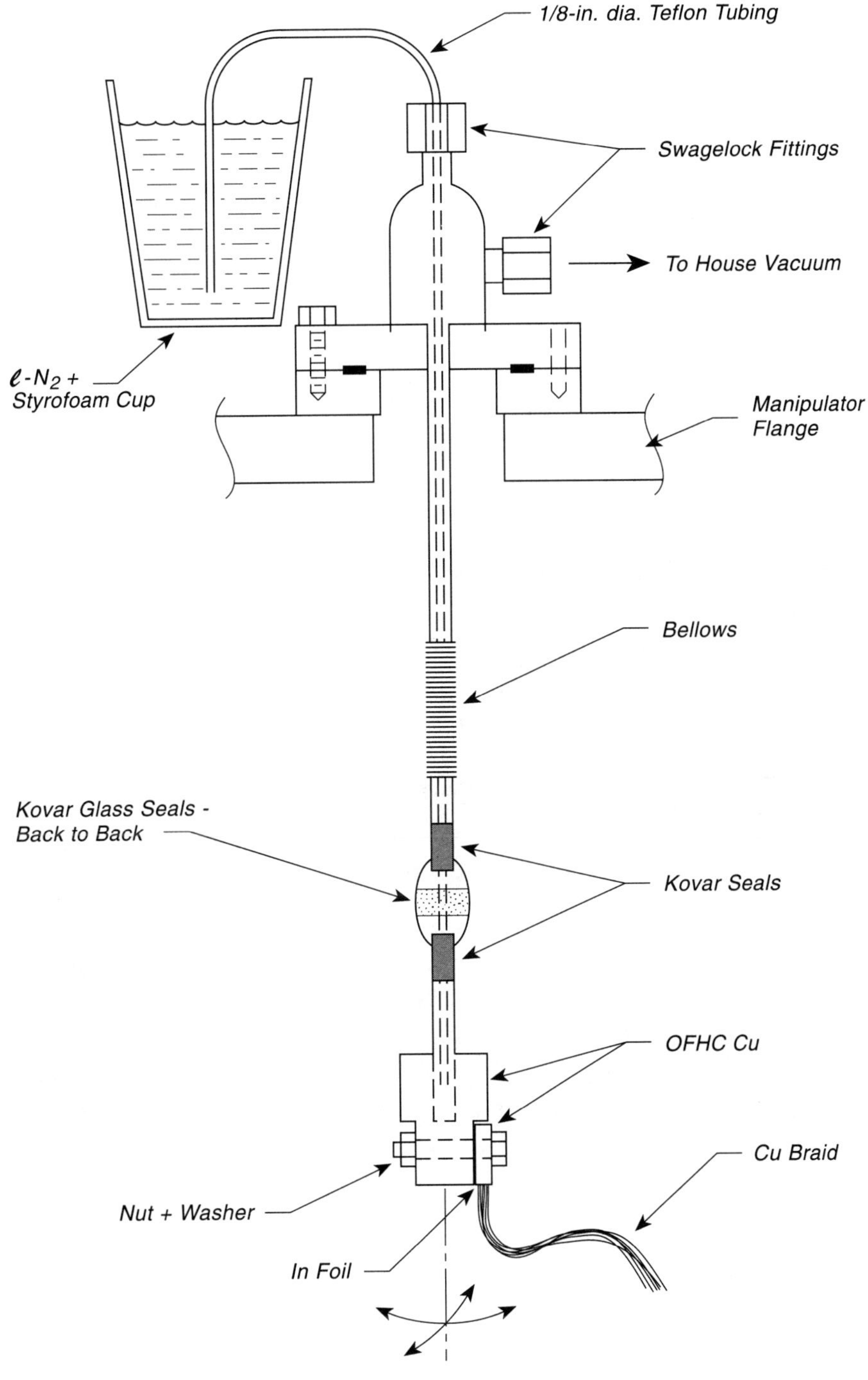


*Fig. 167. Flexible Cooling Reservoir*

# 168

# Enhanced Cooling Using Liquid Nitrogen

A simple method [1] has been devised and tested for cooling below the temperature limit established by liquid nitrogen (77.3 K). The method involves bubbling He gas through the liquid nitrogen (transpiration); evaporation of liquid nitrogen requires heat absorption from the fluid itself, and the temperature falls as a consequence of the evaporation until the triple point (63.1 K) is reached. Thus, for work requiring temperatures slightly below that achieved using liquid nitrogen, this method is convenient. Compared to pumping liquid nitrogen to reduce its equilibrium vapor pressure and to achieve lower temperatures, this method offers the convenience of not requiring high-capacity pumps connected to manipulators by large flexible tubing, a procedure that is often mechanically complex and inconvenient.

A He bubble in equilibrium with $N_2(l)$ will contain both He and $N_2$, so that the partial pressure of $N_2$ will be less than the total pressure within the bubble. Thus as $N_2$ vaporizes within the bubbles, cooling will occur to reduce the temperature below the normal boiling point of $N_2(l)$. Figure 168A shows a demonstration apparatus where this experiment was carried out. A fritted disk bubbler was used in the right-hand Dewar flask, while the thermocouple reference junction was maintained at 77.3 K in the left-hand Dewar. After about 80 min of transpiration, solid nitrogen crystals were observed, and the temperature reached that of the triple point. When the He flow was stopped, the temperature remained stable for some minutes as the solid nitrogen melted at the triple point; heat leakage into the Dewar was responsible for this as well as for the gradual warming beyond 120 min.

Figure 168B shows a similar experiment in a 1.42-m column of liquid nitrogen that comprises the cooling reservoir of a reentrant stainless steel manipulator on an ultrahigh vacuum system. The temperature of a metal single crystal was measured as He bubbled into the bottom of the reservoir from the end of a 0.097-cm-diameter tube. The crystal temperature fell from 86.5 K to 74 K during He flow. Stability of ±0.5 K was achieved, and excellent cooling control may be maintained in this temperature range by adjustment of the He flow rate. Because of the limited amount of $N_2(s)$ produced at the bottom of the reservoir, interruption of the He flow produced an almost immediate rise in the crystal temperature as shown. Because of the length of the column of liquid $N_2$, additional liquid $N_2$ may be added without disturbance, as shown at the time indicated by the dots in Figure 168B. This is a distinct advantage of using a long column of refrigerant where transpiration cooling occurs primarily at the bottom.

## *References*

[1] J. Xu, H.J. Jansch, and J.T. Yates, Jr., J. Vac. Sci. Technol. **A11**, 726 (1993).

***Fig. 168. Cryogenic Trick for Enhanced Cooling with Liquid $N_2$***

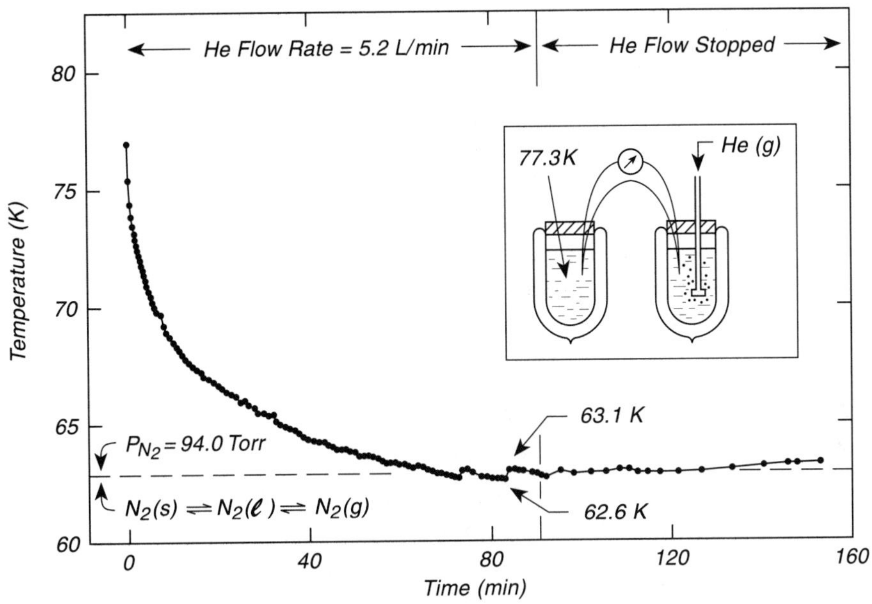


***A. Attainment of $N_2$ Triple Point by He Transpiration***

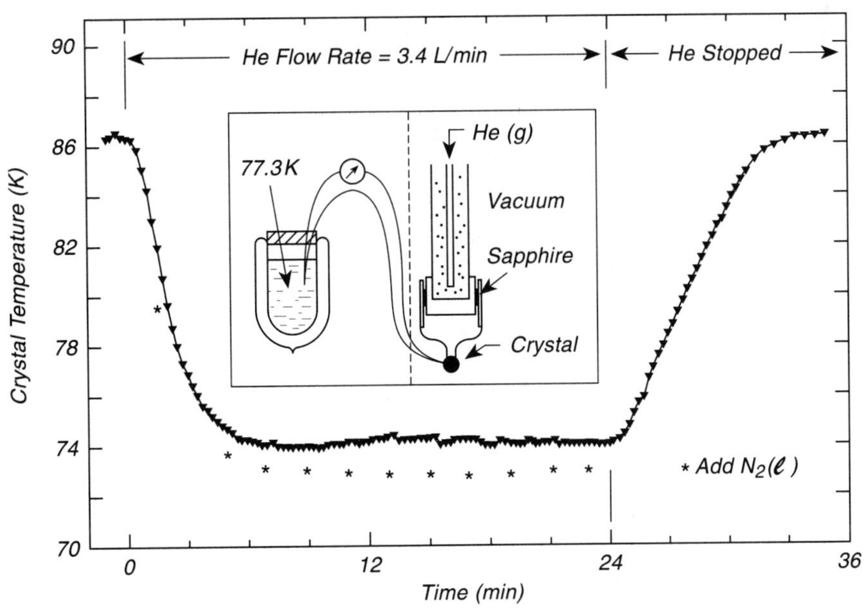


***B. Cooling of Metal Crystal by He Transpiration***

# 169

# Level Alarm-Liquid Nitrogen

There are many designs for level alarms and controllers for liquid-nitrogen Dewars, and Figure 169 shows one of the simplest alarm devices for keeping a small supply Dewar filled [1]. A cork float senses the level of liquid nitrogen in the Dewar, and when the level drops to its lower limit, contact is made between a metal plate and an inverted metal cup at the top of the sealing cork lid. This completes the circuit, causing both an audio and a light signal to be emitted, alerting the operator to refill the Dewar through the refill funnel. The audio oscillator normally used to drive an Auger spectrometer employing a lock-in amplifier can be used.

The Dewar is sealed to the top of the entrance tube using cotton plus ice to make the seal. Simply wet the cotton from a plastic squeeze bottle, admit a little liquid nitrogen to freeze the water, and repair any leaks with more water on the cotton. This type of seal does not put any stress on the glass Dewar during warm-up in contrast to compression seals, which will allow the expanding metal to strain the glass Dewar bottom, sometimes resulting in breakage.

## *References*

[1] Dr. Xing-Cai Guo, Department of Chemical Engineering, Stanford University, Stanford, CA 94305-5025 (private communication).

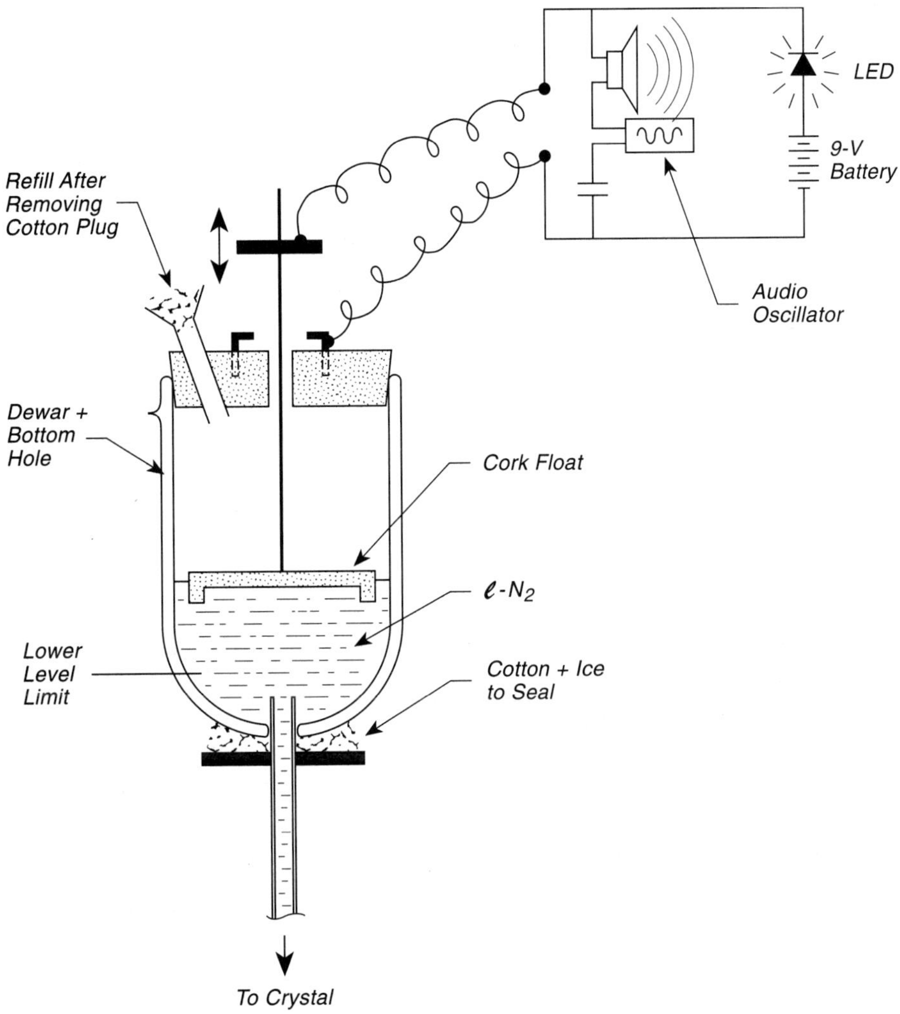


***Fig. 169. Level Alarm—Liquid Nitrogen***

# 170

# Gas Thermal Switch

The ability to turn on and turn off cooling to a sample and its support may be useful in ultrahigh vacuum operations where condensation on the sample/support system must be avoided. This could be the case when the sample and its support are to be immersed in a high-pressure gas in a high-pressure vessel inside the UHV system. The design of a gas thermal switch to vary the cooling rate to a sample and its support is shown here [1].

The principle of the gas thermal switch is to use a heat transfer gas (He(g) is ideal) to conduct heat when cooling is needed, and then to evacuate the He(g) when cooling is not required. As shown in the exploded view of Figure 170A, the OFHC copper cooling stage is maintained at the cryogenic temperature desired by means of contact with a reservoir of the cryogenic fluid ($l$-$N_2$ in this case). Brazed to this heat station is a concentric assembly of copper tubes. The cooling tubes nest into a set of concentric tubes connected to the OFHC copper sample holder that is thermally connected to the sample. Cooling is achieved by introducing about 0.5 Torr of He(g) into the 10-$cm^3$ volume containing the thermal switch. The cooling is interrupted by pumping the He(g) away. Given the geometry of the thermal switch (not exactly specified in the paper), the thermal conductivity varies between about 0.8 and 1.4 W/K with He(g) as a heat transfer medium to about 0.03 W/K with a vacuum of $< 5 \times 10^{-3}$ Torr, which is achieved in 10-s pumping time.

Figure 170B shows the performance of the gas thermal switch. He(g) is admitted with the sample at 300 K and the cooling station at 100 K. Cooling starts immediately with a time constant of about 2 min; after some cooling, the He(g) is evacuated, and cooling slows. Then He(g) is reintroduced, and cooling occurs with a time constant of about 2.6 min. The decrease in cooling rate in the second stage of cooling may be due to the decrease in the thermal conductivity of He as its temperature falls. Cooling to about 102 K is achieved in less than 10 min.

## *References*


[1] O. Sneh and S.M. George, J. Vac. Sci. Technol. **A13**, 493 (1995).

## *Fig. 170. Gas Thermal Switch*

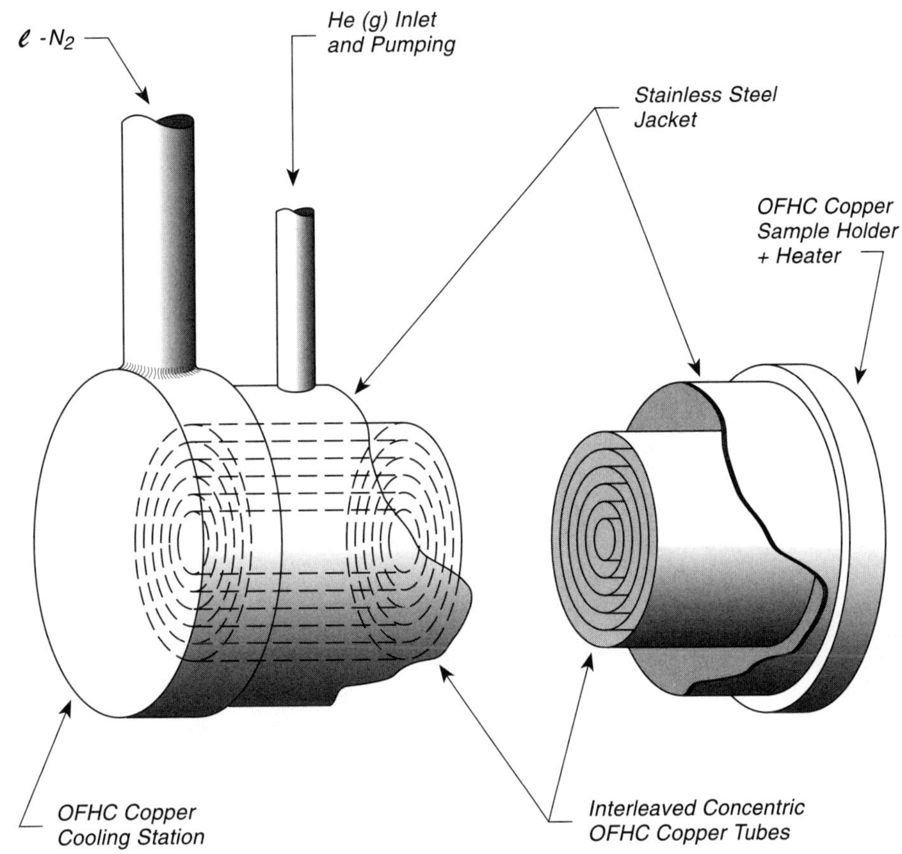


***A. Thermal Switch—Exploded View***

***Fig. 170. Gas Thermal Switch***

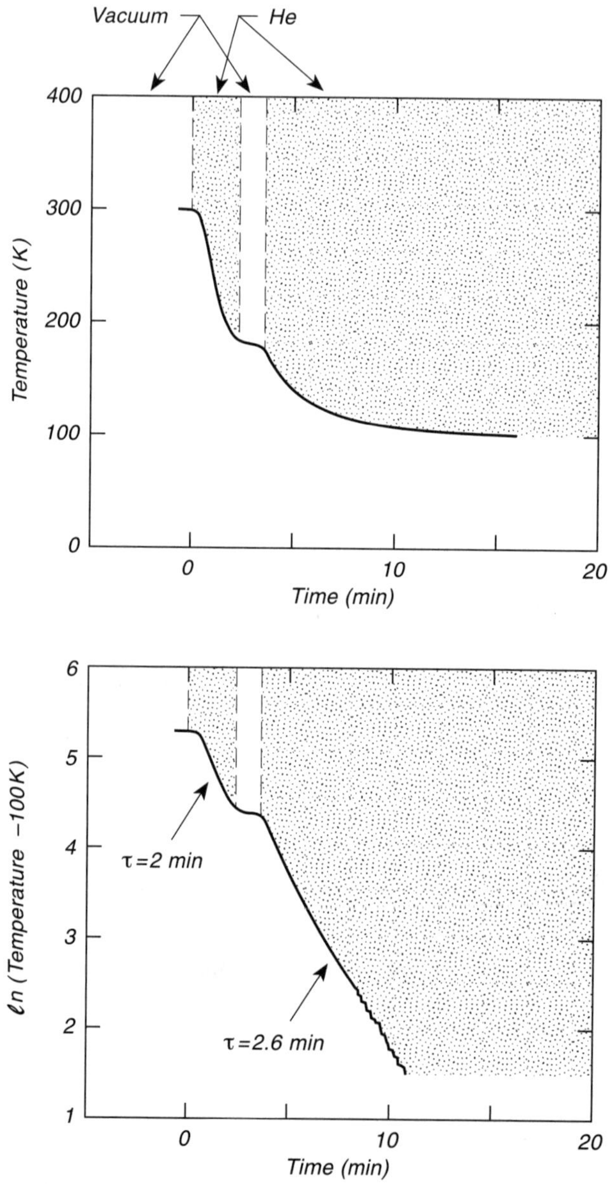


***B. Cooling Performance Using Vacuum and He (g)***

# C. Temperature Measurements

# 171

# Tungsten-Rhenium Thermocouples—Calibration Over a Wide Temperature Range

Tungsten/rhenium thermocouples are often used in ultrahigh vacuum because this class of thermocouples can go to higher temperatures than all other thermocouple alloys [1]. Such thermocouples avoid sample contamination effects owing to metal vaporization, which is common for other thermocouples near the high-temperature end of their operating range. Unfortunately, a factory calibration of tungsten/rhenium thermocouples is not available below 255 K [1], although two calibrations in the range 77–300 K [2], and in the range 4–77 K [3] are published for W/5% Re vs. W/26% Re thermocouples. The calibration data from the three sources have been brought together into a single master calibration that extends from either 4 K or 32 K to 2588 K [4]. The reader is referred to the original paper [4] for calibration tables and for full consideration of the errors present in the master calibration [4]. It should be noted that the variation of thermocouple alloy composition from batch to batch may cause larger errors in the temperature regions not covered by the factory calibration than in those temperature regions which are reported by the manufacturer.

Figure 171A shows the master calibration curve from 4 K to 2588 K, using data from the three sources. This curve is not plotted at high enough resolution to be useful. A tenth-order polynomial function has been fitted to the data, and the function is given in reference [4], as well as at the end of this discussion. The difference between the calibration data and the tenth order function is plotted in Figure 171B, showing that differences are always less than 7 K except for the very low temperature region. The maximum difference above 800 K is near 2 K. For temperatures below 32 K, it is advisable to use the calibration in rcference [3], which is exclusively concerned with this temperature range. The fitting curve of reference [3] cannot be extrapolated with accuracy above 32 K.

The power series for the range 4–2588 K is given below, and tabular solutions to this equation are given in reference [4] at intervals of 0.05 or 0.10 mV.

$$T = \Sigma C_n B^n$$

| | |
|---|---|
| $C_0 = 276.5327901$ | $C_6 = -4.147285358 \times 10^{-3}$ |
| $C_1 = 83.31475475$ | $C_7 = 1.529502993 \times 10^{-4}$ |
| $C_2 = -15.19381797$ | $C_8 = -3.422732361 \times 10^{-6}$ |
| $C_3 = 4.431847913$ | $C_9 = 4.253729476 \times 10^{-8}$ |
| $C_4 = -0.7258516658$ | $C_{10} = -2.251967817 \times 10^{-10}$ |
| $C_5 = 0.06999557967$ | |

An extension of the calibration up to 3273 K has been reported [5].

## *References*

[1] *The Omega Complete Temperature Measurement Handbook and Encyclopedia* (Omega Engineering, Inc., Stamford, CT, 1989) pp. z8–z28.

[2] D.R. Sandstrom and S.P. Withrow, J. Vac. Sci. and Technol. **14,** 748 (1977).

[3] W.J. Mitchell, J. Xie, and W.H. Weinberg, J. Vac. Sci. Technol. **A11,** 3133 (1993).

[4] V.S. Smentkowski and J.T. Yates, Jr., J. Vac. Sci. Technol. **A14,** 260 (1996).

[5] R.R. Asamoto and P.E. Novak, Rev. Sci. Instrum. **38,** 1047 (1967).

**Fig. 171. Tungsten-Rhenium Thermocouples—Calibration over a Wide Temperature Range**

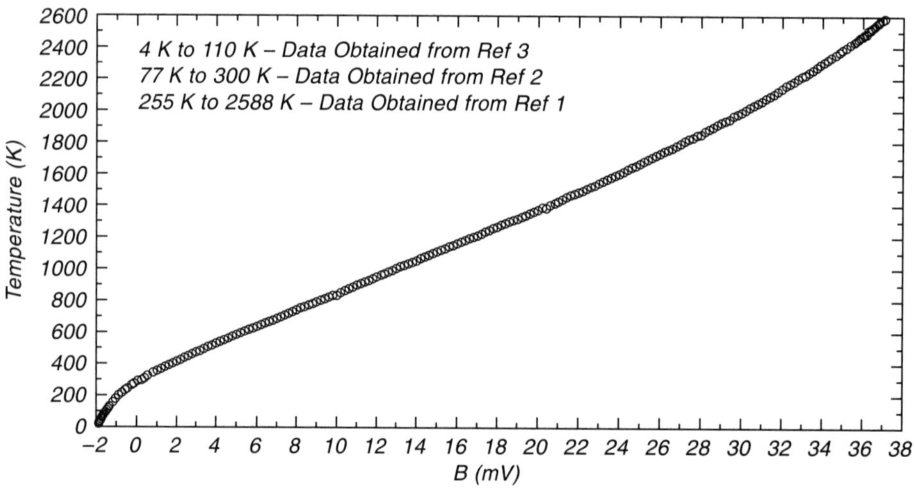


**A. Temperature vs. Potential, W/26% Re vs. W/5% Re Thermocouple**

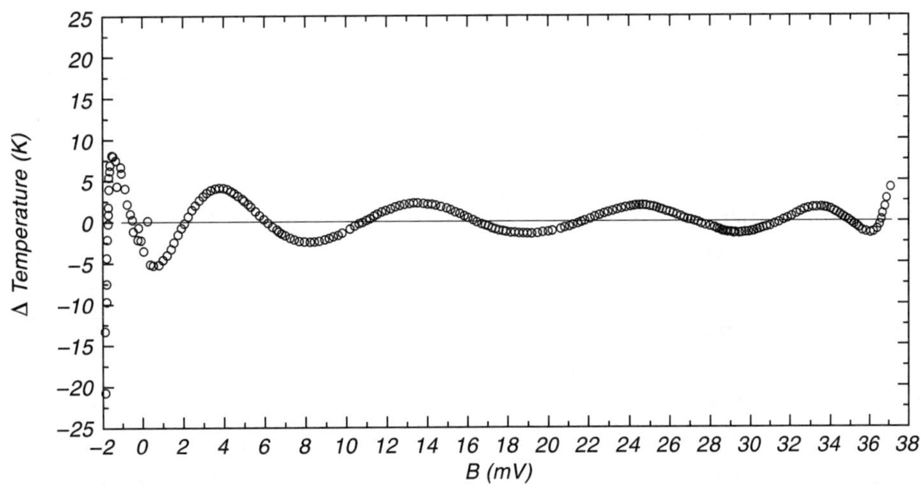


***B. Difference Plot [Measured-Calculated] for Thermocouple—4–2588 K***

# 172

# Measuring Temperatures of Hot Filaments

The temperature of an electrically heated hot filament may be accurately measured by using it as a resistance thermometer. By determining its resistance compared to its resistance at a known temperature, quite accurate temperatures may be measured and maintained.

Figure 172A shows a metal filament containing potential leads across the central third of the filament. The filament is mounted on degassing loops to prevent outgassing effects during heating as described on page 628. A known current, $I$, measured to about four significant figures with a digital ammeter, is passed through the filament using a current-regulated DC power supply. The electrical potential drop, $V_c$, across the central third of the filament, is simultaneously recorded to about four significant figures, using a digital voltmeter. The resistance of the central section of the filament is calculated from Ohm's law, $R_c = V_c/I$, and also the power dissipation from the central section of the filament is calculated, $P_c = IV_c$. If $R_c$ is known at some standard temperature, $T_s$, and if the resistivity, $\rho$, of the filament material is known as a function of temperature (Tables for tungsten are readily available in handbooks), then the temperature of the center section of the filament may be calculated from the ratio $R_c(T)/R_c(T_s) = \rho(T)/\rho(T_s)$, where $T$ is the unknown. In practice, more accurate maintenance of temperature may be achieved if $P_c$ is monitored continuously rather than $R_c$. Typically, from day to day, $P_c$ may be readjusted to match its previous values very accurately, assuring that the original temperature is reestablished. However, in practice, because of slow warming effects in the support system that have a minor effect on $T_c$, the measured value of $P_c$ will drift slightly for some minutes until stability of the hot filament temperature is achieved.

There is one special point that must be emphasized for the most accurate measurements. The potential leads should conduct as little heat as possible away from the center section of the filament. To assure that this is the case, it is wise to etch the potential leads electrochemically to a diameter that is as small as possible, thereby limiting their thermal conductivity. For a well-made system, heating at incandescence will display no optically detectable temperature gradient in the vicinity of the welds of the thinned potential leads to the central section of the filament. However, there will be a gradient of temperature at the welds connecting the filament to the current leads, but the temperature measurement based on the central section of the filament is made far from this region.

For work that does not require extremely high accuracy or reproducibility, one may connect the potential leads outside the vacuum system, as in Figure 172B, making similar measurements as described above. Here, the resistivity ratios will not accurately correspond to the filament temperature ratios because the filament temperature gradient, as well as the resistance of the mounting leads, is being sampled. In practice, measurements made in this fashion are best used to establish the same filament temperature from day to day, and this may be done with surprisingly high accuracy. The potential leads should be hard-soldered to the input current leads in order to avoid the effect of having variable contact points as would occur using clip leads.

One problem that can arise in the use of high-temperature filaments is the evaporation of the filament material, causing the diameter of the filament to decrease, upsetting the temperature calibrations. In addition, filaments containing attached pieces of evaporant material can change resistance over time due to this effect as well as to alloying effects.

A useful nomograph for estimating the power requirements needed to heat tungsten filaments of various diameters is shown in Figure 172C. To determine heating parameters, a straight edge is used across the chart. The example shown in the center of Figure 172C is a guide to tungsten wire heater design, including corrections for tungsten wire cooling at the ends welded to the feedthrough leads.

Fig. 172. Measuring Temperatures of Hot Filaments

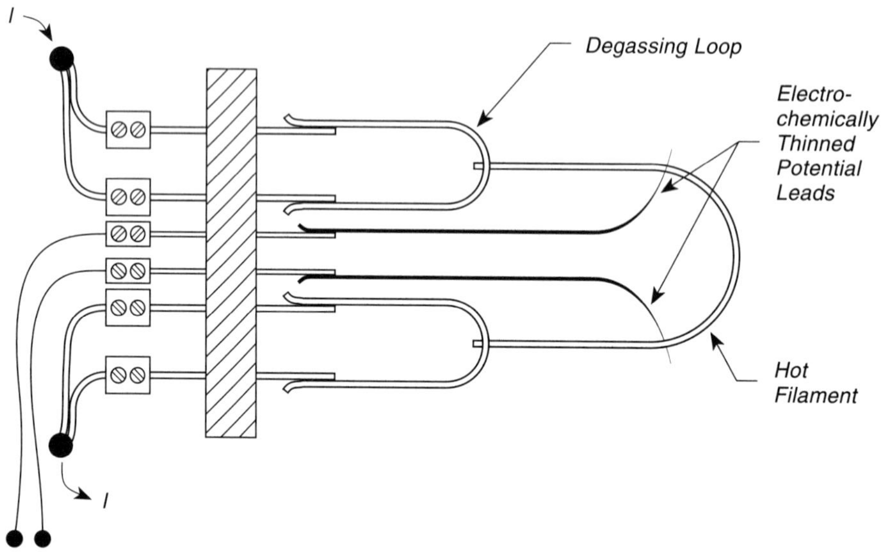


A. Inner-Potential Leads

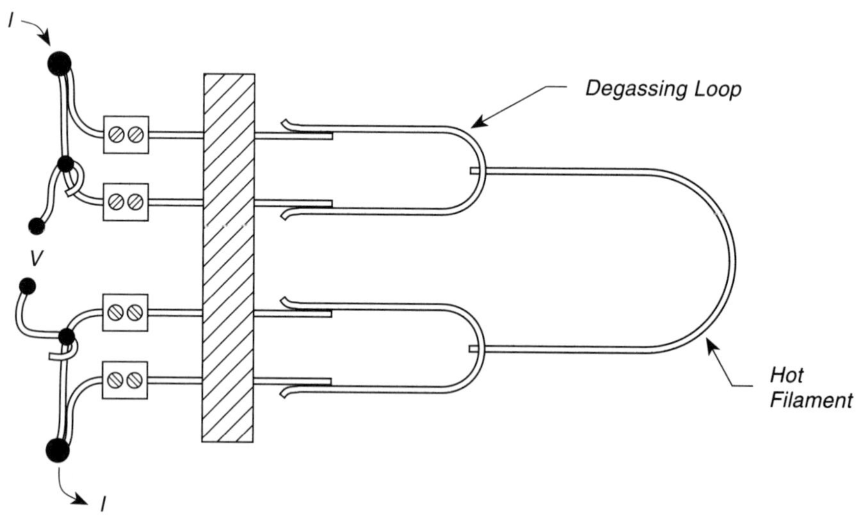


B. Outer-Potential Leads

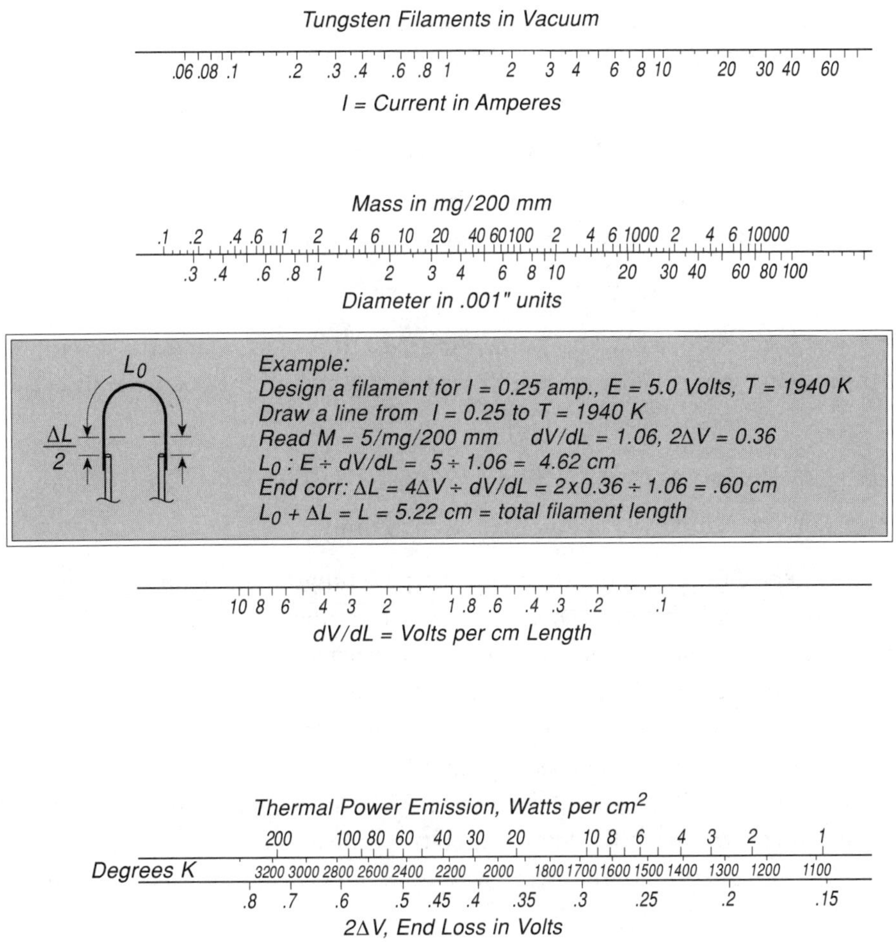


Chart based on Forsythe and Worthing, "The Properties of Tungsten and the Characteristics of Tungsten Lamps" Astrophysical Jour., Vol. 61 pp 146-185 (1925)

**C. Nomograph for Design of Tungsten Wire Heaters**

# 173

# Pyrometric Measurement of Temperatures

The measurement of the brightness temperature of a hot object can be carried out with an optical pyrometer working at a particular wavelength, $\lambda$. Often optical filters in the pyrometer are used to define $\lambda$. Planck's law for a perfect blackbody emitting radiation energy, $W_b(\lambda, T)$, is

$$W_b(\lambda, T) = [2\pi hc^2/\lambda^5]/[\exp(hc/\lambda kT) - 1] \tag{1}$$

where $T$ is the true temperature of the blackbody, $c$ = speed of light, $k$ = Boltzmann's constant, and $h$ = Planck's constant. Since in Equation (1), $hc/\lambda kT >> 1$,

$$W_b(\lambda, T) = [2(hc^2/\lambda^5] \exp - (hc/\lambda kT) \tag{2}$$

When an object is not a blackbody, the surface of the object is a partial inward reflector of the radiant energy being emitted from the object, and the energy emitted, $W_n(\lambda, T)$, is

$$W_n(\lambda, T) = W_b(\lambda, T) \cdot \varepsilon(\lambda, T) \tag{3}$$

where $\varepsilon(\lambda, T)$ is the emissivity, a number less than unity.

In pyrometers of the disappearing-filament type, a tungsten filament, acting as a blackbody, is compared with that of the hot sample by observing a superposition of the hot filament on the hot sample and then finding the pyrometer filament temperature where it becomes invisible against the image of the hot sample. This method is very sensitive, and a skilled operator, working in a darkened room with dark-adapted eyes, can make comparisons to within a few degrees K. Figure 178A shows the image seen through the eyepiece of a disappearing filament pyrometer. The equal-temperature point should be approached alternately from the low-temperature side and from the high-temperature side, and the readings from both directions should be averaged to determine the brightness temperature of the sample.

In the case where the filament and sample brightness temperatures are identical, then

$$W_b(\lambda, T_{pyro}) = W_b(\lambda, T_{sample}) \cdot \varepsilon(\lambda, T) \cdot (1 - R)^2 \tag{4}$$

where $R$ is the reflectivity of each surface of the window material through which the measurement is being made. $R$ is given by

$$R = (n - 1)^2/(n + 1)^2 \tag{5}$$

where $n$ = the refractive index of the window material at $\lambda$.

Using Equations (2) and (3), and solving for $T_{pyro}$, one obtains

$$T_{pyro} = T_{sample}/\{[1 - \lambda k T_{sample}/hc][ln\ \varepsilon(1 - R)^2]\} \quad (6)$$

Both the emissivity, ε, and the window surface reflectivity, $R$, may be determined as correction factors for $T_{sample}$, as shown below.

For the emissivity, an experimentally determined number that is a specific property of the material being measured must be used. Tabulations of the emissivities for various metals are given in reference [1], as well as convenient correction plots to directly convert the brightness temperature to the true temperature for various emissivities.

For the particular case of Si at $\lambda = 0.65$ μm, the emissivity was measured separately by Allen [2] and by Sato [3]. Their data are fit by the following equation [4] in the temperature range 540 K to 1700 K.

$$\varepsilon(0.65\ \mu m, T) = 0.4826 - 2.09 \times 10^{-5}\ T + 0.1583/\{1 + \exp[(T - 1410)/112.9]\} \quad (7)$$

Figure B shows the fit of the empirical Equation (7) to the data. The relationship between $T_{pyro}$ and $T_{silicon}$ can be calculated using Equation (6) with values of $R$ and ε(0.65 μm, $T$) calculated from Equations (5) and (7), respectively. In the case of glass windows, the refractive index for 0.65 μm light is 1.46, and $R = 0.035$.

Figure 173C shows the deviation between the measured brightness temperature with the pyrometer at 0.65 μm, and the true temperature for Si, viewing through a glass window.

Similar considerations apply for pyrometric measurements made by electronic methods in the infrared region, as is common for studies of semiconductor materials, except that a specific emissivity will apply for the particular material and the wavelength of light employed.

## *References*


[1] *Metals Reference Book*, Seventh Edition, edited by C.J. Smithells and E.A. Brandes (Butterworths, London and Boston, 1991).

[2] F.G. Allen, J. Appl. Phys. **28,** 1510 (1957).

[3] T. Sato, Jap. J. of Appl. Phys. **6,** 339 (1967).

[4] H. Nishino, W. Yang, Z. Dohnálek, V.A. Ukraintsev, W.J. Choyke and J.T. Yates, Jr., J. Vac. Sci. Technol. **A15,** 182 (1997).

***Fig. 173. Pyrometric Measurement of Temperatures***

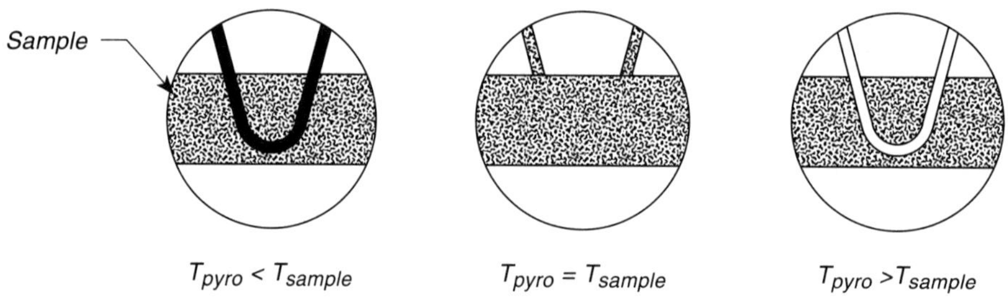


***A. Disappearing Filament Pyrometer - Visual***

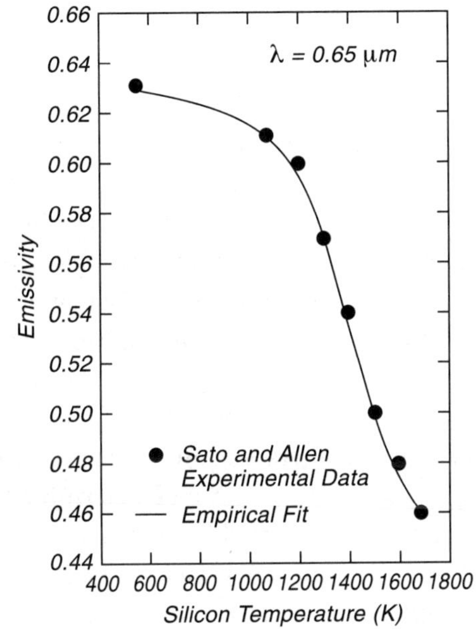


**B. Si Emissivity vs. Temperature**

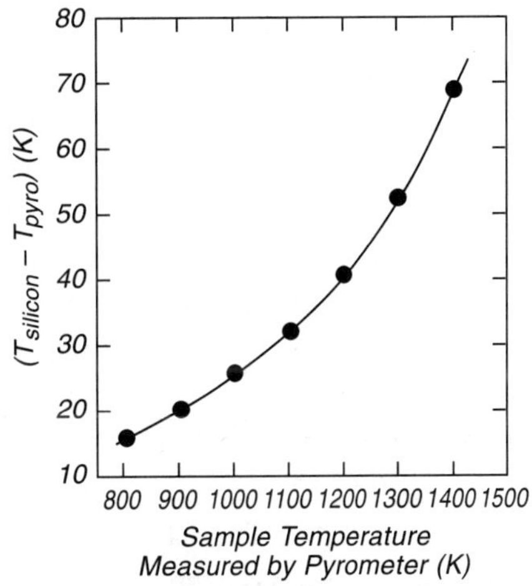


**C. Emissivity Corrections to Brightness Temperature**

# 174

# Pyrometer Measurements Through Windows Coated Inadvertently with Variable-Thickness Thin Films

The deposition of a thin film on an optical window can cause a negative error in optical pyrometer readings, which will increase with the thickness of the deposited film. In order to avoid this effect, an empirical method has been developed and tested against thermocouple readings on a hot sample [1].

Figure 174A shows the basic principle of the method. A movable ionization gauge having an emission regulated thermionic emitter is used as a standard. The temperature of the emitter should remain constant in good vacuum for a standard constant emission current that is set into the controller. When the window is first installed, and when it is free of any thin film deposit, the brightness temperature of the thermionic emitter is measured as a function of the emissivity setting on the pyrometer, and a curve like that of Figure 174B (clear) is constructed. As the window becomes coated with an inadvertent thin film deposit, the brightness temperature of the thermionic emitter filament versus the emissivity setting will decrease, as shown for various time periods in Figure 174B after the thin film has grown to various thicknesses. For any temperature, an emissivity setting returning the filament brightness temperature to the same value can be determined, as shown for example for a brightness temperature of 973 K (dotted line), where emissivity settings of 0.6, 0.5, and 0.3 are needed to bring the pyrometer back to the standard temperature of 973 K. This same emissivity setting is then used when the sample temperature is being studied in the position of the thermionic emitter.

The pyrometer must, of course, be calibrated accurately in the absence of any thin film deposit. Both thermocouple and phase transformation information may be used for this purpose, or in the case of materials with known emissivities, tables of emissivity versus temperature may be used to make the correction from brightness temperature to true temperature. If the sample temperature is considerably lower than the ion gauge filament temperature, it will be necessary to employ an interposed neutral density filter for the thermionic emitter measurements, as shown in Figure 174A. Because the thermionic emitter operates at about 1500 K, a neutral density filter is needed if measurements of sample temperatures considerably below this temperature are needed. In the example given, a neutral filter of optical density 2.0 ($10^{-2}$ transmission) is needed to reduce the apparent filament temperature to about 1000 K for an emissivity setting of 0.6.

Separate calibrations using a clean window that did not become coated with the thin film verified that this procedure was accurate and that the thermionic emitter temperature was constant over a long time [1]. In the case shown here, an IRCON Model 2000 pyrometer, working in the range 0.7 mm to 1.05 μm, was employed [1].

## *References*

[1] S.L. Wright, R.F. Marks, and W.I. Wang, J. Vac. Sci. Technol. **B4,** 505 (1986).

**_Fig. 174. Use of Standard Filament to Calibrate Pyrometer Through Film-Coated Windows_**

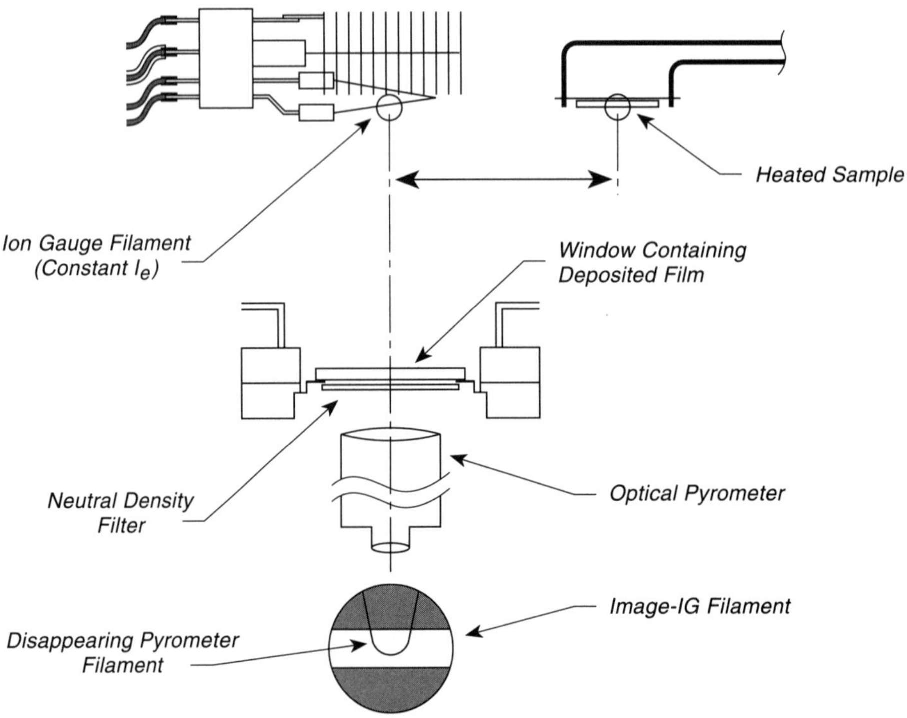


**_A. Ion Gauge Pyrometry Standard_**

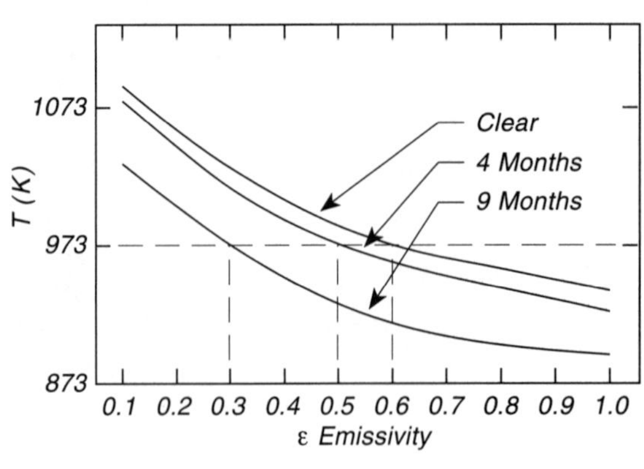


**_B. Typical Temperature vs. Emissivity Setting for Various Window Film Deposits_**

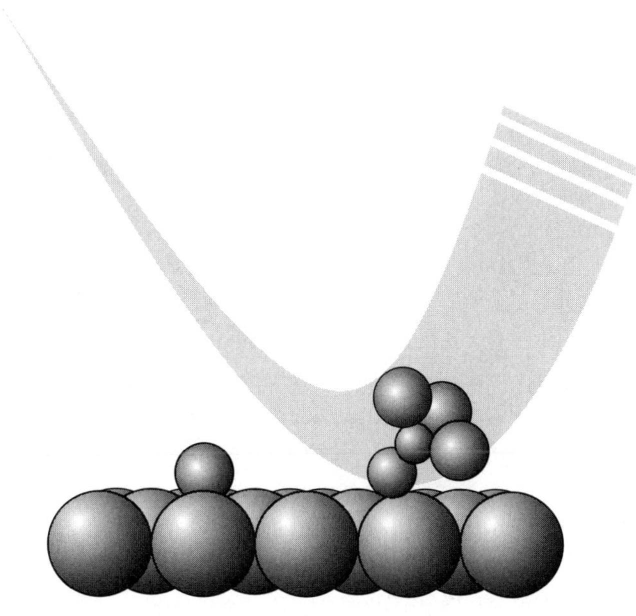

# 175

# Calibration Point for Optical Pyrometer Used for Silicon

The provision of a calibration point for silicon at high temperatures for optical pyrometers may be useful in certain situations. A method developed for doing this in rapid wafer thermal processors [1] can be extended to other situations in the laboratory.

In Figure 175, a silicon wafer, 0.3 mm thick, and covered by a 5000-Å-thick $SiO_2$ film, is coated with a 4000-Å Ge film, and then overcoated with a protective 1000-Å $Si_3N_4$ film. This sandwich structure serves as a useful calibration standard. The optical reflectivity of Ge at 0.6328 μm changes from 49% just before semiconducting Ge melts to 72% for melted Ge, which behaves with metallic properties. The melting point of Ge is 938°C, and melting can be viewed through the $Si_3N_4$ overlayer. As shown in Figure 175A, the sandwich test structure contains a K type thermocouple embedded in the backside of the silicon wafer, and an optical pyrometer to be calibrated at the melting point of Ge. A chopped laser beam is reflected off of the Ge sandwich face, and the reflectivity of the surface is measured as the temperature is raised, using external radiation for heating the wafer. Figure 175B shows the result. At a thermocouple temperature of 938°C, a sharp increase in the reflectivity is observed, followed by a gradual decrease as the temperature continues to rise. On slowly cooling the wafer, the freezing point of the Ge is observed to be depressed to about 904°C as a result of supercooling of the liquid Ge. The melting point of Ge was found to be reproducible to ±1° at 938°C in ten measurements. It was shown that the thermal gradient through the Si crystal was about 2.5°C/mm at the melting point of the Ge. These Ge sandwich structures can therefore be used periodically to calibrate the optical pyrometer emissivity setting (see page 588).

## *References*


[1] J.-M. Dilhac, C. Ganibal, N. Nolhier, and B. Rousset, Rev. Sci. Instrum. **63,** 188 (1992).

## Fig. 175. Pyrometer Calibration at Fixed Point for Silicon

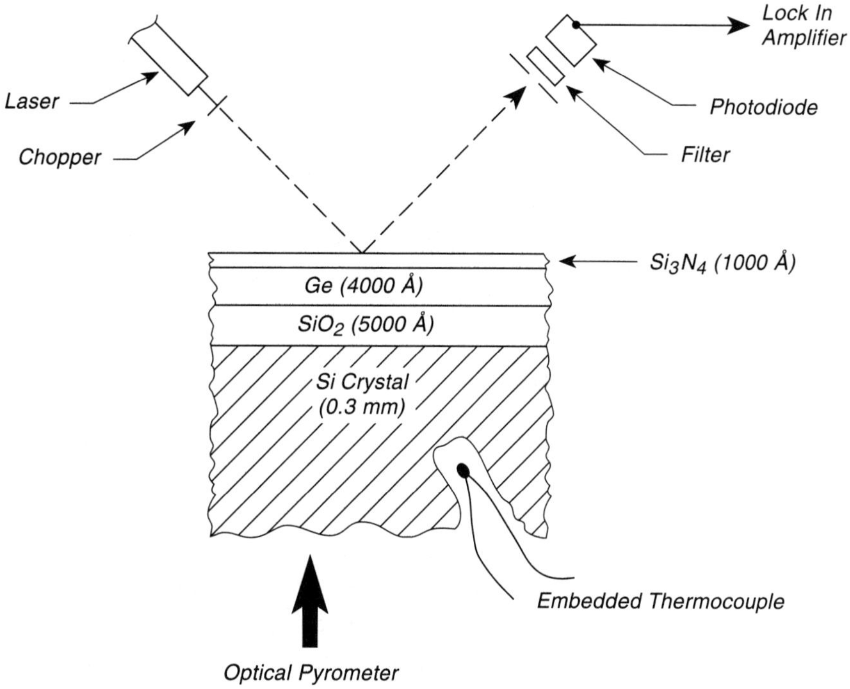


**A. Ge/Si Test Structure**

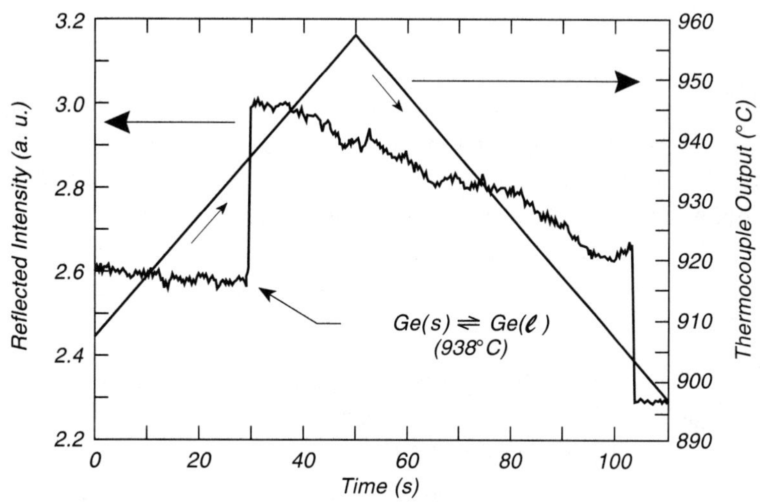


**B. Reflectivity Changes at 0.6328 μm**

# 176

# Control of Silicon Temperature Using Resistivity

Both thermocouples and pyrometers are used to measure and control the temperature of silicon crystals during electrical heating. As discussed elsewhere (page 790), the use of thermocouples for this purpose often produces metal contamination of the silicon crystal. A method is described here involving the measurement of the electrical resistance of the crystal during heating and cooling in order to determine and control the temperature. For purposes of evaluation of the resistance method, the crystal used in these measurements contains a W/26%Re:W/5%Re thermocouple which has a ball end buried inside a laser-drilled hole in the crystal and held in place with a MgO cement (Aremco 571P) [1].

The crystal is mounted on a Ta holder as shown in Figure 146A, page 510. The crystal dimensions are 9.5 x 3 x 0.3 $mm^3$. The silicon is P type, 100 Ω cm, boron doped. The resistance of the crystal was calculated using voltage and current measurements made during heating. Comparisons to the thermocouple and to an optical and an IR pyrometer are described in the original paper [1]. Figure 176A shows the dependence of the logarithm of the measured resistance versus $1/T$. Above about 600 K, the expected linear relationship is observed, where the dominant intrinsic conductivity of the silicon exists. Below 600 K, the dominating conductivity is due to the dopants for this particular crystal. The slope in the linear region corresponds to the quantity $E_G/2k_BT$, and the best value of $E_G$ (the silicon band gap) for this crystal was measured to be in the range 1.22–1.25 ± 0.05 eV, in good agreement with literature values of 1.170–1.205 eV [2–4].

Figure 176B shows the type of temperature control that may be achieved using the resistance and comparing to the temperature measured directly with the thermocouple. A programmed temperature ramp of 3 K/s, interrupted by a constant temperature period, and then a cooling ramp of -3 K/s are shown. The solid line represents the temperature measured with the thermocouple, and the dotted line with the silicon resistance. Good agreement is found with a maximum deviation of 20 K at 1150 K. Control was carried out using LabVIEW [5], which is described on page 498.

The method must be calibrated against a standard at one or more points. This may be done best using a pyrometer and comparing the resistance with the corrected pyrometer reading (see page 588). It would be better to do the calibration against the pyrometer at several points in the intrinsic region in order to make the best calibration, although the knowledge of the bandgap permits single point calibration [1].

## *References*

[1] Z. Dohnálek, S. Mezhenny, I. Lyubinetsky, W.J. Choyke, and J.T. Yates, Jr., J. Vac. Sci. Technol. **A15,** 2766 (1997).
[2] W. Bludau, J. Appl. Phys. **45,** 1846 (1974).
[3] K.L. Shaklee and R.E. Nahory, Phys. Rev. Lett. **24,** 942 (1970).
[4] H.D. Barber, Solid State Electron. **10,** 1039 (1967).
[5] LabVIEW is sold by Analog Devices, One Technology Way, Norwood, MA 02062-9106.

***Fig. 176. Control of Silicon Temperature Using Resistivity***

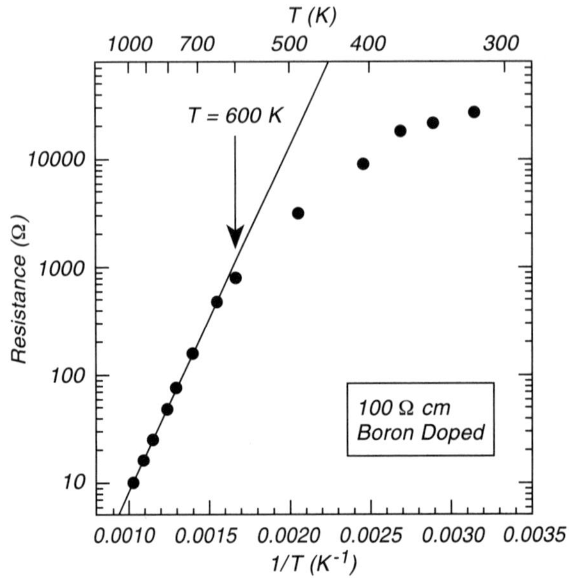


***A. Dependence of Log R vs. Inverse Temperature of Si(100) Crystal***

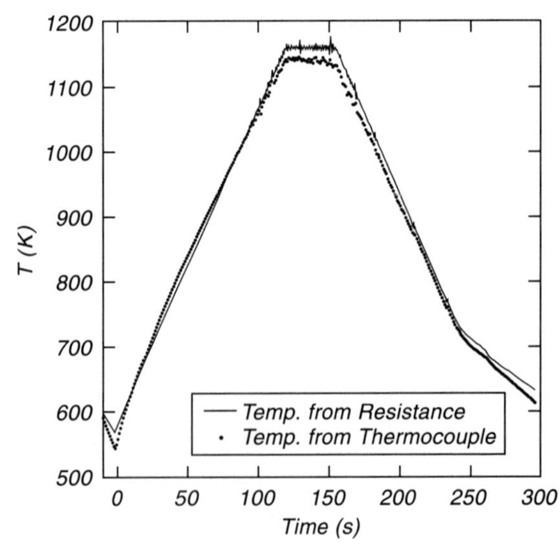


***B. Temperature Programmed Heating and Cooling of Si(100)***

# V. Delivery of Adsorbates to Surfaces


A. Gases . . . . . 603
B. Evaporation Sources . . . . . 627
C. Ions . . . . . 671
D. Active Gases and Species . . . . . 681
E. Electrochemical Sources . . . . . 709
F. Gas Purification and Permeation . . . . . 717

# A. Gases

# 177

# Design and Performance of Microcapillary Array Beam Doser

Beam dosers for quantitative exposure of surfaces to gases have become widely used in surface science research. This is because of the disadvantages in filling an ultrahigh vacuum system with gas in order to study adsorption processes. These disadvantages relate to the large gas loads needed for system dosing as compared to beam dosing. Such large gas loads result in adsorption on the walls of the vacuum chamber, which then contributes to slow achievement of base pressure after adsorption by the sample surface is complete. In addition, large gas loads in a vacuum system can cause displacement effects from the system walls leading to impurity adsorption, as well as regurgitation effects from ion pumps in which impurity gases are produced (see page 118). Also, the use of quantitative and absolute calibration procedures for the gas flux coming from a collimated doser is much superior to dosing with a system dose using an ionization gauge or a mass spectrometer, both of which have serious problems with absolute calibration.

Figure 177A shows a design for a microcapillary array beam doser [1]. A pinhole aperture made of stainless steel serves as a deformable gasket inside a Cajon vacuum fitting [2]. Typical aperture diameters are about 5 μm for convenient dosing fluxes for adsorption studies, as shown on page 610, where calibration methods are described. The flux of gas issuing from the pinhole aperture travels into the doser body, where it strikes a baffle made of metal shim stock, held into place by springy arms that bend against the inside wall of the doser body. This baffle randomizes the gas molecules' motion inside the doser body and behind the microcapillary collimator plate. Collimation of the gas occurs in the high-density capillary array where parallel capillaries of 10 μm diameter and 500–1000 μm length are arranged in a hexagonal close packing arrangement [3]. The microcapillary array disk has a radius, $R_S$, and should be purchased with the capillary axes perpendicular to the array surface, rather than inclined as in the chevron design commonly manufactured [3]. The capillary array is held in place by a metal cap that, with a clearance of several μm, confines the edge of the array in a close-fitting groove on the end of the doser body. Care should be taken not to compress mechanically the very fragile glass capillary array during installation. The capillary array has a relatively high conductance for gases and acts mainly as a collimator in the gas doser.

Figure 177B shows the intercepted flux fraction from the array by the target of radius $R_T$ for various geometries [4]. The most uniform dosing over the target will take place with larger values of $R_S/R_T$, but of course the fraction of the beam intercepted will decrease as shown in Figure 177B. The assumptions used in deriving the functions shown in Figure 177B assume that diffuse scattering occurs on the inside walls of the capillaries under molecular flow conditions where the mean free path is considerably longer than the diameter of the capillaries.

If the capillary array is too close to the target, reflected molecules from the target will have a finite probability of incidence on the crystal for a second encounter when they are reflected back by the capillary array surface.

There are precautionary measures to be considered with glass microcapillary array dosers. They are made of soft glass and therefore contain Si—OH groups on their surface. The hydrogen atoms are labile and will exchange with –OD groups in $D_2O$ or in organic molecules such as $CH_3OD$ [5]. In this case, isotopic exchange overnight can be carried out until the isotopic purity of the transmitted gas is not influenced by isotopic exchange on the inner walls of the capillaries. In addition, reactions of some molecules such as fluorinated ethers with acid sites on the silica surface have been reported [6]. Finally, certain unstable molecules such as $Fe(CO)_5$ have been shown to be reactive with the glass surface [7].

Recently, less finely packed array dosers made of metal have been used with success for uniform gas dosing in directions off the crystal normal [8].

For reactive gases, the use of an inert diluent gas has been described in reference [9] and also on page 138.

Historically, collimated array gas dosers have been used since the 1970s, and some examples are given in references [10–14].

A recent theoretical treatment of the transport of gases through microcapillary array dosers and other tubular devices is found in reference [15], which also contains a bibliography of earlier papers on this subject. A recent analysis of the flux distribution from a tube doser and a capillary array doser has been reported [16].

## *References*


[1] M. Bozack, L. Muehlhoff, J.N. Russell, Jr., W.J. Choyke, and J.T. Yates, Jr., J. Vac. Sci. Technol. **A5,** 1 (1987).

[2] P.L. Hagans, B.M. DeKoven and J.L. Womack, J. Vac. Sci. Technol. A7, 3375 (1989). Laser-drilled micron-size pinholes are produced in disks fitting VCR fittings by Precision Laser Services, 314 E. Wallace St., Ft. Wayne, IN 46803.

[3] Capillary array blanks, made without being coated with active surfaces for use as electron multipliers, are available from Galileo Electro-Optics Corporation, Galileo Park, P.O. Box 550, Sturbridge, MA 01518. Sizes up to 25-mm diameter are available.

[4] A. Winkler and J.T. Yates, Jr., J. Vac. Sci. and Technol. **A6,** 2929 (1988); also see C.T. Campbell and S.M. Valone, J. Vac. Sci. and Technol. **A3,** 408 (1985).

[5] S.M. Gates, J.N. Russell, Jr., and J.T. Yates, Jr., Surf. Sci. **146,** 199 (1984).

[6] P.K. Leavitt and P.A. Thiel, J. Vac. Sci. Technol. **A8,** 148 (1990).

[7] M. Henderson, R.D. Ramsier, and J.T. Yates, Jr., J. Vac. Sci. Technol. **A9,** 2785 (1991).

[8] A.M. Glines, R.N. Carter, and A.B. Anton, Rev. Sci. Instrum. **63,** 1826 (1992).

[9] F.C. Henn, M.E. Bussell, and C.T. Campbell, J. Vac. Sci. Technol. **A9,** 10 (1991).
[10] D.W. Goodman, T.E. Madey, M. Ono, and J.T. Yates, Jr., J. Catalysis **50,** 279 (1977).
[11] D.W. Goodman, J.T. Yates, Jr., and T.E. Madey, Chem. Phys. Lett. **53,** 479 (1978).
[12] S.M. Gates, J.N. Russell, Jr., and J.T. Yates, Jr., Surface Sci. **159,** 233 (1985).
[13] D.E. Ibbotson, T.S. Wittrig, and W.H. Weinberg, Surface Sci. **110,** 294 (1981).
[14] J.E. Deffeyes, A.H. Smith, and P.C. Stair, Surface Sci. **163,** 79 (1985).
[15] D.M. Murphy, J. Vac. Sci. Technol. **A7,** 3075 (1989).
[16] D.E. Kuhl and R.G. Tobin, Rev. Sci. Instrum. **66,** 3016 (1995).

## Fig. 177. Design and Performance of Microcapillary Array Beam Doser

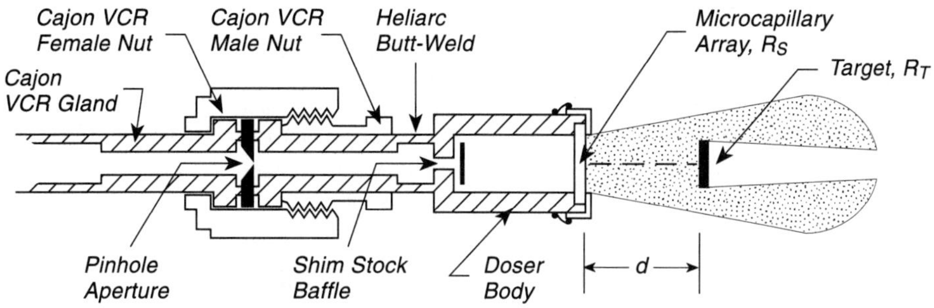


**A. Design of Doser (R = radius)**

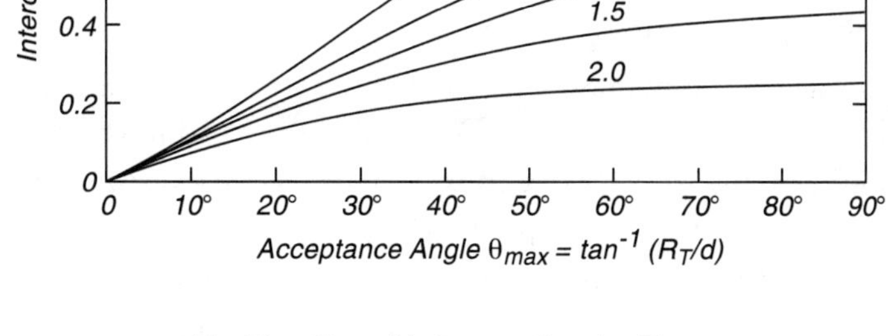


**B. Fractional Interception by Target**

# 178

# Gas Handling System for Array Beam Doser

The effective use of the microcapillary array beam doser shown on page 177 requires a gas-handling system of appropriate design, as shown in Figure 178 [1]. A stainless steel system, pumped by a turbopump, is designed to attach to several glass storage gas bulbs, using Cajon VCR fittings [1].

The storage of gases in glass bulbs at pressures near 1 atm prevents accidental overpressurizing of the gas-handling system by compressed gases in cylinders. Also, during gas transfer to prebaked storage bulbs (for removal of adsorbed $H_2O$ from the glass walls), one has the opportunity to use purification steps for gases such as the freeze-pump-thaw method commonly used to remove noncondensable impurities.

For calibration of the gas doser system, a test gas, such as $N_2$, is admitted to the gas-handling system to a few Torr pressure. The ballast bulb, of exactly known volume, is also filled with this gas. As will be shown on page 610, the conductance of the pinhole aperture may be measured by measuring the time rate of change of pressure in the gas-handling system as the $N_2$ is slowly pumped through the pinhole aperture. To interrupt the flow from the array doser, Valve-1 is closed, and the region behind the doser is rapidly evacuated through Valve-2, keeping the gate valve to the UHV system closed.

Stainless steel systems such as this can be baked with heating tape to 150–200°C while pumping with the turbopump. Because of the low pumping speed through the narrow tubing and the relatively high internal surface area compared to the volume, limiting pressures in the $10^{-8}$ Torr range are usually measured. For gases added to the Torr level, this background gas represents much less than 1 ppm of added impurity, which is usually superior to the initial purity of the gas being studied.

The internal surface of the stainless steel gas-handling system has a significant capacity to react with active gases such as $Cl_2$. Gases of this sort may often be used themselves to passify the stainless walls. This is done by storing the active gas at pressures of several Torr pressure overnight, and then pumping the gas away before adding fresh gas to the system [2]. It may be necessary to do this passivation treatment in several cycles. In other cases, treatment of the gas-handling system with oxygen gas to passivate the inner walls by oxidation has been found to be effective [3].

## *References*


[1] M.J. Bozack, L. Muehlhoff, J.N. Russell, Jr., W.J. Choyke, and J.T. Yates, Jr., J. Vac. Sci. Technol. **A5**, 1 (1987).

[2] A.L. Linsebigler, V.S. Smentkowski, M.D. Ellison, and J.T. Yates, Jr., J. Amer. Chem. Soc. **114**, 465 (1992).

[3] M.A. Henderson, R.D. Ramsier, and J.T. Yates, Jr., J. Vac. Sci. and Technol. **A9**, 2785 (1991).

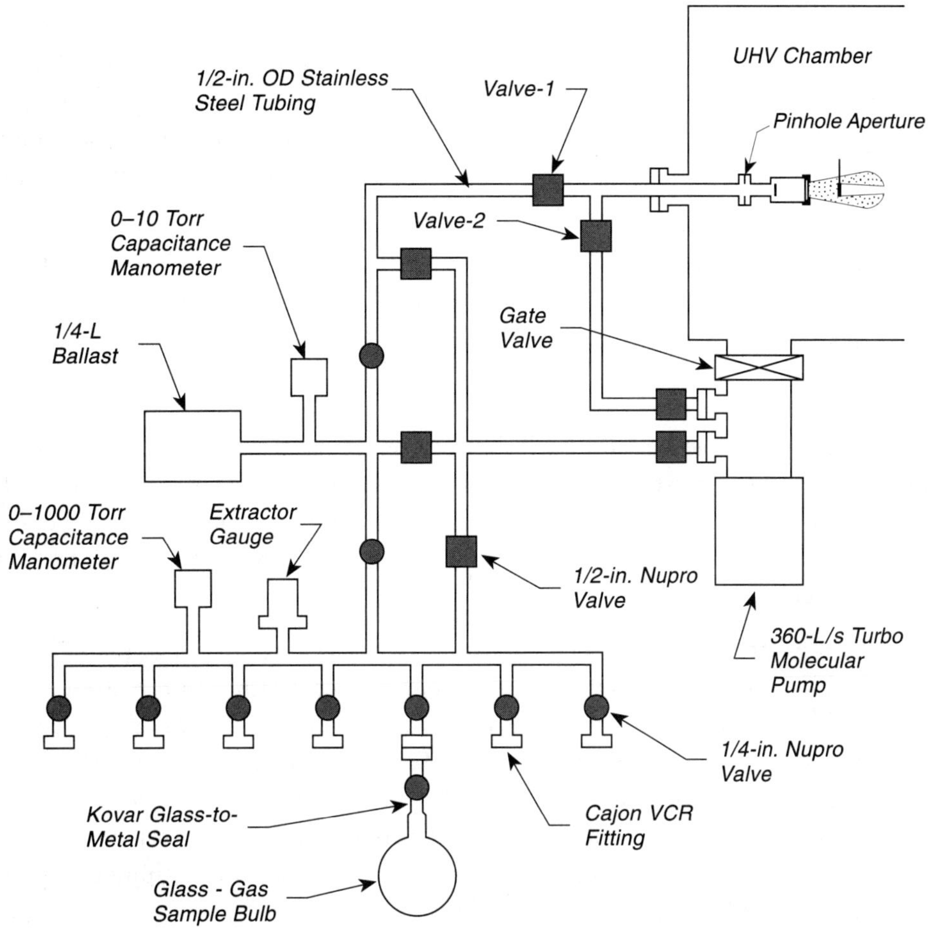


***Fig. 178. Typical Gas Handling System—Beam Doser***

# 179

# Calibration and Use of Array Beam Doser

Figure 179A shows a typical design for an ultrahigh vacuum system equipped to measure the kinetics of adsorption of a gas on an atomically clean crystal surface. A quadrupole mass spectrometer (QMS detector), shielded to receive only a random flux of background gases, and the microcapillary array doser are the principal components of the apparatus for this measurement.

The calibration of the doser was performed with $N_2$ over a period of about $10^5$ s, as shown in Figure 179B. The data, plotted logarithmically, against time, *t*, provide a linear curve whose slope is related to the conductance of the pinhole by Equation (1).

$$dn/Pdt = -V/kT[d\,(ln\,P/P_0)/dt] \tag{1}$$

where *n* number of molecules, *P* = backing pressure, *V* = storage volume, *k* = Boltzmann's constant, and *T* = temperature in Kelvin. For the most accurate measurement of the drop in pressure, the temperature-sensitive capacitance manometer in the gas storage region (see page 608) must be thermally shielded to avoid temperature changes larger than a few tenths of a Kelvin.

Usually measurements of this type have a relative accuracy of better than 5% when gases of various molecular weights are compared to each other in the doser [1]. For molecular flow conditions, where the mean free path of the gas on the high-pressure side of the pinhole is larger than the diameter of the pinhole, the rate of effusion through the pinhole will be proportional to $M^{-1/2}$, where $M$ = the molecular mass of the gas. This condition continues to be approximately true as one enters the transition region in gas density, at higher pressures, where viscous flow is beginning to occur in the pinhole aperture which has finite wall length.

Figure 179C shows measurements of the adsorption of $O_2$ on a Fe(110) surface at 90 K [2]. For this measurement, the shutter is placed between the doser and the crystal, and a steady base pressure of the order of $1 \times 10^{-10}$ Torr or less is established, using the mass spectrometer tuned in this case to 32 amu. After steady conditions are achieved, the doser is turned on with the shutter still in place and the pressure measured by the mass spectrometer rises. Ideally the pressure should level out prior to opening the shutter to allow adsorption from the beam to occur, but since a small flux of random gas scattered from the back of the shutter is incident on the crystal, causing a slow increase in coverage, the waiting time is limited and a compromise must sometimes be reached in which "almost steady state" conditions for $P_{back}$ are established, as shown here. At this point, the shutter is opened, and adsorption occurs until asymptotic behavior of the scattered flux of gas is observed. Then, when the sticking probability has fallen to a low value, placing the shutter into the beam causes no discontinuity to be observed in the mass spectrometer signal.

The data shown in Figure 179C can be converted into sticking coefficient vs. absolute coverage, as shown in Figure 179D, if the fractional interception of the doser beam is calculated (see page 604) (or measured [1]). It is of interest to note that the entire measurement of sticking probability and absolute coverage are determined solely by the accuracy of the flux calibration and the accuracy of the fractional interception of the collimated gas beam by the crystal [3,4]. The mass spectrometer does not have to be calibrated in any absolute sense.

This method is based upon earlier work with variations on the features of the apparatus and the measuring conditions [5–8].

## *References*

[1] C.C. Cheng, R.M. Wallace, P.A. Taylor, W.J. Choyke, and J.T. Yates, Jr., J. Appl. Phys. **67,** 3693 (1990).
[2] V.S. Smentkowski and J.T. Yates, Jr., J. Vac. Sci. Technol. A7, 3325 (1989); Surf. Sci. **232,** 113 (1990).
[3] A. Winkler and J.T. Yates, Jr., J. Vac. Sci. Technol. **A6,** 2929 (1988).
[4] C.T. Campbell and S.M. Valone, J. Vac. Sci. Technol. **13,** 408 (1985).
[5] D.A. King and M.B. Wells, Surf. Sci. **29,** 454 (1972).
[6] T.E. Madey, Surf. Sci. **33,** 355 (1972).
[7] A.A. Bell and R. Gomer, J. Chem. Phys. **44,** 1065 (1966).
[8] C. Wang and R. Gomer, Surf. Sci. **84,** 329 (1979).

## *Fig. 179. Use of Beam Doser for Kinetics and Coverage Measurements*

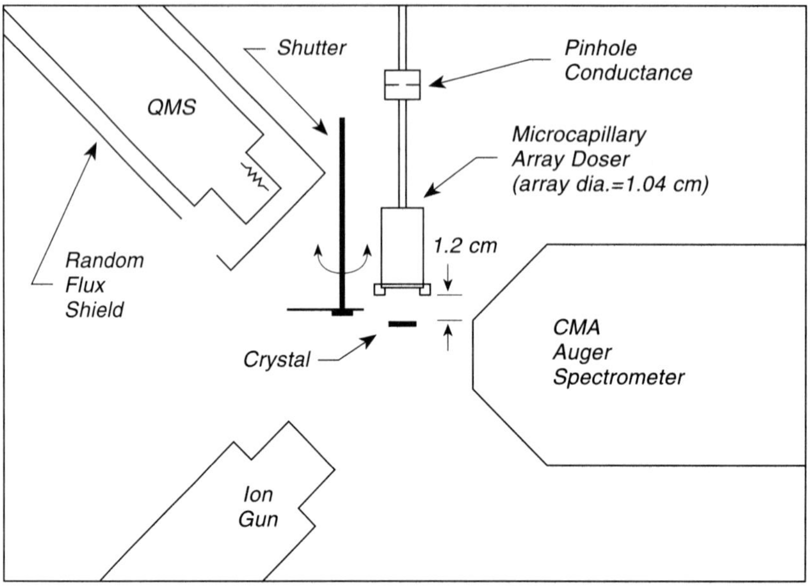


***A. Typical Apparatus - Absolute Coverage and Sticking Probability Measurement***

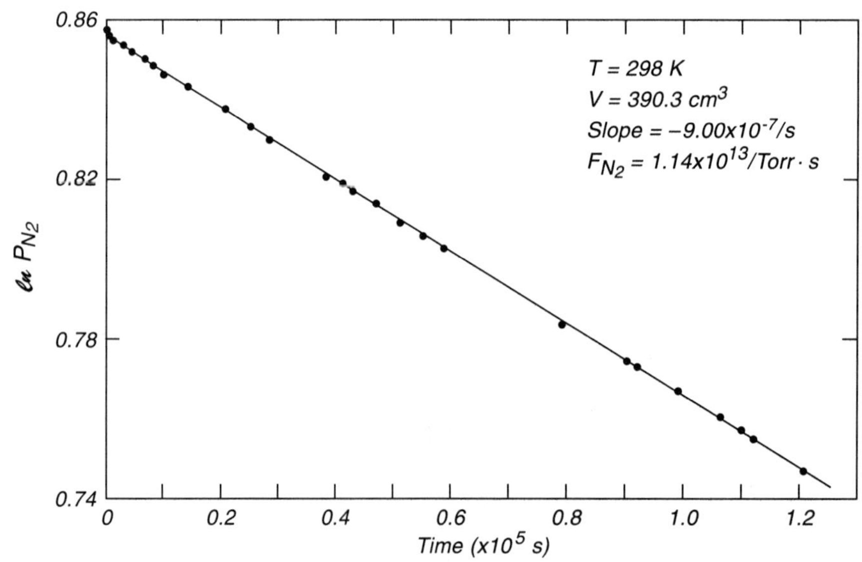


***B. Typical Calibration Curve - Pinhole Conductance***

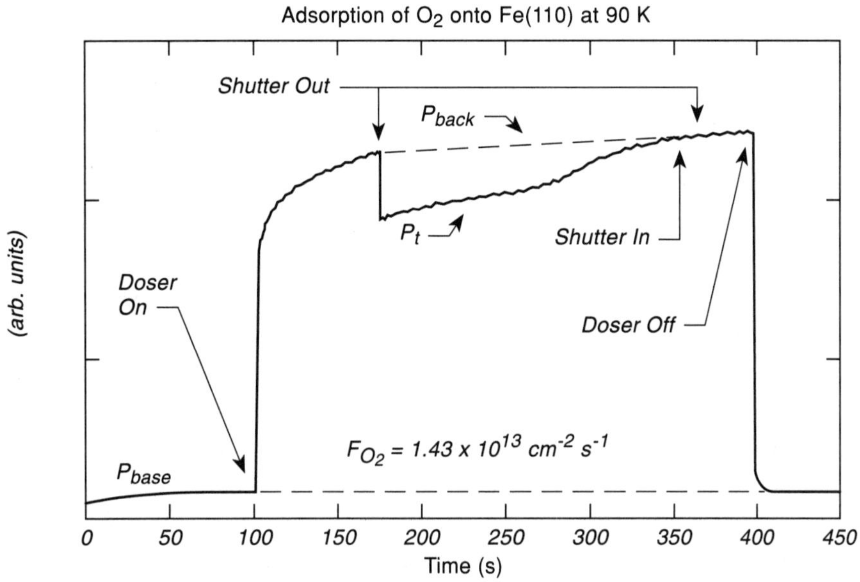


**C. Adsorption Uptake Curve**

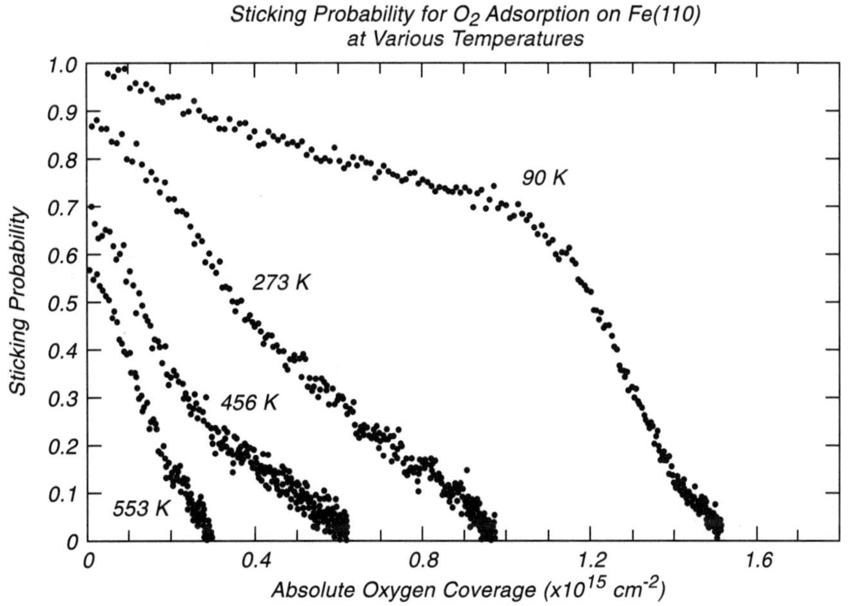


**D. Sticking Probability vs. Absolute Coverage**

# 180

# Beam Doser Shutoff—Condensable Gases

Very simple measures can sometimes be used to interrupt the flow of gases from a gas doser. As indicated on page 620, the use of a small deadspace valve of special design is helpful in defining reproducible and absolute doses of gases into a collimated gas doser system. In some cases, where gases condensable in liquid nitrogen are being used with a gas doser, it is possible to interrupt the flow rapidly by using a cold trap as a pump. In Figure 180, a cold trap made of Pyrex glass is immersed in a Dewar flask containing liquid nitrogen [1]. This cold trap is a very rapid pump for all gases that are condensable at 77 K. Flowing gas through the doser (see page 608) may be turned off rapidly by closing the admission valve and then quickly opening the dump valve. This lowers the gas pressure very rapidly behind the conductance limiting aperture, and turns the doser off. Many doses of gas, and different condensable gases, may be trapped without cleaning the collector tube. For cleaning, the tube is disconnected from the doser line using the VCR fitting, cleaned with solvents, baked briefly in an oven, reconnected, and evacuated using the pump in the gas-handling system. This method is actually preferable to pumping corrosive gases from the trap, after each dose, through a turbopump because of possible failure of the turbopump after exposure over time to corrosive conditions.

## *References*

[1] Author's personal experience.

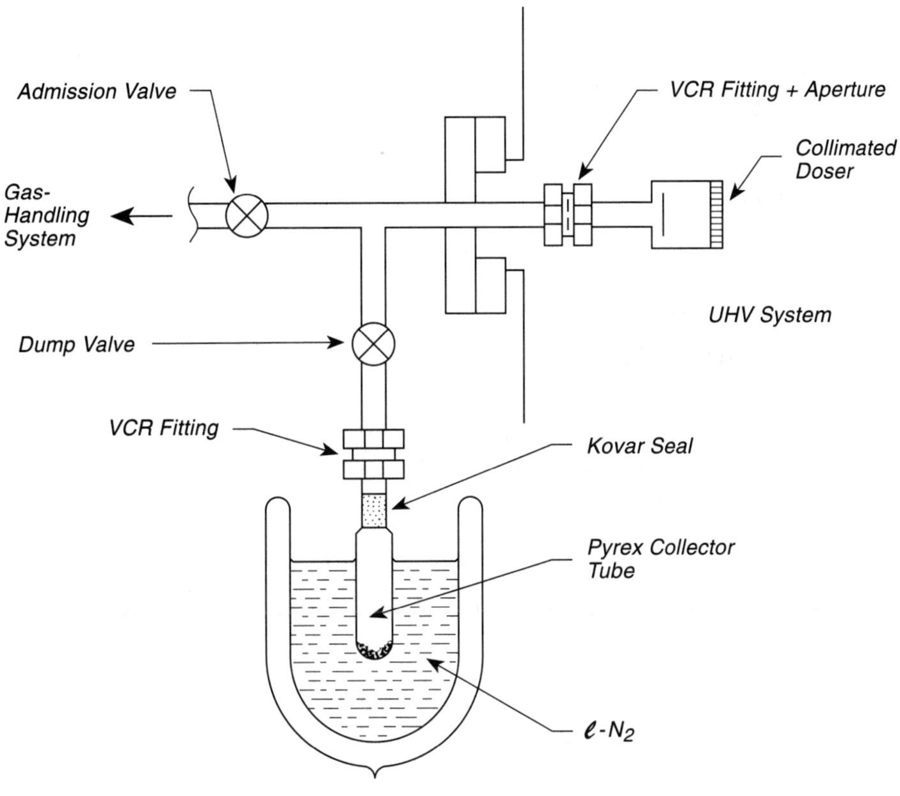


***Fig. 180. Beam Shutoff—Condensable Gases***

# 181

# Neutral Beam Intensity Measurement Using a Stagnation Detector

A stagnation detector may be used to determine the absolute intensity of a molecular beam. In Figure 181, a portion of the beam (having a diameter larger than the orifice), enters a chamber containing an ionization gauge and a calibrated leak. After suitable corrections, the signal indicated by the ionization gauge can be related to the flux of gas in the beam [1].

Under conditions where the beam is shut off, the steady-state pressure, $P_s$, is

$$P_s = Q_o / C_{\text{orf}} \tag{1}$$

where $Q_o$ is the net outgassing rate and $C_{\text{orf}}$ is the conductance of the orifice.

When the valve to the calibrated leak is opened, the pressure increase will be

$$\Delta P_{\text{leak}} = Q_L / C_{\text{orf}} \tag{2}$$

where $Q_L$ is the calibrated leak rate. The conductance of the orifice, which should be sharp-edged is, under molecular flow conditions

$$C_{\text{orf}} = 2.86\ (T/M)^{1/2}\ D^2 \quad \text{L/s} \tag{3}$$

where $D$ is the orifice diameter in cm, $T$ is the absolute temperature, and $M$ is the molecular mass of the gas.

When the molecular beam passes through the calibrated orifice it will cause a pressure rise in the stagnation chamber given by

$$\Delta P_{\text{beam}} = Q_B / C_{\text{orf}} \tag{4}$$

The impingement rate, $I_B$, is related to $\Delta P_{\text{beam}}$ according to Equation (5),

$$I_B = Q_B N_L / \pi D^2 = 0.91\ N_L (T/M)^{1/2}\ \Delta P_{\text{beam}} \tag{5}$$

where $N_L = 3.237 \times 10^{19}$ molecules/Torr-L at 298.15 K.

It is important that the beam diameter be larger than the orifice and that the baffle inside the stagnation detector be arranged to prevent direct entry of the gas from the calibrated leak into the ionization gauge. For reactive gases such as oxygen, the operating temperature of the ionization gauge filament and the emission current should be reduced as much as possible to prevent pumping effects.

The intensity at a sample position that differs from the orifice position is related to the intensity at the orifice by the inverse square law for the distances involved.

A good article detailing the many experimental aspects of the use of supersonic molecular beams in surface chemistry studies may be found in reference [2].

## References


[1] W.G. Dorfeld, J.B. Hudson, and R. Zuhr, Surf. Sci. 57, 460 (1976).

[2] S.T. Ceyer, D.J. Gladstone, M. McGonigal, and M.T. Schulberg in *Investigations of Surfaces and Interfaces—Part A, Physical Methods of Chemistry Series,* 2nd Ed., Vol. IXA, edited by B.W. Rossiter and R.C. Baetzold (Wiley, New York, 1993).

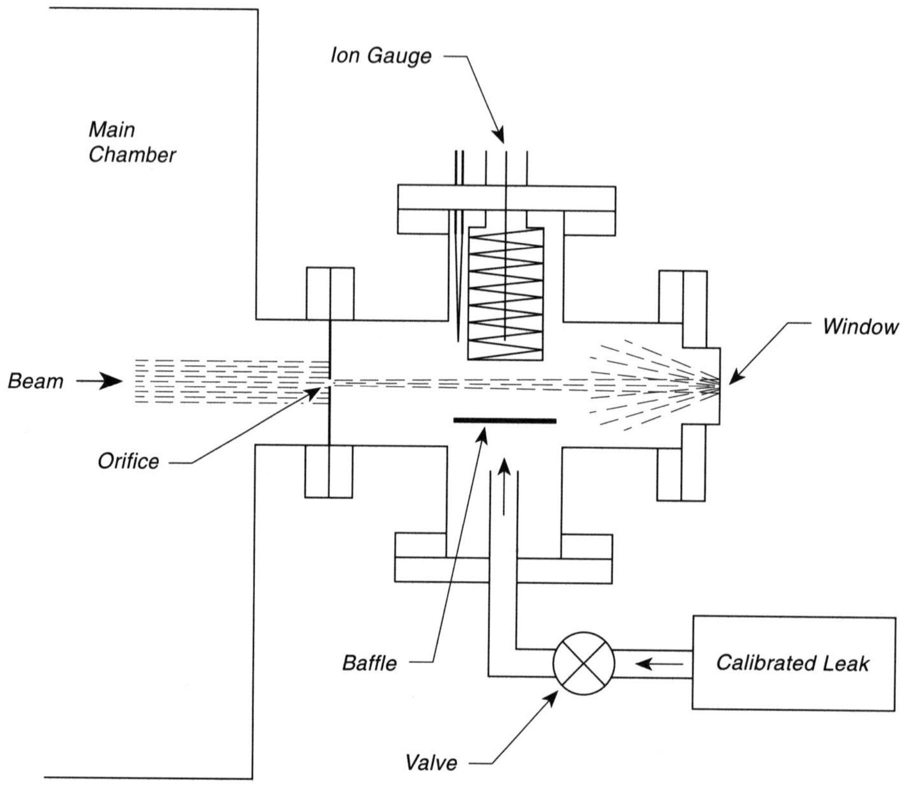


***Fig. 181. Stagnation Detector***

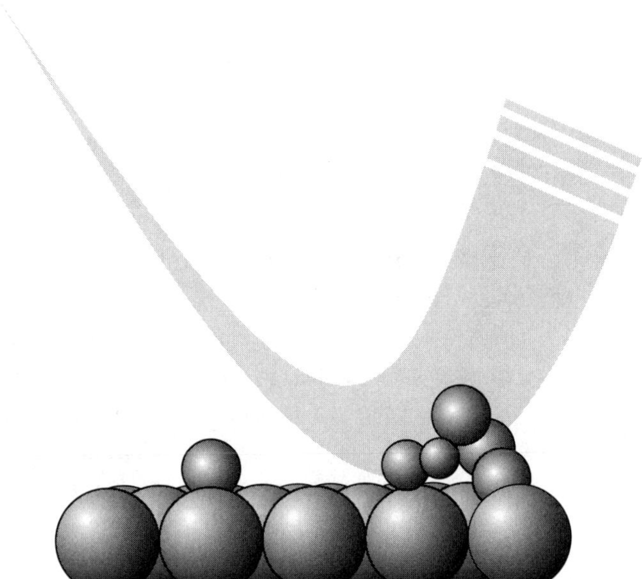

# 182

# Low Deadspace Aperture Valve for Repetitive Gas Dosing

The use of gas dosers for adsorption experiments is highly recommended in order to avoid impurity effects from wall displacement and ion pump regurgitation processes as well as to have quantitative gas dosing capabilities, as discussed on pages 604–615. In the conventional gas dosing design, a rather large volume of gas is held in a storage section of the gas line, and to interrupt dosing, one must pump out this gas or some fraction of it held in a section of the gas storage line. For repetitive gas dosing, this is inconvenient, and in addition for expensive gases it is wasteful. The design shown here [1] avoids this effect by minimizing the deadspace volume so that only a few seconds of gas dosing occurs following closure of the valve. In addition, when a conductance-limiting aperture at the outlet to the small deadspace volume is used, quantitative gas dosing can be done when the valve is opened.

Figure 182A shows a view of the aperture valve assembly. The valve body of a stainless steel diaphragm ultrahigh vacuum valve (Nupro DL Series) is drilled and tapped with 1/16 NPT pipe threads to receive a stainless steel rod also threaded on the upper end. This is done at the correct angle inside the inlet side of the valve. This rod has a central hole, 2 mm in width, bored its length, except for an 0.003-in.-thick membrane at the inner end. The membrane is then laser-drilled with a nominal 60-μm-diameter conical hole [2]. The upper end of the insert contains a screwdriver slot for convenient insertion into the valve; Teflon tape is used to make a gas tight seal. The inlet to the valved assembly (originally the outlet) is connected to a conventional gas-handling system, containing pumping capabilities, a capacitance manometer, and a ballast volume, *V*, used for conductance calibration of the laser-drilled aperture. The outlet of the valve assembly is connected to a small flange that connects to a collimating gas doser (see page 604) inside the ultrahigh vacuum system. With this arrangement, the deadspace when the valve is closed is about 0.2 $cm^3$. Thus, when the dosing is completed, and the valve is closed, evacuation of the deadspace into the ultrahigh vacuum system by way of the doser does not involve the flow of large amounts of gas, even when the backing pressure is near 0.1 Torr. Under these conditions, the number of molecules in the deadspace is about $4 \times 10^{14}$. If 20% of these molecules strike the crystal and the sticking coefficient is 0.1, about 2% of a monolayer will adsorb from the exhaust of the deadspace after the valve is closed.

Figure 182B shows a comparison of the pumping speed through the opened valve for both Ar and $H_2$, and may be used to measure the conductance of the laser-drilled aperture. It is noted that the effusion rate follows a semilogarithmic behavior from backing pressures of 10 Torr and lower, indicating that slight departures from molecular flow conditions do not sensibly disturb the conductance of the aperture from that predicted by the kinetic theory of gases for molecular flow. The superposition of points for the two gases on the curve gives further confidence in the notion that the behavior can be predicted from the molecular flow equations.

Figure 182C shows the response of a mass spectrometer in the vacuum system as the valve is opened and closed for one cycle. A total rise time of about 3 seconds is seen as the valve is opened. Constant Ar flux occurs beyond this time until the valve is closed, where the pumping out of the deadspace volume exhibits a time constant of 1.3 s. Even though these on-and-off periods give relatively slow response times, the doser may still be calibrated to give reproducible doses containing the on-off periods as well as the constant flux period. The use of a smaller laser-drilled aperture is possible, but the time constant for pumping will increase to unacceptable levels. A major feature of the apparatus is that the unit can deliver rather large fluxes at fairly low backing pressures and that it can be quickly cut off without wasting gas. Multiple exposures that are reproducible can be achieved.

A variation on this idea, involving a trapping volume of about 0.4 $cm^3$, has been employed to dump known quantities of gas onto a crystal by way of an effusive doser [3]. The trapping volume contains about $1.4 \times 10^{16}$ molecules/Torr; and the repeatability of the dose, using a capacitance manometer for pressure measurements in the gas line feeding the doser volume, is estimated to be 0.5% standard deviation.

## *References*


[1] S.M. Cohen and M.P. D'Evelyn, J. Vac. Sci. Technol. **A9**, 2414 (1991).

[2] Lenox Laser, 1 Green Glade Ct., Phoenix, MD 21131.

[3] Dr. V.A. Ukrainsev, Texas Instruments, P.O. Box 655012, MS 461, Dallas, TX 75265 (private communication); see also Yu A. Gelman, B.S. Podol'skii, and V.A. Ukraintsev, Pribory I Tekhnika Experimenta, **6**, 191 (1990); I. Harrison, V.A. Ukraintsev, and A.N. Artsyukhovich, SPIE Proceedings, Laser Techniques for Surface Science, edited by H.-L. Dai and S.J. Sibener, **2125**, 285 (1994).

**Fig. 182. Low Deadspace Aperture Dosing Valve**

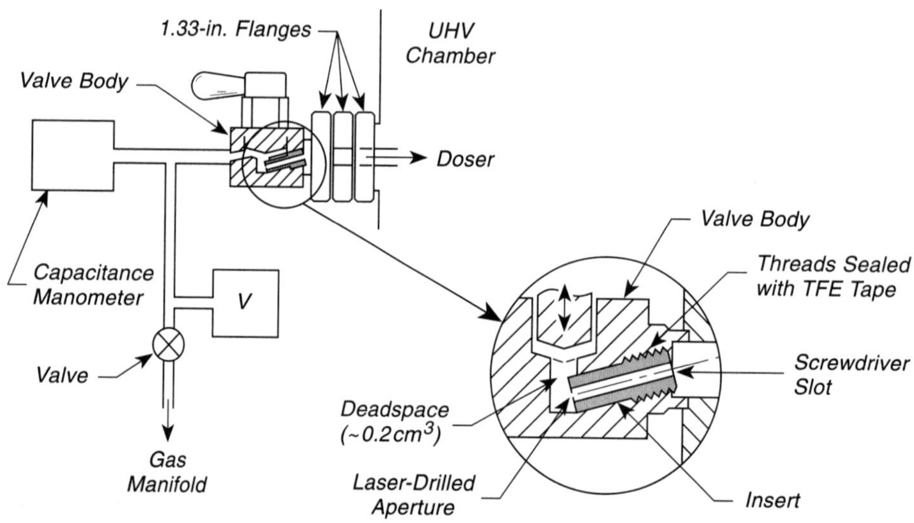


**A. Aperture Valve Assembly**

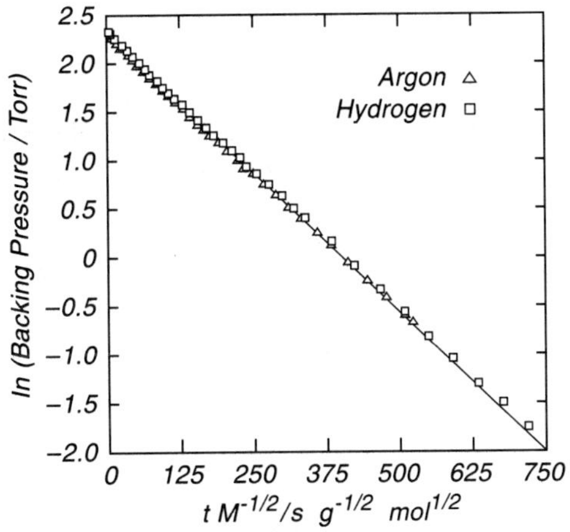


**B. Pumping Behavior—40 μm Aperture**

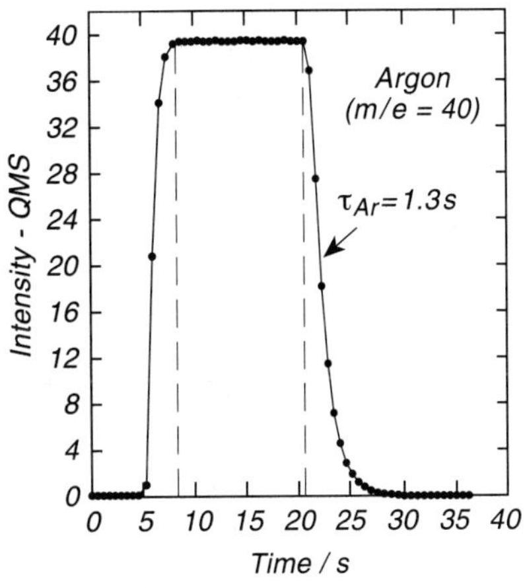


**C. Duty Cycle—Open and Close**

# 183

# Gas Flow Regulation—Squeeze Valve

A very simple method for regulating the flow of gases into an ultrahigh vacuum system with a minimum of contaminants involves the use of a brass squeezing device to restrict the flow through a thick-walled copper tube, as shown in Figure 183A [1]. The copper tubing (3 mm OD x 1 mm ID) is annealed in a flame, then cleaned internally with nitric acid and rinsed with water and alcohol. The brass squeeze device is clamped to the tube, and the flow rate may be adjusted over a wide range using the tightening screws. This action is reversible tens of times. The valve cannot be closed completely. To adjust the flow rate after the valve is set, the pressure in the high-pressure region is varied using the pressure gauge for measurement. To shut down the flow, the upper UHV valve is closed and the region between this valve and the squeeze valve is evacuated to the pump, as may be seen in Figure 183B.

## *References*


[1] Professor G. Scoles, Chemistry Department, Princeton University, Princeton, NJ 08544 (private communication). This method was devised at the FOM Laboratories in Amsterdam 40 years ago and is still in use.

## Fig. 183. Gas Flow Regulation—Squeeze Valve

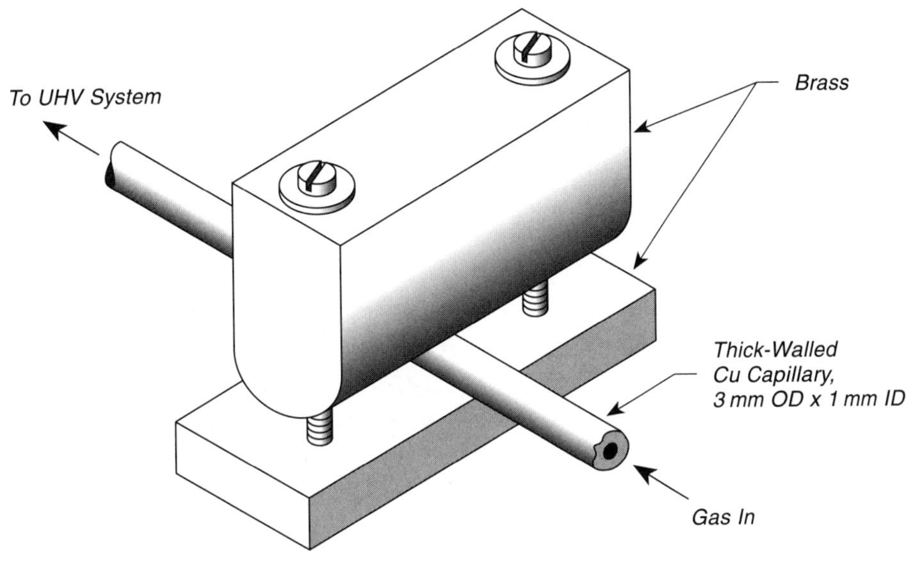


**A. Squeeze Valve**

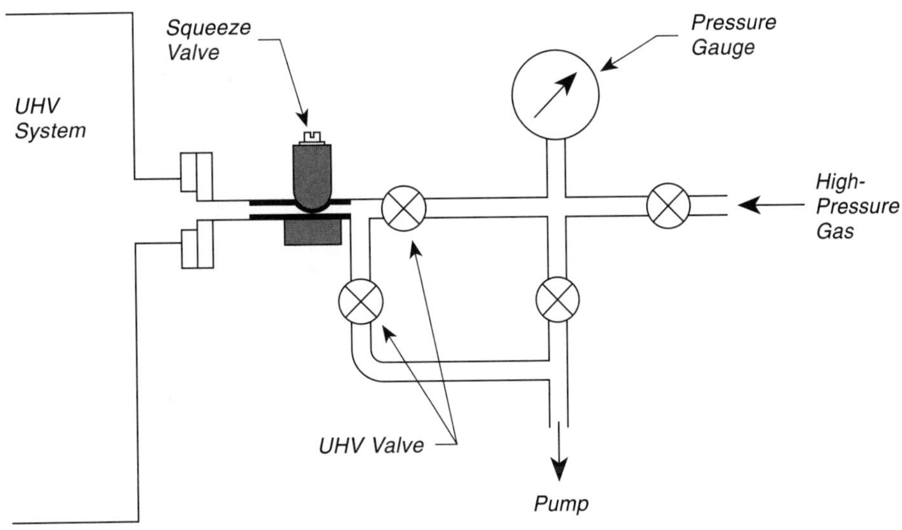


**B. Gas-Handling System**

# B. Evaporation Sources

# 184

# Degassing Evaporation Sources

The deposition of thin films by evaporation constitutes one of the main experimental procedures in surface and materials science. Indeed, the study of chemisorption on clean surfaces has as one of its origins the use of thin metal films as substrates in early studies done in the 1940s.

A major problem in making clean thin films originates from the outgassing of heated surfaces during evaporation, leading to contamination of the thin film by chemisorption from the background vacuum. This may be avoided by careful design in which prior outgassing is achieved for components that will be heated during evaporation. Figure 184 illustrates how this is done, using two principles.

The first principle involves the use of degassing loops to support all filaments that will be heated [1]. In the center of the evaporation shield, the evaporation loop, containing welded pieces of the evaporant at its apex, is mounted on two degassing loops. Knowing the current ($I_1$) required to bring the evaporation loop to its desired temperature for evaporation of the evaporant, one passes current $I_2$ through the degassing loops prior to the evaporation step. The current $I_2$ should be somewhat larger than $I_1/2$ so as to bring the degassing loops to a higher temperature than they will ever attain during evaporation, and the two loops should be outgassed separately so as to avoid passing current through the evaporation loop during outgassing.

The second principle illustrated in Figure 184 is the use of a shield- outgassing filament for heating the shield to temperatures higher than will be achieved while evaporation occurs from the evaporation loop [2]. This is achieved by radiation from the outer shield outgassing filament, where the shield thermocouple may be used to monitor the temperature achieved during outgassing compared to that during evaporation. Note that this shield outgassing filament is also mounted on degassing loops.

Schematic performance curves are shown in Figure 184 to illustrate that extremely rapid employment of a properly outgassed evaporation source is possible compared to a poorly designed source that may outgas over days prior to being useful for evaporation under good vacuum conditions. The shutter should not be opened until an appropriate limiting pressure is achieved during evaporation.

Some evaporation sources will be subject to continual outgassing because of design limitations preventing the use of methods shown here. In that case, reduction of the unwanted gas fluence at the substrate may be achieved by differentially pumping the source and the substrate chambers that are connected by an aperture [3].

The monitoring of the temperature of the evaporation filament may be achieved quite simply by the use of current and voltage drop measurements as described on page 584.

A helpful word about the construction of welded assemblies involving refractory wires such as tungsten and tantalum will be useful. When these wires are fabricated from approximately 0.3-mm-diameter wire for the purposes described here, it is often found that upon heating to high temperatures, the wires will distort geometrically. This distortion is caused by mechanical strain present initially in the welded assembly if the welds and the wire are placed under strain during attempts to make additional welds on the same filament. To avoid this problem, all wires should be formed by bending over gently curved mandrels, and the welding of the preformed filaments should be carried out in such a way that no strain is placed on a weld when subsequent welds are made and that no distortion of the preformed filament shape occurs during final assembly by welding.

## *References*

[1] The use of degassing loops originates from the thin film community but the historical origin of this useful method is not known.
[2] Professor R.J. Behm (personal communication).
[3] W.F. Leonard, Review Sci. Instrum. **40,** 187 (1969).

*Fig. 184. Degassing Evaporation Sources*

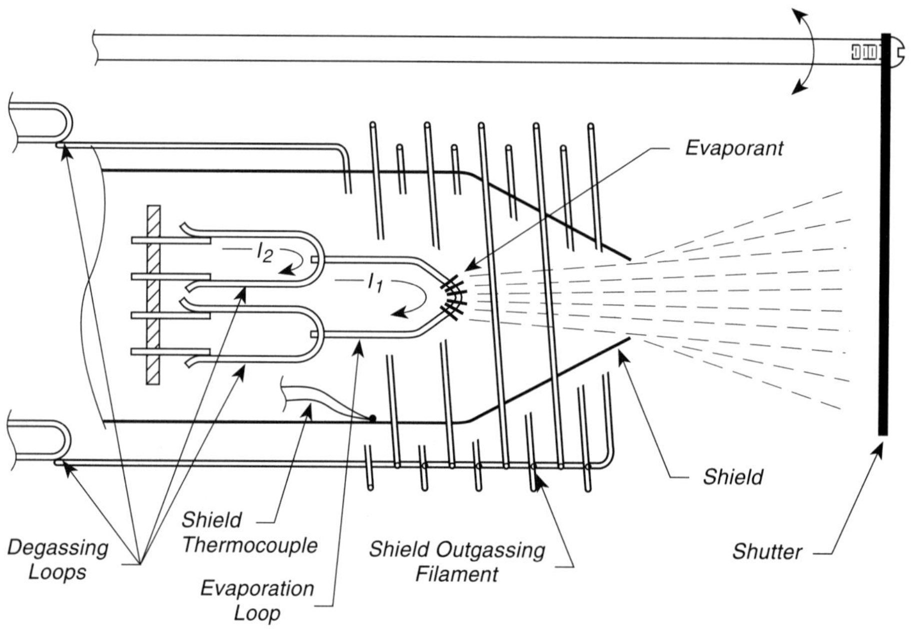

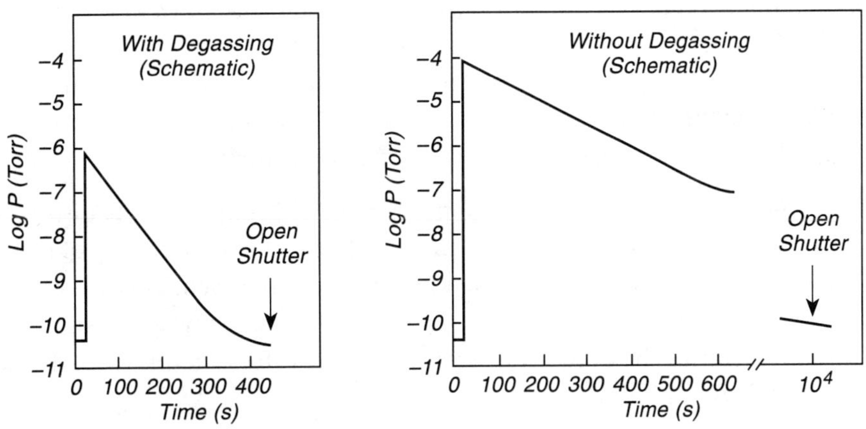
With Degassing
(Schematic)
Log P (Torr)
-4
-5
-6
-7
-8
-9
-10
-11
0 100 200 300 400
Time (s)
Open
Shutter
Without Degassing
(Schematic)
0 100 200 300 400 500 600
10^4

# 185

# Metal Evaporation Sources for Downward Evaporation

Two oven designs for the downward evaporation of metals using radiative heating have been described that operate at very low outgassing rates [1]. The problem with most downward evaporation sources that involve melting of the evaporating metal is that gravity may cause molten metal to fall from the evaporator to the substrate.

The cylindrical oven body is made of tantalum foil, 0.05 mm thick, spot-welded into shape and containing end caps made of the same size material. The oven is then wrapped with several layers of 0.01-mm-thick Ta foil to provide substantial thermal shielding. This heat shielding is essential to permit good thermal efficiency and therefore low outgassing rates. Two 3-mm-diameter Mo rods are used to bring in power to a W coil filament, 1 mm in diameter and 250 mm long. The longer of the two rods is spot-welded to the inside of the Mo oven body to provide mechanical support for the assembly. The second passes through a hole in the top cap and is insulated by a thin tube of pyrolytic boron nitride, pBN. When the oven is operating the resistance of the W coil is about 0.5 Ω. A W25%Re/W3%Re thermocouple monitors the temperature of the top end cap, but does not necessarily read the correct temperature of the metal source.

Figure 185A shows the design for a nonmelting metal (3-mm slab) that is supported with thin W wires on the bottom of the oven. A W shield, placed on the top of the source material reduces upward evaporation, extending the life of the oven. This source has been used for Fe film deposition rates of between 5 and 30 Å/min at power levels of 90 W and above without melting. The oven-to-substrate distance is 10 cm in this case. Mn has also been evaporated with this design.

Figure 185B shows a modified design that permits evaporation of metals that must be melted to achieve the desired rate of film growth. A set of Ta rings (0.01 mm thick) alternate with rings of the evaporating metal (0.1 mm thick). The volume of the metal charge may approach 0.4 $cm^3$. A thicker Ta shield disk (0.05 mm thick) is placed above the stacked rings to direct the metal deposition downward and to shield the substrate from the radiation of the hot W filament. For Cu deposition, a power level of 95 W will give a deposition rate of 10 Å/min, and 145 W gives a rate of about 45 Å/min. The melted metal is held between the thin Ta rings by surface tension. Silver has also been evaporated successfully with this oven design. Metals that alloy with Ta may self-destruct the oven.

The ovens are surrounded by tightly coiled water-cooled Cu tubing that does not touch the oven and that extracts much radiant heat. It is reported that with this design, the total pressure rise is $<5 \times 10^{-11}$ Torr[1].

An excellent review of metal evaporation sources and materials compatibility at high temperatures is given in reference [2].

## *References*

[1] D.A. Steigerwald and W.F. Egelhoff, Jr., J. Vac. Sci. Technol. **A7**, 3123 (1989).

[2] K.J. Ross and B. Sonntag, Rev. Sci. Instrum. **66**, 4409 (1995).

***Fig. 185. Downward Evaporation Source***

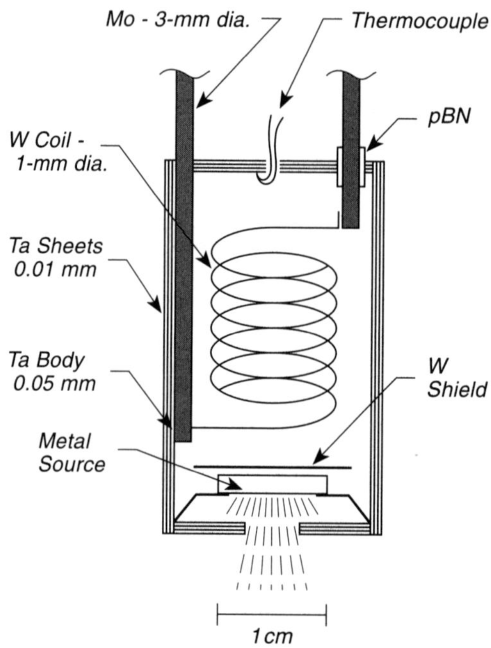


***A. Nonmelting Metal***

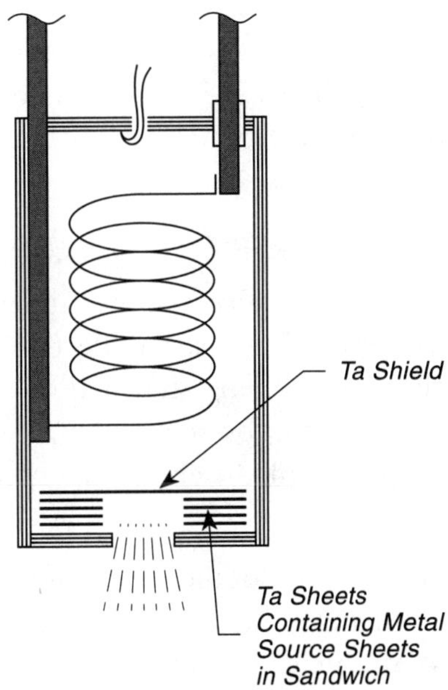


**B. Melting Metal**

# 186

# Ultralow Coverage Metal Evaporation

The use of a quartz thickness monitor (QTM) for absolute measurement of the thickness of deposited films has been discussed elsewhere (page 440). The extension of this technique to very low coverages of deposited material has been made by Chin et al. [1]. Essentially they arrange for the rate of evaporation to be determined by a QTM placed rather close to the evaporation source (7.6 cm), while collecting the deposit on the crystal substrate at a much longer distance ($d_{crystal}$ = 112 cm). The apparatus shown in Figure 186A consists of a set of multiple evaporation sources of similar geometry mounted at a distance, $d_{QTM}$, from the QTM, which is centrally located. The crystal substrate is mounted in an ultrahigh vacuum chamber facing the evaporation source, which is located far from the substrate in a long flight tube as shown in Figure 186B. Considering both the distances and the angles between the normal to the QTM and the crystal, $\alpha_{QTM}$ and $\alpha_{cryst.}$, the equation for the relative rates of deposition is

$$R_{cryst.}/R_{QTM} = (d_{QTM})^2 \cos \alpha_{cryst.} / (d_{cryst.})^2 \cos \alpha_{QTM}$$

In the design and testing of this long-path evaporator, the effect of gravity on the evaporating beam was shown to be negligible as was the effect of beam scattering by residual gases at ultrahigh vacuum. For the geometry shown, it was calculated that because of the lower pumping speed at the evaporation source compared to that at the crystal, the pressure in the evaporator region would be about 40% greater than in the ultrahigh vacuum chamber.

In order to obtain a calibration factor for the apparatus, a second QTM was placed at the crystal position and the ratio of the signals received during evaporation from each of the individual sources was measured. It was found that each of the evaporators gave the designed ratio of about 200 within a 20% error range, whether the evaporator source material is in the form of a bead or a coil of wire. It was also shown that shadows cast by the QTM and its support structure were unimportant at the flight distances employed. Minimum deposition levels of the order of $10^{-4}$ ML are possible with this apparatus. An added advantage of the geometry used is that heating of the crystal substrate by radiation from the sources is negligible.

An excellent review of metal atom evaporation sources and materials compatibility at high temperatures is given in reference [2].

## *References*


[1] K.K. Chin, P. McKernan, and I. Lindau, J. Vac. Sci. Technol. **A4,** 1949 (1986). This paper contains references to tests of this device as well as to related devices.

[2] K.J. Ross and B. Sonntag, Rev. Sci. Instrum. **66,** 4409 (1995).

**Fig. 186. Ultralow Coverage Metal Evaporator**

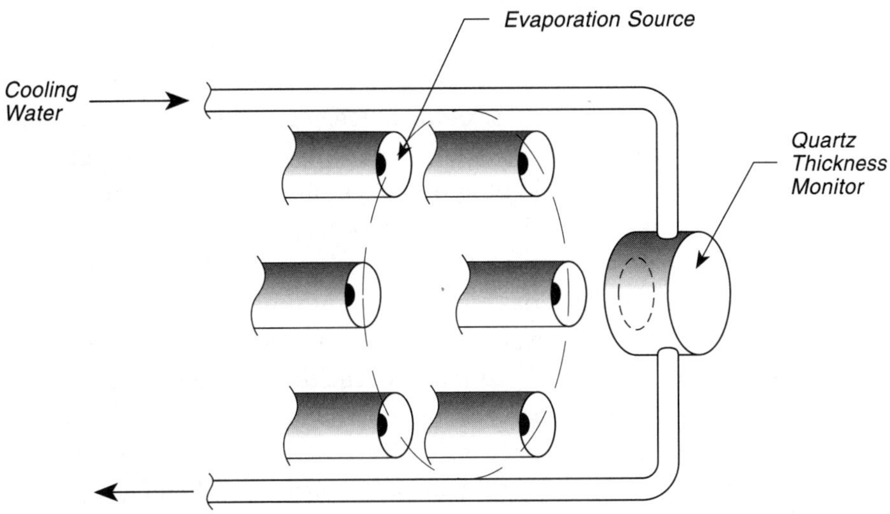


**A. Multiple Evaporator Source**

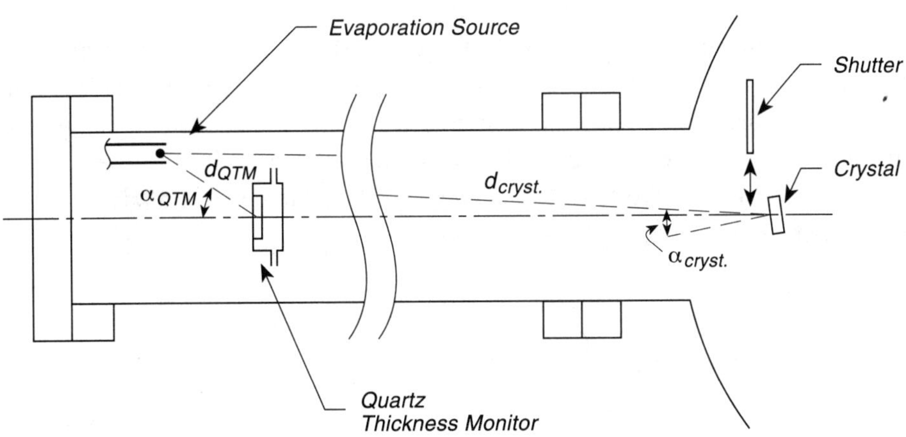


**B. Source Mounted at Long Distance**

# 187

# Congruent Evaporation of Multicomponent Materials

The evaporation of alloys and mixtures will almost always result in composition changes during evaporation because of the differences in vapor pressures of the different components. Thus, flash evaporation is often employed where the entire alloy or mixture is rapidly evaporated to yield films of more uniform composition, equivalent to that of the starting material. Often, the starting material is dispensed as a powder onto a heated surface where evaporation occurs rapidly. This procedure has certain deficiencies, such as the need to powder the material, which might introduce impurities, and the problem with blockage of tubes, etc., with powdered material. Often, too, the powdered material is mechanically ejected, causing film inhomogeneities. The device shown in Figure 187 overcomes these problems by use of a liquid material for the dispenser device [1]. Films of InSb and As/Sb/Se have been made by a predecessor of the method to be described, and films of doped Se have been made by the new method [1].

A quartz capillary tube is partially immersed in a molten alloy of the desired composition, and the liquid is kept slightly above the melting temperature. The molten alloy is forced through the capillary tube by means of an inert gas with a pressure of the order of 10 Torr, and this pressure is dynamically regulated by means of pumping and a gas emission needle valve. The molten source material streams out of the capillary tube into an evaporator heated to a temperature that produces rapid evaporation from its inner surfaces, and is above the melting point of any congruent compound that might form. The evaporator is baffled so that direct line of sight between the capillary end and the substrate is not present. In this arrangement, the molten source material can be thoroughly outgassed prior to film deposition. It is advantageous to heat the top of the capillary and the evaporator to temperatures where the source material already starts to evaporate before leaving the capillary.

## *References*

[1] F. Jansen, J. Vac. Sci. Technol. **21**, 106 (1982).

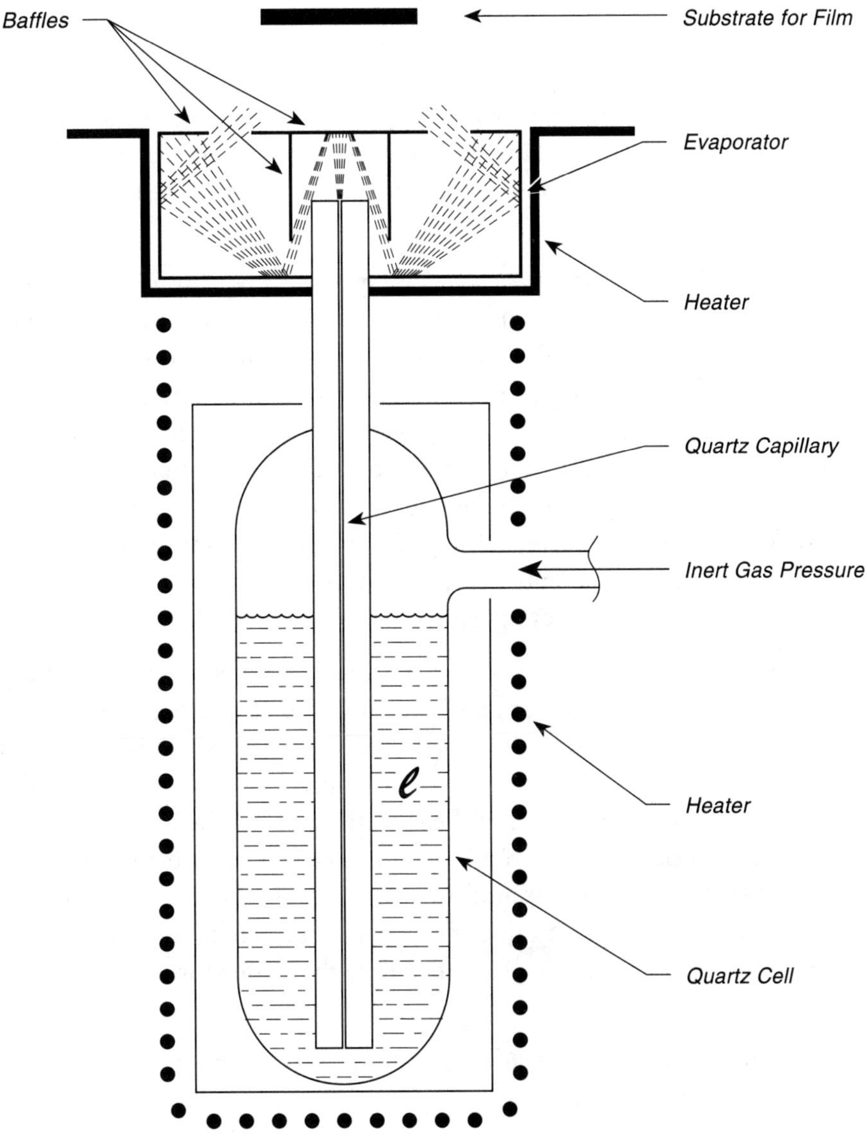


***Fig. 187. Congruent Evaporation Source***

# 188

# Evaporator for Eutectic-Forming Metals

Often, eutectic-forming metals are combined to form evaporation sources for the higher vapor pressure metal. However, at the temperature of vaporization, a lower-melting eutectic alloy may form, causing the mechanical failure of the evaporator device by melting.

A method to avoid this problem has been developed that prevents the formation of the eutectic in sufficient quantities to cause mechanical failure. In the case of a zirconium evaporator, this method has proved satisfactory [1]. Figure 188A shows the basic principle of the method. A tungsten rod serves as a mandrel for the evaporator coil, which consists of smaller-diameter W wire wound together with Zr wire. Ohmic heating of the assembly by passing current through the large Ta connector and support plates raises the Zr and W to the correct evaporation temperature (about 2300 K), using 38–40 A, and permits Zr decomposition at a rate of 0.25–0.35 ML/min on the sample located 3 cm from the evaporation source. A special outgassing procedure was employed for this source [1], which is gassy given the high operating temperature and the method of construction.

The principle of operation may be seen from considerations of Figure 188B. Here, a lower melting eutectic mixture of Zr and W is produced at intermediate atomic fractions of W. In the case of the source described above, one avoids arriving at the eutectic region for a substantial part of the assembly by limiting the amount of Zr available compared to the amount of W. Thus, even if the eutectic forms in the immediate region of a Zr/W interface, diffusion of Zr from this region will of necessity bring one to the left-hand region of the phase diagram and into the composition range where extreme depression of thc W melting point does not occur. This principle is often applied in designs for W/Ni and W/Al [2] evaporators where the same kind eutectic formation occurs, causing difficulties.

An excellent article describing metal evaporation methods and materials compatibility at high temperatures is available [3].

## *References*


[1] P.R. Davis and H.R. Poppa, J. Vac. Sci. Technol. **15,** 1771 (1978).

[2] P.J. Chen, M.L. Colaianni, and J.T. Yates, Jr., Phys. Rev. **B41,** 8025 (1990).

[3] K.J. Ross and B. Sonntag, Rev. Sci. Instrum. **66,** 4409 (1995).

***Fig. 188. Evaporator for Eutectic-Forming Metals***

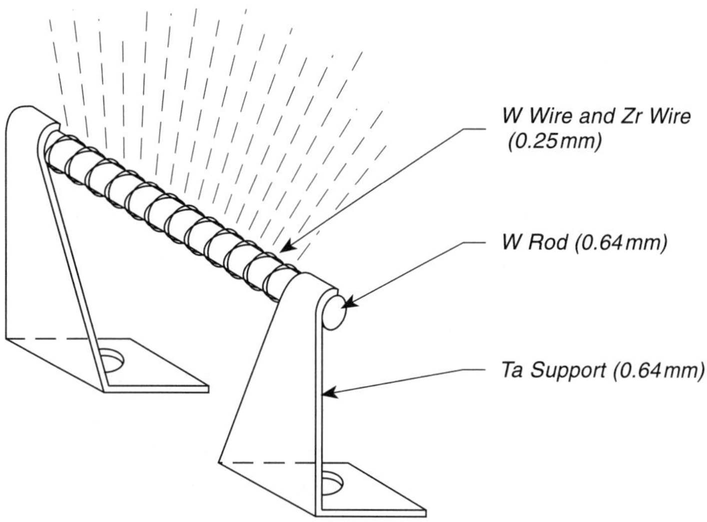


**A. Evaporator**

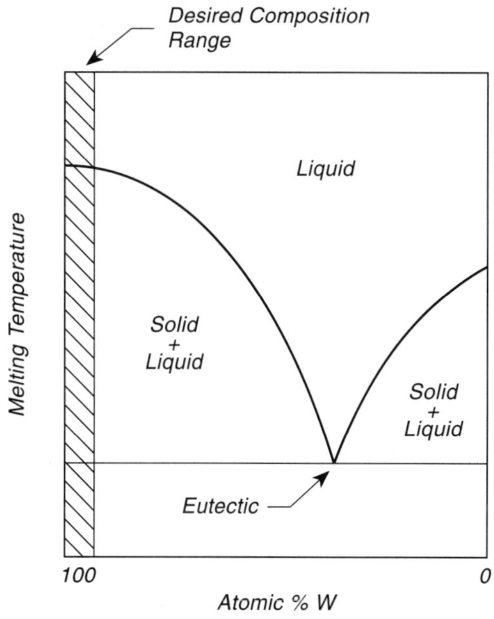


**B. Phase Diagram**

# 189

# Simple Lithium Metal Evaporation Source

The use of alkali metals as adsorbed species and as modifiers of surfaces is well documented in the literature [1], and several alkali metal evaporation sources are used, as indicated on page 694. A particularly nice source for Li has been described and is presented here [2].

A binary alloy of Al and Li, containing 6.5% Li, was used in the form of a disk, 10 mm in diameter and 1.5 mm thick [3]. The material was rinsed in acetone before use and is known to be covered with a thick oxide layer. Two 0.25-mm-diameter Ta wires were looped through two pairs of holes, and a fifth hole was used for attachment of a type K thermocouple, as shown in Figure 189A. Although not used by the developers of this Li source [2], the Ta heating wires should be mounted on degassing loops (see pages 584 and 628) for maximum efficiency in outgassing the support. The evaporation source is surrounded by a stainless steel shield.

Li evolution from the source was monitored with a quadrupole mass spectrometer, and the temperature dependence of Li evolution is shown in Figure 189B. For actual film deposition, a temperature of 695 K was employed, and Li was collected on an atomically clean Al(111) surface on which work-function and X-ray photoelectron spectroscopy measurements were made. From these measurements, it was estimated that approximately 1 ML/min was being deposited with the Al(111) face approximately 1 cm from the source. A 25-ML layer of Li showed no evidence for impurities by X-ray photoelectron spectroscopy. These results indicate that even with the oxide coating, the alloy can supply clean Li by evaporation. The pressure during deposition increased from $1.5 \times 10^{-10}$ Torr to $3.5 \times 10^{-10}$ Torr, and the evolved gas was mainly $H_2$, $CO_2$, and $H_2O$.

A more complex source involving Li/Al alloys is described in reference [4], and the spatial, time-dependent, and temperature-dependent characteristics of this source have been carefully evaluated.

## *References*


[1] *Physics and Chemistry of Alkali Adsorption,* edited by H.P. Bonzel, A.M. Bradshaw, and G. Ertl (Elsevier, Amsterdam, 1989).

[2] F.J. Esposto, K. Griffiths, P.R. Norton, and R.S. Timsit, J. Vac. Sci. Technol. **A12,** 3245 (1994); for details of the surface behavior of Li/Al alloys, see F.J. Esposto, C.-S. Zhang, P.R. Norton, and R.S. Timsit, Surface Sci. **290,** 93 (1993).

[3] Supplied by Alcan, International, Ltd. It is rumored that Canadian beer cans are made of this alloy (personal communication, P.R. Norton).

[4] R.P. Schorn, E. Hintz, S. Musso and B. Schweer, Rev. Sci. Instrum. **60,** 3275 (1989).

***Fig. 189. Li Thin Film Evaporation Source***

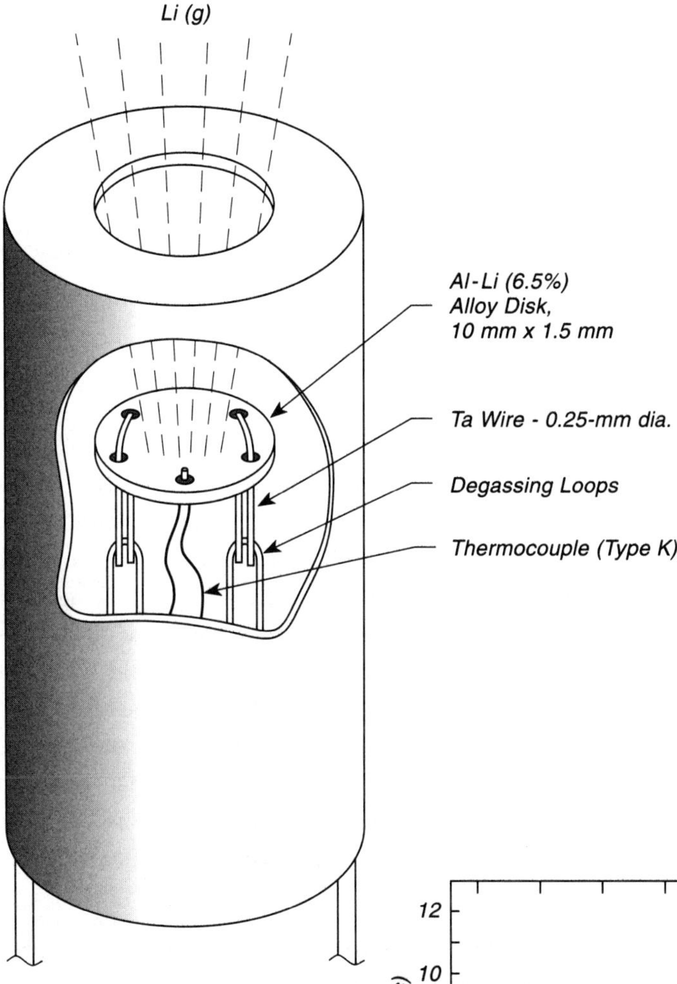


***A. Li Evaporation Source***

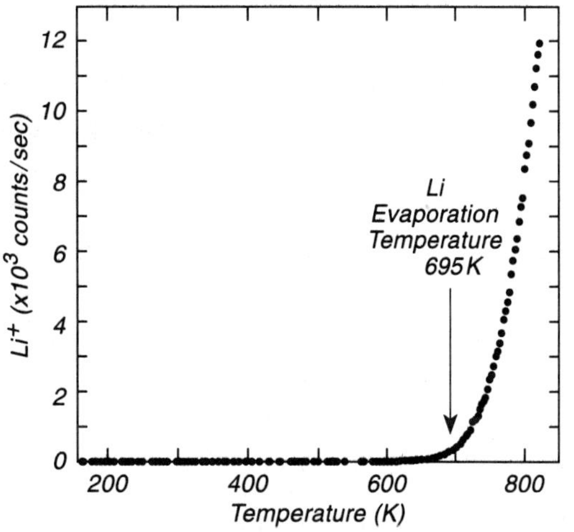


***B. Li$^+$ Mass Spectrometer Signal— T Scan Rate = 0.5 K/s***

# 190

# Aluminum Evaporation Sources—Long Lived

The evaporation of aluminum in ultrahigh vacuum to yield a clean film requires special expertise. The most common design for an evaporator involves the use of W wire as a heater, with Al being evaporated while in contact with the hot W. This is a short-lived source because of alloy formation, producing a low-melting alloy that causes mechanical failure of the W wire. A clever method to avoid alloy formation has been devised, in which a film of Al is oxidized on the W wire to form a barrier that retards alloy effects [1]. A variety of the most recent methods for evaporating Al have been reviewed in a paper which also presents a different solution to the lifetime problem [2].

The evaporation source is made of zero-porosity alumina tubing [DIN 40 685 type AL23-710] [3], 1.1 mm inner diameter and 1.7 mm outer diameter, 6 mm long. Into this was inserted a 6-mm piece of high-purity [99.999%] Al wire, with a 1-mm diameter. A tungsten coil, made of 0.25-mm wire and wrapped around a 1.37-mm-diameter mandrel prior to use, was fastened tightly around the alumina tube by virtue of the size difference between the coil and the tube. This assembly was then spot-welded to two Mo rods of 2-mm diameter. The assembly was rigidly supported between the Mo rods using ceramic locating blocks, not shown in Figure 190. The evaporation source was surrounded by a shield that was supported on the mounting flange, and that contained an 8-mm-diameter hole on the axis of the device. A rotatable shutter was used in front of the aperture to control the Al deposition time and to permit outgassing of the source without deposition.

The source was degassed for hours at a 60% power level compared to the operating conditions used for evaporation. Typical operating conditions were achieved at 2.5 V, 1.4 A, and 3.5 W, under which conditions the alumina tube assembly glowed with a bright orange-yellow color. Final outgassing, with the shutter closed, was carried out for 30 s at the operating temperature. Operation could occur for 5–10 min before the pressure rose to $5 \times 10^{-10}$ Torr. Under these conditions, deposition occurs at a rate of about 0.5 ML/min. Auger spectroscopy and X-ray photoelectron spectroscopy both indicated that impurities on the evaporated Al film were below the detection limit. The source could be operated reproducibly for several months before failure. Since Al melts at the operating temperature, the source should be mounted horizontally.

An improved design for this type of source has recently been reported [4]. Two improvements are incorporated to outgas the source and its supports more thoroughly. First, the heating leads wrapped around the $Al_2O_3$ are mounted on degassing loops as described on pages 584 and 628; secondly, the entire source is surrounded by a Ta radiation shield that can be separately degassed by means of electron bombardment from a W filament mounted inside the shield, somewhat as shown on page 628. After degassing by these methods, the source could be operated for hours in the $10^{-11}$ mbar range [4]. In addition, the outside of the $Al_2O_3$ tube was modified by grinding a helical channel along its length. This reduced the mass of the tube and increased the thermal contact between the heater filament and the tube. Finally a Chromel-Alumel thermocouple was connected to a Pt sphere that itself was attached to the tube by melting a Pt wire with a hydrogen flame and observing that a portion of the molten Pt drops away while another portion wets the alumina [4]. The thermocouple reads a temperature hundreds of degrees Kelvin below the Al temperature but this may still be used to control the Al evaporation rate even when the thermal contact between the heating wire and the alumina tube varies [4].

## *References*


[1] G. Polanski and J.P. Toennies, Surface Sci. **260,** 250 (1992).

[2] W.J. Wytenburg and R.M. Lambert, J. Vac. Sci. Technol. **A10,** 3597 (1992).

[3] Degussa Limited, Wilmslow, Cheshire, England.

[4] H.-G. Keck, E. Kopatzki, and R.J. Behm (private communication).

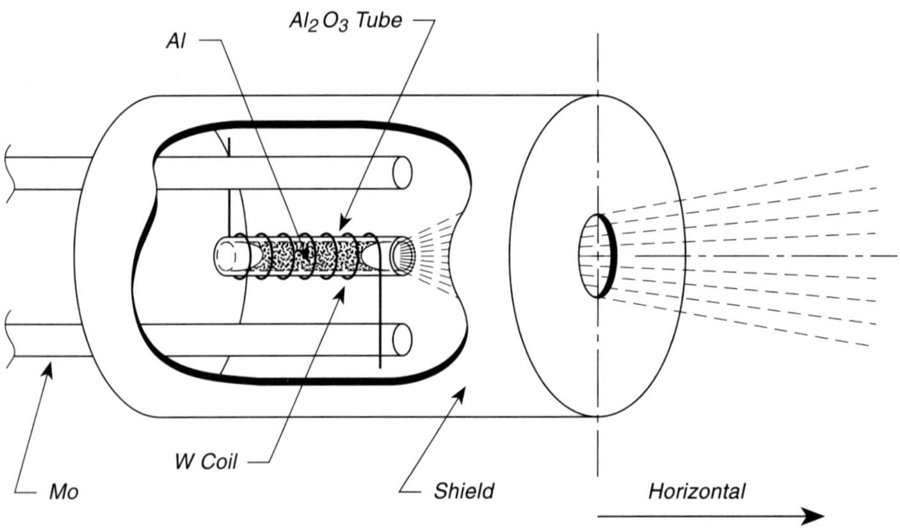


***Fig. 190. Al Evaporation Source—Long-Lived***

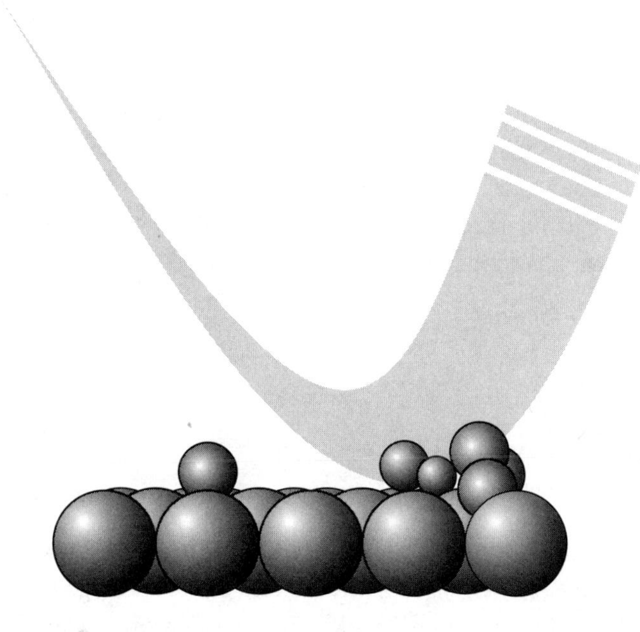

# 191

# Cr Evaporator Sources of Compact Design

Often the methods used for metal film evaporation require large, high-current evaporation devices that are incompatible with experimental apparatus designed to study the films themselves. Both heating and large outgassing effects are associated with many of the high-power evaporation devices. Therefore, very simple designs will often give better results in research experiments that do not require massive film deposition capabilities. The Cr evaporator discussed here is one example of ultimate simplicity which meets these requirements [1].

Two sizes of W wire loops, 0.25 mm and 0.5 mm in diameter, are used. These wires are cleaned with 600 grit sandpaper followed by ultrasonic cleaning in acetone and methanol, and then are bent into a hairpin shape as shown in Figure 191. The hairpin filaments are suspended in an electroplating bath with about 2-mm submersion. The electroplating bath is of the following composition [2]:

100 ml $H_2O$, deionized
30 g $CrO_3$
5 g NaOH (add slowly)
0.6 g $H_2SO_4$
0.1 ml ethanol

Electroplating is carried out with a Pb anode. Currents of about 45 and 90 mA are used for the small and large filaments, respectively. The plating voltage is about 3 V. Electroplating for 18 h typically deposited 75 and 250 mg of Cr on the small and large loops, respectively. Before use, the filaments are rinsed in deionized water and methanol. After installation, degassing is done before bakeout at 5–6 A and 15–16 A for the small and large filaments, respectively. The filaments are outgassed a second time during chamber cooling after bakeout, and may then be used in ultrahigh vacuum without appreciable gas evolution in the $6 \times 10^{-10}$ Torr range.

The small evaporators yielded Cr film growth rates of 4 Å/min at a current of 7 A for a distance of 5 cm. The larger evaporator yielded 4 Å/min at a 10 cm distance at 20-A heating current. These evaporation rates could be increased by a factor of 10–20 by working at higher filament currents. Both filaments produced about 4000 Å of total Cr film thickness before depletion. Auger spectroscopy showed that impurities were not present at the limit of Auger detection if the sources were thoroughly outgassed before use. For early stages of evaporation from a poorly outgassed filament, oxygen impurities were seen.

The films produced are of sufficient purity to be used in studies of the magnetic properties of multilayer structures [3].

## *References*

[1] J.J. McClelland, J. Unguris, R.E. Scholten, and D.T. Pierce, J. Vac. Sci. Technol. **A11**, 2863 (1993).

[2] *Encyclopedia of Chemical Technology,* edited by H.F. Mar, et al. (Wiley, New York, 1979), Vol. 6, p. 104.

[3] J. Unguris, R.J. Celotta, and D.T. Pierce, Phys. Rev. Lett. **67**, 140 (1991).

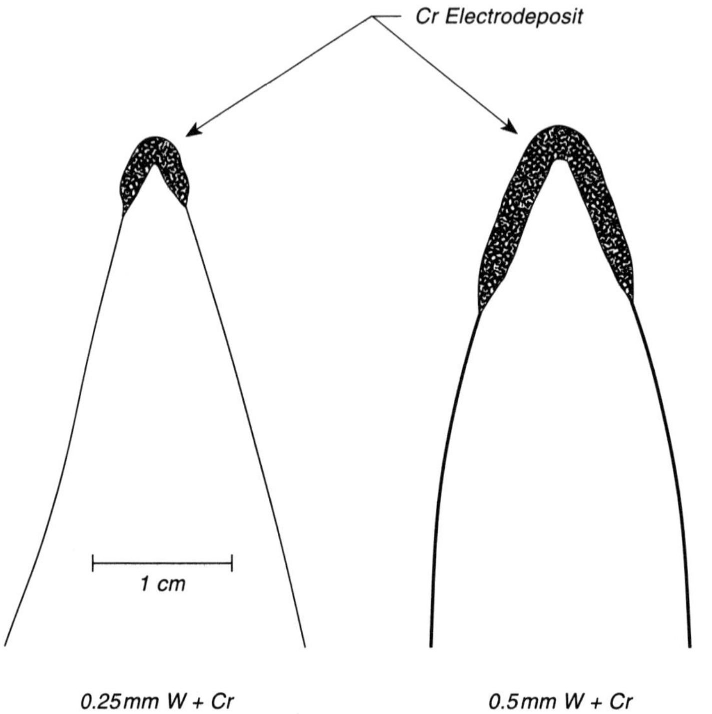


**Fig. 191. Electrodeposited Cr on W Wires for Cr Thin Film Deposition**

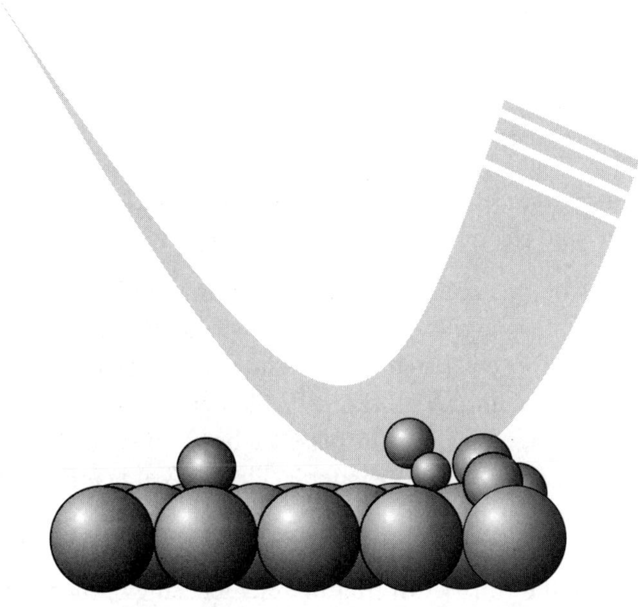

# 192

# Electron Beam Evaporator for Refractory Materials

While many substances can be evaporated using ohmic heating of a filament or a crucible, the successful evaporation of highly refractory materials such as tungsten or molybdenum requires a more intense source of heat, which can be provided by using a high-energy electron beam. A serious problem with conventional electron beam evaporation methods is the emission of both high- and low-energy stray electrons, which cause effects such as—

1. Electron stimulated desorption from walls and the cold shroud which may lead to film contamination.
2. Film damage.
3. Interference with current measuring instruments, such as ionization gauges, mass spectrometers and electron diffraction apparatus.

Of the two types of electron gun evaporators (magnetically and electrostatically focused instruments), the magnetically focused instruments are more susceptible to the difficulties mentioned above [1]. Figure 192A shows a schematic diagram of an electrostatically focused electron gun. A thoriated W filament of circular geometry is located between a shield and a cage, and both are biased at between -5 and -10 kV (*-HV*). The filament is heated by a 12-V, 30-A power supply. Electrons are emitted and follow trajectories to the grounded hearth containing the sample to be evaporated. A focusing mechanism (not shown) can change the relative position between the hearth and the rest of the components of the gun, permitting an exact focusing of the electron beam on the top of the molten metal in the hearth. A second degas mode involves keeping the filament at -3 kV and grounding the remaining components of the electron gun evaporator. Outgassing was reduced by making all components out of refractory materials, such as W, Mo, and Ta. The shield and hearth were made of Mo; the cage was constructed of Ta mesh on a Mo frame with a W top cover containing a central hole. Using a Ta mesh cage prevents heat buildup, and radiation-induced heating of large surfaces leading to outgassing.

The secondary electrons produced in the gun cannot escape from the cage, which is biased at *-HV*. The high-energy electrons have a potential $V$ in the range $-HV < V < -HV + V_f$, where $V_f$ is the potential drop across the emitter filament due to its resistance. In addition, there are high-energy backscattered electrons, which mostly are at a potential of $V$ or a slightly lower potential owing to energy loss upon scattering. Since the potential at the hole in the cage can be positive relative to *-HV* by a few hundred volts, backscattered electrons will escape from it. To eliminate this problem, a tungsten mesh grid with 2-mm spacings was placed over the hole to retard electrons there. In addition, by biasing the filament at -9.9 kV, excellent focusing is achieved and stray electron currents are minimized, as depicted in Figure 192B, where calculated trajectories for cylindrical symmetry are shown.

This evaporation source can operate for refractory metal evaporation at pressures $<2 \times 10^{-9}$ Torr [1]. A cryogenically shrouded source of somewhat similar design has been described [2]. A highly stable electron beam evaporation source has been described in reference [3]. A method involving the measurement of the ion current generated by ionization of evaporating metal atoms and relating this measurement to the rate of metal evaporation is described in reference [4].

An excellent review of metal evaporation methods and materials compatibility at high temperatures is available in reference [5].

## *References*

[1] M. Heilblum, J. Bloch, and J.J. O'Sullivan, J. Vac. Sci. Technol. **A3,** 1885 (1985).
[2] L. Ramberg, E. Flemming, and T.G. Andersson, J. Vac. Sci. Technol. **A4,** 141 (1986).
[3] Y.W. Beag, K. Egawa, and R. Shimizu, Rev. Sci. Instrum. **64,** 3647 (1993).
[4] T. Jones, J. Sawler, and D. Venus, Rev. Sci. Instrum. **64,** 2008 (1993).
[5] K.J. Ross and B. Sonntag, Rev. Sci. Instrum. **66,** 4409 (1995).

**Fig. 192. Electron Beam Evaporation**

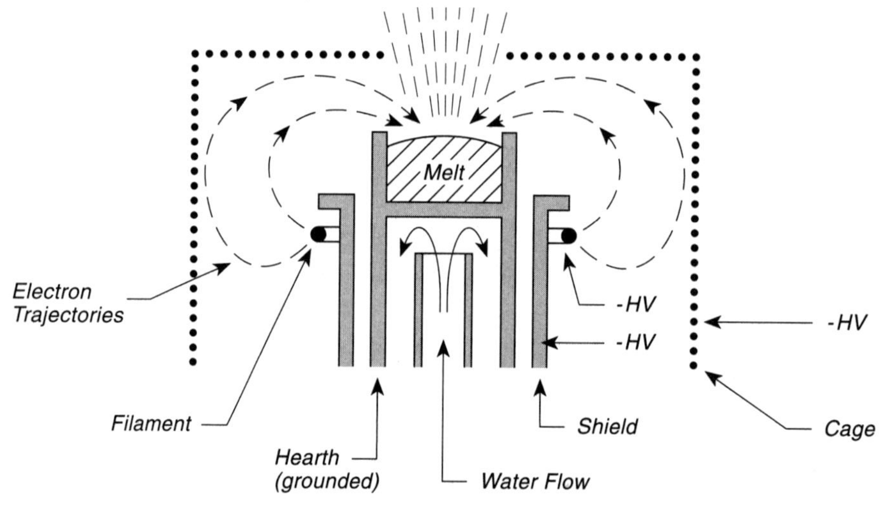


**A. Evaporator**

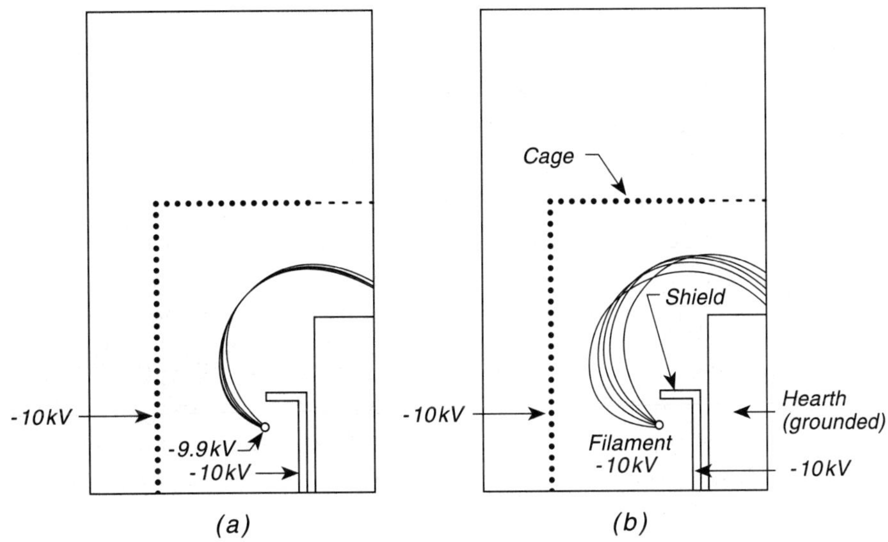


**B. Electron Trajectories**

# 193

# Small Electron Beam Evaporator for Refractory Metals

As discussed previously (page 652), electron beam evaporation is a desirable method for the deposition of thin films of refractory metals that does not involve the problems of ohmic heating of a filament to very high temperatures. The design shown in Figure 193 incorporates simple features [1], combined with low mass components, eliminating problems with large surfaces outgassing during operation. The device is simple to build, does not require cooling water for its components, operates in the $10^{-9}$ Torr range, and produces clean films of refractory metals.

The evaporation device is mounted on a 2.75-in.-diameter flange containing high-voltage feedthroughs. A tungsten emitter filament, about 1 cm in length and 0.012 cm in diameter, is spot-welded on 0.1-cm-diameter Ni rods as shown. It is positioned 1–2 cm from the evaporation source. The source is a 6-cm wire (0.025-cm diameter) made of the refractory metal to be evaporated that is fashioned into a loop on the outer end; the base of this wire is welded to a 0.1-cm-diameter Ni rod connected to a feedthrough pin by a set-screw contact for easy replacement. The source wire is placed on the axis of a Mo wire helix with a diameter of about 1.5 cm, made of 0.025-cm-diameter Mo wire.

For operation, the source wire is set at about +6 kV = $V_a$, and the grid is set at about -3 kV = $V_g$. The filament voltage, $V_f$, is slowly raised using a current supply until the collected current at the evaporator source is in the 5–20-mA range. Next the negative voltage on the grid is empirically adjusted until electron bombardment heating occurs only on the loop of the source wire. Overheating the loop can result in melting which is undesirable. Films with greater than 1000-Å thickness have been evaporated from W, Ta, and Ir at rates in the range 1–10 Å/min.

A variation on this source in which a small Mo crucible replaces the pendant droplet source described here has been designed and tested [2]. The mass of the crucible is kept to a minimum, and the emitter coil is wound so that it encircles the crucible edge and is slightly below the edge [2].

An excellent review of metal evaporation sources and materials compatibility at high temperatures is given in reference [3].

## *References*


[1] J.R. Waldrop and R.W. Grant, J. Vac. Sci. Technol. **A1**, 1553 (1983).

[2] J. Araya-Pochet, G.A. Mulhollan. and J.L. Erskine, Rev. Sci. Instrum. **62**, 2288 (1991).

[3] K.J. Ross and B. Sonntag, Rev. Sci. Instrum. **66**, 4409 (1995).

***Fig. 193. Electron Beam Evaporation***

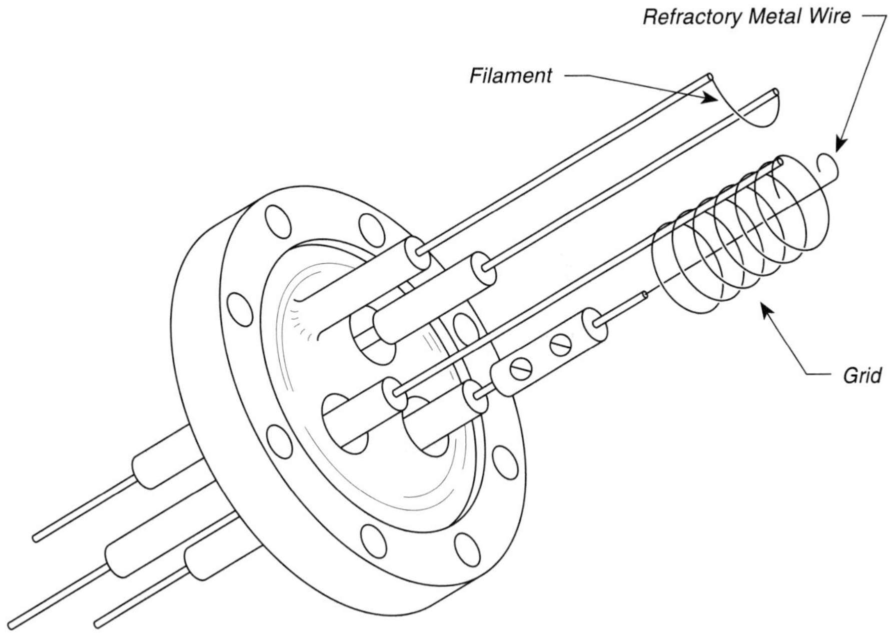


***A. Evaporator Design***

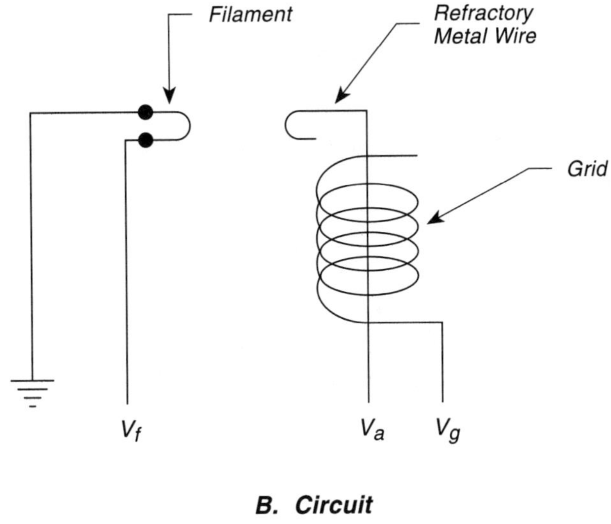


***B. Circuit***

# 194

# Electron Beam Evaporator

A design [1] differing somewhat from those shown elsewhere in this section, which focuses a high-energy electron beam at a point on the end of a source wire for the purpose of evaporating a metal is shown in Figure 194. The metal source wire is 1–2 mm in diameter, mounted on a high-voltage Cu feedthrough that can be translated in the vertical direction. In normal operation, a molten droplet of metal is produced on the end of the wire by electron bombardment. The circular electron emitter filament, made of 0.25-mm-diameter W wire, is looped over itself in a 15-mm diameter circle, and welded to two Ta wire leads (1-mm diameter), which are connected to vacuum feedthroughs as shown. The distance between the plane of the loop and the source tip is typically 6–9 mm. Nominal operating conditions are shown in Figure 194. As the source wire is evaporated, the beam current decreases and may be restored by advancing the source to a position where the current returns to its original value. With a source-to-sample distance of 15 cm, several hundred Å of film may be deposited before source motion is necessary. The filament is shielded from the substrate receiving the evaporated film by the aperture plate, and this reduces radiative warming of the substrate. The source is readily outgassed behind a closed shutter by running with the cooling water removed and at a reduced voltage (0.5–1 kV). During normal operation, with water cooling, the pressure in the UHV system remains in the 1–4 x $10^{-10}$ Torr range. Impurity transport to the film has been measured for Fe and Co source wires using both Auger spectroscopy and X-ray photoelectron spectroscopy, and impurity levels below the detection limit were found.

A design similar to this, but using a very small Mo crucible in the position of the end of the source wire, has been described [2] for the evaporation of low-melting metals under conditions where the low outgassing qualities of this evaporator are retained.

A review of metal evaporation sources is contained in reference [3]. In addition, an excellent review of metal evaporation sources and materials compatibility at high temperatures is given in reference [4].

## *References*


[1] B.T. Jonker, J. Vac. Sci. Technol. **A8,** 3883 (1990).

[2] J. Araya-Pochet, G.A. Mulhollan, and J.L. Erskine, Rev. Sci. Instrum. **62,** 2288 (1991).

[3] L.I. Maissel and R. Glang, *Handbook of Thin Film Technology* (McGraw Hill, NY, 1970), pp. 1-51.

[4] K.J. Ross and B. Sonntag, Rev. Sci. Instrum. **66,** 4409 (1995).

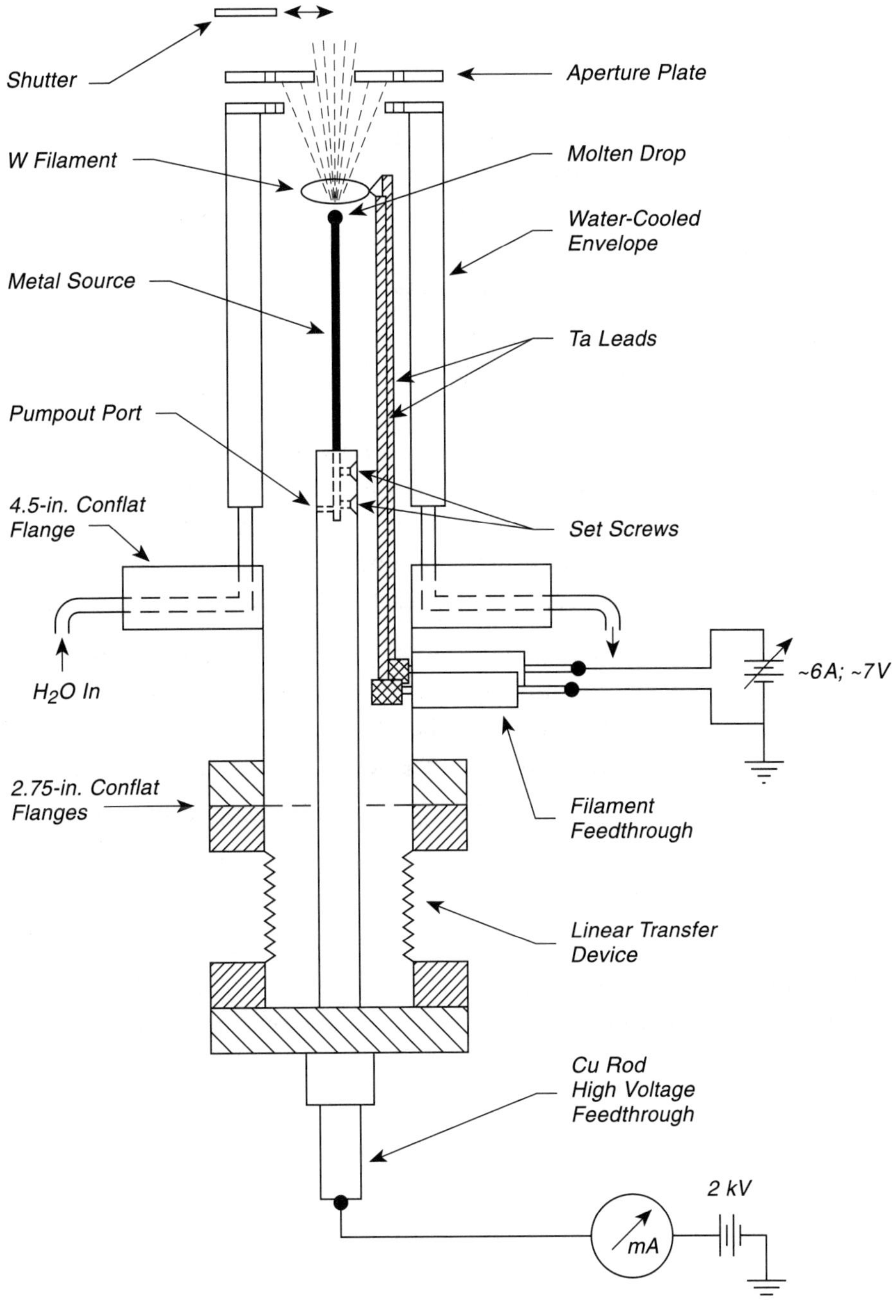


***Fig. 194. Electron Beam Evaporator***

# 195

# Break-Seal Ampoule Doser

Often low-vapor pressure compounds are used as precursors for the growth of compound semiconductor films on a substrate. Many of these compounds react with stainless steel and therefore cannot be transferred through heated tubing. This design places a glass ampoule of the compound close to the substrate so that the vapors do not come into contact with metal prior to deposition [1].

A glass ampoule is filled with the compound in a dry box under an inert atmosphere and the ampoule is sealed. The ampoule is rigidly attached to the end of a transfer rod inside a copper carrier assembly that fits the ampoule (Figure 195). The copper assembly is wound with W wire, sheathed in fiberglass or quartz fiber sleeving, and a thermocouple is placed in contact with the copper carrier for temperature measurements during vaporization from the ampoule. The use of a bellows transfer device is superior to a magnetically driven device, since the heater and thermocouple wires are fixed along the transfer rod during motion and do not have to flex each time the ampoule is moved.

A linear motion device with a stainless steel blade is used to break off the neck of the ampoule when it is desired to start the vaporization process. Once the neck is broken, the ampoule can be moved into the UHV chamber through a gate valve, and the temperature of the doser can be adjusted to produce the desired rate of precursor vaporization.

A cryopump is chosen for the first chamber because it pumps water vapor efficiently and it is not impeded by glass fragments.

## *References*


[1] Dr. M.S. Jackson and Professor J.G. Ekerdt, Department of Chemical Engineering, The University of Texas at Austin, Austin, TX 78712-1062 (private communication).

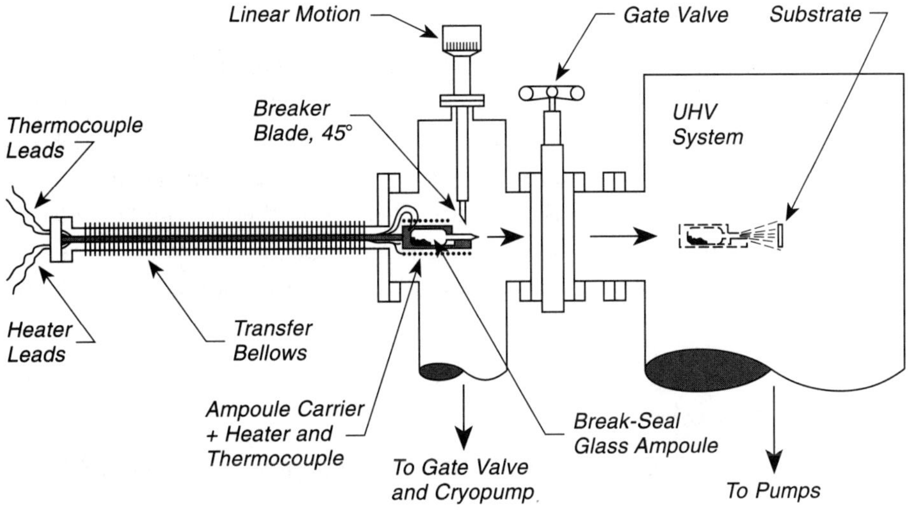


**Fig. 195. Break-Seal Ampoule Doser**

# 196

# Cadmium Sulfide Evaporation Source

Thin films of CdS may be thermally evaporated from CdS powder. The apparatus shown in Figure 196 is made entirely of tantalum, tungsten, and boron nitride [1]. A large area baffled aperture produces a uniform CdS deposit. In addition the baffle shields the substrate from radiation from the indirect heater. The baffle has a 3-mm-wide gap between its outer edge and the Ta shield.

The sublimation source is made of a Ta cup welded to a Ta tube. This slips over a cylindrical shield made of BN, which contains a rigid coil of fairly heavy wound W wire in three strands. The entire assembly may be unloaded by pulling upward on the handle, leaving the W wire exposed. In operation, the CdS nearest to the central heater sublimes first. The source temperature is measured with a Pt/PtRh 13 thermocouple welded to the baffle. At a temperature of 550°C, CdS is deposited at a rate of 7 Å $s^{-1}$ on the substrate at 120°C and 24 cm distant. The operation time at this rate of evaporation is about 4 h before refilling is needed.

By choosing the correct evaporation and source temperatures, the resistivity of the deposited CdS film may be controlled.

A CdS evaporation source using elemental evaporation has been described [2].

## *References*


[1] R.T. Galla, Rev. Sci. Instrum. **36,** 403 (1965).

[2] D. Beecham, Rev. Sci. Instrum. **41,** 1654 (1970).

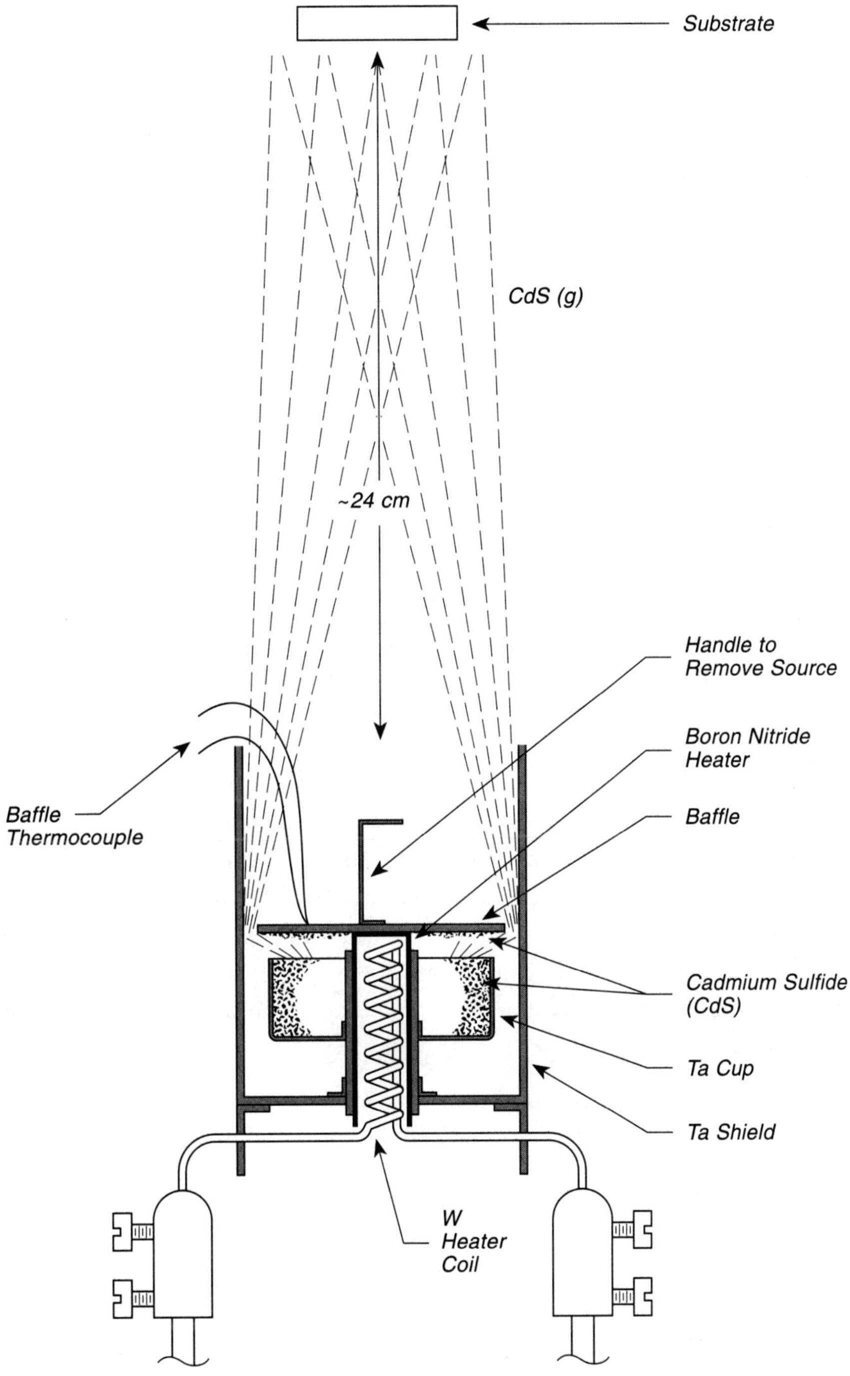


**Fig. 196. Cadmium Sulfide Evaporation Source**

# 197

# Arsenic Atom Source

The use of elements from Groups II, III, IV, V, and VI in the production of semiconductor devices is widespread. Many of the elements sublime as atoms, but in Groups V and VI, sublimation as dimers and tetramers is commonly observed. It is advantageous—for example, in the growth of GaAs surfaces—to employ atomic As, which forms surfaces with fewer defect sites than surfaces grown from $As_2$ or $As_4$ species. The device shown in Figure 197 permits the substantial atomization of $As_2$ and $As_4$ by use of a final heated atomizer stage [1]. The principles embodied here can also be used with other elements where atomization is desired. It should be remembered that the equilibrium between atomic species and higher molecular weight species is favored toward the atomic species by high temperatures and by low pressures.

Figure 197 shows, schematically, the device used to make atomic beams of As containing small fractions of $As_2$ and $As_4$. A sublimation source, loaded with As(s), is connected to a first-stage cracking furnace through an adjustable aperture. Adjustment of this aperture allows the beam pressure of As species to be controlled throughout the apparatus. The subliming species are conducted through the first stage cracking furnace and then exit to a heated plate that serves as the second-stage atomizer. The products from the second stage are significantly enriched in atomic As, and are emitted from a second aperture toward the substrate. The atomizer can be operated in the range above about 1200 K when the local pressure is below about $10^{-3}$ Torr; it can also be operated in the range above 1400 K when the local pressure is about $10^{-2}$ Torr. Many materials can be used for the atomizer plate with various elements to be atomized, including refractory metals such as W, La, Nb, Re, Mo, and Ta, and stainless steel, Ni, graphite, Pt, B, Si, ceramics, and BN. Compound formation between the material of the atomizer plate and the beam is to be avoided.

The preferred range of operation for various materials is given in Table 197, and substantial but not complete atomization will be produced under the listed conditions:

**Table 197. Preferred Range of Operation—Atom Sources**

| Substance | Pressure at atomizer (Torr) | Temperature range (K) |
|---|---|---|
| As | $< 10^{-2}$ | >1400 |
| | $< 10^{-3}$ | >1200 |
| P | $< 10^{-3}$ | > 2000 |
| | $< 10^{-5}$ | >1800 |
| Sb | $< 10^{-2}$ | > 1400 |
| | $< 10^{-4}$ | >1200 |
| S | $< 10^{-2}$ | >2000 |
| | $< 10^{-4}$ | >1700 |
| Se | $< 10^{-2}$ | >1600 |
| | $< 10^{-4}$ | >1300 |
| Te | $< 10^{-2}$ | >1300 |
| | $< 10^{-4}$ | >1100 |

## *References*

[1] Professor Stephen R. Leone, University of Colorado at Boulder, Joint institute for Laboratory Astrophysics, Boulder, CO 80309-0440 (private communication). U.S. Patent No. 5,541,407 Arsenic Atom Source, S.R. Leone, A.L. Alstrin, P.G. Strupp, A.K. Kunz, S.M. Casey, and R.V. Smilgys.

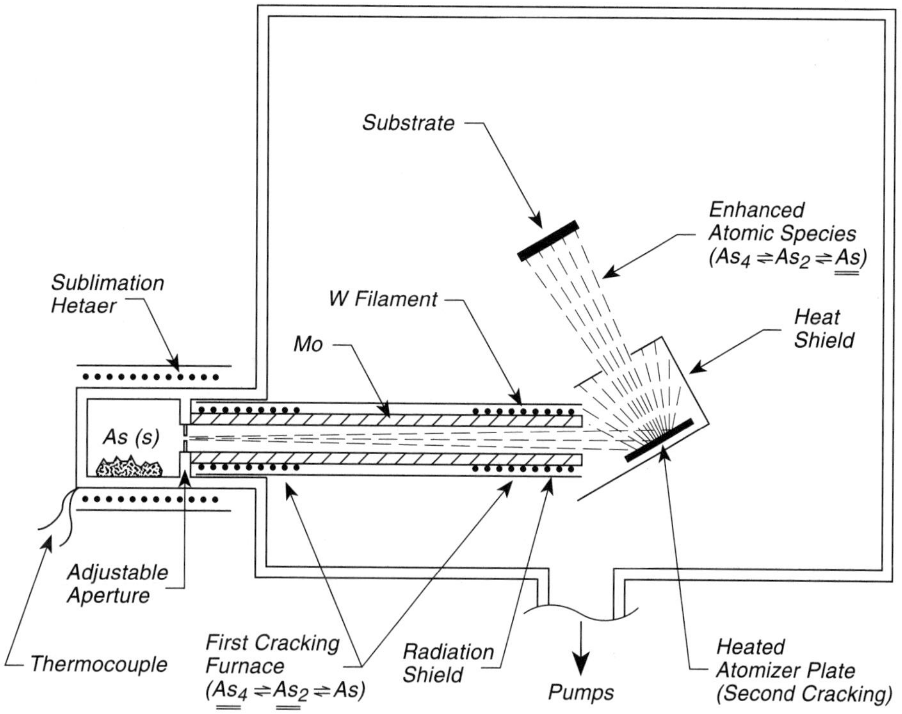


***Fig. 197. Arsenic Atom Source***

# 198

# Vapor Deposition of Organic Thin Films

The evaporation of organic materials and higher-vapor-pressure inorganic materials into an ultrahigh vacuum system may be done with various heated sources. A difficulty arises with large Knudsen cells and metal boat evaporators in that they have a large heat capacity and are difficult therefore to bring to temperature rapidly; also, difficulties in controlling the temperature may be present. A small evaporation source eliminating these problems is shown in Figure 198 [1]. While the device described in the literature was made from the cathode components of a commercial electron gun [1], the basic design principles shown here may be useful in devices built in many ways.

An indirectly heated cathode assembly in a ceramic base, and containing an alumina-coated W spiral heater assembly, is provided with a second closely fitting Ni tube to hold the organic material. In some cases, the organic material is held on a plug of fine quartz wool to prevent it from falling out or bumping during evaporation. A thermocouple is welded to the source tube to monitor the evaporation temperature. A few $mm^3$ of organic material is added, and the source is installed in the vacuum system. One must be careful to determine that the vapor pressure of the material at the base temperature of the vacuum system will not severely degrade the vacuum. Materials such as long chain alkanes, phthalocyanines, and some metal halides have been handled successfully at temperatures up to 500°C. The film deposition rate, controlled by the temperature, ranged from 0.01 to 5 nm/s, typically 0.1 nm/s. The gun can be used for several years.

## *References*


[1] N. Sato, K. Seki, and H. Inokuchi, Rev. Sci. Instrum. **58,** 1112 (1987).

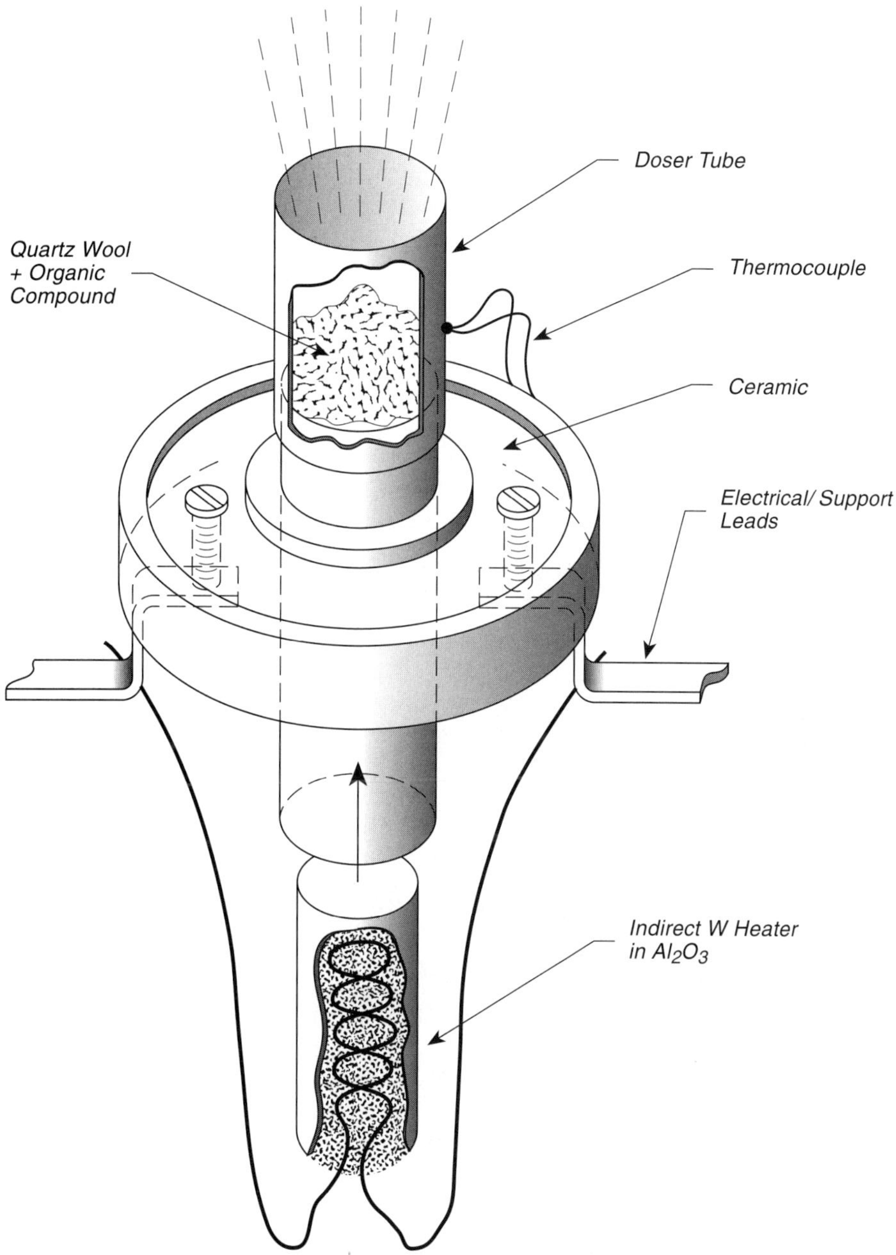


***Fig. 198. Vapor Deposition of Organic Thin Films***

# C. Ions

# 199

# Solid State Cesium Ion Gun Source

The use of Cs as an adsorbate of interest on metals and semiconductors extends back to the early work of Langmuir in this area. Several types of sources have been used, including sources that use elemental liquid Cs [1], and $Cs^+$ surface ionization sources [2,3]. One type of $Cs^+$ source uses Cs vapor that feeds a porous W plug from behind, producing high current density and low beam temperature. This type of source suffers from the need for differential pumping, and the porous W can also clog when the system is vented to air [2,3]. A second type of surface ionization source consists of an aluminosilicate powder deposited on a tungsten base. $Cs^+$ ions are stored in the aluminosilicate and are thermionically emitted [4]. This source requires no differential pumping, but because of the variation in surface potential across its surface, there is a large energy spread of the emitted ions. The source to be described here combines the best features of both of these types of surface ionization sources [4]. Depending upon the energy of the ions desired, different extraction geometries are employed [4].

Figure 199A shows a cross-sectional view of the $Cs^+$ source. A pellet of Cs-M zeolite (cesium mordenite) is made by ion exchange with Na-M zeolite in aqueous solution [5]. Successive exchanges of the $Na^+$ for $Cs^+$ must be carried out, and one must use specially purified CsCl salt (reduced Rb impurity levels) to avoid the incorporation of Rb into the source [6,7]. By heating the exchanged mordenite zeolite to 1050°C, a phase transition from Na-M to Cs-M structure occurs. The powder is then ground to a particle size of 1 μm, placed in a cylindrical die, and heated to 1350°C for 3 h, producing a hard pellet that may be machined slightly with a grinding wheel. The emitting surface of the pellet is then coated with a W film, which provides the surface ionization, as well as a surface of constant potential. The back side of the pellet is coated with a Pt film for electrical contact. When the pellet is heated, the flux of Cs toward the grounded W film can be controlled by the applied bias potential across the pellet. A total charge of 10 C can be extracted from a pellet 0.5 cm in diameter and 1 cm long. Current densities up to 30 $mA/cm^2$ have been extracted.

The time needed to degas the source for operation at less than $10^{-10}$ Torr is about 1 day at 1100°C. When not being used, the source is kept at about 500°C to avoid the need for further outgassing. The advantage of this source is that only $Cs^+$ is emitted, so that measurements of the collected charge can be converted into absolute coverage [4].

In Figure 199A, the W film on the pellet is pressed against an aperture by a spring, and a bias electrode presses against the Pt film on the back of the pellet. Heating by radiation occurs from a 0.25 mm diameter W coil filament, which is coated with alumina powder. Coaxial Ta foils are used as internal shields to reduce the power loss, and 15–18 W brings the pellet to operating temperature (about 1000°C). In this particular design, a Pierce electrode arrangement is used to control the ion focusing for ions in the range 50–500 eV [4]. For ion energies in the range 5–50 eV, a planar diode extraction geometry is used [4], and for ion energies above 500 eV, the Pierce gun is mated to a column containing two Einzel lenses [4]. The performance of the different gun designs is given in the original reference [4], and one example of the Pierce gun performance is shown in Figure 199B. Here, the space charge limitation governs the $V^{3/2}$-dependence of emitted $Cs^+$ current, where $V$ is the ion energy. However, at 920°C, the supply of $Cs^+$ falls near an ion energy of 100 eV, and a larger bias across the pellet restores the performance to that of a space charge limited source.

It has been reported [8] that alkali ion sources involving aluminosilicates in contact with tungsten metal must be "burned in" in ultrahigh vacuum for hours in order to remove the impurity alkalis. Thus, alkalis with low ionization potential will dominate in the ionization process at first, and trace quantities of these impurity alkali atoms will result in a large fractional contamination during first operation of the source. If the source is burned in and then allowed to stand for months in UHV, the impurity ions may reappear in substantial quantity [8]. Thus, it is always wise to use a mass spectrometer for analysis of the alkali ion product from these sources [8]. The application of SIMION ion optical calculations [9] to a large-capacity alkali ion gun has been reported [10].

A careful study of the relative efficiency of zeolite A ion sources for $Li^+$, $Na^+$, $K^+$, and $Cs^+$ has been published [11] and this report contains details of the preparation and evaluation of the various sources. Figure 199C shows the saturation current density for the four zeolite emitters coated on Pt surfaces as a function of emitter temperature, measured in a simple diode arrangement in which the ion emission extrapolated to zero bias was evaluated. An approximately 0.5-$cm^2$ area of Pt mesh emitter was used. A $Rb^+$ ion source, operating with a mixture of $Rb^+$-substituted aluminosilicate zeolite and powdered Mo has been described. Here too a Pierce- type electrostatic focusing structure was used [12].

The investigation of the $Cs^+$ purity from various composition aluminosilicates has been reported [13].

The design of a high-quality ion gun for delivering $Li^+$ ions in the energy range 1–100 eV with an energy spread of only 0.22 eV has been published [14].

## References


[1] J-L. Desplat and C.A. Papageorgopoulos, Surface Sci. **92,** 97 (1980); see also, G.S. Tompa, J.L. Lopes, and G. Wohlrab, Rev. Sci. Instrum. **58,** 1536 (1987); T. Castro and R. Reifenberger Rev. Sci. Instrum. **58,** 289 (1987); R. D. Swenumson and U. Even, Rev. Sci. Instrum. **52,** 559 (1981).

[2] D.G. Welkie, *Secondary Ion Mass Spectrometry SIMS V,* edited by A. Benninghoven et al. (Springer- Verlag, New York, 1986), p. 146.

[3] G.R. Brewer, *Ion Propulsion* (Gordon and Breach, New York, 1970), Chap. 4.

[4] A.E. Souzis, W.E. Carr, S.I. Kim, and M. Seidl, Rev. Sci, Instrum. **61,** 788 (1990).

[5] D.W. Breck, *Zeolite Molecular Sieves* (Wiley, New York, 1977), Chap. 7.
[6] R.E. Weber and L.F. Cordes, Rev. Sci. Instrum. **37,** 112 (1966).
[7] T.E. Madey and J.T. Yates, Jr., J. Vac. Sci. Technol. **8,** 39 (1971).
[8] M.J. Hogan and P.P. Ong, Rev. Sci. Instrum. 61, 1338 (1990).
[9] D.A. Dahl and J.E. Delmore, *The SIMION PC/PS2 User's Manual, Version 3.1,* Informal Report, EGG-CS-7233, Rev. 2 (November 1987). Program available from Idaho National Engineering Laboratory, EG&G Idaho, Inc., P.O. Box 1625, Idaho Falls, ID 83415.
[10] S.T. Beck, D.W. Warner, B.A. Garland, and F.V. Wells, Rev. Sci. Instrum. **60,** 2653 (1989).
[11] Y. Satoh, M. Takebe, and K. Linuma, Rev. Sci. Instrum. **58,** 138 (1987).
[12] D.W. Hughes, R.K. Feeney, and D.N. Hill, Rev. Sci. Instrum. **51,** 1471 (1980).
[13] P.P. Ong and T.L. Tan, Rev. Sci. Instrum. **65,** 3729 (1994).
[14] H.B. Haskell, O. Heinz, and D.C. Lorents, Rev. Sci. Instrum. **37,** 607 (1966).

***Fig. 199. Solid State $Cs^+$ Ion Gun + Other Alkalis***

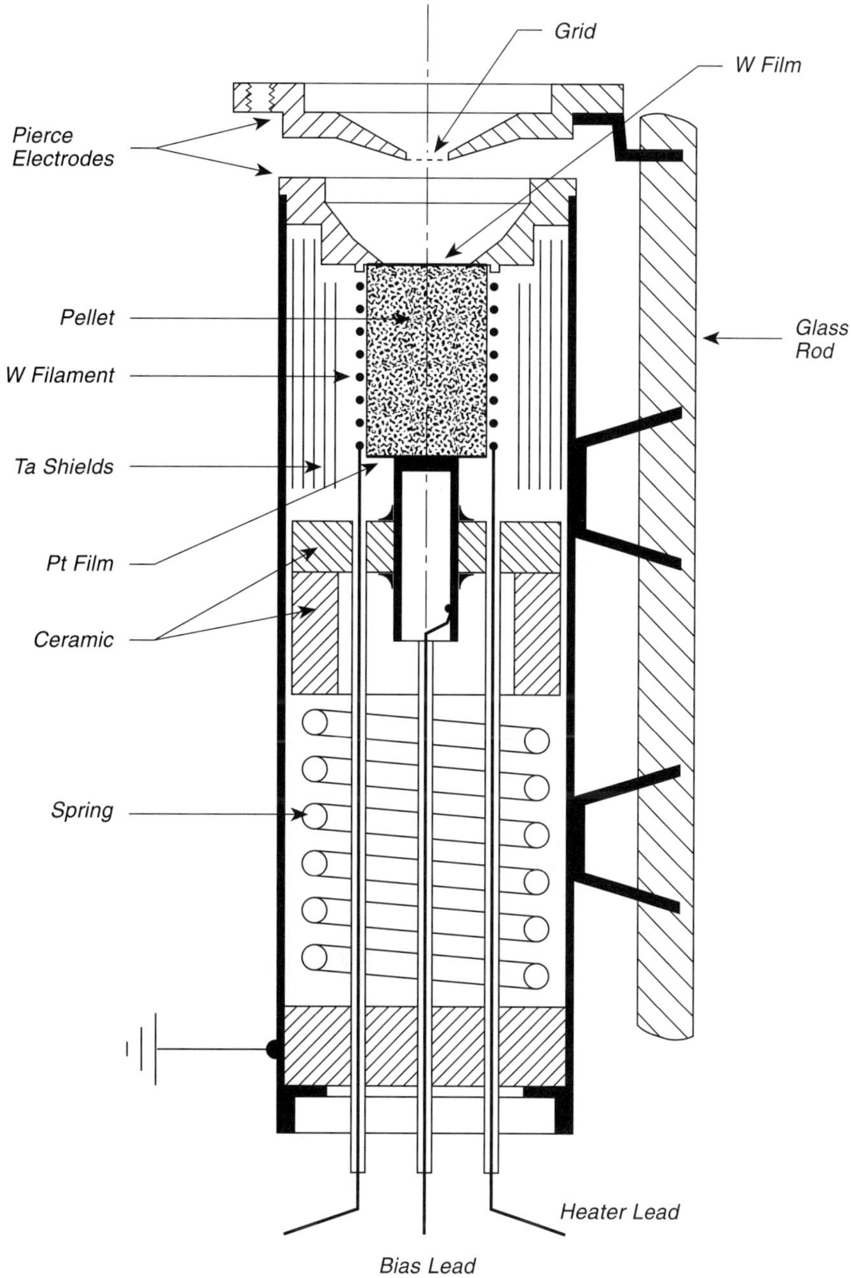


***A. Ion Gun Design***

## *Fig. 199. Solid State $Cs^+$ Ion Gun + Other Alkalis*

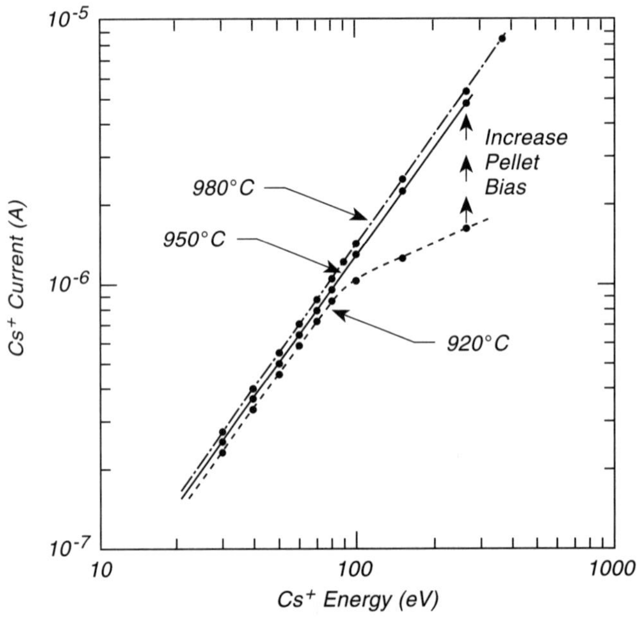


**B. Performance**

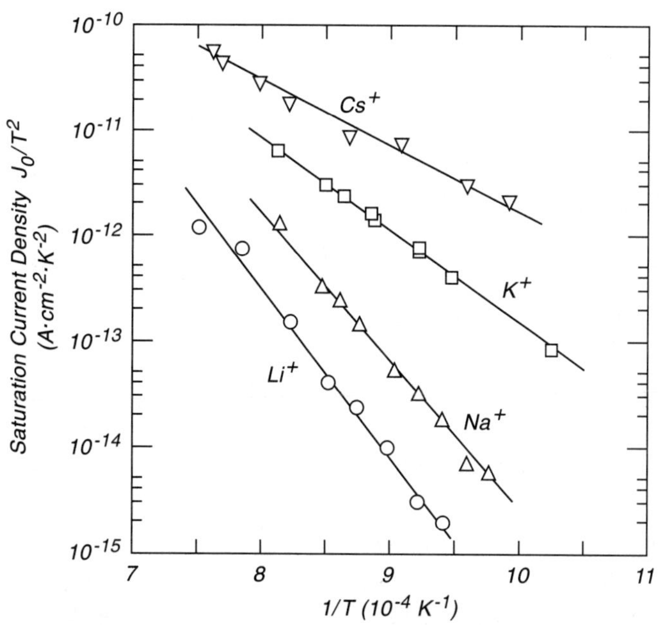


***C. Zeolite Source Performance for Various Alkalis***

# 200

# Alkaline Earth Metal Ion Sources

Although high-temperature sources for production of the alkalis are well known, the use of similar alkaline earth sources is less well established. Recently, a study of the oxides MgO, CaO, SrO, and BaO has been made, and the purity of the emitted $Mg^+$, $Ca^+$, $Sr^+$, and $Ba^+$ ions have been investigated [1].

Filaments (0.25-mm diameter) of either W or Ta were used to support the oxides in front of a quadrupole mass spectrometer as shown in Figure 200A. This was done by making a slurry of the powdered oxide in distilled water and transferring powder to the metal coil. Tungsten wire was used for CaO and MgO, and Ta wire was used for the other oxides, including MgO in later studies, since it was found that MgO reacted with W. Measurements were made of the total ion emission as well as the purity of the emitted ions as a function of the temperature. Temperature measurements were made with a pyrometer.

Figure 200B shows the ion current measured for three of the filaments as the temperature is raised. Currents as high as 1 μA are observed at elevated temperatures, but failure of the filaments occurs rather rapidly at extreme temperatures owing to the reaction of the metal oxide with the refractory metal filament. Figures 200C–E show the impurity behavior for BaO, SrO, and CaO as a function of the temperature. In all cases, alkali metal ion impurities are observed, even though the oxides were of purity of 99.9% or better. This is due to the highly preferred surface ionization of the alkalis as a result of their low first ionization energy. As the filaments are heated to higher temperatures, the alkali ion impurities decrease in yield and rather pure $Ba^+$, $Sr^+$, and $Ca^+$ ions are liberated. The purity of $Mg^+$ from MgO coatings was less than 50% at all temperatures, and this would therefore be an unsuitable source for $Mg^+$.

## *References*

[1] T.L. Tan, Z.L. Zhou, and P.P. Ong, Rev. Sci. Instrum. **66**, 2879 (1995).

***Fig. 200. Alkaline Earth Ion Sources***

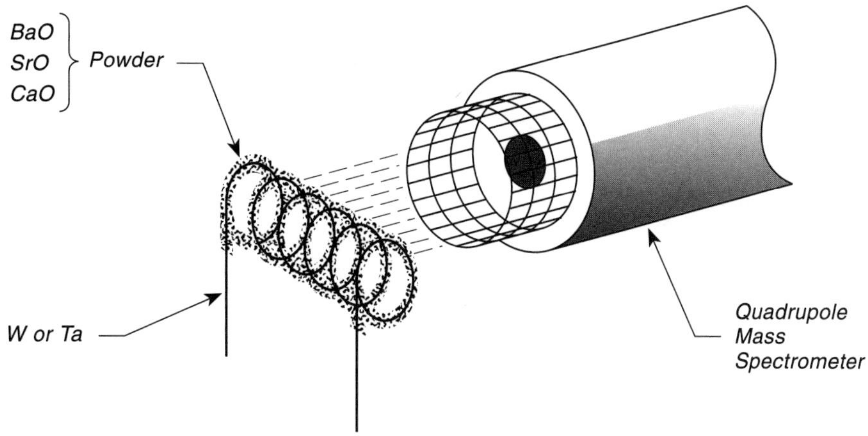


***A. Ion Detection by Mass Spectrometer***

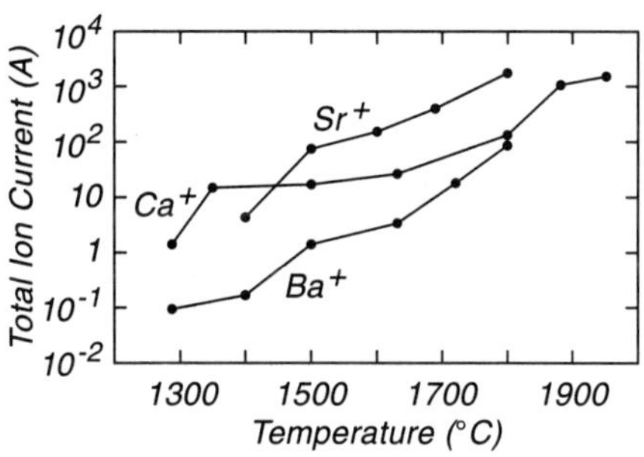


***B. Ion Current vs T***

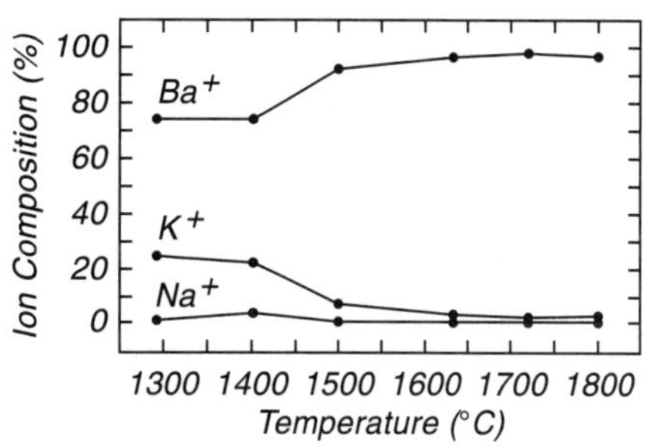


***C. BaO / Ta***

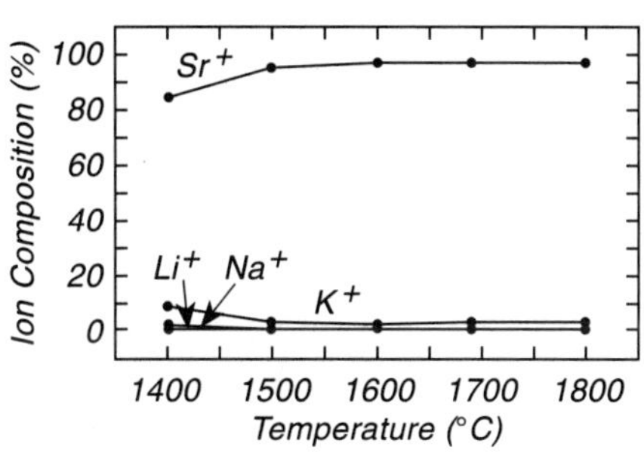


***D. SrO / Ta***

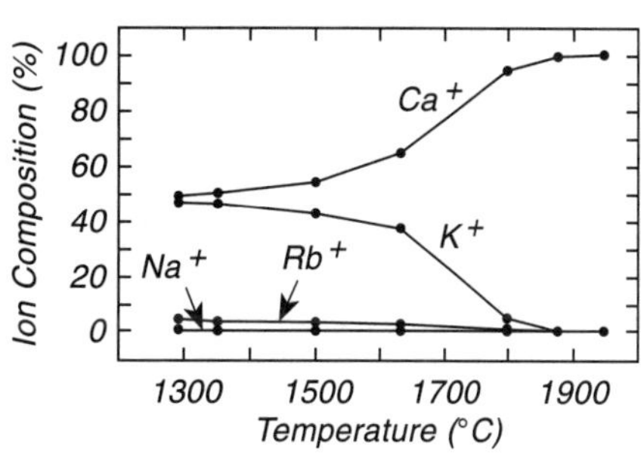


***E. CaO / W***

# D. Active Gases and Species

# 201

# High Flux Atomic Hydrogen Sources—Thermal

Atomic hydrogen is employed for research in the surface science field as a reagent for rapidly producing chemisorbed hydrogen on semiconductor surfaces and also on metal surfaces that do not readily dissociate molecular $H_2$. The surface chemistry of atomic hydrogen has been investigated as early as the work of Langmuir [1,2] who first produced it on a hot tungsten filament. The production of atomic hydrogen has been investigated by a number of workers [3–5]. It has been found that the efficiency of H atom production is proportional to $(P_{H_2})^{1/2}$ at $H_2$ pressures above about $10^{-6}$ Torr where pseudoequilibrium conditions exist. At lower $H_2$ pressures, the rate is proportional to $P_{H_2}$. Under these conditions, the reported efficiency of $H_2$ dissociation is about 0.3 per $H_2$ collision with the tungsten surface at temperatures above 2500 K [3]. It has been observed that the efficiency of dissociation is decreased by impurities often present on a tungsten filament [4].

Two sources that place a relatively high $H_2$ gas density near a heated tungsten surface have recently been described [6,7]. Both of these sources produce large fluxes of atomic hydrogen compared to a bare tungsten filament operating in the vacuum system, if comparisons are made at the same gas throughput to the system's pumps. In Figure 201A a tungsten capillary (0.6-mm ID, 1.6-mm OD) is electron bombarded from a tungsten filament so that the end of the capillary is heated to 1800–1815 K over a 4-mm length. An acceleration voltage of +560 V on the capillary (HV) and a 30 mA emission current achieve this condition. A Cu shield, with a conical hole drilled on axis, serves as a radiation shield. It is cooled by a water circulation tube that is butt-brazed to the solid Cu shield. The sample is placed 10 mm in front of the shield, or 20 mm from the end of the W capillary. The back end of the W capillary is snugly fitted into a ceramic sleeve, which also connects to a 4-mm-diameter stainless tube that delivers $H_2$ gas to the atomizer. Since the gas density is low during operation, no need exists for vacuum tight seals, and the adapter uses only friction fits to the capillary and the tube. The working $H_2$ pressure at the entrance to the W capillary is about $10^{-2}$ Torr. In order to change the position of the H atom doser, it is mounted on a translational device. It is reported that a total atomic H flow of about $3 \times 10^{14}$/s may be achieved under these conditions. The degree of dissociation approaches 50%. Therefore the total gas influx is small, minimizing the background $H_2$ pressure. It is estimated that about 20% of the H atoms are emitted in a central portion of the beam with a cone angle of $\pm 15°$ [8,9].

In Figure 201B, a spiral W filament (1.5-cm diameter, 0.025-cm-diameter wire, 14.5-cm length) [10] is located within a cup-shaped stainless steel doser having a perforated array in its central portion [7]. Hydrogen gas flows into this doser through a tube that is connected to a calibrated conductance aperture (see pages 138–141; 604–615), and the flow rate is adjusted using a capacitance manometer on the high-pressure side of the aperture. A typical flow rate of about 2 x $10^{16}$ $H_2$/s is employed, at a system pressure of 2 x $10^{-7}$ Torr. The incoming $H_2$ encounters a deflector baffle that produces a random flux on the back of the perforated array. The sample is shielded from radiation by the wall of the doser body. In order to reflect the H atoms to the sample (avoiding radiation heating of the sample as well as possible impurities from the hot filament) a cooled Pyrex plate at 120 K is located close to the outlet of the atomic hydrogen source. This reflects atomic H with high efficiency [11,12]. The Pyrex is held onto a stainless steel cooling reservoir with metal clips and is thermally connected to the $l$ - $N_2$ reservoir with a film of In metal.

In studies of the efficiency of the doser [7], it was found that a reduction in efficiency of a factor of 4.8 occurred using the reflector source compared to a line-of-sight geometry. However, experiments working at the same system gas throughput indicate that the doser produces about 50 times more atomic hydrogen than a bare W filament, operating without a doser geometry. Thus, even the factor of 4.8 reduction in efficiency owing to the reflector is compensated by the higher atomic H production rate achieved at the high gas densities in the doser. This source is therefore about a factor of 10 more efficient than a bare W wire in delivering atomic H to the sample for identical $H_2$ gas loads. A monolayer of H atoms (about 4 x $10^{14}$/cm$^2$)was adsorbed by a Si(100) surface at 400 K when 4 x $10^{18}$ $H_2$ molecules had passed through the doser with the W filament operating at 2000 K.

Figure 201C shows the relative efficiency of atomic H production plotted versus the reciprocal temperature (K$^{-1}$). The source produces atomic H at a rate which reaches a maximum near 2000 K. Similar results were reported by George et al. [13].

An interesting study, comparing the pumping speed of atomic H with that of $H_2$ during adsorption on enginering surfaces in the range 2.5–4.3 K, has been published [14]. Interestingly, the pumping speed for $H_2$ is between a factor of 3 and 40 above that of atomic H. This difference in either condensation or adsorption is attributed to the higher internal energy of the atomic H compared to $H_2$.

A hydrogen atom source involving a design similar to that of reference [6] has been carefully evaluated to determine the efficiency of atomic H production [15]. A second similar source has also been evaluated by mass spectrometric methods [16]. In addition, atomic H beams, cooled by collision with surfaces near 6 K, have been produced [17].

A number of detectors for atomic H have been devised and are reviewed briefly in reference [18]. One of these [18] involves the use of a thermistor (primarily NiO) for measurement of an atomic H flux. Electrical conductivity changes, measured with a Wheatstone bridge and phase-sensitive detection, were used for the evaluation of the flux of atomic H made in a microwave discharge. Control experiments eliminated the possibility of effects owing to the influence of heating and radiation on the detector. The mode of operation of the detector is unknown and could be due to local heating as a result of H atom recombination or to electronic surface effects. The lower limit of sensitivity is estimated to be $10^{11}$ H/cm$^2$ s.

An atomic H source involving the use of a W/26 % Re alloy tube ohmically heated to 2400 K has been described. The W/26 % Re alloy does not have the brittle properties of pure W [19].

Many interesting uses for atomic H have been described in the literature and are beyond the scope of this book.

## *References*


[1] I. Langmuir, J. Amer. Chem. Soc. **34,** 1310 (1912).

[2] I. Langmuir, J. Amer. Chem. Soc. **37,** 417 (1915).

[3] J.N. Smith, Jr. and W.L. Fite, J. Chem. Phys. **37,** 898 (1962).

[4] D. Brennan and P.C. Fletcher, Proc. Roy. Soc. London Ser. **A250,** 389 (1959).

[5] T.W. Hickmott, J. Chem. Phys. **32,** 810 (1960).

[6] U. Bischler and E. Bertel, J. Vac. Sci. Technol. **A11,** 458 (1993). W capillary for this source may be obtained from Metallwerke Plansee GmbH, A-6600 Reutte/Tirol, Austria, Tele: 43 5672 70 0; FAX: 43 5672 70 500. The entire source is available from Physikalisches Buro Steinmuller GmbH, Rinnerstr. 16d, A-6074 Rinn/Aldrans, Austria.

[7] K.H. Bornscheuer, S.R. Lucas, W.J. Choyke, W.D. Partlow, and J.T. Yates, Jr., J. Vac. Sci. Technol. **A11,** 2822 (1993).

[8] B.A. Von Gottwald, Vakuum Tech. **22,** 106 (1973).

[9] C.T. Campbell and S.M. Valone, J. Vac. Sci. Technol. **A3,** 408 (1985).

[10] Spiral W filaments in standard or special dimensions may be obtained from CM Furnaces Inc., 103 Dewey Street, Bloomfield, NJ 07003.

[11] H. Wise and B. J. Wood, Adv. At. and Mol. Phys. **3,** 291 (1967).

[12] J.P. Toennies, W. Welz, and G. Wolf, J. Chem. Phys. **71,** 614 (1979)

[13] B.E. Koehler, C.H. Mak, D.A. Arthur, P.A. Coon, and S.M. George, J. Chem. Phys. **89,** 1709 (1988).

[14] V.G. Luppov, M. Mertig, T. Roser, B.S. van Guilder, B. Vuaridel, Yu. M. Melnik, and A.F. Prudkoglyad, Rev. Sci. Instrum. **62,** 2738 (1991).

[15] B. Van Zyl and M.W. Gealy, Rev. Sci. Instrum. **57,** 359 (1986). It is interesting to note that in these studies, the maximum in $H_2$ dissociation probability is not reached until tungsten temperatures above 2400 K are employed. This may be due to the presence of gas phase collisions here, where collisionless conditions generally apply in sources used for surface studies in ultrahigh vacuum.

[16] H. Koschmieder and V. Raible, Rev. Sci. Instrum. **46,** 536 (1975).

[17] A. Hershcovitch, A. Kponou, and T.O. Niinikoski, Rev. Sci. Instrum. **58,** 547 (1987).

[18] K.C. Harvey and C. Fehrenbach, Jr., Rev. Sci. Instrum. **54,** 1117 (1983).

[19] M. Eisenstadt, Rev. Sci. Instrum. **36,** 1878 (1965).

***Fig. 201. High Flux Atomic H Sources***

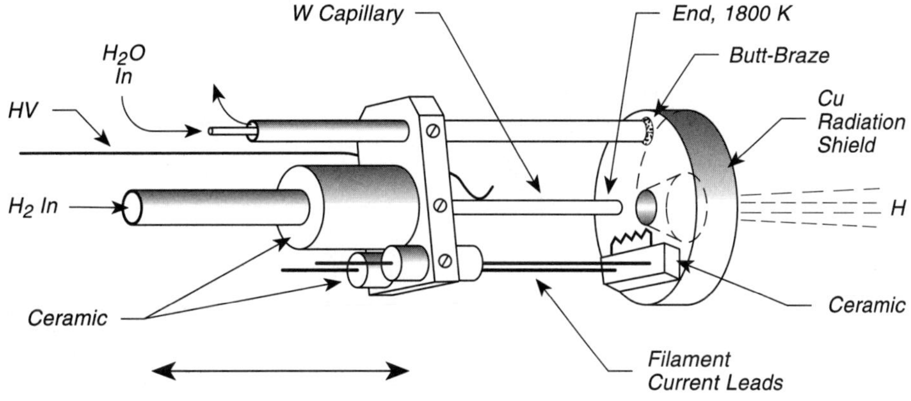


***A. H Atom Source—Hot Tube Design***

***Fig. 201. High Flux Atomic H Sources***

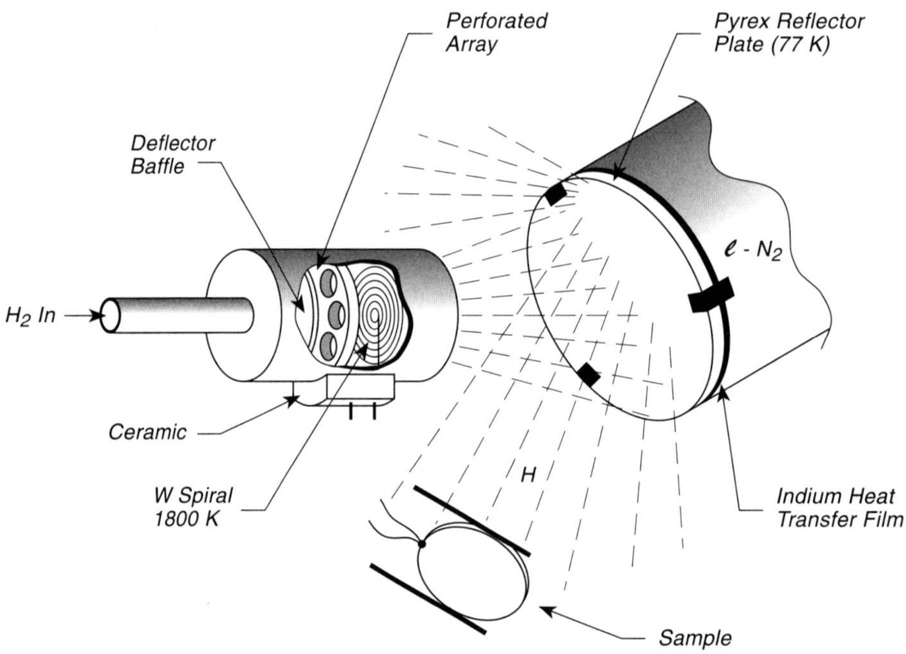


***B. H Atom Source—Spiral Filament Design***

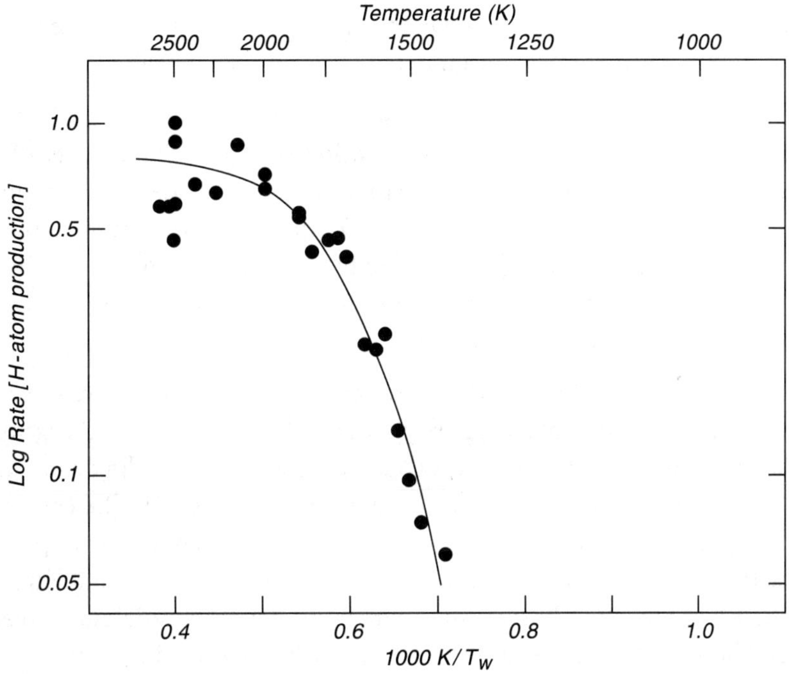


**C. Efficiency of Design B**

# 202

# Atomic Hydrogen Beam Source—Radio Frequency Driven

The use of thermal sources for the production of atomic H has been described on page 682. It is also possible to produce beams of atomic H using microwave discharge sources. These sources are convenient to utilize but one must be cautious that impurities and radiation from the discharge region do not affect the experiments. There are many different RF discharge sources available, as described in reference [1]. A typical source not unlike others that have been made [1] is described below.

Figure 202 shows a diagram of a Pyrex glass source of simple design. It is capable of producing atomic H with up to 95% dissociation of $H_2(g)$. The density of atomic H ranges from 5 x $10^{11}$ to 2 x $10^{13}$ H/cm$^3$. The tube is 18-mm inside diameter and 240 mm long. It is terminated at each end by a 1-mm-ID capillary to prevent the discharge from spreading outside the source, and it is cooled by a water jacket to reduce the temperature and to reduce H atom recombination on the inner walls [2]. This may be responsible for the long life and stability of this design.

The discharge is excited by about 25 W of RF power into a coaxial cavity resonant at 35 MHz. The cavity consists of a copper cylinder, surrounding the discharge tube, and a 12-turn helix of constant pitch. Power is coupled via a single-turn link coupling. Details of cavity design and cavity dimensions have been published [3]. The coaxial cavity is properly impedence-matched to the generator to insure 90% power absorption into the cavity. The RF generator is properly designed to prevent the propagation of unwanted RF signals. The RF oscillator and driver stages are housed within individual screened compartments. The amplifying stages are narrow band with sharply tuned filters in each stage, eliminating harmonics. Notch filters are used in the power input to prevent RF propagation into the laboratory AC supply. Low noise was achieved in auxiliary instruments when the RF supply was operating.

A hydrogen pressure of about 0.15 Torr was used in the source. Great care was taken to clean the discharge tube before use, successively washing with hot chromic acid (in sulfuric acid), water, acetone, HF(aq), and distilled water. A palladium leak (see page 730) was used to purify the $H_2$ before use and back diffusion of contaminants was avoided. The tube operated for about 5000 h without trouble.

## *References*


[1] J. Slevin and W. Stirling, Rev. Sci. Instrum. **52,** 1780 (1981).

[2] B.J. Wood and H. Wise, J. Phys. Chem. **66,** 1049 (1962).

[3] W.W. MacAlpine and R.O. Schildknecht, Proc. IRE, 2099 (1959).

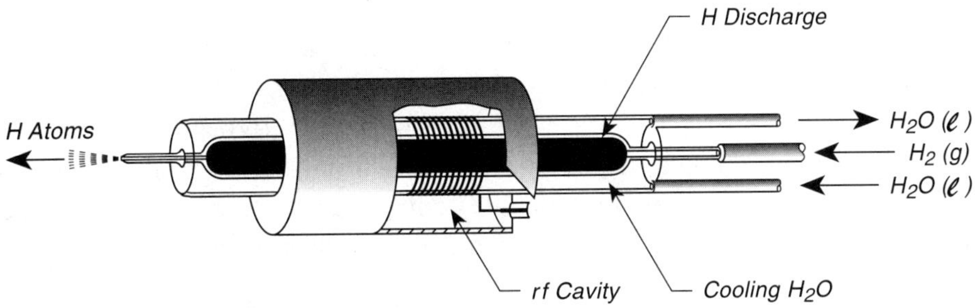


*Fig. 202. H Atom Source*

# 203

# Atomic Fluorine Source

The delivery of reactive atomic species in high purity to surfaces poses one of the difficult experimental problems in adsorption and surface reaction studies. Species such at atomic F are particularly difficult to prepare and transfer without secondary reactions on the walls of the source. An excellent design [1], probably applicable in general to producing a number of reactive species, is shown in Figure 203.

A low pressure 10%$F_2$/90% He mixture is flowed through an alumina tube which passes through an Evanson-type microwave discharge cavity, and the products are then pumped through the apparatus using a rotary pump which is corrosion resistant. The microwave discharge products are admitted to the preparation chamber through a 25-μm-diameter orifice mounted on the end of the outer stainless steel tube. The orifice was laser-drilled (see page 604) in a stainless steel sheet of 25 μm thickness. The alumina tubing inside the source is thought to become passivated with the formation of a nonvolatile layer of $AlF_3$, and is superior to a tube made of silica [1,2]. The microwave discharge source [3] is operated at 30–60 W of power in the total pressure range of 0.05–10 Torr.

The species emitted from the orifice were studied with a modulated quadrupole mass spectrometer, which measured the consumption of $F_2$ ( 38 amu). The results show that essentially complete dissociation occurs below a total pressure of 2 Torr, as shown in Figure 203B. When working in the molecular flow regime below a pressure of 5.7 Torr, the angular distribution is described as having a $\cos^2$ distribution, and for sample sizes of the order of 0.5 cm at source-to-sample distances of about 1 cm, the fluence is rather uniform. It is calculated that at a total pressure of 0.7 Torr, the source delivers F atoms at a flux of about $1 \times 10^{14}$/cm$^2$ s in the beam with a random flux from the source and the chamber being at about 5% of this beam flux.

Microwave sources are widely used in molecular beam studies [2,4], but this source is one of the first such sources to be successfully employed in surface studies [5]. It is also useful for generating chlorine atoms. For chlorine, a quartz tube may be used instead of an alumina tube [6].

It should be noted that large gas loads accompany the use of this source, and that a rapidly pumped preparation chamber, isolated from the ultrahigh vacuum surface analysis chamber, is one way to reduce the gas load in the analysis chamber when the source is being used.

## *References*

[1] C.D. Stinespring, A. Freedman, and C.E. Colb, J. Vac. Sci. Technol. **A4,** 1946 (1986).
[2] C.E. Kolb and M. Kaufman, J. Phys. Chem. **76,** 947 (1972).

[3] Microwave sources are available from Opthos Instruments, Inc., 17805 Caddy Drive, Rockville, MD 20855.

[4] S. Sibener, R.J. Buss, C.Y. Ng, and Y.T. Lee, Rev. Sci. Instrum. **51,** 167 (1980).

[5] C.D. Stinespring and A. Freedman, Appl. Phys. Lett. **48,** 718 (1986); A. Freedman, J. Appl. Phys. **75,** 3112 (1994); G. N. Robinson, A. Freedman, and R. L. Graham, Langmuir **11,** 2600 (1995).

[6] A. Freedman and C.D. Stinespring, J. Phys. Chem. **96,** 2253 (1992).

**Fig. 203. Microwave Source of F Atoms**

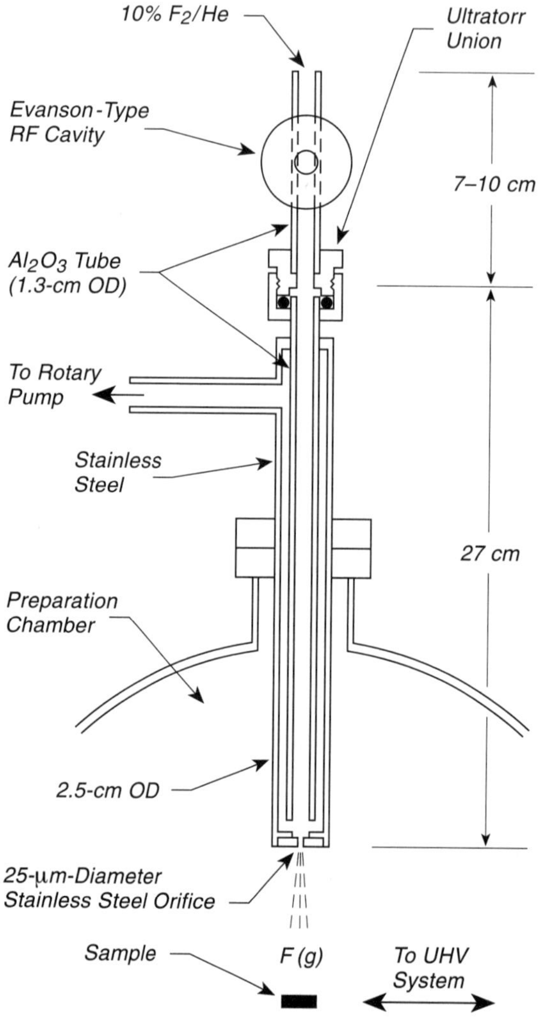


**A. Source**

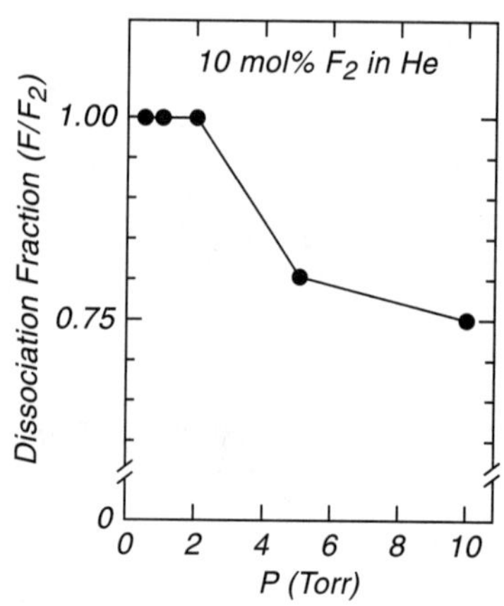


**B. Performance vs. Pressure**

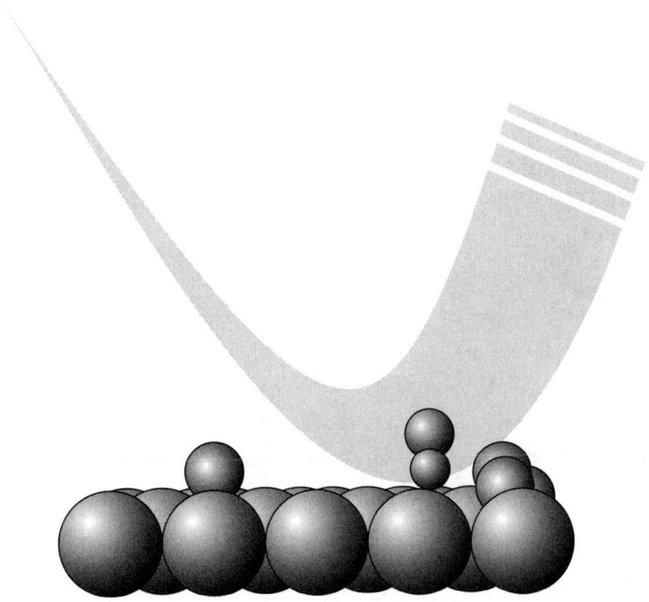

# 204

# Alkali Metal Sources

The alkali metals form one of the most interesting class of adsorbates on the surfaces of both metals and semiconductors [1]. As a result they have been widely studied since the pioneering work of Langmuir in the 1930s [2]. The alkali metals are also used in the production of low-work-function photocathode surfaces, and in other applications in the field of physical electronics. As a result, excellent commercial sources of alkalis for research are available from SAES getters [3].

Figure 204A shows a view of a typical SAES alkali getter source; these are available to deliver Li, Na, K, Rb, and Cs. A powdered mixture of the alkali metal chromate, $M_2CrO_4$, and a reducing agent (St 101 alloy—84 % Zr and 16 % Al) is held in a nichrome boat with a very narrow slit running along its length, as shown. A nichrome retaining wire is squeezed along the backside of the slit to prevent powdered material from being ejected from the slit. Butting pieces of two designs are described, which block the ends of the boat and which serve as connectors for electrical heating. As current is passed, the temperature rises into the 550–850°C region, as indicated in Figure 204B, and the solid state reduction reaction liberates the alkali metal which escapes from the slit. The Zr also acts as a getter for impurity gases such as $H_2$ and CO. The typical operating temperature for deposition must be adjusted on the basis of experience, but it is usually preferable to use lower rates for controlled deposition experiments. The typical amount of $H_2$ emitted from one of the sources is about 1 Torr-cm$^3$ per cm of getter length, and degassing is achieved by initially working at 500°C. Various geometrical shapes and sizes are available.

The solid state oxidation-reduction reaction between the reducing agents, Zr and Al in the metal alloy reactant can be written as shown in Figure 204C, but the exact details of the reaction are probably not well worked out. An excess of the Zr-Al St 101 alloy is employed so as to have material for the gettering of active gases during the reaction.

In addition to alkali metal getters, SAES also makes alkaline-earth Ba getter sources of a different configuration from the alkali metal sources (Type St 15A/20HY/400 and St 15/20KLS/200). The reactant consists of 45% $BaAl_4$ alloy mixed with 55% Ni powder. This composition, on being heated to about 850°C, reacts exothermically instantaneously releasing about 25%–30% of its Ba. Continued heating by RF methods releases the remainder of the Ba. These sources are used to process vacuum tubes by throwing a heavy getter film of Ba, which is active for CO [4], $H_2$ and other gas adsorption. An SAES source of Ba geometrically similar to the alkali sources is the type St 14/Fr, which can be ohmically heated as is done with the alkali sources.

A description of an alternate design for a Cs ion source, involving diffusion of Cs through a heated porous W plug, is given in reference [5].

Alkali metal sources may also be based on the direct evaporation of the metal, and many such sources have been described. A novel source, which delivers large quantities of Na as an atomic beam, has been described; it employs a metal mesh wick to return excess Na to the evaporator section. This source also works under ultrahigh vacuum conditions [6]. In addition, zeolite sources containing alkalis are often employed, and one of these is discussed on page 672.

## *References*

[1] H. Bonzel, A.M. Bradshaw, and G. Ertl, *Physics and Chemistry of Alkali Metal Adsorption* (Elsevier, Amsterdam, 1989).
[2] J.B. Taylor and I. Langmuir, Phys. Rev. **44,** 423 (1933).
[3] Getter sources may be obtained from SAES Getters S.p.A., Head Office, Via Gallarate, 215, 20151 Milano, Italy. USA Offices: SAES Getters/U.S.A. Inc., 1122 E. Cheyenne Mtn. Blvd., Colorado Springs, CO 80906.
[4] F. Ricca and P. Della Porta, Vacuum **10,** 215 (1960).
[5] D.L. Schaefer, Rev. Sci. Instrum. **41,** 274 (1970).
[6] L.V. Hau, J.A. Golovchenko, and M.M. Burns, Rev. Sci. Instrum. **65,** 3746 (1994).

**Fig. 204. Alkali Metal Sources**

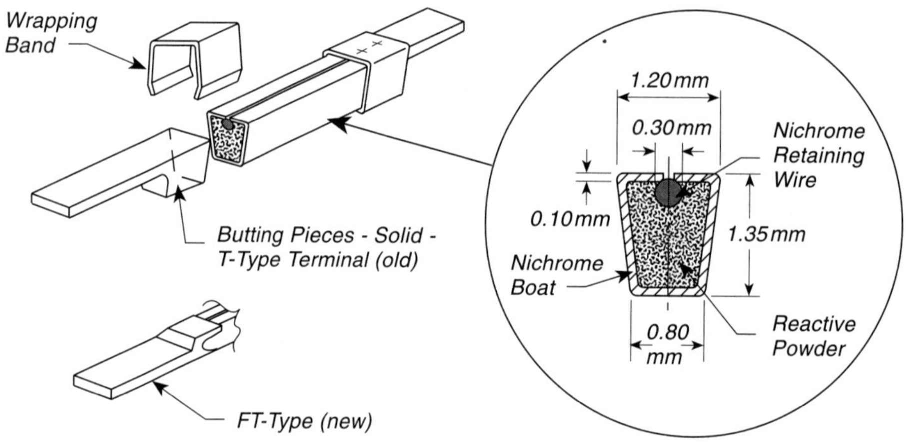


**A. Structure of SAES Alkali Metal Dispenser**

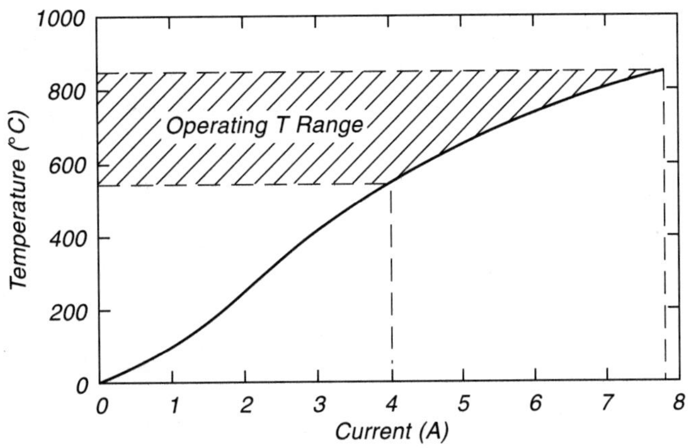


***B. Typical Temperature—Current Performance***

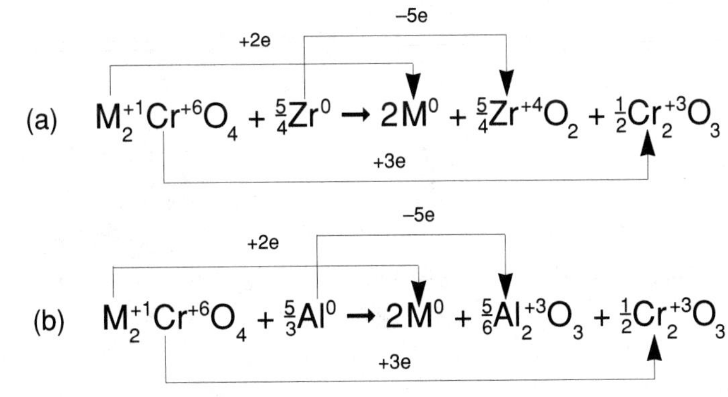


***C. Solid State Reaction—Using Zr or Al***

# 205

# Thermal Source of Methyl Free Radicals

The involvement of methyl radicals in the surface chemistry of hydrocarbon reactions is often encountered. However, because of the difficulty of producing clean sources of $CH_3$, this radical species has been neglected in adsorption studies. Pyrolysis of molecules containing methyl groups has been the preferred method for the generation of methyl radicals [1–5], but most methods also generate other reactive species that can cause interference in studies of the methyl radical surface chemistry. The source described here avoids this problem by using azomethane as the source of the $CH_3$ and by working at molecular flow conditions where high pressure gas phase processes to produce other highly reactive molecules are avoided to a high degree [6]. The pyrolysis reaction

$$CH_3NNCH_3 \longrightarrow 2CH_3 + N_2 \qquad (1)$$

is a fairly clean reaction, and if gas phase reactions of the methyl radicals are avoided, then only surface reactions in the source can contribute impurities.

Figure 205 shows a schematic design of the pyrolysis apparatus. Azomethane is allowed to pass through a quartz tube that has its central section heated by means of a closely wound Ta wire, 0.25-mm diameter, which is cemented to the outside of the pyrolysis tube using a ceramic cement (Aremco Type 576). The quartz tube is connected to the source of azomethane using a graphite ferrule [7] in a swagelock reducing union. The thermocouple embedded into the heating section allows the temperature to be adjusted to the optimum one for complete conversion of azomethane to pyrolysis products, which occurs near 1200 K. A cooling section beyond the pyrolysis region allows the $CH_3$ radicals to become accommodated at a lower temperature than their production temperature. Various lengths of the three zones shown in Figure 205B were studied [6]. A typical mass spectrum for azomethane and for the completely pyrolyzed product is shown in Figure 205C. In addition to $CH_3$, measurable fractions of $CH_4$, $C_2H_6$, and $H_2$ are produced along with $N_2$ at 1173 K and a chamber pressure of $5 \times 10^{-7}$ Torr.

The azomethane was synthesized using the method of Renaud and Leitch [8] and was stored in a stainless steel container over a period of several months without decomposition. Ar was mixed with the azomethane as an internal standard for quantitative flow measurements [6]. A quadrupole mass spectrometer was used to acquire the mass spectra shown, working inside a stainless steel chamber pumped by a turbomolecular pump [6].

## *References*

[1] F. Kalos and A.E. Grosser, Rev. Sci. Instrum. **40,** 804 (1969).

[2] K.N. Walzl, C.F. Koerting, I.M. Xavier, and A. Kuppermann, J. Chem. Phys. **86,** 88 (1987).

[3] F.A. Elder, C. Giese, B. Steiner, and M. Inghram, J. Chem. Phys. **36,** 3292 (1962).

[4] P. Chen, S.D. Colson, W.A. Chupta, and J.A. Berson, J. Phys. Chem. **90,** 2319 (1986).

[5] J. Dyke, N. Jonathan, E. Lee, and A. Morris, J. Chem. Soc, Faraday II **72,** 1385 (1976).

[6] X-D. Peng, R. Viswanathan, G.H. Smudde, Jr., and P.C. Stair, Rev. Sci. Instrum. **63,** 3930 (1992).

[7] Alltech Assoc., Inc. 2051 Waukegan Rd., Deerfield, IL 60015.

[8] R. Renaud and L.C. Leitch, Can. J. Chem. **32,** 545 (1954). Proper caution should be employed in making azomethane. The precursors are poisonous, and there is a danger of explosion in glass apparatus, especially in the presence of ground glass. Special azomethane production kits are available.

**Fig. 205. Methyl Radical Source**

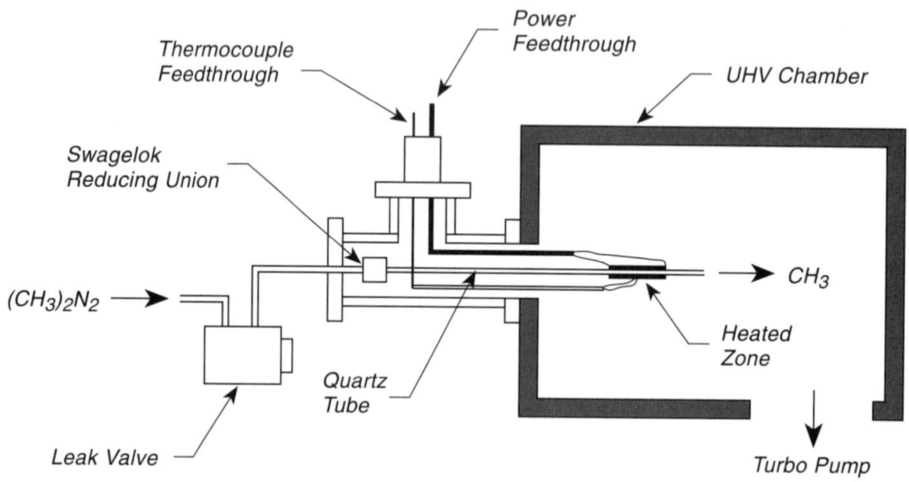


**A. Methyl Radical Source**

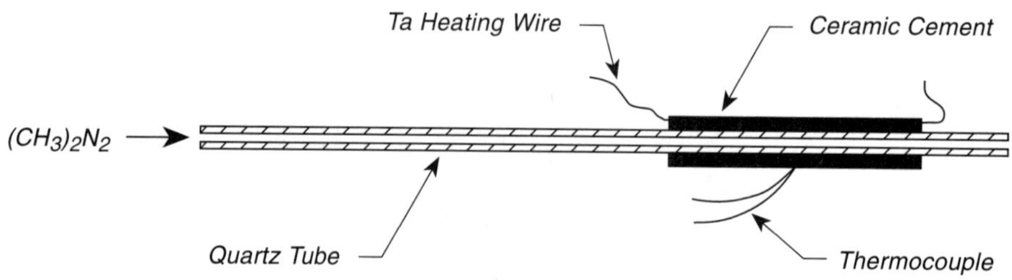


**B. Pyrolysis Tube**

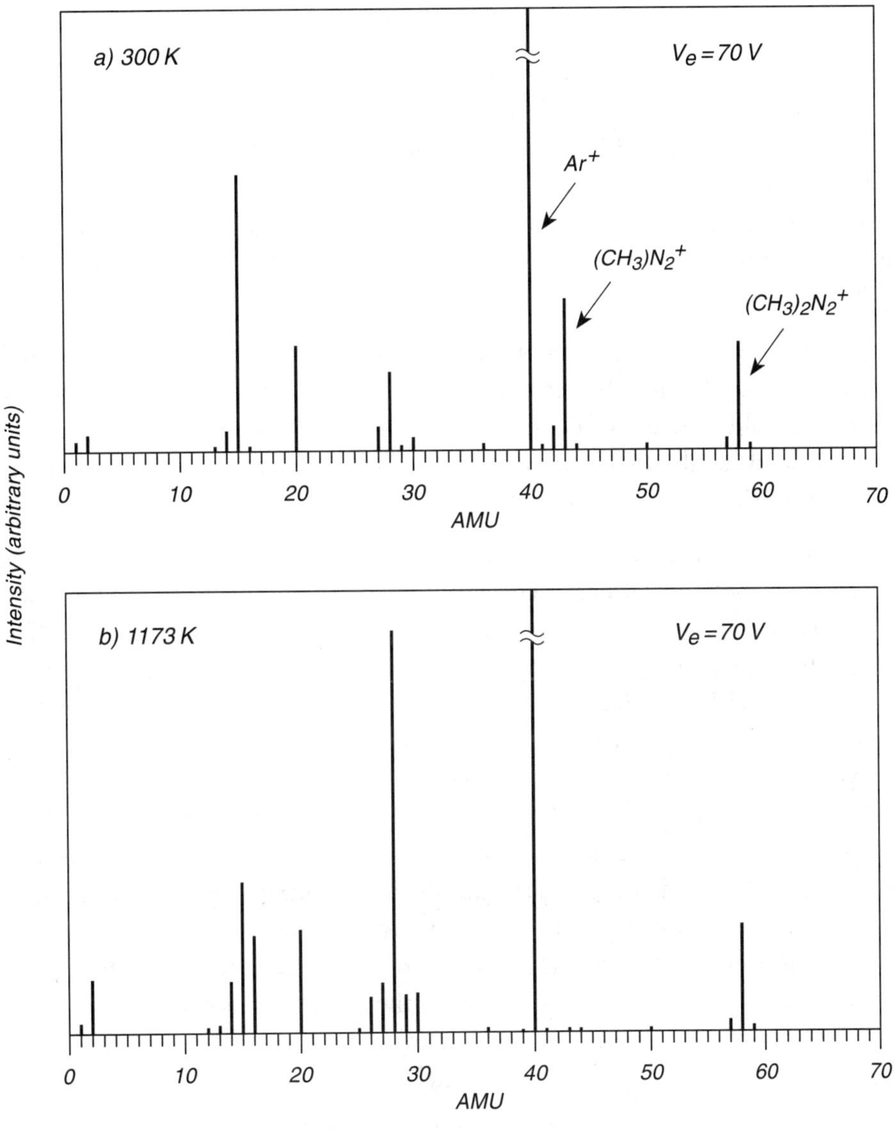


**C. Mass Spectrum of Gas from Source**

# 206

# Production, Storage, and Use of Ozone

The ozone molecule, $O_3$, is a powerful oxidizer, useful in situations where oxidation of surfaces is needed under mild conditions. It has been used, for example, in the growth of $Ba_2YCu_3O_{7-\delta}$ thin films [1]. However, when stored as the liquid at cryogenic temperatures, $O_3$ is dangerous and may explode. The apparatus described here avoids this problem by storing $O_3$ by adsorption in the pores of silica gel at temperatures slightly above its boiling point [2]. The rate of decomposition under these storage conditions is about 5% per week.

Ozone is toxic at a concentration of around 1 ppm by volume. The odor threshold in air is about 0.01–0.02 ppm. The threshold limit value (TLV) is 0.1 ppm, which is above the odor threshold. The use of ozone must involve safe storage, use of inert surfaces such as passivated stainless steel, quartz, Pyrex glass, Teflon, or passivated aluminum, and constant monitoring with an ozone detector [3].

Figure 206 shows an oxygen cylinder containing dry oxygen that feeds a commercial ozone generator (which produces a corona discharge). The rate of production of $O_3$ is such that a concentration of $O_3$ of 8 weight % is produced steadily at 1 atm with an $O_2$ feed rate of 1.7 L atm/min. This corresponds to the production of about 11 g of $O_3$/h. The $O_3$ is trapped at 195 K on 300 g of silica gel [4] up to levels of about 7 weight %, so that in 2 h of operation this limit will be reached. Excess oxygen is vented to the atmosphere with Valve A open and all other valves (B–D) closed. The gel should display a deep blue color at this loading level. To transfer small quantities of $O_3$ into the vacuum chamber, Valve A is closed, and Valves B and C are opened. Under these conditions, a steady pressure of $O_3$ will be evolved and pumped into the rotary pump, with a pressure of about $10^{-2}$ Torr, as monitored by the capacitance manometer. Valve D may then be opened to admit $O_3$ to the vacuum chamber. This procedure has the advantage that the flowing $O_3$ is continuously produced by vaporization from the silica gel, maintaining purity during the exposure of samples in the vacuum system to the tiny fraction of $O_3$ that arrives through the doser tube, compared to that which is exhausted.

The system is equipped with $O_3$ destruct tubes containing copper sponge that catalytically decomposes $O_3$. These are also available commercially. The rotary pump oil should be non-hydrocarbon-based, using either Fomblin [5] or Krytox [6]. Teflon O-rings should be used in all locations, and not rubber O-rings, since ozone is well known to destroy rubber by attacking carbon-carbon double bonds. Extreme care should be taken to keep the dry ice + acetone trap cold at all times when $O_3$ is trapped on the silica gel.

An alternate method [7] employs a cryopump controlled by a thermocouple feedback to boil off $O_3$ from liquid $O_3$ at a precisely controlled temperature. The $O_3$ is condensed in a vacuum-jacketed reservoir whose temperature is precisely adjustable through the operation of a cryopump. These authors provide a rupture disk for this $O_3$ storage system in case of an explosion, which is a possibility for liquid $O_3$ stored in this manner.

A simple electrical discharge apparatus for producing 100% pure $O_3$ in a nonflow system, and for continued repurification, has been described. Less than 100 $cm^3$ of pure oxygen at atmospheric pressure is converted in a batch operation, and purification occurs by pumping traces of residual $O_2$ away at 77 K [8].

The calibration of a mass spectrometer for $O_3$ and the measurement of its cracking pattern in a magnetic mass spectrometer has been carried out carefully, using a 5.3-L storage volume (Teflon coated) to receive the $O_3$ from a cold trap after pumping off impurity $O_2$, etc. The storage volume was the source for a molecular beam of $O_3$ that entered the ionizer of the mass spectrometer. It was found that the half-life of $O_3$ at a pressure of 0.1 Torr in the storage volume was about 12 h at room temperature [9].

## *References*


[1] a) D.D. Berkley, A.M. Goldman, I. Maps, B.R.Johnson, J. Morton, and T. Wang, Rev. Sci. Instrum. **60,** 3769 (1989); b) D.D. Berkley, B.R. Johnson, N. Anand, K.M. Beauchamp, L.E. Conroy, A.M. Goldman, K. Mauersberger, M.L. Mecartney, J. Morton, M. Tuominen, and Y.-J. Zhang, Appl. Phys. Lett. **53,** 1973 (1988).

[2] E. Coleman, T. Siegrist, D.A. Mixon, P.L. Trevor, and D.J. Trevor, J. Vac. Sci. Technol. **A9,** 2408 (1991); see also G.A. Cook, A.D. Kiffer, C.V. Klumpp, A.H. Malik, and L.A. Spence, *Advances in Chemistry Series 21* (Amer. Chem. Soc., Washington, DC, 1959) p. 44.

[3] *Matheson Gas Data Book,* 6th ed. (Matheson Gas Products, Secaucus, NJ, 1980), pp. 574-577.

[4] Aldrich #24,982-3; CAS (7631-86-9). Sold by Aldrich Chemical Co., P.O. Box 2060, Milwaukee, WI 53201.

[5] Fomblin is a registered trademark of Montedison Specialty Chemicals, Milano, Italy.

[6] Krytox is a registered trademark of E.I. dupont De Nemours and Co., Inc., Wilmington, DE.

[7] S. Hosokawa and S. Ichimura, Rev. Sci. Instrum. **62,** 1614 (1991).

[8] M. Griggs and S. Kaye, Rev. Sci. Instrum. **39,** 1685 (1968).

[9] S. Anderson and K. Mauersberger, Rev. Sci. Instrum. **52,** 1025 (1981).

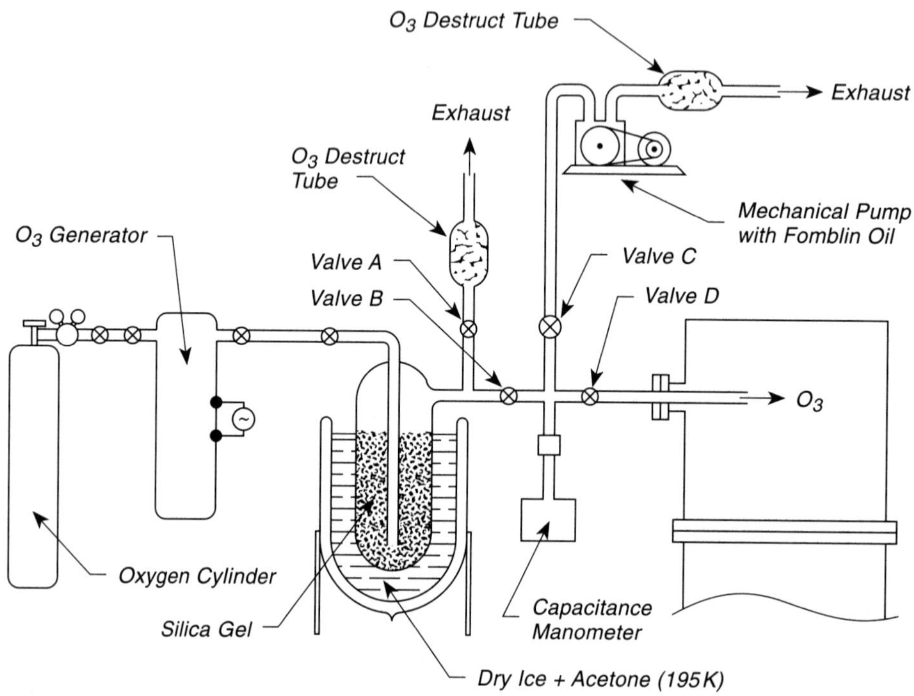


***Fig. 206. System for Production, Storage, and Use of Ozone***

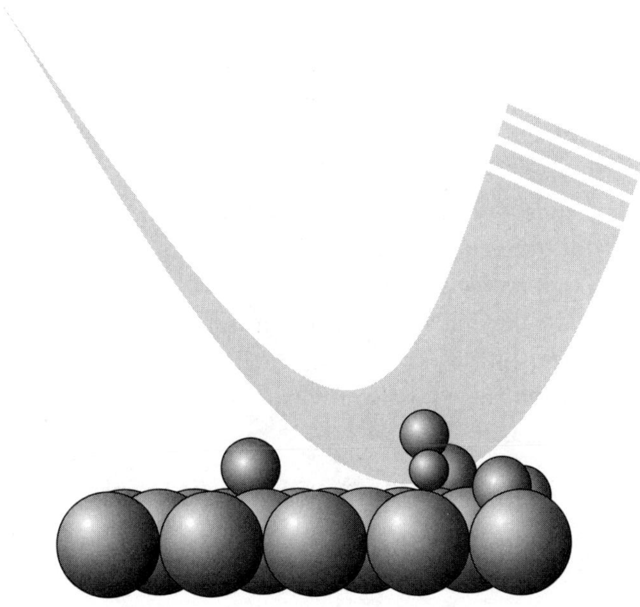

# 207

# Detection of Ozone or Atomic Oxygen by Oxidation of Silver Films

Silver reacts only slowly with molecular oxygen, while its reactivity with $O_3$ is very rapid. In the apparatus shown in Figure 207A, $O_3$, originating from liquid $O_3$ vaporization in an $O_3$ cryogenic storage system, is incident on a Ag film (500 Å thickness) evaporated onto a quartz crystal microbalance [1] (see pages 702 and 440). The rate of increase of mass is proportional to the $O_3$ flux in the initial stages of oxidation, and then falls off to a lower rate as bulk diffusion effects begin to control the growth kinetics as shown in Figure 207B. The $O_3$ angular distribution from the delivery tube (stainless steel) was also measured by moving the detector off axis, maintaining a constant tube-to-detector distance of 6.3 cm, and a constant $O_3$ flow rate of 1.9 standard $cm^3\ s^{-1}$. The profile of $O_3$ is Gaussian (Fig. 207C), and when Ar diluent was used, the flux of $O_3$ decreased and the angular profile was broadened somewhat. It was estimated that the detection efficiency of the Ag film was 26% for $O_3$, assuming that one O atom from $O_3$ oxidizes the Ag, and that molecular $O_2$ is evolved. This efficiency assumes that the $O_3$ is 100% pure when it exits the delivery tube, and if this is not so, then the detection efficiency is higher. The near 100 % purity of the distilled $O_3$ was chemically checked using reaction with KI solution [2], and also by its vapor pressure. $O_3$ fluxes in the range $10^{14}$–$10^{16}\ cm^{-2}\ s^{-1}$ were measured.

The fate of the $O_3$ in the film growth system was studied by placing $O_3$ detectors in various positions outside the beam. It was found that almost no $O_3$ was detected, suggesting that efficient wall decomposition reactions occur, converting $O_3$ to $O_2$.

A similar study, involving atomic O made in an RF discharge, was made using Ag films on a quartz crystal microbalance [3].

## References


[1] J.T. Kucera, J.D. Perkins, K. Uwai, J.M. Graybeal, and T.P. Orlando, Rev. Sci. Instrum. **62**, 1630 (1991).

[2] D.H. Byers and B.E. Saltzman, *Advances in Chemistry Series*, Vol. 21 (American Chemical Society, Washington, DC, 1959), p. 93.

[3] V. Matijasevic, E.L. Garwin, and R.H. Hammond, Rev. Sci. Instrum. **61**, 1747 (1990).

***Fig. 207. Detection of Ozone or Atomic Oxygen***

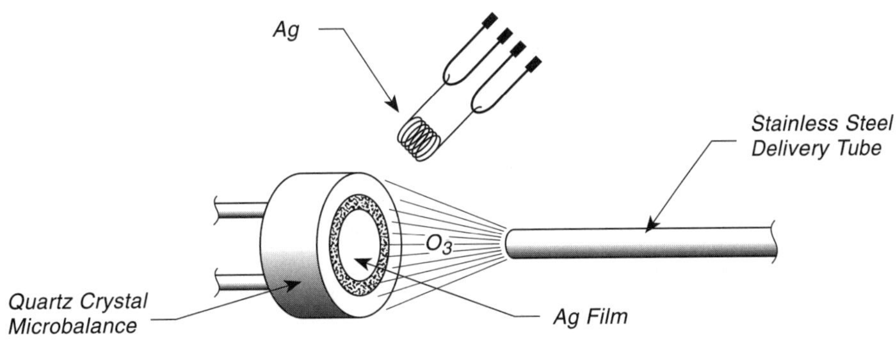


***A. Geometry for $O_3$ Detection***

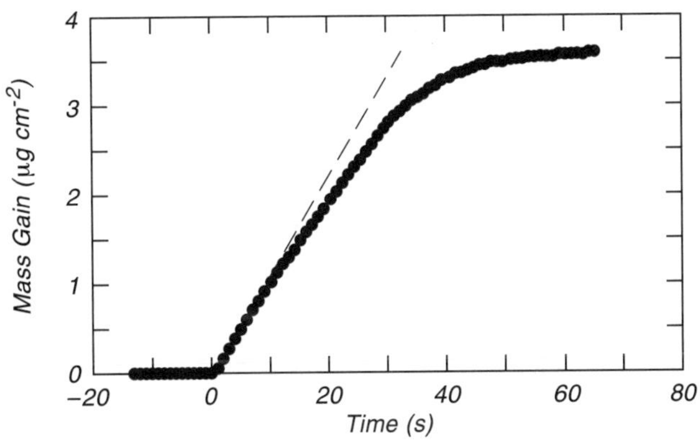


***B. Mass Change During $O_3$+Ag Reaction***

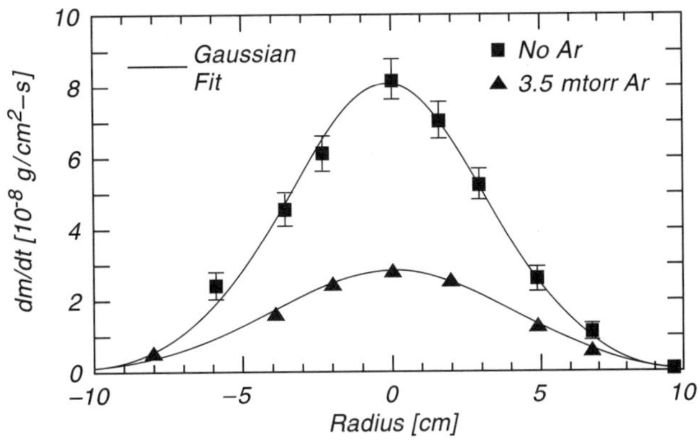


***C. Angular Distribution of $O_3$***

# E. Electrochemical Sources

# 208

# Electrochemical Sulfur Source

A solid-state electrochemical cell is especially convenient as a source for sulfur vapor deposition in surface science experiments. The cell was originally designed by Wagner [1] in the 1950s, and variations on its design have been in use since [2–6]. In Figure 208A, a cross-sectional view of the cell is shown. It consists of a glass tube, 2.6 mm in inside diameter at the outlet on the top, which has been made by a glassblower to have a slightly larger diameter on its bottom end. The waist joint in the center serves as a stop to hold the contents under compressive force, as will be described. A pellet of AgI fits tightly into the glass tube and is contacted by a silver foil cathode electrode on its bottom that is welded to a Pt wire. The Pt wire is directed through the center of a glass capillary piston and through a spring device that exerts a compression force on the piston from the bottom. Above the AgI pellet is a $Ag_2S$ pellet, compressed against a Pt ring that is welded to a Pt wire anode that passes up through the collimating tube. The assembly is contained inside a metal tube that may be heated by ohmic heating from a resistance wire insulated from the metal tube and wrapped around it. At a cell temperature of 485 K and a potential drop of 0.15 V, a maximum flux of 4–5 x $10^{12}$ $S_2$ molecules/s was generated. The literature indicates that $S_2$ is the major gas species generated under these conditions [5].

The measured angular distribution of S on a Pt crystal is shown in Figure 208B, but unfortunately the doser-crystal distance was not reported for these measurements [2]. A Gaussian distribution is observed with Auger spectroscopy, where data measured in both the *x*- and *y*-direction are superimposed on the same graph.

The electrochemical cell operates in ultrahigh vacuum with minimal outgassing and may be operated with pressures being maintained in the $10^{-10}$-Torr range.

## *References*


[1] C. Wagner, J. Chem. Phys. **21,** 1819 (1953).

[2] W. Heegemann, K.H. Meister, E. Bechtold, and K. Hayek, Surf. Sci. **49,** 161 (1975).

[3] T.E. Gallon, I.G. Higginbotham, M. Prutton, and H. Tokutaka, Surf. Sci. **21,** 224 (1970).

[4] H. Tokutaka, M. Prutton, I.G. Higginbotham, and T.E. Gallon, Surf. Sci. **21,** 233 (1970).

[5] J.C. Greenbank and B.B. Argent, Trans. Far. Soc. **61,** 655 (1965); D. Detry, J. Drowart, P. Goldfinger, H. Keller, and H. Rickert, Z. Physik. Chem. **NF 55,** 314 (1967).

[6] N.D. Spencer, P.J. Goddard, P.W. Davies, M. Kitson, and R.M. Lambert, J. Vac. Sci. Technol. **A1,** 1554 (1983).

## *Fig. 208. Electrochemical S Source*

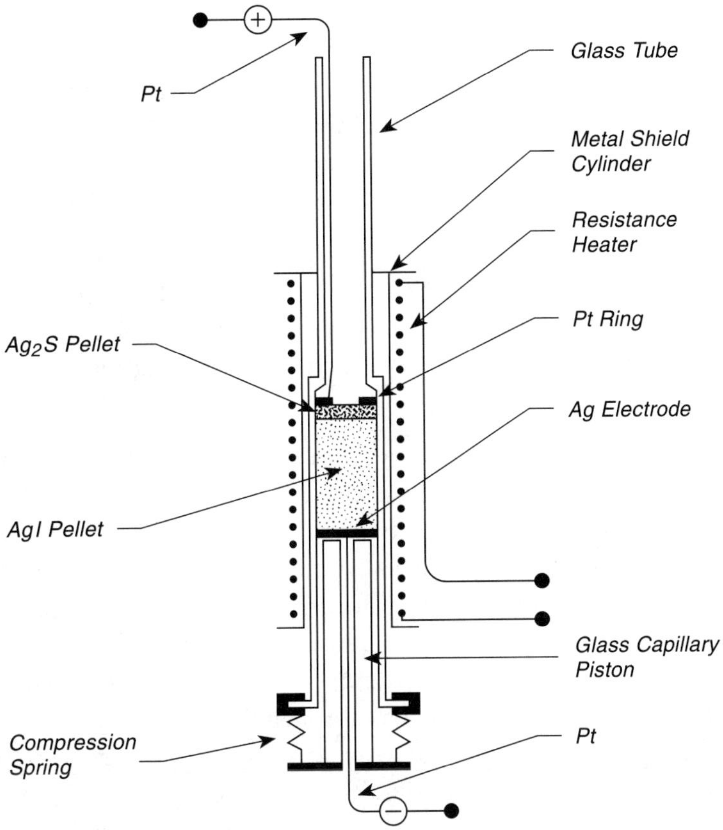


**A. Electrochemical Cell**

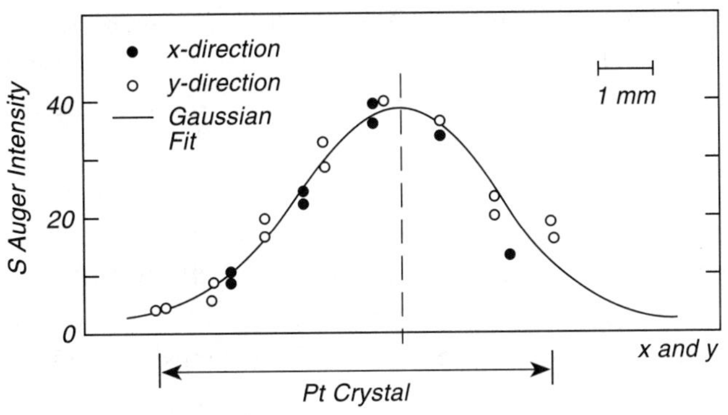


**B. Spatial S Distribution**

# 209

# Electrochemical Halogen Sources

Halogens are important surface species in surface chemistry and their delivery to surfaces in a controlled manner forms an important research technology. This discussion concerns the delivery of iodine gas by means of electrochemical generation [1], but others have used similar methods for both chlorine and bromine [2–4].

The source shown in Figure 209A is a silver iodide pellet. At moderate temperatures, an electrochemical reaction, $2AgI(s) \rightarrow 2Ag(s) + I_2(g)$, occurs in the pellet as a result of a potential applied between two electrodes, relying on the ionic mobility in the heated pellet. The mobile defect concentration may be increased by mixing about 5% $CdI_2$ with the AgI before casting the pellet.

The current flow through the heated pellet can be regulated using a constant current DC source, and a voltage of about 25 V will cause a current of about 25 μA to flow without external heating. The cells display extremely non-ohmic behavior, especially during start-up. In order to increase the efficiency of the source, it is housed in a Pyrex heater tube coated on the outside with a semiconducting $SnO_2$ thin film (see page 190) having a resistance of about 10–100 Ω/cm of tube. Alternatively, the Pyrex tube can be heated using a spiral nichrome wire wrapped around the tube [5]. The temperature of the tube and the pellet is adjusted by passing current through the $SnO_2$ film, which has a band of Aquadag colloidal graphite on each end to act as a contact between the conductive film and two clamps. At a temperature of 390 K, an applied voltage of 20 V produces a stable cell current of about 100 μA. The cell walls may adsorb $I_2$ during initial operation, but then stable operating conditions are achieved. Faraday's laws may be used to roughly estimate the flux of iodine from the source.

Figure 209B shows the output of the electrochemical cell as a function of the current as measured with a quadrupole mass spectrometer in line of sight, and on axis with the source. The system pressure was simultaneously measured with an ionization gauge that did not have line of sight to the source, and it is believed that it measures mainly $I_2$ gas. The mass spectrum indicates that $I^+$ at 127 amu and $I_2^+$ at 254 amu are the main ions generated in the mass spectrometer from $I_2(g)$, although some $H_2$ and $Cl_2$ are also produced. If the pellet is heated to higher temperatures, near 430 K, the conductivity may be increased to give higher $I_2$ fluxes, but deposition of $CdI_2$ also occurs slowly. It is also reported that a baffle at room temperature may be used to avoid deposition of undesirable substances, and that gaseous iodine can still be deposited on the sample with this arrangement.

To make the $AgI/CdI_2$ button, 7 g of the mixed salt is melted and poured onto the Pt/Rh mesh that is held on a glass plate under a cylindrical mold made of Pyrex. The Ag coil is inserted into the top of the melt before solidification. If the Pt/Rh mesh is heated to redness in a Bunsen burner flame prior to being covered with the liquid melt, and if it is held flat on the plate, the melt will cover only the exposed wire and will not submerge the electrode. Alternatively, red-hot Pt/Rh can be fused into the cold pellet [2]. The pellet can easily be removed from the Pyrex mold after cooling for insertion into the long support-heater tubing, which acts as a beam collimator in use.

Similar devices have been made for producing all of the halogens [2], and their performance is similar to that reported here. It is reported that after many hours of operation, a dead short often occurs through the pellet. This indicates that a metallic Ag dendrite has grown inside the pellet; brief passage of a high current through the heated pellet may break the dendrite. In addition, for the $Cl_2$ cell used at excessive temperatures (> 420 K) significant AgCl(g) was evolved. The halogen sources may be voltage modulated in the frequency range of 0.5–22.5 Hz, and the modulated halogen beam produced can then be used in kinetic studies on single crystal surfaces [3], and useful waveforms may be employed.

The production of pure $F_2$ by electrochemical means has been partially successful [4], although the high reactivity of fluorine with materials has caused problems. A $CaF_2$ pellet, held by springy contacts between a Pt plate and a Pt gauze, was employed for electrochemical generation at temperatures of 420–450 K. The gas evolved at the anode was focused onto the sample with a short tube, coated with $CaF_2$. Small amounts of $SiF_4$ were detected if glass tubing was used. The source delivers small amounts of platinum fluorides and HF. Studies with this source of the adsorption of $F_2$ on Pt showed that $PtF_2$ is evolved above about 500 K from Pt at low coverages, and that $PtF_4$ is evolved at high coverages, in addition to F(g) at 890 K [6]. The HF yield from the electrochemical cell may be reduced to a few percent of the $F_2$ yield by outgassing and electrolysis. Currents between 2 and 50 μA were used. During $F_2$ dosing , the total pressure of the vacuum system was below $1 \times 10^{-9}$ mbar.

Halogen atom sources have been described and one example is a pulsed source involving $Cl_2$ dissociation in a microwave cavity. One interesting feature of this work is the study of $2P_{3/2}$- and $2P_{1/2}$-Cl atom losses on Teflon walls [7].

A source of halogen ions produced by surface ionization on a $LaB_6$ surface of low work function has been described [8].

## *References*

[1] A. Mowbray and R.G. Jones, J. Vac. Sci. Technol. **A7,** 3373 (1989).

[2] N.D. Spencer, P.J. Goddard, P.W. Davies, M. Kitson, and R.M. Lambert, J. Vac. Sci. Technol. **A1,** 1554 (1983). This paper contains references to a number of arrangements to the electrochemical generation of halogens, etc.

[3] W.T. Tysoe, N.D. Spencer, and R.M. Lambert, Surface Sci. **120,** 413 (1982).

[4] R.G. Jones, Prog. in Surf. Sci. **27,** 25 (1988). A general review of halogen adsorption on surfaces.

[5] Professor Philip R. Watson, Department of Chemistry, Oregon State University, Corvallis, Oregon 97331-4003 (private communication).
[6] E. Bechtold, Appl. Surf. Sci. **7**, 231 (1981).
[7] W. Müller-Markgraf and M.J. Rossi, Rev. Sci. Instrum. **61**, 1217 (1990).
[8] N. Kashihira, E. Vietzke, and G. Zellermann, Rev. Sci. Instrum. **48**, 171 (1977).

***Fig. 209. Typical Electrochemical Halogen Source***

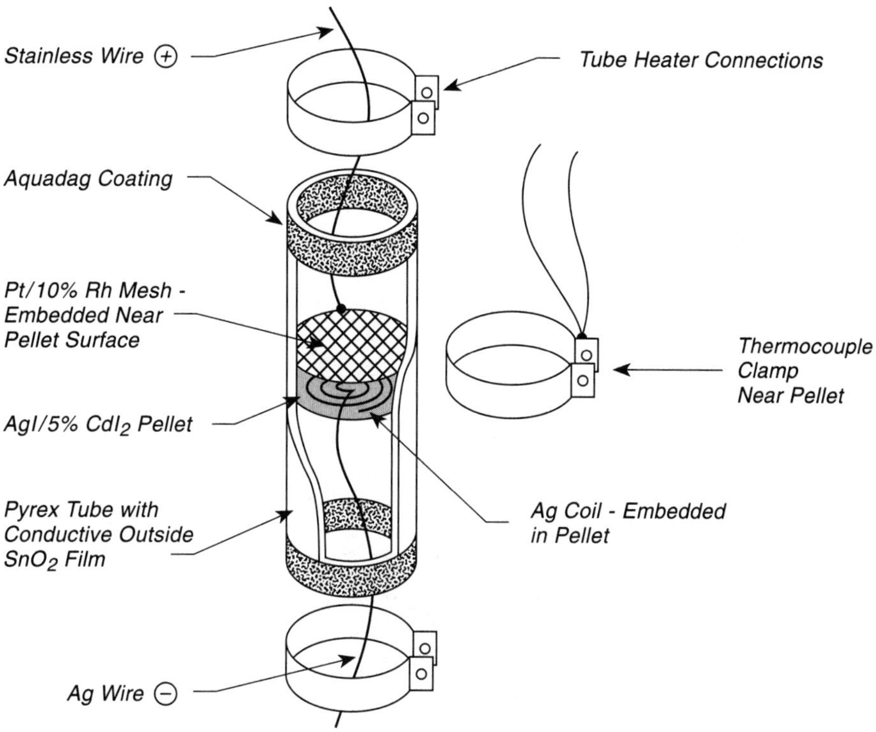


***A. Cell***

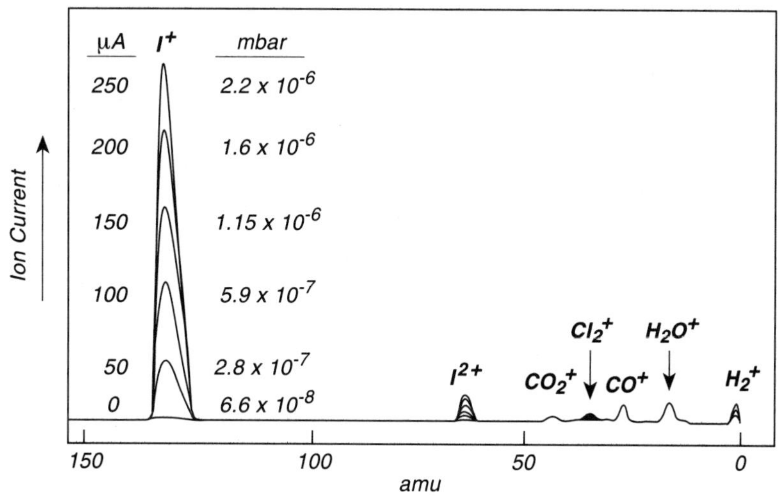


***B. Mass Spectrum at 150°C***

# F. Gas Purification

# 210

# Gas Purification for Experiments at High Pressures

The use of high-pressure gases for experiments involving single crystal surfaces always involves the problem of the preferential adsorption on the crystal of trace impurities present in the gas. At high pressures, even 1 ppm of adsorbable impurity can result in the selective adsorption of a monolayer or more of impurity. This problem may be eliminated by several methods, depending upon the gases and the impurities involved.

For gases with low condensation temperatures, storage over previously baked (573 K) zeolite pellets [1] at liquid nitrogen temperature is a favored method for the adsorption of trace impurities with high condensation temperatures [2]. This is a favored method for the additional purification of 99.9999% pure methane, for example [2]. In addition, carbon monoxide almost always contains traces of $Ni(CO)_4$ and $Fe(CO)_5$ as impurities produced within the storage cylinder or in the stainless steel gas handling system, and these metal carbonyls can be removed by storage of the CO over a wide pore zeolite at liquid nitrogen temperature. See page 726 for a discussion of the use of cryogenic storage and distillation at liquid hydrogen temperature. Alternatively, metal carbonyls can be removed by passing CO through a copper coil heated to 100°C, which is sufficient to decompose the carbonyls, and then through a trap cooled with liquid nitrogen. It is important to avoid the use of stainless steel tubing following the purification of the CO [3].

Adsorption of active chemisorbing gases in the presence of inactive gas is often effectively carried out by means of a Ti getter film which is freshly deposited inside a storage vessel under vacuum [3–5]. See page 722 for the use of a Ti getter film for the final stages of purification of He(g) [5]. An apparatus involving the use of zeolite trapping followed by storage over a Ti film is shown in Figure 210A. A commercial titanium sublimation pump cartridge is employed in a stainless steel storage vessel, and gas storage takes place at 77 K.

Catalytic chemists have for years used a method involving gas exposure to a high area catalyst prior to admission to the reactor chamber. This method has also been applied to Ni single crystals being studied under high pressures of methane [6]. Figure 210B shows the configuration of the gas purification apparatus used in these studies. A U-tube is filled with a Ni catalyst (in this case), which consists of Ni particles supported on a high surface area oxide support. The particle size ranges from 0.3–0.8 mm diameter. Both ends of the U-tube are plugged with quartz wool to prevent transfer of dust away from the catalyst. The catalyst is activated by reduction in flowing $H_2$ at 800 K, and then is evacuated at 800 K to remove $H_2$. Either methane or nitrogen may be cleaned of impurities such as CO, $O_2$, $CO_2$, and hydrocarbons (excluding $CH_4$) using this apparatus.

The cleanup of $H_2$ may be achieved using a supported Cu catalyst in the same manner. The chemisorption of $H_2$ by Cu is suppressed by working at liquid nitrogen temperature. The activation of the Cu catalyst is first carried out using $H_2$, and one must be careful to initially use a low pressure of $H_2$, since the exothermic reduction of copper oxides may cause a runaway condition. Reduction temperatures of 650 K are recommended [7,8].

## *References*


[1] One may use Varian Vacsorb molecular sieve, catalog No. 944-0000, obtainable from Varian Vacuum Products, 121 Hartwell Avenue, Lexington, MA 02173.

[2] T.P. Beebe, D.W. Goodman, B.D. Kay, and J.T. Yates, Jr., J. Chem. Phys. **87,** 2305 (1987).

[3] Dr. Jorg Wambach, Danish Technical University, Department of Physics, Building 307, DK-2800 Lyngby, Denmark (private communication).

[4] L. Hanley, Z. Xu, and J.T. Yates, Jr., Surface Sci. **248,** 265 (1991).

[5] J.T. Yates, Jr., J.J. Zinck, and W.H. Weinberg, Rev. Sci. Inst. **50,** 132 (1979).

[6] I. Chorkendorff, Technical University of Denmark, Building 307, DK-2800, Lyngby, Denmark (private communication). See I. Chorkendorff, I. Alstrup, and S. Ullmann, Surface Sci. **227,** 291 (1990).

[7] P.A. Taylor, P.B. Rasmussen, C.V. Ovesen, P. Stoltze, and I. Chorkendorff, Surface Sci. **261,** 191 (1992).

[8] P.B. Rasmussen, M. Kazuta, and I. Chorkendorff, Surface Sci. **318,** 267 (1994).

***Fig. 210. Gas Purification for Experiments at High Pressure***

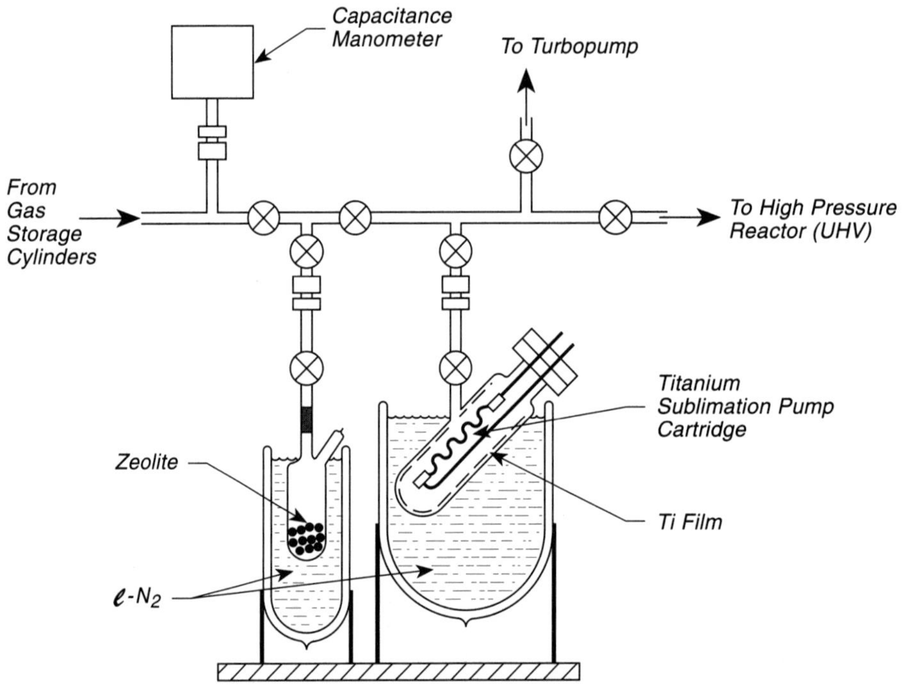


***A. Gas Purification by Adsorption***

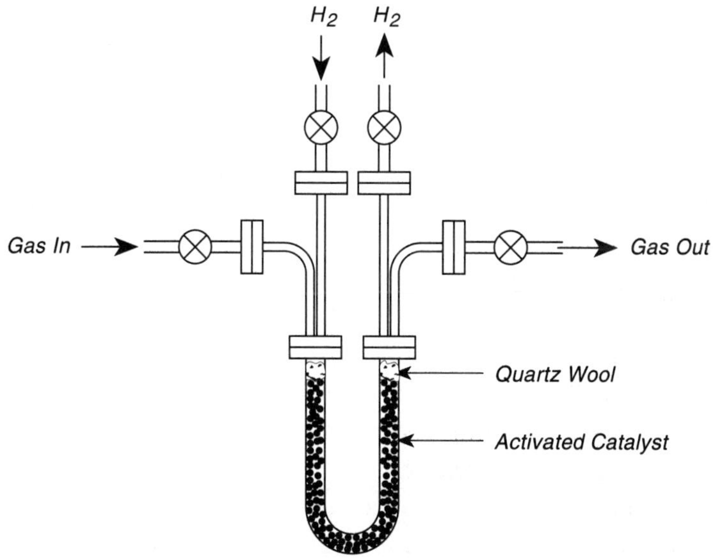


**B. Gas Purification over Catalyst (High Area)**

# 211

# Production and Purity Measurement for Extremely Pure He

The production of He with extremely high purity for use in ultrahigh vacuum systems is a challenge requiring special gas handling techniques. It must be remembered that hydrogen is a major impurity in stainless steel ultrahigh vacuum systems, and that to avoid hydrogen impurities, glass vacuum systems must therefore be employed. The bakeout treatment of a glass vacuum system differs from a metal system, and bakeout temperatures of the order of 600 K must be used [1].

A number of methods for producing pure He have been reported in the literature. The use of superfluid He leaks [2] and diffusion through Vycor glass [3,4] have been used by others, but the diffusion rate for $H_2$ through hot Vycor is about 0.1 of that for He [2], so that other methods must be used to achieve large factor reductions in the hydrogen impurity level.

The apparatus [5] shown in Figure 211 was mainly constructed of Pyrex glass with bakeable metal valves, and it was processed by bakeout to achieve a base pressure of 1 x $10^{-10}$ Torr. The pumping speed in this apparatus is only about 2 L/s. Operation of the purification system begins with the deposition of Ti getter films on the inner walls of the two bulbs, $G_1$ and $G_2$, by heating W loops to which Ti wire segments have been spot welded. The surface area of these films is each about 300 $cm^2$. The Vycor leak is outgassed at 1025 K until the outgassing cannot be detected using the Bayard–Alpert gauge at 1 x $10^{-10}$ Torr pressure. Valve $V_4$ is then closed and He is effused through the Vycor leak at 975 K. The He used on the high pressure side of the Vycor leak is specified to be 99.999% pure, and it is admitted to approximately 1 atm pressure, into a gas handling system initially at $10^{-7}$ or $10^{-8}$ Torr. Typical He effusion rates are approximately 6 x $10^{14}$ He atoms/s. Storage of He takes place overnight at several Torr pressure in the closed system over the freshly evaporated Ti getter film, $G_1$.

The Rh(111) single crystal was used as an integrating detector of $H_2$ in the purified He, using temperature-programmed desorption (TPD) with the mass spectrometer (QMS) as the detector. Purified He was admitted to less than 1 Torr pressure, with Valve $V_2$ open to the system, and with a freshly deposited Ti film in getter bulb, $G_2$, in order to continue getter purification of the He during exposure to the clean Rh(111) single crystal. An exposure of 1000 s was used in the He purity test, leading to exposures of from 100 to 900 Torr s to the He. Following pumping with the Hg diffusion pump, followed by use of the ion pump, TPD of both $H_2$ and CO was carried out, and compared to blank experiments without He. A least squares fit to the data gave 5.4 x $10^{10}$ H/$cm^2$-Torr s (He) for the contamination rate with the CO contamination rate about an order of magnitude smaller. It was shown that the rate of $H_2$ adsorption was not limited by gas diffusion kinetics at the He pressures used. Assuming a sticking coefficient of unity for the $H_2$, these measurements indicate that $P(H_2)/P(He) = 1.9 \times 10^{-11}$, which is a factor of about $10^6$ improvement over the original high purity He [5]. Flowing He has been similarly purified using hot Ti sponge in the flow lines [6].

Superleaks for the control of He flow into a vacuum system have been described [7]. They are made of porous Vycor glass [8] with approximately 40-Å-diameter pores occupying about 30% of the volume. By using various diameters of the Vycor, the He leak rate can be adjusted. A leak made of 1-mm-diameter porous Vycor, 8 mm long, exhibits a leak rate for He of $4 \times 10^{-6}$ std. $cm^3/s$ at 300 K at 1 atm presure. The leaks may be used also by immersion in liquid He [7].

## *References*

[1] P.A. Redhead, J.P. Hobson, and E.V. Kornelsen, *The Physical Basis of Ultrahigh Vacuum* (American Institute of Physics, New York, NY, 1993). pp. 417–430.

[2] M.A. Biondi, Rev. Sci. Instrum. **22,** 535 (1951).

[3] F.J. Norton, J. Appl. Phys. **28,** 34 (1957).

[4] J.R. Young and N.R. Whetten, *Transactions of 8th Vacuum Symposium and 2nd International Congress* (American Vacuum Society, New York, NY, 1961), p. 625.

[5] J.T. Yates, Jr., J.J. Zinck, and W.H. Weinberg, Rev. Sci Instrum. **50,** 132 (1979).

[6] G.B. Hoflund and R.P. Merrill, J. Vacuum Sci. and Technol. **A1,** 1560 (1983).

[7] M.F. Wilson, D.O. Edwards, and J.T. Tough, Rev. Sci. Instrum. **39,** 134 (1968).

[8] Vycor glass is made by Leaching Pyrex glass in alkali solution. It is available from major scientific glass manufacturers.

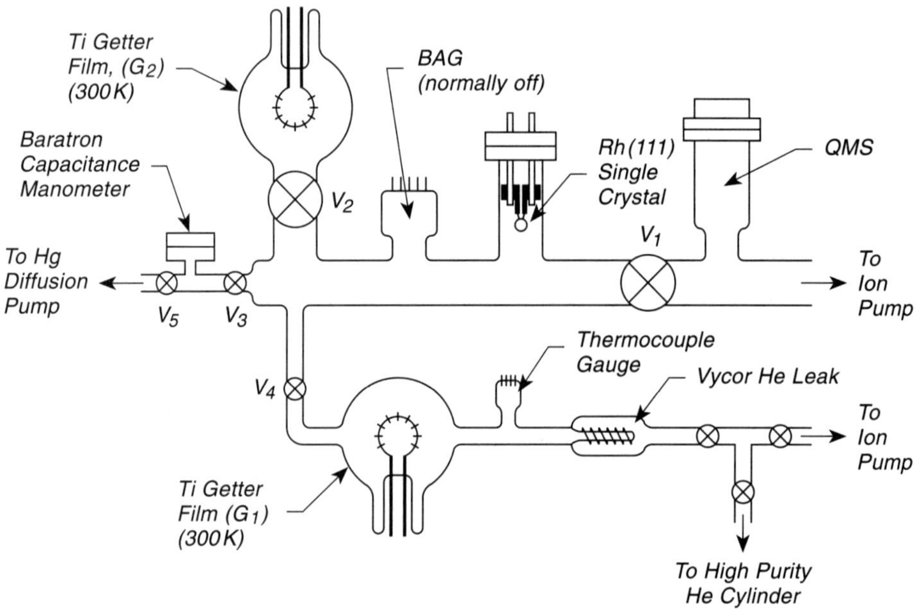


***Fig. 211. Production and Purity Measurement for Extremely Pure He***

# 212

# Purification of Gases by Cryogenic or Gettering Methods

The production of high-purity gases for laboratory use in ultrahigh vacuum applications has been discussed previously in the case of He (see page 722). Here we present other methods for the purification of gases such as He, $H_2$, Ar, $N_2$, and $O_2$[1].

As shown in Figure 212A, a combination of zeolite adsorption and cryogenic impurity condensation was employed for both He and $H_2$ purification. The apparatus, made of glass, consists of a conventional gas handling system connected to a cryogenic purification unit. The zeolite trap and the glass wool trap, along with the collection tube and associated glass lines, were baked out to achieve ultrahigh vacuum. The cryostat was then filled with either *l*-$H_2$ or *l*-He, which causes the pressure in the tube to drop to $<10^{-12}$ Torr. Gas from the gas handling system was then slowly admitted to the cold system, and condensation of highly pure gas was observed to occur, drop by drop, off of a tungsten field emission tip located in the collection tube, which was also a field emission microscope. This indicates that the glass wool section of the apparatus was condensing impurities preferentially from the incoming $H_2$ or He gas, and that gas phase transport of purified gas occurred from the glass wool section to the collection tube. The purity of the collected gas was measured by evaporation of the liquified gas, which filled the collection tube, and observing that the W field emission tip was still clean after immersion and re-evacuation. Concentrations of chemisorbing impurities were estimated to be $<5 \times 10^{12}/\text{cm}^3$ of liquid, considering the diffusion lengths in the liquid. This corresponds to an atomic fraction of impurities of the order of $10^{-14}$ in the purified gas.

In the case of Ar, $N_2$, and $O_2$ that cannot be purified by the method shown in Figure 212A, because of the high freezing point of these gases, a second method of purification was devised, as shown in Figure 212B [1]. Here, the glass vacuum system was baked to 430°C to achieve ultrahigh vacuum, and a Ta or Ni film was evaporated on the walls of the 2-L vessel, after seal-off from the high vacuum line using a torch seal-off. Ar or $N_2$ was then admitted into the system, and condensed into the trap at *l*-$H_2$ temperature, after which the connection to the gas source was sealed off. Then the condensed gas was allowed to evaporate and condense several times to facilitate gas contact with the getter film, since diffusion through the gas will be slow at the ~1 atm pressure employed. The gas could be condensed into the collection tube using liquid $N_2$-$O_2$ mixtures whose temperature could be adjusted by using different compositions, and monitored with an immersed thermocouple. By reestablishing ultrahigh vacuum by condensing the gas into the trap within the cryostat, field emission measurements showed that no detectable adsorption had occurred on the field emission tip for the case of Ar. For $N_2$ purification, a Ni getter film was employed, but irreversible chemisorption of $N_2$ prevented the use of field emission to detect impurities. For $O_2$, admission of tank gas through *l*-$N_2$ traps was used without a getter film, and field emission measurements were also not possible because of oxygen chemisorption

The power of metal getter films for maintaining very low vacuum conditions in sealed-off glass vacuum systems is clear from observations made with field emission tips sealed off in the presence of getter films. In one instance, almost no contamination of the tip was observed in one year, indicating a pressure of the order of $10^{-16}$ Torr or lower [2].

## *References*


[1] B. Halpern and R. Gomer, J. Chem. Phys. **51,** 1031 (1969).

[2] Personal observations.

Fig. 212. Gas Purification Apparatus

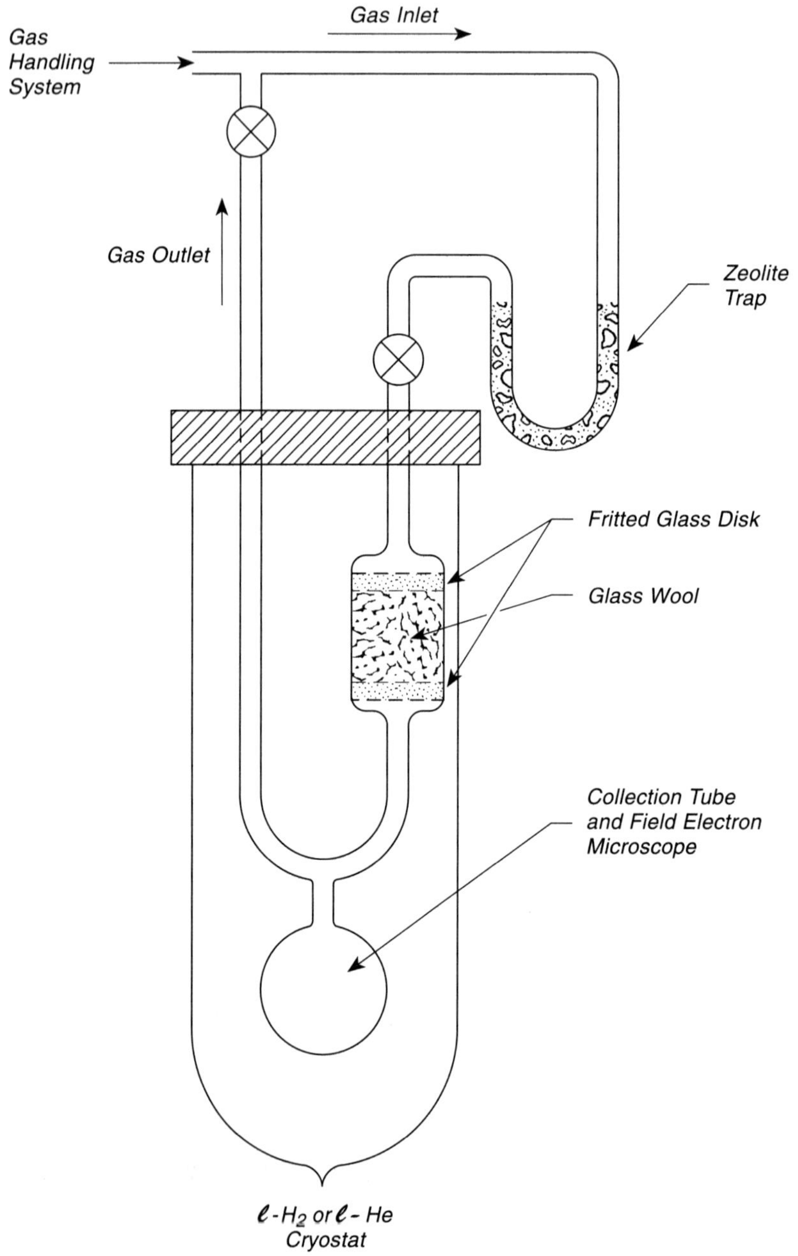


A. He and $H_2$ Purification

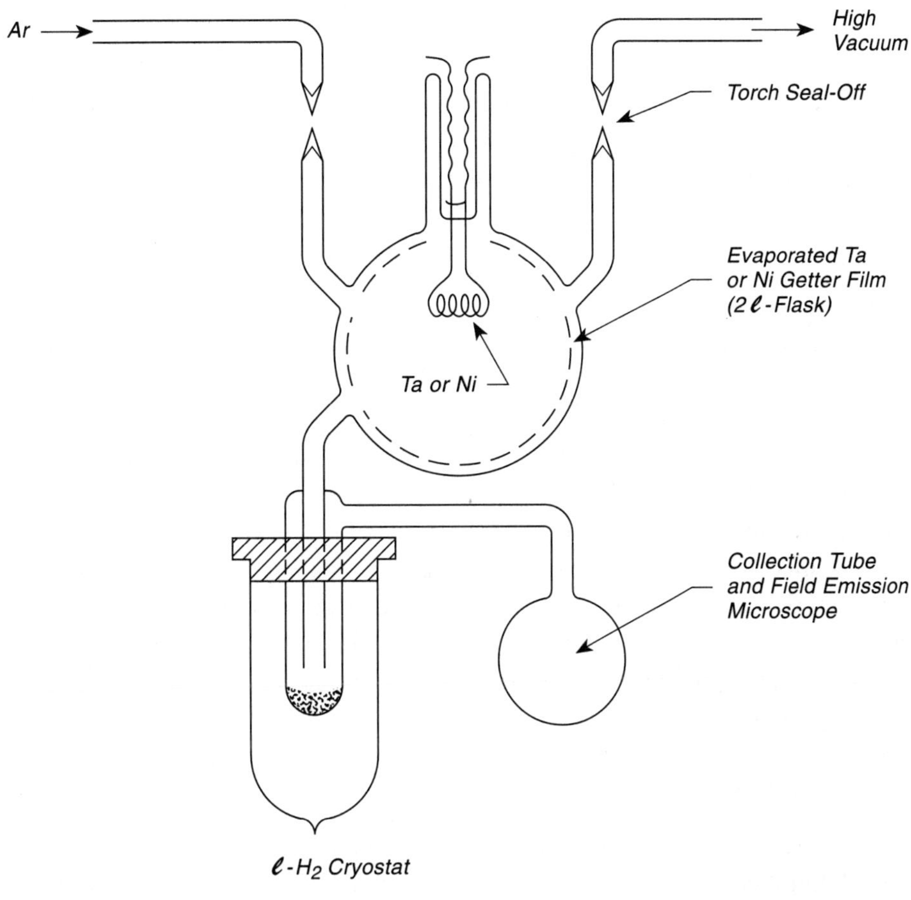


**B. Ar and $N_2$ Purification**

# 213

# Gas Purification by Permeation Through Metals

For some applications the supply of extremely pure gases into a vacuum system is needed, and conventional cylinder sources are inadequately pure for this purpose. Thus, for years, chemists have used permeation through high-temperature membranes for the removal of impurities from the gas. In addition to supplying pure gas, the permeation technique also affords easy control of flow rate by the adjustment of the temperature.

In Figure 213A, either hydrogen or oxygen permeation sources are shown [1]. For hydrogen, pure Ni permeation tubing is used [2]; for oxygen, pure silver permeation tubing is used [2]. The permeation tubes are 10–13 cm in length. The thin permeation tubes are silver brazed (in hydrogen) to the ends of 0.25-in.-diameter copper tubes mounted on ceramic insulator feedthroughs welded into a flange. Passing current through the device provides the heating necessary to allow gas permeation to occur. The purity of gases produced in this manner has been studied by Whetten and Young [3] and Young [4], and comparisons are made in the book by Redhead [5]. The worker should remember that while high gas purity can be achieved by permeation methods, it is also likely that the gas in a UHV system will become impure owing to displacement from the walls of the chamber or by interaction with hot filaments [5]. Studies of Ag permeation sources for the production of pure $O_2$ have shown that long periods of operation of the source must occur before gases like $H_2S$, $SO_2$, CO, and $CO_2$ are eliminated to the percent level [6]. It may be best to condition a Ag source on a separate vacuum system to avoid the evolution of the gases into the main system [6].

In Figure 213B, a glass device involving a permeation thimble is shown. For hydrogen, both pure Ni thimbles and pure Pd thimbles have been used. For oxygen, pure silver thimbles are employed. The thimble, brazed to a Kovar seal, is centrally mounted in a glass envelope containing a helical refractory metal heating filament that heats the thimble by radiation. Permeation of purified gas from the outer envelope to the inside of the thimble takes place. In both experimental arrangements, shown in Figure 213A,B, it is appropriate to flow the source gas slowly through the apparatus to replenish gas at the source.

A solid state source of oxygen, followed by permeation through a Ag tube wall, is described on page 734.

The use of a pure iron permeation source for purification of hydrogen and deuterium has been reported in reference [7].

## *References*

[1] L.C. Burton, J. Vac. Sci. Technol. **8,** 758 (1971).

[2] Type "A" Ni tubing was used; fine (99.99%) silver tubing was used. Uniform Tubes, Inc., 7th Avenue, Collegeville, PA 19426.

[3] N.R. Whetten and J.R. Young, Rev. Sci. Instrum. **30,** 472 (1959).

[4] J.R. Young, Rev. Sci. Instrum. **34,** 891 (1963).

[5] P.A. Redhead, J.P. Hobson, and E.V. Kornelsen, *The Physical Basis of Ultrahigh Vacuum* (originally published by Chapman and Hall, 1968; reprinted American Institute of Physics, New York, 1993).

[6] L.C. Beavis, Rev. Sci. Instrum. **43,** 122 (1972); E.G. Burroughs, Rev. Sci. Instrum. **40,** 33 (1969).

[7] M. Arbab and J.B. Hudson, Appl. Surf. Sci. **29,** 1, (1987); see also, Surf. Sci. **206,** 316 (1988); Surf. Sci. **209,** 183 (1989).

***Fig. 213. Gas Purification by Permeation Through Metals***

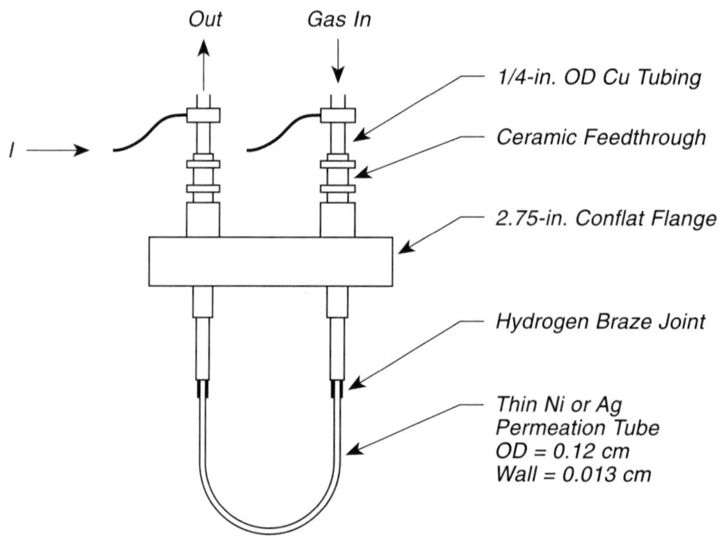


***A. $H_2$ or $O_2$ Permeation Source—Ohmic Heating***

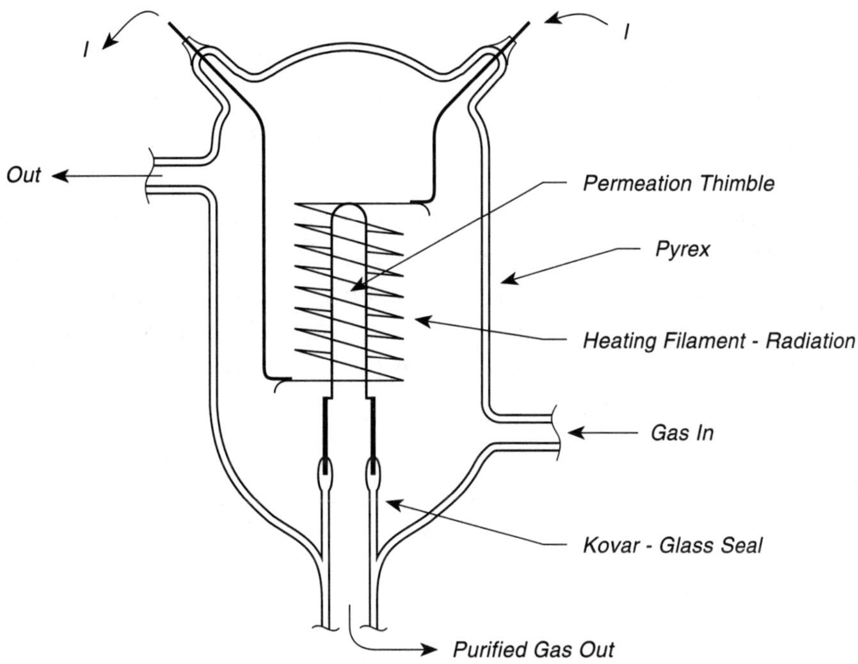


***B. Permeation Thimble—Radiation Heating***

# 214

# $O_2$ Source—Solid State

Various permeation sources, in which gaseous $O_2$ is purified by diffusion through a selective metal, are described on page 730. These sources all involve the use of oxygen gas, supplied as a gas, to drive the diffusion/purification process. A clever modification of this idea involves the use of a solid that liberates $O_2$ at the diffusion/purification temperature, supplying a long-lived source of purified oxygen without the need for a vacuum system to introduce $O_2$ gas into the permeation source [1].

Figure 214 shows a Ag tube device packed with $MnO_2$ powder. The tube is sealed by electron beam welding and then welded to two Ag tabs as shown. The electron beam welding, being done under vacuum, automatically removes most of the entrained air in the powdered $MnO_2$.

Passing electrical current through the Ag tube raises its temperature first to a degassing temperature of about 450°C; when the source is resistively heated to about 550°C, the $MnO_2$ begins to decompose, liberating $O_2$ inside the tube that then diffuses through the Ag, becoming highly purified. The rate of production of purified $O_2$ increases as the temperature is raised, and gas pressures from about $10^{-9}$ Torr to $10^{-2}$ Torr can be sustained in pumped UHV systems using this source.

## *References*


[1] D.L. Schaefer, Rev. Sci. Instrum. **41**, 274 (1970).

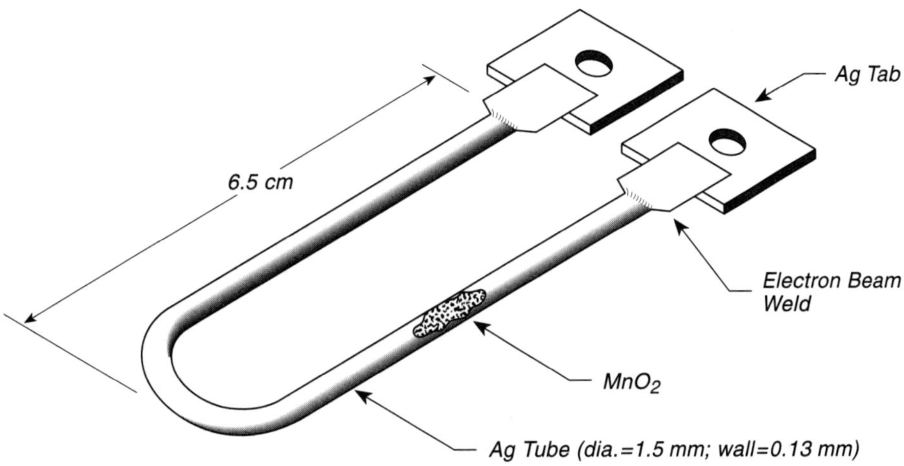


***Fig. 214. $O_2$ Source—Solid State***

# 215

# Reusable Sample Holder for Hydrogen Permeation Studies

The permeation of hydrogen gas through pure metals and alloys is of technological and scientific interest. One of the methods used to make permeation measurements is to electron-beam-weld the metal sample diaphragm to a cylindrical tube end [1]. Another method is to use a sample with machined knife edges that contact a deformable gasket [2]. Elastomer gasket or O-ring sealing is only useful at room temperature. The method to be described here differs from the earlier methods, is simple, and the apparatus is reusable [3].

Two standard Conflat flanges with a standard Cu gasket are used. The sample, in the form of a circular disk, is mounted inside the Cu gasket, and two Al gaskets are placed on either side of the sample. The Cu gasket, when bolted in place, protects the sample from the outside environment. The Al gaskets are plastically deformed between the sample and the inner region of the flanges, and their thickness should be properly adjusted to a dimension leading to compression and sealing. In cases where an incomplete seal at the Al gasket was found, this was attributed to work hardening of the gasket during manufacture. This could be corrected by annealing the assembled sample holder for 1–3 h at 573 K, then retightening the bolts. The sample holder was used up to 623 K, and is shown in Figure 215.

An entirely different method for measuring hydrogen diffusivity in metals, involving electrical measurements as hydrogen diffuses along the length of a foil, is described in reference [3].

## *References*


[1] M. Kanno, M. Yamawaki, and T. Namba, *Proceedings of the 2nd Japan Institute of Metals International Symposium, Hydrogen in Metals,* Supplement to Trans. JIM 21, 105 (1980).

[2] E.H. Van Deventer and V.A. Maroni, J. Nucl. Mater. **113**, 65 (1983).

[3] D.J. Pine and R.M. Cotts, Rev. Sci. Instrum. **55**, 614 (1984).

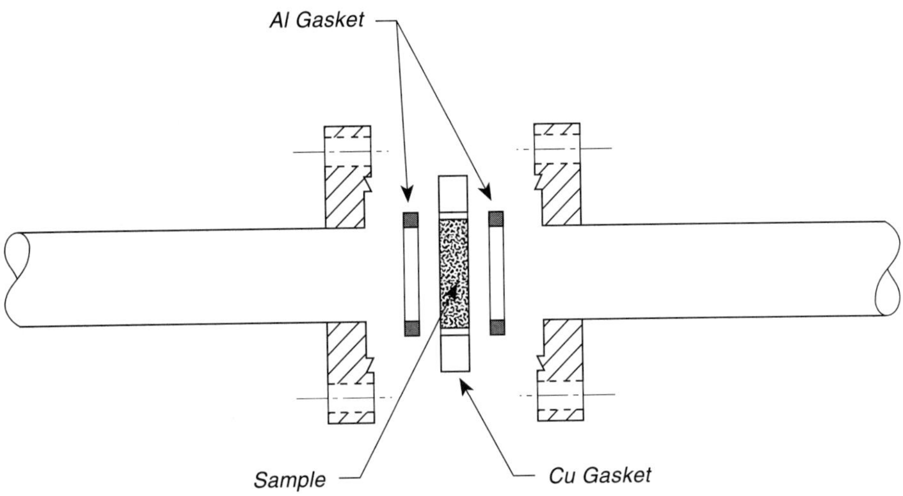


**Fig. 215. Reusable Sample Holder—Hydrogen Permeation**

# VI. UHV Windows

A. Spectroscopic Windows .............................. 741
B. Observation Windows............................... 767

# B. Spectroscopic Windows

# 216

# Mounting Infrared Windows on UHV Systems

For the transmission of infrared radiation to and from an ultrahigh vacuum system for the purpose of carrying out infrared spectroscopic measurements, it is essential to employ windows that are highly transparent in the desired spectral region. Often, alkali halide crystals are used for this purpose. Alkali halides have a small linear thermal expansion coefficient relative to metals (about 0.1 of the metal) and thus undergo high stresses under elevated temperature conditions when rigid contact is made between the window material and the metal. Several window-mounting designs that place elastomer seals between the window and the metal have been reported [1–5]. In each of these designs the elastomer yields by means of shear strains during bakeout relieving the strain in the alkali halide window and thereby avoiding fracture. The common elastomer in all of the designs is Viton (see page 166 for studies of gas diffusion through Viton), and this material will stand repeated bakeout at 200°C. Two of the designs, shown in Figures 216A,B, involve differential pumping to avoid leakage and/or diffusion of atmospheric gases through the Viton seals. The designs shown in Figures 216C,D do not employ differential pumping. The design of Figure 216C places an Al gasket between two Viton O-rings, eliminating leakage or diffusion through the outer O-ring. The design of Figure 216D eliminates differential pumping or seals between the Viton gaskets.

Figure 216A is the original differentially pumped design by Hollins and Pritchard [1], and has proven effective in many different laboratories [2]. The NaCl or KBr window is retained in a stainless steel flange by an outer clamping flange. The inner space between the Viton O-rings is differentially pumped to pressures of the order of 0.1 Torr by a rotary pump, and it is reported that this pumping is essential for the best system pressure. On the ultrahigh vacuum side of the window, the O-ring is compressed outside of a loosely fitting retainer ring that keeps it from moving inward and also avoids local trapping of gas introduced during assembly. If the differentially pumped space is let up to air before bakeout, the system pressure responds immediately; however, after bakeout, the O-ring seals mate more effectively to the metal and window, and only a small degradation of vacuum occurs if air is introduced into the differential volume. The indicated base pressure after bakeout is about $5 \times 10^{-11}$ Torr.

Figure 216B shows a modification of the original design in which no extension beyond the system outer flange occurs (zero profile) [3]. Here, a modified Conflat flange, equipped with differential pumping, receives an O-ring plus a metal ring retainer to hold the O-ring in a circular geometry. An IR window is seated on this O-ring, and a double O-ring plus retainer ring is compressed on the outside of the window by a threaded thrust ring that screws into the flange opening. The inner O-ring (1/16-in.) of the pair seals against the window, while a slightly larger diameter 1/16-in. O-ring seals against a ledge in the flange. The atmospheric pressure forces the inner O-ring against the outer rim of its groove in the retainer ring. This device has been employed with KRS-5 and CsI window material, and is routinely baked at 100–110°C. Once the seal is made, no effect of pumping out the differential space is observed, and pressures of the order of $5 \times 10^{-11}$ Torr are reported.

Figure 216C shows a third method for sealing IR windows to an ultrahigh vacuum system [4]. The components are constructed of KF 50 flanges, which are commercially available. The outer section has its groove widened in order to hold the O-ring. The inner section has been modified by welding tubulation containing an O-ring groove into a hole cut into the KF 50 flange. A specially designed Al gasket is placed between the inner and outer flanges before assembly. An IR window is then pressed between the inner and outer O-ring seals and compression by three separate clamps results in gasket deformation, leading to a vacuum-tight seal between the two O-rings, and removing the need for differential pumping. Bakeout of the window was limited to 150°C and the O-rings were observed not to stick to the sealing surfaces. The leak rate was estimated to be smaller than $1 \times 10^{-12}$ Pa $m^3$ $s^{-1}$.

Figure 216D shows a double seal made of two flat Viton gaskets that seal against the IR window [5]. A special thrust ring is bolted onto a modified Conflat flange using six bolts and a small volume at the edge of the IR window is captured by both gaskets. Bakeout temperatures up to 150°C have been used, and the base pressure is reported to be $2 \times 10^{-10}$ mbar.

All four of these designs must be carefully made to avoid metal contact with the brittle window material upon compression of the O-ring seals.

A design similar to that of Figures 216A,B has been reported [6] for sealing IR windows to a glass system.

For some applications, $CaF_2$ single crystal windows [7] and MgO windows [8] may be attached to flanges for IR studies, and these seals require neither differential pumping nor O-ring-sealing technology. A review of this method may be found in reference [9], and an RF heating method for making these seals is given in reference [10]. Irtran 4 windows (ZnSe, polycrystalline) have been sealed to Voscovit 501 alloy using Pyroceram 7575 sealant [11].

The use of a lead gasket in conjunction with sealing a NaCl window to a stainless steel flange has been reported to be successful for ultrahigh vacuum applications. A specially shaped flange sealing surface and a series of conical washers acting as compression springs are involved to compensate for differential thermal expansion. These windows assemblies are bakable to 275°C, and exhibit a leak plus outgas rate of $<3 \times 10^{-10}$ atm $cm^3/s$ [12].

A simple method involving the sealing of Irtran 2 and Irtran 6 [13] into rotatable Conflat flanges has been given in reference [14]. Here, a silicone cement (Gevac [15]) is used to seal the window to the rotatable insert, and a ring of Al is placed between the outer flange and the insert in order to apply pressure directly over the underlying knife edge to prevent flexure of the flange during tightening. This design is superior to that of reference [16], which involves flexure of the nonrotatable flange employed.

## *References*


[1] P. Hollins and J. Pritchard, J. Vac. Sci. Technol. **17,** 665 (1980). This report also contains references to earlier window sealing methods.

[2] Z. Xu and J.T. Yates, Jr., J. Vac. Sci. Technol. **A8,** 3666 (1990).

[3] R.G. Tobin , C. Chung, and J.S. Luo, J. Vac. Sci. Technol. **A12,** 264 (1994).

[4] C. Damerow and W. Erley, J. Vac. Sci. Technol. **A5,** 2974 (1987).

[5] A. Bradshaw, private communication. The gaskets may be obtained from Busak and Shamban GmbH & Co., Handwerkstr. 5-7, 70565 Stuttgart, Germany.

[6] R. Raval, S. Munro, A. Coats, and G. Beange, J. Vac. Sci. Technol. **A10,** 3593 (1992).

[7] $CaF_2$ windows, sealed directly to Conflat flanges through a thin silver/silver chloride intermediate sealing assembly, may be obtained commercially from Harshaw Crystal and Electronic Products, Solon Technologies, 6801 Cochran Road, Solon, OH 44139. These windows are unsatisfactory for use with ultraviolet light because of the sensitivity of the AgCl cement to photodecomposition. A description of the sealing process is given in D.D. Berger, Rev. Sci. Instrum. **53,** 1619 (1982); R.W. Roberts, J.F. Harrod, and H.A. Poran, Rev. Sci. Instrum. **38,** 1105 (1967); B. Caras, Rev. Sci. Instrum. **39,** 1441 (1968); Y.-C. Chiu, Rev. Sci. Instrum. **41,** 639 (1970).

[8] MgO windows have been sealed to Pyrex glass sleeves using gold/silver chloride sealing with the Pyrex being sealed by a Kovar graded seal to stainless steel flanges. For a diagram of this method, see J.T. Yates, Jr., and D.A. King, Surface Sci. **30,** 601 (1972); also see R.W. Roberts, J.F. Harrod, and H.A. Poran, Rev. Sci. Instrum. **38,** 1105 (1967).

[9] Y.-C. Chiu Rev. Sci. Instrum. **41,** 639 (1970).

[10] M. Aslam, Rev. Sci. Instrum. **42,** 1892 (1971).

[11] P.E. Wierenga, G.J. Mollenhorst, and A.T.B. Wikkerink, Rev. Sci. Instrum. **49,** 408 (1978).

[12] T.J. Manuccia, J.R. Peele, and C.E. Geosling, Rev. Sci. Instrum. **52,** 1857 (1981).

[13] Irtran optics are available from Eastman Kodak Co., 5/81 RC MC 02010, Rochester, NY 14650-2010.

[14] M. Kottke and R.G. Greenler, Rev. Sci. Instrum. **42,** 1235 (1971).

[15] A silicon resin, type KL5, like that used in ref. [14] can be obtained from Kurt J. Lesker Company, 1515 Worthington Avenue, Clairton, PA 15025-9911.

[16] L.J. Schkolnick, Rev. Sci. Instrum. **39,** 122 (1968).

**Fig. 216. Infrared Window Mountings**

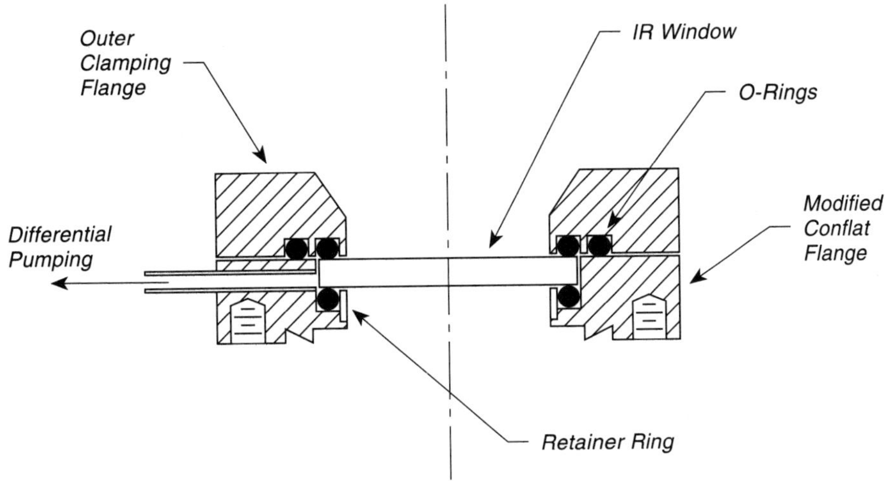


**A. Original Design—Outer Clamping Flange**

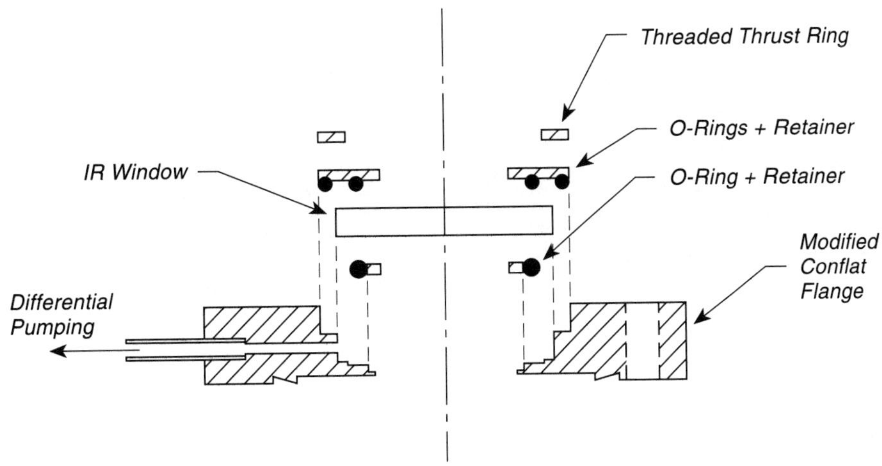


**B. Modified Flange Design—Zero Profile**

**Fig. 216. Infrared Window Mountings**

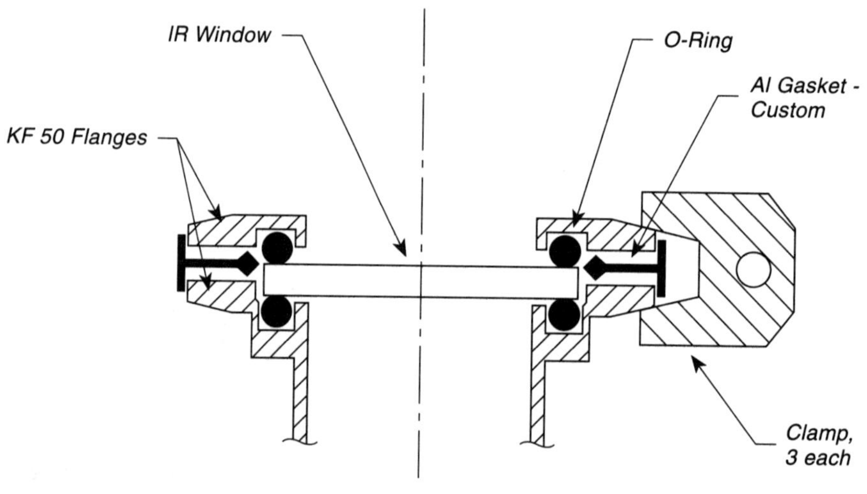


**C. Modified Design—No Differential Pumping**

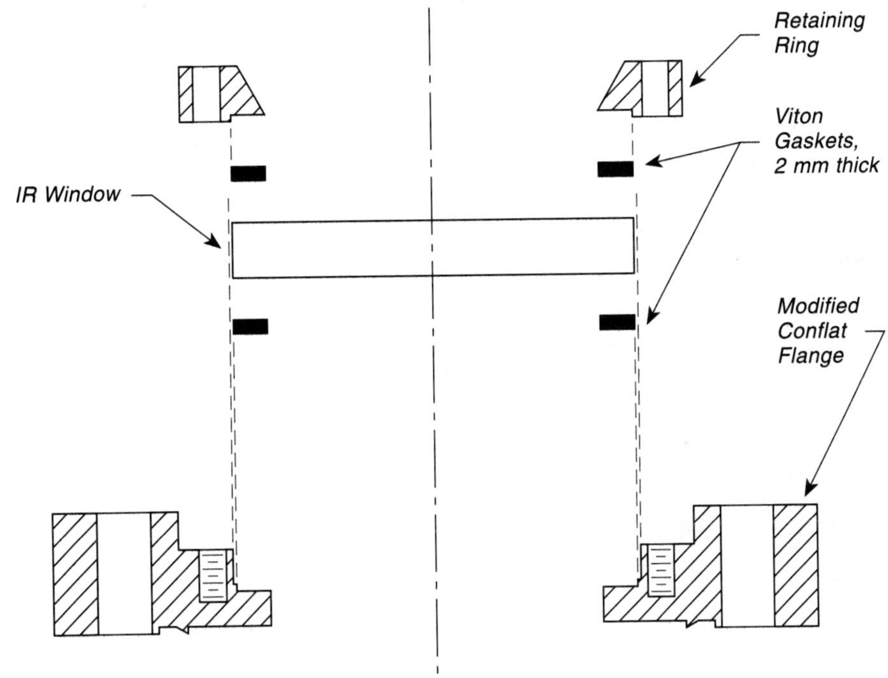


***D. Viton Gasket Design—No Differential Pumping***

# 217

# Windows for Vacuum UV Transmission into Ultrahigh Vacuum Systems

The Viton O-ring technology for sealing windows such as KBr, $CaF_2$, and MgO to ultrahigh vacuum systems involve severe limitations on the bakeout temperatures that may be employed (see page 742). Other sealing methods involving, for example, fused AgCl cements, are susceptible to leak formation upon exposure to ultraviolet light, and also are incompatible with many metals because of corrosion effects. A new method, which works for both $CaF_2$ and $MgF_2$ windows, is described below [1].

Figure 217 shows a stainless steel window flange welded to a tubulation that connects to the ultrahigh vacuum system. A 0.65-cm-wide step is provided to support the window, which is 1 mm smaller in diameter than the outer diameter of the step. An outer window retaining ring presses the window against the seal, and the pressure at the seal can be adjusted using bolts which compress springs around the perimeter. A soft aluminum foil cushion is used between the retaining ring and the outer part of the window.

The vacuum seal is made by building up a Vacseal [2] bead on both the window and the step surface of the flange as shown in Figure 217. The Vacseal bead is not considered to be a leak sealer between the window and the flange, but a deformable gasket material. The following recipe is recommended for the production of a high temperature seal which is He leak tight:

1. Clean and degrease the window and the flange.
2. Apply Vacseal as shown to the step and window surfaces as shown, keeping both surfaces horizontal. A 1.5–2.5-mm-thick bead is formed on each surface.
3. Place flange and window into an oven at room temperature and heat to 125°C for 5 h. Then, following this precuring procedure, allow to cool to room temperature. There should be no bubbles in the beads.
4. Invert window and carefully place on the flange. The two beads should coalesce into a single seal about 2–2.5 mm in thickness without voids or bubbles.
5. Gently tighten the retainer ring but do not displace the Vacseal sealing material.
6. Cure at 150°C for 15 h. Cool to room temperature and inspect for cracks or voids. The sealant should be amber in color. The spring-loaded bolts may be tightened slightly in this step.
7. Place the seal into the oven at 200°C for 8 h. Cool and leak check.

If voids or cracks appear during the above procedure, start over. Methyl ethyl ketone (MEK) may be used to dissolve the cured Vacseal. Ethyl alcohol will not dissolve the seal and may be used to clean the window without causing a leak to form. One must be careful to not apply mechanical stresses to the cured window assembly. The seals are bakeable to 330–400°C and are leak-tight for periods in excess of 6 months [1]. This sealing technique has been used on windows of 7.5-cm diameter [1]. The optical distortion of the widows is small, and thermal cycling may be carried out.

Another method, useful for mounting quartz windows on Al flanges, has been reported [3]. Here, an indium seal is employed and the window may be baked to 120°C, which is the bakeout temperature needed for Al ultrahigh vacuum systems.

## *References*

[1] C.H. Muller, III, M.W. Barrett, and D.D. Lowenthal, Rev. Sci. Instrum. **59**, 1425 (1988).

[2] Vacseal is sold as a viscous liquid in 3 oz. bottles. Space Environment Laboratories, Box 1061, Boulder, CO 80306.

[3] H. Saeki, J. Ikeda, and H. Ishimaru, Vacuum **39**, 563 (1989).

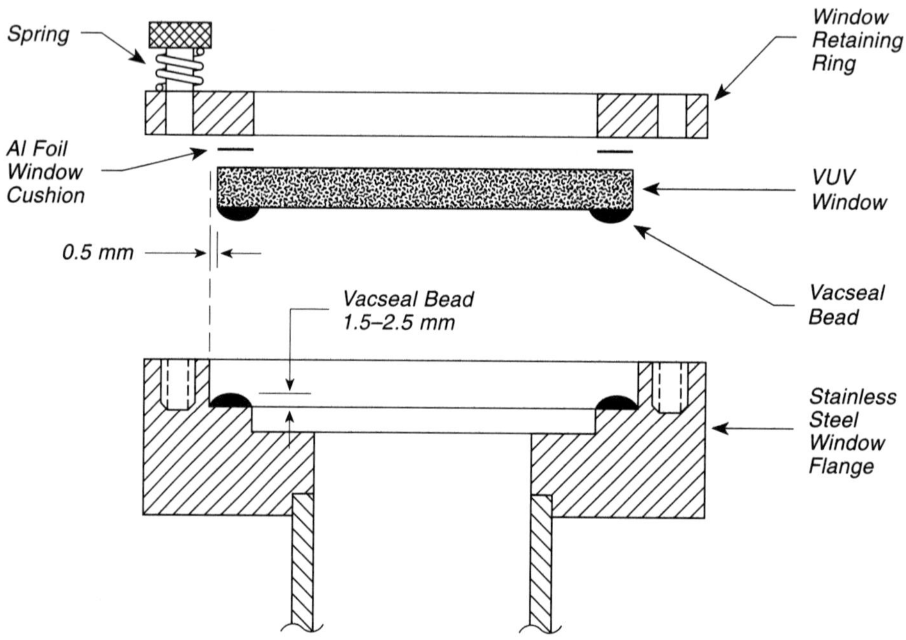


***Fig. 217. Vacuum UV Window—UHV***

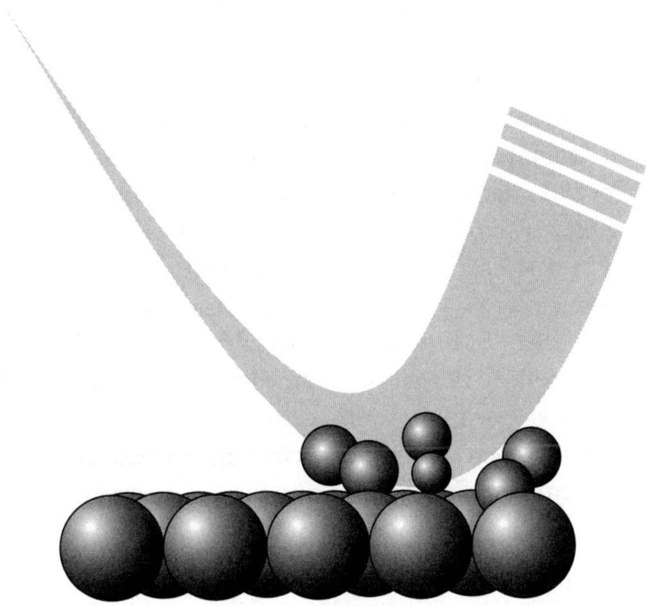

# 218

# Construction of Be Windows for Ultrahigh Vacuum Use

Thin Be windows are often employed for transmission of X-rays into UHV systems. Although electron-beam-welded windows are commercially available [1], brazing onto stainless steel Conflat flanges is sometimes used [2,3]. The procedure for doing this is reported here [3].

Figure 218A schematically shows the mechanical procedure employed for making an ultrahigh vacuum brazed seal to type 304 stainless steel. Beryllium foil, 0.25–0.50 mm thick, must be specified to be of vacuum quality [4]. Such foils may be used for spans up to 20 cm in length. The stainless steel surface is abraded with 250 grit sandpaper, and all sharp edges on the vacuum side are rounded. A brazed region 6–9 mm wide is employed. The Be foil is shaped in exterior dimensions by etching, using common fingernail polish to protect the regions to be preserved. The protective layer can later be removed using acetone solvent. The acid etching mixture is specified in Table 218A below. The braze material is 99.99% Ag foil of 0.075-mm thickness, cut into a ring. The silver should be cleaned with silver polish or by means of a chemical bath specified in Table 218B, in which metallic Al is used to reduce oxidized and sulfided Ag layers. The cleaned components are loaded into a stainless steel cylinder and the lower flange is tack welded to the cylinder. The stacked assembly is covered with quartz cloth and loaded with a weight as shown in Figure 218A. The assembly is fired in a vacuum furnace according to the procedure outlined in Table 218C.

He leak detection indicates no leaks at the sensitivity of the detector, $10^{-11}$ mbar L $s^{-1}$.

The transmission of X-rays through Be foils of different thicknesses is shown in Figure 218B.

**Table 218A. Be Etching solution (0.5 mm/min)**

| | |
|---|---|
| $HNO_3$ | 2% by volume |
| $H_2SO_4$ | 2% by volume |
| HF(aq) | 2% by volume |
| $H_2O$ | Balance |

**Table 218B. Silver Cleaning**

| | |
|---|---|
| $H_2O$ | 1 L, boiling in beaker |
| $NaHCO_3$ | 15 cm$^3$ of powder |
| NaCl | 15 cm$^3$ of powder |
| Add large piece of Al foil | |
| Add Ag ring—boil 1–2 min. | |
| Remove Ag ring—clean in $H_2O$ | |

**Table 218C. Brazing Procedure**

| |
|---|
| Evacuate to $10^{-4}$ Torr |
| Preheat to 850°C, 45 min |
| Stabilize at 900°C, 25 min |
| Braze at 960°C, 10 min. |
| Cool |
| Vent furnace, leak check Be seal |

## *References*


[1] Electrofusion Corp., 34325 Ardenwood Blvd., Fremont, CA 94536.
[2] P.L. Hartman, Rev. Sci. Instrum. **57**, 2564 (1986).
[3] G. Navrotski, J. Vac. Sci. Technol., **A12**, 262 (1994).
[4] Brush-Wellman, Elmore, OH 43416.

***Fig. 218. Beryllium Window Fabrication***

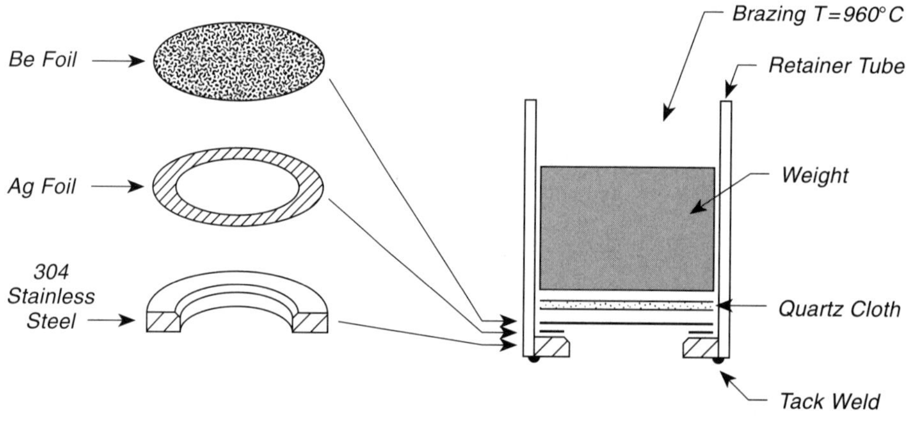


***A. Brazing Be Windows to Stainless Steel***

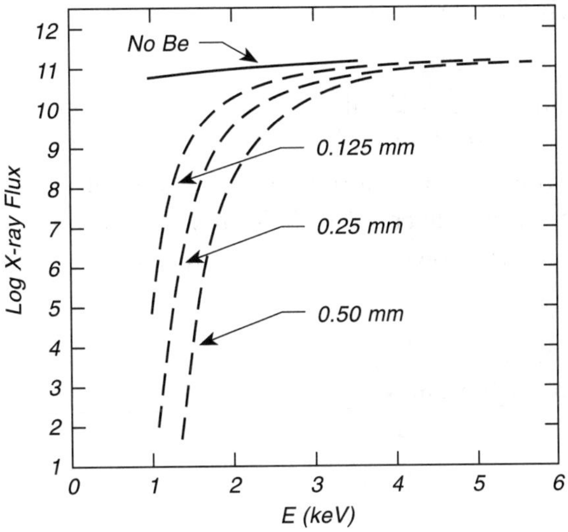


**B. X-ray Transmission through Be**

# 219

# Bakeable Aluminum Ultrahigh Vacuum Window

Low atomic number elements are often used for windows to transmit X-rays or high-energy particle beams. A very simple bakeable aluminum foil window mounting design is shown in Figure 219 [1]. The Al foil window, cut into a circle, is mounted between two Cu gaskets that are used to make vacuum seals to two Conflat flanges. The window material is Alcoa 3003 aluminum with an H19 temper. The only special machining required is to put a radius, *R*, onto the inside edge of the Cu gasket placed on the vacuum side of the window. The radius *R* should equal the thickness of the window material, *T*.

Both 0.05-mm and 0.13-mm Al foil of 3.81-cm diameter and 10-cm diameter exhibit a leak rate for He of $<5 \times 10^{-10}$ std. $cm^3\ s^{-1}$. Foils of 0.13-mm thickness have been baked at 260°C overnight with one side at atmospheric pressure without developing leaks; foils of 0.05-mm thickness have been baked at 180°C overnight without developing leaks. This sealing technique has been shown to be unsuccessful with Be windows.

## *References*


[1] J. Edgecumbe, Rev. Sci. Instrum. 37, 1419 (1966).

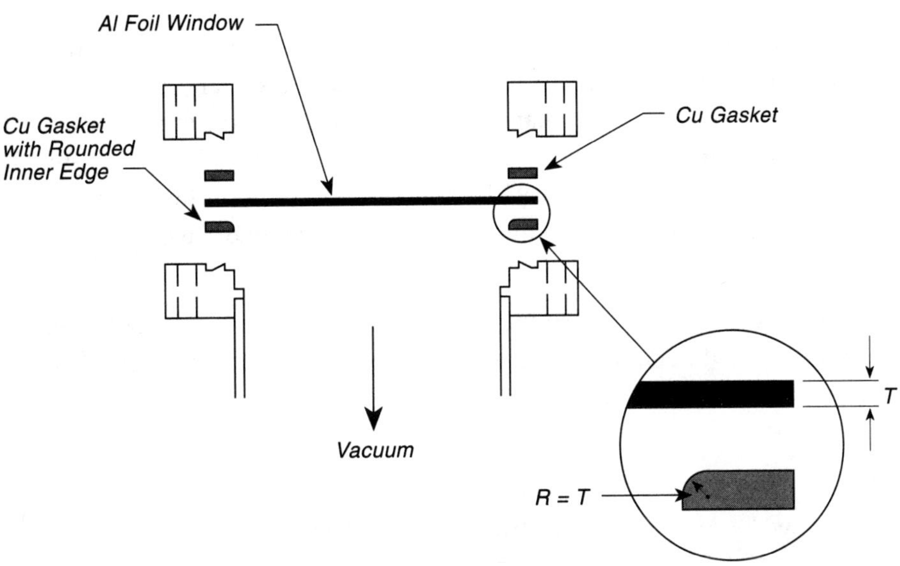


***Fig. 219. Conflat Flange Seal for Al Foil Window***

# 220

# Metal Thin Foil Windows—Welding

Very thin metal foils may be welded to a stainless steel flange using electron beam welding in vacuum. However, there is danger of damaging the foil because of the lack of heat conduction away from the fusion region. A method [1] for successful welding is shown in Figure 220.

The foil, which may be of Cu, Ni, or stainless steel, is several microns in thickness. It is placed on the flange and a backup washer of 304 stainless steel (0.05–0.13 mm thick, made by photoetching from sheet material) is placed on top of the foil, as shown. A hold-down fixture and clamps are used. The electron beam (120 kV, 250 μA, 0.13-mm diameter) is directed as shown in Figure 220, and the foil is fused to the flange, after which the washer may be removed. A rotating table, with a velocity of 76 cm/min at the weld point is used.

If the foil initially contains wrinkles, it will flatten as a result of welding at the circumference. This is thought to be due to radial forces exerted at the circumference during cooling of the weld puddle.

## *References*

[1] P. Fowler and J.V. Minges, Rev. Sci. Instrum. **40,** 841 (1969).

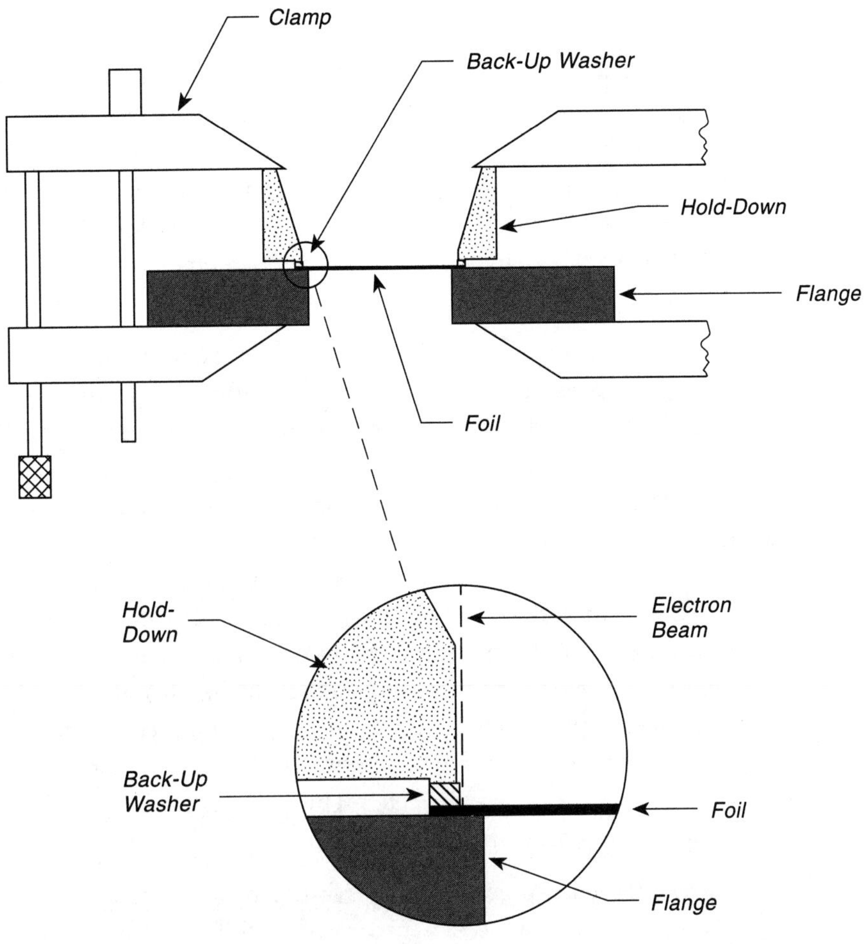


***Fig. 220. Metal Thin Film Windows—Welding***

# 221

# Preparation of Large-Area Si Single Crystal Windows

The preparation of large windows of uniform thickness made of single crystal silicon could be useful in many ways, for example, in X-ray lithography, in energy-loss particle detectors, in Schottky diodes, and Josephson junctions, in ion beam analysis, and in in situ transmission electron microscopy. A variety of thinning methods have been used [1], and the description here is concerned with preparing such windows in a thickness range from 1000 to 20,000 Å [1]. The method has been demonstrated for crystals of (100), (111), and (110) orientation [1].

Figure 221A shows the basic sequence of steps employed. Device-grade silicon wafers with a high-quality surface finish (e.g., Syton polish), of either *p*-type or *n*-type with 1–10 Ω-cm resistivity are subjected to boron diffusion in a furnace operating at 1050°C with a planar BN diffusion source and nitrogen gas as the carrier gas. Roughly speaking, a diffusion time of 15 min will give a B penetration depth that is appropriate for making a window of 2000–2500-Å thickness. The diffusion time increases quadratically as the desired window thickness. The boron glass deposited on the surface is later removed by dipping into boiling $HNO_3$ (aq) followed by HF (aq), causing controlled crystal thinning. The silicon wafer is then cleaved into smaller pieces by a diamond scriber. Flat quartz discs are then used to support the wafer sections, using a paraffin dam around the perimeter of the Si sample. The wafer with the polished side down is then placed on the dam, and sealed on the edge by pressing down a piece of Teflon tape [2] onto the top of the wafer. The tape has a hole cut into it for the window etch. This protects the shielded regions of the wafer from etch solutions, as shown in Figure 221B.

Two etch procedures are used. A preetch, which is fast, removes the majority of the silicon, at a rate of 25 μm/min. This consists of a mixture of the following solutions: HF(aq) (48% concentration); $HNO_3$ (70.4% concentration), and glacial acetic acid, $CH_3COOH$ (100% concentration) in the volumetric ratio of 2:5:2. The etch rate can be increased or decreased by adjusting the HF concentration.

For the preetch procedure, as shown in Figure 221C, the quartz-Si assembly is placed in a Teflon or polypropylene beaker (250 ml), which may be rotated and which contains a gas jet bubbler (plastic) supplying oxygen or nitrogen gas for agitation. The rotation of the beaker plus agitation by bubbling often gives uniform etching conditions, and also detaches bubbles that may form on the Si surface. Thinning to about 0.0025-cm thickness leads to a transparent film that will appear to be reddish by light transmission, indicating that the preetching should be immediately stopped. The Teflon tape is then carefully removed using a razor blade working from the outside, or using boiling trichloroethylene to dissolve the adhesive on the tape. Care must be taken not to disturb the fragile Si window.

The selective, slow etching procedure is done with a mixture of pyrocatechol ($C_6H_4(OH)_2$), ethylene diamine ($NH_2(CH_2)_2NH_2$)-(EDA), and water in the ratio 3 g:17 ml:8 ml. This etching solution does not attack Si if the B doping exceeds 7 x $10^{19}/cm^3$. Etching rates for Si(111), Si(100), and Si (110) are found to be 3, 50, and 30 μm/h, respectively. The selective etch is carried out as shown in Figure 221D, by leaning the quartz-Si assembly against a glass slide and operating at 110°C. A curved glass cover (watch glass) should be placed over the boiling solution to condense the distillate and return it to the main etch solution. The EDA is the etching agent.

With Si(111) crystals, the EDA etching rate is slow, and lateral etching around the edge of the crystal will dominate, giving a noncircular boundary. This can be prevented by arranging in the preetch to have nonuniform etching by lowering the gas flow rate which leads to preferential etching in the middle of the window. Under these conditions, the EDA etching will start at the thinnest central region and will proceed laterally toward the edge, etching the non-(111) planes which exist there. The etching is stopped before the etching reaches the window boundary. This process leads to oval/round etched shapes which are mechanically more stable than etch configurations with sharp corners.

Modification of the highly doped windows is possible after etching. The B concentration may be lowered [3], and the window may probably be used as a seed to grow Si layers by CVD processes [1]. The thin Si windows can be made amorphous by ion implantation up to 1/3 of the thickness [1]. The window can be heated to 1000°C without breaking. Layers of $SiO_2$ and metal silicides can also be grown at temperatures up to 850°C [4,5].

## *References*

[1] N.W. Cheung, Rev. Sci. Instrum. **51**, 1212 (1980).

[2] Obtained from Connecticut Hard Rubber Co., P.O. Box 2560, Meriden, CT 06450.

[3] C.L. Huang and T. van Duzer, IEEE Trans. Electron Devices, ED-23, 579 (1976).

[4] N.W. Cheung, L.C. Feldman, P.J. Silverman, and I. Stensgaard, Appl. Phys. Lett. **35**, 859 (1979).

[5] N.W. Cheung, M-A. Nicolet, and J.W. Mayer, in *Proceedings of the Symposium on Thin Film Phenomena- Interfaces and Interactions,* Vol. 80.2, edited by J.E.E. Baglin and J.M. Poare (The Electrochemical Society, Princeton, NJ, 1980).

**Fig. 221. Making Thin Si Windows by Etching**

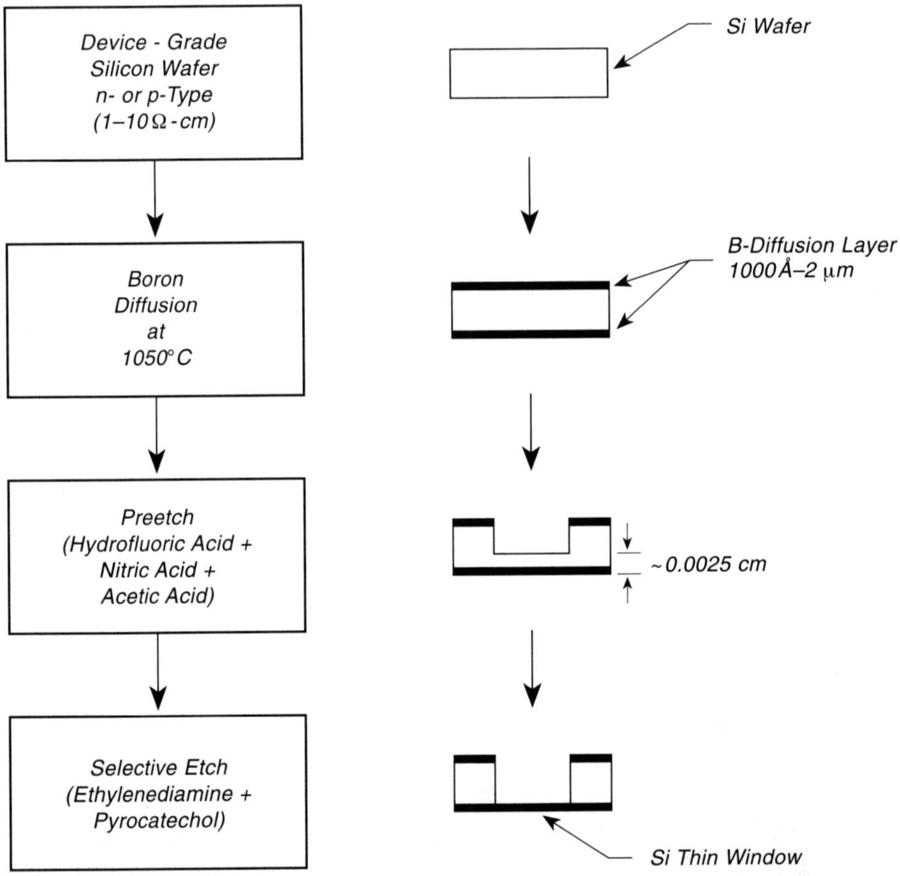


**A. Sequence of Steps—Schematic**

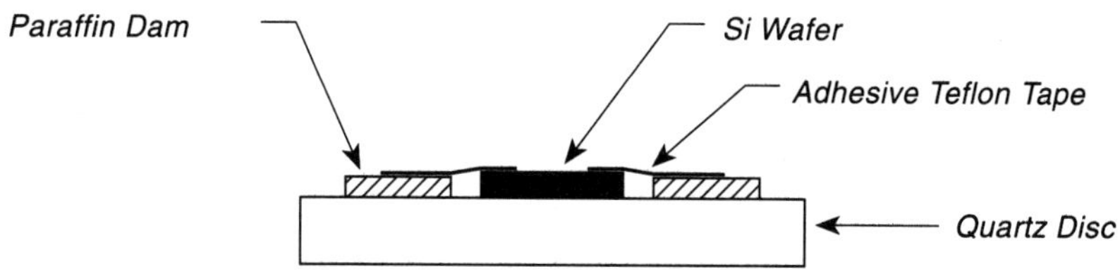


**B. Structure Before Preetch**

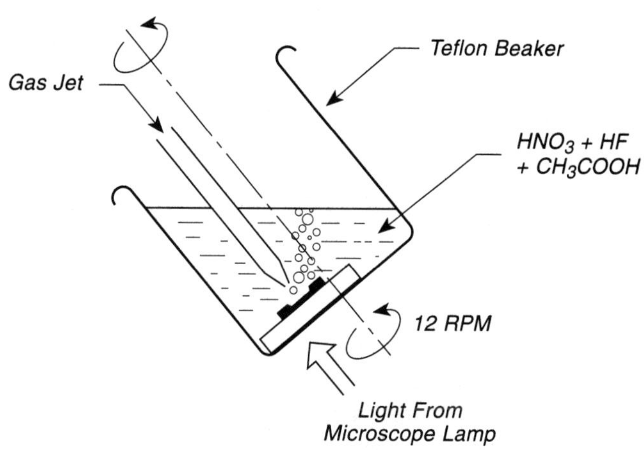


**C. Preetch**

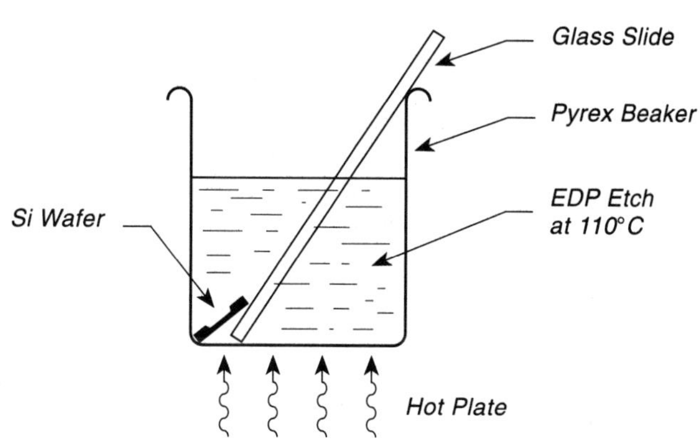


**D. Selective Etch**

# 222

# Strain-Free Optical Window Mounting

The effect of birefringence in strained optical windows can be a significant perturbation in ellipsometric measurements made inside of an ultrahigh vacuum system. This strain can be caused by the distortion of a flange during tightening of the mounting bolts that can be transmitted to the window. A re-entrant design such as that shown in Figure 222 can eliminate this type of strain [1].

To construct this mounting, the glassblower will start with a Kovar-to-glass seal that has been welded into the flange. This seal first has glass matched to the thermal expansion coefficient of Kovar alloy, and then a graded seal leading to Pyrex on the outer end of the seal. The Pyrex is fashioned into an S-shaped reentrant seal, with the window mounting flared and polished flat to receive the window and its sealant. The length of glass in the S-section is sufficient to eliminate the transmission of stress to the window when the flange is tightened.

Glass-to-metal seals other than Kovar are sometimes used in these applications [2].

## *References*


[1] Professor David Waldeck, Department of Chemistry, University of Pittsburgh, Pittsburgh, PA 15260 (private communication). The design originated with Professor C.B. Harris' group at the University of California, Berkeley, CA.

[2] Glass-to-metal sealing methods are described in E.L. Wheeler, *Scientific Glassblowing* (Interscience [Wiley], NY, 1958), p. 157; W.E. Barr and V.J. Anhorn, *Scientific and Industrial Glass Blowing and Laboratory Techniques* (Instrument Publishing Co., Pittsburgh, 1947), p. 116.

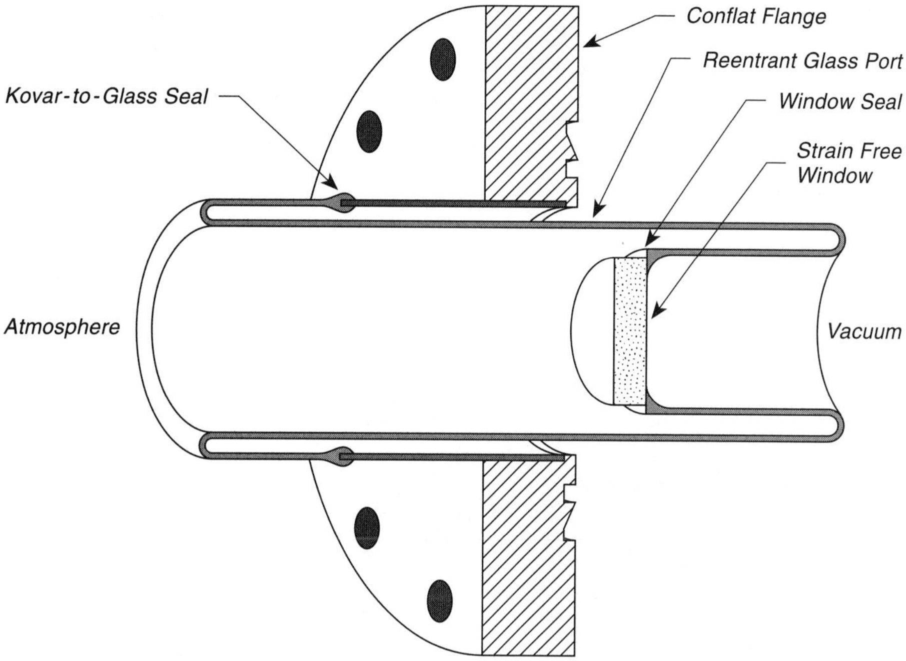


**Fig. 222. Strain-Free Optical Window Mounting—Section**

# B. Observation Windows

# 223

# Shielded Observation Window

In ultrahigh vacuum evaporation of materials to produce thin films, the accumulation of deposits on observation windows can be detrimental to temperature measurements using an optical pyrometer. A calibration procedure using a constant temperature internal source has been described earlier (see page 592), and this method will work for thin film deposits. However, a preferable procedure involves shuttering the window so that short observations can be made; then the window can be protected during most of the evaporation period. An internal rotatable shutter provides one way to do this, and the design shown in Figure 223 shows another method [1].

A Pyrex window is sealed onto a glass-metal seal that involves either Kovar alloy or stainless steel tubing mated to the glass. The tubing is welded to a formed metal bellows connected to a flange as shown. A retention mechanism, held in place by a wing nut, allows the bellows to be bent so that the window is out of line of sight from the evaporation source during most of the evaporation period. For observation, the retention mechanism is removed and the window then aligns with the source. One should use formed bellows, not welded bellows, for constructing this device.

Alternatively, a disposable window, inserted into the chamber, can be used to protect observation windows [2].

## *References*


[1] A. Chambers, J. Vac. Sci. Technol. A7, 259 (1989).

[2] Dr. W.A. Egelhoff, National Institute of Standards and Technology, Gaithersburg, Md. 20899-0001 (private communication).

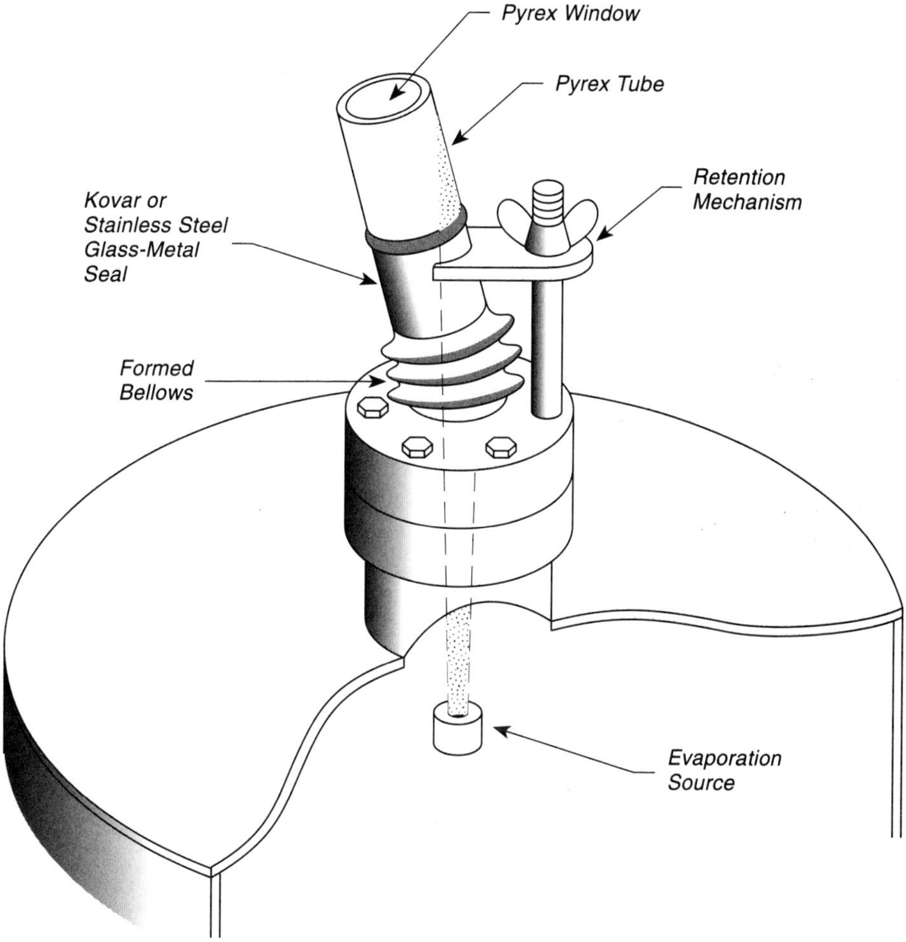


***Fig. 223. Shielded Observation Window***

# 224

# Heated Internal Window for Prevention of Film Deposition

External windows used for observation of evaporation sources may be protected from film deposition internally by using a heated window to intercept the evaporating material. If the temperature of the internal window is sufficiently high to cause evaporation, optical measurements may be made through the two windows without interference from film deposition.

Figure 224 shows schematically an arrangement for heating a quartz internal window to temperatures sufficiently high the prevent As deposition during MBE growth of GaAs, for example [1]. The inner quartz window is heated ohmically through a Ti film deposited around the edge of the quartz disk on both faces to a thickness of about 1.5 μm. The electrical resistance of the film, when contacted at points opposite each other, is about 2 Ω at room temperature and about 10 Ω at the operating temperature. Electrical contact to the Ti film is made using W clips made from 0.25-mm-diameter wire and protective Ta foil pads prevent abrasion of the Ti film by the W clips. Electrical heating at a power level of about 20 W maintains the temperature around the edge of the quartz window at 550–600°C, and the temperature in the center is estimated to be 350–400°C, based on a calculation for radial heat transfer [2]. It was reported that evaporation of over 1000 g of As over 1 year was carried out using one protective quartz window, and it was found that the optical transmission of the internal window only decreased by 5%, corresponding to a temperature error of only 2° for pyrometry measurements at 600°C. For GaAs deposition, it was found that the heated window did not introduce impurities into the films produced by molecular beam epitaxy.

Alternatively, the use of disposable internal glass windows has been employed successfully for this purpose [3].

## *References*


[1] D.E. Mars and J.N. Miller, J. Vac. Sci. Technol. **B7**, 682 (1989).

[2] D.E. Mars and J.N. Miller, J. Vac. Sci. Technol. **B4**, 571 (1986).

[3] Dr. W.F. Egelhoff, National Institute of Standards and Technology, Gaithersburg, MD 20899-0001 (private communication).

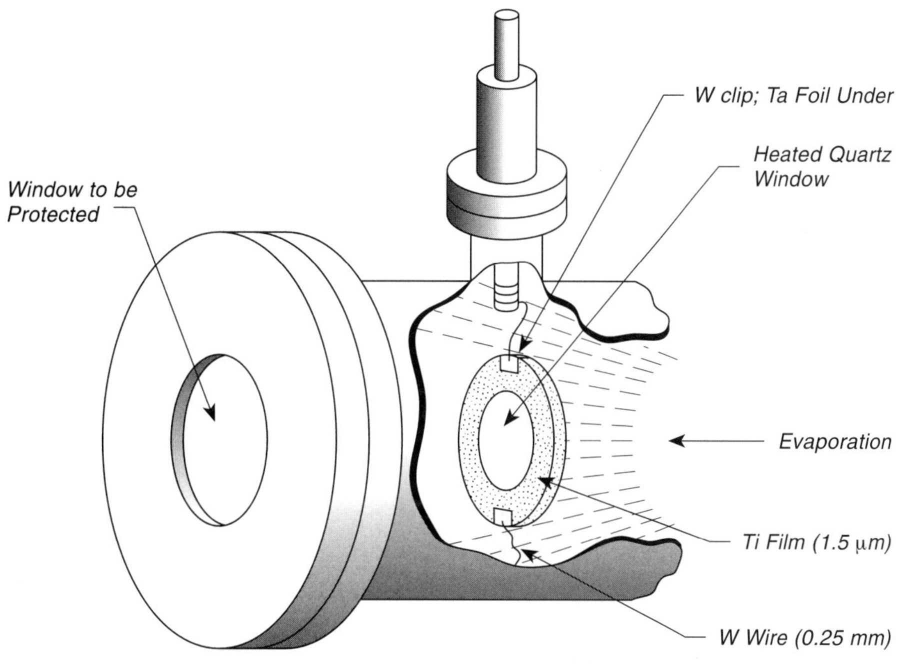


***Fig. 224. Heated Windows***

# 225

# UHV Glass Window Seal—Low Optical Distortion

Glass seals in Conflat flanges are generally made by a brazing technique that distorts the window, limiting its usage for certain optical studies. In addition, when optical coatings are applied to flange-sealed glass windows, the high temperature of the coating process may further distort the window. A number of alternative window seals have been described [1–5] but often require considerable modification of the standard flanges.

In Figure 225, a modified Conflat flange and gasket are employed to seal BK 7 [6] glass into an ultrahigh vacuum chamber [7]. The window material can range from high precision optical flats to window glass. These glass blanks can be optically coated if desired. A commercial OFHC Cu gasket is machined to have its own knife edge. The window seals the gasket by deforming the Cu knife edge. The dimensions shown are somewhat arbitrary. Pressure is applied to the window seal using a rotatable bolt ring which may be modified to increase the area which applies force to the window. Twenty sheets of 0.001-in.-thick Al foil are used as a cushion between the retaining ring and the glass. Alternatively either Cu or Pb wire may be used for the cushion.

The bolts are tightened uniformly and slowly until no leak is observed with a He leak detector. For a 4.5-in.-diameter flange, torques of about 3 ft lb/bolt are employed for sealing. The technique forms seals with less than $10^{-10}$ atm $cm^3$ $s^{-1}$ He leak rates. Baking to 250°C may open leaks, and these can be eliminated by retightening the flange bolts. A test for stress-induced birefringence showed that only 0.01 % of 851-nm light passed through two crossed linear polarizers with a window mounted between them.

It is important to keep the Cu knife edge at least 0.1 in. from the edge of the glass to prevent breakage, and this may necessitate offsetting the Cu knife edge from the underlying stainless steel knife edge, as shown in Figure 225 [7].

## *References*


[1] T.J. Manuccia, J.R. Peele, and C.E. Geosling, Rev. Sci. Instrum. **52,** 1857 (1981).

[2] C. Damerow and W. Erley, J. Vac. Sci. Technol. **A5,** 2974 (1987).

[3] C.H. Muller III, M.W. Barrett, and D.D. Lowenthal, Rev. Sci. Instrum. **59,** 1425 (1988).

[4] H. Saeki, J. Ikeda, and H. Ishimru, Vacuum **6,** 563 (1989).

[5] M.T. Duignan, D.J. Garhardt, and W.T. Whitney, Rev. Sci. Instrum. **60,** 3537 (1989).

[6] BK 7 blanks are supplied by Newport Optical Glass Inc., Costa Mesa, CA, and by Glassfab Inc., P.O. Box 1880, Rochester, NY 14603.

[7] A. Noble and M. Kasevich, Rev. Sci. Instrum. **65,** 3042 (1994).

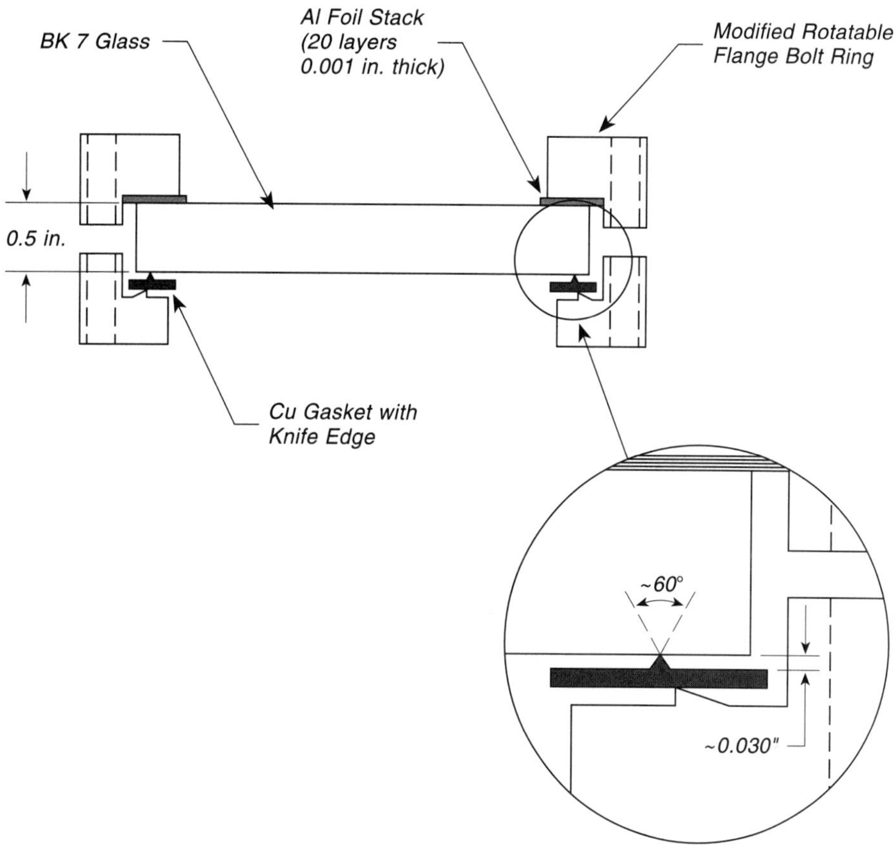


**Fig. 225. UHV Glass Window Seal—Low Optical Distortion**

# 226

# Zero-Length Window Assembly

Several methods for mounting windows on ultrahigh vacuum systems are available, but the method shown here is the most simple, requiring only standard components. As shown in Figure 226, a window is retained between two Viton O-ring seals, using a rotatable outer flange without its normal insert [1]. The assembly is mounted onto a normal Conflat flange attached to the vacuum system. The inner seal may be either a O-ring or a special Viton gasket manufactured by Varian for use in place of copper gaskets. The upper bakeout temperature for the Viton is 150°C. The general design is also applicable to larger flanges.

## *References*


[1] L.A. Biagi, J. Vac. Sci. Technol. **8**, 464 (1971).

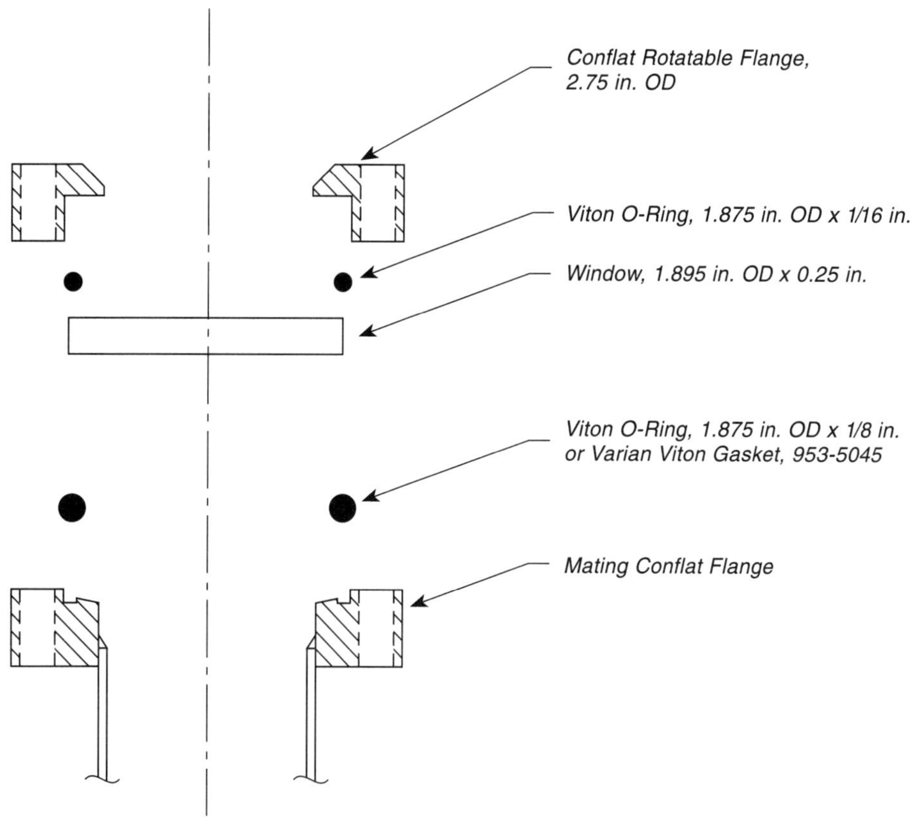


**Fig. 226. Zero Length Window**

# VII. Surface Preparation Methods

A. Cleaning Metal and Semiconductor Crystals—Examples . . 779

# A. Cleaning Metal and Semiconductor Crystals—Examples

# 227

# Surface Segregation and the Cleaning of Metal Single Crystals

The preparation of clean single crystals of the metals is fundamentally important in surface science research, since surface properties are often highly dependent on poisons and promoters that are present. It is commonly known that bulk impurities present in zone-refined single crystals at the ppm level can often segregate to the surface upon heating. Based on the phase diagram of two component alloys, Burton and Machlin [1] gave an easy method to predict whether a solute would segregate to the surface upon heating, as described below.

In Figure 227A, two typical phase diagrams are shown for alloys, where the region between the solidus (dashed curve) and the liquidus (solid curve) is a two-phase region. In phase diagram (a), at some constant % solute, if the solid is warmed to the melting point on the dashed curve, a liquid depleted in solute will be formed. In contrast, in phase diagram (b), melting of the alloy solid will produce a liquid phase that is enriched in solute. The authors point out that many of those qualities that distinguish a liquid from a solid—namely, lower symmetry and lower coordination—also distinguish the surface from the bulk, so that the tendency for the liquid to be enriched or depleted of the solute on warming may apply also to the surface of an alloy. This expectation was confirmed for 16 alloys, where alloys with almost coincident liquidus and solidus curves were categorized as type (b′), and behaved as type (a) systems, and exhibited no surface segregation.

If methods can be found to stabilize a solute on a surface by reaction to form a stable compound or to remove the solute from the surface as a volatile species, then it is possible to deplete the bulk phase of this solute. This is commonly achieved by using oxygen as a reactant, producing a volatile oxide that enhances the concentration gradient of the solute, allowing more solute to diffuse to the surface where it is removed by reaction. In combination with sputtering, surface segregation is effective in purifying the bulk (or at least the near surface region) so that bulk impurities are not available at the surface during an experiment.

A description of the surface cleaning methods employed for a total of 74 elements has been given in a very useful review by Musket et al. [2]

Several examples of the use of chemical methods for cleaning surfaces in this manner are presented below to illustrate the general principles involved and the types of measurements used to follow the cleaning process.

## Platinum

Elements such as Si, Ca, Al, S, C, and P are often found in zone-refined Pt crystals, and an excellent review of and insight into this problem has been provided in the work of Bonzel et al. [3]. Si segregates strongly to the surface of Pt, but this was not appreciated until the early 1980s because the prominent Si (92 eV) feature is almost coincident with a Pt(94eV) feature. This may be seen in Auger spectra in Figure 227B, where heavy Si segregation is observed under equilibrium conditions at 873 K, and less heavy segregation is observed at higher temperatures. Spectrum (a) is of the freshly sputtered surface (Si free) and the small Pt Auger feature at about 94 eV is evident. Studies of the effect of $O_2$ on the rate of Si segregation to the surface show that at $10^{-7}$ mbar and at Pt temperatures of 873 K, no enhancement of the Si segregation occurs in the presence of $O_2$, as shown in Figure 227C. Studies at higher temperature [3] indicate that bulk $SiO_2$ will form for $T > 1000$ K. It is pointed out that the Auger intensity of the O(KLL) feature, associated with Si atoms, is the best monitor of the amount of Si present after oxidation. Platinum oxides are thermodynamically unstable in vacuum at the temperatures discussed here. The properties of the oxide on silicon-contaminated Pt first led investigators to postulate the presence of a "platinum oxide" with unexpected thermodynamic stability on Pt surfaces [3].

## Palladium

Similar results to those found for Pt are reported for Pd [4]. Here, too, Si-stabilized oxides are formed under conditions where PdO would be thermodynamically unstable, as shown in Figure 227D. The oxidation occurred at 1073–1273 K in $10^{-6}$ mbar $O_2$ for spectra (a) and (b). Argon ion sputtering was used prior to acquiring spectrum (c).

## Rhodium

A common impurity in single crystal Rh is boron, which is difficult to remove by sputtering [5,6]. It has been found that heavy oxidation is effective in removing boron by the formation of a boron oxide species that is best detected by high-resolution electron spectroscopy (HREELS) [7]. The interference between Rh and B Auger spectral features makes Auger spectroscopy impractical for seeing small amounts of B on Rh. Figure 227E shows an HREELS spectrum of the B-O vibrational modes obtained when Auger spectroscopy was ineffective in detecting B.

## *References*

[1] J.J. Burton and E S. Machlin, Phys. Rev. Lett. **37,** 1433 (1976).

[2] R.G. Musket, W. McLean, C.A. Colmenares, D.M. Makowiecki, and W.J. Siekhaus, Appl. Surf. Sci. **10,** 143 (1982).

[3] H.P. Bonzel, A.M. Franken and G. Pirug, Surf. Sci. **104,** 625 (1981); earlier important studies of this problem include M.J. Cardillo and G.E. Becker, Surf. Sci. **99,** 269 (1980); H. Niehus and G. Comsa, Surf. Sci. **102,** L14 (1981).

[4] S.D. Bader, L. Richter and T.W. Orent, Surf. Sci. **115,** 501 (1982).

[5] D.G. Castner and G.A. Somorjai, Appl. Surf. Sci. **6,** 29 (1980).

[6] P.A. Thiel, J.T. Yates, Jr., and W.H. Weinberg, Surf. Sci. **82,** 22 (1979).

[7] S. Semancik, G.L. Haller, and J.T. Yates, Jr., Appl. Surf. Sci. **10,** 546 (1982).

***Fig. 227. Surface Segregation and Cleaning of Metal Surfaces Chemically***

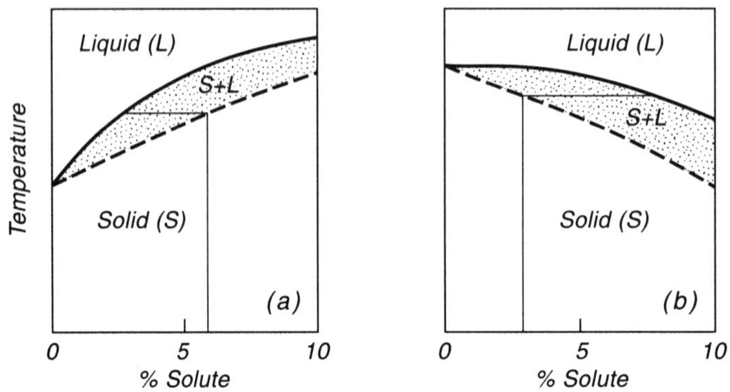


***A. Typical Phase Diagrams for Alloys***

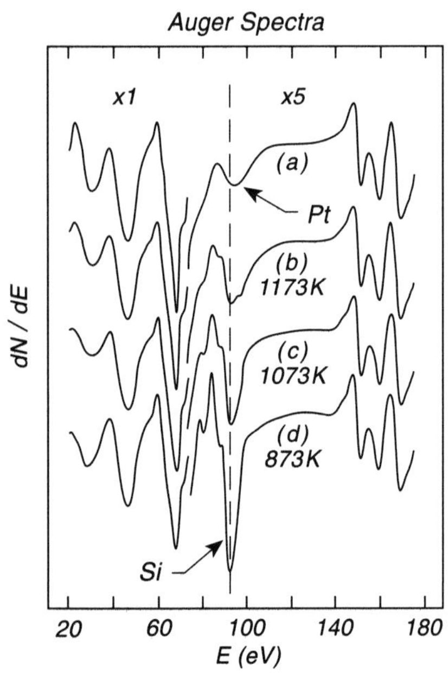


***B. Si Segregation on Pt***

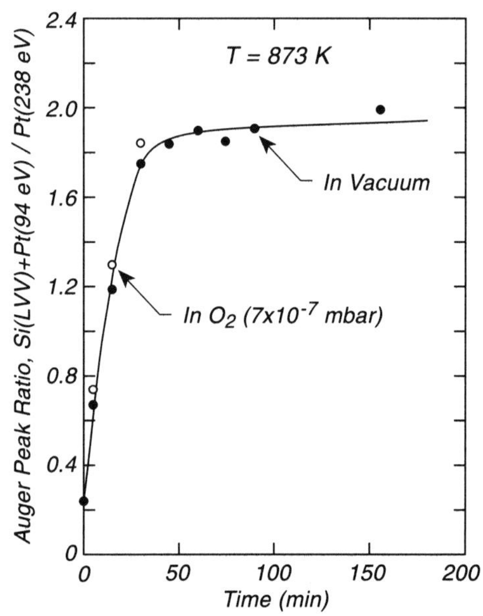


***C. Si Segregation in Vacuum and in $O_2$ on Pt***

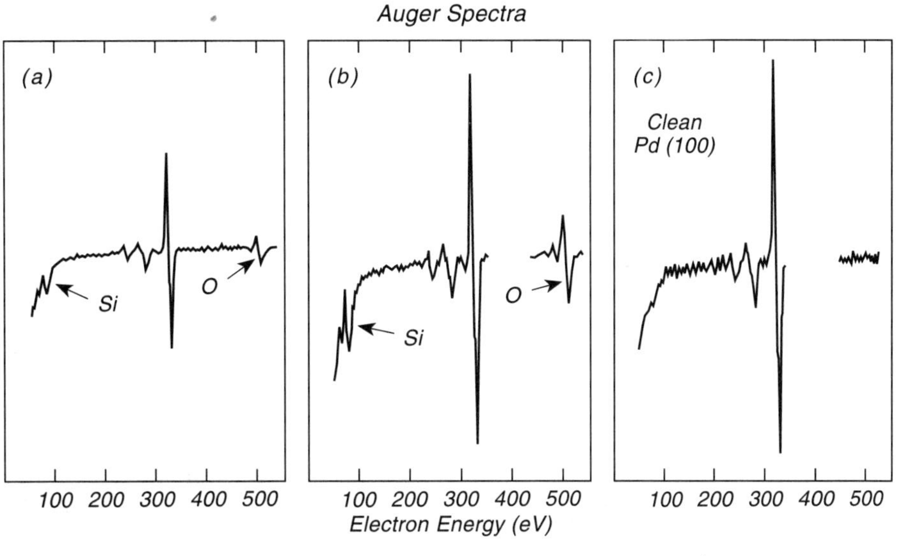


**D. Silicon Segregation on Pd**

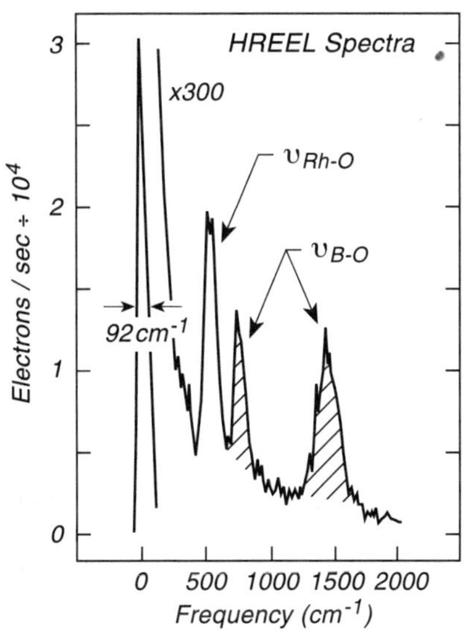


**E. B-O Vibrational Modes on Rh(111) After Reaction at 970 K in $O_2$**

# 228

# Chemical Method to Measure and Remove Carbon from Pd

The high thermodynamic energy of NO compared to $O_2$ may be used to oxidize impurities from single crystal surfaces as well as from the walls of ultrahigh vacuum systems as was described on page 136.

In a special case, involving carbon impurities purposely introduced to a Pd single crystal, this effect was employed not only to clean the Pd crystal of carbon but also to provide a rather sensitive method for showing that carbon had been removed [1]. Carbon on Pd is difficult to measure by Auger spectroscopy because of the overlap of the C(KLL) Auger line with the Pd(MNN) line. Figure 228A shows a series of Auger spectra for increasing exposure to $C_2H_4$ using quantitative exposure conditions. A careful examination of the Auger spectra shows that small changes occur, and a ratio measurement shown in Figure 228B indicates that the (C + Pd)/Pd Auger ratio does increase slightly following $C_2H_4$ exposure, but the poor signal-to-noise ratio on this number makes a quantitative measurement impossible.

The carbon may be quantitatively reacted away using NO as an oxidizing agent in much the same fashion as was done by others using $O_2$ [2] to oxidize the C as well as to measure the presence of C using measurements of the CO and $CO_2$ formation in thermal desorption. In the case of the NO being employed as an oxidizing agent, using isotopically labeled $^{15}NO$ permits the discrimination of $^{15}N_2$ (30 amu) product from CO (28 amu product) and hence to supply a quantitative basis for measuring the coverage of carbon from the $^{15}N_2$ formation. Figure 228C shows the thermal desorption spectra for 30 amu species produced after treating carbon-covered Pd with $^{15}NO$. Successive experiments were done with higher levels of carbon, deposited by quantitative doses of $C_2H_4$. The cross-hatched portion of the thermal desorption spectra is due to $^{15}N_2$ from the reaction $2C(a) + 2NO(a) \rightarrow 2CO + N_2$. The sensitivity of the method for low carbon levels may be judged by the fact that $C_2H_4$ exposures much less that $1 \times 10^{14}$ cm$^{-2}$ give measurable $^{15}N_2$ yields in the high-temperature peak. The low-temperature desorption peak, also from $^{15}N_2$, originates from the normal dissociation of NO on the Pd surface.

Using this measurement method for determining the presence of small amounts of carbon on Pd, a cleaning procedure involving exposure to $3 \times 10^{16}$ NO cm$^{-2}$ at 510 K followed by heating in vacuum to 850 K was devised. This method gave reproducibly clean Pd, which yielded the three overlapping $^{15}N_2$ desorption curves shown at the bottom of Figure 228C for three sequential $^{15}NO$ adsorption experiments on the carbon-free crystal.

## *References*

[1] R.D. Ramsier, K.-W Lee, and J.T. Yates, Jr., J. Vac. Sci. Technol. **A13**, 188 (1995).
[2] H. Conrad, G. Ertl, J. Küppers, and E.E. Latta, Surf. Sci. **65**, 235, (1977); ibid 245, (1977).

***Fig. 228. Chemical Method to Measure and Remove C from Pd***

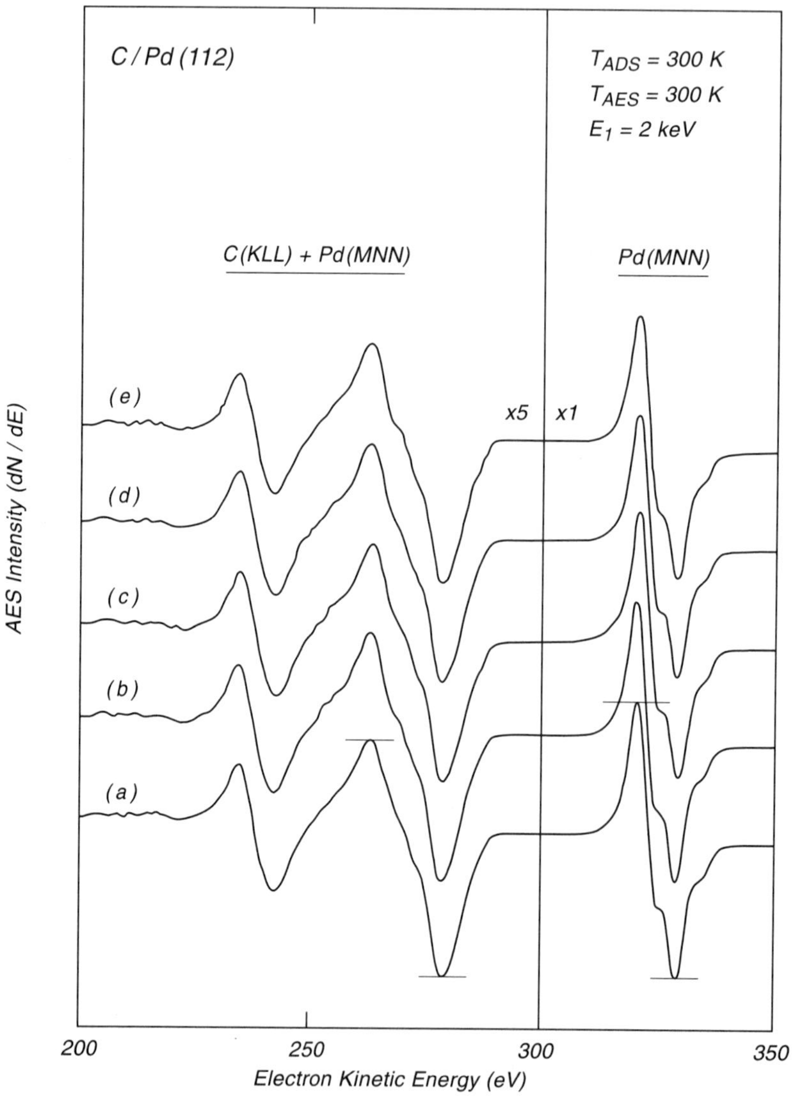


**A. Pd Auger Spectra for Increasing C Coverage**

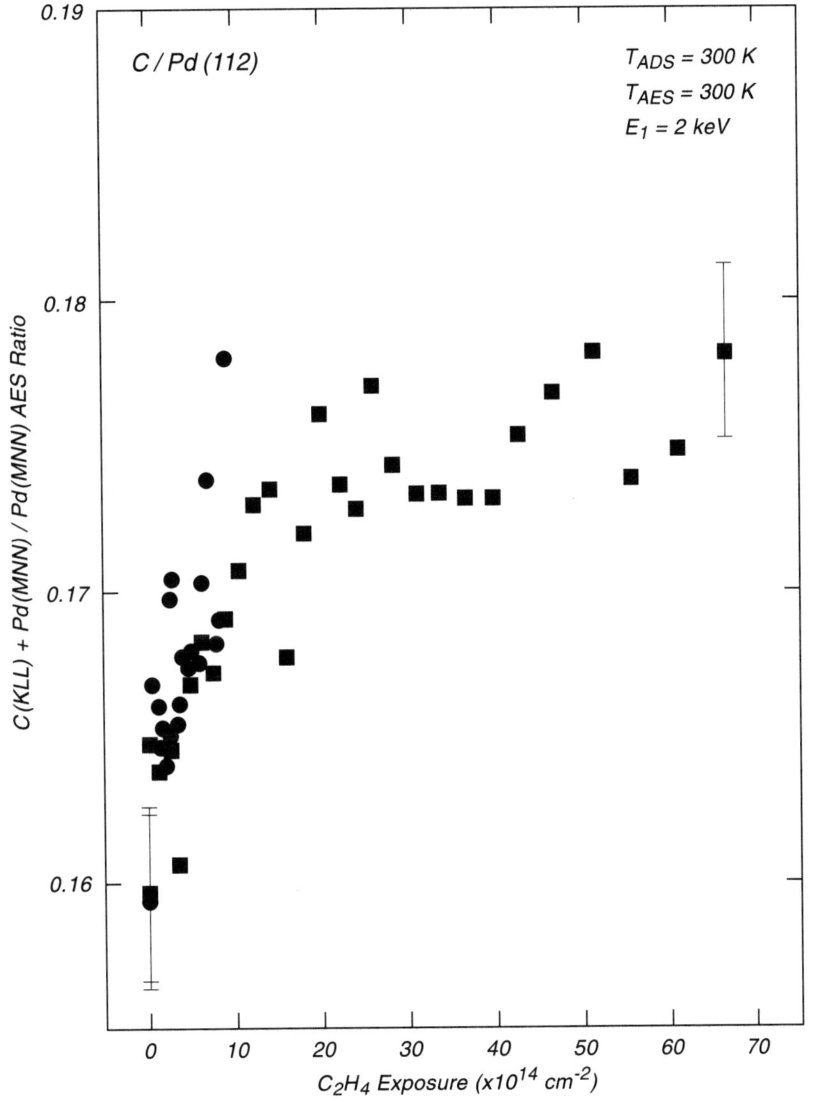


**B. Auger Ratio for Increasing C Coverage**

***Fig. 228. Chemical Method to Measure and Remove C from Pd***

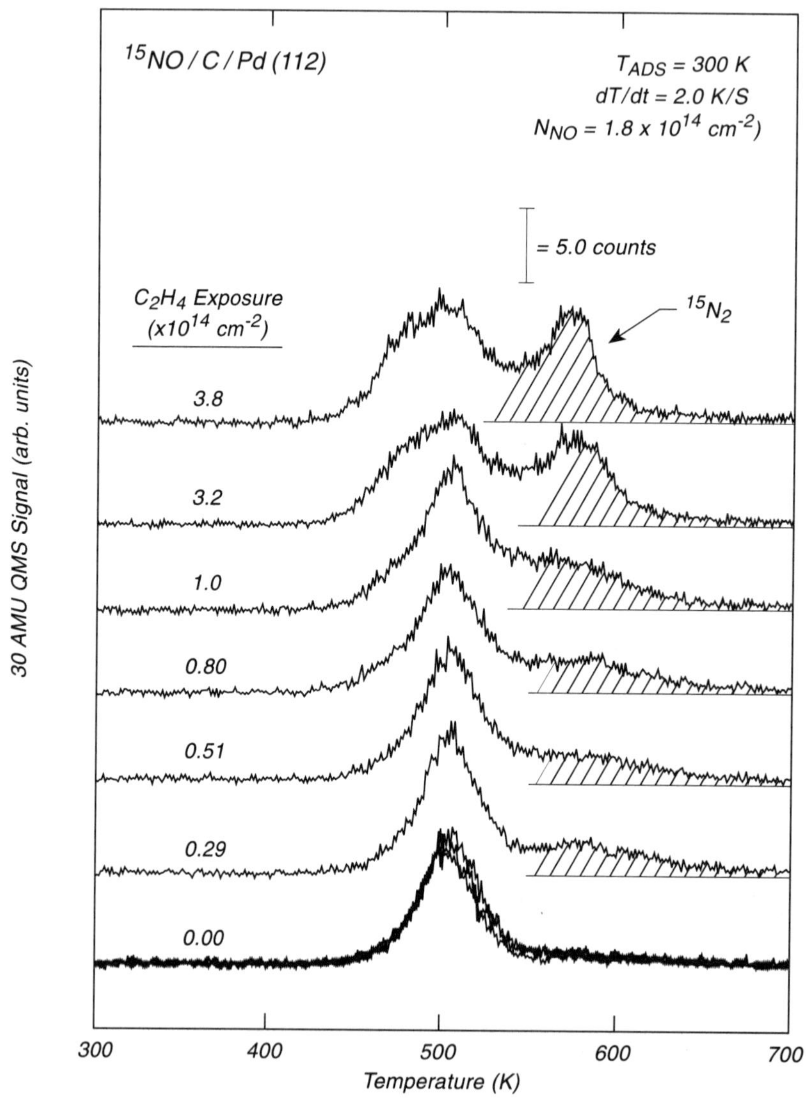


**C.** $2C + 2\,^{15}NO \rightarrow {}^{15}N_2 + 2CO$

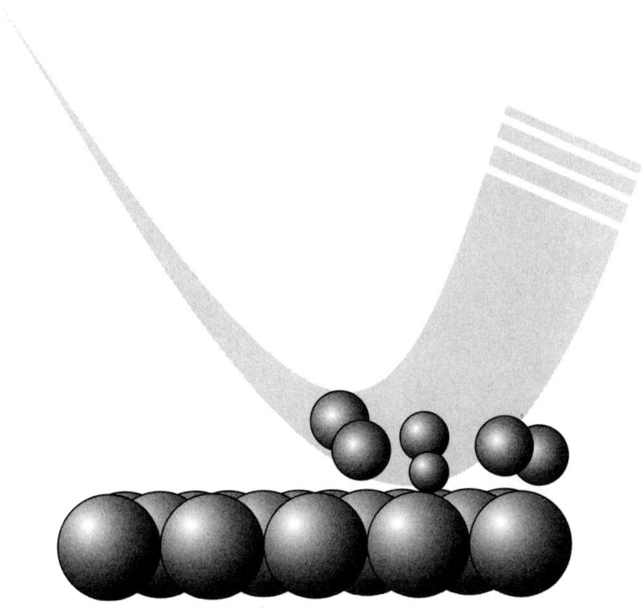

# 229

# Surface Cleaning Procedures—Silicon

The production of metal-free silicon surfaces is a common starting point in much surface science work dealing with this element. The metals nickel and iron are common impurities originating from the vacuum system environment owing to various types of sources. Their presence causes significant reconstruction effects on Si(100) [1–3] and on Si(111) [1,4,5]. The STM permits the observation of reconstruction effects at high sensitivity [6], and recently STM has been used to compare the statistical distribution of defects with the measured Auger intensity from impurity Ni in the surface region [6]. The transport of Ni through the silicon lattice is extremely rapid [7], causing Ni contamination effects to spread throughout the crystal from their origin during ordinary thermal processing.

The fundamental unit of the reconstructed Si(100) surface owing to very low levels of Ni contamination is the "split-off-dimer" structure, shown in Figure 229A. Four silicon dimer sites are affected and produce a characteristic STM image of two neighboring dark dimer sites, a neighbor bright dimer site, and a neighbor dark dimer site. From detailed Auger spectroscopy studies, about one-half of the detected Ni atoms are responsible for the split-off dimer sites [6]. As the concentration of impurity Ni increases, the split-off dimer structures organize themselves to produce vacancy channels having (2 x $n$) overlayer periodicities. Ultimately, high corrugation levels of Si(100) with 50-Å roughness are produced. Figure 229B shows the correlation between the statistical counting of split-off-dimer structures and other methods of surface analysis. A linear relationship is observed between the density of the defects and the surface analysis for Ni .

The origin of the Ni impurities may be from several sources listed below:

1. Ni bulk contamination in Ta used in contact with silicon: The bulk analysis of high-purity Ta may give 3 ppm of Ni. The typical amount of Ta in the vicinity of the Si could therefore supply hundreds of monolayers of Ni to the Si.
2. Ni surface contamination of Ta following machining: The transfer of Ni from alloy tools used to manufacture Ta parts may be a source of significant Ni levels.
3. Ni contamination from thermocouple alloys: Alloys such as chromel-alumel (type K) are about 90% Ni. At temperatures of 1450–1500 K, the vapor pressure of Ni will be in the $10^{-4}$–$10^{-5}$ Torr range, and even protection of the Si by Ta foil envelopes surrounding the thermocouple junction will not prevent Ni contamination from this source. It is recommended that Ta-protected tungsten/rhenium thermocouples be used.
4. Metals deposited by sputtering. Almost all workers using the STM as a detector of surface cleanliness and structure on Si(100) have discarded the use of ion bom-

bardment as a way of cleaning the surface. Metal atoms produced from sputtering effects inside the ion gun, or from sputtering of metal structures in the neighborhood of the Si sample, are probably transferred to the Si surface.

A simple method that seems to eliminate Ni contamination in Ta components consists of boiling for 5–10 min in concentrated HCl (<5 ppb Ni), and then boiling in deionized water. This procedure has been shown to reduce the Ni concentration in the Si surface to below the Auger sensitivity (signal to noise = 0.0003) level, giving Si crystals that are long-lived and almost Ni-free as measured by STM [6].

The cleaning procedure for Si(100) wafer strips is then to heat to 900–950 K overnight to outgas and then to anneal to 1420–1450 K for 30–60 s, cool quickly to 1000–1100 K, hold at temperature to 120 s, and finally cool down to 300 K at a slow rate of 2 K/s. Alternating current heating is used.

## *References*

[1] K. Kato, T. Ide, S. Miura, A. Tamura, and T. Ichinokawa, Surf. Sci. **194,** L87 (1988).
[2] H. Niehus, U.K. Kohler, M. Copel and J.E. Demuth, J. Microscopy **152,** 735 (1988).
[3] X.F. Lin, K.J. Wan, and J. Nogami, Phys. Rev. **B47,** 13491 (1993).
[4] A.E. Dolbak, B.Z. Olshanetsky, S.I. Stenin, S.A. Teys, and T.A. Gavrilova, Surf. Sci. **218,** 37 (1989).
[5] R.M. Wallace, C.C. Cheng, P.A. Taylor, W.J. Choyke, and J.T. Yates, Jr., Appl. Surf. Sci. **45,** 201 (1990).
[6] V. Ukrainsev and J.T. Yates, Jr., Surf. Sci. **346,** 31 (1996).
[7] E.R. Weber and N. Wiehl, Mat. Res. Soc. Symp. Proc. **14,** 19 (1983).

***Fig. 229. Surface Cleaning Procedures—Silicon***

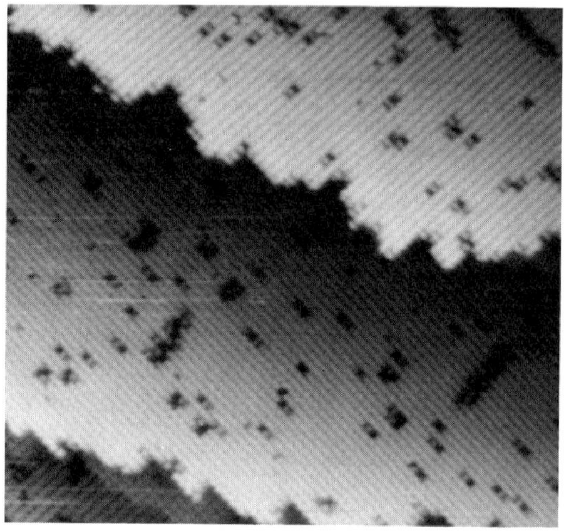

***A. STM Image of Si(100) Showing "Split-Off-Dimer" Defects***

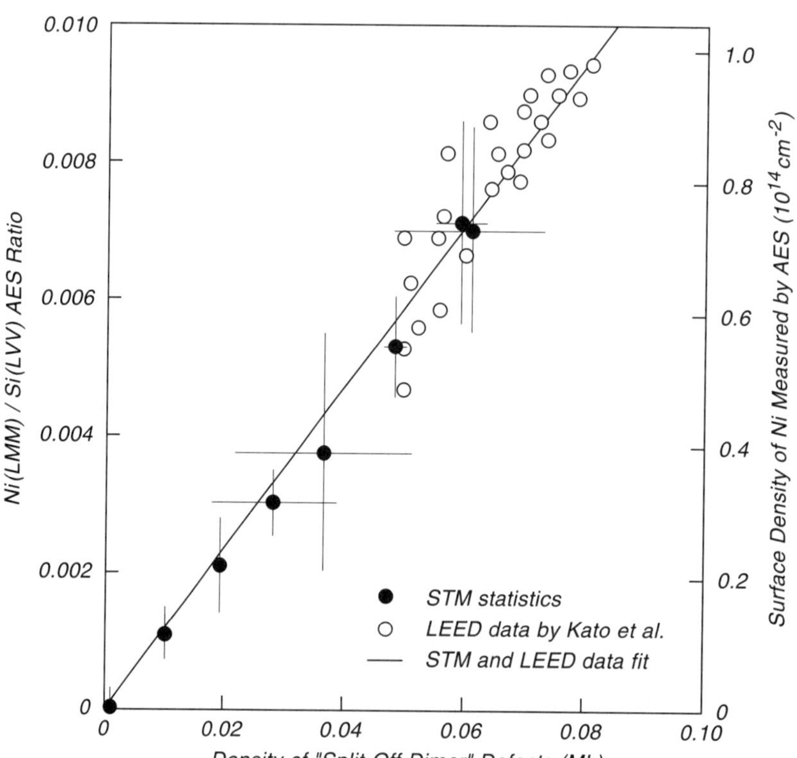


***B. Correlation of STM Defect and AES and LEED Measurements***

# VIII. High-Area Solids

A. Infrared Cells . . . . . . . . . . . . . . . . . . . . . . . . . . . . . . . . . . . . 795
B. Adsorption/Desorption—Thermal . . . . . . . . . . . . . . . . . . . . 807

# A. Infrared Cells

# 230

# Infrared Cell for Adsorption Studies on Supported Catalysts

The study of supported catalysts using transmission IR spectroscopy is a well-established technique employed widely throughout the field. Over 50 cell designs have been devised and reported in the literature [1]. The cell described in Figure 230 has special features of importance for some types of applications, especially where a simple design for use at cryogenic temperatures is desired. A unique feature of this cell is the capability to study the adsorption of gases on the support and on the support plus the active catalytic material in the same experiments [2].

The cell consists of two 2.75-in. diameter Conflat flanges containing IR transparent (down to 1000 $cm^{-1}$) windows made of $CaF_2$ single crystals, and joined to the flange through semiflexible Ag mounting rings, using fused AgCl cement. These window flanges are sealed with Cu gaskets to a double-sided central flange that contains the support for the sample. The sample support is a Cu ring (OFHC Cu) that is brazed with Ag braze (no Cd alloy) to a thin stainless steel tube that exits through two thin-walled support tubes at the edge of the central flange. A mixture of liquid nitrogen and nitrogen gas is passed through the supporting tubes to cool the sample to near 100 K, and the temperature of the support ring is monitored by a thermocouple entering the central flange at another position. Temperature control may be achieved in the low-temperature region by control of the flow rate of the refrigerant, and at high temperatures electrical heating by an external resistive heater may be controlled. The cell is evacuated through a stainless steel valve that leads to the gas-handling and vacuum system. Temperature gradients across the $CaF_2$ sample holding disk (to be described below) that is mounted in the Cu ring are negligible, as shown by experiments with multiple thermocouples cemented to the $CaF_2$ plate [3].

The sample of a high area oxide support containing metal particles is prepared by spraying a slurry of the support material and a salt of the metal onto a $CaF_2$ single crystal disk held on a hot plate at about 70°C. Here a 9:1 acetone:water solution is the solvent for the metal salt and a carrier for the oxide support. Flash evaporation of the solvent occurs during the spraying, and the oxide sticks to the disk and retains the ions from the salt. In some cases it will be necessary to lightly abrade the $CaF_2$ disk to permit better adhesion. The disk is masked so that one-half is sprayed with the catalyst/support mixture while the other half is sprayed with the support alone. This permits adsorption processes on the support to be studied independently under the identical conditions present at the supported catalyst. Typical deposited surface densities are at the 0.01 $g/cm^2$ level. The disk containing the adjacent sprayed sections is transferred to the cell and clipped into the Cu ring, and the cell is assembled, using annealed Cu gaskets (see page 150) and gently tightening alternate bolts a little at a time to prevent excessive flexure of the window flanges and fracture of the $CaF_2$ windows. The catalyst is then prepared for study by appropriate heating and reduction conditions using hydrogen, and heating the cell in a small oven that surrounds it. The $CaF_2$ windows should be carefully protected against thermal shock in use.

After reduction and evacuation to remove hydrogen, the sample is ready for chemisorption experiments. Figure 230B shows an example of the chemisorption of $N_2$ on Rh metal that is supported on $Al_2O_3$. At the temperature of the experiment, $N_2$ is also physically adsorbed on the $Al_2O_3$. The spectrum on the left was measured on the right-hand side of the sample and shows only physisorbed $N_2$. The spectrum on the right, measured on the left-hand side of the cell, shows the presence of both physisorbed $N_2$ on the $Al_2O_3$ support and chemisorbed $N_2$ on the metallic Rh catalyst surface. The cell is translated from side to side inside the spectrometer using a simple linear bearing track arrangement in order to measure each half of the sample.

This cell has been used extensively where a simple design was satisfactory. A more complex design, having additional capabilities, is shown in reference [1] and discussed on page 800. The $CaF_2$ windows mounted in Conflat flanges are commercially available [4].

## *References*

[1] P. Basu, T. Ballinger and J.T. Yates, Jr., Rev. Sci. Instrum. **59,** 1321 (1988).

[2] H.P. Wang and J.T. Yates, Jr., J. Phys. Chem. **88,** 852 (1984).

[3] T.P. Beebe, P. Gelin, and J.T. Yates, Jr., Surface Sci. **148,** 526 (1984).

[4] Windows of $CaF_2$ single crystal mounted in Conflat flanges may be obtained from Harshaw Crystal and Electronic Products, Solon Technologies, 6801 Cochran Rd., Solon, OH 44139.

Fig. 230. IR Cell for Adsorption Studies on Supported Catalysts

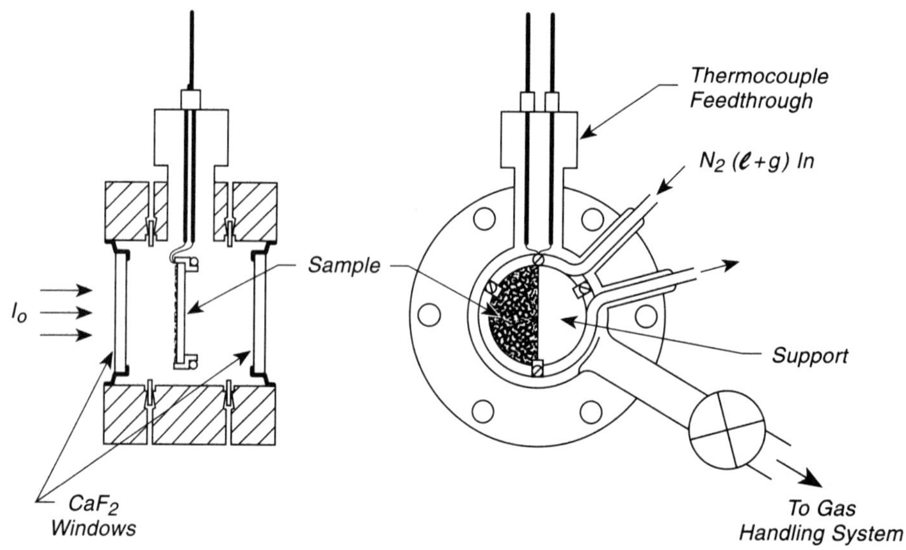


A. IR Cell

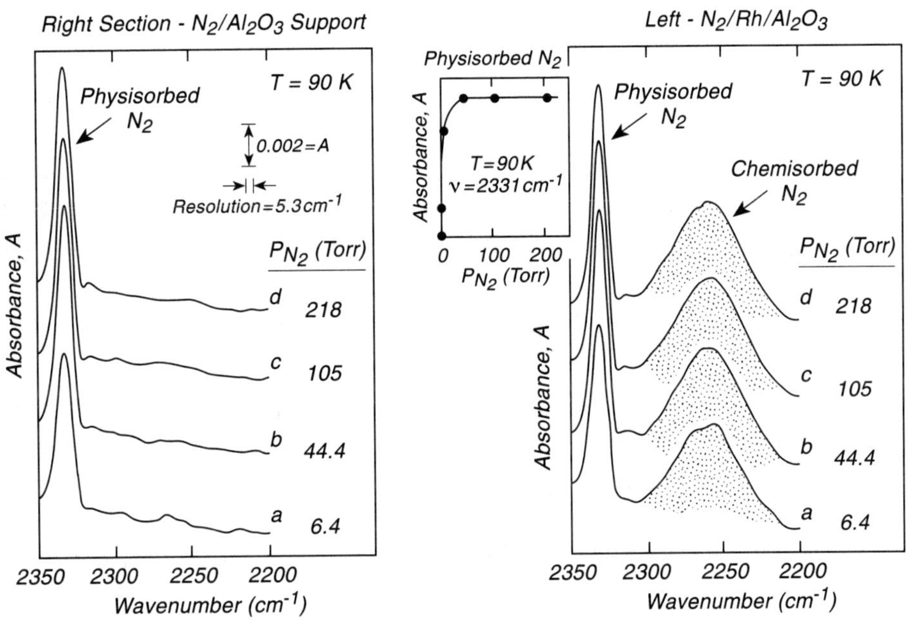


***B. Transmission IR Spectra***

# 231

# Wide-Temperature-Range IR Cell for High-Area Solids

Over 50 cell designs for use in the study of high area surfaces by infrared spectroscopy have been reported since 1967, as reviewed in reference [1]. The cell shown in Figure 231 possesses unique features, which, in combination, are not found in other designs reported. These features include ultrahigh vacuum operation, wide temperature range extending from about 100 K to above 1500 K without use of an external furnace, accurate temperature control to 1 K, linear temperature programmability up to 8 K/s, and very rapid sample cooling rates [1].

The cell design in Figure 231A comprises a six-way stainless cube to which windows and gas delivery and pumping tubes may be attached by means of 2.75-in. Conflat flanges. The sample itself is held on a tungsten grid that is clamped between high thermal conductivity Ni clamps, mounted on an electrical feedthrough that exits from the bottom of a reentrant Dewar made of stainless steel. The feedthrough also contains matched thermocouple leads that are attached to a thermocouple welded to the top center of the grid. The entire assembly is oriented using a rotatable flange and is inserted into the six-way cube, along with other desired attachments.

The grid material is made of thin tungsten foil that has precise square holes photo-etched in 0.0125–0.05-mm-thick material and exhibiting 58% optical transparency in the IR [1]. Each of the square holes is 0.22 mm in width, surrounded by 0.05-mm boundaries. The uniform geometry is advantageous in controlling the uniformity of powder deposition, as will be shown below. Gold grids of the same geometry have also been employed in this cell.

Two methods may be used for the deposition of high-area solids on the W grid. As shown in Figure 231C, Step 1, a spraying technique [1] like that described on page 190 may be used in which the powdered solid is suspended in a volatile solvent and sprayed onto the heated grid. Heating to about 70°C is achieved using temporary power leads to the grid and its Ni clamp support, causing flash evaporation of the solvent. The grid is rigidly held in a T-shaped Macor jig-mask that is designed to maintain correct spacing between the Ni clamps so that they may be slid over the power feedthrough leads on the bottom of the reentrant Dewar without flexure of the grid containing the powder deposit as shown in Step 2. The Macor jig is then removed, as shown in Step 3. The powder first sticks to the boundaries of the 0.22-mm-wide openings and then as the deposition continues, the powder uniformly fills the holes. Oxide powder densities up to about 0.01 $g/cm^2$ may be achieved in the grid openings.

A second method for placing powder into the grid openings is shown in Figure 231C [3]. Here, the unmounted grid is held between two sheets of glossy weighing paper and an excess of powder is sprinkled over the grid. Two stainless steel dies are used to sandwich the grid and its protective paper, and hydraulic pressure of about 12,000 psi is applied. This causes the powder to enter the grid openings and to be rigidly held there. Upon removal of the glossy paper, a razor blade is used to scrape off the excess powder from the top of the grid, and the grid is then mounted in the Ni clamps for loading into the IR cell. The powder material in the grid openings will withstand gentle bending of the grid. Powder densities of the order of 0.004 g/cm$^2$ may be achieved. The spectroscopic, gas transport, and thermal behavior of the pressed powders is very similar to that of the sprayed deposits [3].

The thermal properties of the grid during electrical heating to a stable temperature by a feedback controller have been characterized as shown in Figures 231D,E [1]. In Figure 231D, four thermocouples were attached as shown in the diagram. The open circle and black triangle points were used to measure temperature gradients along a vertical line in the center of the grid. The three open point measurements were made to determine the temperature gradient along a horizontal line through the center of the grid. In the central 4-mm-wide region of the grid that is sampled by the IR beam, no vertical gradient in temperature is observed. A small dropoff in temperature is seen in the central region as one moves horizontally owing to heat conduction away from the grid by the Ni clamps. The ability to temperature-program the grid and the supported powder to elevated temperatures is illustrated in Figure 231E. Here, almost linear heating rates between 1.9 K/s and 8.4 K/s are demonstrated, with the controller being programmed to stop the heating ramp at about 750 K in this case. Rapid cooling in 10–20 s from 900 K to the base temperature of 325 or 150 K may be achieved [1].

Other cells for heating catalysts to high temperatures and cooling to low temperatures have been described. One of these is used for EXAFS measurements and works in a range from about 77 K to 700 K [4]. Another, involving the use of a pressed sample in a copper grid, has been described and characterized for thermal behavior [5].

Cells in which the sample can be heated provide an ideal situation for the study of infrared emission spectroscopy. While many IR emission studies have been done on planar surfaces containing coatings, a few studies of high area surfaces as IR emission sources have been carried out too [6,7]. One must be careful to rule out artifacts in this type of measurement [7].

## *References*

[1] P. Basu, T.H. Ballinger, and J.T. Yates, Jr., Rev. Sci. Instrum. **59,** 1321 (1988).

[2] Custom fabricated grids made of W or other metals are available from Buckbee Mears, 245 E. Sixth Street, St. Paul, MN 55101.

[3] T.H. Ballinger, J.C.S. Wong, and J.T. Yates, Jr., Langmuir **8,** 1676 (1992).

[4] F.W.H. Kampers, T.M.J. Maas, J. van Grondelle, P. Brinkgreve, and D.C. Konigsberger, Rev. Sci. Instrum. **60**, 2635 (1989).

[5] M. Hughes, L. Bourget, and R.G. Greenler, J. Cat. **139**, 691 (1933).

[6] D. Kember, D.H. Chenery, N. Sheppard, and J. Fell, Spectrochimica Acta **35A**, 455 (1979).

[7] D. Kember and N. Sheppard, J. Chem. Soc. Faraday Trans. **2**, 77, 1321 (1981).

## Fig. 231. Wide-Temperature-Range IR Cell for High Area Solids

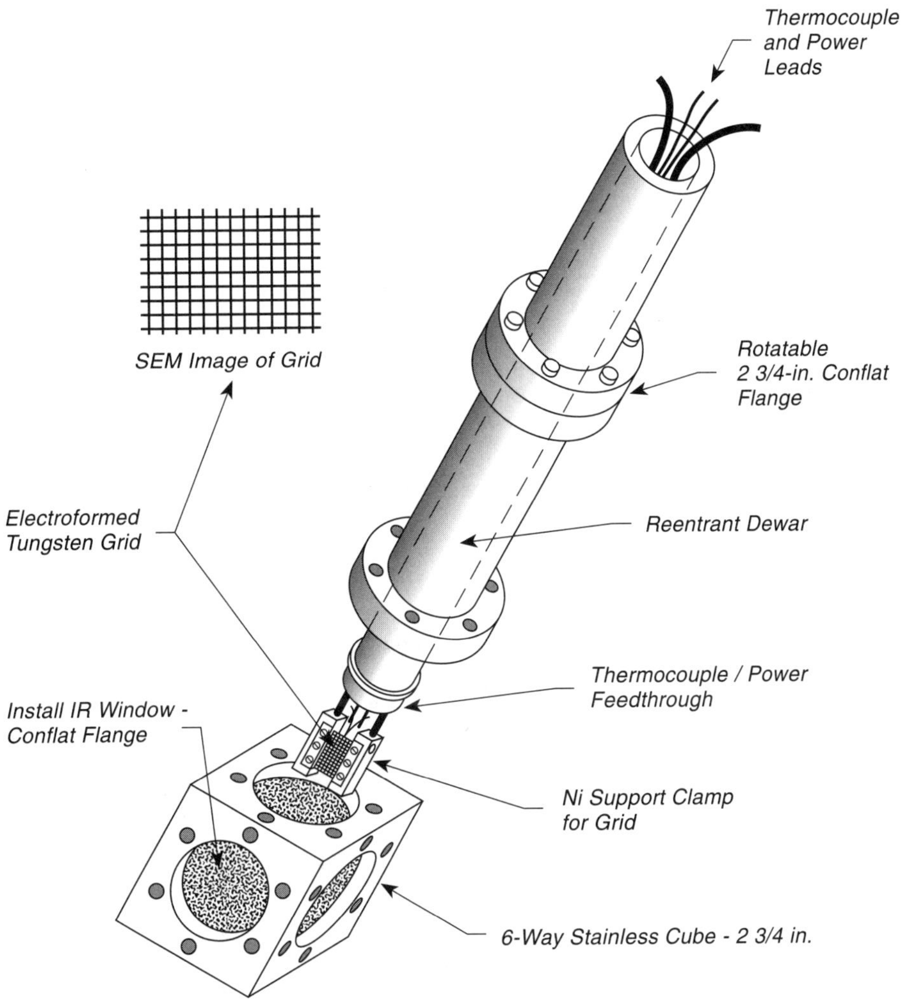


**A. Cell + Sample Holder**

## Fig. 231. Wide-Temperature-Range IR Cell for High Area Solids

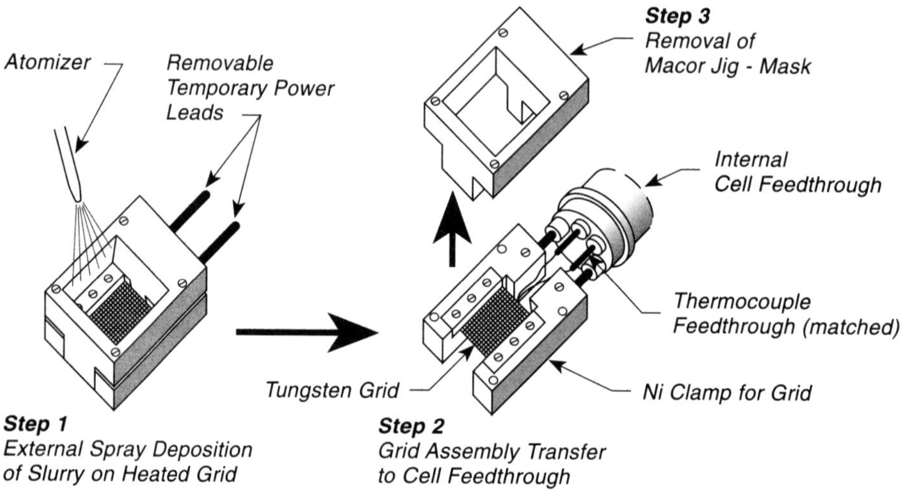


**B. Spraying Technique for Sample Preparation**

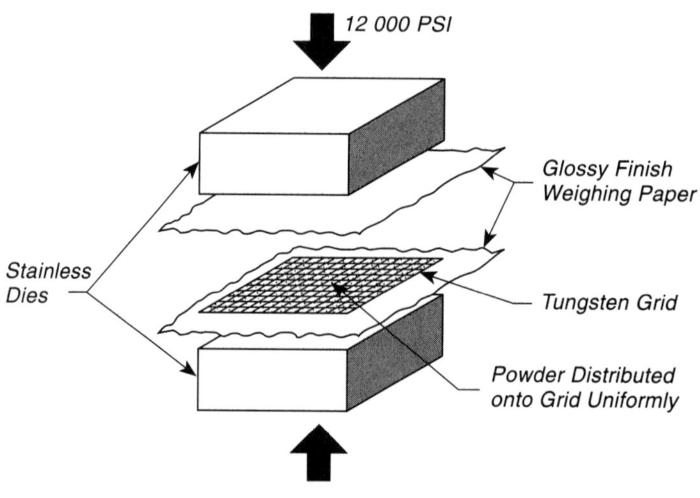


**C. Hydraulic Pressing Technique**

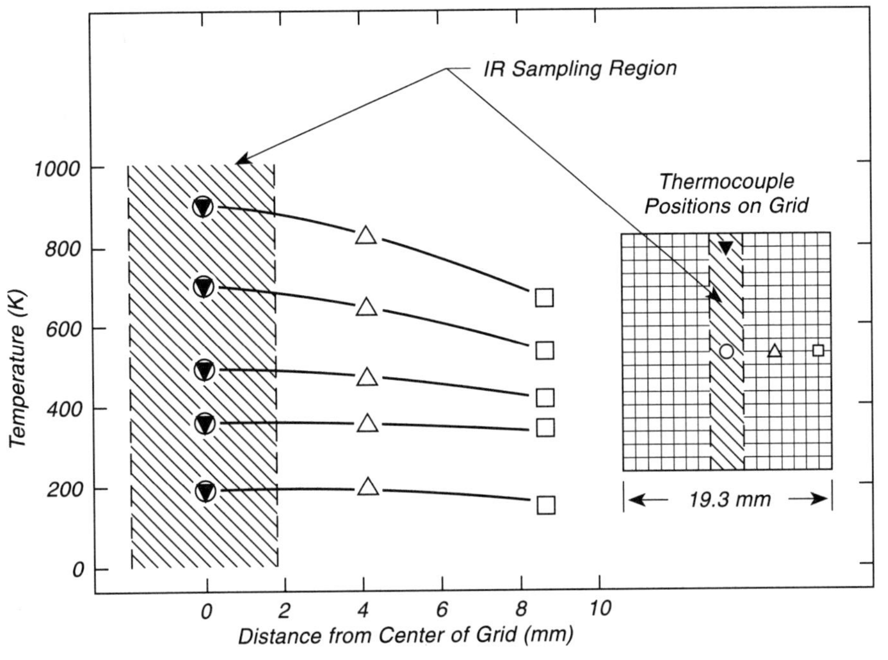


**D. Grid Temperature Distribution at Steady State**

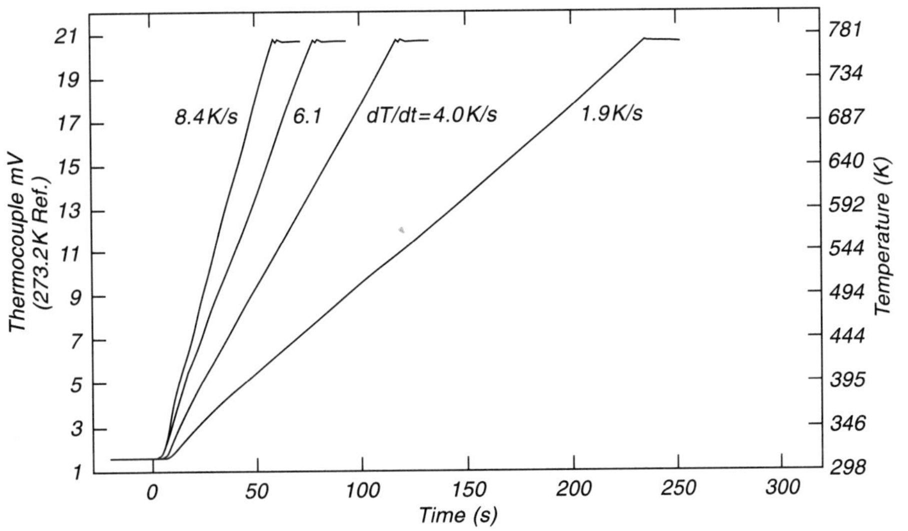


**E. Temperature Programming Grid**

# B. Adsorption/Desorption—Thermal

# 232

# Adsorption and Desorption from High-Area Solids

The study of the surface behavior of high area solids can be carried out using extensions of the fundamental measurement methods developed for studies of low area single crystal surfaces. Among the most widely used methods in single crystal studies is temperature-programmed desorption (TPD), often called thermal desorption spectroscopy (TDS) [1]. The TPD method, with appropriate calibration, can be used to measure absolute surface coverages as well as the kinetics of desorption as a function of surface coverage on both single crystals and high-area solids.

Figure 232A shows a thermal desorption spectrometer that was used for the study of ethylene adsorption on a high-area (423 $m^2/g$) NaX zeolite [2a]. The sample, of mass 10 μg, was deposited on a Ta disk by use of a micropipet that deposited known volumes of a slurry containing the suspended zeolite in suspension. Upon evaporation of the solvent, it was found that the zeolite particles were separated from each other as they adhered to the Ta surface. Thus, interparticle diffusion effects during adsorption and desorption were eliminated by the high particle dispersion, although pore diffusion in the zeolite particles was clearly evident in the measurements [2a].

The high-area sample is first cleaned by heating in vacuum using a temperature programmer that operates in a feedback circuit involving the thermocouple. Gases are adsorbed in the pumped system, or for high exposures, with the gate valve closed.

The sample is held in front of a 2-μm sampling aperture that may be opened and closed using an O-ring seal on the end of a linear motion device. With the gate valve closed, this permits the sampling of gases near the sample surface at high pressures as would be necessary for the study of catalytic reactions. However, with the gate valve open after adsorption, sampling of desorbing gas from the powdered sample is possible in the pumped system through direct transmission from the sample to the mass spectrometer. The mass spectrometer and pumping speed of the system may be separately calibrated using an absolute flow of gas into the pumped system through a calibrated leak (see page 610). Knowing the mass of the sample and its surface area per unit mass, the absolute coverage of adsorbate may be determined by thermal desorption measurements, integrating the pressure signal over the time duration of the desorption peak.

Figure 232B shows an example of the uptake of ethylene versus exposure on NaX zeolite at 114 K. At low exposures, adsorption is postulated to occur on the outer surfaces of the isolated zeolite particles [2]; at exposures above about 200 L (1 L = 1 x $10^{-6}$ Torr s), adsorption occurs into the pores of the zeolite, and the efficiency of adsorption increases significantly as indicated by the lower curve of Figure 232B. Over the entire upper range of exposures, a typical sigmoid uptake curve is found, as shown in the upper curve of Figure 232B. The inset shows an analysis that may be used to predict the saturation coverage of the adsorbate, which occurs in the limit of $1/\varepsilon = 0$, where $\varepsilon$ is the gas exposure.

Extensive work on the analysis of TPD spectra from high-area solids has been reported and appropriate references are listed in reference [2a]. Noteworthy is a study by Demmin and Gorte [3]. A major consideration is the effect of interparticle diffusion on the observed desorption kinetics, and the method shown here, involving very high particle dispersion, avoids this effect, while also offering sensitivity at the $10^{-4}$ ML level for powders with very high specific surface areas. Small corrections for the adsorption on the Ta support may be easily applied.

Similar methods have been applied to larger samples (0.1 g) of powdered material in the form of compressed pellets, but the influence of interparticle diffusion is not eliminated in these studies [4].

A recent review of temperature-programmed desorption from high area solids may be found in reference [5].

## *References*


[1] J.T. Yates, Jr., Methods of Exp. Phys. **22,** 425 (1985).

[2] a) P. Basu and J.T. Yates, Jr., Surf. Sci. **177,** 291 (1986); U.S. Patents 4,877,584 and 4,663,297; b) M. Kiskinova, G.L. Griffin, and J.T. Yates, Jr., J. Catal. **71,** 278 (1981).

[3] R.A. Demmin and R. Gorte, J. Catal. **90,** 32 (1984).

[4] G.D. Moggridge, J.P.-S. Badyal, and R.M. Lambert, J. Vac. Sci. Technol. **A8,** 3874 (1990).

[5] R.J. Gorte, Catal. Today **28,** 405 (1996).

***Fig. 232. Adsorption / Desorption on Powdered Solids***

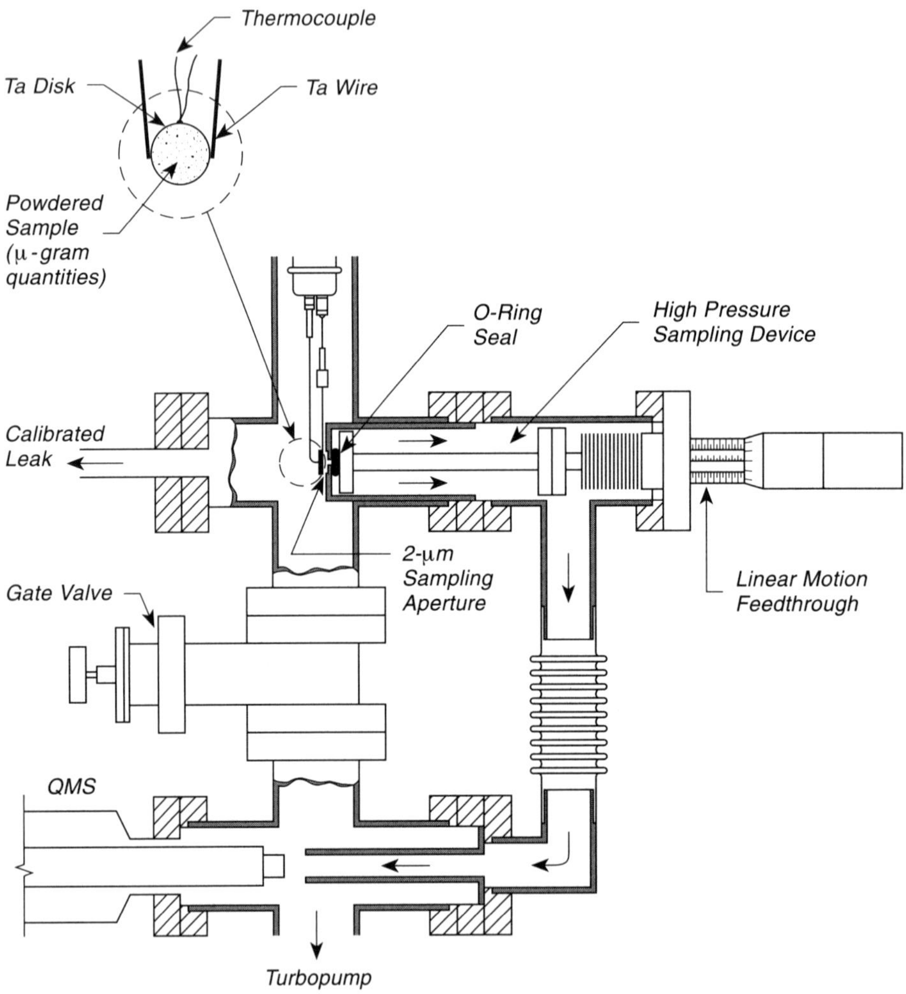


***A. Thermal Desorption Mass Spectrometer***

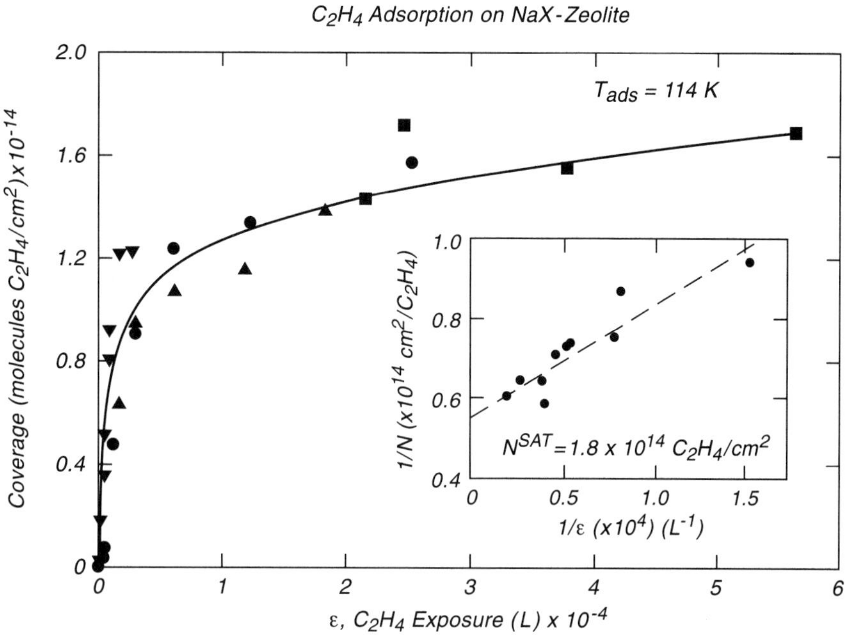


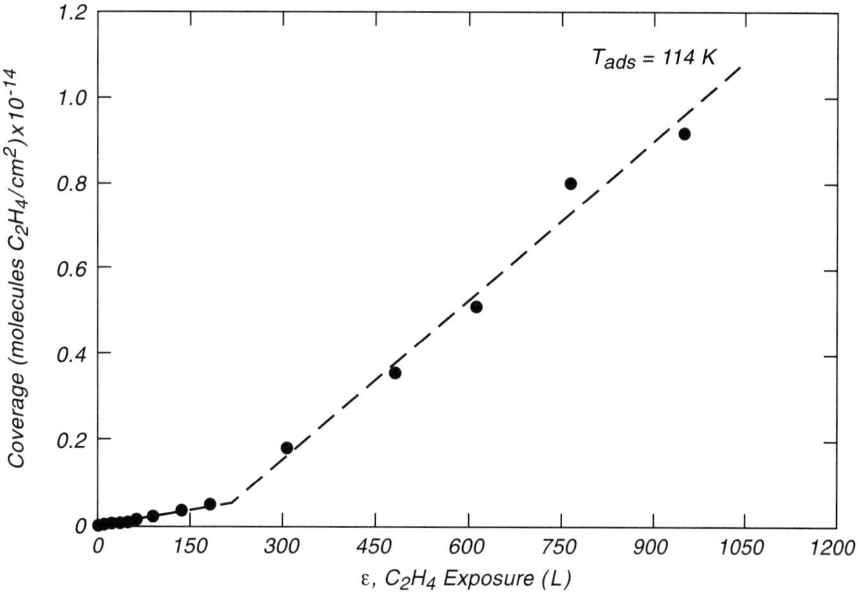


***B. Typical Adsorption Uptake Data***

# 233

# Microcalorimetric Studies of Adsorption Heat on Powders

The direct measurement of the energy released during chemisorption provides a measurement of the binding energy of adsorbed molecules that depends only of the first law of thermodynamics. A method for making these measurements on thin single crystal surfaces has been given on page 476.

In Figure 233A, the cell used for calorimetric measurement on high-area metal catalysts is shown [1]. A metallic Ni catalyst, prepared from $Ni(II)(NO_3)_2 \cdot 6H_2O$ powder by reduction in $H_2$ at 623 K, is stored under He in a sealed-off Pyrex NMR tube, and this tube is located in one arm of the calorimeter cell. The metal weighs between 0.5 and 1.0 g and has a surface area of 1.7 $m^2$/g and a CO capacity of 22 μmol/g. The leak rate of the closed cell was measured to be about $10^{-6}$ μmol/min in a 70-$cm^3$ volume.

Calorimetric measurements are made on the catalyst by first breaking the glass NMR tube using the linear motion feedthrough that presses down on the inner tube. Then small amounts of CO (or other gases) are admitted and the thermal response of the calorimeter is measured, comparing the sample cell to the reference cell in two identical calorimeters. The uptake of gas is measured using standard volumetric techniques and a capacitance manometer with a sensitivity of $0.5 \times 10^{-4}$ Torr.

The calorimeter is a commercial device, a Setaram C-80 heat-flux calorimeter.

Typical results are shown in Figure 233B, and are compared with measurements made on a Ni(100) single crystal [2]. The coverage scales are normalized to the saturation coverage in both cases. It may be seen that the agreement is excellent, especially for the heat of adsorption at zero coverage. These results suggest that the chemical reduction technique used for preparation of the powdered Ni results is a clean surface that behaves in a fashion similar to atomically clean Ni surfaces. In reference [2], the range of the initial adsorption heat was 8%, comparing three low index planes of Ni.

A second example of the use of microcalorimetry for powdered samples may be found in reference [3]. Here, the sample, weighing about 0.5 g, is spread over the flat bottom of the calorimeter (6.3-$cm^2$ geometrical area). This permits rapid equilibration between the gas and the adsorbed layer, and also rapid thermal equilibrium.

## *References*


[1] B.E. Spiewak, P. Livin, R.D. Cartright, and J.A. Dumesic, J. Phys. Chem. **100,** 17260 (1996).

[2] J.T. Stuckless, N. Al-Sarraf, C. Wartnaby, and D.A. King, J. Chem. Phys. **99,** 2202 (1993).

[3] D.J. Parrillo, C. Lee, and R.J. Gorte, App. Catal. A: General **110,** 67 (1994); see also D.J. Parrillo and R.J. Gorte, Catal. Lett. **17,** 16 (1992); D.J. Parrillo and R.J. Gorte, J. Phys. Chem. **97,** 8786 (1993).

**Fig. 233. Microcalorimeter—Adsorption Heats on Powders**

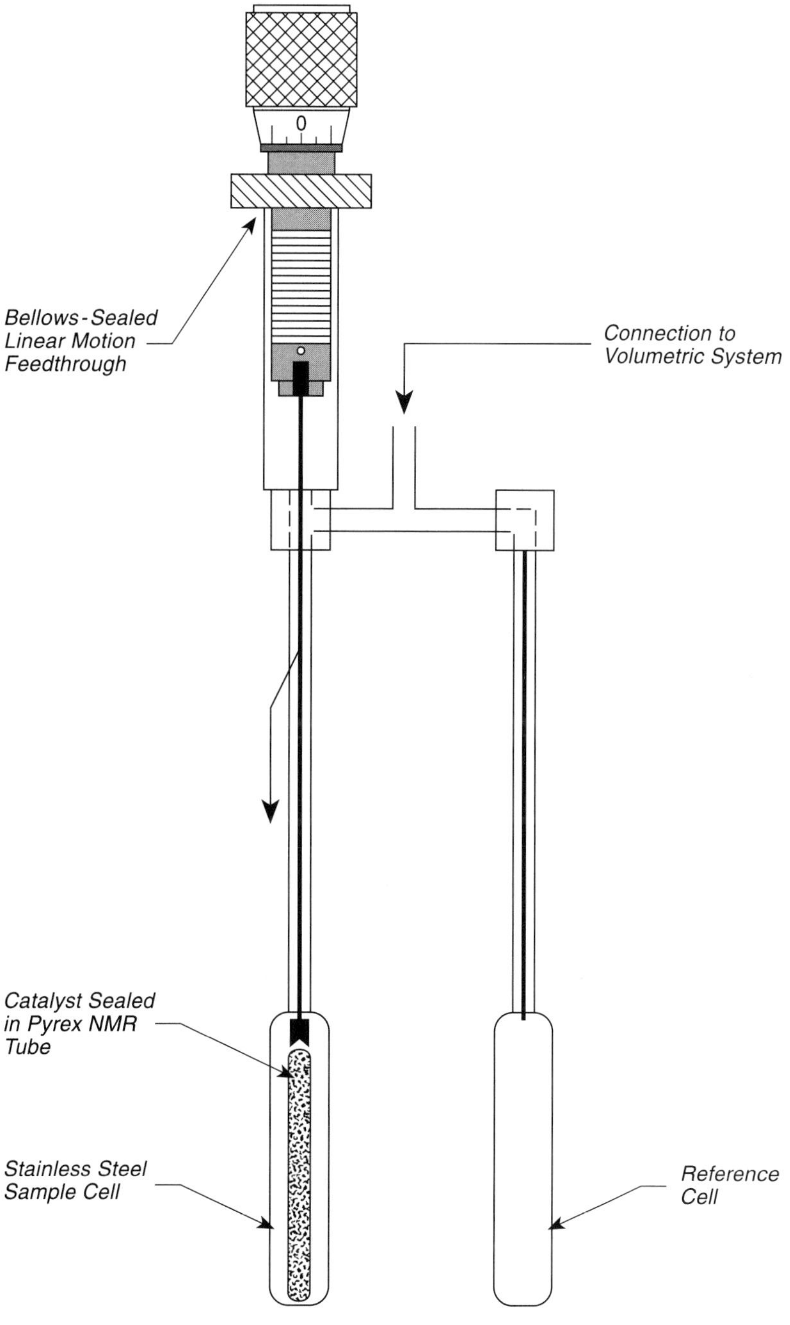


**A. Calorimeter Cells**

**Fig. 233. Microcalorimeter—Adsorption Heats on Powders**

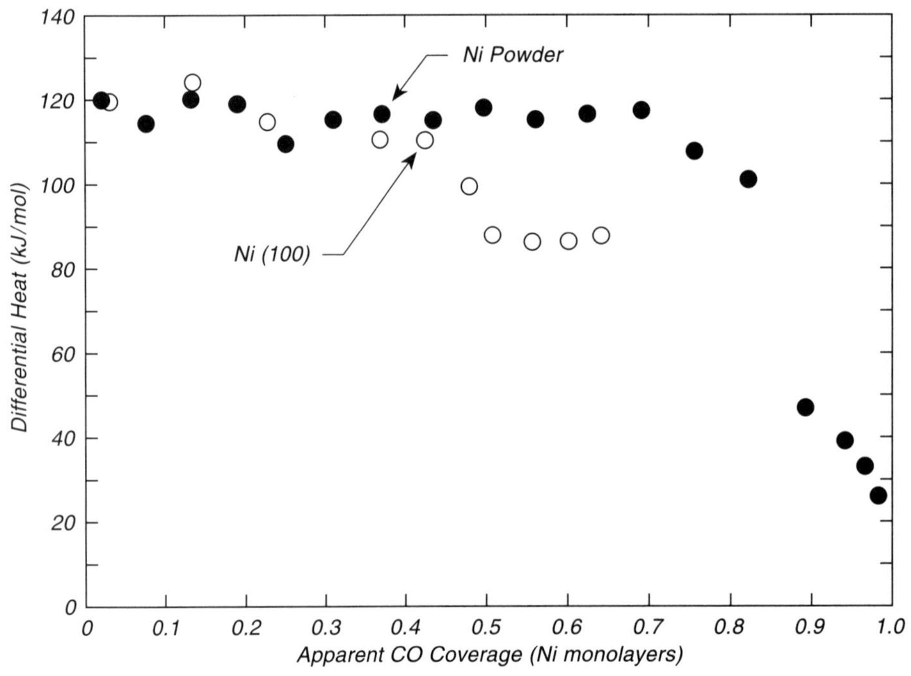


**B. CO Chemisorption on Ni**

# IX. Safety

A. Protecting the Vacuum System . . . . . . . . . . . . . . . . . . . . . . . 817
B. Protecting Personnel . . . . . . . . . . . . . . . . . . . . . . . . . . . . . . 831

# A. Protecting the Vacuum System

# 234

# Delay Circuit for Turbopumping Protection Against Vacuum Loss from Power Interruptions

Power interruptions can play havoc with UHV systems operating with a turbopump, and particularly when these systems are being baked out. A power failure has the effect of venting a turbopump, using the commercial vent valve designed to prevent oil backstreaming from the forepump into the turbopump and the vacuum chamber during a power interruption. A partial solution to avoid massive air entry into the UHV system is to employ a pneumatic gate valve between the turbopump and the vacuum system. However, there are two defects in this protection system: 1) the time of closure of the gate valve is too long to prevent at least partial loss of vacuum in the chamber; and, 2) when the power returns, the gate valve will quickly open, exposing the chamber to a turbopump which is at or near atmospheric pressure. A circuit to prevent these effects is described below [1]. There is a commercial device that will do almost the same thing electrically [2], and a pneumatic device also for the same purpose [3].

The first problem mentioned above is solved by venting the turbopump only after the gate valve is completely closed. This is done by supplying short-term emergency power to the turbopump vent valve for the necessary period of time (approximately 8 s is used). The second problem is solved by controlling the pneumatic gate valve with a relay [4] that latches in the off position upon power failure. This prevents the valve from reopening when the power returns until the relay is overridden.

Figure 234 shows the circuit, costing less than \$100, which accomplishes both objectives. The nondelayed socket supplies power to the gate valve relay. A line switch between the nondelayed socket and the gate valve supplies manual control of the gate valve. The nondelayed socket power is designed to latch in the off position in the case of a power interruption, and pushing the reset button reactivates power at this socket when the main power is restored.

The delayed socket supplies emergency power to the turbopump vent valve. It is powered by a battery-driven backup AC power supply [5]. When main power is interrupted, emergency power is supplied and then shuts off, venting the turbopump. For momentary main power interruptions lasting less than 8 s, the turbopump vent valve will not open because power is supplied from the backup supply.

The control system does not assume that when power returns the turbopump restarts and can be automatically opened to the vacuum system, unlike other systems [2,3]. It is deemed best to leave the system shut down after power interruptions, awaiting manual restarting.

## *References*

[1] J.A. Polta and P.A. Thiel, J. Vac. Sci. Technol. **A5,** 386 (1987).
[2] Balzers Part No. TCV 102. Obtained from Balzers, 8 Sagamore Park, Hudson, NH 03051.
[3] J.P. Saint-Germain, G. Abel, and B.L. Stansfield, J. Vac. Sci. Technol. **A4,** 2391 (1986).
[4] Time delay CNS-35-76 from Newark Electronics, 4801 N. Ravenswood Avenue, Chicago, IL 60640-4496.
[5] Available from Tripp Lite, 500 N. Orleans Street, Chicago, IL 60610.

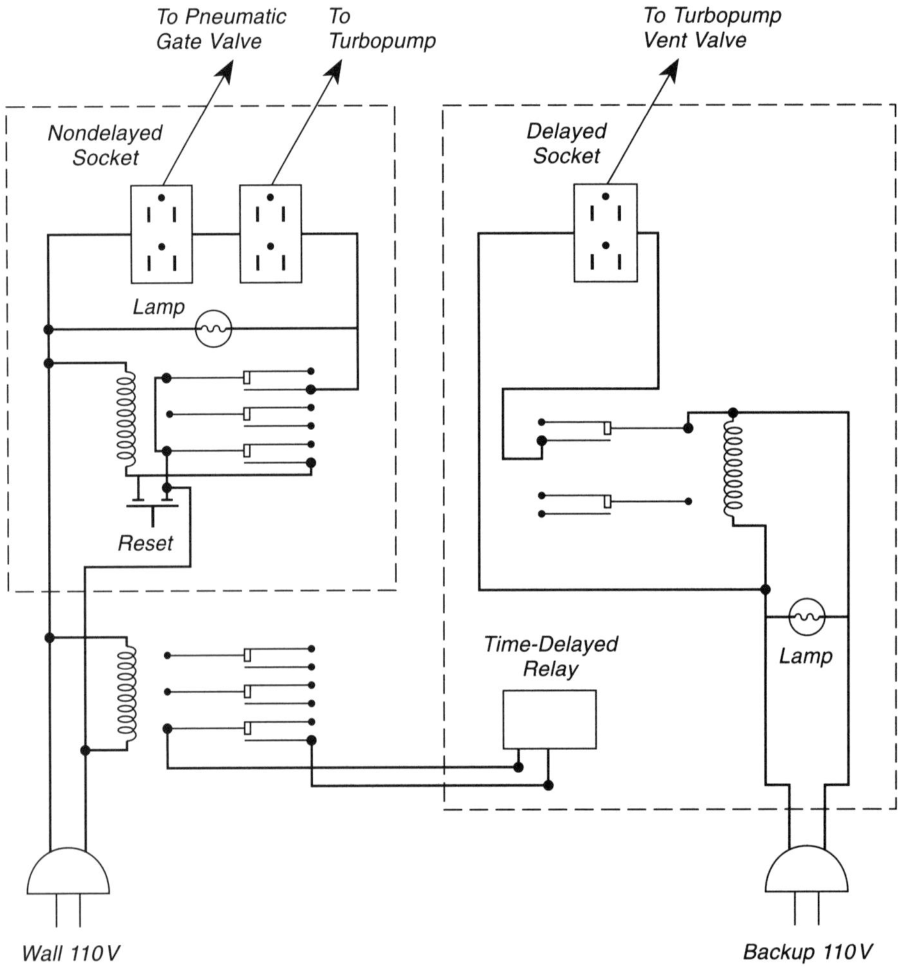


**Fig. 234. Delay Circuit for Turbopumping Protection Against Vacuum Loss from Power Interruptions**

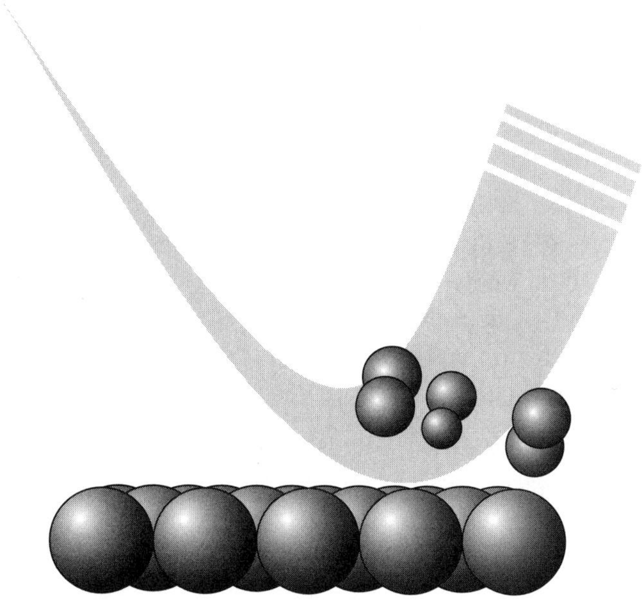

# 235

# Fast-Closing Beam Valve for UHV Chamber Protection

It is often necessary to protect an ultrahigh vacuum system from vacuum failure in an auxiliary system, such as an attached beam source system. In this particular case, since the beam is of small cross section, the valve can be very small. The valve must be electrically actuated by relays connected to pressure gauges in both systems.

Figure 235 shows a modified solenoid actuated valve [1] that is inexpensive, fast, and ultrahigh-vacuum-compatible. The solenoid driving the valve is encased in a housing that is sealed by O-rings so that gas evolution from the heated coil will not degrade the vacuum. The housing is tightened into place by using a nut that engages a threaded rod at the top of the assembly.

When the current is on, the wedged-shaped Cu seal is drawn up into the solenoid; interruption of vacuum results in turning the solenoid off, and the Cu seal is dropped into the sealing position, where it is held in place by a spring and by gravity. The seal is made against a Viton O-ring in the lower portion of the valve. Copper cooling coils are used to dissipate heat from the solenoid. Electrical connections are brought out from the solenoid housing through a Conflat flange, not shown.

The leak rate through the closed valve is $4 \times 10^{-5}$ Torr L/s, and is too large to be used to protect an ultrahigh vacuum system without a pumping section between the valve and the system. In the case of the use of this valve, a differentially pumped chopper region is present and serves this purpose [2].

A similar principle has been used to insert a large gate valve into an ultrahigh vacuum system [3].

## *References*


[1] Vacoa model SOV-050-00-SS, Vacuum Accessories Corporation of America, 390 Central Avenue, Bohemia, NY 11716.

[2] T. Engel and D.N. Braid, Rev. Sci. Instrum. 57, 301 (1986).

[3] J.J. Zinck and W.H. Weinberg, Rev. Sci. Instrum. 55, 810 (1984).

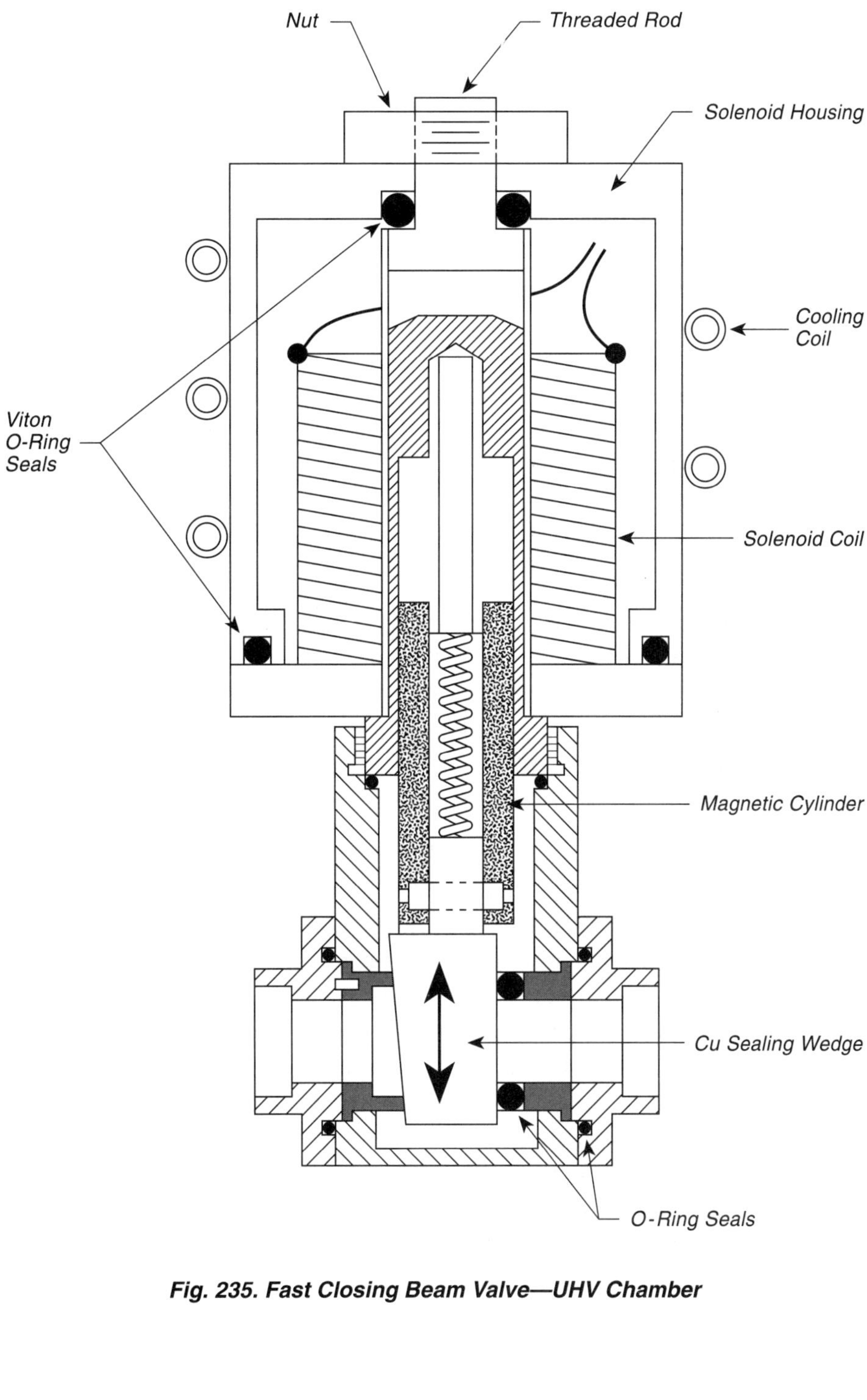


*Fig. 235. Fast Closing Beam Valve—UHV Chamber*

# 236

# Safety System for Oil Diffusion-Pumped UHV Systems

Modern oil diffusion pumps are now used to achieve ultrahigh vacuum, and these pumps exhibit some advantages over more expensive ion pumps or turbomolecular pumps. Among the advantages are lower cost, the ability to pump all gases, the absence of magnetic fields, and the absence of vibrations. Among the disadvantages are the possibilities of system contamination if failure of various components occurs [1]. The failure items are loss of vacuum, loss of power, loss of cooling water, and loss of liquid nitrogen cooling in the ultrahigh vacuum trap. A pneumatically actuated valve between the pumps and the ultrahigh vacuum system is the key element that closes to protect the system if some failure should occur.

A schematic drawing of an oil diffusion pumped ultrahigh vacuum system is shown in Figure 236A. It consists of a standard bakeable chamber with an ionization gauge and a titanium sublimation pump (TSP). The pneumatically actuated gate valve separates the system from the external diffusion pump and liquid nitrogen trap. The oil diffusion pump contains special perfluoro polyether pump oil (Fomblin type) or polyphenyl ether (Santovac-5) pump oil used for achieving the best pressure. Silicone diffusion pump oils should be avoided in all UHV apparatus. The oil diffusion pump is protected from the oils of the mechanical pump by a bakeable molecular sieve trap, and the foreline pressure is monitored by a thermocouple gauge or other pressure gauge.

The simple relay circuit shown in Figure 236B is designed to protect the system from the various failure possibilities [1]. Digital electronic components were avoided to reduce the circuit sensitivity to electrical noise. The control items in the vacuum system are: 1) foreline thermocouple gauge maximum pressure set point—to J1; 2) water flow sensor-to J2; 3) ionization gauge maximum pressure setpoint—to J3; 4) liquid nitrogen filler with low level sensor—to J4; 5) pneumatically actuated system valve—to J5. The relay, R1, is a double-pole, single-throw relay and is turned on by closing the power switch, SW1, and then depressing the momentary contact switch, SW2. This sets R1 if the foreline pressure is below the thermocouple gauge set point and if there is sufficient water flowing through the diffusion pump, closing J2. The switch SW3, which controls the relay, R2, may then be used to switch on the diffusion pump. The system valve, SW4, can then be actuated if continuity exists through the right-hand side of the circuit, meaning that the liquid nitrogen trap and the ion gauge are operating in the proper range. If one is in the initial pumpdown period, the pressure will be too high at the ionization gauge, and this gauge and the thermocouple gauge can be bypassed temporarily by closing switch SW5.

Table 236 shows the various failure conditions and the components' condition when these failures occur.

**Table 236. Interlock Response to Failure Conditions**

| Failure | Diffusion Pump Power | System Valve |
|---|---|---|
| Chamber pressure too high | On | Closed |
| Liquid $N_2$ level too low | On | Closed |
| Foreline pressure too high | Off | Closed |
| Water failure[a] | Off | Closed |
| Electrical failure | Off | Closed |

[a]An additional bimetallic thermal switch control is available on most diffusion pumps that interrupts the heating power if the water goes off.

## *References*

[1] L.H. Dubois, J. Vac. Sci. Technol. **A6,** 162 (1988).

**Fig. 236. Safety System for Oil-Diffusion-Pumped UHV Systems**

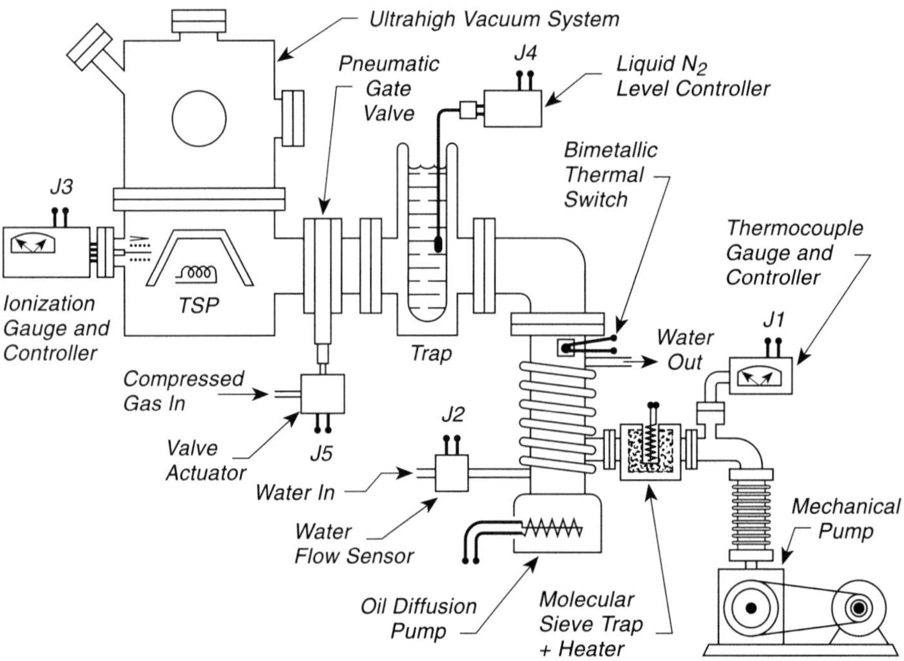


**A. Schematic Oil Diffusion Pumped UHV System**

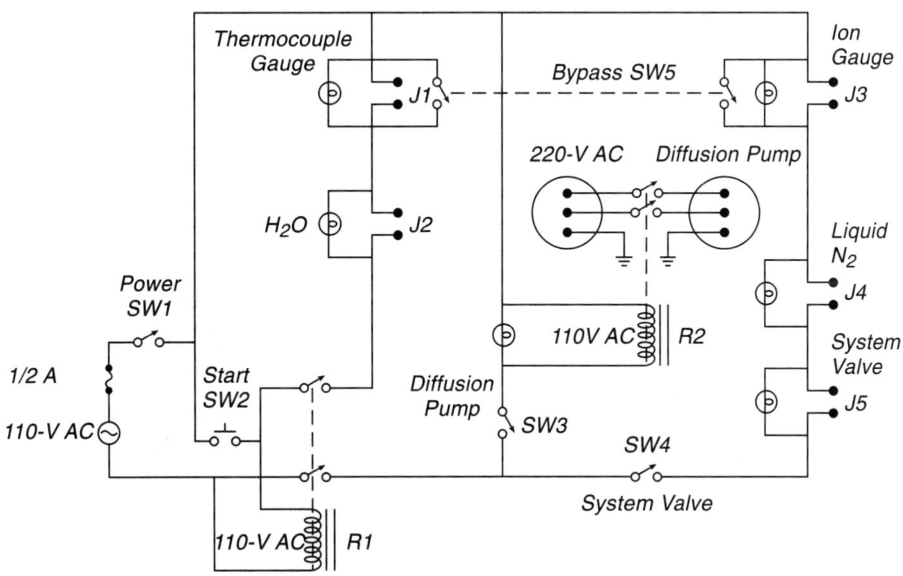


***B. Safety System Schematic***

# 237

# Folding Linear Magnetic Translator

Rather long linear and rotary magnetic translators are often used in ultrahigh vacuum systems, and these are always in danger of being hit by laboratory personnel as they pass by. The design shown in Figure 237 prevents this by folding the device housing containing the retracted magnetic translator/rotor.

Figure 237A shows a side view of the device. The welded bellows is folded upward through a 90° angle by a hinge and locked into place. In the horizontal position the bellows is held in a slightly compressed configuration by the gimbal subassembly. The outer end of the outer cylinder of the translator has a threaded rod attached on axis and this is retained in a "scissors" subassembly that permits accurate vertical positioning of the end of the translator. The bench holding the device is attached to the system lower flange by a rigid yoke assembly.

Figure 237B shows the gimbal subassembly and the two rotational adjustments that are possible as well as a lateral translation for alignment. A double-sided flange is used in order to be able to disassemble either the bellows or the translator section independently. Washers, used as spacers under the gimbal, permit vertical adjustment.

A critical design consideration is the appropriate size of the welded bellows. Figure 237C shows the dimensional parameters associated with a welded bellows-namely, the relaxed or free length (FL) at one atmosphere, the extended length (EL), and the compressed length (CL). These parameters are specified by the bellows manufacturer. In the bent condition, the inner radius of the bellows is defined as the working length (WL) in Figure 237D. The outer length of the bent bellows is therefore WL + $\theta r$, and the condition in Equation (1) must be met:

$$\mathrm{WL} + \theta r < \mathrm{EL} \tag{1}$$

In order to achieve the condition in Equation (1), the bellows must be compressed during installation by using the gimbal's subassembly, such that WL < FL. Note that welded bellows must be used for this device, since preformed bellows will not have the range of extension or the flexibility needed.

## *References*


[1] J.T. Fitch, J.J. Sumakeris, and G. Lucovsky, J. Vac. Sci. Technol. **A11**, 2871 (1993).

## Fig. 237. Folding Linear Magnetic Translator

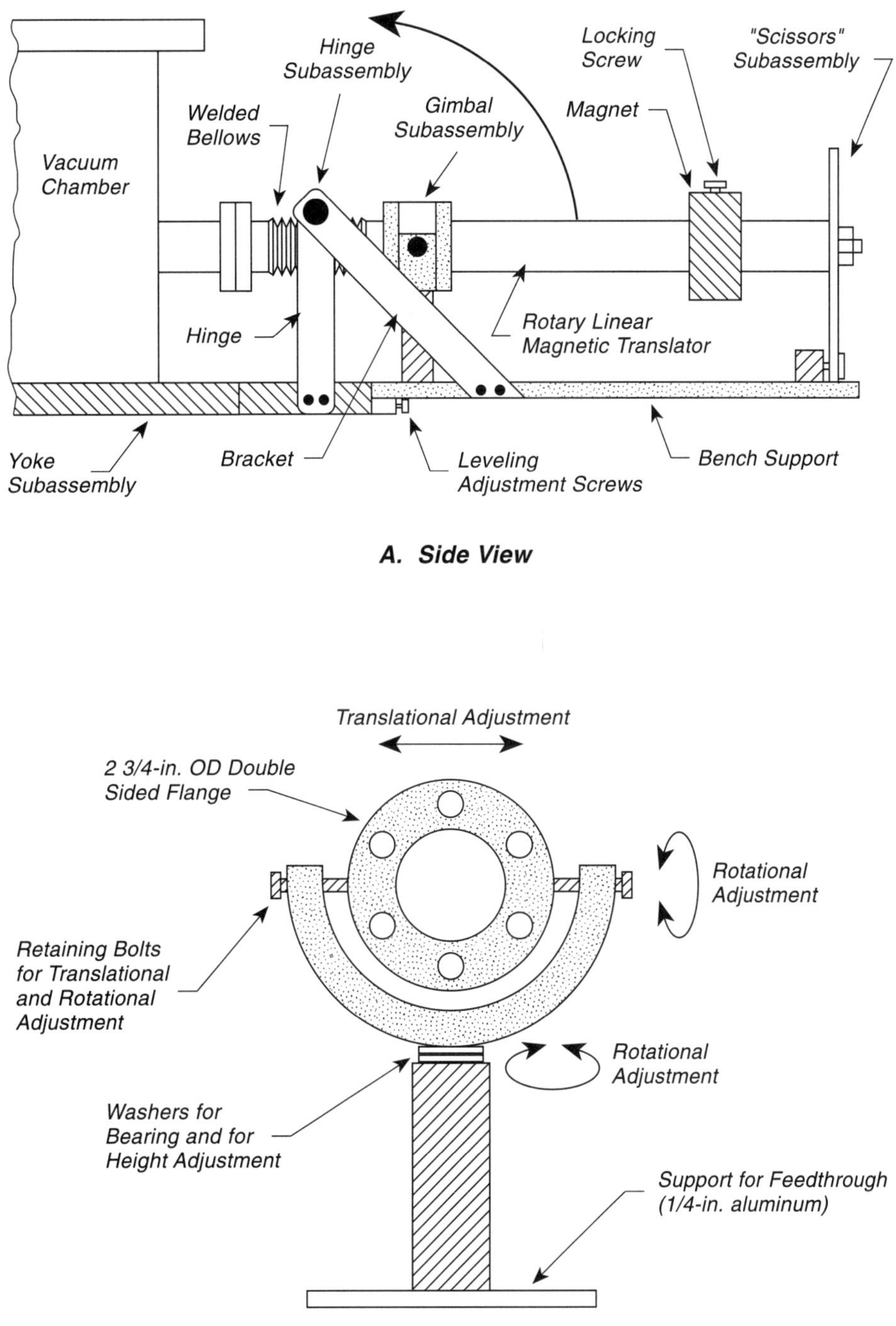


**A. Side View**

**B. Gimbal Subassembly**

**Fig. 237. Folding Linear Magnetic Translator**

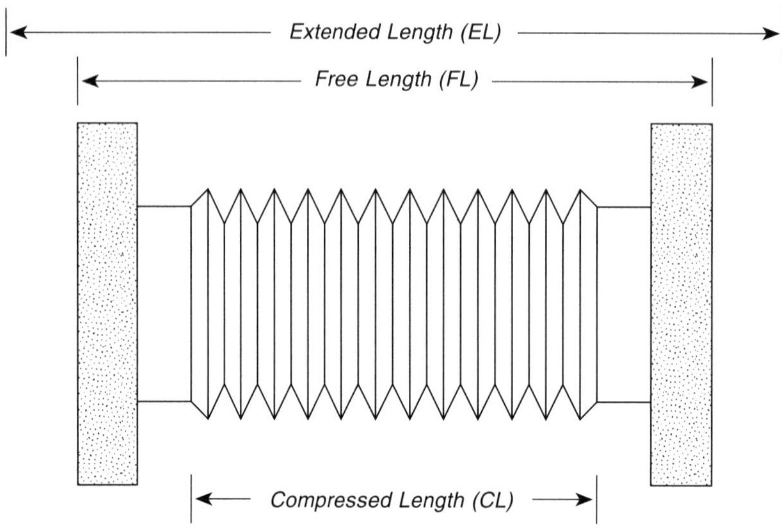


**C. Welded Bellows—Dimensional Definitions**

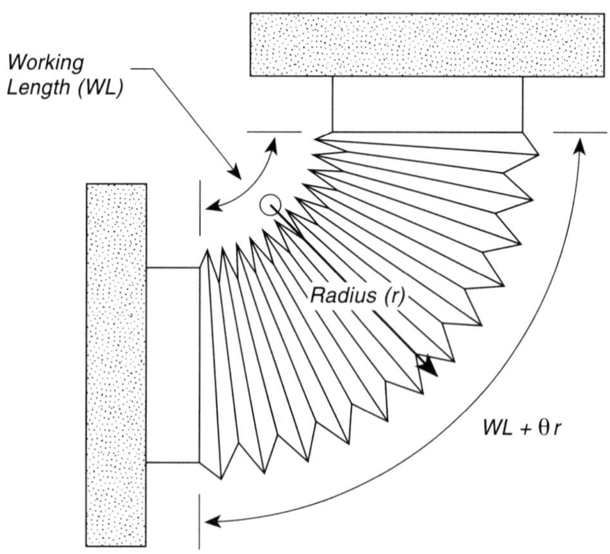


**D. Bellows in Bent Configuration**

# B. Protecting Personnel

# 238

# Electrical Shocks in the Laboratory

The quantitative effects of electrical current flowing through the human body are listed in the Table 238 below [1]. It is the current flow through the body that is the fundamental parameter involved in electrical shock, and the level of danger with a high-voltage power supply is related to the current rating of the supply. Safety with electrical devices involves minimizing accidental current flow through the body, especially through the heart. One may have to work with one hand in a pocket when there is danger of shock from high-voltage wires, but of course this is not an adequate protection against shock. Electrical shielding of leads and terminals is an obvious precaution that must be taken.

**Table 238. Quantitative Effects of Electric Current on Man**

| | DC | | AC—60 Hz | | AC—10,000 Hz | |
|---|---|---|---|---|---|---|
| Effect | Men (mA) | Women (mA) | Men (mA) | Women (mA) | Men (mA) | Women (mA) |
| Slight sensation on hand | 1 | 0.6 | 0.4 | 0.3 | 7 | 5 |
| Perception threshold | 5.2 | 3.5 | 1.1 | 0.7 | 12 | 8 |
| Nonpainful shock—muscular control not lost | 9 | 6 | 1.8 | 1.2 | 17 | 11 |
| Painful shock—muscular control not lost | 62 | 41 | 9 | 6 | 55 | 37 |
| Painful shock—let-go control threshold | 76 | 51 | 16 | 10.5 | 75 | 50 |
| Painful and severe shock—muscular contractions; breathing difficult | 90 | 60 | 23 | 15 | 94 | 63 |
| Possible ventricular fibrillation | | | | | | |
| 3-s shocks | 500 | 500 | 100 | 100 | | |
| *T*-s shocks | | | $165T^{-1/2}$ | | $165T^{-1/2}$ | |

The avoidance of electrical shocks in the laboratory is carried out with an understanding of the meaning of the electrical ground and how to connect electrical equipment to ground. Figure 238A shows a three-wire wall outlet and a polarized male plug. The green ground wire is a non-current-carrying voltage reference line that is connected to the earth somewhere near the laboratory building. The white wire, which is neutral, is connected to the ground wire in the vicinity of the wall outlet. At the point of connection, the electrical potential of the white wire will also be at ground potential. However, as one moves from this connection point, there will be a change in the potential of the white wire owing to the IR drop caused by the current flow through the white wire. One should avoid connecting the white wire to the ground wire in the laboratory. The black wire is the hot wire, and it remains isolated from both the ground and the neutral wire. The wall outlet contains three contacts—the large slot for the neutral wire, the small slot for the hot wire, and the larger hole for the ground connection.

Figure 238B shows two electronic units connected to AC power through three-wire power leads. Usually the chassis of each of the units will be grounded internally to the green grounding wire. However, to make doubly sure that the chassis of each is grounded, it is good practice to connect each chassis to a separate external ground buss by means of the chassis ground connection on each unit. The reason for this is that if the hot wire or any internal high-voltage source should accidentally touch the chassis from the inside, the grounding of the chassis will protect the laboratory worker from accidentally contacting a high electrical potential.

## *References*

[1] Dr. W.T. Peria, Department of Electrical Engineering, Institute of Technology, 4-174 EE/CSci Building, 200 Union Street S.E., Minneapolis, MN 55455 (private communication). The table is taken from the *Physical Electronics Laboratory Manual,* ca. 1965.

**Fig. 238. Electrical Shocks in the Laboratory**

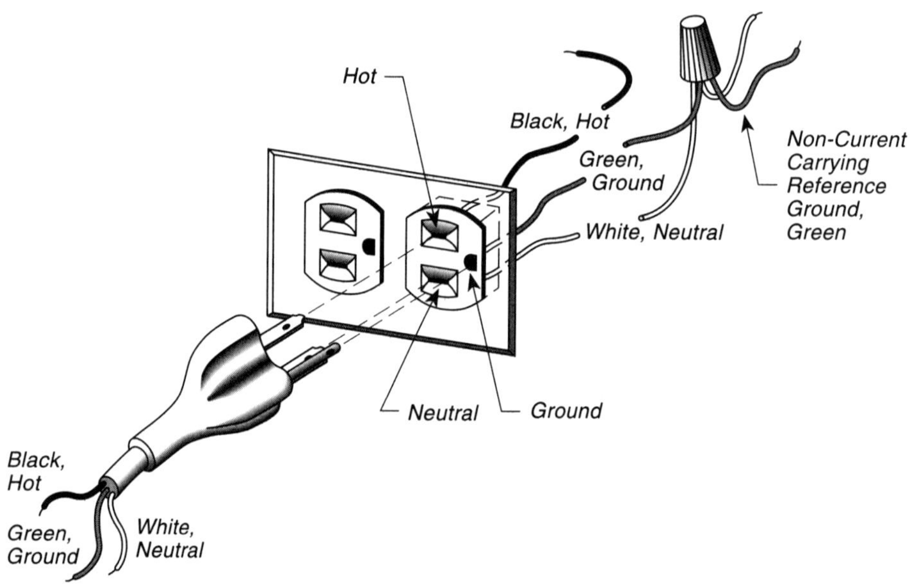


**A. 110V AC Wall Outlet + Polarized Plug**

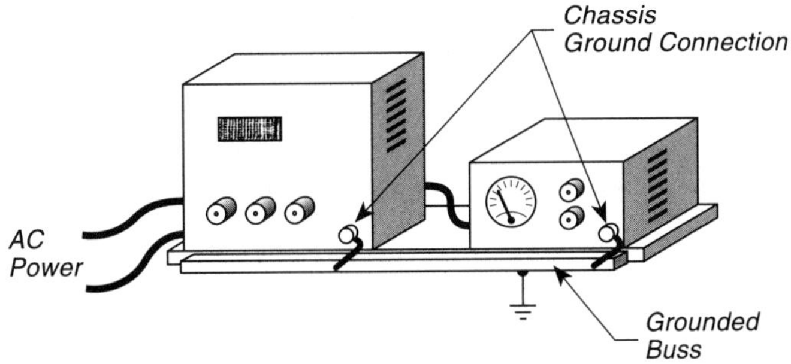


***B. Coupled Chassis Ground***

# 239

# Accidental Electrical Charging From Ionization Gauge

Accidental activation of the electron bombardment degassing capability of an ionization gauge in the presence of gas pressures that will support an electrical discharge can lead to potentially lethal electrostatic potentials on isolated components of an ultrahigh vacuum system. Figure 239 shows a schematic diagram of an ionization gauge operating at pressures in the $10^{-3}$ Torr range. A plasma is produced when the gauge is accidentally switched to the degas mode as a positive potential in the kilovolt range is applied to the grid for normal electron bombardment cleaning. An electrically isolated feedthrough inside the vacuum system is immersed in the plasma and comes to the plasma potential which is near the potential of the grid. Contact with this feedthrough electrode external to the vacuum system can result in exposure to high electrical potentials, and one example of this effect leading to death has been reported. A safe operating procedure to avoid this type of effect is never to attempt to outgas an ionization gauge when the pressure is high, which would be a pointless thing to do anyway.

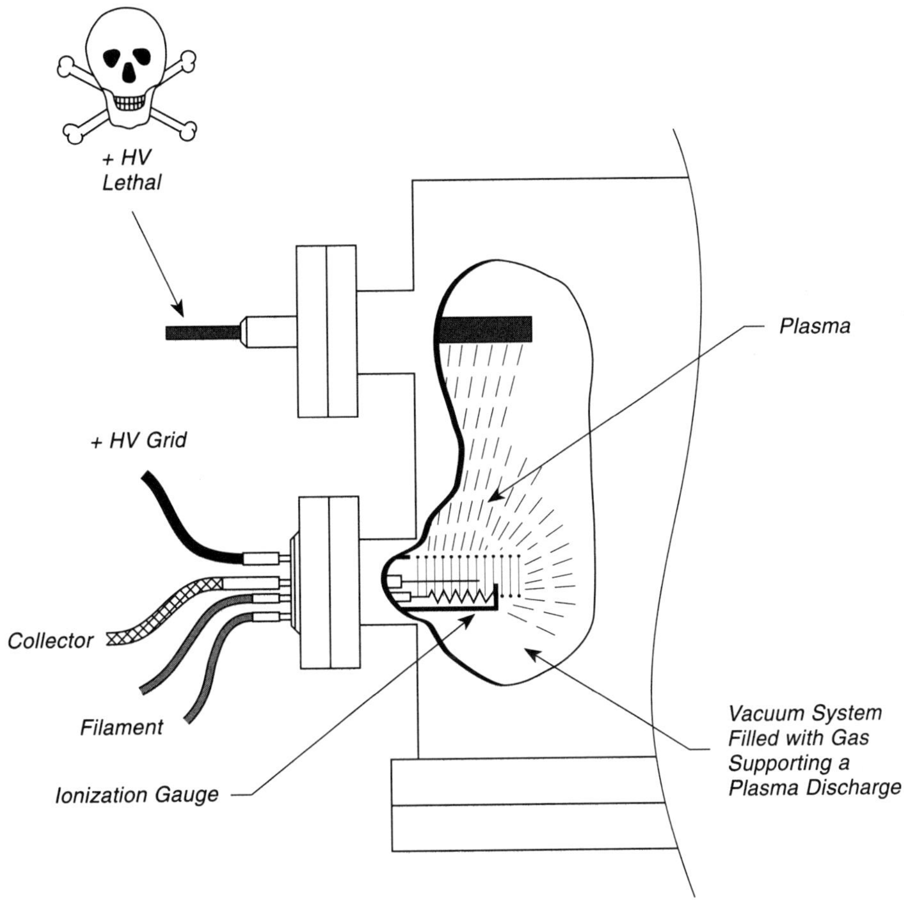


***Fig. 239. Accidental Electrical Charging from Ionization Gauge***

# Appendixes

# Appendix A

# Useful Books—Experimental Surface Science Methods


Anderson, R.L., *Revealing Microstructures in Metals,* Scientific Paper 425-C000-P2 (Westinghouse Research Laboratories, Pittsburgh, PA, 1961).

Baria, D.N. and Bautista, R.G., *Compilation of Experimental Work on Thermal Emittances of Materials, Part I—Metals; Part II—Non-Metals; Part III—Alloys* (Ames Laboratory, USAEC, Iowa State University, Ames, IA 50010, 1974). Available from National Technical Information Service, Dept. A, Springfield, VA 22151.

Barr, W.E. and Anhorn, V.J., *Scientific and Industrial Glassblowing and Laboratory Techniques* (Instrument Publishing Co., Pittsburgh, PA, 1947).

Beavis, L.C., Harwood, V.J. and Thomas, M.T., *Vacuum Hazards Manual,* AVS Monograph Series M-1, 2nd Ed. (American Vacuum Society, New York, NY 10017, 1979).

Berman, A., *Total Pressure Measurements in Vacuum Technology* (Academic, New York, 1985).

Braker, W. and Mossman, A.L., *Matheson Gas Data Book* (Matheson Gas Products, Secaucas, NJ 07094, 1980).

Breck, D.W., *Zeolite Molecular Sieves* (Wiley, New York, NY, 1977).

Bunshah, R.F., Ed., *Handbook of Deposition Technologies for Films and Coatings,* 2nd Ed. (Noyes, Park Ridge, NJ, 1994).

Dawson, P.H., *Quadrupole Mass Spectrometry and Its Applications* (American Institute of Physics, New York, NY 10017-3483, 1995).

Drinkwine, M.S. and Lichtman, D., *Partial Pressure Analyzers and Analysis,* AVS Monograph Series, M-3 (American Vacuum Society, New York, NY 10017, 1979).

Dushman, S., *Scientific Foundation of Vacuum Technique,* 2nd Ed. (Wiley, New York, NY, 1962).

Ertl, G. and Küppers, J., *Low Energy Electrons and Surface Chemistry,* 1st Ed., (Verlag Chemie, Weinheim, 1974); 2nd Ed., completely revised (VCH Verlagsgesellschaft, mbH, Weinheim, 1985).

Fomenko, V.S., *Handbook of Thermionic Properties* (Translated), Edited by Samsonov, G.V. (Plenum, New York, NY, 1966).

French, T.E. (Revised by Vierck, C.J.), *A Manual of Engineering Drawing,* 7th Ed. (McGraw Hill, New York, NY, 1947).

Gomer, R., *Field Emission and Field Ionization* (American Institute of Physics, New York, NY 10017-3483, 1993).

Gordon, A.J. and Ford, R.A., *The Chemist's Companion—A Handbook of Data, Techniques, and References* (Wiley, New York, NY, 1972).

Hablanian, M.H., *Diffusion Pumps,* AVS Monograph Series, M-5 (American Vacuum Society, New York, NY 10017, 1983).

Kennard, E.H., *Kinetic Theory of Gases* (McGraw-Hill, New York, NY, 1938).

Kohl, W.H., *Handbook of Materials and Techniques for Vacuum Devices* (American Institute of Physics, New York, NY 10017-3483, 1995).

Leco Corporation, *Metal Etching and Polishing* (St. Joseph, MI 49085, 1977).

Leybold-Heraeus, *Vacuum Technology Its Foundations, Formulae, and Tables* (Leybold-Heraeus, Pittsburgh, PA 15632).

Loeb, L.B., *The Kinetic Theory of Gases,* 3rd Ed., (Dover, NY, 1961).

Lunn, G. and Sansone, E.B., *Destruction of Hazardous Materials in the Laboratory* (Wiley, New York, NY, 1990).

Madey, T.E., Ed., *Vacuum Science and Technology, History of Vacuum Science and Technology,* Vol. 1 (American Vacuum Society, AIP Press, New York, NY, 1984).

Maissel, L.I. and Glong, R., *Handbook of Thin Film Technology* (McGraw-Hill, New York, NY, 1970).

Mark, H.F., Ed., et al., *Encyclopedia of Chemical Technology* (Wiley, New York, NY, 1979).

Moore, J.H., Davis, C.C. and Coplan, M.A., *Building Scientific Apparatus* (Addison-Wesley Reading, MA, 1983).

Müller, E.W. and Tsong, T.T., *Field Ion Microscopy* (Elsevier, New York, NY, 1969).

O'Hanlon, J.F., *A User's Guide to Vacuum Technology* (Wiley, New York, NY, 1980).

*The Omega Complete Temperature Measurement Handbook and Encyclopedia,* Omega Engineering, Inc., P.O. Box 4047, One Omega Drive, Stamford, CT 06907.

Pierce, J.R., *Theory and Design of Electron Beams* (D. van Nostrand Co., Inc., New York, NY, 1949); 2nd Ed. (1954).

Redhead, P.A., Hobson, J.P. and Kornelsen, E.V., *The Physical Basis of Ultrahigh Vacuum* (American Institute of Physics, New York, NY 10017-3483, 1993).

Redhead, P.A., Ed., *Vacuum Science and Technology, History of Vacuum Science and Technology,* Vol. 2 (American Vacuum Society, AIP Press, New York, NY, 1994).

Roberts, R.W. and Vanderslice, T.A., *Ultrahigh Vacuum and Its Applications* (Prentice-Hall, Engelwood Cliffs, NJ, 1963).

Robinson, N.W., *The Physical Principles of Ultrahigh Vacuum Systems and Equipment* (Chapman and Hall, London, 1968).

Rosebury, F., *Handbook of Electron Tube and Vacuum Techniques* (AIP, American Institute of Physics, New York, NY, 1993).

Roth, A., *Vacuum Sealing Techniques* (Pergamon, Oxford, UK, 1966).

Roth, A., *Vacuum Technology,* 2nd Ed. (North Holland, Amsterdam, 1982).

Sala, A., *Radiant Properties of Materials* (Elsevier, Amsterdam, 1986).

Smithells, C.J. and Brandes, E.A., Eds., *Metals Reference Book,* 7th Ed. (Butterworths, London and Boston, 1991).

Spangenberg, K.R., *Vacuum Tubes* (McGraw-Hill, New York, NY, 1948).

Strong, J., Neher, H.V., Whitford, A.E., Cartwright, C.H. and Hayward, R., *Procedures in Experimental Physics* (Prentice-Hall, New York, NY, 1946).

Thompkins, H.G., *The Fundamentals of Vacuum Technology,* AVS Monograph Series, M-6 (American Vacuum Society, New York, NY 10017, 1991).

Thompkins, H.G., *Pumps Used in Vacuum Technology,* AVS Monograph Series, M-9 (American Vacuum Society, New York, NY 10017, 1991).

Thompkins, H.G., *Vacuum Gauging and Control,* AVS Monograph Series, M-12 (American Vacuum Society, New York, NY 10017, 1994).

Wheeler, E.L., *Scientific Glassblowing* (Wiley, New York, NY, 1958).

Wilson, E.B., *An Introduction to Scientific Research,* 1st Ed. (McGraw-Hill, New York, NY, 1952).

Wilson, N.G. and Beavis, L.C., *Handbook of Vacuum Leak Detection,* AVS Monograph Series, M-2 (American Vacuum Society, New York, NY 10017, 1979).

Wood, E.A., *Crystal Orientation Manual* (Columbia University Press, New York, NY, 1963).

Wutz, M., Adam, H. Walcher, W., *Theory and Practice of Vacuum Technology* (Friedr. Vieweg und Sohn, Braunschweig/Wiesbaden, Germany, 1989).

# Appendix B

# Manufacturers

## United States, UK, and Canadian Manufacturers

| Company | Products |
|---|---|
| 12R<br>PO Box 159<br>Cheltenham, PA 19012<br>Tel: 215-379-3333<br>Fax: 215-663-8847 | Tools for scientists |
| II-VI<br>375 Saxonburg Blvd.<br>Saxonburg, PA 16056<br>Tel: 412-352-4455<br>Fax: 412-352-4980 | Infrared materials |
| Accuratus<br>RD4 Brass Castle Rd.<br>Washington, NJ 07882<br>Tel: 908-689-0880<br>Fax: 908-689-8796 | Macor machineable glass; ceramics |
| Ace Glass, Inc.<br>1430 Northwest Blvd.<br>PO Box 688<br>Vineland, NJ 08360<br>Tel: 800-223-4524<br>Fax: 800-543-6752<br>or<br>63941 S. Hancock St.<br>PO Box 996<br>Louisville, KY 40201<br>Tel: 800-626-5381<br>Fax: 800-635-4698 | Glass products |
| Ace Plastic Co.<br>494 Wortman Ave.<br>Brooklyn, NY 11208-5428<br>Tel: 212-523-5500<br>Fax: 407-659-3816 | Plastic balls for crystal models |

| | |
|---|---|
| ACF Metals<br>2239 E. Kleindale Rd.<br>Tucson AZ 85719-2440<br>Tel: 520-325-9557<br>Fax: 520-32509493 | Metal foils, carbon foils |
| Acheson Colloids<br>1607 Washington Ave.<br>PO Drawer 611747<br>Port Huron, MI 48061-1747<br>Tel: 800-255-1908<br>Fax: 313-984-1446 | Aquadag—colloidal graphite |
| Ad-Vance Magnetics, Inc.<br>625 Monroe St.<br>PO Box 69<br>Rochester, IN 46975<br>Tel: 219-223-3158<br>Fax: 219-223-2524 | Magnetic shields |
| Advanced Ceramics Corp.<br>P.O. Box 94924<br>Cleveland, OH 44101-4924<br>or<br>11907 Madison Ave.<br>Lakewood, OH 44107-5026<br>Tel: 800-822-4322<br>Fax: 216-529-3954 | Graphite monochromators |
| Advent Associates, Ltd.<br>6663 Woodwell St.<br>Pittsburgh, PA 15217<br>Tel: 412-422-8100<br>Fax: 412-422-8101 | Ion guns, vacuum components |
| Advent Research Materials LTD<br>Blyth Road Industrial Estate<br>Halesworth Suffolk<br>England 1P19 8DD<br>Tel: 44 1986 874555<br>Fax: 44 1986874557 | Metals, alloys, foils, sheets, wires, rods, tubes |
| AESAR<br>Johnson Matthey, Inc.<br>PO Box 1087<br>Seabrook, NH 03874<br>Tel: 1-800-343-1990 | Single crystals |

| | |
|---|---|
| AIRCO<br>Lansbrook Rd.<br>Acton, MA 01720<br>Tel: 508-263-7769<br>Fax: 508-263-6014 | Specialty gases |
| Aithaca<br>50 Charles Lindbergh Blvd.<br>Suite 400<br>Uniondale, NY 11553<br>Tel: 516-229-2330<br>Fax: 516-227-2350 | Single crystals |
| Alberox<br>Industrial Park<br>New Bedford, MA 02745-1289<br>Tel: 508-995-1725<br>Fax: 508-995-6954 | Feedthroughs |
| Alcatel<br>Phil Danielson<br>Danielson Associates, Inc.<br>PO Box 157<br>51471/2 Main St.<br>Donners Grove, IL 60515<br>Tel: 612-370-1279 | Pumps, mechanical |
| Aldrich Chemical Co.<br>PO Box 2060<br>Milwaukee, WI 53201<br>Tel: 414-273-3850<br>Fax: 414-273-4979 | Chemicals |
| Alfa<br>Johnson Matthey<br>30 Bond St.<br>Ward Hill, MA 01835-8099<br>Tel: 800-343-0660<br>Fax: 800-322-4757 | Chemicals, high purity electronic grade |
| Allen Datagraph, Inc.<br>Industrial Way<br>Salem, NH 03079<br>Tel: 800-258-6360<br>Fax: 603-893-9042 | X-Y recorders/plotters |

| | |
|---|---|
| Alltech Assoc. Inc.<br>2051 Waukegan Rd.<br>Deerfield, IL 60015<br>Tel: 800-255-8324<br>Fax: 708-948-1078 | Graphite fittings for swagelok fittings |
| American Laboratory<br>PO Box 3220<br>Lowell, MA 01853-9878 | Annual buyer's guide edition |
| Analog Devices<br>One Technology Way<br>Norwood, MA 02062-9106<br>Tel: 617-329-4700<br>Fax: 617-326-8703 | Computer chips and boards |
| APD Cryogenics, Inc.<br>1833 Vultee St.<br>Allentown, PA 18103-4783<br>Tel: 610-791-6700<br>Fax: 610-791-0440 | Cryopumps; Helitron He coolers |
| Aremco<br>23 Snowden Ave.<br>PO Box 429<br>Ossining, NY 10562-0429<br>Tel: 914-762-0685<br>Fax: 914-762-1663 | Single crystals |
| Automation Gages, Inc.<br>850 Hudson Ave.<br>Rochester, NY 14621<br>Tel: 1-800-922-0329<br>Fax: 716-338-7370 | Translation stages |
| Baikowski International Corp.<br>6006 Old Pineville Rd.<br>Charlotte, NC 28210<br>Tel: 704-523-1707 | Alumina polishing powder |
| Baker Process Equipment<br>PO Box 898<br>308-Moon Clinton Rd.<br>Coraopolis, PA 15108<br>Tel: 412-264-8271<br>Fax: 412-264-8758 | Filters, dryers, purifiers |

| | |
|---|---|
| Balzers<br>8 Sayamore Park Rd.<br>Hudson, NH 03051<br>Tel: 603-595-3250<br>Fax: 603-595-3200 | Mass spectrometers |
| Barnes/Spectra Tech.<br>652 Glenbrook Rd.<br>Stamford, CT 06906<br>Tel: 800-243-9186 | IR spectrometers |
| Bearings, Inc.<br>5536 Baum Blvd.<br>Pittsburgh, PA 15232<br>Tel: 412-681-4900 | Ball bearing power transmission units |
| Berghof/America<br>PO Box 6029<br>Concord, CA 94524<br>Tel: 800-544-5004<br>Fax: 510-827-1189 | Teflon products |
| Bertan<br>121 New South Rd.<br>Hicksville, NY 11801<br>Tel: 516-433-3110<br>Fax: 516-935-1766 | High-voltage power supplies |
| Bid Service<br>PO Box 128<br>Bradley Beach, NJ 07720<br>Tel: 908-775-8300<br>Fax: 908-774-1443 | Surplus vacuum and semiconductor processing equipment |
| Blake Industries, Inc.<br>660 Jerusalem Rd.<br>Scotch Plains, NJ 07076<br>Tel: 908-233-7240 | X-ray instruments |
| Bomco Inc.<br>125 Gloucester Ave.<br>Gloucester, MA 01930-2294<br>Tel: 508-283-9000<br>Fax: 508-283-2882 | Quartz-Mo seals |
| Bristol Organics LTD<br>Sharpness Docks<br>Berkeley<br>Glos GL 13 9UG<br>UK<br>Tel: 0453-811400 | Fluorine compounds (organic) |

| | |
|---|---|
| Brush-Wellman<br>Elmore, OH 43416<br>Tel: 419-862-2745 | Beryllium |
| Buckbee Mears Corp.<br>245 East 6th St.<br>St. Paul, MN 55101-1959<br>Tel: 612-228-6400<br>Fax: 612-228-6572 | Photolithograph W grid |
| Buehler Analyst<br>41 Waukegan Rd.<br>PO Box 1<br>Lake Bluff, IL 60044<br>Tel: 847-295-6500<br>Fax: 847-295-7979 | Crystal cutting, polishing, fiber optics, optical instrumentation |
| C&H Sales<br>2176 E. Colorado Blvd.<br>PO Box 5356<br>Pasadena, CA 91117-9988<br>Tel: 800-325-9465<br>Fax: 818-796-4875 | Surplus equipment |
| C.M. Furnaces, Inc.<br>103 Dewey St.<br>Bloomfield, NJ 07003<br>Tel: 201-338-6500<br>Fax: 201-338-1625 | Evaporation sources, filaments |
| California Fine Wire Co.<br>PO Box 446<br>Grover, CA 93483-0446<br>Tel: 805-489-5144<br>Fax: 805-489-5352 | Wire, thermocouple wire, fine wire |
| Cambridge Isotopes<br>50 Frontage Rd.<br>Andover, MA 01810-5413<br>Tel: 800-322-1174<br>Fax: 508-749-2768 | Isotopes, stable |
| Canberra Industries, Inc.<br>Nuclear Products Group<br>800 Research Pkwy.<br>Meriden, CT 06450<br>Tel: 203-238-2351<br>Fax: 203-235-1347 | Counting electronics |

| | |
|---|---|
| Ceramaseal<br>PO Box 260, Rte. 20<br>New Lebanon, NY 12125<br>Tel: 518-794-7800<br>Fax: 518-794-8080 | Feedthroughs |
| Cercoa<br>ECS Co. 55 Webster Ave.<br>New Rochelle, NY 10801<br>Tel: 914-636-4325<br>Fax: 914-636-3701 | Ceramic materials |
| Chemvac Associates<br>808 High Point Ave.<br>Virginia Beach, VA 23451<br>Tel: 804-422-9823 | Vacuum joints |
| Chromalox<br>641 Alpha Dr.<br>Pittsburgh, PA 15238<br>Tel: 800-443-2640<br>Fax: 412-967-5148 | Heaters |
| Cliftronics, Inc.<br>2 South St.<br>Clifton Springs, NY 14432<br>Tel: 315-462-9471 | Electronics equipment and supplies, electron guns |
| Cole-Palmer<br>7425 North Oak Park Ave.<br>Chicago, IL 60648<br>Tel: 800-323-4340 | Laboratory supplies—general |
| Comstock<br>1005 Alvin Weinberg Dr.<br>Oak Ridge, TN 37830<br>Tel: 615-483-7690<br>Fax: 615-481-3884 | Electron energy analyzers, electron guns |
| Connecticut Hard Rubber Co.<br>PO Box 2560<br>Meriden, CT 06450<br>Tel: 203-634-0555 | Teflon tape |
| Connecticut Technology<br>PO Box 524<br>Stratford, CT 06497<br>Tel: 203-375-5015<br>Fax: 203-375-0541 | Langmuir-Blodgett film deposition |

| | |
|---|---|
| Cooke Vacuum Products<br>13 Merritt St.<br>South Norwalk, CT 06854<br>Tel: 203-853-9500<br>Fax: 203-858-9553 | Vacuum components |
| Copper & Brass Sales<br>17401 Ten Mile<br>East Detroit, MI 48021<br>Tel: 800-926-2600<br>Fax: 313-775-5078 | UHV soft aluminized bronze |
| Cory Lab Inc.<br>823 5th St.<br>Menominee, MI 49858<br>Tel: 906-863-9336 | Sapphire/metal feedthroughs |
| Cotronics<br>3379 Shore Pkwy.<br>Brooklyn, NY 11235<br>Tel: 718-646-7996<br>Fax: 718-646-3028 | Ceramic materials |
| CREE<br>2810 Meridian Pkwy.<br>Durham, NC 27713<br>Tel: 919-361-5709<br>Fax: 919-361-4630 | Silicon carbide |
| Cryofab<br>PO Box 485<br>540 N. Michigan Ave.<br>Kenilworth, NJ 07033<br>Tel: 207-686-3636<br>Fax: 908-686-9536 | Cryogenic equipment, dewars, etc. |
| Cryomech Inc.<br>1630 Erie Blvd. E.<br>Syracuse, NY 13210<br>Tel: 315-475-9692<br>Fax: 315-422-1202 | Cryogenic equipment |
| Custom Cable Corp.<br>242 Butler St.<br>PO Box 1050<br>Westbury, NY 11590-3193<br>Tel: 800-242-2253<br>Fax: 516-334-3989 | Electric equipment and supplies, coaxial cable |

| | |
|---|---|
| CVC<br>525 Lee Rd.<br>PO Box 1886<br>Rochester, NY 14603<br>Tel: 716-458-2550 | Vacuum products |
| CVT Engineering Ltd.<br>110 Shawmut Rd.<br>Canton, MA 02021 | Vacuum transfer devices |
| Dabilier<br>207 Radley Rd.<br>Abingdon Oxon<br>OX14 3XA<br>England<br>Tel: 0235-28271 | Ion sources |
| Daedal, Inc.<br>1140 Sandy Hill Rd.<br>Harrison City, PA 15636<br>Tel: 800-245-6903<br>Fax: 412-861-3330 | Positioning equipment |
| DEI<br>2301 Research Blvd.<br>Suite 105<br>Fort Collins, CO 80526<br>Tel: 970-493-1901<br>Fax: 970-493-1903 | Pulse generators (high voltage) |
| DoAll Northern IL. Co.<br>236 Laurel Ave.<br>Des Plaines, IL 60016<br>Tel: 800-955-8191<br>Fax: 708-824-4340 | Machine tools, precision gauges, cutting tools |
| Dow Corning<br>PO Box 0994<br>Midland, MI 48640-0994<br>Tel: 517-496-4000 | Silicone products |
| Duniway Stockroom Corp.<br>1600 N. Shoreline Blvd.<br>Mountain View, CA 94043<br>Tel: 800-446-8811<br>Fax: 415-965-0764 | Rebuilt ion pumps and vacuum components |

| | |
|---|---|
| Dupont<br>1007 Market St.<br>N-2426-A<br>Wilmington, DE 19898<br>Tel: 800-527-2601<br>Fax: 302-773-2291 | Polymers |
| E. McGrath, Inc.<br>35 Osborne St.<br>Salem, MA 01970-2599<br>Tel: 508-744-3546<br>Fax: 508-741-4020 | Used and new surplus equipment |
| Ealing Corp.<br>89 Doug Brown Way<br>Holliston, MA 01746<br>Tel: 800-343-4912<br>Fax: 508-429-7893 | Optical components |
| East Coast Sales<br>11 Business Park Dr.<br>Armonk, NY 10504<br>Tel: 914-273-1400<br>Fax: 914-273-5866 | Tools—semiconductor |
| Eastman Kodak Co.<br>5/81 RC MC 02010<br>Rochester, NY 14650-2010<br>Tel: 716-722-4385 ext. 1704<br>Fax: 716-477-4947 | Irtran optics |
| Edmond Scientific<br>A-415 Edscorp. Bldg.<br>Barrington, NJ 08007<br>Tel: 609-573-6250<br>Fax: 609-573-6295 | Optics, etc. |
| Edwards High Vacuum<br>301 Ballardvale St.<br>Wilmington, MA 01887<br>Tel: 508-658-5410<br>Fax: 508-658-7969 | Vacuum components |
| Electrofusion Corp.<br>34325 Ardenwood Blvd.<br>Fremont, CA 94536<br>Tel: 415-795-1100 | Beryllium windows |

| | |
|---|---|
| Electronic Space Products International<br>1050 Bensen Way<br>Ashland, OR 97520<br>Tel: 800-638-2581 | |
| Energy Beam Sciences<br>11 Bowles Rd.<br>PO Box 468<br>Agawam, MA 01001<br>Tel: 800-992-9037<br>Fax: 413-789-2786 | $LaB_6$ cathodes |
| Energy Research & Generation, Inc.<br>900 Stanford Ave.<br>Oakland, CA 94608<br>Tel: 510-658-9785<br>Fax: 510-658-7428 | Aluminum foams |
| Engelhard Industries Division<br>Engelhard Corp.<br>Electronic Materials Systems<br>1 West Central Ave.<br>East Newark, NJ 07029<br>Tel: 201-268-7800 | Liquid bright gold |
| EPI MBE Products Group<br>1290 Hammond Rd.<br>St. Paul, MN 55110<br>Tel: 612-653-0488<br>Fax: 612-653-0725 | MBE products |
| Esco<br>171 Oakridge Rd.<br>Oakridge, NJ 07438<br>Tel: 800-327-1399<br>Fax: 201-697-3011 | Optical materials—sapphire |
| Extrel Products<br>575 Epsilon Dr.<br>Box 11512<br>Pittsburgh, PA 15238-2838<br>Tel: 412-963-7530<br>Fax: 412-963-6578 | Quadrupole mass spectrometers |
| Fairfield Synthesis<br>PO Box 20<br>Blythewood, SC 29016<br>Tel: 803-754-3856<br>Fax: 803-754-8833 | Custom organic synthesis |

| | |
|---|---|
| Farchan<br>2603 NW 74th Place<br>Gainesville, FL 32653<br>Tel: 352-378-5864 | Organic chemicals |
| FEI<br>PO Box 654<br>McMinnville, OR 97128<br>Tel: 503-648-7075 | Electron guns |
| Fil-Tech<br>1143 Diamond St.<br>San Francisco, CA 94114 | Vacuum supply filaments |
| Fine Science Tools<br>373 Vintage Park Dr.<br>Foster City, CA 94404<br>Tel: 800-521-2109<br>Fax: 415-349-3729<br>e-mail: info@finescience.com | Surgical tools |
| Fluorocarbon Corp.<br>10871 Kyle St.<br>PO Box 520<br>Los Alamitas, CA 90720<br>Tel: 213-594-0941 | Teflon |
| Fluorochem Ltd.<br>Wesley St.<br>Old Glossop<br>Derbyshire SK13 9RY<br>England<br>Tel: 04574-68921 | Molecular formula index |
| Fluoroware<br>102 Jonathan Blvd. N.<br>Chaska, MN 55318<br>Tel: 612-448-3131<br>Fax: 612-448-5576 | Teflon—semiconductor tools, handling equipment |
| Foametal, Inc.<br>37645 Vine St.<br>Willoughby, OH 44904<br>Tel: 216-951-9081<br>Fax: 216-951-9360 | Metal foams |
| Friedfich & Diownick, Inc.<br>2127 Wheaton Ave.<br>PO Box 230<br>Millville, NJ 08332<br>Tel: 609-825-0305<br>Fax: 609-327-4299 | Glass capillary |

| | |
|---|---|
| Furon<br>4402 Corporate Center Dr.<br>Los Alamitos, CA 90720<br>Tel: 714-995-1818<br>Fax: 714-761-1270 | Gaskets, seals |
| Fyrepel<br>PO Box 518<br>Newark, OH 43055<br>Tel: 614-344-0391<br>Fax: 614-344-6025 | Aluminized fiberglass sheet |
| Galileo<br>Galileo Park<br>PO Box 550<br>Sturbridge, MA 01566<br>Tel: 800-648-1800<br>Fax: 508-347-9191 | Channeltron and channel plate multipliers |
| Gallard/Schlesinger<br>584 Mineola Ave.<br>Carle Place, NY 11514<br>Tel: 516-333-5600<br>Fax: 516-333-5628 | Chemicals, metals |
| Gast Manufacturing Corp.<br>2300 NM 63 #139<br>Benton Harbor, MI 49022<br>Tel: 616-926-6171 | Industrial equipment and supplies, carbon vane vacuum pump |
| Glassfab, Inc.<br>PO Box 1880<br>Rochester, NY 14603<br>Tel: 716-262-4000<br>Fax: 716-454-4305 | |
| Goodfellow Metals<br>800 Lancaster Ave.<br>Berwyn, PA 19302-1780<br>Tel: 610-640-1612<br>Fax: 800-283-2020 | Metal foils |
| Hamamatsu<br>420 South Ave.<br>Middlesex, NJ 08846<br>Tel: 908-469-6640 | Photomultipliers |

| | |
|---|---|
| Hamilton Technology, Inc.<br>101 N. Queen St.<br>PO Box 4787<br>Lancaster, PA 17604<br>Tel: 717-299-2581<br>Fax: 717-397-8510 | HAVAR foil |
| Hanovia Division of Engelhard Industries<br>1 West Central Ave.<br>East Newark, NJ 07029<br>Tel: 201-268-7800 | Organometallic paints |
| Hansen R.G. & Associates<br>631 Chapala St.<br>Santa Barbara, CA 93101<br>Tel: 805-564-3388<br>Fax: 805-963-0733 | Laboratory cryogenic systems and components |
| Harrick Scientific Corp.<br>Croton Dam Rd.<br>Box 867<br>Ossining, NY 10562<br>Tel: 914-762-0020 | IR-VIS-UV accessories, spectrometers |
| Harshaw Optical<br>6801 Cochran Rd.<br>Solon, OH 44139<br>Tel: 800-472-5656<br>Fax: 216-349-6581 | IR and UV windows |
| Helicoflex<br>400 Myrtle Ave.<br>Boonton, NJ 07005<br>Tel: 201-334-8100 | Vacuum fittings |
| Helicoflex Co.<br>118 W. 12th St.<br>Parkerburg, WV 26101<br>Tel: 304-428-1644<br>Fax: 304-428-3490<br>e-mail hxrpvseals@aol.com | Seals |
| Helicoflex Co.<br>PO Box 9889<br>Columbia, SC 29290<br>Tel: 803-783-1880<br>Fax: 803-783-4279 | Seals |

| | |
|---|---|
| Hex<br>Metal Specialties Inc.<br>515 Commerce Dr.<br>Fairfield, CT 06430<br>Tel: 203-384-0331<br>Fax: 203-368-4082 | Single crystals |
| High Vacuum Apparatus<br>7055 Engle Rd.<br>Suite 6-607<br>Middleburg Hts., OH 44130<br>Tel: 216-826-0559<br>Fax: 216-826-1171 | Gate valves |
| Honeywell<br>Industrial Automation & Controls<br>1100 Virginia Dr.<br>Washington, PA 19034 | Electronics |
| Honeywell<br>Micro Switch Div.<br>11 W. Spring St.<br>Freeport, IL 61032-4353<br>Tel: 800-537-6945<br>Fax: 815-235-5574 | Electronics |
| IBM Instruments, Inc.<br>Orchard Park<br>PO Box 332<br>Danbury, CT 06810 | FTIR spectrometers |
| ICON<br>19 Ox Bow Ln.<br>Summit, NJ 07901<br>Tel: 908-273-0449<br>Fax: 908-273-0449 | Isotopes |
| Indium Corporation of America<br>PO Box 269<br>1676 Lincoln Ave.<br>Utica, NY 13503-0269<br>Tel: 800-448-9240<br>Fax: 800-221-5759 | Indium |
| Infrared Laboratories, Inc.<br>1808 E. 17th St.<br>Tucson, AZ 85719-6505<br>Tel: 520-622-7074<br>Fax: 520-623-0765 | Dewars, He detectors |

| | |
|---|---|
| Insaco<br>1365 Canary Rd.<br>PO Box 9006<br>Quakertown, PA 18951-9006<br>Tel: 215-536-3500<br>Fax: 215-536-7750 | Sapphire fabrication |
| Instrument Specialties Co.<br>PO Box A<br>Delaware Water Gap, PA 18327<br>Tel: 717-424-8510<br>Fax: 717-424-6213 | Springs |
| Instruments SA, Inc.<br>3880 Park Ave.<br>Edison, NJ 08820-3097<br>Tel: 800-438-7739<br>Fax: 908-549-5125 | Monochromators, optical |
| Insulator Seal, Inc.<br>23874-B Cabot Blvd.<br>Hayward, CA 94545-1661<br>Tel: 800-548-9509<br>Fax: 510-887-7475<br>e-mail: sales@isi-seal.com | Feedthroughs—ceramic/metal |
| Intek Services, Inc.<br>7 McMillan Way<br>Suite 101<br>Newark, DE 19713<br>Tel: 302-366-8530 | Teflon suspension |
| International Crystal Lab<br>11 Erie St.<br>Garfield, NJ 07026<br>Tel: 201-478-8944<br>Fax: 201-478-4201 | IR accessories |
| IonSpec<br>One Longstreet<br>Irvine, CA 92714<br>Tel: 714-857-0793 | Fourier transform mass spectrometry |
| Isotec, Inc.<br>3858 Benner Rd.<br>Miamisburg, OH 45432-4304<br>Tel: 513-859-1808<br>Fax: 513-859-4878<br>e-mail: isosales@isotec.com | Isotopic gases |

| | |
|---|---|
| Israelachvili Instrument<br>2233 Foothill Ln.<br>Santa Barbara, CA 93105<br>Tel: 805-899-9292<br>Fax: 805-899-9190 | Surface forces instruments |
| J&H Berge, Inc.<br>4111 S. Clinton Ave.<br>So. Plainfield, NJ 07080-0310<br>Tel: 1-800-684-1234 | Ball bearings, ultrasonic (nalgene) cleaners |
| Jensen Tools<br>7815 S. 46th St.<br>Phoenix, AZ 85044-5399<br>Tel: 800-426-1194<br>Fax: 800-366-9662 | Tools |
| JML Direct<br>690 Portland Ave.<br>Rochester, NY 14621-5196<br>Tel: 716-342-9482<br>Fax: 716-342-6125 | Optical components |
| Keithley<br>28775 Aurora Rd.<br>Cleveland, OH 44139-1891<br>Tel: 216-248-0400<br>Fax: 216-248-6168 | Electronics—measurement |
| Kepco<br>131-38 Sanford Ave.<br>Flushing, NY 11352<br>Tel: 718-461-7000<br>Fax: 718-767-1102 | Power supplies |
| Kimball Physics Inc.<br>311 Kimball Hill Rd.<br>Wilton, NH 03086<br>Tel: 603-878-1616<br>Fax: 603-878-3700 | Electron and ion guns, $LaB_6$ emitters |
| Kulite Semiconductor Products, Inc.<br>1 Willow Tree Rd.<br>Leonia, NJ 07605<br>Tel: 201-461-0900<br>Fax: 201-461-0990 | Strain gauges |
| Lab Safety Supply<br>PO Box 1368<br>Janesville, WI 53547-1368<br>Tel: 800-356-0783<br>Fax: 800-543-9910 | Safety lab equipment |

| | |
|---|---|
| Lake Shore Cryotonics, Inc.<br>64 Walnut St.<br>Westerville, OH 43081-2399<br>Tel: 614-891-2243<br>Fax: 614-891-1392 | Cryogenics, cryogenic thermometry |
| Lamba<br>515 Broad Hollow Rd.<br>Melville, NY 11747-3700<br>Tel: 516-694-4200 | Power supplies |
| Lasertechnics<br>5500 Wilshire Ave., NE<br>Albuquerque, NM 87113<br>Tel: 505-822-1123<br>Fax: 505-821-2213 | Pulsed valve |
| LDS Vacuum Products, Inc.<br>980 Sunshine Ln.<br>Aitamonte Springs, FL 32714<br>Tel: 407-862-4643<br>Fax: 407-862-8723 | Vacuum supplies |
| Lebow Co.<br>5960 Mandarin Ave.<br>Goleta, CA 93117<br>Tel: 805-964-7117<br>Fax: 805-964-7117 | |
| Leisk<br>Albert Dr., Burgess Hill<br>West Sussex, RH15 9NX<br>England<br>Tel: 044-46-45821 2 | Flanges and gaskets |
| Lenox Laser<br>1 Green Glade St.<br>Phoenix, MD 21131<br>Tel: 301-592-3106 | Laser beam machining, laser repair |
| Lesker, Kurt J. Co.<br>1515 Worthington Ave.<br>Clairton, PA 15025<br>Tel: 412-233-4200<br>Fax: 412-233-4275 | Vacuum components |
| Linseis<br>PO Box 666<br>Princeton, NJ 08550<br>Tel: 609-799-6282<br>Fax: 209-799-7739 | X-Y recorders |

| | |
|---|---|
| Love Controls Corp.<br>475 Wheeling Rd.<br>Wheeling, IL 60090<br>Tel: 800-828-4588<br>Fax: 708-541-7366 | Temperature controllers |
| M-G Industries<br>175 Meister Ave.<br>PO Box 5328<br>North Branch, NJ 08876<br>Tel: 610-695-7400<br>Fax: 610-695-7600 | Gases |
| Magnetic Shielding Corp.<br>750 N. Thomas Dr.<br>Bensenville, IL 60106<br>Tel: 708-766-7800<br>Fax: 708-766-2813 | Magnetic shields |
| Mark V Laboratory<br>18 Kripes Rd.<br>East Granby, CT 06026<br>Tel: 800-243-9776<br>Fax: 203-653-4087 | Metallographic equipment |
| Markson<br>PO Box 1359<br>Hillsboro, OR 97123-9981<br>Tel: 800-528-5114<br>Fax: 800-858-2243 | Laboratory supplies/general |
| Materials Research Corp.<br>542 Rte. 303<br>Orangeburg, NY 10962<br>Tel: 914-359-4200<br>Fax: 914-359-3453 | High-purity materials for thin film deposition |
| McAllister<br>W. 280-T Prairie Ave.<br>Cover D'Alend, ID 83814<br>Tel: 208-772-9527 | STM, Kelvin probe, vacuum components |
| MDC<br>2384 Cabot Blvd.<br>Hayward, CA 94545-1651<br>Tel: 800-443-8817<br>Fax: 510-887-0626 | Vacuum components |

| | |
|---|---|
| Melles Grot<br>1770 Kettering St.<br>Irvine, CA 92714<br>Tel: 800-835-2626<br>Fax 714-261-7589 | Optical items |
| Metal Bellows<br>1075 Providence Highway<br>Sharon, MA 02067<br>Tel: 617-784-1400<br>Fax: 617-784-1405 | Bellows, etc. |
| Metal Crystals & Oxides, Ltd.<br>Button End, Harston<br>CB2 5WX<br>Cambridge, England<br>Tel: 0223-872022 | Metals (pure), metal foils, single crystal metals |
| Metaleurop<br>111 Richmond St. West<br>Suite 418<br>Toronto, Ontario Canada<br>Tel: 416-366-3954<br>Fax: 416-366-0586 | Single crystals |
| Meyer Tool & Mfg., Inc.<br>4601 Southwest Highway<br>Oak Lawn, IL 60453<br>Tel: 708-425-9080<br>Fax: 708-425-2612 | Vacuum chambers—fabrication |
| Micromeritics<br>One Micrometitics Dr.<br>Norcross, GA 30093-1877<br>Tel: 770-662-3633<br>Fax: 770-662-3696 | Particle size and surface area measurement instruments |
| Midac<br>17911 Fifth Ave.<br>Irvine, CA 92714<br>Tel: 714-660-8558<br>Fax: 714-660-9334 | IR emission |
| Miller-Stephenson Chemical Co.<br>Beckus Ave.<br>Danbury, CT 06810<br>Tel: 203-743-4447<br>Fax: 203-791-8702 | Fluorocarbon lubricants |

| | |
|---|---|
| Minnesota Valley Engineering, Inc.<br>407 Seventh St. NW<br>New Prague, MN 56071<br>Tel: 612-758-4484 | Vacuum insulated piping systems |
| MKS<br>Six Shattuch Rd.<br>Andover, MA 01810-2449<br>Tel: 800-227-8766<br>Fax: 508-975-0093 | Capacitance manometers |
| MMR Technologies<br>1400 N. Shoreline Blvd. A5<br>Mountain View, CA 94043-1312<br>Tel: 415-962-9620<br>Fax: 415-962-9647 | Cryogenic-closed cycle refrigerators |
| Monocrystals<br>24400 Highland Rd.<br>Richmond Heights, OH 44143<br>Tel: 216-531-9820<br>Fax: 216-531-9789 | Single crystals |
| MSA<br>PO Box 426<br>Pittsburgh, PA 15230<br>Tel: 800-672-2222<br>Fax: 412-967-3451 | Air pollution sensors |
| MSC<br>151 Sunnyside Blvd.<br>Plainview, NY 11803<br>Tel: 800-753-7937<br>Fax: 516-349-0265 | Machinist's tools |
| MTS Systems Corp.<br>Box 24012<br>Minneapolis, MN 55424<br>Tel: 612-937-4000<br>Fax: 612-937-4515 | |
| Munns Mfg., Inc.<br>5885 La Ribera St.<br>Livermore, CA 94550<br>Tel: 510-447-0310<br>Fax: 510-447-0110 | Vacuum equipment (UHV) |
| MVE<br>8011 34th Ave., S.<br>Bloomington, MN 55425-1636<br>Tel: 800-247-4446<br>Fax: 612-853-9661 | Vacuum-insulated products |

| | |
|---|---|
| National Instruments<br>6504 Bridge Point Pkwy.<br>Austin, TX 78730-5039<br>Tel: 512-794-0100<br>Fax: 512-794-8411<br>Website: www.natinst.com | LabVIEW, other instrumentation for data analyses and acquisition |
| Neslab Instruments<br>PO Box 1178<br>Portsmouth, NH 03802-1178<br>Tel: 603-436-9444<br>Fax: 603-436-8411 | Cooling units |
| New Focus, Inc.<br>340 Pioneer Way<br>Mountain View, CA 94041<br>Tel: 415-961-2108<br>Fax: 415-961-6072 | Optical components |
| Newark Electronics<br>4801 N. Ravenswood Ave.<br>Chicago, IL 60640-4496<br>Tel: 312-784-5100<br>Fax: 312-907-5217 | Electronics distributor |
| Newport Corp.<br>1791 Deere Ave.<br>Irvine, CA 92714<br>Tel: 800-222-6440<br>Fax: 714-253-1680 | Pulsed valves, laser optics supports |
| Newport Optical<br>Costa Mesa, CA | Optical components |
| Nolan Supply<br>PO Box 6289<br>111-115 Leo Ave.<br>Syracuse, NY 13217<br>Tel: 800-736-2204<br>Fax: 315-463-0316 | Tools |
| Omega<br>PO Box 4047<br>One Omega Dr.<br>Stamford, CT 06907<br>Tel: 800-826-6342<br>Fax: 203-359-7700 | Thermocouples, etc. |

| | |
|---|---|
| Opthos Instruments Inc.<br>17805 Caddy Dr.<br>Rockville, MD 20855<br>Tel: 301-926-0589<br>Fax: 301-963-3060 | RF sources |
| Oriel<br>250 Long Beach Blvd.<br>PO Box 872<br>Stratford, CT 06497-0872<br>Tel: 203-377-8282<br>Fax: 203-378-2457 | Micrometer positioners |
| Ortec<br>PO Box 1397<br>Easley, SC 29640<br>Tel: 803-859-1471<br>Fax: 803-859-8580 | Electronics |
| Osaka Vacuum<br>5547 Beacon St.<br>Pittsburgh, PA 15217<br>Tel: 412-422-8100<br>Fax: 412-422-8101 | Turbo pumps |
| Park Scientific Instruments<br>1171 Borregas Ave.<br>Sunnyvale, CA 94089-1304<br>Tel: 800-776-1602<br>Fax: 408-747-1601 | STM |
| PCR<br>PO Box 14166<br>Gainesville, FL 32802<br>Tel: 904-376-8246<br>Fax: 904-371-6246 | Fluorinated compounds, silanes |
| Perkin Elmer<br>6509 Flying Cloud Dr.<br>Eden Prairie, MN 55344<br>Tel: 800-237-3603<br>Fax: 612-828-6322 | Vacuum equipment and electron spectrometers |
| Petrarch Systems Inc.<br>Bartram Rd.<br>Bristol, PA 19007<br>Tel: 215-781-9255 | Silicon compounds |

| | |
|---|---|
| Pfizer, Inc.<br>235 E. 42nd St.<br>New York, NY 10017<br>Tel: 212-573-2323<br>Fax: 212-573-1166 | Chemicals, pyrolytic graphite |
| Phoenix Equipment Co.<br>100 Fernwood Ave.<br>Rochester, NY 14621-5684<br>Tel: 716-266-5550<br>Fax: 716-266-1598 | Used equipment—scientific |
| PIC Design<br>PO Box 1004<br>Benson Rd.<br>Middlebury, CT 06762<br>Tel: 800-243-6125<br>Fax: 203-758-8271 | Gears |
| Pittsburgh Corning<br>800 Presque Isle Dr.<br>Pittsburgh, PA 15239<br>Tel: 412-327-6100<br>Fax: 412-327-5890 | Insulation/fiberglass |
| Polysciences<br>400 Valley Rd.<br>Warrington, PA 18976-2590<br>Tel: 800-523-2575<br>Fax: 800-343-3291<br>e-mail: polysci@tiggar.jvnc.com | Metals |
| Polytec PI Inc.<br>3001 Redhill Ave.<br>Costa Mesa, CA 92626<br>Tel: 714-850-1835<br>Fax: 714-850-1831 | Piezoelectric translators |
| Praxair<br>3101 Preble Ave, NE<br>Pittsburgh, PA 15233<br>Tel: 412-766-6900<br>Fax: 412-766-9532 | Specialty gases |
| Precision Laser Services Inc.<br>314 E. Wallace<br>Fort Wayne, IN 46803<br>Tel: 219-744-4375<br>Fax: 219-744-5666 | Drilling, welding, cutting |

| | |
|---|---|
| Progressive Technology Inc.<br>12122 Dry Creek Rd.<br>#1021<br>Auburn, CA 45602<br>Tel: 916-823-5047<br>Fax: 916-823-3275 | Ceramic tubing, washers, bolts, nuts, and balls |
| Pyrometer Instrument Co.<br>234 Industrial Pkwy.<br>PO Box 70<br>Northvale, NJ 07647<br>Tel: 201-768-2000<br>Fax: 201-768-2570<br>e-mail: pyroinfo@pyrometer.com | Optical pyrometers |
| Qualitech Gas Products, Inc.<br>110 Nichol Ave.<br>McKees Rocks, PA 15136<br>Tel: 412-771-7660 | Gases |
| Quantar Technology, Inc.<br>3004 Mission St.<br>Santa Cruz, CA 95060-5700<br>Tel: 408-429-5227<br>Fax: 408-429-5131 | |
| Queensgate Instruments, Inc.<br>Suite 600<br>90 Merrick Ave.<br>East Meadow, NY 11554<br>Tel: 516-357-3900<br>Fax: 516-357-9686 | Nano positioning systems |
| R.G. Hansen & Associates<br>631 Chapala St.<br>Santa Barbara, CA 93101<br>Tel: 805-564-3388<br>Fax: 805-963-0733 | Cryogenic equipment |
| R.M. Jordan Co.<br>990 Golden Gate Terrace<br>Grass Valley, CA 95945<br>Tel: 916-272-4580<br>Fax: 916-272-2955 | Reconditioned quadrupole mass spectrometers, etc. |
| Rama Corporation<br>600 W. Esplanade Ave.<br>San Jacinto, CA 92583<br>Tel: 909-654-7351<br>Fax: 909-654-3748 | Heaters—bakeout |

| | |
|---|---|
| Raychem<br>300 Constitution Dr.<br>Menlo Park, CA 94025<br>Tel: 415-361-3860 | Electronic equipment and supplies, SME alloys |
| RHK Technology<br>209 Voorheis Rd.<br>Pontiac, MI 48053<br>Tel: 313-858-8958 | High-resolution electron energy loss spectrometer |
| S&J Trading, Inc.<br>PO Box 337<br>Glenoaks, NY 11004 | Mica |
| SAES Gettters, Inc.<br>1122 East Cheyenne Mountain Blvd.<br>Colorado Springs, CO 80906<br>Tel: 719-576-3200<br>Fax: 719-576-5025 | Alkali getters |
| Sargeant-Welch<br>PO Box 5229<br>Buffalo Grove, IL 60089-5229<br>Tel: 800-727-4368<br>Fax: 800-676-2540<br>e-mail: sarwel@sargentwelch.com | Vacuum components |
| Saville & Co.<br>414 Rodi Rd.<br>PO Box 27286<br>Pittsburgh, PA 15235<br>Tel: 412-731-6401<br>Fax: 412-731-4098 | Shim stock |
| Scientific Gas Products<br>2330 Hamilton Blvd.<br>S. Plainfield, NJ 07080-9986<br>Tel: 201-754-7700 | Gases |
| Scientific Instrumentation Services, Inc.<br>1027 Old York Rd.<br>Ringoes, NJ 08551-1039<br>Tel: 908-788-5550<br>Fax: 908-806-6631 | |
| Scientific Instruments<br>Rd. 3, Box 593<br>Ringoes, NJ 08551-9675<br>Tel: 908-788-5550 | Cryogenics, closed cycle refrigerators, low temperature thermocouples (Cr/Fe:Au) |

| | |
|---|---|
| Scott Specialty Gases<br>6141 Easton Rd.<br>PO Box 310<br>Plumsteadville, PA 18949-0310<br>Tel: 800-220-4436<br>Fax: 215-766-0320 | Gases |
| Sensote, Inc.<br>1202 Chesapeake Ave.<br>Columbus, OH 43212<br>Tel: 800-848-6564 x880<br>Fax: 614-486-0506 | |
| Setra<br>45 Nagog Park<br>Acton, MA 01720<br>Tel: 508-263-1400 or 800-257-3872<br>Fax: 508-264-0292 | Pressure gauges, digital capacitance manometers |
| Shipley Company, Inc.<br>2300 Washington St.<br>Newton, MA 02162<br>Tel: 617-969-5500 | Photo resist coating materials |
| Sigmund Cohn Corp.<br>121 S. Columbus Ave.<br>Mt. Vernon, NY 10553<br>Tel: 914-664-5300<br>Fax: 914-664-5377 | Very fine wire, precision fabricated metals |
| Small Parts<br>13980 NW 58th Ct.<br>Box 4650<br>Miami Lakes, FL 33014-0650<br>Tel: 800-220-4242<br>Fax: 800-423-9009 | Parts, small nuts, bolts, gears, washers, miscellaneous parts—stainless steel PO mechanical components |
| South Bay Technology<br>1120 Via Callejon<br>San Clemente, CA 92672<br>Tel: 714-492-2600<br>Fax: 714-492-1499 | Crystal cutting equipment |
| Space Environment Laboratories<br>Box 1061<br>Boulder, CO 80306<br>Tel: 303-443-4090 | Vacseal leak sealer |

| | |
|---|---|
| Specialty Minerals, Inc.<br>640 N. 13th St.<br>Easton, PA 18042<br>Tel: 610-250-3321<br>Fax: 610-250-3325 | Pyrolytic graphite |
| Spectra Tech<br>2 Res. Dr.<br>PO Box 869<br>Shelton, CT 06484-0869<br>Tel: 800-243-9186<br>Fax: 203-926-8909 | Ion sources, heaters |
| Spectra-Mat, Inc.<br>100 Westgate Dr.<br>Watsonville, CA 95076<br>Tel: 408-722-4116<br>Fax: 408-722-4172 | Ion sources |
| Starrett<br>29 McKean St.<br>Pittsburgh, PA 15219<br>Tel: 412-431-8700<br>Fax: 412-431-5150 | Tool chests, precision tools |
| Stavac Associates<br>114 Black Thorn Dr.<br>Butler, PA 16001<br>Tel: 412-586-2100<br>Fax: 412-586-5927 | Ion sources |
| Strem<br>7 Mulken Way<br>Dexter Industrial Park<br>Newburyport, MA 01950-4098<br>Tel: 508-462-3191<br>Fax: 508-465-3104 | Chemicals, custom synthesis |
| Sumitomo Electric<br>551 Madison Ave.<br>New York, NY 10022<br>Tel: 212-308-6444 | GaAs |
| Surface Interface, Inc.<br>110 Pioneer Way Suite<br>Mountain View, CA 94041<br>Tel: 415-965-8205<br>Fax: 415-965-8207 | Manipulators, UHV banana plugs |

| | |
|---|---|
| Swagelok<br>31400 Aurora Rd.<br>Solon, OH 44139<br>Tel: 216-439-5934<br>Fax: 216-349-5843 | Vacuum fittings |
| Swiss Jewel Co.<br>613 Lafayette Bldg.<br>Philadelphia, PA 19106<br>Tel: 215-925-2867<br>Fax: 215-922-3055 | Sapphire-fabricated bearings, etc. |
| T.E. Brown & Co.<br>Button End<br>Harston<br>Cambridge England<br>Tel: 0223-872072 | Spark cutter |
| Technical Devices, Inc.<br>5721 Prescott Rd.<br>Prescott, AZ 86301<br>Tel: 602-772-7500 | Wire bonding tool |
| Technitool<br>5 Apollo Rd.<br>Box 368<br>Plymouth Meeting, PA 19462<br>Tel: 610-941-2400<br>Fax: 610-828-5623 | Tools, specialized |
| Terranova<br>5841 Bell Rd.<br>Auburn, CA 95602<br>Tel: 800-519-2634<br>Fax: 916-269-2877 | Pressure measurement, vacuum valves |
| Test Lab Co.<br>1066 Linda Vista Ave.<br>Mountain View, CA 94043<br>Tel: 800-42-5835 | Reconditioned test equipment |
| Thermionics<br>22815 Sutro St.<br>PO Box 3711<br>Hayward, CA 94540<br>Tel: 510-538-3304<br>Fax: 510-538-2889 | UHV apparatus and accessories |

| | |
|---|---|
| Thorlabs Inc.<br>435 Rte. 206<br>Newton, NJ 07860-0366<br>Tel: 201-579-7227<br>Fax: 201-383-8406 | Optical components |
| Time Motion Tools<br>12778 Brookprinter Pl.<br>Poway, CA 92064<br>Tel: 619-679-0303 | Tools |
| Topac, Inc.<br>99 Derby St.<br>Suite 303<br>Hingham, MA 02043<br>Tel: 617-740-8778<br>Fax: 617-740-8779<br>e-mail: topac@wing.net | Channel multipliers |
| Tripp Lite<br>500 N. Orleans St.<br>Chicago, IL 60610<br>Tel: 312-329-1777<br>Fax: 312-644-6505 | Electronics |
| Tube Sales<br>Interstate Industrial Park<br>Bellmaur, NJ 08031<br>Tel: 215-923-5176 | Stainless steel tubing |
| Tucker Electronic<br>PO Box 551419<br>Dallas, TX 75355-1419<br>Tel: 800-527-4642<br>Fax: 214-348-0367<br>e-mail: sales@tucker.com | Electronics |
| Tylan General<br>1110 Shepard Dr.<br>Blue Bell, PA 19422<br>Tel: 610-272-1933<br>Fax: 603-894-4909 | Vacuum components |
| U-C Components, Inc.<br>410 Logue Ave.<br>Mountain View, CA 94043-4019<br>Tel: 415-964-3827<br>Fax: 415-964-0216 | Vented screws |

| | |
|---|---|
| Uniform Tubes, Inc.<br>7th Ave.<br>Collegeville, PA 19426<br>Tel: 215-539-0700 | Metal tubing, specialty |
| Unimetrics Corp.<br>501 Earl Rd.<br>Shorewood, IL 60436<br>Tel: 800-854-6931<br>Fax: 815-741-0799 | Micro syringes and accessories |
| Unique Wire Weaving Co., Inc.<br>762 Ramsey Ave.<br>Hillside, NJ 07205-1094<br>Tel: 800-445-6912<br>Fax: 908-688-4601 | Stainless mesh |
| United Chemical Technologies, Inc.<br>2731 Bartram Rd.<br>Bristol, PA 19007<br>Tel: 215-781-9255<br>Fax: 215-785-1226 | Silanes and silicones |
| United Detector<br>12515 Chadron Ave.<br>Hawthorne, CA 90250<br>Tel: 310-978-1150<br>Fax: 310-644-1727 | Electronic equipment and supplies |
| UTI<br>1313 Fifth St. SE<br>Suite 130<br>Minneapolis, MN 55414<br>Tel: 612-378-1393<br>Fax: 612-378-2287 | Mass spectrometers |
| Vacoa<br>10-7 Technology Dr.<br>E. Setauket, NY 11733<br>Tel: 516-751-7115<br>Fax: 516-751-7154 | Vacuum valves and accessories |
| Vacuum General<br>9577 Chesapeake Dr.<br>San Diego, CA 92123<br>Tel: 619-571-1222<br>Fax: 619-576-1703 | Capacitance gauges |

| | |
|---|---|
| Vacuum Science Workshop, Ltd.<br>5045 N. 30th St.<br>Colorado Springs, CO 80919<br>Tel: 719-531-7711<br>Fax: 719-531-7714 | Electron spectrometers |
| Varflex<br>512 W. Court St.<br>Rome, NY 13440<br>Tel: 315-336-4400<br>Fax: 315-336-0005 | Glass braid |
| Varian<br>121 Hartwell Ave.<br>Lexington, MA 02173-3133<br>Tel: 1-800-882-7426<br>Fax: 617-860-5437 | Turbopumps, vacuum products, feedthroughs |
| Vat<br>600 West Cummings Park<br>Woburn, MA 01801<br>Tel: 617-935-1446<br>Fax: 617-935-3940 | Gate valves |
| Velmex, Inc.<br>7550 State Routes 5 & 20<br>Bloomfield, NY 14469-9389<br>Tel: 800-642-6446<br>Fax: 716-657-6153 | Unislide assemblies |
| Victor Equipment Co.<br>101 S. Hanley Rd.<br>St. Louis, MO 63105<br>Tel: 314-721-5573<br>Fax: 314-721-4822 | Gas apparatus |
| Virginia Semiconductor, Inc.<br>1501 Powhatan St.<br>Fredericksburg, VA 22401<br>Tel: 540-373-2900<br>Fax: 540-371-0371 | Silicon membranes, silicon crystals |
| Voltaix<br>PO Box 5357<br>197 Meister Ave.<br>North Branch, NJ 08876<br>Tel: 908-231-9060<br>Fax: 908-231-9063 | Gases-semiconductor process |

| | |
|---|---|
| W.R. Grace<br>7221 W. Parkland Court<br>Milwaukee, WI 53223<br>Tel: 414-354-4400<br>Fax: 414-354-5088 | Silica, powdered; silica gel |
| Watlow<br>2987 Babcock Blvd.<br>Pittsburgh, PA 15237<br>Tel: 412-311-5004<br>Fax: 412-322-1322 | Heating elements—bakeout |
| Wavemat<br>44191 Plymouth Oaks Blvd.<br>Suite 100<br>Plymounth, MI 48170 | Microwave plasma sources |
| Wiley<br>1245 South Sixth St.<br>Coshocton, OH 43812<br>Tel: 614-622-0755<br>Fax: 614-622-3231 | Organic chemicals |
| Wilmad Glass Co.<br>Rt. 40 & Oak Rd.<br>Buena, NJ 08310<br>Tel: 800-297-8165<br>Fax: 800-297-8174 | Glass tubing-capillary |
| Winfred M. Berg Inc.<br>499 Ocean Ave.<br>East Rockaway, NY 11518<br>Tel: 516-599-5010 | Precision mechanical components |
| YSI<br>Yellow Springs Instrument Co,. Inc.<br>1725 Branum Ln.<br>Yellow Springs, OH 45387<br>Tel: 513-767-7241<br>Fax: 513-767-9353 | Temperature measurement |

## German and European Manufacturers

*All addresses with D postal codes are located in Germany*

| Company | Products |
|---|---|
| Advent Research Materials Ltd<br>Blyth Road Industrial Estate<br>Halesworth, Suffolk<br>England IP 19 8DD<br>Tel: 44-1986 874555<br>Fax: 44-1986-874557 | Pure materials, metals, foils, wires, powders |
| Alcatel Hochvakuumtechnik GmbH<br>Am Kreuzeck 10<br>D-97877 Wertheim<br>Tel: 09342-8700<br>Fax: 09342-87030 | Vacuum components, quadrupole mass spectrometer, pumps |
| Alfa Produkte<br>Zeppelinstr. 7<br>D-76185 Karlsruhe<br>Tel: 0721-853061<br>Fax: 0721-849674 | Pure metals, metal products, noble metal technology, single crystals, wires, foils, powder |
| Araldit AtI<br>Ciba-Geigy GmbH<br>Oeflinger Str.<br>D-79664 Wehr (Boswn) Germany | |
| Aurel GmbH<br>Malteser Str. 443C<br>D-86899 Landsberg | Si-crystals |
| Balzers Hochvakuum GmbH<br>Otto-von-Guericke-Ring 7<br>D-65205 Wiesbaden<br>Tel: 06122-706-0<br>Fax: 06122-70610 | Vacuum components |
| Bender & Hobein GmbH<br>Fraunhofer Str. 7<br>D-85737 Ismaning<br>Tel: 089-9965480<br>Fax: 089-9965490 | "Surfrasil" silconizing agent, glass coating |
| Berger Lahr GmbH<br>Breslauer Str. 7<br>D-77933 Lahr<br>Tel: 07821-5820<br>Fax: 07821-582313 | Synchronous and step motors of all sizes, positioning systems |

| | |
|---|---|
| or<br>Zuständiger Kundendienst<br>SIG Positec GmbH<br>Büro Mitte<br>Maybachstr. 14<br>D-63456 Hanau<br>Tel: 06181-66580<br>Fax: 06181-665818 | |
| Best<br>Ventil & Fitting GmbH<br>Robert-Bosch-Str. 20<br>D-63477 Maintal-Dörnigheim<br>Tel: 06181-46041<br>Fax: 06181-432161 | Swagelok, Whitey, Cajon, Nupro, Sno-trik, valves, glass-metal connections |
| Busak & Shamban GmbH & Co.<br>Handwerkstr. 5-7<br>D-70565 Stuttgart 80<br>Tel: 0711 7864-0<br>Fax: 0711-7803171 | Gaskets |
| Caburn-MDC<br>Ostendstr. 1<br>D-12459 Berlin<br>Tel: 030-6953-9840<br>Fax: 030-635-3786 | UHV components, feedthroughs, coaxial cables |
| Caddock Europe<br>Leopoldstr. 28a/2<br>D-80802 München<br>Tel: 089-390799<br>Fax: 089-391230<br>or<br>Caddock Electronics Inc.<br>1717 Chicago Ave.<br>Riverside, CA 92507-2364<br>Tel: 001-714-788-1700<br>Fax: 001-714-369-1151 | High precision resistors, high voltage resistors up to 10000 MOhm |
| Carl Roth GmbH+Co<br>Postfach 21 11 62<br>D-76161 Karlsruhe<br>Tel: 0721-56060<br>Fax: 0721-560649 | Chemicals, laboratory equipment |

| | |
|---|---|
| CETEC Spezialarmaturen<br>Ernst H.F. Hiller<br>Vahrer Str. 248 d<br>D-28329 Bremen<br>Tel: 0421-461855<br>Fax: 0421-462385 | UHV components |
| CKL Vacuumtechnik Vertriebs GmbH<br>Unterfeldring 17<br>D-85256 Vierkirchen<br>Tel: 08139-890<br>Fax: 08139-8943 | UHV components, pumps, sputter guns |
| CompoTherm<br>Oberflächentechnik GmbH<br>Siemenstr. 16-18<br>D-28857 Syke<br>Tel: 04242-593-12<br>Fax: 04242-593-13 | High-speed pyrometers |
| Conatex<br>Postfach 1220<br>D-66592 St. Wendel/Saar<br>Tel: 06851-2071<br>Fax: 06851-2076 | Technical equipment for scientists: thermometers, thermocouples |
| Cryo Vac<br>Dr. Johannes Klug, Handelsvertretung<br>Hasbacher Str. 1a<br>D-51503 Rösrath<br>Tel: 02205-85454<br>Fax: 02205-5797 | Cryo technology, cryo components and materials, insulation, low temperature glue |
| Cryophysics<br>Landwehrstr. 48<br>D-64293 Darmstadt<br>Tel: 06151-86281<br>Fax: 06151-84481 | Modulators, crystals, harmonic generating, polarizers, Q-Switches, lasermetrics, LN2 cryostats, super insulation |
| Cryoson GmbH<br>Postfach 1104<br>D-63821 Schöllkrippen<br>Tel: 06024-1233<br>Fax: 06024-7879 | Steel dewars, LN2-containers |
| Deutsche Air Liquide—Kryotechnik<br>König-Adolf-Str. 3<br>D-65191 Wiesbaden<br>Tel: 0611-543048<br>Fax: 0611-547264 | LN2 dewars, cryo technology, LHe containers |

| | |
|---|---|
| Dr. Sjuts Optotechnik<br>Max-Planck-Str. 1<br>D-37191 Katlenburg-Lindau<br>Tel: 05556-1011<br>Fax: 05556-4587 | Electron multipliers, channeltrons |
| EA-Elektro-Automatik<br>Helmholtzstr. 33-35<br>D-31747 Viersen<br>Tel: 02162-3785-0<br>Fax: 02162-16230 | High current power supplies |
| EBARA Germany GmbH<br>Donaustr. 7<br>D-63452 Hanau<br>Tel: 06181-1876-0<br>Fax: 06181-1876-40 | Turbo and cryo pumps |
| EMCO Elektronik GmbH<br>Einsteinstr. 35<br>D-82152 Martinsried<br>Fax: 089-85-97785 | RF power amplifiers, micro wave components |
| EMV Elektronische Meßgeräte<br>Vertriebs GmbH<br>Wallbergstr. 7<br>D-82024 Taufkirchen<br>Tel: 089-6128054<br>Fax: 089-6128593 | Electronic measuring instruments |
| ENI Germany<br>A Division of Astec Germany GmbH<br>Holderäckerstr. 15<br>D-70839 Gerlingen<br>Tel: 07156-21095<br>Fax: 07156-49372 | RF amplifiers, power amplifiers |
| Epoxy Produkte GmbH<br>Guntherstr. 31<br>Postfach 1207<br>D-64658 Fürth/Odenwald<br>Tel: 06253-4546<br>Fax: 06253-21687 | Solvent free epoxy products conducting cement (for $LiTaO_3$ contacts) |
| Farnell Advance Electronic GmbH<br>Dieselstr. 21<br>D-63533 Mainhausen<br>Tel: 06182-21091-93<br>Fax: 06182-28819 | Power supplies, measuring instruments |

| | |
|---|---|
| FAST ComTec GmbH<br>Grünwalder Weg 28a<br>D-82041 Oberhaching<br>Tel: 089-613-1081<br>Fax: 089-613-6171 | Transient recorder, PC-boards, NIM boards, multichannel scaler, discriminators, timer-counters |
| FHR Anlagenbau, GmbH<br>Bergener Ring 41<br>D-01458 Ottendorf-Okrilla/Dresden<br>Tel: 035205-520-0<br>Fax: 035205-520-40 | UHV, HV, sputter instruments and supplies, pure gases, valves |
| FISONS Instruments<br>Vertriebs-GmbH<br>Peter-Sander-Str. 43<br>D-55252 Mainz-Kastel<br>Fax: 06134-289904 | UHV components, surface science analysis instruments |
| FOM<br>Kruislan 407<br>1098 SJ Amsterdam<br>The Netherlands<br>Tel: 31-20-6081301<br>Fax: 31-20-6684106 | Single crystal preparation |
| Friatec<br>(ehemals Friedrichsfeld GmbH)<br>Postfach: 710261<br>D-68222 Mannheim<br>Tel: 0621-4860<br>Fax: 0621-477999 | Ceramic parts, UHV feedthroughs |
| Fritz Eichenauer GmbH<br>Industriestr. 1<br>D-76770 Hatzenbühl<br>Tel: 07275-7020<br>Fax: 07275-702200 | Heating elements |
| FuG Elektronik GmbH<br>Florianstr. 2<br>D-83024 Rosenheim<br>Tel: 08031-81093<br>Fax: 08031-81099 | High-precision power supplies, high-voltage power supplies and resistors |
| General Scanning GmbH<br>Münchener Str. 2<br>D-82152 Planegg<br>Tel: 089-8595031<br>Fax: 089-8599487 | Galvo drives, scanners |

| | |
|---|---|
| Goodfellow GmbH<br>Postfach 5427<br>D-65729 Eschborn<br>Tel: 06196-703441<br>Fax: 06196-482954 | Metals in all forms: foils, crystals, tubes, powder |
| Gossen GmbH Mess- und Regeltechnik<br>Postfach 1780<br>D-91007 Erlangen<br>Tel: 09131-8271<br>Fax: 09131-28895 | General measuring instruments, controlling instruments |
| Gräff<br>Bonner Str. 54<br>D-53842 Troisdorf<br>Tel: 02241-41031<br>Fax: 02241-404002 | High-precision thermometers, thermocouple instruments |
| Hahn & Bischof<br>Donarweg 3<br>D-81739 München<br>Tel: 089-601-4409<br>Fax: 089-601-2748 | UHV-, MBE-instruments, analysis instruments, nano crystal fabrication |
| Hameg Instruments<br>Keisterbacher Str. 17<br>D-60528 Frankfurt<br>Tel: 069-67805-0<br>Fax: 069-67805-13 | Oscilloscopes, signal generators, multimeters |
| Hela GmbH<br>Postfach 411011<br>D-12120 Berlin<br>Tel: 030-7927080 | Electrical wires, polymide cables, kapton wire |
| Heraeus Sensor GmbH<br>Im Stift 6-8<br>D-58119 Hagen<br>Tel: 02334-8040<br>Fax: 02334-80432 | Temperature measuring program, thermocouples, high-precision resistors, temperature probes |
| Heraeus Wittmann GmbH<br>Englerstr. 11<br>D-69126 Heidelberg<br>Tel: 06221-30430 | Silicon-heating elements, heating ribbons |

| | |
|---|---|
| Hiden Analytical<br>Vertretung für Deutschland<br>VACUA GmbH<br>Niedmannweg 13<br>D-82431 Ried/Kochel<br>Tel: 088-57693010<br>Fax: 088-579771 | Quadrupole mass spectrometer, ion sources for surface analysis |
| HMS-Electronik Hans M Strassner GmbH<br>Am Arenzberg 42<br>D-51381 Leverkusen<br>Tel: 02171-3814<br>Fax: 02171-33852 | Extremely low-noise pre-amplifiers, lock-in amplifiers |
| Hoke GmbH<br>Hanauer Landstr. 11-13<br>D-60134 Frankfurt<br>Tel: 069-4050090<br>Fax: 069-4050094 | Pure gases and gas handling supply |
| Hositrad/Holland<br>P.O.B. 114<br>3870 CC Hoevelaken<br>Teh Netherlands<br>Tel: 03495-37210<br>Fax: 03495-35274 | CVD, UHV parts, manipulators, loadlocks, Ti Pumps. Repair of ion pumps (Varian and Leybold) |
| Hositrad/Deutschland<br>Lindnergasse 2<br>D-93074 Regensburg<br>Tel: 0941-55827<br>Fax: 0941-562296 | |
| HP Hewlett-Packard GmbH<br>Vertriebszentrum Bad Homburg<br>Hewlett-Packard-Str.<br>D-61352 Bad Homburg v.d.H.<br>Tel: 06172-16-0<br>Fax: 06172-16-1309 | Synthesizer, function generators, oscilloscopes, RF components, multimeters, power supplies |
| HP Direct<br>Dornierstr. 7<br>D-71034 Böblingen<br>Tel: 07031-140<br>Fax: 07031-142100 | |

Ina Wälzlager Schaeffler KG
Industriestr. 1-3
D-91074 Herzogenaurach
Tel: 09132-820
Fax: 09242-5349

Needle bearings, linear bearings, linear motions, planetary gearings

Intervac Henning GmbH
Gemeindedingerstr. 7
D-97892 Kreuzwertheim
Tel: 09342-22417
Fax: 09342-7058

Vacuum components supply

Isopad GmbH
Engelerstr. 11
D-69126 Heidelberg
Tel: 06221-30430
Fax: 06221-304343

Heating supplies: heating ribbons, collars, tubes, etc.

Joachim Schwarz
Hindenburgstr. 12
D-76332 Bad Herrenalb
Tel: 07083-51833
Fax: 07083-51925

UHV components

Joisten & Kettenbaum
Asselborner Weg 14-16
D-51429 Bergisch Gladbach
Tel: 02204-839-0
Fax: 02204-839-60

Tools for surface treatment, polishing supply, diamond paste

JUST Vakuum Technik
Lehnstr. 34
D-66869 Kusel
Tel: 06381-6060
Fax: 06381-6040

Vacuum components, pumps, quadrupole mass spectrometers

Karlsruher Glastechnisches Werk
Schieder GmbH
Spezialfabrik für Isoliergefäße
Gablonzer Str. 6
D-76185 Karlsruhe
Tel: 0721-95897-0
Fax: 0721-95897-77

Glass instruments, cryostats for $LN_2$ and LHe. Specialize in insulation dewars

Keithley Instruments GmbH
Landsberger Str. 65
D-82110 Germering
Tel: 089-8493070
Fax: 089-84930759

High precision measuring instruments

Klaus Schaefer GmbH
Postfach 1660
D-63206 Langen
Tel: 06103-79085
Fax: 06103-71799

Marketing of products: Crowcon, Extrel, General Vacuum, Walker Scientific, Ceramaseal, Cryomagnetics, Leak Sealant, IMAG

KMF Laborchemie GmbH
Postfach 900129
D-51111 Köln
Tel: 02241-96850
Fax: 02241-67353

"Surfrasil" siliconizing agent, glass coating

Knuerr-Heinzinger elecronic GmbH
Anto-Jakob-Str. 4
D-83026 Rosenheim
Tel: 08031-44040
Fax: 08031-440444

High-precision power supplies. High voltage and high-current power supplies

Kristallhandel Kelpin
Im Schilling 18
D-69181 Leimen
Tel: 06224-72558
Fax: 06224-77189

Metal single crystals, electro-optical and acusto-optical crystals

Kurt J. Lesker Company
Karl-Heinz-Beckurts-Str. 13
D-52428 Jülich
Tel: 0130-812843
Fax: 0130-812842

Vacuum components

LeCroy GmbH
Mannheimer Str. 177
D-69123 Heidelberg
Tel: 06221-831001
Fax: 06221-834655

Digital oscilloscopes, CAMAC modules, transient recorders

Leybold Vakuum GmbH
Bonner Str. 498
D-50968 Köln
Tel: 0221-347-0
Fax: 0221-347-1250
or
Postfach 510760
D-50943 Köln

Ultra high vacuum parts and analysis instruments

Linde AG
Kostheimer Landstrae 25
D-55246 Mainz-Kostheim
Tel: 06134-208-31
Fax: 06134-208-25

Gases and gas handling supplies

| | |
|---|---|
| M.K. Juchheim GmbH<br>Moltkestr. 13-31<br>D-36039 Fulda<br>Tel: 0661-60030<br>Fax: 0661-6003-500 | Thermometer, process controlling, thermocouples, time switches |
| MaTecK Material-Technologie & Kristalle<br>Dr. H. Schlich<br>Karl-Heinz-Beckurts-Str. 13<br>D-52428 Jülich<br>Tel: 02461-690740<br>Fax: 02461-690749 | Single crystals, wafers, pure materials and alloys, high-precision polishing service |
| Mawi-therm Temperatur-Prozeβtechnik<br>Hofstr. 23<br>D-40789 Monheim<br>Tel: 02173-51094<br>Fax: 02173-58623 | Thermometer, temperature controller infrared thermometer, pyrometer, thermoelements |
| Melja Diagnostik GmbH<br>Lilienthalstr. 7-25<br>D-34123 Kassel<br>Tel: 0561-57-15-77<br>Fax: 0561-57-16-36 | "Surfrasil" siliconizing agent, glass coating |
| Menz & Könecke GmbH<br>D-41379 Brüggen<br>Tel: 02163-5940<br>Fax: 02163-5210 | Electrical heating supplies for baking out |
| Merck<br>Frankfurterstr. 250<br>D-64271 Darmstadt<br>Tel: 06151-723368<br>Fax: 06151-722000 | Chemicals and metals. Also inexpensive vacuum parts and lubricants of different companies: Alcatel., Busch, Kinney, Tietschle, CTI, Edwards, Leybold, Welch, Sargent, Stokes, Balzers, etc. Precision Plus Vacuum Parts, Inc. |
| Messer Griesheim GmbH<br>Postfach 210553<br>In der Steinwiese 5<br>D-57074 Siegen<br>Tel; 0271-69040<br>Fax: 0271-6904160 | Gases and gas handling supply, liquid nitrogen and helium |
| Metal Crystals & Oxides LTD<br>Unit B5 Button End Industrial Estate<br>Harston<br>Cambridge<br>England CB2 5NX<br>Tel: 1223-872072<br>Fax: 1223-872517 | Single crystals, special materials |

| | |
|---|---|
| Metrofunk Kabel-Union GmbH<br>Postfach 410109<br>D-12111 Berlin<br>Tel: 030-8314052<br>Fax: 030-7925343 | Electrical wires, polymide cables, kapton wire |
| MKS Instruments Deutschland GmbH<br>Schatzbogen 43<br>D-81829 München<br>Tel: 089-420008-0<br>Fax: 089-424106 | Vacuum components, gas analyzer, capacitance manometer, gas flow controllers |
| Moorfield Associaes<br>Moorfield House<br>Plumley Moor Rd.<br>Plumley<br>Cheshier WA 16 9RS<br>UK<br>Tel: 0565-722609 | VSW instruments, HREELS IB200 spectrometer. Used UHV analysis instruments |
| Niederlassung Hanau<br>Wilhelm-Rohn-Str. 25<br>D-63405 Hanau<br>Tel: 06181-34-1202<br>Fax: 06181-34-1212 | |
| neoLAB<br>Rischerstr. 7<br>D-69123 Heidelberg<br>Tel: 06221-844219<br>Fax: 06221-844233 | General laboratory equipment |
| Normag<br>Labor- und Verfahrenstechnik GmbH<br>Postfach 1269<br>Feldstr. 1<br>D-65719 Hofheim a. Ts.<br>Tel: 06192-2099-0<br>Fax: 06192-209939 | Special glass instruments, glass grinding spindle valves |
| Omicron Vakuumphysik GmbH<br>Idsteiner Str. 78<br>D-62232 Taunusstein<br>Tel: 06128-73044<br>Fax: 06128-73596 | Rear view LEED, STM, AFM, UPS He-source, evaporation sources |
| Oxford Instruments<br>Kreuzberger Ring 36<br>D-65205 Wiesbaden<br>Tel: 0611-7640<br>Fax: 0611-764175 | High precision superconducting magnets, cryotechnology, low temperature STM |

| | |
|---|---|
| Paskovsky Meßelektronik GmbH<br>Postfach 1259<br>D-63502 Langenselbold<br>Tel: 06184-3042<br>Fax: 06184-62610 | Amplifier and linearization modules for thermocouples |
| Perkin Elmer<br>Bruckmannring 40<br>D-85764 Oberschleissheim<br>Tel: 089-315-717-0<br>Fax: 089-315-3117 | UHV surface analysis equipment: XPS, AES, ESCA, UPS, ISS, SAM, SIMS, LEED, HEED, electron, photon, and ion sources, hemispherical or cylindrical analyzers, manipulators |
| Phytron-Elektronik GmbH<br>Industriestr. 14<br>D-82194 Gröbenzell<br>Tel: 08142-503-0<br>Fax: 08142-503-190 | Step motors, special clutches |
| Piezosystem Jena<br>Wildenbruchstr. 15<br>D-07745 Jena<br>Tel: 03641-675-390<br>Fax: 03641-675-393 | Piezo, high-precision positioning systems |
| Pink Vakuumtechnik<br>Am Kessler 6<br>Postfach 1509<br>D-97877 Wertheim<br>Tel: 09342-872-0<br>Fax: 09342-872-20 | Vacuum systems, special manufacturing |
| Polytec GmbH<br>Polytec Platz 5-7<br>D-76337 Waldbronn<br>Tel: 07243-6040<br>Fax: 07243-69944 | Inorganic high-temperature cements |
| Rohde und Schwarz<br>Vertriebs-GmbH<br>Mühldofst.r 15<br>D-81671 München<br>Tel: 089-41290<br>Fax: 089-41293208 | Synthesizer, function generators, signal generators, RF components, oscilloscopes |
| Rolf Müller<br>Juethorn Str. 72<br>D-22043 Hamburg<br>Tel: 040-652-65602<br>Fax: 040-570-3000 | Thermo electrical components, thermocouples, miniaturized welding tools |

|  |  |
|---|---|
| Rosenkranz Elektronik<br>Gro-Gerauer Weg 55<br>D-64295 Darmstadt<br>Tel: 06151-3998-0<br>Fax: 06151-399818 | Used electronic instruments, oscilloscopes, function generators multimeters |
| SAES GETTERS<br>Postfach 420552<br>Gerolsteiner Str. 1<br>D-50899 Köln<br>Tel: 021-443086<br>Fax: 0221-441095 | Alkali-metal dispensers, NEG pumps, getter material |
| SASKIA Hochvakuum—und Labortechnik<br>Am Vogelherd 3<br>D-98693 Ilmenau<br>Tel: 03677-6040<br>Fax: 03677-604110 | Vacuum systems, pumps, gauges |
| Scandivac<br>Nordic Vacuum Engineering AB<br>Industrivägen 16<br>S-59041 Rimforsa<br>Sweden<br>Tel: 46-494-20530<br>Fax: 46-494-21197 | Thermionics vacuum parts, UHV-rotation tables for manipulators, electron guns, pumps, valves |
| Schlumberger Technologie GmbH<br>Ingolstädter Str. 62a<br>D-80939-München<br>Tel: 089-318890<br>Fax: 089-35908 | CAMAC and NIM modules, crates, Si-detectors, digital voltmeters |
| Schulz-Electronic GmbH<br>Postfach 11 01 18<br>D-76486 Baden-Baden<br>Tel: 07223-58054<br>Fax: 07223-58093 | Power supplies |
| SI Spectroscopy Instruments GmbH<br>Rudolf-Diesel-Str. 7a<br>Postfach 1326<br>D-82239 Gilching<br>Tel: 08105-5011<br>Fax: 08105-5577 | SRS Stanford Research Systems: lock-in amplifier, signal generators, timer counter, boxcar averager, chopper |
| Sigma-Aldrich Chemie GmbH<br>Postfach<br>D-82039 Steinheim<br>Tel: 0130-5155<br>Fax: 0130-6490 | Chemical materials and equipment, glass instruments |

| | |
|---|---|
| Specs GmbH<br>Volastr. 5<br>D-13355 Berlin<br>Tel: 030-4633031<br>Fax: 030-4642083 | Surface science analysis instruments, electron, photon and ion sources |
| Struck<br>Bäckerbarg 6<br>D-22889 Tangsted/Hamburg<br>Tel: 04109-55-0<br>Fax: 04109-5533 | Transient recorders, CAMAC modules |
| Struers GmbH<br>Albert Einstein Str. 5<br>D-40699 Erkrath<br>Tel: 0211-20030<br>Fax: 0211-200399 | Metallographic products, grinding and polishing machines, polishing supplies, diamond paste |
| tectra GmbH<br>Reuterweg 65<br>D-60323 Frankfurt/M<br>Tel: 069-720040<br>Fax: 069-720400 | Vacuum components, all kinds of surface science equipment and analysis instruments |
| Tektronix GmbH<br>Sedanstr. 13-17<br>D-50668 Köln<br>Tel: 0221-72220<br>Fax: 0221-722362 | Digital and analog oscilloscopes, spectrum analyzers |
| Telemeter Electronic<br>Posthof 4<br>D-86609 Donauwörth<br>Tel: 0906-706-930<br>Fax: 0906-21706 | RF amplifier, signal generators, microwave components |
| Thermal Quarz-Schmeize GmbH<br>Vor dem Haßel 8<br>D-21438 Brackel<br>Tel: 04185-4035<br>Fax: 04185-5309 | Various products made of glass |
| VAB Vakuum Anlagen Bau GmbH<br>Sandhöhe 11<br>D-25337 Elmshorn<br>Tel: 04121-73095 | UHV systems and components, Ti sublimation pumps, valves, manipulators |
| Vacom<br>Leutragraben 8a<br>D-07743 Jena<br>Tel: 03641-443994<br>Fax: 03641-443993 | Vacuum components, pumps, valves, gauges, feed throughs |

| | |
|---|---|
| Vakuumschmelze GmbH<br>Grüner Weg 37<br>Postfach 2253<br>D-63456 Hanau<br>Tel: 06181-380<br>Fax: 06181-382645 | Permanent magnets |
| Varian GmbH - Vacuum Products<br>Kuehnstr. 71d<br>D-22045 Hamburg<br>Tel: 040-6696031<br>Fax: 040-6682282 | Vacuum components, electron guns, surface science analysis instruments |
| VAT<br>Bretonischer Ring 7<br>D-85630 Grasbrunn-Neukeferloh<br>Tel: 089-465015<br>Fax: 089-463765 | Valves, bellows, vacuum parts |
| Victoreen Inc.<br>6000 Cochran Rd.<br>Cleveland, OH 44139<br>Tel: 001-216-248-9300<br>Fax: 001-216-248-9301<br>or<br>Compotron<br>Geretsrieder Str. 4<br>D-84478 Waldkraiburg<br>Tel: 08638-3051<br>Fax: 08638-2781 | High-precision, high-voltage resistors |
| VSE Vakuumtechnik GmbH<br>Rottmosserstr. 16<br>D-83543 Rott a. Inn<br>Tel: 08039-4454<br>Fax: 08039-4457 | Metal valves, leak valves, vacuum components |
| VSI Vakuum Science Instruments GmbH<br>Bahnhofstr. 2<br>D-65307 Bad Schwalbach<br>Tel: 06124-70830<br>Fax: 06124-708333 | Analyzer for UPS, AES, ISS, HREELS-spectrometer Video-LEED, electron guns, vacuum components |
| Wacker-Chemie GmbH<br>Hans-Seidel-Platz 4<br>D-81737 München<br>Tel: 089-627901<br>Fax: 089-62791770 | Si-single crystals |

| | |
|---|---|
| Wallace & Tiernan<br>Postfach 1563<br>D-89305 Günzburg<br>Tel: 08221-904-0<br>Fax: 08221-904-203 | Pennwalt manometer, aneroid-absolute manometer |
| Wavetek Electronics GmbH<br>Freisinger Str. 34<br>D-85737 Ismaning<br>Tel: 089-9609490<br>Fax: 089-967170 | Function generators, pulse generators, microwave components |
| Well-Diamantdrahtsägen<br>Oppauerstr. 37<br>D-68305 Mannheim<br>Tel: 0621-741990<br>Fax: 0621-745897 | Diamond wire saws, crystal cutting tools |
| Witzenmann GmbH<br>Östliche Karl-Friedrich-Str. 134<br>D-75175 Pforzheim<br>Tel: 07231-5810<br>Fax: 07231-581820 | Bellows, diaphragm bellows |
| Zinsser Analytical GmbH<br>Eschborner Landstr. 135<br>D-60489 Frankfurt<br>Tel: 069-7891060<br>Fax: 069-78910680 | Pure metals, alloys, chemicals |

## Japanese Manufacturers

| Company | Products |
|---|---|
| Anelva<br>5-8-1 Yotsuya<br>Fuchu-city<br>Tokyo 183<br>Japan<br>Tel: 31 423 34 0221<br>Fax: 81 423 60 2277 | Gauges, ion pumps, mechanical pumps, vacuum flanges, valves, etc. |
| Daikin<br>2-4-12 Nakazaki-Nishi<br>Kita-ku<br>Osaka 530<br>Japan<br>Tel: 81 6 374 9306<br>Fax: 81 6 374 9296 | Cryopumps, cryo-refrigerator |

| | |
|---|---|
| Denka<br>141 Yurakucho<br>Chiyoda-ku<br>Tokyo 100<br>Japan<br>Tel: 81 3 3507 5257<br>Fax: 81 3 3507 5078 | Ceramics, $LaB_6$ cathodes |
| Hamamatsu Photonics<br>3-8-21 Toranomon<br>Minato-ku<br>Tokyo 105<br>Japan<br>Tel: 81 3 3436 0491<br>Fax: 81 3 3433 6997<br>or<br>314-5 Toyooka-mura<br>Iwata-gun<br>Shizuoka 438-01<br>Japan<br>Tel: 81 539 62 5245<br>Fax: 81 539 62 2205 | CCD camera, microchannel plate, multi-capillary plate, etc. |
| Irie Kiken Co., Ltd.<br>4-11-7 Ginza<br>Chuou-ku<br>Tokyo 104<br>Japan<br>Tel: 81 3 3542 4692<br>Fax: 81 3 5565 7064 | Flexible tubes, quick-coupling(NW) components |

# Index

'f' *denotes Figure;* 'n' *denotes Footnote*

Absolute adsorption
  coverage measurement, 358–361
  King and Wells method, 358
  TPD method, 808
Absolute coverage, 440
Absolute sticking coefficient
  at desorption temperature, 362–363
  measurement, 358–361
Adsorption
  metal fatigue and, 146
  gas purification and, 720f
  powders, 812
  supported catalysts, 796
Adsorption/desorption from high area solids, 808–811
Alkali metal source, 694–695
Aperture, adjustable iris, 56–57
Aquadag, 16, 81, 190, 292, 538
Argon purification, 729f
Atomic arsenic, 664–667
Atomic fluorine, 690–693
Atomic hydrogen
  detectors, 683
  high-flux thermal source, 682–686
  radiofrequency driven source, 688–689
Atomic oxygen detection, 706–707
Atomization
  arsenic, 664
  fluorine, 690
  hydrogen, 682, 688
Au UPS spectrum, 342f
Auger electron spectroscopy (AES), 289
  calibration with quartz microbalance, 441
  lubrication and, 488
Azimuthal sample motion, 4, 6f
  cryogenic manipulators for, 558–559
  strain-free mounting for, 506, 507f
Azomethane
  explosion danger, 699
  methyl radical source, 698

Bakeout
  eliminated, 76
  glass systems, 722
  heaters for, 132
  in metal UHV systems, 132–134
  to UHV condition, 114
  Viton seal and, 166
  wall oxidation and, 136
Bayard–Alpert gauge
  contamination of, 203
  filament replacement in, 218–219
  high sensitivity of, 368
  ion gun based on, 314–317
Bearings
  high-speed rotary shafts, 54–55
  internal ball, 14f
  lifetime, 46
  rotation translation, 49f
  self lubricating, 48–49
  steel ball, 54
  Teflon, 54
  tungsten carbide ball, 48
  vacuum ball, 10
Bellows
  folding, 830f
  hypalon/nylon, 86
  temporary leak sealing of, 164–165
Beryllium window construction, 752
Birefringence, 764
Bond-Barrel holder, 230–233
Books, experimental methods, 841
Boron as impurity, 781
Boron nitride
  diffusion, 760
  lubricant, 48
  shield, 662
Brightness temperature, 588

Calibrated effusive molecular beam measurements, 358–361
Calorimeter, Setaram C-80, 813f
Calorimetric heats of adsorption, 476–477, 812–814

Capacitance
crystal positioning using, 16–17
measuring circuit, 19f
Capacitance manometer, 610, 621
Capillaries, cleaning, 139–140
Carbon monoxide. *See* CO
Cataphoretic deposition
of $LaB_6$, 206–208
of $ThO_2$, 202
Cathodes
$ThO_2/LaB_6$ coated, 202, 206–208, 276
$CeB_6$ coated, 211
$ThO_2$ coated Ir, 320
CEMA. *See* Chevron electron multichannel plate array
Cement
AgCl, 748, 796
MgO (Aremco 571P), 598
semiconductor crystal mounting, 528, 529f
Vacseal, 526, 748
$ZrO_2$ (Aremco Ultratem 516), 528
Ceramic insulator deterioration, 106
Cesium ion gun design, 672–675
Channeltron detector, 372–375
Charged particle oscillator, He(I) and He(II) ultraviolet source, 344
Chemisorption
CO on Ni, 812, 814f
IR transmission spectra, 805f
$N_2$ on Rh, 797
$O_2$ on Pd, 348–355
Chevron electron multichannel plate array (CEMA) detector, 334
Cleaning crystals
metal, 780
palladium, 784
silicon, 790
CMA Auger spectrometer
Pt deposition measurement, 440–445
photoelectron spectrum measurement, 338
CO, carbon monoxide
desorption from polycrystalline W, 366
purification, 718
Coefficient of friction, 486–489
Conductive coatings, 190–192
Conflat flange
aluminum, 154
for Al foil window, 757f
sexless, 150
Cooling reservoir, 550, 558
flexible, 570, 571f
for horizontal manipulator, 562, 563f
Cooling samples
contact with thermal reservoir, 548–549
crystal holder for, 544–546
gas thermal switch, 576–578
on manipulators, thermal conductivity issues, 550–553
using direct contact, 560–561
variable rate, 576–578
Cryogenic connections, 564–565
Crystal heating
electron bombardment, 516–518
interrupted ohmic heating, 512–515
Crystal cleaving, 144–145
Crystal manipulator, 4–7
Crystal mounting
non-weldable, 504–505f
strain free for azimuthal motion, 506–507
Crystal positioning, 16–17
Crystal temperature measurement, 517
Cu, copper
braid, 94, 436, 564
for cooling tank, 560, 561f
gasket removal device, 156–157
improved heat conductivity, 564
oxidation of, 544
reservoir, 554, 557f, 558
*See also* OFHC Cu
Cu-Be cup, 558

Dark current, 372
Debye temperature, 480–483
Desorption kinetics
angular resolved, 386–389
from single crystal faces, 366
Detectors
MgCdTe (MCT) infrared, 456, 472, 478f
InSb infrared, 472
ion-pumped infrared, 473f
MCP-type, 331, 413
resistive anode, 330
silicon surface barrier, 436, 438f
stagnation, 477, 616, 618
Deuterium kinetic isotope effect, 399
Diamondlike carbon coating, 151
Diamond, heating behavior, 521, 523f
Differential pumping method, 8–9, 344
Diode ion pump performance in He pumping, 122–123
Dynode multiplier, regeneration of, 373
Dynode rejuvenation, 302–303

EDA. *See* ethylene diamine
Einzel lens, 276, 324, 673
Electrical connections
band clamp, 110, 111f
cold-formed wire connector, 104–105
leakage, 106
phosphor bronze spring, 108
sliding metal contact, 108–109
tungsten-to-tungsten wrap, 102–103
ultrahigh resistance vacuum feedthroughs, 106–107
Electrical isolation of UHV components, 226–227
Electrical shock
avoidance, 834f
effect on man, 832
from ionization gauge, 836–837

Electrodeposition, Cr on W wire, 648–650
Electron beam
  angular divergence measurement, 294–296
  damage from stray electrons, 367
  evaporator, 652, 659f
  point focused, 658
  position monitor, 328–329
  small evaporator design, 656–657
Electron bombardment heating, experimental difficulties, 394
Electron energy analyzers, 288–291
Electron gun
  beam properties, 278–279
  cathode, 216
  design and behavior, 276–277
  electrostatically focused, 652, 654f
  emission current regulation, 277
  low-energy design, 284f
  low energy/broad beam, 280–287
  pulsed, 277, 334, 336f
  work function measurement, 416–419
Electron multipliers
  channeltron-type, 372–375
  Cu/Be-type rejuvenation, 302–303
Electron reflection coefficient measurement, 420–423
Electron spectrometer calibration, 292–293
Electrons, specularly reflected, 300
Electroplating, 526
Emissivity, 589
Epoxy adhesive, 266
ESDIAD
  pattern for CO on Cu(110), 330, 333f
  planar grid apparatus, 331
  pulse counting analyzer, 330–333
  *See also* LEED
Etching
  large thin window fabrication procedure, 760–763
  STM tips, 252, 254–255
Ethylene adsorption/desorption study, zeolite surface, 808–811
Ethylene diamine (EDA) etching agent, 760
Eutectics, Zr and W, 641f
Evanston-type microwave discharge cavity, 690
Evaporation source
  aluminum, 644–646
  cadmium sulfide, 662–663
  congruent, 638–639
  lithium, 642–643
  *See also* Metal evaporation
EXAFS measurement, 801

Faraday cup, 244
  absolute detection efficiency using, 331
  modified for electron current measurement, 300–301
  movable, 480–482
  multiple, 276, 294–296
$Fe(CO)_5$ decomposition, 138
Feedthrough
  high-speed rotary, 46–47
  ultrahigh resistance sapphire-metal, 106
Fermi level determination, 292
Filament temperature measurement, 584–587
  W filament nomograph, 587
Films
  BeO on dynodes, 302
  electrically conductive $SnO_2$, 190
  for $O_2$ and $O_3$ detection, 706–707
  $Si_3N_4$, 596
  thickness measurement, 448
Flanges
  Conflat, 8
  KF 50, 743
Formic acid, isotopically labeled, 383
Friction measurement, 486–489

Ga, gallium, metallic joint, 560, 566
GaAs, gallium arsenide
  indirect heating of, 524–525
  internal reflection IR spectroscopy and, 470–471
  MBE growth of, 956
Galling,
  prevention, 81
  protection against, 50
Gas chromatography sensitivity enhancement, 408–409
Gas dosers, 138–141, 604–615
  designs, 141f
  repetitive dosing, 620
Gas flow regulation, 624
Gas purification
  apparatus, 728f
  at high pressures, 718–719
  cryogenic or gettering method, 726
  extremely pure He, 722–724
  metal permeation method, 730–733
  nitrogen, 719, 726
Gaskets, 149–160
  annealed copper, 150
  copper, 84, 96
  cryogenic, 154–155
  deformable, 736, 748
  gold wire, 151, 152f
  OFHC Cu type, 158
  removal device, 156–157
  silver-plated copper, 150
  titanium nitride-coated copper, 151
  *See also* Seals
Gas thermal switch, 576–578
$Gd_2(MoO_4)_3$. *See* GMO
Ge, germanium, optical pyrometer calibration, 596–597
Geiger Müller counter
  for $^{14}C$ on Pt(111), 436
  for inverse photoemission intensity, 304–305
Getter films
  He storage, 722

maintaining vacuum and, 727
Ti, 386, 718
TSP pump, 118
GMO [$Gd_2(MoO_4)_3$], in measuring ion-beam size, 326, 327f
Graphite, electron spectrometer calibration, 292
Grids
conical grid fabrication, 186–187
hemispherical, 180–183, 331
in electron gun, 280
IR cell sample support, 800–805
large flat mesh, 184–185
tungsten, 800, 803f–805f

Halogens,
electrochemical source, 712–715
microwave discharge source, 690, 713
He, helium
cryocooler, 144
extremely pure, 722–724
leak detection probe, 162–163
leaktight seal, 748
production and purity of, 722–724
pumping tests, 122
purification apparatus, 728f
transpiration, use for lowering $l$-$N_2$ temperature, 572–573
He(I)/He(II)
photoemission, source, 338–342
UV resonance line source, 344–347
Heaters
button type, 520f
paint-on type, 538, 539f
pyrolytic graphite, 536–537
tungsten filaments, 540
Heating
compound semiconductors, 524
conduction, 522f
design for insulator crystals, 526–527
indirect, 520–523
irregular shapes, paint-on type, 538
ohmic, 86, 94, 523f
radiation and electron bombardment, 394, 518f, 522f
silicon crystals, 508–511
temperature programmed, 394, 494–503, 598–600
High area solids
adsorption and desorption on, 808–811
microcalorimetric studies, 812
IR cell for, 800–805
High-pressure reactor
gold-plated pan bronze, 67
sampling with gas chromatography 408
High-pressure sample transfer cell, 86–89
Hot filament resistance thermometer, 380, 584–587
HREELS, high-resolution electron energy loss spectroscopy, 289, 513
for boron detection on Rh, 781
Hydrogen
as impurity, 722
atomic, 682–686, 688–689
desorption from Ni(111), 386, 389f
dissolved in stainless steel, 136
permeation studies, 736–737
purification, 719

Infrared cells
adsorption studies, 796
high surface area study, 800
wide temperature range, 800, 803f
Infrared reflection-absorption spectroscopy (IRAS), 456–466
Fourier transform spectrometers (FTIR) and, 456–459, 466
measurement at high pressure, 462–465
Infrared spectroscopy
continuously pumped detector, 472–473
differential pumping window seals and, 745f
sealing windows to glass, 743
semiconductors and, 466
window mountings, 742–747
Insulators, electrical
$Al_2O_3$, 106
Kapton, 16, 550
Macor, 304
sapphire, 106, 550
Internal reflection IR spectroscopy
gallium arsenide, 470–471
*p* and *s* polarization vector diagram, 468f
silicon and, 466–469
UHV optical system, 469f
Inverse photoemission, 304–305
Ion beam
angular distribution, 330, 334
polishing, 238
STM tip sharpening, 256–257
Ion beam sources
axial magnetic field and, 322f
Coultron GD-2, 324
low energy, magnetic focused, 324–325
measuring beam dimensions, 326–327
Ion bombardment
evaporation, 450
neutrals desorbed by, 334
sources, 310–313
Ion gun
based on Bayard–Alpert gauge, 314–317
cesium, 672
cold cathode, 318–319
design of, 672, 675
low gas emission, 308–309
modified, 309
Pierce, 673
pulsed, 334

Ion pumps
cleaning by chemical etching, 126–127
methane liberation from, 118
Ion sources
alkali, 672
alkaline earth, 678
broad beam, 320–322
Ionization gauge
electrical shock from, 836
filament replacement, 218, 219f
geometry in evaporation rate measurement, 447f
metal film evaporation rate measurement, 446–447
movable, 592
IRAS. *See* Infrared reflection-absorption spectroscopy
$IrO_2$ cathode insulating layer, 320
Isosteric heat measurements, 465f, 476

JASON program linear Poisson equation, magnetic shielding, 222
Joule-Thompson refrigerator, 174

Kapton insulator, 16, 226
Kapton/sapphire, comparative cooling rates, 550–553
Kaufman ion sources, 320
Kelvin probe
Au electrode, 432
effect of stray capacitance in , 429
electromagnetic driven, 430f
high-temperature flow reactor, 432, 433f
piezodriven, 426f, 428
reproducibility, 429
thermal desorption spectroscopy and, 428
work function in semiconductors, 425
Kinetic isotope effect, 404
for deuterium, 399
Kinetics of adsorption, 358, 362
Knudsen cell, 668

$LaB_6$ thermionic emitter material, 210
LabVIEW program, 394, 498, 598
controller logic, 500f
Laser-induced fluorescence (LIF), 334
Laser-ionized neutrals time of flight detection, 334–336
Leaks
calibrated, 808
check groove, 156, 157f
detection probe, 162–163
liquid sealers, 164
temporary sealing of bellows, 164–169
LEED, low energy electron diffraction
angle of incidence measurement in, 244–245
beam intensity measurement, 298–299
Kr isotherms using, 145f
movable Faraday cup in, 480, 482f
pattern, for Cu(110), 330, 333f
retarding diode method and, 413
LEED/ESDIAD analyzer, 330–333
LIF. *See* Laser-induced fluorescence
Linear motion
actuator for, 25f
inexpensive device for, 26–27
of translating Auger spectrometer, 28–29
platform (LMP), 58–59
solenoid driven, 24–25
Liquid nitrogen
enhanced cooling of, 572
level alarm, 574, 575f
trap, 128–129
triple point, 572, 573f
Low-energy electron diffraction. *See* LEED
Lubricants
Aquadag. *See also* Aquadag
boron nitride, 48
degradation, 54
fluorocarbon, 266
for heavy sliding loads, 50–52
graphite, 9, 151
molybdenum sulfide, 9, 56
Santovac 5 pump oil, 9
TiC/Ag, 50
Lubrication
coefficient of friction and, 488
of metal/metal interfaces, 50

Macor insulator, 214, 304
Magnetic eddy current drive, 42
Manipulator
horizontal cooling of, 562–563
rotary high precision, 20
cooling samples on, 550–553
multiple motion, 4–7, 30–31
vacuum-jacketed cryogenic, 554–557
Manufacturers, equipment and components, 844
Mass spectrometers
calibration, 382, 384f
differentially pumped, 118
magic angle thermal desorption, 390–393
sensitivity comparison with bare ion gauge, 368
sensitivity ratios for, 382
Ta filament mass scale calibration, 380, 381f
thermal desorption, 810f
unshielded, 390
UTI-100C, 383
*See also* Quadrupole mass spectrometers
Mass spectrometry, 366–384
MCP. *See* Microchannel plate detector
Mechanical action on samples
crystal cleaver, 144–145
fatigue of Al(111) crystal, 146
mica cleavage device, 144
piezoelectric fatigue apparatus, 146–147
tungsten carbide blade, 144
Metal evaporation
alloys and mixtures, 638

chromium evaporator, 648
degassing, 628
downward evaporation, 632
eutectic mixtures, 640
flash evaporation, 638
measurement of rate, 446
multiple evaporator, 637f
ultralow coverage, 636
Metal fatigue, 146
Metal foam, vacuum applications, 124–125
Metal foil
aluminum, 756
beryllium, 752
HAVAR, 266
stretching, 266–267
tantalum/platinum sandwich weld, 261, 263f
wrinkle free, 266–267
Metal single crystal cleaning, 779–792
Methanol
reducing agent in spotwelding, 260, 262f
reaction on Ni(111), 398–403
Methyl chloride thermal desorption spectra, 118
Methyl radical source, 698
Microcalorimetric studies, 812–814
Microcapillary array beam doser, 604–615
calibration and use of, 610–613
design, 604, 607f
fractional interception, 611
gas handling for, 608, 609f
glass, 138
kinetic studies, 610, 613f
precautions, 605
shutoff with condensable gas, 614
Microchannel plate detector (MCP), 330–333
Motion in vacuum, 4–63
Motional freedom
cooling braid and, 566, 568f
cryogenic manipulator, 554
shape memory effect and, 38–40
six degrees of, 4–5
*See also* Rotary motion
Motors
high speed, 44–45
very high speed, 47f
Mu metal, 428
Multicapillary array collimator, 308

$N_2$, nitrogen, polarizability from work function measurement, 416
Neutral beam intensity measurement, 616
Ni permeation source, hydrogen purification, 730
NO, nitric oxide
impurity oxidation, 784
oxidative wall passivation, 136–137

Observation windows
film deposition prevention, 770–771
low optical distortion glass, 772–773
shielded, 768–769
strain-free mounting, 764–765
zero length assembly, 774–775
OFHC Cu, oxygen free high conductivity copper, 158
cooling braid, 564, 709
cooling reservoir, 570
crystal holder, 504
hollow bolt seal, 160f
sample cooling, 544, 550
thermal shield, 60
Optical pyrometer
calibration point for silicon, 596–597, 598
disappearing filament type, 588–590, 594f
radiant energy equation, 588
measurement through coated window, 592–595
Optical windows. *See* Observation windows
Outgassing
degassing loops, 584–587, 628–631
from motors, 44
prevention, 584
studies, 166
Oxygen
cathode oxidation
chemisorbed, 349, 352, 355f
detection of atomic oxygen, 706
$MnO_2$ solid state source, 734–735
purification, 719
Ozone
detection of, 706
mass spectrometer cracking pattern, 703
production, storage, use, 702–704

Palladium
carbon removal from, 784–788
surface segregation and, 781
Palladium leak, hydrogen purification with, 688
Palladium permeation source, hydrogen purification, 730
Pancake radiation heater, 80
Passivation, surface
oxidative, 608
Teflon coating, 140
Permeation
atmospheric, 166
gas purification by, 730–733
reusable samples, 736–737
Phosphor screen
beam dimension measurement, 294
CEMA detector and, 334, 336f
dusting method for coating, 198–199
electrostatic deposition of, 195, 198
LEED apparatus, 190, 331
sedimentation deposition of, 194–197
spot photometer and, 298
Photochemical surface processes, 348–355
Photochemistry

of chemisorbed $O_2$, 349, 351, 355f
polarized UV light in, 352–355
Photoemission studies, hollow cathode resonance lamp for, 338–342
Pierce ion gun, 673
Piezoceramic foil, in Kelvin probe, 424–427
Piezoelectric actuators
drive, 60, 146
fatigue on polycrystalline Ag, 146, 147f
Kelvin probe actuator, 432
tube, 60
Planck's law, 588
Polarized UV light for photochemistry, 352–355
Polycrystalline metals
Ag, 146
Ni, 425, 427f
W, 366
Pulse counting, 330–333
Pump oil
Fomblin, 824
Santovac-5, 824
Pumps
aspirator/sorption, 114–115
diode ion, 118, 122–123
ion, 96, 114, 122, 308, 344, 367
oil contamination elimination, 114
oil diffusion, 344
regurgitation effect, 118
safety system, 824–827
triode ion, 122
turbomolecular, 67, 344
zeolite sorption, 114, 174
*See also* Titanium sublimation pump
Pumping
differential, 8–9, 67, 344, 347f
getter film, 118
parallel, 120
pressure equalizing device, 120–121
tests, 123f
two-stage differential, 347f
vibration-free ballast-type, 116, 117f
Pyrolytic graphite
chemical vapor deposited, 536
paint-on heater, 536–537
Pyrometer. *See* Optical pyrometer

QTM. *See* Quartz thickness monitor
Quadrupole mass spectrometers (QMS)
channeltron detector, 372–375
enhanced signal-to-noise ratio in, 372–375
floating grid, 376, 379f
ozone and atomic oxygen detection, 706
shielded, 366–371, 610
spurious electron emission elimination, 376–379
*See also* Mass spectrometers
Quartz crystals
oscillator, 457
thin film measurement on, 448–449
*See also* Quartz thickness monitor
Quartz microbalance. *See* Quartz thickness monitor
Quartz prism UV polarizer, 348, 352
Quartz thickness monitor (QTM), 440–449
calibration of, 441
magnetic drive, 42
very low coverage measurement, 636
Quartz flat partial reflector, 348

Radioactive adsorbates, 436–438
Rapid wafer thermal processors, 596
Re ribbon, $ThO_2$ deposition on, 206
Resonance enhanced multiphoton ionization (REMPI), 334
Retarding diode method, 412–415
Richardson equation, 202
RHEED intensity measurement, 446
Rotary motion, 4–23
arcsec resolution, 20–23
axial, 8
bearings, 55f
cryogenic azimuthal, 558–559
electrical readout for, 10
fine angular adjustment of, 23f
heavy-duty turntable, 36–37
high-speed feedthrough, 46–47
magnetic shutter, 42–43
optimization of angular resolution, 32
rapid, 10
through small angles, 32–35
turntable, 36
two-axis device, 34f
ultrahigh vacuum and, 8–15

SAES alkali metal sources, 694
Safety, 831–837
Sample cooling, 566–568
Sample transfer, 65–97
at low temperature, 80–83
from high pressure to UHV, 66–74
from UHV to external cell, 76–79
high-pressure cell, 86–89
internal cup high-pressure cell, 84–85
Lebold-Heraeus device, 67
longitudinal with disconnect, 92f
magnetic rod, 90
movable crystal, 69f
rod-through-chamber, 67, 73f, 79f
rotation to multiple ports, 76
transfer rod, 86
translatable reactor cup, 70f
Sample transfer with disconnect
inexpensive, 94, 95f
in standard UHV system, 90
in STM systems, 90
Sapphire

crystal holder, 504
electrical leakage prevention, 106–107
rod, 432
support bearing, 506
thermal conductivity of, 550
UV grade window, 348
Scanning kinetic spectroscopy (SKS), 398–404
Scanning tunneling microscope. *See* STM
Seals
compression, 198–199
doubly differentially pumped, 76, 457
for infrared windows, 742
gold wire cryogenic-type, 154, 155f
Helicoflex O-ring, 172
high force, 68
high temperature, 8, 748
indium, 749
Kovar glass, 764
metal to Teflon, 68
rotary design, 9, 11f, 12f
sliding rod, 72f
spring-loaded Teflon, 9, 13, 86, 96
UHV glass window, strain-free, 764
Vacseal leak sealer, windows sealer, 748
Wilson-type, 8
*See also* Gaskets
Sealants, use of, 44, 748
Semiconductor crystals
cement mounting for, 528–529
cleaning, 779–788
interferometry and temperature measurement, 509
low stress mount for heating/cooling, 530–531
SiC, 470
Shape memory effect (SME)
clamp, 40f
NiTi alloy, 38
shutter, 40f
small motions with, 38–40
Shelton diode for work function measurement, 420–423
Shielding, 222–227
ad-mu 80, 223
electrical, 226–227
magnetic in UHV, 222–224
Shutter device, rotatable, 42–43
Silica gel
Davisil, 156n
ozone storage, 702
Silicon
crystal heating, 508–511
crystal mounting, 510f
emissivity of, 589–591
etching, 760
split-off-dimer defects, Ni induced, 792f
surface cleaning, 790–792
temperature control using resistivity, 598
Silver films, for ozone and atomic oxygen detection, 706
Silver permeation source, oxygen purification, 730
SIMION electrostatic lens and analysis program, 280
ion optical calculations, 673
Single crystals
alignment, 240–243
Bond-Barrel holder, 230, 233f
$CaF_2$ disk, 796
curvature of, in grinding and polishing, 231
cutting goniometer, 236–237
desorption kinetics on, 366
electron beam angle of incidence on, 246, 247f
fabrication, 229–239
ion beam polishing, 238–239
orientation, grinding and polishing, 230–234
photochemical processes on, 348
radiation heating of through glass windows, 532–535
reconstruction effects, 790
use as calorimeter, 352, 354f
UV irradiation of, 348
*See also* Metal single crystals
Single crystal surface
chemisorption and reaction on, 436
Kelvin probe desorption measurement, 425
kinetics on, 404
synthesis reactions on, 404
SKS. *See* Scanning kinetic spectroscopy
SME. *See* Shape memory effect
Sources
alkali metal, 694
alkaline earth ion, 678
atomic arsenic, 664
atomic fluorine, 690–693
BN diffusion, 760
cesium, 672–677
electrochemical cell, 710, 711f
He(I), He(II), 346f
high-flux atomic hydrogen, 682–686
ion bombardment, 310–313
Kaufman ion source, 320
methyl free radicals, 698
UV from high-pressure Hg arc, 348
*See also* Atomic hydrogen; Thermionic emitters; Vacuum ultraviolet sources
Spot photometer construction and use, 298–299
Spot-welding
difficult junctions, 260–263
slotted crystals, 261
*See also* Welding
Springs
chromel wire, 530
Cu-Be, 110, 558
phosphor bronze, 108
Ta wire, 108
Sputtering
axial, to sharpen tips, 256
cleaning by, 310
coating tube inner walls, 452–453
contamination effects, 310

crystal polishing, 238
deposition rates, 318
for hybrid thin film deposition, 450–451
ion gun for, 308–324
neutral species, 391
TiN, 151
Stagnation detector, 616
Sticking coefficient, 382, 620
measurement, 476–477
Sticking probability versus absolute coverage, 613f
equation, 362
STM, scanning tunneling microscopy, 250, 790
repeatable sample positioning, 60–63
STM tip
approach mechanism for, 58
etching for field ion microscopy, 252f
fabrication, 250–257
reproducible etching, 254–255
sharpening, 250
single crystal metal, 256–257
Strain gauges, 486, 488
Sulfur source, 710
Supersonic
molecular beams, 617
nozzles, 268–269
Supported catalysts, 796
Surface barrier detector, 436–438
Surface contaminants, 376
Surface Debye temperature, 480–483
Surface poisoning, 399
Surface reaction kinetics at high pressure, 68
Surface reaction processes, 398
Surface segregation, 780
Surface studies
radiotracer techniques for, 436–438
electrochemistry apparatus, 176f

Tantalum tube, hydrogen embrittlement prevention, 452
TDS. *See* Thermal desorption spectroscopy; TPD
Teflon
compression fitting, 174
impurities from, 86
liners, 67
Temperature measurement
hot filaments, 584–587
thermocouples, 580–583
pyrometric, 588–597
W filament nomograph, 587
Temperature-programmed desorption (TPD), 366–371, 404–405f, 494
Temperature-programmed reaction spectroscopy (TPRS), 399, 404–406
Temperature programming
metal crystals, 494–497
metal and semiconductor crystals, 498–502
Thermal braid cooling, 561, 566–569
Thermal desorption, 386–397
angular distribution, 386–389
apparatus for, 388f
CO on Ru(001), 429, 431f
$H_2$ on Ni(111), 386, 389f
Thermal desorption spectroscopy (TDS), 366, 394–397. *See also* TPD
Thermionic emitters
brightness temperature, 592
carburization, 203
cathode design, 217f
electron gun, 276
emitter evaporation rates, 210, 213f
filament replacement in, 218–219
high-temperature operation, 216–217
$LaB_6$ coated, 206, 208f
lanthanum hexaboride, 210–213
low work function coatings, 216
mounting, 214–215
shielded and open emitters, 215f
shielded, 314
spiral, 214
thoriated, 202–205
Thermocouples
attachment, 508, 510–511
avoiding excess local heating, 394
calibration, 509, 580–583
cemented, 521
chromel-alumel, 508, 645
contamination, 598
welding to metal single crystal, 260
Thermocouple gauge, sealing, 158, 160f
Thin film deposition, 440–453
absolute surface coverages, 440–445
hybrid methods, 450–451
measurement, thickness limitations, 448
on tube inner walls, 452–453
organic vapor deposition cell, 668, 669f
stable/metastable films, 450
Tilt motion, 4, 7f
Titanium sublimation pump (TSP), 118, 174, 308, 367
bakeout and, 133
Topping model, 416, 419f
TPD, temperature-programmed desorption, 366, 394–397, 722, 808–811
TPRS. *See* temperature-programmed reaction spectroscopy
Translator, folding linear magnetic, 828
Traps
liquid nitrogen, 128–129
metal foam, 124–125
water-cooled Cu, 338
zeolite, 338
Tribiology, review, 50
Tribometer, 486–489
TSP. *See* Titanium sublimation pump
Tungsten

coils, 102
crystal holder, 504
grid, 800, 803f–805f
hairpin filament, 440
high-temperature operation, 540–541
wire, fabrication, 102
wire heaters, nomograph for design, 587f
Tungsten ribbon attachment to power leads, 100–101
Turbopump power interrupt protection, 818–820

Ultra high vacuum systems (UHV)
access port, 96–97
aluminum, 172–173
bakeout, 132
chamber protection, 822
common impurities in, 790
differential pumping in, 8
grating rotation, 32–35
stainless steel wall passivation, 136–141
surface electrochemistry apparatus, 174–176
Ultrathin metal single crystals, 476, 478f
Ultraviolet light (UV)
lamp for photochemistry on surfaces, 348
polarized, 352
Ultraviolet photoelectron spectroscopy (UPS), 292
gold (Au) spectrum, 342f

Vacuum ultraviolet sources (VUV), 338–347
Valves
fast closing, 822–823
low deadspace aperture, 620, 622f
small gate, 76
squeeze, 624–625
straight through, 338
polyfluoroalkoxy (PFA), 138
Vapor deposition, organics, 668
Variable aperture, flange mounted, 56–57
Viton
atmospheric permeation, 166–169
bakeout limitation, 748
outgassing, 67, 457
Vycor permeation source, helium purification, 722

Wagner electrochemical cell, 710
Wall passivation, 136–141
Wall reactions
minimization, 138–140
reduction with gold film, 66
Wehnelt cylinder, 276
Welding
for thermal contact, 564
metal foil window, 758
sandwich weld, 261, 263f
*See also* Spot-welding
Wien filter, 324
Wilson seal, 8
Wilson differential pumping method, 8
Windows
alkali halide, 742
bakeable aluminum, 756
bakeout limitations, 748
beryllium, 752, 755f
$CaF_2$, 304, 457, 743
disposable, 768
for vacuum UV transmission, 748–750
glass seals for, 772–773
heated, 770–771
IR transparent, 457, 796
Itran, 743
large-area Si crystal preparation, 760
metal thin foil welding, 758
mounting, 742
$SrF_2$, 304
thin film coated, 592
UV grade sapphire, 348
UV transparent, 304
X-ray transmission, 752, 755–756
zero length assembly, 774
*See also* Observation windows
Wire
ceramic-bead insulated, 108
polycrystalline, 256
single crystal, 256
straightening and braiding, 270–271
tantalum, 86, 380
tungsten, 86, 102, 587f
Work function
changes, retarding diode method, 412–415
difference for two surfaces, 420
electron gun measurement, 416–419
electron reflection coefficient measurements, 420–423
measurement, 428
minimization of differences along a surface, 282
$SnO_2$, 190
tungsten, 202

X-ray standing wave spectroscopy (XSW), 20

Zeolite
ethylene adsorption study on, 808
for gas storage, 718
trap, 125f, 726
Zone electropolishing, 250-252

# Experimental Innovations in Surface Science

### *How you can contribute to Future Editions*

If you have an innovation that could be published in future editions, please fill in this card and return it to Springer-Verlag New York, Inc. You will then be contacted by Dr. John T. Yates, Jr. for more details.

**Brief description of Innovation**

________________________________________

________________________________________

________________________________________

________________________________________

Name:________________________________

Title:________________________________

Affiliation:____________________________

________________________________________

Address:______________________________

________________________________________

City:_________________________________

State:__________________ Mailing Code:____________

Country:______________________________

Phone:_______________________________

Fax:_________________________________

E-Mail:______________________________

FOLD ON DOTTED LINE LEAVING
"BUSINESS REPLY MAIL" PANEL EXPOSED

CLOSE WITH TAPE

NO POSTAGE
NECESSARY
IF MAILED
IN THE
UNITED STATES

# BUSINESS REPLY MAIL

FIRST CLASS MAIL PERMIT NO. 5863 NEW YORK, NY

POSTAGE WILL BE PAID BY ADDRESSEE

ATTN: MARIA TAYLOR
**SPRINGER-VERLAG NEW YORK, INC.**
175 FIFTH AVENUE
NEW YORK NY 10160-0266